R.H. TUTTLE

Developments in Primatology: Progress and Prospects

Series Editor
Louise Barrett
Lethbridge, Alberta, Canada

More information about this series at http://www.springer.com/series/5852

Tracy L. Kivell • Pierre Lemelin
Brian G. Richmond • Daniel Schmitt
Editors

The Evolution of the Primate Hand

Anatomical, Developmental, Functional, and Paleontological Evidence

Springer

Editors
Tracy L. Kivell
Animal Postcranial Evolution (APE) Lab
Skeletal Biology Research Centre
School of Anthropology and Conservation
University of Kent
Canterbury, UK

Department of Human Evolution
Max Planck Institute for Evolutionary Anthropology
Leipzig, Germany

Brian G. Richmond
Division of Anthropology
American Museum of Natural History
New York, NY, USA

Department of Human Evolution
Max Planck Institute for Evolutionary Anthropology
Leipzig, Germany

Pierre Lemelin
Division of Anatomy
Department of Surgery
Faculty of Medicine and Dentistry, University of Alberta
Edmonton, AB, Canada

Daniel Schmitt
Department of Evolutionary Anthropology
Duke University
Durham, NC, USA

ISSN 1574-3489 ISSN 1574-3497 (electronic)
Developments in Primatology: Progress and Prospects
ISBN 978-1-4939-3644-1 ISBN 978-1-4939-3646-5 (eBook)
DOI 10.1007/978-1-4939-3646-5

Library of Congress Control Number: 2016935857

Printed on acid-free paper

This Springer imprint is published by Springer Nature
The registered company is Springer Science+Business Media LLC New York

Foreword

Rarely are we privileged to witness the appearance of a book that strikes out in a completely new, groundbreaking direction and will accelerate a major field of research. The new direction of this book on the evolution of the primate hand is toward a comprehensive, highly informed, critical review of the subject in chapters, not by a single author but by a team of experts actively involved in the varied fields that contribute to the understanding of the subject. The team approach ensures uniform high quality of the review together with a shared underlying theme (the variability of primate hands upon a shared primitive pattern). Acceleration of research on the evolution of the hand will be the inevitable result of the team's having laid a deep, broad foundation of knowledge and ideas for the production of new investigations.

This highly readable collection of chapters will be a welcomed resource to all who are interested in the human hand, its functions, and its origins. There is a fascination with hands and their expression of human behavior, manifest in representations ranging from the walls of prehistoric caves to the Social Programs Bas-relief created by Robert Graham at the Roosevelt Memorial on the Washington D.C. mall. Hand surgeons marvel to me at the dexterity and adaptability of the human hand to its varied roles, and teachers lament the trend away from "hands-on" activities that enhance learning in K-12 science courses and even in medical anatomy laboratories. It seems to them almost as if our hand was freed from locomotion only to become captive to computer keyboards! All these readers will find that most of our remarkable manipulative capabilities originated early in our primate ancestry and may be understood in the context of nonhuman primate locomotor and manipulative behavior, including the constant interaction of touch and proprioceptive cues in learning about the physical and social environment.

For the first time we find in a single book descriptions of all currently known human and nonhuman primate fossil hand bones, together with detailed descriptive and quantitative data on living primate musculoskeletal hand anatomy, development, and uses that inform functional and phylogenetic analyses of the fossils. In addition, we are introduced to the most recent developments in approaches to fine-tuning these analyses. Three chapters review experiments involving new techniques for imaging, motion capture, and hand pressure recording, along with sophisticated modeling of joint movements and stresses in locomotor and manipulative behavior. Readers learn about the wide variety and impressive flexibility of primate hand postures during locomotion as well as manipulation and are encouraged to expand studies to include active haptic sensing of shapes, weights, and textures of objects with which primates interact in their environment. A chapter on hand development brings the reader up to date on the current understanding of genetic and developmental factors in phenotypic variation and addresses potential for and constraints on phylogenetic change in the hands of primates. Especially welcome in a new book on primate hand evolution are chapters on the comparative morphology of hand integument and on neural control of the hand.

Most impressive is the tremendous effort made by the authors not only to review the literature critically in each area but also to summarize extensive, detailed information in tables and to provide drawings and photographs that are informative and relevant to the text. Careful attention is given to the definitions and uses of terms for anatomical features, grips, thumb and finger movements, and taxonomic categories, which should at last reduce confusion in the literature. Here is a springboard for new research that will enable us to communicate knowledgeably and effectively about how our future findings relate to the findings of our current and former colleagues.

In their suggestions for future directions in research, the authors echo persistent calls in the literature for more fossils, especially for associated elements of the hand and evidence from the early evolutionary stages of the genus *Homo*. They strongly emphasize in addition the need for comparative studies and functional analyses of morphological variability within and among a much larger range of primate species. However, their book also reveals the large store of data already available for functional and phylogenetic analysis of living and fossil primate hands, and the challenge now will be to keep the book up to date as the future research they propose comes to fruition. It is a great personal pleasure to introduce the book, which will be an invaluable resource for all whose work in evolutionary biology and human health care focuses on the fascinating diversity of primate hands. The editors deserve congratulations for their powerful concept and for their monumental achievement.

Mary W. Marzke
School of Human Evolution and Social Change
Arizona State University
Tempe, AZ, USA

Contents

Contributors

Sergio Almécija Center for the Advanced Study of Human Paleobiology, Department of Anthropology, The George Washington University, Washington, DC, USA

Department of Anatomical Sciences, Stony Brook University School of Medicine, Stony Brook, NY, USA

Institut Català de Paleontologia Miquel Crusafont, Universitat Autònoma de Barcelona, Barcelona, Spain

David R. Begun Department of Anthropology, University of Toronto, Toronto, ON, Canada

Jonathan I. Bloch Florida Museum of Natural History, University of Florida, Gainesville, FL, USA

Doug M. Boyer Department of Evolutionary Anthropology, Duke University, Durham, NC, USA

Stephen G.B. Chester Department of Anthropology and Archaeology, Brooklyn College, CUNY, Brooklyn, NY, USA

Department of Anthropology, Graduate Center of the City University of New York, New York, NY, USA

New York Consortium in Evolutionary Primatology, New York, NY, USA

Jessica Crast Yerkes National Primate Research Center, Emory University, Atlanta, GA, USA

Rui Diogo Department of Anatomy, College of Medicine, Howard University, Washington, DC, USA

Dorothy M. Fragaszy Department of Psychology, University of Georgia, Athens, GA, USA

Laurie R. Godfrey Department of Anthropology, University of Massachusetts, Amherst, MA, USA

Marc Godinot UMR 7207 CR2P, École Pratique des Hautes Études, Paris, France

Michael C. Granatosky Department of Evolutionary Anthropology, Duke University, Durham, NC, USA

Terry Harrison Center for the Study of Human Origins, Department of Anthropology, New York University, New York, NY, USA

William D. Hopkins Neuroscience Institute and Language Research Center, Georgia State University, Atlanta, GA, USA

Division of Developmental and Cognitive Neuroscience, Yerkes National Primate Research Center, Atlanta, GA, USA

William L. Jungers Department of Anatomical Sciences, Stony Brook University, Stony Brook, NY, USA

Association Vahatra, BP 3972, Antananarivo 101, Madagascar, Stony Brook University, Stony Brook, NY, USA

Amanda K. Kingston Interdepartmental Doctoral Program in Anthropological Sciences, Stony Brook University, Stony Brook, NY, USA

Tracy L. Kivell Animal Postcranial Evolution (APE) Lab, Skeletal Biology Research Centre, School of Anthropology and Conservation, University of Kent, Canterbury, UK

Department of Human Evolution, Max Planck Institute for Evolutionary Anthropology, Leipzig, Germany

Pierre Lemelin Division of Anatomy, Department of Surgery, Faculty of Medicine and Dentistry, University of Alberta, Edmonton, AB, Canada

Stephanie A. Maiolino Interdepartmental Doctoral Program in Anthropological Sciences, Stony Brook University, Stony Brook, NY, USA

Department of Pathology and Anatomical Sciences, University of Missouri School of Medicine, Columbia, MO, USA

Masato Nakatsukasa Laboratory of Physical Anthropology, Kyoto University, Kyoto, Japan

Caley M. Orr Department of Cell and Developmental Biology, University of Colorado School of Medicine, Aurora, CO, USA

Kelly Ostrofsky Center for the Advanced Study of Human Paleobiology, Department of Anthropology, The George Washington University, Washington, DC, USA

Biren A. Patel Department of Cell and Neurobiology, Keck School of Medicine, University of Southern California, Los Angeles, CA, USA

Thomas R. Rein Department of Anthropology, Central Connecticut State University, New Britain, CT, USA

Brian G. Richmond Division of Anthropology, American Museum of Natural History, New York, NY, USA

Department of Human Evolution, Max Planck Institute for Evolutionary Anthropology, Leipzig, Germany

Neil T. Roach Division of Anthropology, American Museum of Natural History, New York, NY, USA

Department of Human Evolutionary Biology, Harvard University, Boston, MA, USA

Campbell Rolian Department of Comparative Biology and Experimental Medicine, Faculty of Veterinary Medicine, University of Calgary, Calgary, AB, Canada

Daniel Schmitt Department of Evolutionary Anthropology, Duke University, Durham, NC, USA

Chet C. Sherwood Department of Anthropology and Center for the Advanced Study of Human Paleobiology, The George Washington University, Washington, DC, USA

Erik Trinkaus Department of Anthropology, Washington University, Saint Louis, MO, USA

Evie E. Vereecke Jan Palfijn Anatomy Lab, Department of Development and Regeneration, University of Leuven, Leuven, Belgium

Andrey Verendeev Department of Anthropology and Center for the Advanced Study of Human Paleobiology, The George Washington University, Washington, DC, USA

Erin Marie Williams-Hatala Department of Biology, Chatham University, Pittsburgh, PA, USA

Roshna E. Wunderlich Department of Biology, James Madison University, Harrisonburg, VA, USA

Gabriel S. Yapuncich Department of Evolutionary Anthropology, Duke University, Durham, NC, USA

Angel Zeininger Department of Evolutionary Anthropology, Duke University, Durham, NC, USA

Chapter 1
Introduction

Tracy L. Kivell, Pierre Lemelin, Brian G. Richmond, and Daniel Schmitt

> *"There is evidently something extraordinarily primitive about the hand that has been preserved and passed on to Man; but like the primitive rotating forearm, this primitive, simple and unspecialized five-fingered hand is full of possibilities."*
>
> *(Wood Jones 1916: 20)*

Since Darwin (1871) first discussed it in the *Descent of Man*, scientists and lay persons alike intuitively recognize that the hand has played a key role in primate and human evolution. Frederic Wood Jones and John Russell Napier were two of the leading thinkers who helped establish the foundations of biological anthropology in the twentieth century, and both conducted pioneering research demonstrating the importance of comparative and functional anatomy of the hand in primate and human evolution. One theme that unified their research was the radical notion that the human hand, rather than being a specialized organ, is instead primitive in many aspects, retaining features found in earlier primates and other pentadactyl (five digits) mammals.

T.L. Kivell (✉)
Animal Postcranial Evolution (APE) Lab, Skeletal Biology Research Centre, School of Anthropology and Conservation, University of Kent, Canterbury, UK

Department of Human Evolution, Max Planck Institute for Evolutionary Anthropology, Leipzig, Germany
e-mail: t.l.kivell@kent.ac.uk

P. Lemelin
Division of Anatomy, Department of Surgery, Faculty of Medicine and Dentistry, University of Alberta, Edmonton, AB, Canada

B.G. Richmond
Department of Human Evolution, Max Planck Institute for Evolutionary Anthropology, Leipzig, Germany

Division of Anthropology, American Museum of Natural History, New York, NY, USA

D. Schmitt
Department of Evolutionary Anthropology, Duke University, Durham, NC, USA

T.L. Kivell et al. (eds.), *The Evolution of the Primate Hand*, Developments in Primatology: Progress and Prospects, DOI 10.1007/978-1-4939-3646-5_1

Labeling the primate hand as "primitive" can seem counterintuitive given the remarkable dexterity typical of primates and especially humans. In addition, there is considerable diversity in primate hand form and use that allows the exploitation of a wide range of substrates and foods. However, when examining the diversity of hand morphology across primate clades in comparison with other mammals, an astonishing number of primitive qualities are preserved, even in those primates with extraordinary specializations for movement and food procurement (e.g., aye-ayes, lorises, spider monkeys). It is precisely this elaboration on a primitive and versatile *bauplan* that facilitated the evolution of many morphologies and behavioral abilities in living and extinct primates. Reaching an understanding of how this combination of primitive and novel traits in the primate hand develops, functions, and ultimately evolved is inherently a multidisciplinary problem that requires a variety of approaches, methods, and expertise.

Our aim with this edited volume is to provide an all-in-one resource that captures this diverse perspective needed to approach a richer understanding of the primate hand. We asked our contributors to explore the diversity in primate hand anatomy and function in light of development, biomechanics, and evolution from a broader mammalian perspective, highlighting both the primitiveness and specializations of the primate hand. We also asked our contributors to address these topics in a straightforward, accessible language with data-rich tables and illustrations that will serve as a comprehensive guide for any researcher interested in the primate hand. We are delighted with the chapters that our contributors produced.

Many other books have been written on the human and nonhuman primate hand that have laid the foundation for this volume and greatly improved our understanding of the primate hand from different research perspectives. For example, Napier's *Hands* (revised in 1993 by Russell Tuttle) is a classic, particularly for nonspecialists. With this volume, however, we aim to provide an up-to-date and much more in-depth review of the primate hand than Napier's book offers. Lewis' (1989) *Functional Morphology of the Evolving Hand and Foot* has been an invaluable and detailed resource for researchers interested in primate hand anatomy, particularly with its broad comparative perspective on hand anatomy of other mammals. Our book builds on Lewis (1989) and Napier (1993) with the aim of providing user-friendly anatomical and functional descriptions, aided in particular by the wealth of biomechanical and behavioral research that have been conducted in the decades since Lewis' book was published. Preuschoft and Chivers' (1993) *Hands of Primates* is another valuable resource for researchers interested in primate hand use, function, and development. Again, we aim to complement this volume by providing more review-oriented (rather than a focus on specific research questions) chapters that summarize the most up-to-date research on hand anatomy, biomechanics, and evolution across all primate clades. In short, we hope to build upon the foundation created by these, and other seminal books, to provide researchers with an easy-to-understand, comprehensive, and current summary of what we know (and do not know) about primate hand anatomy, development, function, and evolution, current methods, and future directions of research. We cover all primate clades, from strepsirrhines to hominoids, and from the earliest primate fossils and close relatives to the evolution of modern *Homo sapiens*.

We have divided the book into three sections. The first is **Anatomical and Developmental Evidence**, in which we review the history of research on primate hand anatomy (Chap. 2), the skeletal morphology of the primate wrist (Chap. 3) and digital rays (Chap. 4), the development (Chap. 5) and neural control (Chap. 6) of the primate hand, and primate hand musculature (Chap. 7) and integument (Chap. 8). The second section is **Biomechanical, Experimental, and Behavioral Evidence**, including recent engineering and imaging methods for exploring the functional morphology of the primate hand (Chap. 9), experimental research on primate hand use and function (Chap. 10), the biomechanics of the human hand (Chap. 11), grasping function in primates (Chap. 12), and hand posture during positional behavior (Chap. 13). The third section, **Paleontological Evidence**, reviews the fossil record of hand morphology, together with its functional and evolutionary significance, of the earliest fossil primates (Chap. 14), subfossil lemurs (Chap. 15), non-hominoid anthropoids (Chap. 16), Miocene hominoids (Chap. 17), early hominins (Chap. 18), and Pleistocene *Homo* (Chap. 19).

Across the chapters, different authors have addressed a variety of specific questions and provided their perspectives, but all explore the main themes described above to provide an overarching "primitive primate hand" thread to the book. Each chapter provides (1) an in-depth review and critical account of the available literature, (2) a balanced interpretation of the evidence from a variety of perspectives, and (3) prospects for future research questions. In order to make this a useful resource for researchers at all levels, the basic structure of each chapter is the same, so that information can be easily consulted from chapter to chapter. An extensive reference list is provided at the end of each chapter so the reader has additional resources to address more specific questions or to find specific data.

Together, the chapters of this book demonstrate how the primate hand combines both primitive and novel morphology, both general function with specialization, and both a remarkable degree of diversity within some clades and yet general similarity across many others. When we first undertook this initiative, we hoped to produce a book that each of the coeditors wished they had had when they began their doctoral dissertations on comparative anatomy, functional morphology, and evolution of the primate forelimb and hand. We are delighted and grateful that all of our contributors have gone above and beyond our expectations and allowed us to reach this goal.

References

Darwin C (1871) Sexual selection and the descent of man. John Murray, London
Lewis OJ (1989) Functional morphology of the evolving hand and foot. Clarendon Press, Oxford
Napier JR (1993) Hands (revised by R.H. Tuttle). Princeton Science Library, Princeton
Preuschoft H, Chivers DJ (eds) (1993) Hands of primates. Springer-Verlag, Vienna
Wood Jones F (1916) Arboreal man. Edward Arnold, London

Chapter 2
On Primitiveness, Prehensility, and Opposability of the Primate Hand: The Contributions of Frederic Wood Jones and John Russell Napier

Pierre Lemelin and Daniel Schmitt

1 Introduction

Humans are profoundly invested in their unique place in the natural world. People want to believe that we represent a peak of evolution. Popular science and literature feed that conceit. Many people would describe us as the smartest, most creative, and most dexterous animal on earth. If you ask people what makes us special, many will answer with a list of attributes that are associated with our hands. We gesture, paint, play music, write, and carve objects. We do all that, and more, with our amazingly nimble hands. This perception of the human hand as unique and more capable than that of any other primates is held not only by laypeople but also by many anatomists and most clinicians who write and comment on hand anatomy and function. In addition to our dexterous hand, many people would go on to say that one of the things that separates humans from other primates is the opposable thumb, at which point they usually touch their index finger and thumb together.

All this, of course, is relativistic. To use the popular adage, "history is written by the victors." In this case, it is written by those who can write. We see our hand as special because we can and are writing the story of our own evolution. But if chimps could compose a chapter about their own hands, they might write: "We climb, termite fish, groom, swing in the trees, crack nuts, and even knuckle-walk with

Both authors contributed equally on writing this chapter.

P. Lemelin (✉)
Division of Anatomy, Department of Surgery, Faculty of Medicine and Dentistry, University of Alberta, 5-05A Medical Sciences Building, Edmonton, AB, Canada, T6G 2H7
e-mail: plemelin@ualberta.ca

D. Schmitt
Department of Evolutionary Anthropology, Duke University, Box 90383, Durham, NC 27708, USA

T.L. Kivell et al. (eds.), *The Evolution of the Primate Hand*, Developments in Primatology: Progress and Prospects, DOI 10.1007/978-1-4939-3646-5_2

those long hands. And even though our thumbs are short, they are still very mobile and useful. In this way, the chimpanzee hand is amazing and special."

This idea that the human hand is a specialized organ dates back to at least Darwin (1859: 434) who classified the human hand as "formed for grasping." Darwin (1871) further emphasized the role of the human hand in facilitating throwing weapons, an idea reinforced by Dart (1959) in his model of early hominin weapon use (see Young 2003 for a review). The idea of the human hand playing an integral role to our evolutionary success in making tools for hunting and protection has held a prominent place in our public understanding of human evolution (e.g., Le Gros Clark 1967). The assumption that the human hand is a remarkable departure from anything that came before is deeply embedded in our thinking and has influenced almost all of our discussions of human evolution. Recent findings of *Australopithecus sediba* and *Homo naledi* focus on the hands (and especially the relative thumb length) in an effort to determine whether these early hominins had undergone two fundamental shifts in human evolution: a shift to committed terrestriality and to toolmaking, both of which are thought to be reflected in the hand (Kivell et al. 2011, 2015; see Chaps. 11, 12, 18, and 19).

But at least twice in the past 100 years, two prominent anatomists—Frederic Wood Jones and John Russell Napier—emphasized the primitiveness of the human hand and its resemblance to those of other pentadactyl mammals (i.e., mammals with five digits or rays). They also defined the specificities of the primate and human hand vis-à-vis other mammals in terms of prehensile and opposable functions. Both were clinicians who shifted their careers to include anthropology and evolution. They were connected by their relationship to Wilfrid Le Gros Clark, a friend to Wood Jones and mentor to Napier (see Le Gros Clark 1955; Day 1988). Both made lasting contributions in their seminal books on the hand (Wood Jones 1920 [revised and reissued in 1942]; Napier 1980 [revised and reissued in 1993]). Being almost 40 years younger than Wood Jones, Napier also benefited from the rapidly growing fields of hand surgery, primate biology, and paleoanthropology in the 1950s and 1960s, fields in which he made significant impacts. Moreover, Napier had more popular appeal with his book *Hands* and his articles in *Scientific American* compared to Wood Jones—a more polarizing figure with his unorthodox views of evolutionary theory. Nonetheless, Wood Jones paved the way with his views on the evolution of the hand, some of which were later adopted by Napier. Here we briefly present some of these contributions organized in three major themes shared by their work: primitiveness, prehensility, and opposability. Those themes are recurrent throughout the chapters that follow in this volume and represent the intellectual foundations on which this book is constructed.

2 Primitiveness of the Primate Hand

At the turn of the last century, Frederic Wood Jones began publishing a series of influential books in anatomy and physical anthropology (Wood Jones 1916, 1920 [revised and reissued in 1942], 1929, 1944). His iconic 1916 book—Arboreal Man—laid out a holistic view of early primate and human locomotor evolution. In this book, and those that followed, Wood Jones covered a wide range of topics in primate and human anatomy, often in an evolutionary context. Wood Jones is well known (and often dismissed) for his "tarsian hypothesis" of the importance of orthograde in human evolution, an idea that derives from his own connections with Sir Arthur Keith and Grafton Elliot Smith. However, this idea is not irrelevant to his overall conception of the hand as primitive. Wood Jones embraced the idea that the human lineage split very early from tarsiers and anthropoids and that orthograde was a primitive feature that laid the foundation for the origin of human bipedalism. At the time, little was known about primate phylogeny and evolutionary history. This idea, in combination with his detailed studies of anatomy, led to an important and foundational point, which resonates with all the chapters of this volume: the primate hand is "primitive." Wood Jones stressed the similarities between primates and other pentadactyl mammals, noting the "minimal departure" of the primate forelimb and hand from the mammalian "archetype" (Wood Jones 1916: 24). Napier (1962b) made a similar argument in *Scientific American* and again in *Hands* (1980), saying that:

> Man's hand shows an extraordinary degree of primitiveness, an astounding conclusion when one thinks of its specialised movements, its acute sensitivity, its precision, subtlety and expressiveness....There is an explanation of this apparent paradox between specialised and primitive. The hand itself is derived from yeoman stock but the factor that places it among the nobles is, as it were, its connections—its connections with the higher centers of the brain. (Napier 1980: 24–25)

This raises the question as to what Wood Jones and Napier precisely meant by "primitive." What counts as primitive is not simple to define. For example, it would be easy to say that our body plan is primitive because we have four limbs like those of early mammals, or even reptiles and amphibians. This is true, but does not reveal much about our history. One could also argue that aspects of our anatomy are "primitive" because they preserve the same neuromotor patterns observed across many different lineages (see Smith 1994). Still, a claim of primitive retention needs a clear and precise definition. On this, Wood Jones (1916: 20) offered some clarity:

> By a primitive hand we mean a very definite thing, and one essential in the make-up of this hand is the possession of five separate, and fairly equally developed digits.

Here Wood Jones contended that having five distinct and equally developed digits is what makes the hands of all primates the same and primitive. It is easy to see that a hand with five digits of subequal length does separate primates from hoofed mammals with reduced numbers of digits, sloths with tightly bound digits in the shape of a hook, whales with a hand in the form of a fin that includes extra phalanges, and bats with digits of hugely different proportions sporting a patagium. These animals have far more specialized hands compared to primates and early fossil mammals (see Ji et al. 2002). However, this definition masks the considerable variation that exists in the lengths and proportions of the digits among primates (e.g., Jouffroy et al. 1993; Lemelin and Jungers 2007; Fig. 2.1; see also Chap. 4). More importantly, it is also an unspecific definition that does not include the behavioral aspects of the hand in primates, including humans, and what appears to be greater prehension, dexterity, and neural control compared to most other mammals, a point later taken up by Napier that is the topic of the next section.

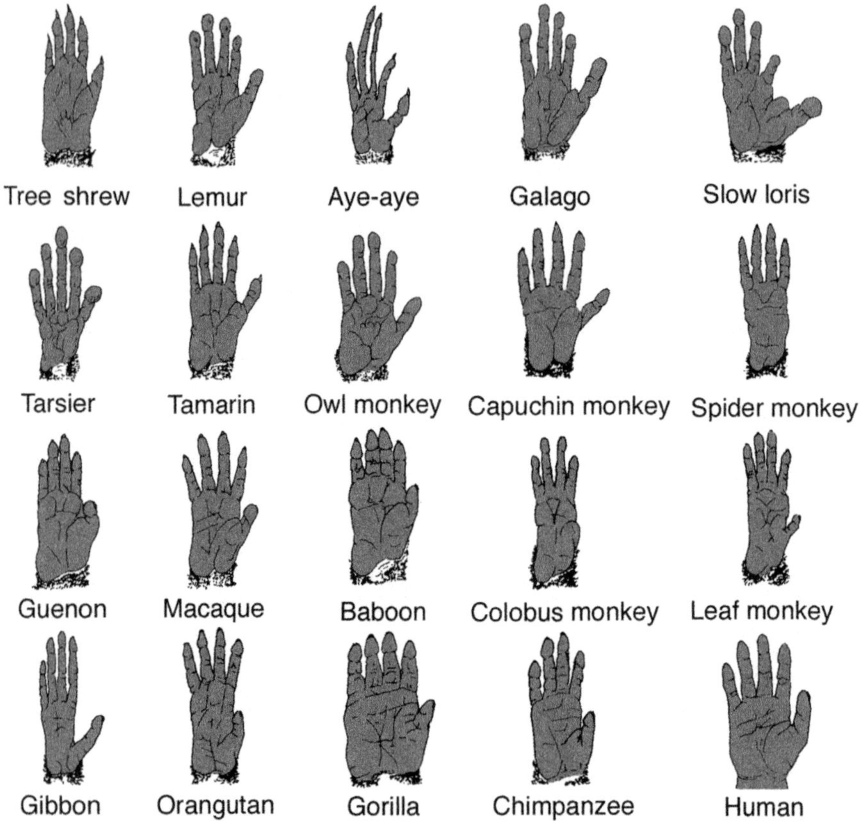

Fig. 2.1 Diversity in shape and proportions of the hand of primates and tree shrew (adapted from Schultz 1969). Note the reduction of the index finger in the slow loris or thumb of the spider monkey and colobus monkey compared to the elongation digits 3 and 4 in the aye-aye (see Chap. 4 for more details). Hands not to scale

3 Prehensility and Opposability of the Primate Hand

Aside from primitiveness of the hand, Wood Jones (1916) was also concerned with its prehensile functions. He saw a "progression" toward dexterous grasping with the adoption of an arboreal lifestyle. In his view, an arboreal environment favored segregation of the functional roles of the forelimb and hand versus those of the hind limb and foot: manipulation versus weight support. This process, which he termed "emancipation of the forelimb," led to the evolution of the hand as "a free organ full of possibilities" (Wood Jones 1916: 17). In other words, a versatile hand allows for grasping of branches in "arboreal progression" and collecting and manipulating of food to become a "hand-feeder" (Wood Jones 1916: 22). Napier borrowed the same concept of "emancipation of the hand" when discussing functional changes observed in human children as they become committed bipeds (see pp. 87–88 in Napier 1980).

Wood Jones was the leading scholar on the comparative and functional anatomy of the hand when Napier began publishing his own work on the topic in the early 1950s. The contributions of John Napier with regard to advancing our understanding of the functional roles of the hand and its evolutionary history cannot be overstated. In our minds, he remains the most influential scholar on the study of hand function, locomotion, and evolution in living and fossil primates. His contributions are numerous, thorough, accessible, and provocative (e.g., Napier 1952, 1955, 1956, 1960, 1961, 1962a, b, 1963, 1967; Napier and Davis 1959; Day and Napier 1961, 1963; Napier and Napier 1967; Napier and Walker 1967). At heart, Napier was a scientist concerned about sorting out proximate and ultimate causes in order to explain the diversity of primate anatomy and behavior. With his wife Prue Napier, he classified the entire order of primates on the basis of locomotor categories and body metrics such as the intermembral index in the classic *Handbook of Living Primates* (Napier and Napier 1967), so that functional associations could be established (e.g., long-legged primates leap, long-armed primates brachiate). Although there are difficulties associated with summing up locomotor variation into discrete categories such as "semibrachiation" (see Mittermeier and Fleagle 1976) or interpreting intermembral index variation because of confounding allometric factors (see Jungers 1984), such functional associations are still very useful to infer behavior in fossil primates (e.g., Jungers 1980, 1982, 2009; Ishida et al. 2004).

Napier adopted a similar classification when studying the prehensile functions of the hand in humans and other primates. His 1956 classification of prehensile movements of the human hand represents such an example. Prehensile movements are defined as "movements in which an object is seized and held partly or wholly within the compass of the hand," and their fundamental requirements are "that the object, whether it is fixed or freely movable, should be held securely" (Napier 1956: 902). In his classic comparative study of primate and mammal hands, Napier (1961: 116–117) provided a more comprehensive definition of a hand capable of prehensility:

> A convergent hand can be termed prehensile when the digits approximate in such a manner that an object may be grasped and held securely *by one hand* against external influences (e.g. gravity) that may be tending to displace it.

According to Napier, stability of the object held by the hand is paramount and can be achieved using two different grips: *power* and *precision* grips (Napier 1956; see Chaps. 11 and 12). Fifty years later, these grip categories are still widely in use among anatomists, anthropologists, clinicians, and other students of the primate hand, although they have been refined and redefined in significant ways (e.g., Marzke 1997; Gumert et al. 2009; Marzke and Pouydebat 2009). In Napier's mind, the anatomical and functional distinctions between power and precision grips are critical, with the position of the thumb being a key difference. The power grip allows an object to "be held in a clamp formed by the partly flexed fingers and the palm, counter pressure being applied by the thumb lying more or less in the plane of the palm" (Napier 1956: 903). In contrast, the precision grip involves an object to be "pinched between the flexor aspects of the fingers and the opposing thumb" (Napier 1956: 903). During a precision grip, the thumb is held in *opposition*, which Napier (1955: 362) defines as a:

> …movement which results in the pulp surface of the thumb becoming diametrically opposed to the pulp surface of one or other of the remaining digits for the purpose of prehension.

Always concerned about defining concepts with anatomical and functional meaning, Napier (1961: 119) offered a definition of opposition that can be found in many anatomy textbooks today:

> …opposition is a compound movement of abduction, flexion and medial rotation occurring at the carpo-metacarpal articulation of the pollex.

Napier (1960: 654) also made a very important point concerning the different perceptions of the term "opposition" among anatomists and zoologists, a reflection of his personal training as a physician and his newly found interest in comparative primate biology:

> The term opposition has no uniform connotation among human anatomists and zoologists and it is not surprising that confusion exists. To a human anatomist opposition has a special meaning and signifies the movement of the thumb as a whole in relation to one of the remaining digits (see footnote p. 650); to a zoologist the term has a more general import; in a static sense it implies that the thumb is set at an angle or even opposite to the remaining digits as in certain birds and reptiles (schizodactyly) or, in a dynamic sense, is capable of being moved to such a position by muscular action, as in most Primates…To many anatomists, opposition is a "hallmark of mankind", to many zoologists, it is simply a function of the Primate hand.

Alongside his efforts to clarify the concept of opposition, Napier was also preoccupied by the identification of anatomical correlates of opposability in primates. He identified several of those traits, including sellar-shaped joint surfaces between the trapezium and pollical metacarpal (Napier 1955, 1961) and the presence of specific hand muscles (Day and Napier 1963). He categorized the hand of primates with those attributes—Old World monkeys and humans in particular, with the exclusion of other hominoids—as "truly opposable" (Day and Napier 1963: 132). In other primates for which opposability can be achieved by means other than "carpometacarpal opposition, the term *pseudo-opposability* is suggested" (Napier 1961: 120). Like semibrachiation, the term pseudo-opposable underestimates variation in hand anatomy and behavior among primates. For example, capuchin monkeys—categorized by Napier (1961) as "pseudo-opposable"—turn out to have sellar-shaped trapeziometacarpal joints (Rose 1992; see Chap. 3) and regularly use precision grips involving the thumb and index finger (Costello and Fragaszy 1988). Moreover, sellar-shaped trapeziometacarpal joints may be primitive for mammals (Lewis 1977), and hand muscles thought to define "true opposability" (Day and Napier 1963) are found in virtually all primates (Diogo et al. 2012; see Chap. 7). Again, these examples underscore the challenges Napier faced when trying to summarize primate hand diversity into discrete categories and sorting out the anatomical correlates of various prehensile behaviors of the hand. The same challenges are still present today as researchers do not always agree on a common language when, for example, describing "thumb opposition" in living primates (see Chap. 12) or interpreting the evolution of "opposable" functions and capabilities of the thumb in early fossil hominins (see Chap. 18).

4 Final Remarks on Contributions by Frederic Wood Jones and John Napier

Both Wood Jones and Napier had revolutionary and disruptive ideas. They turned the idea of uniqueness on its head and forced anthropologists to rethink the concept of the human hand as being "special." Wood Jones reminded anthropologists of the primitive nature of the primate hand. Napier picked up on this theme and defined the concepts of hand prehension and opposability that hold sway today. Decades of research that followed have only reinforced the notion that primates have relatively "simple" hands. They have five rays with roughly equal lengths (although there is substantial variation across primates in terms of hand proportions). Primates have fewer hand muscles compared to many nonprimate mammals and even to lizards (see Chap. 7). The human hand seems unexceptional in having roughly equal length digits and fewer short muscles in the palm compared to most other primates (although the thumb is longer and more muscles attach on it; see Chaps. 4 and 7). The final conclusion then is that humans and other primates are not far separated from other pentadactyl mammals in terms of hand anatomy. Prehension, precision, and dexterity, however, are functions that appear to set primates, and especially humans, apart (see Chaps. 6, 11, and 12).

At the end of the day, readers have to consider for themselves how "special" the human hand is. Some will conclude that the human hand is fundamentally primitive and all changes from this primitive condition are relatively small. Others will focus on the opposability of the human thumb and see it as a releaser for technology and art and come to the conclusion that we have an exceptionally derived hand. The thought-provoking ideas of Wood Jones and Napier represent the foundations of the research on the evolution of our own hands with which we write this final sentence.

Acknowledgments For over 25 years, Mike Rose—a student of John Napier—has been a mentor and an inspiration to us both through his writings and personal discussions. His insights on primate limb anatomy and locomotor behavior have been critical throughout our careers and in writing this chapter. We are also very grateful to Tracy Kivell and Brian Richmond for comments that improved this chapter immeasurably.

References

Costello MB, Fragaszy DM (1988) Prehension in *Cebus* and *Saimiri*: I. Grip type and hand preference. Am J Primatol 15:235–245

Dart R (1959) Adventures with the missing link. Harper and Brothers, New York

Darwin C (1859) On the origins of species by means of natural selection, or the preservation favoured races in the struggle for life. John Murray, London

Darwin C (1871) The descent of man, and selection in relation to sex. John Murray, London

Day MH (1988) In memoriam Professor John Russell Napier, M.R.C.S., L.R.C.P., D.Sc. J Anat 159:227–229

Day MH, Napier JR (1961) The two heads of flexor pollicis brevis. J Anat 95:123–130

Day MH, Napier J (1963) The functional significance of the deep head of flexor pollicis brevis in primates. Folia Primatol 1:122–134
Diogo R, Richmond BG, Wood B (2012) Evolution and homologies of primate and modern human hand and forearm muscles, with notes on thumb movements and tool use. J Hum Evol 63:64–78
Gumert MD, Kluck M, Malaivijitnond S (2009) The physical characteristics and usage patterns of stone axe and pounding hammers used by long-tailed macaques in the Andaman Sea region of Thailand. Am J Primatol 71:594–608
Ishida H, Kunimatsu Y, Takano T, Nakano Y, Nakatsukasa M (2004) *Nacholapithecus* skeleton from the Middle Miocene of Kenya. J Hum Evol 46:69–103
Ji Q, Luo Z-X, Yuan C-X, Wible JR, Zhang J-P, Georgi JA (2002) The earliest known eutherian mammal. Nature 416:816–822
Jouffroy FK, Godinot M, Nakano Y (1993) Biometrical characteristics of primate hands. In: Preuschoft H, Chivers DJ (eds) Hands of primates. Springer-Verlag, Vienna, pp 133–171
Jungers WL (1980) Adaptive diversity in subfossil Malagasy prosimians. Z Morphol Anthropol 71:177–186
Jungers WL (1982) Lucy's limbs: allometry and locomotion in *Australopithecus afarensis*. Nature 297:676–678
Jungers WL (1984) Body size and scaling of limb proportions in primates. In: Jungers WL (ed) Size and scaling in primate biology. Plenum Press, New York, pp 345–381
Jungers WL (2009) Interlimb proportions in humans and fossil hominins: variability and scaling. In: Grine FE, Fleagle JG, Leakey RE (eds) The first humans: origin and early evolution of the genus *Homo*. Springer, New York, pp 93–207
Kivell TL, Kibii JM, Churchill SE, Schmid P, Berger LR (2011) *Australopithecus sediba* hand demonstrates mosaic evolution of locomotor and manipulative abilities. Science 333:1411–1417
Kivell TL, Dean AS, Tocheri MW, Orr CM, Schmid P, Hawks J, Berger LR, Churchill SE (2015) The hand of *Homo naledi*. Nat Commun 6:8431
Le Gros Clark WE (1955) Frederic Wood Jones. 1879–1954. Biogr Mems Fell R Soc 1:118–134
Le Gros Clark WE (1967) Man-apes or ape-men? The story of discoveries in Africa. Holt, Rinehart and Winston, New York
Lemelin P, Jungers WL (2007) Body size and scaling of the hands and feet of prosimian primates. Am J Phys Anthropol 133:828–840
Lewis OJ (1977) Joint remodelling and the evolution of the human hand. J Anat 123:157–201
Marzke MW (1997) Precision grips, hand morphology, and tools. Am J Phys Anthropol 102:91–110
Marzke MW, Pouydebat E (2009) Comments on E. Pouydebat, P. Gorce, Y. Coppens, V. Bels, 2009. Biomechanical study of grasping according to the volume of object: human versus non-human primates. J Biomech 42:2628–2629
Mittermeier RA, Fleagle JG (1976) The locomotor and postural repertoires of *Ateles geoffroyi* and *Colobus guereza*, and a reevaluation of the locomotor category semibrachiation. Am J Phys Anthropol 45:235–256
Napier JR (1952) The attachments and function of the abductor pollicis brevis. J Anat 86:335–341
Napier JR (1955) The form and function of the carpo-metacarpal joint of the thumb. J Anat 89:362–369
Napier JR (1956) The prehensile movements of the human hand. J Bone Joint Surg [Br] 38B:902–913
Napier JR (1960) Studies of the hands of living primates. Proc Zool Soc Lond 134:647–657
Napier JR (1961) Prehensility and opposability in the hands of primates. Symp Zool Soc Lond 5:115–132
Napier J (1962a) Fossil hand bones from Olduvai Gorge. Nature 196:409–411
Napier J (1962b) The evolution of the hand. Sci Am 207:56–62
Napier J (1963) The locomotor functions of hominids. In: Washburn SL (ed) Classification and human evolution. Aldine, Chicago, pp 178–189
Napier JR (1967) Evolutionary aspects of primate locomotion. Am J Phys Anthropol 27:333–341
Napier J (1980) Hands. Pantheon Books, New York
Napier J (1993) Hands. Revised edition by Russell H. Tuttle. Princeton University Press, Princeton
Napier JR, Davis PR (1959) The fore-limb skeleton and associated remains of *Proconsul africanus*. Fossil Mammals of Africa, No. 16. British Museum (Natural History), London

Napier JR, Napier PH (1967) A handbook of living primates. Academic Press, New York
Napier JR, Walker AC (1967) Vertical clinging and leaping—a newly recognized category of locomotor behaviour of primates. Folia Primatol 6:204–219
Rose MD (1992) Kinematics of the trapezium-1st metacarpal joint in extant anthropoids and Miocene hominoids. J Hum Evol 22:255–266
Schultz AH (1969) The life of primates. Weidenfeld and Nicolson, London
Smith KK (1994) Are neuromotor systems conserved in evolution? Brain Behav Evol 43:293–305
Wood Jones F (1916) Arboreal man. Edward Arnold, London
Wood Jones F (1920) The principles of anatomy as seen in the hand. J. and A. Churchill, London
Wood Jones F (1929) Man's place among the mammals. Edward Arnold, London
Wood Jones F (1942) The principles of anatomy as seen in the hand, 2nd edn. Ballière, Tindall, and Cox, London
Wood Jones F (1944) Structure and function as seen in the foot. Ballière, Tindall, and Cox, London
Young RW (2003) Evolution of the human hand: the role of throwing and clubbing. J Anat 202:165–174

Part I
Anatomical and Developmental Evidence

Chapter 3
The Primate Wrist

Tracy L. Kivell

1 Introduction

"Carpus" is derived from the Greek word *karphoo*, meaning "to shrink together". This is an appropriate name as the carpus, or wrist, is arguably one of the most complex joint systems in the mammalian body, incorporating some 15–17 bones interconnected by at least 20 articulations and bound together by numerous ligaments and tendons. Wood Jones (1942) considered learning the identity and laterality of the human carpal bones to be minutiae not worth the time of modern-day medical students. However, the carpal bones together function to transfer loads between the hand and forearm (radius and ulna) and permit the mobility of the hand in multiple planes. The study of variation in carpal morphology across primates since Owen (1866), Mivart (1867, 1869) and Leboucq's (1884) first comparative descriptions not only has provided unique insight into primate wrist evolution, hand use and hand mobility but also has played an important role in hypotheses regarding primate origins (e.g., Godinot and Beard 1991; Boyer et al. 2013), hominoid origins (e.g., Cartmill and Milton 1977; Beard et al. 1986) and particularly human evolutionary history (e.g., Marzke 1971; Begun 1992; Richmond et al. 2001; Tocheri et al. 2008; Kivell and Schmitt 2009). A history of detailed morphological descriptions by a select few (e.g., Lewis 1989 and references therein; Ziemer 1978; Sarmiento 1988; Hamrick 1996a, b, 1997; Richmond et al. 2001; Daver et al. 2012) and recent advancements in 3D (Tocheri 2007; Tocheri et al. 2003, 2005; Orr et al. 2013) and in vivo/in vitro imaging (e.g., Neu et al. 2001; Crisco et al. 2005; Moritomo et al. 2006; Pillai et al. 2007; Orr et al. 2010; see Chap. 9) have provided insight into the

T.L. Kivell (✉)
Animal Postcranial Evolution (APE) Lab, Skeletal Biology Research Center, School of Anthropology and Conservation, University of Kent, Canterbury, Kent CT2 7NR, UK

Department of Human Evolution, Max Planck Institute for Evolutionary Anthropology, Deutscher Platz 6, Leipzig 04103, Germany
e-mail: t.l.kivell@kent.ac.uk

T.L. Kivell et al. (eds.), *The Evolution of the Primate Hand*, Developments in Primatology: Progress and Prospects, DOI 10.1007/978-1-4939-3646-5_3

complexities of carpal movement and a better understanding of the implications of what subtle variation in carpal morphology may mean with regard to overall wrist function. Thus, the tiny, irregular-shaped bones of the wrist often considered a tedious nightmare by biological anthropology or medical students hold important insight into our own evolution.

This chapter will review the functional morphology of the carpus across major primate clades (strepsirrhines, New and Old World monkeys and hominoids, including humans), with reference to morphology in other, closely related mammals. Much of this review is based on the tome of work by Lewis (1965; 1969; 1970; 1971a, b; 1972a, b; 1974; 1977; 1985a, b; Lewis et al. 1970), which is summarized in Lewis (1989). Although many researchers have disagreed with Lewis's functional and evolutionary interpretations (e.g., Jenkins and Fleagle 1975; Cartmill and Milton 1977; Sarmiento 1988; Hamrick 1997; Orr et al. 2010), his detailed comparative morphological descriptions of the primate wrist (and hand) have provided an invaluable foundation for much of the work that has been done since on the extant and fossil primate wrist.

The bones of the primate carpus can be organized into four main joint complexes: (1) **antebrachiocarpal** (between the forearm and carpus), (2) **radial carpometacarpal** (between scaphoid/os centrale, trapezium, trapezoid and first and second metacarpals), (3) **midcarpal** (between the proximal and distal carpal rows) and (4) **ulnar carpometacarpal joints** (between the trapezoid, capitate, hamate and second to fifth metacarpals). This chapter is organized by joint complex, with variation in carpal morphology across primates depicted graphically rather than described. Given the complexity of carpal shapes, the function of the multiple intercarpal joints, and the morphological variation across primates, this chapter is by no means exhaustive. Furthermore, this chapter focuses on the bony morphology only and generally ignores soft tissues, as the network of interosseous ligaments that is critical for stabilization of the carpus is too complex to discuss in detail here. Readers interested in more detailed functional morphology (both bony and soft tissue) are referred to Lewis (1989) for a comprehensive review of the primate carpus across all clades with comparisons to other mammals; Hamrick (1996a, b, 1997) for strepsirrhines; O'Connor (1975), Ziemer (1978), Youlatos (1996) and Daver et al. (2012) for Old and New World monkeys; and Corruccini (1978), Sarmiento (1988), Richmond et al. (2001), Begun (2004), Richmond (2006), Tocheri (2007) and Orr et al (2010) for hominoids and references therein.

2 The Primitive Primate Carpus

In most primates, the carpus is composed of nine bones, which have been given various names since they were first named by Lyser in 1653 (the most common alternative names are listed below; see also Playfair McMurrich 1914). The carpals can be divided into three functional columns (most often used in reference to humans only; Taleisnik 1985; Fisk 1981; Feipel et al. 1994) or in two radioulnar

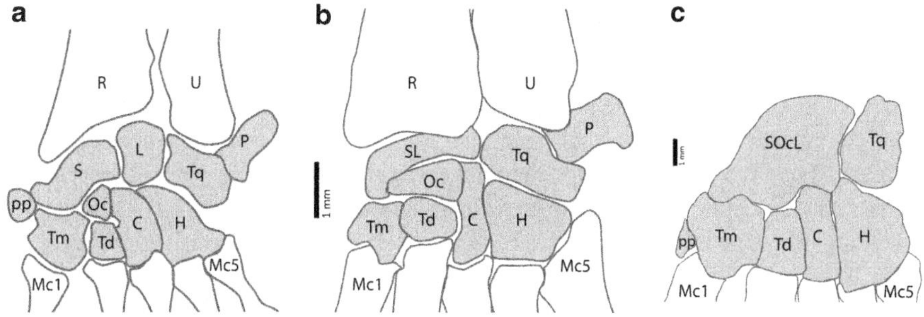

Fig. 3.1 The non-primate mammalian carpus. (**a**) A hypothetical generalized ancestral mammal, redrawn from Lewis (1989); (**b**) a tree shrew (*Tupaia tana*); and (**c**) a colugo (*Cynocephalus volans*), both adapted from Stafford and Thorington (1998). The primate carpus is most similar to the hypothesized ancestral mammalian condition. Mammals closely related to primates show more carpal fusions (i.e., a more derived carpus) than most primates; tree shrews have a fused scaphoid-lunate (SL) and colugos have a fused scaphoid-os centrale-lunate (SOcL). Note that the prepollex is missing in (**b**) and the pisiform is missing in (**c**). Abbreviations: *R* radius, *U* ulna, *S* scaphoid, *Oc* os centrale, *L* lunate, *Tq* triquetrum, *P* pisiform, *pp* prepollex, *Tm* trapezium, *Td* trapezoid, *C* capitate, *H* hamate, *Mc1* first metacarpal, *Mc5* fifth metacarpal

rows; the latter is more common in comparative primate and mammalian studies (e.g., Lewis 1989; Stafford and Thorington 1998) and is used here. The proximal row is comprised of (from radial to ulnar) the **scaphoid** (or radiale), **os centrale**, **lunate** (semilunar or intermedium), **triquetrum** (cuneiforme or ulnare) and **pisiform**. The distal row is made up of the **trapezium** (greater multangular), **trapezoid** (lesser multangular), **capitate** (os magnum) and **hamate** (unciforme) (Fig. 3.1). In humans, African apes and some strepsirrhines, the os centrale is fused to the scaphoid, and thus the carpus is composed of only eight bones in these taxa (see below and Kivell and Begun 2007). The retention of eight or nine carpal bones in primates represents a primitive pattern compared with many other mammals. A reduction in the number of carpal bones—either via fusion or loss of the bone—is common in marsupials, cetaceans, carnivores, rodents, bats, tree shrews and dermopterans (colugos or "flying lemurs") (Flower 1885; Yalden 1970, 1971; Stafford and Thorington 1998). For example, among the taxa that are most closely related to Primates, Tupaiidae (tree shrews) and Rodentia (e.g., squirrels, mice) have a fused scaphoid and lunate (i.e., scapholunate), and Dermoptera show further fusion with the os centrale (i.e., scaphocentralolunate) (Stafford and Thorington 1998; Fig. 3.1) [For a discussion of the homology of different carpal elements throughout tetrapod evolution, see Čihák (1972) and Lewis (1989).] Given the diversity of locomotor, postural and manipulative behaviors, typical of the primate clade, the retention of more separate elements within the carpus may allow for more versatility in wrist function, which is particularly useful for navigating arboreal environments. For example, increased arboreality has been suggested as the functional explanation for why pen-tailed tree shrews (*Ptilocercus*) retain nine carpals compared with other tree shrews, which have seven (Stafford and Thorington 1998).

3 Primate Carpal Ossification

Chapter 5 focuses on how the bones of the wrist and hand develop up to the point of ossification. The degree of carpal (and hand bone) ossification is commonly used to estimate skeletal maturity and age in humans (Greulich and Pyle 1959; Tanner et al. 1983), while variation in skeletal growth in general has been used as a proxy for assessing differences in life history across primates (e.g., Cheverud 1981; Glassman 1983; Winkler 1996; Zihlman et al. 2007). Within primates, however, there is strong variation in both the timing and sequence of carpal ossification (Table 3.1). The capitate, hamate and triquetrum are typically among the first carpal bones to ossify across primates, while the pisiform is usually among the last. In humans, the capitate and hamate begin ossifying between 2 and 5 months postnatally (Scheuer and Black 2000). In contrast, the capitate and hamate begin ossifying prenatally in other apes (*Pan*, *Pongo*, *Hylobates*) (Schultz 1944; Nissen and Riesen 1949; Winkler 1996; Marzke et al. 1987), and in Old and New World monkeys, most carpal ossification centers are present at birth (Phillips 1976; Sirianni and Swindler 1985; Galliari 1988). In humans, the carpus is fully ossified by 12.5 years in females and 15 years in males, while most carpals in great apes are fully ossified by approximately 10–12 years of age (when the third molar is freshly erupted, but not in occlusion) (Kivell 2007). Winkler (1996) found a positive relationship between the individual body mass and number of carpals present at birth in *Pongo*, which may help to explain some of the variation in carpal ossification. However, there is a great deal of variation in timing and sequence of carpal ossification, both intra- and interspecifically (Newell-Morris et al. 1980; Winkler, 1996; Kivell 2007).

4 General Carpal Function

Compared with most other mammals, primates have a diverse repertoire of positional behaviors, and, particularly in arboreal environments, the wrist and hand must deal with a variety of irregular and discontinuous supports. Primates are capable of using a wide range of hand postures to accommodate variation in substrate size and orientation, which require compromises in carpal joint mobility and stability and diverse mechanical demands on carpal morphology (e.g., Yalden 1972; Jenkins and Fleagle 1975; Fleagle and Meldrum 1988; Lewis 1989; Hamrick 1996a; Daver et al. 2012; see also Chaps. 12 and 13). For these reasons, primates retain the versatility of a primitive mammalian carpal *Bauplan*, but also show variations in carpal morphology that reflect differences in the functional demands placed on the wrist and hand during locomotion and manipulation.

Most primates are pronograde quadrupeds; thus, the wrist assumes an extended and pronated (i.e., palmigrade or digitigrade) posture during the support phase of quadrupedal walking or running (e.g., Jenkins and Fleagle 1975; O'Connor 1975; Whitehead 1993; Schmitt 1994; Hamrick 1996a; Lemelin and Schmitt 1998; Patel 2010; Patel and Wunderlich 2010). During terrestrial quadrupedalism, the wrist and

Table 3.1 Most common carpal ossification sequence in different primates and other mammals

Taxon	Carpal sequence	Reference
Homo	**CAP**, **HAM**, **TRIQ**, LUN, TRPZM, TRPZD, SCAPH, PISI	Scheuer and Black (2000)
Pan	**CAP**, **HAM**, TRIQ, TRPZM, LUN, SCAPH, PISI, TRPZD	Nissen and Riesen (1949), Marzke et al. (1987), Winkler (1996)
Gorilla	(**CAP**, **HAM**) (TRPZM, **TRIQ**) SCAPH (TRPZD, LUN, PISI)	Noback (1930)
Pongo	**CAP**, **HAM**, SCAPH, TRPZM, LUN, (TRPZD, CENT) TRIQ, PISI	Winkler (1996)
Hylobates	(**CAP**, **HAM**) (SCAPH, TRPZM) (**TRIQ**, LUN, TRPZD, PISI, CENT)	Schultz (1944)
Macaca nemestrina	**CAP**, **HAM**, **TRIQ**, PISI, SCAPH, TRPZM, TRPZD, LUN, CENT	Newell-Morris et al. (1980), Sirianni and Swindler (1985)
Macaca mulatta	**TRIQ**, **HAM**, SCAPH, **CAP**, PISI (LUN, TRPZM) TRPZD, CENT	van Wagenen and Asling (1964), Michejda and Bacher (1980)
Saimiri boliviensis	**CAP**, **TRIQ**, **HAM**, TRPZM, LUN, CENT, TRPZD, PISI, SCAPH	Galliari (1988)
Cebus albifrons	**TRIQ**, **HAM**, CENT, **CAP**, TRPZD, SCAPH, PISI, LUN, TRPZD	Thurm et al. (1975)
Cebus apella	(**TRIQ**, SCAPH) **HAM** (**CAP**, LUN, TRPZM, TRPZD, PISI, CENT)	Watts (1990)
Callithrix jacchus	(CENT, TRPZM, **CAP**, **HAM**) (SCAPH, LUN, **TRIQ**, TRPZD) PISI	Phillips (1976)
Tarsius	(**CAP**, **HAM**) (PISI, TRPZM, **TRIQ**) (SCAPH, LUN, TRPZD, CENT)	Kindahl (1944)
Rattus	(SCAPH, LUN) (**TRIQ**, CENT) **HAM**, CENT, **CAP**, TRPZM, TRPZD	Strong (1925)
Canis	PISI, LUN, **HAM**, TRPZM, **CAP**, TRPZD, SCAPH, CENT, **TRIQ**	Curgy (1965)
Felis	PISI, CENT, **HAM**, LUN, **CAP**, TRPZM, SCAPH, TRPZD, **TRIQ**	Curgy (1965)

The capitate (CAP), hamate (HAM) and triquetrum (TRIQ) are typically the first carpal bones to ossify in primates and are shown in bold to help visualize this compared with other mammals. Carpal bones in parentheses reflect ossification at the same time or in an unknown sequence. *LUN* lunate, *TRPZM* trapezium, *TRPZD* trapezoid, *SCAPH* scaph, *CENT* os centrale, *PISI* pisiform hand tend to be more in line with the forearm (i.e., more neutral posture) (Lemelin and Schmitt 1998), while during arboreal quadrupedalism, the hand is more ulnarly deviated. These hand postures generally hold true of quadrupedal primates with mesaxonic (i.e., third digit is the longest) and ectaxonic (i.e., fourth digit is the longest) hands, though there are several exceptions (e.g., callitrichids and spider monkeys, see Lemelin and Schmitt 1998). Thus, quadrupedal primates share several morphological features related to stabilizing the wrist during compression in an extended, pronated posture. The articulation between the radius and carpus is relatively flat to resist unidirectional weight-bearing loads (Jenkins and Fleagle 1975; Sarmiento 1988; Hamrick 1996a). Furthermore, the antebrachiocarpal and midcarpal joints are close packed—meaning the joint surfaces are in maximum congruency—in an extended, pronated and ulnarly deviated posture. This means that the antebrachiocarpal and midcarpal wrist joints are in their most stable position during the support phase of the typical quadrupedal hand posture. In this close-packed position, radioulnar deviation and rotation are not possible and the wrist can only flex (O'Connor 1975).

Primates that regularly engage in vertical clinging or suspensory behaviors use a variety of hand postures that place different functional demands on the wrist than the typical quadrupedal primate. Vertical clinging strepsirrhines (e.g., *Propithecus*, *Avahi*, *Lepilemur*) use a flexed and partly supinated posture during vertical clinging and suspension (Hamrick 1996a, b). Their wrists, in turn, show morphological features, such as a deeply curved radiocarpal joint or dorsally constricted embrasure between the capitate and trapezoid (see below), that allow more mobility than quadrupeds, but also stabilize the wrist during flexion at the antebrachiocarpal joint and supination at the midcarpal joint (Hamrick 1996a, b).

During suspension or brachiation (i.e., *Ateles*, *Lagothrix* and hylobatids), the grasping fingers are fixed to the substrate while the body rotates below (e.g., approximately 90° during brachiation in spider monkeys) through the swing (Jenkins 1981). The wrist accommodates most of this rotation (i.e., supination; ~70° vs. ~20° by supination of the radius) by having a midcarpal configuration that acts as a highly mobile ball-and-socket joint. The wrist is capable of high degree of supination, but has limited mobility in the opposite direction (hyperpronation, beyond a neutral pronated posture) (Jenkins 1981). This motion and mobility is essentially the opposite of what we see in quadrupedal, palmigrade taxa (e.g., *Macaca*), which are capable of very limited supination, but a high degree of hyperpronation (Jenkins 1981).

Variation in carpal morphology is largely responsible for differences in the range of motion at the wrist and hand (Fig. 3.2). For example, in most primates (Old and New World monkeys and most strepsirrhines), the radius and ulna both articulate with the carpus, making the antebrachiocarpal joint relatively stable, and therefore ulnar deviation occurs mainly at the midcarpal joint (Jouffroy and Medina 2002; Daver et al. 2012; see below). In contrast, in hominoids (including humans), for which contact of the ulna with the carpus has been lost, ulnar deviation occurs primarily at the antebrachiocarpal joint (Jouffroy and Medina 2002; Crisco et al. 2005). Furthermore, the degree of curvature of the facets (e.g., the proximal facets of the capitate and hamate) and relative size of the articular areas (e.g., dorsally extended proximal facets of the capitate and hamate in some primate taxa) is a good indication of the range of movement at a particular joint (Sarmiento 1988; Hamrick 1996b; Richmond 2006). However, it is important to note that the interosseous ligamentous

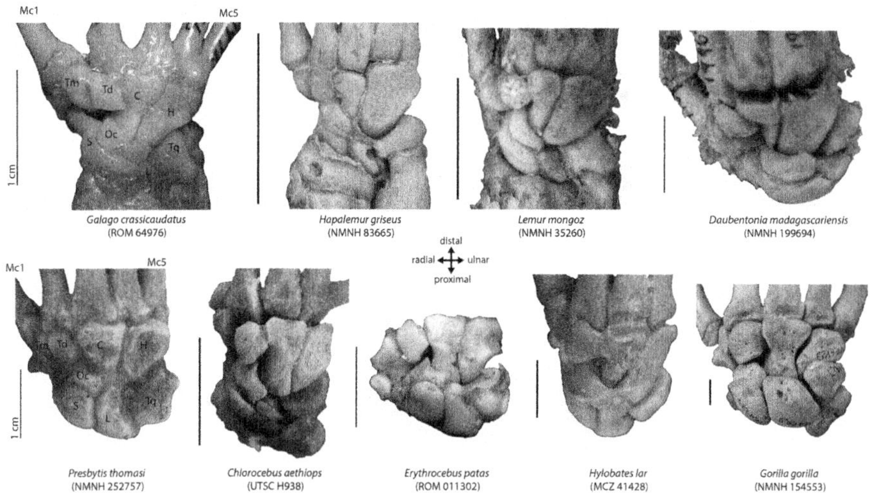

Fig. 3.2 Dorsal view of articulated wrists in a sample of primates. Note the relative variation in size between the capitate and hamate. In strepsirrhines and hylobatids, the hamate is much larger than the capitate. In strepsirrhines, the os centrale often articulates with the hamate, cutting off the articulation between the capitate and lunate. In contrast, the capitate of most catarrhines is equal to or larger in size than the hamate, and the hamate does not articulate with the os centrale or scaphoid. Also, note the large size of the triquetrum in most strepsirrhines and Old World monkeys, compared with hominoids. The pisiform is missing for most specimens. All wrists are shown from the right side and scaled to roughly the same size (scale represents 1 cm for each taxon) See Fig. 3.1 for abbreviations

network also plays a critical role in carpal mobility/stability (Martin et al. 1998). As such, in vivo mobility can be more limited than might be predicted from bony morphology alone (Richmond 2006; see below).

Given the complexity of the wrist, understanding the movement or kinematics of specific joints or carpal bones is particularly challenging (see Chap. 9). Kinematic studies of humans are most common. They show that the distal carpal row functions essentially as a single unit during wrist motion, but the carpals of the proximal row have more functional independence from one another because they are more loosely tethered by ligaments (Garcia-Elias et al. 1994; Wolfe et al. 2000; Moojen et al. 2003). However, for decades, much of what we understood about the nonhuman primate carpal movement stemmed largely from two in vivo cineradiographic studies: one of a juvenile chimpanzee knuckle-walking (Jenkins and Fleagle 1975) and another of spider monkeys brachiating (Jenkins 1981). Most movements of the wrist and hand require simultaneously combining the flexion or extension with radial deviation or ulnar deviation, making the kinematics of particular bones difficult to visualize and understand. Furthermore, since most of the extrinsic forearm muscles bridge, rather than insert onto, the wrist, the carpals move largely via indirect forces from the activation of muscles inserting on other bones of the hand and forces on the metacarpals and phalanges (Jouffroy and Medina 2002). Finally, the ligaments play an important but rather poorly understood functional role within the wrist. For example, when the wrist is not in an close-packed extended and pronated posture, such as during suspension, vertical clinging or climbing, the articular surfaces may not be in close contact, and thus the ligaments help resist tensile stress and allow individual carpal elements to move while maintaining overall integrity of the wrist (Lovejoy et al. 2001; Jouffroy and Medina 2002). New in vivo cineradiography imaging techniques (e.g., Crisco et al. 2005, see below) or 3D computed tomography (CT) of cadavers that provide 3D models or movies of the movement, such as those provided by Orr et al. (2010), represent some of the best ways to better understand the complexity of movement in the wrist (see Chap. 9).

5 The Antebrachiocarpal Joint

The **antebrachiocarpal joint** refers to the articulation between the forearm and wrist (Fig. 3.3). In most non-hominoid primates, this joint is composed of a radial and ulnar portion, both of which are weight bearing. In all primates, the radial portion is formed by the articulation between the radioulnarly and dorso-palmarly concave distal radius and the correspondingly convex articular surfaces of the scaphoid and lunate (Figs. 3.4 and 3.5). However, modifications to the ulnar portion of the antebrachiocarpal joint, particularly in hominoids and

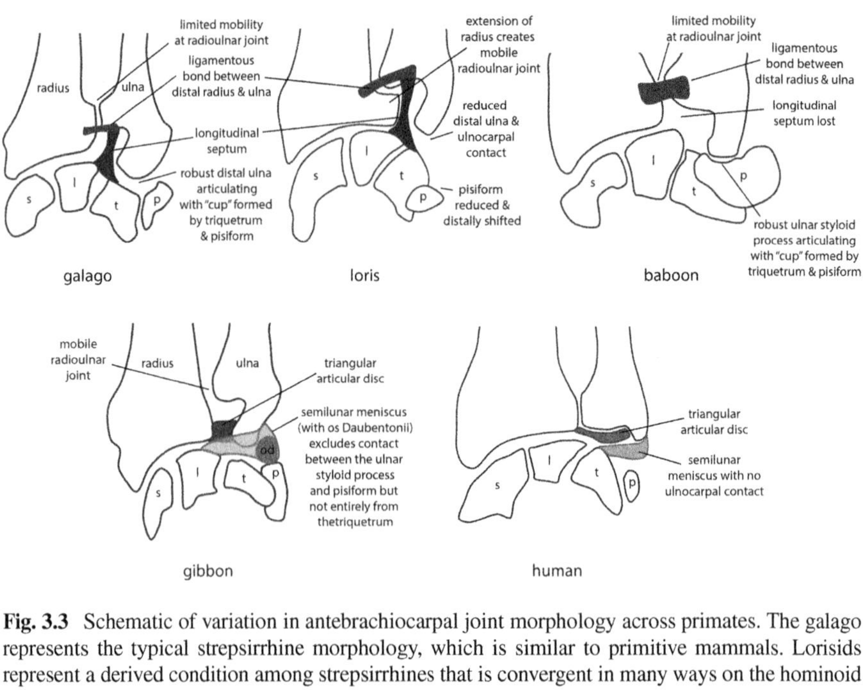

Fig. 3.3 Schematic of variation in antebrachiocarpal joint morphology across primates. The galago represents the typical strepsirrhine morphology, which is similar to primitive mammals. Lorisids represent a derived condition among strepsirrhines that is convergent in many ways on the hominoid morphology. The baboon is representative of the typical Old World monkey morphology (many New World monkeys still retain a longitudinal septum). The gibbon demonstrates the derived hominoid morphology, including the triangular articular disc and semilunar meniscus that partially (gibbons and chimpanzees) or fully blocks (orangutan, gorilla, human) contact between the ulna and carpus. The os Daubentonii (od) is only consistently found in hylobatids. In the humans, the ulnar styloid process and semilunar meniscus are further reduced than that of other hominoids. See also Lewis et al. (1970)

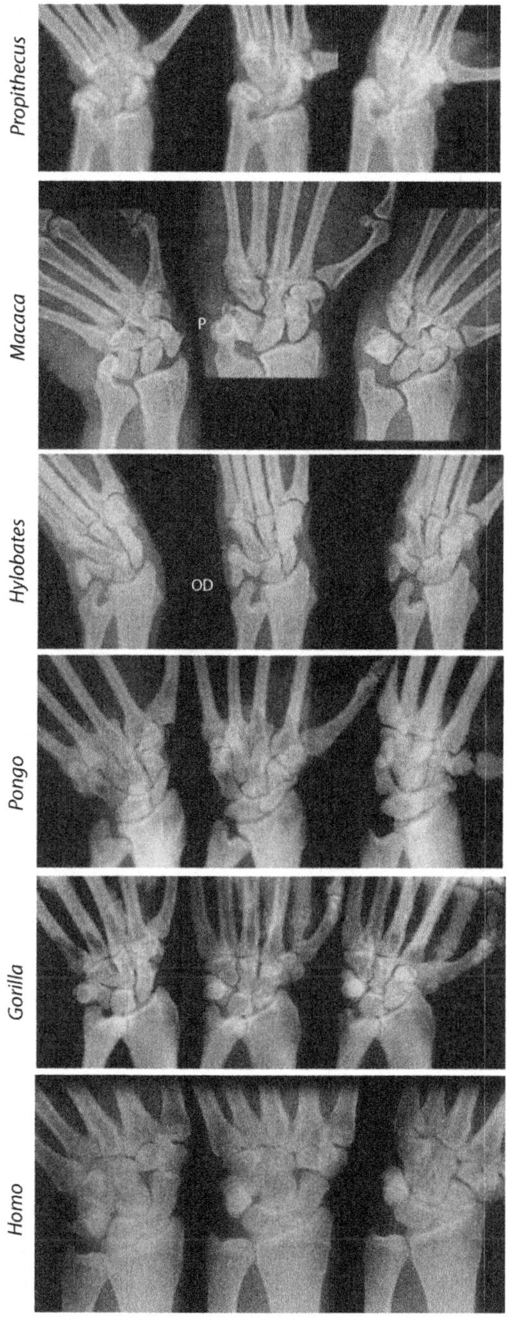

Fig. 3.4 Carpal movement at the antebrachiocarpal and midcarpal joints. Radiographs showing the wrist in ulnar deviation (*left*), neutral posture (*middle*) and radial deviation (*right*), adapted from Jouffroy and Medina (2002). Movement during radioulnar deviation stems primarily from the midcarpal joint in *Propithecus*, *Macaca* and *Hylobates*, because there is full or at least partial (in the case of *Hylobates*) contact between the ulnar styloid process and the carpus. In contrast, in *Pongo*, *Gorilla* and humans, there is greater movement at the antebrachiocarpal due to loss of the ulnocarpal articulation. However, note that there is greater ulnar-radial deviation in this particular *Macaca* specimen [species not provided by Jouffroy and Medina (2002)] compared with humans, despite differences in the antebrachiocarpal articulation. *P* pisiform, *OD* os Daubentonii. For additional informative radiographic images of hominoid and macaque wrists, see Jenkins and Fleagle (1975)

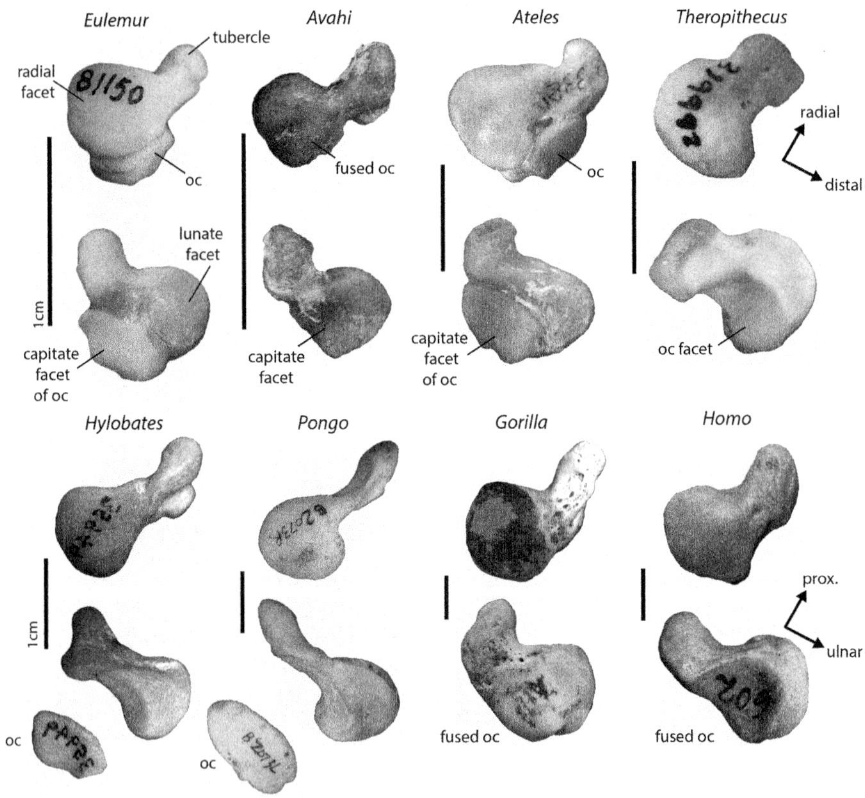

Fig. 3.5 Variation in primate scaphoid and os centrale morphology. *Top* row for each taxon shows the roughly proximoradial view of the scaphoid, featuring the radial facet. *Bottom* row for each taxon shows the distomedial view of the scaphoid [and os centrale (oc)], featuring the lunate and os centrale/capitate articular areas. In *Eulemur* and *Ateles*, the os centrale is independent, but still articulated with the scaphoid via its strong ligamentous attachment. In *Avahi*, *Gorilla* and *Homo*, the os centrale is fused early in ontogeny to become part of the scaphoid. All specimens are shown as the right side and scaled to approximately the same size (scale represents 1 cm for each taxon)

lorisids, are arguably the most significant evolutionary changes in the primate wrist compared with other mammals (Mivart 1867; Lewis 1969, 1985a, 1989; Cartmill and Milton 1977).

Most strepsirrhines (e.g., *Lemur*, *Galago*, but see lorisids below) have an ulnocarpal (and antebrachiocarpal overall) joint that is similar to the typical mammalian pattern (Cartmill and Milton 1977; Lewis 1989). The pisiform is elongated and projects proximally to act as a supporting "heel" of the hand (Cartmill and Milton 1977; Lewis 1989). The pisiform and triquetrum combine to form a cup-shaped facet that articulates with a spindle-shaped extension of the distal ulna (homologous to the narrower and more projecting ulnar styloid process in hominoids). Like other mammals, most strepsirrhines retain a thick, fibrous longitudinal septum that divides the antebrachiocarpal joint into two compartments: an ulnocarpal portion and radiocarpal portion (Fig. 3.3). This septum links several ligaments together: the lunatotriquetrum ligament distally, the ulnocarpal ligament palmarly and a proximal ligament that unites the distal portions of the radius and ulna (Cartmill and Milton 1977; Lewis 1989). Thus, the septum prevents the proximal carpal row from sliding either radially or ulnarly across the distal radius, thereby limiting radioulnar deviation (Cartmill and Milton 1977; Hamrick 1996a). Therefore, in taxa with a longitudinal septum, the majority of radioulnar deviation occurs at the midcarpal joint (Hamrick 1996a). There is no true synovial joint between the radius and ulna (Lewis 1989). Instead, the radial portion of the distal ulna is more like a small projection, jutting out radially to articulate with the radius, which limits pronation and supination more so than in hominoids or lorisids (Cartmill and Milton 1977; see below) (Fig. 3.3).

The antebrachiocarpal joint of most Old and New World monkeys (e.g., *Colobus*, *Cebus*, *Alouatta*; but see *Ateles* below), particularly palmigrade monkeys, is overall quite similar to the general mammalian and strepsirrhine morphology described above (Lewis 1971b, 1989; Youlatos 1996; Daver et al. 2012; Fig. 3.3). The distal articulation between the radius and ulna is typically a syndesmosis, with a firm ligamentous bond and minimal mobility (Lewis 1965; Cartmill and Milton 1977), although some taxa have an incipient synovial articulation with slightly more mobility (e.g., *Cercopithecus*, *Ateles* and *Lagothrix*; Lewis 1989). The pisiform is rod-like and robust and projects proximally into the heel of the hand. The distal end of the ulna articulates with the concave facet formed by the triquetrum and pisiform and is weight bearing as in strepsirrhines. However, the projecting articulating portion of the ulna has a constricted neck and thus resembles more the ulnar styloid process of hominoids than the morphology of strepsirrhines (O'Connor 1975; Lewis 1989; Hamrick 1996a). Furthermore, there is usually no longitudinal septum separating the radiocarpal and ulnocarpal compartments of the antebrachiocarpal joint in Old World monkeys, though a septum is still found in most New World monkeys (Lewis 1989; Youlatos 1996; Daver et al. 2012; but see Cartmill and Milton 1977).

In hominoids, the distal articulation between the radius and the head of the ulna is a fully elaborated synovial joint, which provides greater mobility (pronation and supination) of the wrist and hand than most other primates. A **triangular articular disc**, which is strongly connected to the palmar ulnocarpal ligament running from the ulnar styloid process to the lunate, separates the ulnar head joint cavity from the remainder of the antebrachiocarpal joint (Fig. 3.3). The pisiform is smaller compared with strepsirrhines and monkeys and projects palmarly and distally into the palm, rather than proximally (except in hylobatids), and does not articulate with the ulna. The ulnar styloid process is reduced, and, instead, a fibrocartilaginous, intra-articular meniscus fills this space and wraps around the ulnar side of the antebrachiocarpal joint from the lunate palmarly to the radius dorsally (Cartmill and Milton 1977; Lewis 1989). Hylobatids are distinct in the presence of an ossified sesamoid-like bone (a lunula), called an os Daubentonii, within the thick, ulnar portion of meniscus [although a small lunula can occasionally be found in *Gorilla*; Lewis (1989); see Sarmiento (1988) for a different interpretation]. The presence of a small aperture in the meniscus in hylobatids and sometimes in *Pan* allows the ulnar styloid process to articulate with the triquetrum. However, in *Gorilla* and *Pongo*, the meniscus blends with the triangular articular disc, thus completely excluding the ulnar styloid process from articulating with the carpus (Lewis 1989). In humans, the ulnocarpal joint is further modified, such that there is no longer a discrete meniscus, but instead it merges with the remainder of the proximal articular surface of the radiocarpal joint (Lewis et al. 1970; Cartmill and Milton 1977; Lewis 1989). Altogether, the derived hominoid morphology allows the radius and ulna to still be held together, but to rotate freely around each other (Sarmiento 1985). Thus, in hominoids (and lorisids; see below), the antebrachiocarpal joint largely consists of a radiocarpal articulation only, which is why this joint is often referred to simply as the **radiocarpal joint** in humans and other apes.

Lorises and spider monkeys, as well as sloths (Mendel 1979), show some convergent morphology with hominoids related to greater mobility, particularly ulnar deviation and rotation, of the wrist needed for the climbing, bridging or suspensory behaviors common to all of these taxa. Lorises (e.g., *Loris*, *Nycticebus*, *Arctocebus*) have a derived antebrachiocarpal joint compared with other strepsirrhines (Nayak 1933; Cartmill and Milton 1977). The pisiform is slightly smaller and displaced distally (along with the triquetrum) and does not articulate with the ulna. The distal end of the ulna has a narrow projection, which is similar in shape to the ulnar styloid process of hominoids (Cartmill and Milton 1977). The radioulnar articulation is also derived such that the radial portion of the distal ulna no longer projects radially, but instead articulates with an ulnarly extended "shelf" of the radius (Fig. 3.3). This configuration expands the articulation with the radius, creating an "ulnar head" morphology, similar to that of hominoids, enhancing pronation and supination (Cartmill and Milton 1977). However, a longitudinal septum is still present and is similar in morphology to that of other strepsirrhines, and there is no meniscus like in hominoids (Cartmill and Milton 1977).

Spider monkeys have an intermediate morphology; they have also lost the articulation between the ulna and pisiform and have a small, distopalmarly positioned pisiform compared with other New World monkeys (e.g., *Alouatta*, *Lagothrix*) (Youlatos 1996; but see Lewis 1971b). This morphology is consistent with their increased suspensory, climbing and clambering locomotion (Cant et al. 2001). However, spider monkeys also retain a large distal ulna that articulates with the triquetrum, providing support on the ulnar carpus during pronograde quadrupedalism.

The functional implications of the derived morphology of the antebrachiocarpal joint in hominoids, and the convergent development of some of these features in lorisids and spider monkeys, were traditionally thought to allow for greater supination and ulnar deviation at the wrist, common to the habitual wrist postures used during climbing, vertical clinging or suspension. However, Jouffroy and Medina (2002) show that some taxa (e.g., *Macaca*[1]) with a fully elaborated ulnocarpal articulation have greater ulnar deviation than those without (e.g., humans), highlighting the

[1] Note that the particular *Macaca* specimen depicted in Fig. 3.6 and adapted from Jouffroy and Medina (2002) seems to display an unusual carpal placement that may suggest a greater degree of ulnar deviation (measured as 56°) than is typical for *Macaca*. Jouffroy and Medina (2002) do not

importance of ligaments and tendons in overall wrist mobility. That being said, loss of ulnocarpal articulation does allow for greater mobility at the antebrachiocarpal joint, such that radioulnar deviation in humans and great apes derives primarily from the antebrachiocarpal joint, while Old and New World monkeys, strepsirrhines and hylobatids have greater midcarpal mobility (Jouffroy and Medina 2002; Fig. 3.4).

The other half of the antebrachiocarpal joint—the **radiocarpal articulation**—has received comparatively much less attention, likely because the morphology does not vary substantially across primates (e.g., Jenkins and Fleagle 1975; Ziemer 1978; Lewis 1989). For example, slow-climbing strepsirrhines have more radioulnarly curved radiocarpal (and midcarpal) joint surfaces compared with vertical clinging and arboreal quadrupedal strepsirrhines (Hamrick 1996b). However, the latter two locomotor groups do not differ significantly in radiocarpal curvatures despite loading their forelimbs quite differently (Hamrick 1996b). Most of the discussion about the radiocarpal articulation has focused on extension-limiting mechanisms in terrestrial taxa, such as African apes and digitigrade monkeys (e.g., Jenkins and Fleagle 1975; Corruccini 1978; Zylstra 1999; Richmond and Strait 2000; Richmond et al. 2001; Begun 2004). In African apes, the radiocarpal joint is stabilized in the weight-bearing, slightly extended wrist posture in part by two osteological features of the distal radius: (1) a distal extension of the dorsal margin—called the dorsal ridge (Richmond and Strait 2000)—that buttresses the scaphoid as it rotates during extension and (2) a large scaphoid notch along the dorsolateral margin that contacts a concavity on the dorsal surface of the scaphoid that limits further extension [Tuttle 1967, 1969; see Richmond and Strait (2000) for images]. In addition, the scaphoid has a larger articulation with the radius than that of the lunate (the opposite condition to that of *Pongo*; Zylstra 1999), and the scaphoid and lunate articular surfaces share a similar distoulnar orientation (i.e., they are roughly coplanar). Together, these features are thought to better resist stress during weight bearing, particularly on the radial side of the wrist, and prevent the wrist joint from collapsing into extension (Richmond and Strait 2000; Richmond et al. 2001; Begun 2004).

Terrestrial digitigrade Old World monkeys, such as baboons and patas monkeys, also have limited extension (and ulnar deviation) at the antebrachiocarpal joint compared with palmigrade monkeys (Tuttle 1969; Lemelin and Schmitt 1998; Richmond 2006). Like African apes, they have a similar projecting dorsal ridge of the distal radius (Richmond and Strait 2000). However, the scaphoid notch is much larger and deeper, thus allowing for a greater degree of extension before the radius contacts the scaphoid (Whitehead 1993; Richmond and Strait 2000; Richmond et al. 2001). These more terrestrial primates also have a meniscus in between the dorsal articular areas of the radius and scaphoid that further helps to limit extension (Daver et al. 2012).

In contrast, suspensory apes typically have a much smaller dorsal projection of the distal radius, a smaller scaphoid notch and scaphoid-lunate articular surfaces that are more angled relative to each other, all of which contribute to a much greater range of extension at the antebrachiocarpal joint (Tuttle 1967, 1969; Richmond 2006).

---

provide information on the species or sample size. For comparison, Richmond (2006), using different methods, reports 45° ulnar deviation in *Erythrocebus* and 61° ulnar deviation in *Papio*.

6 Scaphoid-Os Centrale Fusion

Scaphoid-os centrale fusion is one of the most discussed features of the primate wrist (e.g., Mivart 1867; Schultz 1936; Marzke 1971; Jenkins and Fleagle 1975; Sarmiento 1988; Begun 1992, 2004; Gebo 1996; Richmond et al. 2001; Fig. 3.5). Many of the initial morphological descriptions of the primate carpus in the late nineteenth century have all discussed fusion of the scaphoid with the os centrale (Lucae 1865; Mivart 1867, 1869; Giebel 1879; see Kivell and Begun 2007 for historical review). For example, Huxley (1863) noted that African apes and humans have eight carpals, compared with nine in most other primates. In spite of all of these observations, there has been confusion over which taxa have consistent scaphoid-os centrale fusion (*versus* fusion later in life due to, for example, ossified ligaments or osteophytic growth) and the functional reasons behind this fusion (e.g., Schultz 1936; Yalden 1972; Jouffroy 1975; Tuttle 1975; Sarmiento 1985; Schwartz and Yamada 1998; Whitehead 1993; Hamrick 1996a; Richmond et al. 2001; Begun 2004).

Kivell and Begun (2007) undertook the first systematic study of scaphoid-os centrale fusion across a broad sample of primates and found consistent fusion (>95 %) in *Pan*, *Gorilla* and humans that occurred early in ontogeny, compared with rare fusion in Asian apes (~7 %) that occurred only in adulthood. In the smaller samples of strepsirrhines they examined, they found consistent scaphoid-os centrale fusion in two extant species of Indriidae (*Indri* and *Avahi*), one species of Megaladapidae (*Lepilemur*) and one species of the subfossil Palaeopropithecidae (*Babakotia*) (see Chap. 15). Wood Jones (1942) related scaphoid-os centrale fusion to the importance of the index finger and the need for stability at the base of the second digit in African apes and, particularly, humans. Most researchers today, however, suggest that scaphoid-os centrale fusion in African apes is a functional adaptation to the increased shear stress on this joint during knuckle-walking (Marzke 1971; Tuttle 1975; Corruccini 1978; Sarmiento 1994; Gebo 1996; Richmond et al. 2001; Begun 2004) and that its presence in humans is due to phylogenetic "lag" (Richmond et al. 2001) or is an exaptation to shear stress during power-grip postures (Marzke 1971). Fusion in particular species of strepsirrhines has been considered advantageous for having a large and divergent thumb (Begun 2004) along with increased loading of the radial side during vertical climbing and quadrupedalism (Sarmiento 1994). However, there are several reasons why such functional hypotheses are not consistent with the sporadic occurrence of fusion across strepsirrhines (e.g., the absence of fusion in lorises, who have the most divergent thumbs among strepsirrhines and engage in substantial quadrumanous climbing), and there is a strong heterochronic and genetic underpinning to fusion in hominines that can make functional explanations more challenging (Kivell and Begun 2007). Furthermore, roughly half the individuals (58 % of $n=12$) of the highly suspensory subfossil lemur *Palaeopropithecus* show fusion, suggesting that we have much to learn about the interplay of genetics, development and function when it comes to this feature (Kivell and Begun 2007).

7 The Pisiform: It Is Not a Sesamoid!

The pisiform is often described as a sesamoid bone and distinguished from the remaining "true" carpal bones (e.g., Flower 1885; Belliappa and Burke 1992; Scheuer and Black 2000). This description reflects the bias towards human morphology, where the pisiform is tiny and "pea-shaped" (hence, the name "pisiform") bone, developing from a single ossification center within the tendon of the flexor carpi ulnaris muscle (FCU) (Scheuer and Black 2000) and articulating solely with the triquetrum (Fig. 3.3). However, in other primates and mammals, the pisiform is a much larger, elongated bone that articulates with the distal ulna and, in some taxa, the hamate (Gillies 1929; Etter 1974; Cartmill and Milton 1977) or even the radius (*Daubentonia*; Mivart 1867; Flower 1885; Nayak 1933). The pisiform develops from two ossification centers (Eckstein 1944; Lewis 1989; Jouffroy 1991) that are divided by a palmar epiphyseal or growth plate (Harris 1944; Kjosness et al. 2014). Thus, the primate pisiform does not follow a sesamoid developmental pattern, and, because it articulates with more than one bone, it cannot be considered a true sesamoid (Gillies 1929, but see Flower 1885).

The large pisiform of most nonhuman primates functions as a "heel" for the hand. It is usually considered not to be weight bearing (Lewis 1989; Whitehead 1993), although this should be experimentally tested (see Patel and Wunderlich 2010). The pisiform projects proximopalmarly to provide a bony origin for some of the hypothenar muscular and forms a cup-like articulation (with the triquetrum) for the distal ulna, which together stabilize the ulnar side of the wrist during the compressive loading of quadrupedal locomotion (Lewis 1989; Youlatos 1996). The pisiform defines the medial "wall" of the carpal tunnel with the hamate's hamulus and serves as the attachment for the FCU and the abductor digiti minimi tendons (Diogo and Wood 2011 and references therein, but see Jouffroy 1975; see Chap. 7).

An elongated pisiform appears to be more functionally important for increasing the moment arm of the FCU rather than carpal tunnel depth (Sarmiento 1988; Lewis 1989; Hamrick 1997). For example, in humans, the planar joint between the pisiform and triquetrum allows for approximately 1 cm of movement, and thus contraction of the FCU is necessary to stabilize the pisiform for effective action of the abductor digiti minimi muscle (Brand and Hollister 1993; Marzke et al. 1998). The pisiform is especially long and palmarly projecting in pronograde quadrupedal taxa where the FCU acts to flex the wrist from an extended posture (Sarmiento 1988; Whitehead 1993; Hamrick 1997; Patel et al. 2012). In arboreal climbing primates, particularly lorises, spider monkeys and hominoids, the pisiform is relatively short (compared with the hamate hamulus) and more distopalmarly positioned in the palm (Lewis 1989; Youlatos 1996; Hamrick 1997). This morphology is thought to reflect a decreased commitment to quadrupedalism (i.e., less wrist flexion from an extended wrist posture) (Hamrick 1997) and, along with other derived features of the ulnocarpal region (see above), enhance pronation-supination and ulnar deviation (Sarmiento 1988). In African apes, it has also been proposed that the elongated pisiform plays a role in forelimb propulsion during knuckle-walking, especially at high speeds (Sarmiento 1985).

8 Radial Carpometacarpal Joints

The radial carpometacarpal (CM) joints include the complex articulations between the scaphoid, trapezium, trapezoid and the first and second metacarpals. In primates, the distolateral portion of the scaphoid body, often including its tubercle, and os centrale (or the os centrale portion of the scaphoid when fused) articulate with the trapezium and trapezoid (Fig. 3.6). The trapezoid also articulates with the radiodorsal portion of the capitate (except in the gorilla for which there is often no articulation at all; Lewis 1989; Tocheri et al. 2005). This region is further complicated by the tendon of flexor carpi radialis, which runs palmarly around the base of the trapezium's tubercle, and the trapezium also serves as the origin for some the thenar muscles (see Chap. 7). The first metacarpal articulates solely with the trapezium (the first CM joint), and the second metacarpal articulates proximally with the trapezoid, laterally with the trapezium and medially with the capitate. The **prepollex**, meaning "before the thumb", is greatly reduced in primates compared with other mammals and appears, if at all, at the base of the thumb, articulating with (or fused to, which is common in gorillas) to the scaphoid tubercle and trapezium (Lewis 1989). The prepollex is thought to have little influence on the mechanics of the radial CM joints, though it can serve as an attachment site for the abductor pollicis longus and abductor pollicis brevis muscles (Howell and Straus 1933). Within the radial CM joint, the trapezium-Mc1 joint (Tm-Mc1) has received the most attention, both with regard to primate

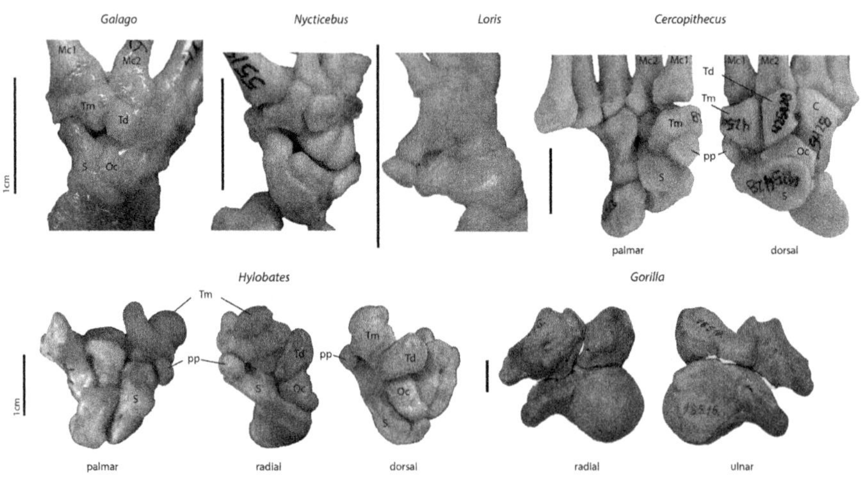

Fig. 3.6 Radial carpometacarpal joints. *Top* row showing the articulated radial carpometacarpal joints in some strepsirrhines (radial view) and *Cercopithecus* (palmar and dorsal views). *Bottom* row showing articulated radial carpal joints in *Hylobates* (palmar, radial and dorsal views), and re-articulated carpals in *Gorilla* (radial and ulnar views). Carpal bones included in radial carpal joints are labelled. Abbreviations: *Tm* trapezium, *Td* trapezoid, *S* scaphoid, *Oc* os centrale, *pp* prepollex, *C* capitate, *Mc1* first metacarpal, *Mc2* second metacarpal. All specimens are shown as the right side and scaled to approximately the same size (scale represents 1 cm for each taxon)

morphology and in clinical studies (e.g., Haines 1944; Napier 1955; Tuttle 1969; Kuczynski 1974; Cooney and Chao 1977; Lewis 1977, 1989; Rafferty 1990; Rose 1992).

Primates differ from tree shrews and other mammals in having a thumb complex that is relatively independent and more divergent from the remainder of the hand, which is likely related to more efficient manual grasping (Altner 1971; Boyer et al. 2013; see Chap. 14). The opposable thumb is a functionally important feature of many, but not all, primate hands. Napier was the first to put forth classifications of thumb opposability in primates (Napier 1955, 1961; Napier and Napier 1967). Napier (1961: 119) defined opposition as "a compound movement of abduction, flexion and medial rotation" that is made possible by a saddle-shaped Tm-Mc1 articulation. Thus, "true opposability" generally applies to the thumbs of catarrhines (e.g., Napier 1961; Rose 1992). In contrast, New World monkeys and strepsirrhines are often considered to have "pseudo-opposable" thumbs because the Tm-Mc1 joint is cylindrical or relatively flat, rather than saddle-shaped (Napier 1961; Jouffroy and Lessertisseur 1979; Ziemer 1978). Thus, the joint acts more like a hinge and does not allow rotation (Napier 1961). Although these taxa can converge their thumb towards the fingers, this ability is due partly to a deep carpal arch (i.e., such that the trapezium is more in-turned relative to the other carpals) and flexion-extension, with only limited abduction-adduction, at the Tm-Mc1 joint (Napier 1961).

However, the terms "opposable" and "pseudo-opposable" are poorly defined and understood. Many since Napier have shown that a saddle-shaped Tm-Mc1 joint (implying both flexion-extension and abduction-adduction movements are possible) is common in most platyrrhines (Rafferty 1990), many strepsirrhines (Etter 1974; Jouffroy and Lessertisseur 1979; Boyer et al. 2013) and some marsupials and carnivorans (Haines 1958; Lewis 1977, 1989). Indeed, Lewis (1977) suggested that a saddle-shaped joint Tm-Mc1 may be the primitive condition for mammals. The more limited opposability in platyrrhines (and likely strepsirrhines) stems from having more congruent axes of the saddle joint, such that conjunct axial rotation during flexion of the thumb is limited or absent (Rafferty 1990; Rose 1992). In addition, there are certainly muscular and neurological adaptations (e.g., more efficient digital coordination) that can have a strong influence on opposability, but may not be reflected in the bony morphology (Costello and Fragaszy 1988; Spinozzi et al. 2004; see Chap. 12), further complicating how we define these terms functionally and how we might identify such abilities in the fossil record.

Across cercopithecoids, the Tm-Mc1 joint is generally saddle-shaped with little variation in bony morphology (Rafferty 1990). Even in colobines that have a diminutive thumb, the Tm-Mc1 joint is saddle-shaped, although the curvature of the concavity is less developed than that of the convexity at this joint (Rafferty 1990). In contrast to cercopithecoids, there is much more variability in the New World monkey trapezium-Mc1 joint (Mivart 1867; Lewis 1977; Rafferty 1990). For example, in callitrichids, one of the few primates considered to have little thumb opposability, the trapezium's Mc1 facet is usually flat and is positioned distally and palmarly such that the Mc1 is less divergent and in the same plane as the other metacarpals (and even articulates with the Mc2) (Rafferty 1990; Boyer et al. 2013). Capuchin monkeys, for which some species are known for their dexterity (i.e., *Sapajus*), have a saddle-shaped Tm-Mc1 facet with a deeper concavity-convexity that is more similar to that of catarrhines than other ceboids (e.g., *Aotus*). Squirrel monkeys (*Saimiri*) have a small and relatively flat trapezium-Mc1 facet (Mivart 1867; Lewis 1977; Rafferty 1990), while still other species (i.e., *Pithecia*, *Chiropotes*) have a pronounced groove (on the trapezium) and keel (on the Mc1) across the dorsopalmar surface of the Tm-Mc1 joint that makes the joint surfaces highly congruent and limited in their mobility (Rafferty 1990). In *Ateles*, in which the thumb is extremely reduced, the trapezium shows no reduction relative to other carpals, but the Tm-Mc1 facet is small and flat (as in *Lagothrix* and *Alouatta*) (Rafferty 1990; contra Lewis 1977; Ziemer 1978).

Finally, all great apes have a well-developed saddle-shaped Tm-Mc1 joint and differ from non-hominoid primates in having much greater abduction-adduction mobility (Rose 1992). Hylobatids are unusual among primates in that the Mc1 facet on the trapezium is a convex half sphere, rather than saddle-shaped, and fits with a concave trapezium facet on the Mc1 (Lorenz 1971; Lewis 1977, 1989; Rafferty 1990; Fig. 3.6). This morphology creates a distinctive ball-and-socket Tm-Mc1 joint that allows for greater mobility in hylobatids compare with other primates, although the large tubercle of the trapezium limits abduction and flexion of the thumb (Lorenz 1971; Lewis 1977, 1989; Rafferty 1990).

The articulations between the scaphoid/os centrale-trapezium-trapezoid (STT) and the second metacarpal have received much less attention than the Tm-Mc1 joint (Kauer 1986; Moritomo et al. 2000a, b; Tocheri et al. 2003, 2005; Begun 2004; Sonenblum et al. 2004; Tocheri 2007). Most studies limit discussion of this region to the variation in the capitate-trapezoid embrasure (i.e., v-shaped gap) for the os centrale (or scaphoid, when the os centrale is fused) during midcarpal joint rotation (e.g., Jenkins 1981; Lewis 1989; Hamrick 1996a; Schwartz and Yamada 1998; see below). For example, in arboreal quadrupedal strepsirrhines, the embrasure is widest on the dorsal surface, which facilitates palmar rotation (pronation) of the proximal carpals into the capitate-trapezoid embrasure. In contrast, in vertical clinging and leaping strepsirrhines, the embrasure is wider on the palmar side (and constricted dorsally), thus facilitating dorsal rotation (supination) of the proximal carpals at the midcarpal joint (Hamrick 1996a). Hylobatids and *Ateles* also have a dorsally constricted capitate-trapezoid embrasure, allowing for a greater range of midcarpal supination (Jenkins 1981; see below). These differences in morphology stem largely from variation in the shape of the capitate and variation in how the trapezoid is oriented within the carpus. For example, the scaphoid-trapezoid articulation is more proximodistally oriented in vertical clinging and suspensory primates to enhance mobility, while in quadrupeds the scaphoid-trapezoid articulation is more radioulnarly oriented to better resist compression (Hamrick 1996a; Richmond et al. 2001; Begun 2004). However, there is substantial morphological variation in the orientation of the scaphoid-trapezoid articulation that does not always correlate well with locomotor strategy [e.g., Figs. 3.2 and 3.6; see also Fig. 7 in Hamrick (1996a) and Fig. 9 in Richmond et al. (2001)].

The articulations across the STT joint have only been well studied in extant hominids by Tocheri and colleagues (Tocheri et al. 2003, 2005; Tocheri 2007). The trapezoid in most primates can be described as wedge-shaped with a narrow palmar non-articular surface and broad dorsal non-articular surface. In most Old World monkeys and African apes, the Mc2 articulation is strongly keeled (rather than relatively flat as in Asian apes) to provide a more stable CM joint (Fig. 3.2, see below). Furthermore, the human trapezoid is derived in having a radioulnarly-expanded non-articular palmar surface, which gives it its characteristic "boot-shape" appearance and promotes reorientation of the radial carpals and thumb into a more supinated position (Lewis 1989; Tocheri et al. 2003, 2005). In most primates, the scaphoid's articulation with the trapezoid is larger than that with the trapezium, while the opposite pattern is true for humans (Marzke et al. 1992; Tocheri et al. 2005). Humans also have a larger trapezoid-capitate articulation that is positioned more palmarly rather than dorsally as in other primates. These differences in the radial CM joints across apes are functionally consistent with how the hands are used during positional behavior. In knuckle-walking or suspension, loading is transmitted distoproximally through the Mc2-trapezoid-scaphoid. In contrast, forceful precision and power grips load this region of the human hand more transversely (radioulnarly), mainly via the Mc1-trapezium-scaphoid (Lewis 1989; Tocheri et al. 2005; Tocheri 2007). Thus, reorientation of the radial CM articulations in humans is thought to better accommodate compressive loading from the thumb towards the palm (Lewis 1989; Tocheri et al. 2003; Tocheri 2007).

9 The Midcarpal Joints

The **midcarpal joint** refers to the articulation between the proximal and distal rows of the carpus. It is a complex joint that varies strongly in the relative contributions and orientations of particular bones and joint surfaces, but generally can be described broadly as a ball-and-socket-type articulation: the mainly distally oriented facets of the scaphoid/os centrale, lunate and triquetrum in the proximal carpal row form a radioulnarly and dorsopalmarly concave "socket", and the proximal facets of the capitate and hamate in the distal carpal row form the correspondingly convex "ball" (Figs. 3.2, 3.4, 3.7, 3.8). The capitate and hamate are firmly bound together by interosseous ligaments and essentially function as a unit in all primates. In most strepsirrhines and tarsiers, the hamate is much larger than the capitate; the os centrale is ulnarly extended such that it articulates with the hamate, cutting off the articulation between capitate and lunate (Jouffroy 1975; Godinot and Beard 1991; Schwartz and Yamada 1998; Stafford and Thorington 1998; Fig. 3.2). This is thought to be functionally related to the frequent ulnar deviation at the midcarpal joint common in strepsirrhines (Hamrick 1997; Lemelin and Schmitt 1998). In contrast, in most anthropoids (although hylobatids are a notable exception), the capitate is larger than the hamate, such that the midcarpal articular configuration is primarily between the capitate-os centrale/scaphoid-lunate and between the hamate-triquetrum (Figs. 3.2 and 3.4). The radial articulations between

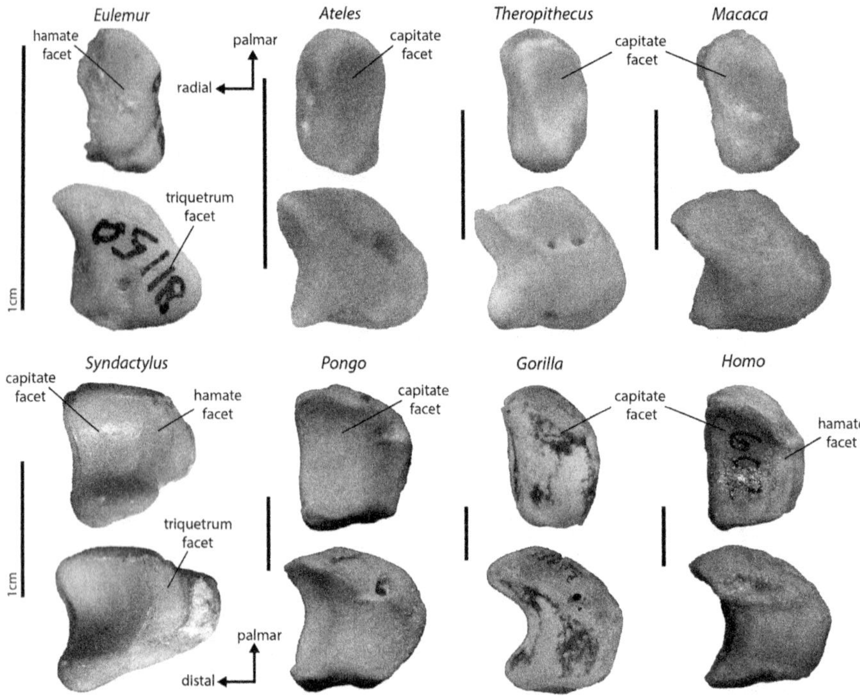

Fig. 3.7 Variation in primate lunate morphology. *Top* row for each taxon shows the distal view of the lunate, featuring the midcarpal articulation with either the hamate, capitate or both. *Bottom* row for each taxon shows the ulnar view of the lunate, featuring the triquetrum facet. All specimens are shown as the right side and scaled to approximately the same size (scale represents 1 cm for each taxon)

the scaphoid, trapezium and trapezoid discussed above also can be strictly considered the "midcarpal joint", but are usually considered as a somewhat separate functional unit and not included in discussions of mobility/stability of the primate midcarpal joint (e.g., Lewis 1985a, b; Lemelin and Schmitt 1998; Begun 2004; Richmond 2006; but see Jenkins 1981 and Richmond et al. 2001).

The midcarpal joint permits movement of the hand relative to the forearm in three planes: flexion/extension, radial/ulnar deviation and pronation/supination. Combinations of movements in these planes in which the proximal carpal row moves into a stabilized or "close-packed" position on the distal carpal row is referred to as a "screw-clamp" mechanism (MacConaill 1941; Lewis 1989; Orr et al. 2010; Fig. 3.9). In quadrupedal primates and humans, movement at the midcarpal joint into a close-packed position is achieved via a combination of extension, ulnar deviation and pronation (i.e., rotation) (Lewis 1989). This combined movement appears to be primitive, shared with other mammals such as opossums (Jenkins 1971; Lewis 1989). Supination at the midcarpal joint in quadrupedal primates is relatively limited (Jenkins 1981). In contrast, the screw-clamp mechanism in suspensory primates seems not to be as effective as in quadrupeds (at least in the

Fig. 3.8 Variation in primate triquetrum morphology. *Top* row for each taxon (except for *Alouatta*) shows the radial view of the triquetrum, featuring the lunate facet. *Bottom* row for each taxon shows the distoradial view of the triquetrum, featuring the hamate facet. Only distoradial views are shown for *Alouatta* and *Lagothrix*. All specimens are shown as the right side and scaled to approximately the same size (scale represents 1 cm for each taxon)

taxa that it has been studied in vitro); close packing occurs at a much higher degree of extension in *Pongo* compared with *Pan* (Orr et al. 2010; see Chap. 9). In brachiators (*Ateles*, *Lagothrix* and hylobatids), the midcarpal joint undergoes a high degree of supination, with limited mobility in terms of pronation (Jenkins 1981); the opposite pattern of quadrupeds (Fig. 3.10).

The degree of midcarpal joint curvature is correlated with the range of radioulnar deviation and midcarpal rotation of the hand (Jenkins and Fleagle 1975; Jenkins 1981; Sarmiento 1988; Hamrick 1996a, b). Arboreal or terrestrial quadrupedal primates have flatter midcarpal joint surfaces (i.e., larger radius of curvature) than those of suspensory (e.g., Asian apes, spider monkeys) or climbing/bridging primates (e.g., lorisids). The scaphoid-os centrale-capitate and triquetrum-hamate articulations are more proximally oriented compared with suspensory primates, and in terrestrially-adapted quadrupedal taxa [e.g., baboons, patas monkeys, African apes (but also humans)], the proximal capitate and hamate facets are also radioulnarly broader (relative to carpal length) (Yalden 1972; O'Connor 1975; Jenkins 1981; Lewis 1985b; Sarmiento 1988; Richmond 2006; Lemelin et al. 2008).

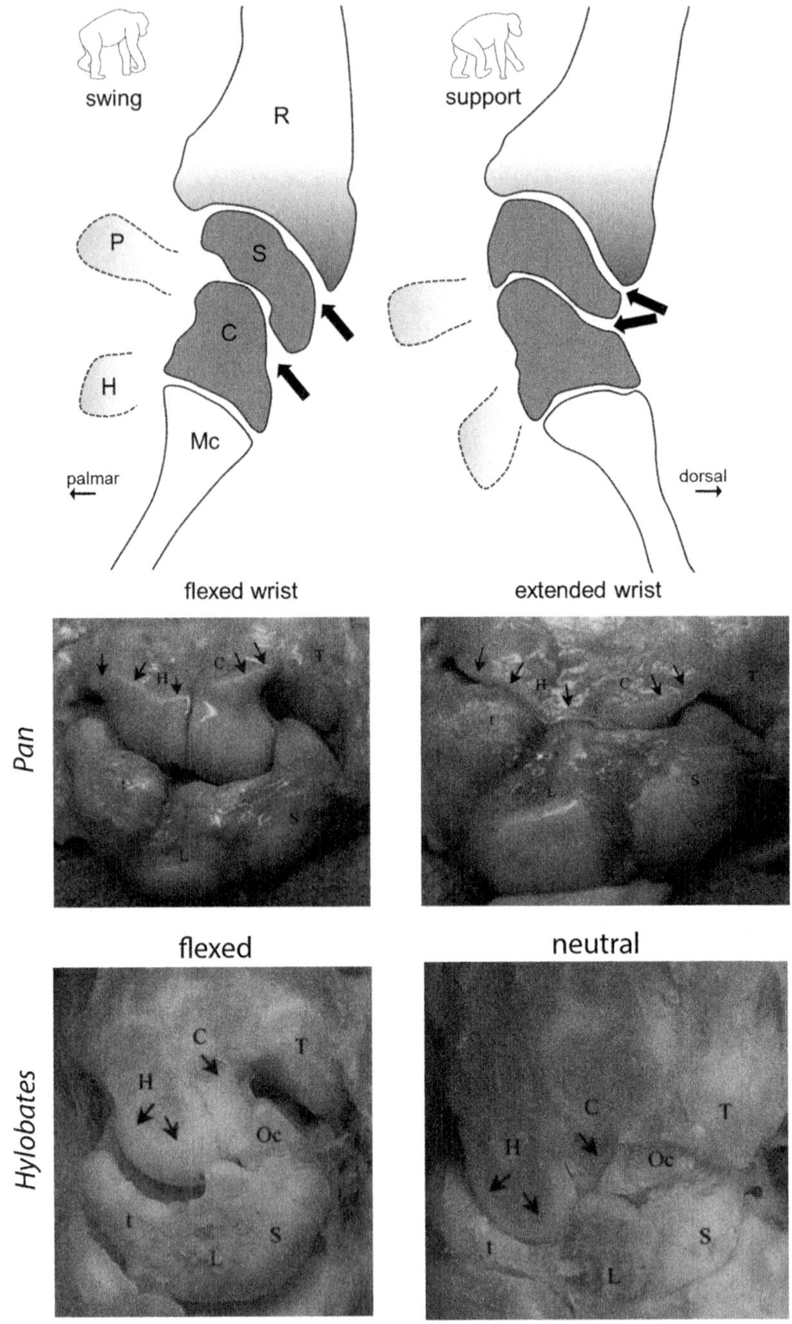

Fig. 3.9 "Close packing" of the midcarpal joint. Schematic *above* shows wrist posture during knuckle-walking swing phase and support phase. In swing phase the wrist is flexed, and there is space between the articulations of the radius-scaphoid-capitate when viewed dorsally (*block arrows*). In the support or weight-bearing phase, the wrist is extended the radius-scaphoid-capitate achieve a "close-packed" articulation. The palmarly projecting pisiform (P) and hamate hamulus (H) are also depicted. *Middle*, dorsal views of a *Pan* cadaveric wrist shown in a flexed and extended "close-packed" postures. *Below*, dorsal views of *Hylobates* cadaveric wrist in a flexed and neutral posture. *R* radius, *S* scaphoid, *C* capitate, *Mc* metacarpal. All images adapted from Richmond et al. (2001)

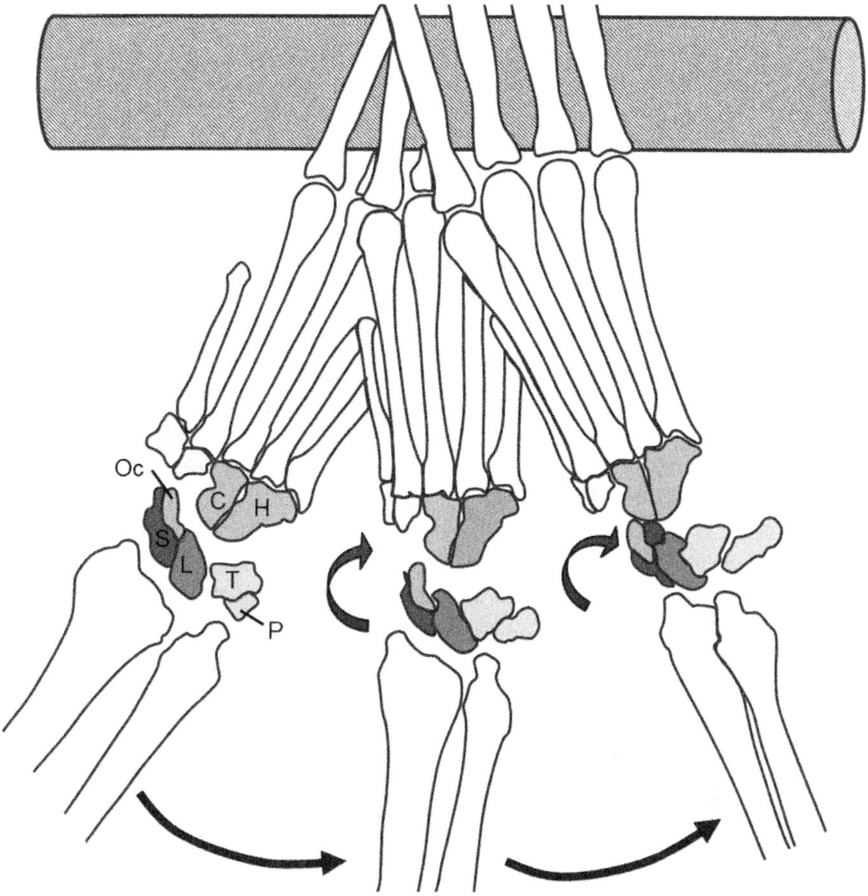

Fig. 3.10 Rotation at the midcarpal joint during brachiation in a spider monkey. "Exploded" dorsal view of the hand at the beginning (*left*), middle and end (*right*) of swing, demonstrating how the proximal carpal row, with the forearm, supinates around the "ball" formed by the capitate and hamate proximal facets. Image adapted from Jenkins (1981)

Together, this morphology limits radioulnar deviation and supination, creating greater stability during the extended wrist postures typically used during palmigrade or digitigrade locomotion. However, it is interesting to note that baboons and patas monkeys have more distally extended dorsal joint surfaces on the capitate and hamate than in hylobatids and yet have much more limited extension, showing that bony morphology is not always necessarily indicative of mobility (Lovejoy et al. 2001; Richmond 2006); instead, joint curvature appears to be more closely linked to mobility (see Chap. 9).

During brachiation the majority of rotation in the hand occurs at the midcarpal joint (Jenkins 1981; Fig. 3.10). Thus, brachiators such as spider monkeys and hylobatids, as well as suspensory primates (e.g., orangutans), share similar, highly curved midcarpal joint morphology. The scaphoid-os centrale-capitate articulation is more radially oriented, and the hamate-triquetrum articulation is more ulnarly oriented, which allows for considerable radioulnar deviation. The more distally extended (both dorsally and palmarly) proximal facets on the capitate and hamate allow for increased flexion-extension (Richmond 2006). Together, this morphology allows for a much larger degree of supination at the midcarpal joint than is typically found in pronograde quadrupedal primates (Fig. 3.10).

Lorisids converge on the suspensory morphotype with proximal facets of capitate and hamate that are more curved and a hamate-triquetrum facet that is oriented more dorsally compared with other quadrupedal strepsirrhines (Lewis 1985a; Hamrick 1996b). This morphology, in addition to the reduced ulnar-carpal articulation described above, promotes pronation and extreme ulnar deviation of the hand (Hamrick 1996b; Lewis 1985a), as well as supination (Lemelin and Schmitt 1998), required for frequent climbing and bridging.

There has been much discussion about the midcarpal joint of African apes and humans (e.g., Tuttle 1967; Jenkins and Fleagle 1975; Corruccini et al 1975; Corruccini 1978; Sarmiento 1994; Dainton and Macho 1999; Richmond and Strait 2000; Richmond et al. 2001; Begun 2004; Kivell and Schmitt 2009; Williams 2010). Because the triquetrum is reduced in size and the os centrale is fused, the scaphoid's capitate facet (the os centrale portion) contributes more to the midcarpal joint than in other anthropoids, in which the os centrale, lunate and triquetrum contribute roughly equally to the "socket" of the midcarpal joint (Richmond et al. 2001; Begun 2004; Figs. 3.2, 3.5, 3.8). There are several aspects of the African ape midcarpal joint that are considered advantageous for limiting extension and making the wrist more stable during the knuckle-walking [the retention of some of these features in humans may suggest hominins evolved from a knuckle-walking ancestor; see Richmond et al. (2001) for a review]. On the capitate, the concave distal portion of the scaphoid facet is expanded in African apes compared with the solely convex articular surface in Asian apes or the smaller concave portion in macaques (Jenkins and Fleagle 1975). The convex-concave midcarpal articulation on the capitate contributes to the "waisting" or narrowing of the capitate body, which allows the scaphoid to wedge firmly into the capitate-trapezoid embrasure during extension (Figs. 3.2 and 3.9). The hamate-triquetrum facet is described as a spiraling, concavo-convex articulation, with the most distal portion of this articulation facing nearly proximally, to provide additional stability. Dorsal ridges at the most distal extent of the capitate and hamate midcarpal articulations also help to limit extension and provide stability during compression in a slightly extended wrist posture typical during knuckle-walking (Jenkins and Fleagle 1975; Richmond et al. 2001). Many or all of these traits have been considered specific adaptations to knuckle-walking (e.g., Tuttle 1967, 1969; Corruccini 1978; Zylstra 1999; Begun 2004; Richmond et al. 2001). However, many of these features (e.g., spiral triquetrum-hamate facet and dorsal ridges) are also found in Old World monkeys and, thus, likely reflect quadrupedal adaptations more generally (Jenkins and Fleagle 1975; Richmond et al. 2001; Richmond 2006, but see Lewis 1989). Furthermore, there is substantial variation in degree of expression or even presence of these "knuckle-walking" features across African apes (Sarmiento 1994; Richmond 2006; Kivell and Schmitt 2009). For example, capitate waisting and dorsal ridges of the capitate and hamate proximal facets are less pronounced in *Gorilla* than in *Pan*, despite their more frequent terrestrial knuckle-walking (Richmond 2006; Kivell and Schmitt 2009).

10 The Ulnar Carpometacarpal Joints

The ulnar carpometacarpal (CM) joints refer to the articulations between the trapezoid, capitate and hamate and ulnar metacarpals (Mc2-Mc5) (see also Sect. 8 above for discussion on the trapezoid-Mc2 articulation) (Figs. 3.11 and 3.12). These articulations take on various forms depending on the taxon, but are usually much more stable, planar joints in contrast to the mobile, often saddle-shaped, first CM joint.

Primates retain a primitive mammalian condition of a "stepped" configuration across the ulnar CM joints: the Mc2 extends proximally on its ulnar side to articulate primarily with the trapezoid, but also the capitate and Mc3 ulnarly and the trapezium radially (Lewis 1989; Figs. 3.1 and 3.2). The Mc3 also extends slightly proximally to articulate with hamate ulnarly. The Mc5-hamate articulation is usually oriented more ulnarly than the remaining ulnar CM articulations. There is a complex network of CM and intermetacarpal ligaments that help stabilize the

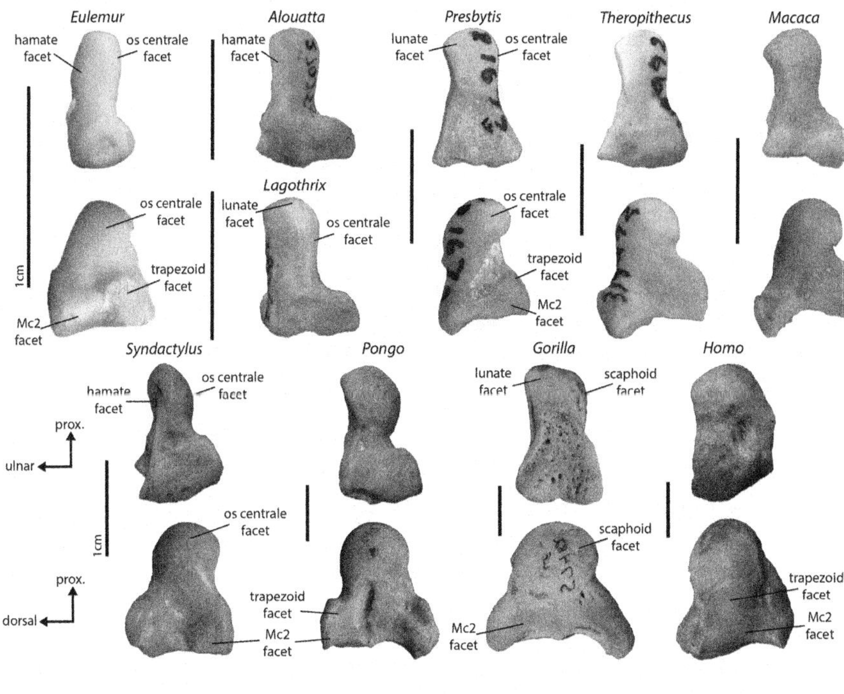

Fig. 3.11 Variation in primate capitate morphology. *Top* row for each taxon (except for *Lagothrix*) shows the dorsal view of the capitate, featuring the dorsal portion of the proximal facet. *Bottom* row for each taxon shows the radial view of the capitate, featuring the os centrale facet (or scaphoid facet in *Gorilla* and *Homo*), second metacarpal (Mc2) facet and trapezoid facet. Note that the capitate-trapezoid articulation in *Gorilla* is variable, ranging from absent to being palmarly positioned, like in humans (Lewis 1989). Only dorsal views are shown for *Alouatta* and *Lagothrix*. All specimens are shown as the right side and scaled to approximately the same size (scale represents 1 cm for each taxon)

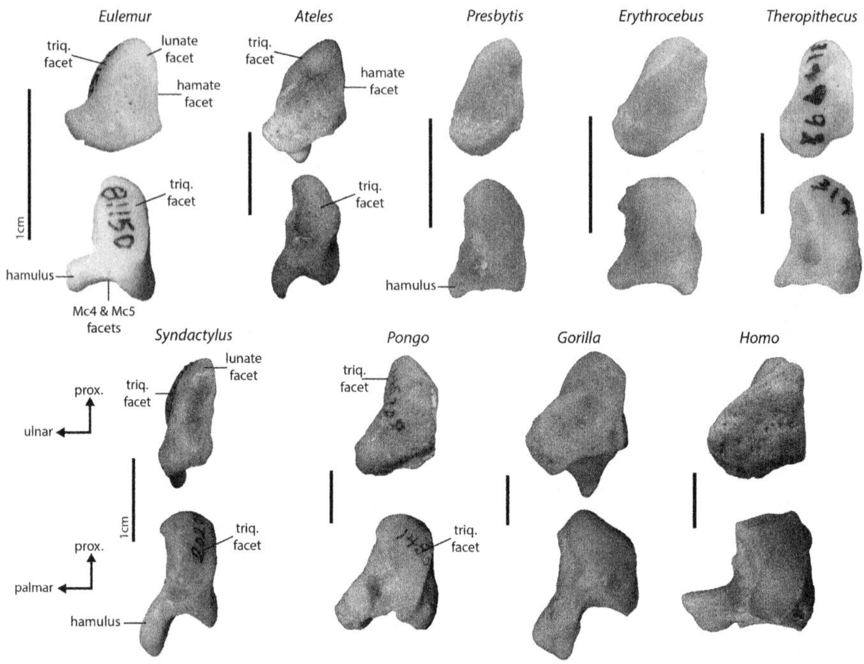

Fig. 3.12 Variation in primate hamate morphology. *Top* row for each taxon shows the dorsal view of the hamate. *Bottom* row for each taxon shows the ulnar view of the hamate, featuring the triquetrum facet and variation in the size and orientation of the hamulus. Note that the proximal hamate in *Eulemur* and *Symphalangus* also articulates with the lunate (because the capitate is comparatively small) at the midcarpal joint, while in most other anthropoids the hamate articulates only with the triquetrum (apart from occasional hamate-lunate articulation in African apes and humans; see Marzke et al. 1994). The distal articular surface of the hamate articulates with the Mc4 and Mc5 in all taxa. All specimens are shown as the right side and scaled to approximately the same size (scale represents 1 cm for each taxon)

joints between the distal carpals and Mc2-Mc5 (while the Tm-Mc1 is notably separated from this ligamentous network, emphasizing its distinctive functional role for movement of the thumb) (Lewis 1989). Tree shrews, some insectivorous eutherians (e.g., tenrecs) and some marsupials (e.g., opossums) also retain a stepped configuration of the ulnar CM joints, but show more derived osteological and soft tissues morphologies than most primates (Lewis 1989). For example, in some marsupials, a convex Mc2-capitate articulation and loss of the CM ligament permits more mobility of the Mc2, while the extensor carpi ulnaris tendon not only attaches on the Mc5 (as in primates) but also crosses the palm all the way to the Mc2 (Lewis 1989), possibly enhancing the mobility of the radial metacarpals at their base. Furthermore, while many primates (particularly strepsirrhines and hominoids) have a well-developed and projecting hamate hamulus, the hamate of tree shrews and squirrels has only a small palmar protuberance (Lewis 1989; Hamrick 1997, see his Fig. 3; Fig. 3.12).

Most strepsirrhines display the stepped configuration of ulnar CM joints described above; however, in some taxa (e.g., *Lemur*), the articulation between the hamate and Mc3 is reduced (Fig. 3.2). The capitate and hamate metacarpal articulations are generally planar, with limited mobility. The hamate hamulus is more well developed than in Old and New World monkeys (Hamrick 1997), creating a deeper carpal tunnel (in conjunction with a well-developed scaphoid tubercle) (Fig. 3.12).

Old and New World monkeys share similar ulnar CM articulations as in strepsirrhines, except that an articulation between the hamate and Mc3 is typically lacking and the Mc4 also articulates with the capitate (Lewis 1989; Marzke et al. 1994; Figs. 3.2 and 3.4). The metacarpal articulations of the capitate and hamate tend to be smooth and dorsopalmarly concave to match dorsopalmarly convex facets on the proximal metacarpals. As such, there is some degree of movement at these joints. For example, the Mc3 is capable of some flexion and combined extension-supination movement (Marzke and Marzke 1987). Furthermore, when the Mc4 and Mc5 extend, this movement is combined with slight pronation. These combined movements bring the metacarpals inline transversely and provide stability during the extended posture of palmigrade or digitigrade locomotion (Marzke 1983; Marzke and Marzke 1987). However, there is subtle variation in the morphology of these CM joints across cercopithecoids, which translates into slight variations in mobility. For example, *Papio* has a more complex concavo-convex capitate-Mc3 articulation, reminiscent of the morphology found in *Pan* (see below), which provides greater stability during digitigrady (Marzke and Marzke 1987). Although the hamate hamulus is usually small (i.e., beak-like process) as in other mammals (Lewis 1989; Hamrick 1997), the metacarpal articular surfaces extend onto the hamulus. The short hamulus of *Papio* reflects a shallow carpal tunnel compared to strepsirrhines and hominoids (Hamrick 1997), likely reflecting a de-emphasis of the digital flexor musculature used during climbing or suspensory activities.

Hominoids also display a stepped configuration, although the Mc4-capitate articulation is reduced relative to Old and New World monkeys, articulating only at the dorsal corner, if at all (Lewis 1989; Marzke et al. 1994). The Mc4 and Mc5 facets extend onto a well-developed hamulus in all hominoids (except humans). However, the hamulus varies in its angulation: Asian apes tend to have a hamulus that is more distally extended; the *Pan* hamulus is best described as being equally distally and palmarly extended; the *Gorilla* and human hamuli primarily project palmarly [see Orr et al. (2013) for a quantification of hamulus shape and angle]. A well-developed hamulus is most clearly associated with a deep carpal tunnel and a strong digital flexor musculature (Hamrick 1997; Ward et al. 1999; Ward 2002). There have been various functional explanations provided for a more distally projecting *versus* a more palmarly projecting hamulus. A more distally extended hamulus, like that of *Pan*, obstructs the amount of flexion at the Mc5-hamate joint (Ward et al. 1999; Ward 2002), while a more palmarly projecting hamulus would increase the mechanical advantage of the opponens digiti minimi and flexor digiti minimi muscles (Ward et al. 1999). It has also been suggested that variation in the orientation of the hamulus would enhance the ability of the FCU to act as a wrist adductor or wrist flexor, respectively (Sarmiento 1988; Ward et al. 1999). However, Ward (2002) noted importantly that the FCU attaches (via the pisohamate ligament) to the base of the hamulus, not its tip, and thus its extension and orientation likely have little effect on FCU function.

In Asian apes, the capitate and hamate metacarpal articulations are smooth and slightly dorsopalmarly concave, and the Mc5 articulation is more proximally oriented, falling more in line with the remaining ulnar CM facets compared with other haplorhines. In contrast, the African ape capitate and hamate metacarpal articular areas, particularly in *Pan*, are concavo-convex, creating a more complex and stable articulation that limits sliding and rotation at these joints (Marzke and Marzke 1987; Begun 2004). The trapezoid-Mc2 articulation has also been described as more keeled than that of other primates (Begun 2004), although this is debatable as hylobatids and many Old World monkeys also show similar keeling. Altogether, the complexity of the ulnar CM articulations in African apes has been functionally associated with increasing stability needed to resist compressive or shear forces during knuckle-walking (Richmond et al. 2001; Begun 2004).

Humans arguably have the most derived ulnar CM condition among primates. In humans, the trapezoid-Mc2 articulation is oriented in more of a radioulnar plane, rather than sagittal (proximodistal) plane, as in other apes (Lewis 1989; Tocheri et al. 2005; Tocheri 2007). The proximal Mc3 displays a distinct styloid process at its radiodorsal corner, and the corresponding portion of the capitate is bevelled. The styloid process is thought to result from a separate ossification center fusing, via an embryonic migration, to the base of the Mc3 instead of the dorsoradial corner of the capitate as in other primates (Marzke and Marzke 1987; Lewis 1989; Lovejoy et al. 2009). A separate "os styloideum" is found in about 6 % of humans (O'Rahilly 1953) and rarely in other primates (Marzke and Marzke 1987). Because a styloid process is found consistently only in humans and Neanderthals (i.e., committed tool-using hominins), it is thought to help stabilize the intercarpal joints during forceful and complex manipulative tasks and is associated with the suite of changes that evolved in the human radial CM joints described above (Marzke and Marzke 1987, 2000; Marzke 2013).

The human hamate-Mc4/Mc5 articulation is much flatter than the more complex articulation of African apes, with only a slight dorsopalmar concavity on the hamate. However, the hamate facet for the Mc5 is often described as saddle-shaped, with a slightly radioulnar convexity as well (Lewis 1989; Marzke and Marzke 2000). The Mc4 and Mc5 facets do not extend onto the hamulus. Furthermore, the hamate facet of the Mc5 is proportionately larger than that of the Mc4, while the opposite is true in African apes (Marzke et al. 1992; Orr et al. 2013). Together, the relatively enlarged and saddle-shaped Mc5 facet have been hypothesized to be related to greater loading of the ulnar digits and allowing slight rotation during flexion of the fifth digit during forceful precision and power-squeeze grips (Marzke et al. 1992). However, the relatively larger Mc5 facet in humans likely reflects a reduction in the Mc4 facet rather than enlargement of the Mc5 itself. Orr et al. (2013) suggested that reduction of the Mc4 is related biomechanically to reducing obstruction for Mc5 movement, more specifically, for increasing rotation of the Mc5 during flexion as it opposes the thumb.

11 Conclusions

Just as Wood Jones (1916) and Napier (1961) described the human hand as generalized and primitive compared with most mammals, the primate carpus also can be considered as such. The retention of eight or nine wrist bones, like the retention of five digits, is primitive among mammals and differs from the more derived occurrence of carpal fusions in many mammal orders, including those closely related to primates. But the primitiveness of the primate carpus stems from more than just having retained independent carpal bones; the conjunct movement of the carpal bones (i.e., close packing in extension, ulnar deviation and pronation) is a primitive condition among mammals as well. Even the saddle-shaped trapezium-Mc1 joint that permits the opposability of thumb—a defining feature of the primate hand—is considered primitive among mammals. Primates likely retain this primitiveness because it allows for a greater degree of versatility in wrist (and hand) function necessary for the complex, three-dimensional locomotor and manipulative environments they inhabit.

That being said, there have been some changes in wrist morphology from this primitive condition across primates: some subtle and some not so subtle. The more subtle differences in carpal structure, such as relative sizes of particular carpal bones, slight changes in orientation or size of facets, or development of ridges or more complex articular morphology, translate into slight variations in mobility that are generally consistent with differences in locomotor behavior and habitual hand use. Not-so-subtle changes include convergent changes in the antebrachiocarpal joint, such that hominoids, lorisids and, to a lesser extent, spider monkeys have reduced or no contact between the ulna and carpus. Similar derived morphology across other mammals is only known in highly suspensory two-toed sloths (Mendel 1979). Scaphoid-os centrale fusion in African apes, humans and a few strepsirrhine taxa is also a derived feature of the carpus, reminiscent of carpal fusions that are found in closely related mammals, though the development, functional or phylogenetic reasons for this fusion across different primate clades is still unclear.

It is interesting to note that across primates, taxa with extremely specialized hands, such as aye-ayes (elongated third and fourth digits), colobines and spider monkeys (reduced thumbs) or lorisids (highly divergent thumb and reduced index finger for pincer-like grasping) could be described as showing relatively limited change in their carpal morphology. In other words, the derived changes appear more so in the digits rather than the carpus. Somewhat ironically, the taxon with the most derived changes to the carpus is arguably humans, which shows scaphoid-os centrale fusion, reorientation of the radial carpals, development of a styloid process and extreme reduction in the pisiform (see Chap. 18). However, despite these changes, Wood Jones and Napier's description of the human hand as primitive is still a valid assessment for the human and nonhuman primate carpus within the broader context of mammals.

12 Future Directions

There is still much to be learned about the basic morphology of the carpal bones, particularly the lunate, triquetrum and trapezoid, that remains relatively understudied across all primates compared with other carpals. There has been a historical focus on the great ape carpus given their close phylogenetic relationship to humans. Still, we have a comparatively poor understanding about the intraspecific variation within each taxon, or how subspecies might differ based on variation in ecology or frequency of locomotor behaviors (e.g., Tocheri et al. 2011). A common caveat of functional analyses is that morphological variability *within* extant species may hinder our ability to use living species as models for the functional interpretation of morphology in fossils (Marzke et al. 1994). Given the particular importance African ape morphology plays in the functional interpretation of fossil hominin and hominoid morphology (e.g., Beard et al. 1986; Ward et al. 1999; Tocheri et al. 2007; Kivell and Begun 2009; Lovejoy et al. 2009; Begun and Kivell 2011; Kivell 2011; Kivell et al. 2011; see Chap. 18), understanding both intraspecific and interspecific variation in these taxa is especially important.

Furthermore, it is commonly stated or assumed that Old World monkey carpal morphology is generally similar across the clade (e.g., Corruccini 1978; Lewis 1989; Rafferty 1990). Further research is needed to determine if this representation holds true for all Old World monkey wrists, given the extensive variation in (1) arboreal/terrestrial locomotor behaviors (including high frequencies of bridging and suspensory behaviors in some taxa), (2) ecology (e.g., highly terrestrial macaques *versus* cliff-climbing langurs), (3) hand postures [e.g., palmigrade, graspwalk or digitigrade (Hunt et al. 1996)] or (4) autapomorphic hand morphologies (e.g., reduction in the thumb in colobines) throughout the clade. This would provide a better comparative context for understanding the variation in hominoid morphology, particularly when Miocene fossils are considered (Chap. 17), and the seemingly greater variation in New World monkey carpal morphology (Rafferty 1990), although the latter is also relatively understudied.

Four recent methodological advancements (see Chap. 9) are ideal for analyzing the complex and irregular shape of carpal bones. Firstly, 3D analysis of external morphology (e.g., via surface scanning) has proven to be a much more objective and informative method for quantifying and comparing morphology across taxa than traditional 2D measures or qualitative descriptions (e.g., Tocheri et al. 2003, 2005; Orr et al. 2013). Secondly, analyses of the internal bone structure, including cortical and trabecular bone distribution, may help to reveal variation in how individual carpal bones are habitually loaded during life. Initial analyses of trabecular bone in the primate carpal bones using traditional volume-of-interest-based methods have proven functionally uninformative (Schilling et al. 2014). However, new methods of analyzing the distribution of cortical thickness and trabecular structure throughout the bone promise to reveal greater functional information (Gross et al. 2014; Skinner et al. 2015).

Thirdly, although single-plane cineradiography (e.g., Jenkins and Fleagle 1975) offers in vivo information on carpal movement during "natural" behavior, in vitro 3D kinematic studies, like that of Orr et al.(2010) are able to provide a much greater understanding of the complexity of carpal movement, particularly of individual bones. Finally, MRI or X-ray Reconstruction of Moving Morphology (X-ROMM) 3D imaging technology allows one to visualize and quantify skeletal movement in vivo. This has been applied successfully to the human carpus (e.g., Crisco et al. 2005; Moritomo et al. 2006; Pillai et al. 2007). Ideally, application of such methods to nonhuman primates (although not without its ethical challenges) would provide the much-needed information on individual carpal movements and range of motion during "natural" locomotor or manipulative behaviors (compared to "imposed" postures on cadaveric or sedated specimens). Application of these methods, particularly to understudied Old and New World monkeys and most strepsirrhines, will greatly improve our understanding of the morphology and subsequent functional interpretations of both extant and fossil carpal bones.

Acknowledgments This review is the result of the knowledge, guidance, encouragement, expertise, generosity and patience of many, many people. Among them, I am grateful to David Begun with whom my interest in the wrist and hand initially developed, Daniel Schmitt and Roshna Wunderlich who taught and inspired me think about this anatomy in a completely new way, Lee Berger and Steve Churchill whose amazing fossil discoveries gave me the opportunity to move beyond the wrist, and Matthew Skinner, Jean-Jacques Hublin, Richard Lazenby and Dieter Pahr, who gave me the opportunity to look "inside" these bones for the first time. I thank the many museum curators whose generosity allowed me access to collections in their care. I am also grateful for the many insightful discussions that I have had over the years with fellow researchers of the hand, including (but not limited to) Caley Orr, Matt Tocheri, Biren Patel, Nick Stephens, Erin Marie Williams-Hatala, Mary Marzke, Campbell Rolian, Carol Ward, Sergio Almécija, and especially Pierre Lemelin, Daniel Schmitt and Brian Richmond, whose expertise and constructive comments greatly improved this chapter in particular. This work was funded by the Natural Sciences and Engineering Research Council of Canada, General Motors Women in Science and Mathematics Award, The University of Toronto, The Max Planck Society, and the European Research Council Starting Grant #336301.

References

Altner G (1971) Histologische und vergleichend-anatomische untersuchungen zur ontogenie und phylogenie des handskeletts von *Tupaia glis* und *Microcebus murinus*. Folia Primatol 17:1–106

Beard KC, Teaford MF, Walker A (1986) New wrist bones of *Proconsul africanus* and *P. nyanzae* from Rusinga Island, Kenya. Folia Primatol 47:97–118

Begun DR (1992) Miocene fossil hominids and the chimp-human clade. Science 257:1929–1933

Begun DR (2004) Knuckle-walking and the origin of human bipedalism. In: Meldrum DJ, Hilton CE (eds) From biped to strider: the emergence of modern human walking, running, resource transport. Kluwer Academic/Plenum Publishers, New York, pp 9–33

Begun DR, Kivell TL (2011) Knuckle-walking in *Sivapithecus*? The combined effects of homology and homoplasy with possible implications for pongine dispersals. J Hum Evol 60:158–170

Belliappa PP, Burke FD (1992) Excision of the pisiform in piso-triquetral osteoarthritis. J Hand Surg Br 17:133–136

Boyer DM, Yapuncich GS, Chester SGB, Block JI, Godinot M (2013) Hands of early primates. Yearb Phys Anthropol 57:33–78
Brand PW, Hollister A (1993) Clinical mechanics of the hand. Mosby Year Book, Chicago
Cant JGH, Youlatos D, Rose MD (2001) Locomotor behavior of *Lagothrix lagothricha* and *Ateles belzebuth* in Yasuni National Park, Ecuador: general patterns and nonsuspensory modes. J Hum Evol 41:141–166
Cartmill M, Milton K (1977) The lorisiform wrist joint and the evolution of "brachiating" adaptations in the Hominoidea. Am J Phys Anthropol 47:249–272
Čihák R (1972) Ontogenesis of the skeleton and intrinsic muscles of the human hand and foot. Ergeb Anat Entwicklungsgesch 46:1–194
Cheverud JM (1981) Epiphyseal union and dental eruption in *Macaca mulatta*. Am J Phys Anthropol 56:157–167
Cooney WPI, Chao EY (1977) Biomechanical analysis of static forces in the thumb during hand function. J Bone Joint Surg Am 59:27–36
Corruccini RS (1978) Comparative osteometrics of the hominoid wrist joint, with special reference to knuckle-walking. J Hum Evol 7:307–321
Corruccini RS, Ciochon RL, McHenry HM (1975) Osteometric shape relationships in the wrist joint of some anthropoids. Folia Primatol 24:250–274
Costello MB, Fragaszy DM (1988) Prehension in *Cebus* and *Saimiri*: I. Grip type and hand preference. Am J Primatol 15:235–245
Crisco JJ, Coburn JC, Moore DC, Akelman E, Weiss A-PC, Wolfe SW (2005) *In vivo* radiocarpal kinematics and the dart thrower's motion. J Bone Joint Surg Am 87A:2729–2740
Curgy J-J (1965) Apparition et soudure des points d'ossification des membres chez les mammifères. Mém Mus Natl d'Histoire Nat Sér A Zool 32:173–307
Dainton M, Macho GA (1999) Did knuckle-walking evolve twice? J Hum Evol 36:171–194
Daver G, Berillon G, Grimaud-Hervé D (2012) Carpal kinematics in quadrupedal monkeys: towards a better understanding of wrist morphology and function. J Anat 220:42–56
Diogo R, Wood B (2011) Soft-tissue anatomy of the primates: phylogenetic analyses based on the muscles of the head, neck, pectoral region and upper limb, with notes on the evolution of these muscles. J Anat 219:273–359
Eckstein FM (1944) The pisiform bone. Nature 154:182
Etter H-UF (1974) Morphologisch- und metrisch-vergleichende Untersuchung am Handskelett rezenter Primaten Teil II. Gegenbaurs Morphol Jahrb 120:153–171
Feipel V, Rooze M, Louryan S, Lemort M (1994) Functional anatomy of the carpus in flexion and extension and in radial and ulnar deviations: an *in vivo* two- and three-dimensional CT study. In: Schuind F, An K-N, Cooney WP III, Garcia-Elias M (eds) Advances in the biomechanics of the hand and wrist. Plenum Press, New York, pp 255–270
Fisk G (1981) Biomechanics of the wrist joint. In: Tubiana R (ed) The hand, vol 1. W.B. Saunders, Philadelphia, pp 191–201
Fleagle JG, Meldrum DJ (1988) Locomotor behavior and skeletal morphology of two sympatric pithecine monkeys, *Pithecia pithecia* and *Chiropotes satanas*. Am J Primatol 16:227–249
Flower WH (1885) An introduction to the osteology of the mammalia. Macmillan, London
Galliari CA (1988) A study of postnatal appendicular skeletal maturation in captive-born squirrel monkeys (*Saimiri boliviensis*). Am J Primatol 16:51–61
Garcia-Elias M, Smith DK, Ruby LK, Horii E, An K-N, Linscheid RL, Chao EY, Cooney WP (1994) Normal and abnormal carpal kinematics. In: Schuind F, An K-N, Cooney WP, Garcia-Elias M (eds) Advances in the biomechanics of the hand and wrist. Plenum, New York, pp 247–253
Gebo DL (1996) Climbing, brachiation and terrestrial quadrupedalism: historical precursors of hominid bipedalism. Am J Phys Anthropol 101:55–92
Giebel CG (1879) Saügethiere: mammalia. In: Bronn HG (ed) Klassen und Ordnungen des Thierreichs wissenschaftlich dargestellt in Wort und Bild 6. C.F. Winter'sche Verlagshandlung, Leipzig, pp 516–541
Gillies CD (1929) The origin of the os pisiforme. J Anat 63:380–383

Glassman DM (1983) Growth and development in the saddle-back tamarin: the sequence and timing of dental eruption and epiphyseal union. Am J Primatol 5:51–59

Godinot M, Beard KC (1991) Fossil primate hands: a review and an evolutionary inquiry emphasizing early forms. Hum Evol 6:307–354

Greulich WW, Pyle SI (1959) Atlas of skeletal development of the hand and wrist. Stanford University Press, Stanford

Gross T, Kivell TL, Skinner MM, Nguyen NH, Pahr DH (2014) A CT-image-based framework for the holistic analysis of cortical and trabecular bone morphology. Palaeontol Electron 17:1–13

Haines RM (1944) The mechanism of rotation at the first carpo-metacarpal joint. J Anat 78:44–46

Haines RW (1958) Arboreal or terrestrial ancestry of placental mammals. Q Rev Biol 33:1–23

Hamrick MW (1996a) Functional morphology of the lemuriform wrist joints and the relationship between wrist morphology and positional behavior in arboreal primates. Am J Phys Anthropol 99:319–344

Hamrick MW (1996b) Articular size and curvature as determinants of carpal joint mobility and stability in strepsirrhine primates. J Morphol 230:113–127

Hamrick MW (1997) Functional osteology of the primate carpus with special reference to strepsirhini. Am J Phys Anthropol 104:105–116

Harris HA (1944) The pisiform bone. Nature 153:715

Howell AB, Straus WL Jr (1933) The muscular system. In: Hartman GG, Straus WL Jr (eds) The anatomy of the rhesus monkey (*Macaca mulatta*). Hafner, New York, pp 89–175

Hunt KD, Cant JGH, Gebo DL, Rose MD, Walker SE, Youlatos D (1996) Standardized descriptions of primate locomotor and postural modes. Primates 37:363–387

Huxley TH (1863) Evidence as to man's place in nature. Williams and Norgate, London

Jenkins FA Jr (1971) Limb posture and locomotion in the Virginia opossum (*Didelphis marsupialis*) and in other non-cursorial mammals. J Zool 165:303–315

Jenkins FA Jr (1981) Wrist rotation in primates: a critical adaptation for brachiators. Symp Zool Soc Lond 48:429–451

Jenkins FA Jr, Fleagle JG (1975) Knuckle-walking and the functional anatomy of the wrists in living apes. In: Tuttle RH (ed) Primate functional morphology and evolution. Mouton, The Hague, pp 213–231

Jouffroy FK (1975) Osteology and myology of the lemuriform postcranial skeleton. In: Tattersall I, Sussman RW (eds) Lemur biology. Plenum Press, New York, pp 149–192

Jouffroy FK (1991) La "main sans talon" du primate bipède. In: Coppens Y (ed) Origine(s) de la Bipédie chez les Hominidés. Center National de la Recherche Scientifique, Paris, pp 21–35

Jouffroy FK, Lessertisseur J (1979) Relationships between limb morphology and locomotor adaptations among prosimians: an osteometric study. In: Morbeck ME, Preuschoft H, Gomberg N (eds) Environment, behavior, and morphology: dynamic interactions in primates. Gustav Fischer, New York, pp 143–181

Jouffroy FK, Medina MF (2002) Radio-ulnar deviation of the primate carpus: an x-ray study. Z Morphol Anthropol 83:275–289

Kauer JMG (1986) The mechanism of the carpal joint. Clin Orthop Relat Res 202:16–26

Kindahl M (1944) On the development of the hand and foot in *Tarsius tarsius* and *Microcebus myoxinus*. Acta Zool 25:49–58

Kivell TL (2007) Ontogeny of the hominoid midcarpal joint and implications for the origin of human bipedalism. Ph.D. dissertation, University of Toronto

Kivell TL (2011) A comparative analysis of the hominin triquetrum (SKX 3498) from Swartkrans, South Africa. S Afr J Sci 107:60–69

Kivell TL, Begun DR (2007) Frequency and timing of scaphoid-centrale fusion in hominoids. J Hum Evol 52:321–340

Kivell TL, Begun DR (2009) New primate carpal bones from Rudabánya (late Miocene, Hungary): taxonomic and functional implications. J Hum Evol 57:697–709

Kivell TL, Schmitt D (2009) Independent evolution of knuckle-walking in African apes shows that humans did not evolve from a knuckle-walking ancestor. Proc Natl Acad Sci U S A 106:14241–14246

Kivell TL, Kibii JM, Churchill SE, Schmid P, Berger LR (2011) *Australopithecus sediba* hand demonstrates mosaic evolution of locomotor and manipulative abilities. Science 333:1411–1417

Kjosness KM, Hines JE, Lovejoy CO, Reno PL (2014) The pisiform growth plate is lost in humans and supports a role for *Hox* in growth plate formation. J Anat 225:527–538
Kuczynski K (1974) Carpometacarpal joint of the human thumb. J Anat 118:119–126
Leboucq H (1884) Recherches sur la morphologie du carpe chez les mammifères. Arch Biol 5:35–102
Lemelin P, Schmitt D (1998) The relation between hand morphology and quadrupedalism in primates. Am J Phys Anthropol 105:185–197
Lemelin P, Hamrick MW, Richmond BG, Godfrey LR, Jungers WL, Burney DA (2008) New hand bones of *Hadropithecus stenognathus*: implications for the paleobiology of the Archaeolemuridae. J Hum Evol 54:405–413
Lewis OJ (1965) Evolutionary change in the primate wrist and inferior radioulnar joints. Anat Rec 151:275–286
Lewis OJ (1969) The hominoid wrist joint. Am J Phys Anthropol 30:251–268
Lewis OJ (1970) The development of the human wrist joint during the fetal period. Anat Rec 166:499–516
Lewis OJ (1971a) Brachiation and the early evolution of the Hominoidea. Nature 230:577–578
Lewis OJ (1971b) The contrasting morphology found in the wrist joint of semibrachiating monkeys and brachiating apes. Folia Primatol 16:248–256
Lewis OJ (1972a) Osteological features characterizing the wrists of monkeys and apes, with a reconsideration of this region in *Dryopithecus* (*Proconsul*) *africanus*. Am J Phys Anthropol 36:45–58
Lewis OJ (1972b) Evolution of the hominoid wrist. In: Tuttle RH (ed) The functional and evolutionary biology of primates. Aldine-Atherton, Chicago, pp 201–222
Lewis OJ (1974) The wrist articulations of the Anthropoidea. In: Jenkins FA Jr (ed) Primate locomotion. Academic Press, New York, pp 143–169
Lewis OJ (1977) Joint remodeling and the evolution of the human hand. J Anat 123:157–201
Lewis OJ (1985a) Derived morphology of the wrist articulations and theories of hominoid evolution. Part I The lorisine joints. J Anat 140:447–460
Lewis OJ (1985b) Derived morphology of the wrist articulations and theories of hominoid evolution. Part II The midcarpal joints of higher primates. J Anat 142:151–172
Lewis OJ (1989) Functional morphology of the evolving hand and foot. Clarendon Press, Oxford
Lewis OJ, Hamshere RJ, Bucknill TM (1970) The anatomy of the wrist joint. J Anat 106:539–552
Lorenz R (1971) The functional interpretation of the thumb in the Hylobatidae. Proc. 3rd Int. Congr. Primat., Zurich 1970, vol 1, pp 130–136
Lovejoy CO, Heiple KG, Meindl RS (2001) Palaeoanthroplogy: did our ancestors knuckle-walk? Nature 410:325–326
Lovejoy CO, Simpson SW, White TD, Asfaw B, Suwa G (2009) Careful climbing in the Miocene: the forelimbs of *Ardipithecus ramidus* and humans are primitive. Science 326:70e-1-8
Lucae JCG (1865) Die Hand und der Fuss. Ein Beitrag zur vergleichenden Osteologie der Menschen, Affen und Beutelthiere. Abhandlungen Senckenbergischen Naturforschende Gesellschaft 5:275–332
Lyser M (1653) *Culter anatomicus*. Copenhagen
MacConaill MA (1941) The mechanical anatomy of the carpus and its bearing on some surgical problems. J Anat 75:166–175
Martin RB, Burr DB, Sharkey N (1998) Skeletal tissue mechanics. Springer, New York
Marzke MW (1971) Origin of the human hand. Am J Phys Anthropol 34:61–84
Marzke MW (1983) Joint function and grips of the *Australopithecus afarensis* hand, with special reference to the region of the capitate. J Hum Evol 12:197–211
Marzke MW (2013) Tool making, hand morphology and fossil hominins. Phil Trans R Soc B 368:1–8
Marzke MW, Marzke RF (1987) The third metacarpal styloid process in humans: origin and functions. Am J Phys Anthropol 73:415–431
Marzke MW, Marzke RF (2000) Evolution of the human hand: approaches to acquiring, analysing and interpreting the anatomical evidence. J Anat 197:121–140
Marzke MW, Morbeck ME, Alongi C, Fritz P (1987) Ossification of the hand and wrist in the chimpanzee. Am J Primatol 12:359 (abstract)

Marzke MW, Wullstein KL, Viegas SF (1992) Evolution of the power ("squeeze") grip and its morphological correlates in hominids. Am J Phys Anthropol 89:283–298

Marzke MW, Wullstein KL, Viegas S (1994) Variability at the carpometacarpal and midcarpal joints involving the fourth metacarpal, hamate and lunate in Catarrhini. Am J Phys Anthropol 93:229–240

Marzke MW, Toth N, Schick K, Reece S, Steinberg B, Hunt K, Linscheid RL, An K-N (1998) EMG study of hand muscle recruitment during hard hammer percussion manufacture of Oldowan tools. Am J Phys Anthropol 105:315–332

Mendel FC (1979) The wrist joint of two-toed sloths and its relevance to brachiating adaptations in the Hominoidea. J Morphol 162:413–424

Michejda M, Bacher J (1980) Skeletal age as a determinant of gestation in *Macaca mulatta:* a radiographic study. J Med Primatol 10:293–301

Mivart SG (1867) On the appendicular skeleton of the Primates. Phil Trans R Soc Lond 157:299–429

Mivart SG (1869) Contributions towards a more complete knowledge of the skeleton of the Primates. Part I The appendicular skeleton of Simia. Trans Zool Soc Lond 6:175–225

Moojen TM, Snel JG, Ritt MJ, Venema HW, Kauer JM, Bos KE (2003) *In vivo* analysis of carpal kinematics and comparative review of the literature. J Hand Surg 28A:81–87

Moritomo H, Viegas SF, Elder K, Nakamura K, DaSilva MF, Patterson RM (2000a) The scaphotrapezio-trapezoidal joint. Part 2: a kinematic study. J Hand Surg Am 25A:911–920

Moritomo H, Viegas SF, Nakamura K, DaSilva MF, Patterson RM (2000b) The scaphotrapezio-trapezoidal joint Part 1: an anatomic and radiographic study. J Hand Surg Am 25A:899–910

Moritomo H, Murase T, Goto A, Oka K, Sugamoto K, Yoshikawa H (2006) *In vivo* three-dimensional kinematics of the midcarpal joint of the wrist. J Bone Joint Surg 88A:611–621

Napier JR (1955) The form and function of the carpometacarpal joint of the thumb. J Anat 89:362–369

Napier JR (1961) Prehensility and opposability in the hands of primates. Symp Zool Soc Lond 5:115–132

Napier JR, Napier PH (1967) A handbook of living primates. Academic Press, New York

Nayak UV (1933) The articulations of the carpus in *Chiromys madagascarensis* with reference to certain other lemurs. J Anat 68:109–115

Neu CP, Crisco JJ, Wolfe SW (2001) *In vivo* kinematic behavior of the radio-capitate joint during wrist flexion-extension and radio-ulnar deviation. J Biomech 34:1429–1438

Newell-Morris L, Tarrant LH, Farenbach CE, Sackett GP (1980) Ossification in the hand of the pigtail macaque (*Macaca nemestrina*) 2: order of appearance of centers and variability in sequence. Am J Phys Anthropol 53:423–439

Nissen HW, Riesen AH (1949) Onset of ossification in the epiphyses and short bones of the extremities in chimpanzee. Growth 13:45–70

Noback CV (1930) Digital epiphyses and carpal bones in the growing infant female gorilla with sitting height and estimated age. Zoologica 11:117–151

O'Connor BL (1975) Functional morphology of cercopithecoid wrist and inferior radioulnar joints, and their bearing on some problems in evolution of Hominoidea. Am J Phys Anthropol 43:113–121

O'Rahilly R (1953) A survey of carpal and tarsal anomalies. J Bone Joint Surg Am 35:626–642

Orr CM, Leventhal EL, Chivers SF, Marzke MW, Wolfe SW, Crisco JJ (2010) Studying primate carpal kinematics in three dimensions using a computed-tomography-based markerless registration method. Anat Rec 293:692–709

Orr CM, Tocheri MW, Burnett SE, Due Awe R, Saptomo EW, Sutkina T, Jamiko, Wasisto S, Morwood MJ, Jungers WL (2013) New wrist bones of *Homo floresiensis* from Liang Bua (Flores, Indonesia). J Hum Evol 64:109–129

Owen R (1866) Osteological contributions to the natural history of the anthropoid apes (no. VII): comparison of the bones of the limbs of the *Troglodytes Gorilla*, *Troglodytes niger*, and of different varieties of the human race and on the general characters of the skeleton in *Gorilla*. Trans Zool Soc Lond 5:8–27

Patel BA (2010) The interplay between speed, kinetics, and hand postures during primate terrestrial locomotion. Am J Phys Anthropol 141:222–234

Patel BA, Wunderlich RE (2010) Dynamic pressure patterns in the hands of olive baboons (*Papio anubis*) during terrestrial locomotion: implications for cercopithecoid primate hand morphology. Anat Rec 293:710–718
Patel BA, Larson SG, Stern JT Jr (2012) Electromyography of wrist and finger flexor muscles in olive baboons (*Papio anubis*). J Exp Biol 215:115–123
Phillips IR (1976) Skeletal development in the foetal and neonatal marmoset (*Callithrix jacchus*). Lab Anim 10:317–333
Pillai RR, Thoomukuntla B, Ateshian GA, Fischer KJ (2007) MRI-based modelling for evaluation of *in vivo* contact mechanics in the human wrist during active light grasp. J Biomech 40:2781–2787
Playfair McMurrich J (1914) The nomenclature of the carpal bones. Anat Rec 8:173–182
Rafferty KL (1990) The functional and phylogenetic significance of the carpometacarpal joint of the thumb in anthropoid primates. M.A. dissertation, New York University
Richmond BG (2006) Functional morphology of the midcarpal joint in knuckle-walkers and terrestrial quadrupeds. In: Ishida H, Pickford M, Tuttle R, Ogihara N, Nakatsukasa M (eds) Human origins and environmental backgrounds. Springer, New York, pp 105–122
Richmond BG, Begun DR, Strait DS (2001) Origin of human bipedalism: the knuckle-walking hypothesis revisited. Yearb Phys Anthropol 44:70–105
Richmond BG, Strait DS (2000) Evidence that humans evolved from a knuckle-walking ancestor. Nature 404:382–385
Rose MD (1992) Kinematics of the trapezium-1st metacarpal joint in extant anthropoids and Miocene hominoids. J Hum Evol 22:255–266
Sarmiento EE (1985) Functional differences in the skeleton of wild and captive orang-utans and their adaptive significance. Ph.D. dissertation, New York University
Sarmiento EE (1988) Anatomy of the hominoid wrist joint: its evolutionary and functional implications. Int J Primatol 9:281–345
Sarmiento EE (1994) Terrestrial traits in the hands and feet of gorillas. Am Mus Novit 309:1–56
Scheuer L, Black S (eds) (2000) Developmental juvenile osteology. Academic Press, San Diego
Schilling A-M, Tofanelli S, Hublin J-J, Kivell TL (2014) Trabecular bone structure in the primate wrist. J Morphol 275:572–585
Schmitt D (1994) Forelimb mechanics as a function of substrate type during quadrupedalism in two anthropoid primates. J Hum Evol 26:441–457
Schultz AH (1936) Characters common to higher primates and characters specific to man. Q Rev Biol 11:259–283
Schultz AH (1944) Age changes and variability in gibbons. Am J Phys Anthropol 2:1–129
Schwartz JH, Yamada TK (1998) Carpal anatomy and primate relationships. Anthropol Sci 106(Suppl):47–65
Sirianni JE, Swindler DR (1985) Growth and development of the pigtailed macaque. CRC Press, Boca Raton
Skinner MM, Stephens NB, Tsegai ZJ, Foote AC, Nguyen NH, Gross T, Pahr DH, Hublin J-J, Kivell TL (2015) Human-like hand use in *Australopithecus africanus*. Science 347:395–399
Sonenblum SE, Crisco JJ, Kang L, Akelman E (2004) *In vivo* motion of the scaphotrapeziotrapezoidal (STT) joint. J Biomech 37:645–652
Spinozzi G, Truppa V, Laganà T (2004) Grasping behavior in tufted capuchin monkeys (*Cebus apella*): grip types and manual laterality for picking up a small food item. Am J Phys Anthropol 125:30–41
Stafford BJ, Thorington RW Jr (1998) Carpal development and morphology in archontan mammals. J Morphol 235:135–155
Strong RM (1925) The order, time, and rate of ossification of the albino rat (*Mus norvegicus albinus*) skeleton. Am J Anat 36:313–355
Taleisnik J (1985) The wrist. Churchill Livingstone, New York
Tanner JM, Whitehouse RH, Cameron N, Marshall WA, Healey MJR, Goldstein H (1983) Assessment of skeletal maturity and prediction of adult height, 2nd edn. Academic Press, London
Thurm DA, Samonds KW, Fleagle JG (1975) An atlas for the skeletal maturation of the cebus monkey; the first year. Harvard School of Public Health, Department of Nutrition, Boston

Tocheri MW (2007) Three-dimensional riddles of the radial wrist: derived carpal and carpometacarpal joint morphology in the genus *Homo* and the implications for understanding the evolution of stone tool-related behaviors in hominins. Ph.D. dissertation, Arizona State University

Tocheri MW, Marzke MW, Liu D, Bae M, Jones GP, Williams RC, Razdan A (2003) Functional capabilities of modern and fossil hominid hands: three-dimensional analysis of trapezia. Am J Phys Anthropol 122:101–112

Tocheri MW, Razdan A, Williams RC, Marzke MW (2005) A 3D quantitative comparison of trapezium and trapezoid relative articular and nonarticular surface areas in modern humans and great apes. J Hum Evol 49:570–586

Tocheri MW, Orr CM, Jacofsky MC, Marzke MW (2008) The evolutionary history of the hominin hand since the last common ancestor of Pan and Homo. J Anat 212:544–562.

Tocheri MW, Orr CM, Larson SG, Sutikna T, Jatmiko, Saptomo EW, Awe Due R, Djubiantono T, Morwood MJ, Jungers WL (2007) The primitive wrist of *Homo floresiensis* and its implications for hominin evolution. Science 317:1743–1745

Tocheri MW, Solhan CR, Orr CM, Femiani J, Frohlich B, Groves CP, Harcourt-Smith WE, Richmond BG, Shoelson B, Jungers WL (2011) Ecological divergence and medial cuneiform morphology in gorillas. J Hum Evol 60:171–184

Tuttle RH (1967) Knuckle-walking and the evolution of hominoid hands. Am J Phys Anthropol 26:171–206

Tuttle RH (1969) Quantitative and functional studies on the hands of Anthropoidea; I The Hominoidea. J Morphol 128:309–364

Tuttle RH (1975) Parallelism, brachiation, and hominoid phylogeny. In: Luckett WP, Szalay FS (eds) Phylogeny of the primates: a multidisciplinary approach. Plenum Press, New York, pp 447–480

van Wagenen G, Asling CW (1964) Ossification in the fetal monkey (*Macaca mulatta*). Am J Anat 114:107–132

Ward CV (2002) Interpreting the posture and locomotion of *Australopithecus afarensis*: where do we stand? Yearb Phys Anthropol 45:185–215

Ward CV, Leakey MG, Brown B, Brown F, Harris J, Walker A (1999) South Turkwel: a new Pliocene hominid site in Kenya. J Hum Evol 36:69–95

Watts ES (1990) A comparative study of neonatal skeletal development in *Cebus* and other primates. Folia Primatol 54:217–224

Whitehead PF (1993) Aspects of the anthropoid wrist and hand. In: Gebo DL (ed) Postcranial adaptation in nonhuman primates. Northern Illinois University Press, DeKalb, pp 96–120

Williams SA (2010) Morphological integration and the evolution of knuckle-walking. J Hum Evol 58:432–440

Winkler LA (1996) Appearance of ossification centers of the lower arm, wrist, lower leg, and ankle in immature orangutans and chimpanzees with an assessment of the relationship of ossification to dental development. Am J Phys Anthropol 99:191–203

Wolfe SW, Neu CP, Crisco JJ III (2000) *In vivo* scaphoid, lunate and capitate kinematics in wrist flexion and extension. J Hand Surg 25A:860–869

Wood Jones F (1916) Arboreal man. Edward Arnold, London

Wood Jones F (1942) The principles of anatomy as seen in the Hand, 2nd edn. Baillière, Tindall and Cox, London

Yalden DW (1970) The functional morphology of the carpal bones in carnivores. Acta Anat 77:481–500

Yalden DW (1971) The functional morphology of the carpus in ungulate mammals. Acta Anat 78:461–487

Yalden DW (1972) The form and function of the carpal bones in some arboreally adapted mammals. Acta Anat 82:383–406

Youlatos D (1996) Atelines, apes and wrist joints. Folia Primatol 67:193–198

Ziemer LK (1978) Functional morphology of forelimb joints in the woolly monkey *Lagothrix lagothricha*. Contrib Primatol 14:1–130

Zihlman AL, Bolter DR, Boesch C (2007) Skeletal and dental growth and development in chimpanzees of the Taï National Park, Côte D'Ivoire. J Zool 273:63–73

Zylstra M (1999) Functional morphology of the hominoid forelimb: implications for knuckle-walking and the origin of hominid bipedalism. Ph.D. dissertation, University of Toronto

Chapter 4
Morphological Diversity in the Digital Rays of Primate Hands

Biren A. Patel and Stephanie A. Maiolino

1 Introduction

The primate forelimb consists of three segments distal to the pectoral girdle. The most proximal is the brachium (the arm), which contains the single humerus. The middle segment is the antebrachium (the forearm), which contains two bones, the radius and ulna. The distal segment is the manus (hand), which contains several bones within the carpus (wrist), the metacarpus (palm), and the digits (fingers). Collectively, the **metacarpus** and **digits** form the digital rays of the hand. The hand, including the carpus, can account for 20–42 % of total forelimb length across primates (Table 4.1). The **rays** are numbered 1–5 from radial to ulnar. The first ray is the **thumb** (pollex), and this is followed by the **index** (demonstratorius or pointer), **middle** (medius), **ring** (annularis), and **little** (minimus) **fingers** according to the conventional terminology used for the human hand. Historically, each digit has been referred to by various names, but typical among nonhuman primates the thumb is referred to as the pollex and the remaining rays as ulnar rays 2 through 5. The interested reader is encouraged to see Napier (1980) for a more colorful and detailed history of hand ray nomenclature. The focus of this chapter is on the digital rays of the hand.

B.A. Patel (✉)
Department of Cell and Neurobiology, Keck School of Medicine,
University of Southern California, 1333 San Pablo Street, BMT 404,
Los Angeles, CA 90033, USA
e-mail: birenpat@usc.edu

S.A. Maiolino
Interdepartmental Doctoral Program in Anthropological Sciences,
Stony Brook University, Stony Brook, NY, USA

Department of Pathology and Anatomical Sciences,
University of Missouri School of Medicine, Columbia, MO 65211, USA

T.L. Kivell et al. (eds.), *The Evolution of the Primate Hand*, Developments in Primatology: Progress and Prospects, DOI 10.1007/978-1-4939-3646-5_4

Table 4.1 Relative hand length in primates

	Napier and Napier (1967)[a]		Jouffroy et al. (1993)[b]		
Genus	*N*	Mean (%)	*N*	Mean (%)	SD
Alouatta	5	27	8	27.5	1.1
Aotus	4	30	8	29.2	0.8
Arctocebus	2	24	14	21.2	1.6
Ateles	6	27	9	27.2	1.1
Avahi	2	34	6	34.0	1.5
Brachyteles	–	–	1	27.3	–
Callicebus	2	28	3	30.5	0.6
Callimico	2	28	–	–	–
Callithrix	14	29	6	30.4	1.7
Cebuella	1	30	–	–	–
Cebus	5	27	11	27.5	0.7
Cercocebus	3	26	–	–	–
Cercopithecus	13	26	11	26.5	1.5
Cheirogaleus	–	–	8	30.6	1.4
Chiropotes	1	29	2	27.8	0.5
Colobus	3	30	10	29.1	1.1
Daubentonia	4	42	9	41.2	1.0
Erythrocebus	3	23	–	–	–
Euoticus	–	–	20	33.3	0.8
Galago[c]	6	30	55	30.6	2.5
Gorilla	10	25	6	25.2	1.6
Hapalemur	1	30	9	30.4	1.7
Homo	–	–	3	24.3	0.5
Hylobates	5	26	8	25.1	1.0
Indri	2	33	7	34.9	0.9
Lagothrix	2	26	6	26.4	0.5
Lemur[d]	3	29	38	27.8	1.1
Leontopithecus	3	31	7	32.2	1.7
Lepilemur	1	32	11	32.6	0.9
Loris	6	22	10	20.5	1.4
Macaca	18	27	12	27.8	0.8
Mandrillus	3	27	–	–	–
Microcebus	1	32	16	28.8	1.0
Miopithecus	–	–	7	27.4	0.5
Nasalis	4	23	–	–	–
Nycticebus	4	23	22	25.5	1.8
Pan	9	30	14	29.3	0.9
Papio	7	23	10	24.3	1.6
Perodicticus	4	29	7	28.2	1.5
Phaner	–	–	2	32.0	1.7

(continued)

Table 4.1 (continued)

	Napier and Napier (1967)[a]		Jouffroy et al. (1993)[b]		
Genus	*N*	Mean (%)	*N*	Mean (%)	SD
Pithecia	4	28	–	–	–
Pongo	6	28	9	27.9	1.2
Presbytis	13	29	10	28.8	1.0
Propithecus	2	33	6	33.2	1.2
Pygathrix	1	27	–	–	–
Rhinopithecus	1	25	2	26.7	0.5
Saguinus	11	31	–	–	–
Saimiri	6	28	8	27.2	1.0
Semnopithecus	–	–	5	28.8	1.0
Symphalangus	–	–	3	26.2	1.4
Tarsius	3	40	11	37.7	1.4
Trachypithecus	–	–	2	29.2	0.2
Varecia	–	–	9	30.1	1.6

[a]Napier and Napier (1967) calculated relative hand length as the sum of the length of the hand (carpus length in line with the third digit + metacarpal 3 length + proximal phalanx 3 length + intermediate phalanx 3 length + distal phalanx 3 length) divided by the sum of the length of the forelimb (humerus length + radius length + hand length) multiplied by 100

[b]Jouffroy et al. (1993) calculated relative hand length as the sum of the length of the hand (carpus length + metacarpal length + proximal phalanx length + intermediate phalanx length + distal phalanx length of the longest ray) divided by the sum of the length of the forelimb (humerus length + radius length + hand length) multiplied by 100

[c]These data likely include species of both *Galago*, *Galagoides*, and *Otolemur*

[d]These data likely include species of both *Lemur* and *Eulemur*

After briefly reviewing the evolutionary history of the primate hand, followed by a discussion on intrinsic hand bone proportions, the rest of the chapter emphasizes the general osteology of the digital rays. The most proximal elements, the **metacarpals**, are discussed first, followed by the **proximal** and **intermediate phalanges**, and finally the **distal phalanges**. For each bony element, the typical morphology is described using human morphology as a reference followed by a brief review of the diversity of nonhuman primates. When available, the general bony configuration of closely related non-primate taxa (i.e., euarchontans) is also reviewed to demonstrate either similarity to or divergence from the average primate morphology. Figure 4.1 illustrates the major characters discussed in the text and shows the broad diversity in the morphology of the primate hand skeleton. In most cases, the comparative literature on primate hand bones has focused on hominoids and cercopithecoids (and will likely continue to do so for obvious reasons; but see Boyer et al. 2013). Therefore, most of the citations in this chapter reflect this bias. Where clear gaps in the literature exist, new preliminary descriptions are presented.

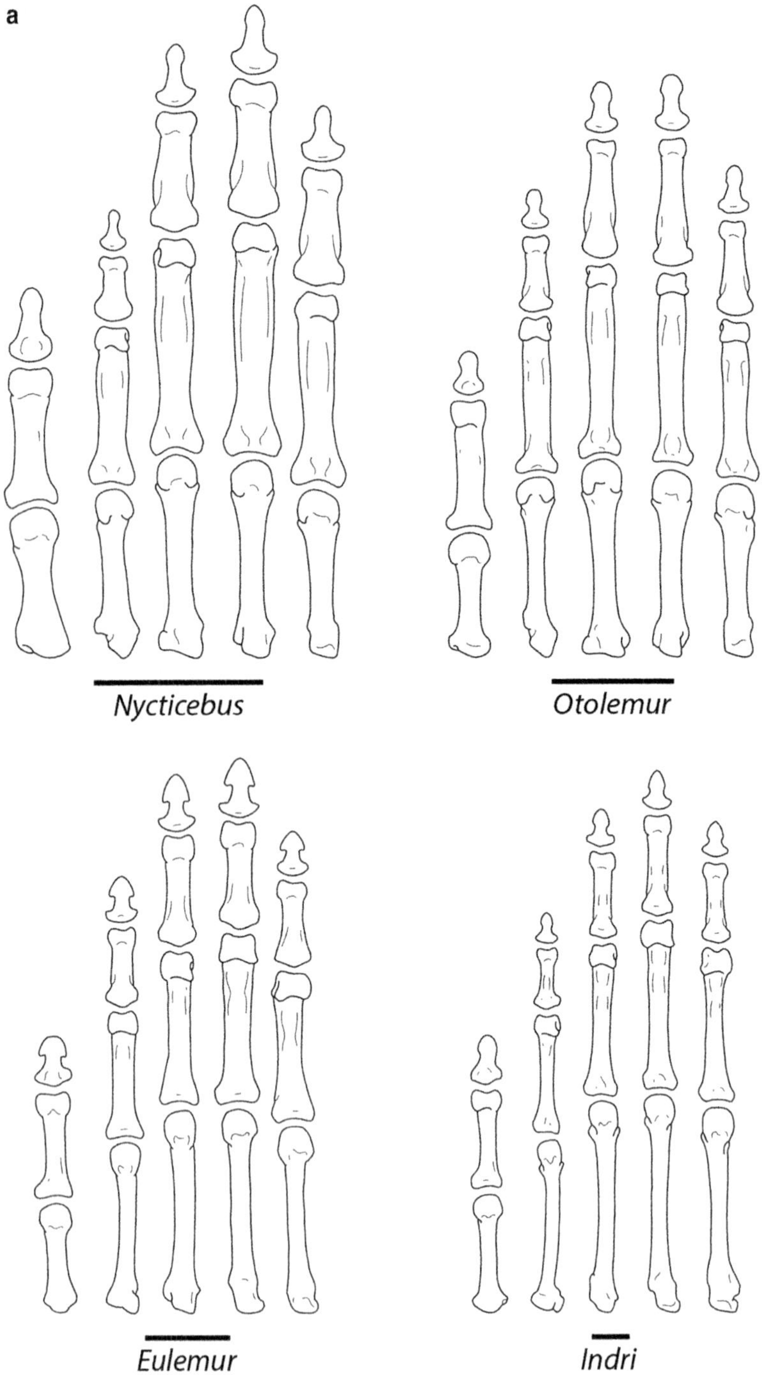

Fig. 4.1 **(a–e)** Articulated metacarpals and phalanges of the left hand (palmar view) showing major differences in ray morphology and proportions among extant primates. Each hand is scaled to the same length (based on the third metacarpal). Scale bars are 1 cm

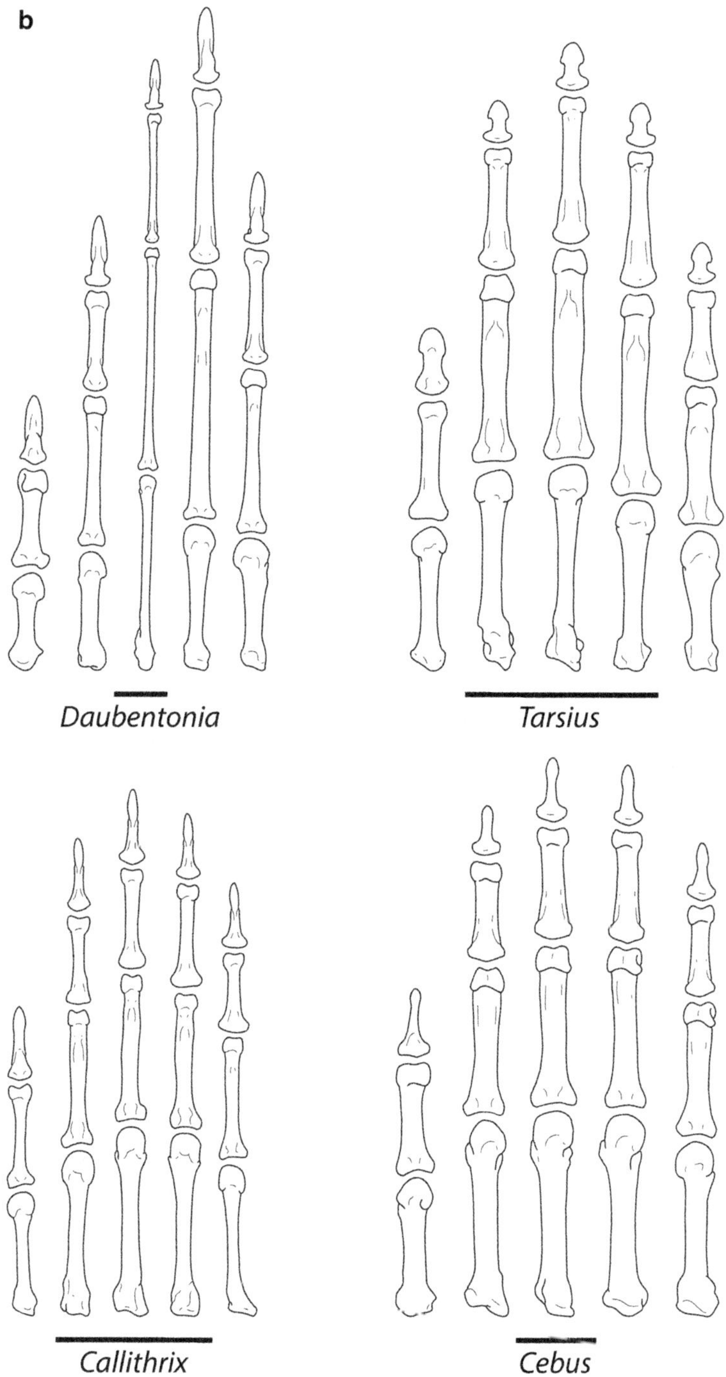

Fig. 4.1 (continued)

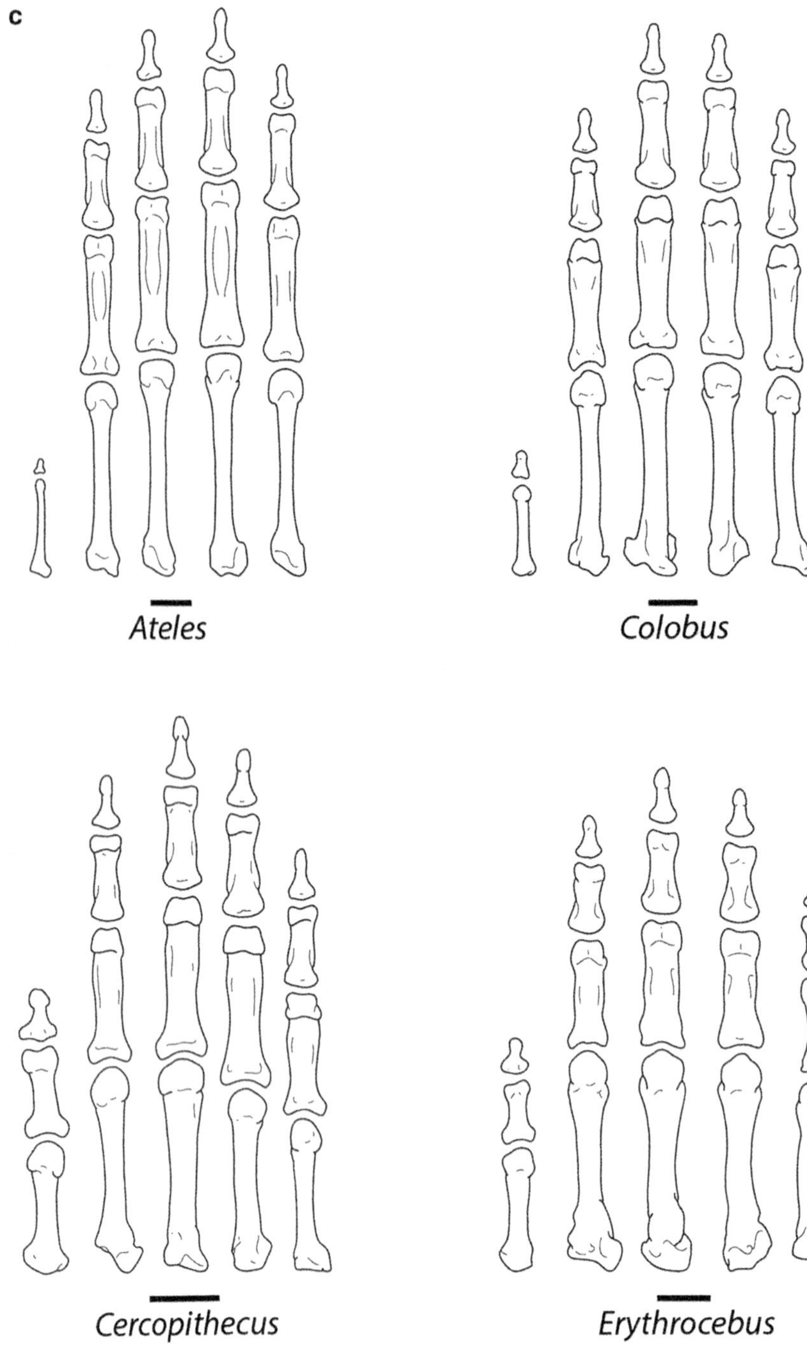

Fig. 4.1 (continued)

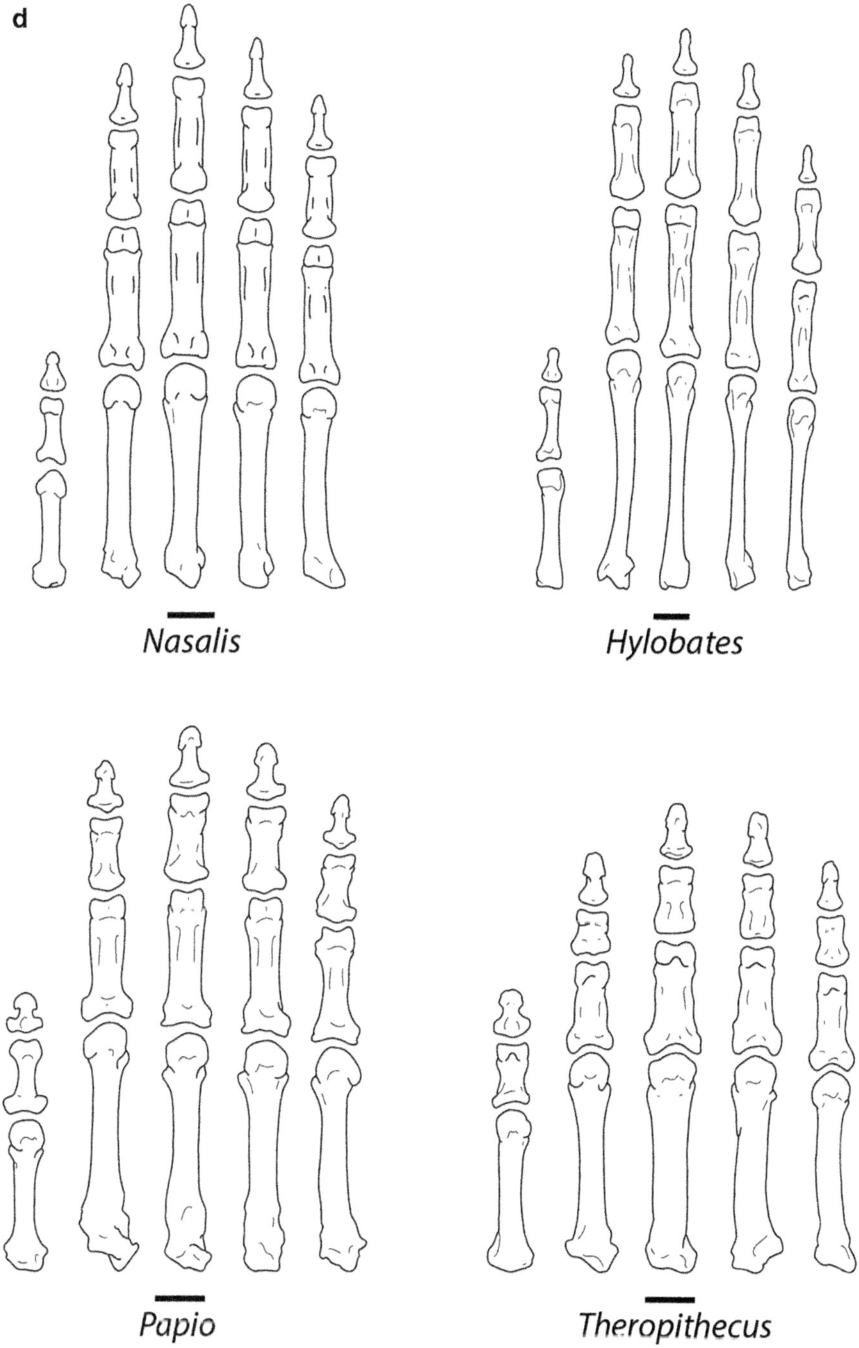

Fig. 4.1 (continued)

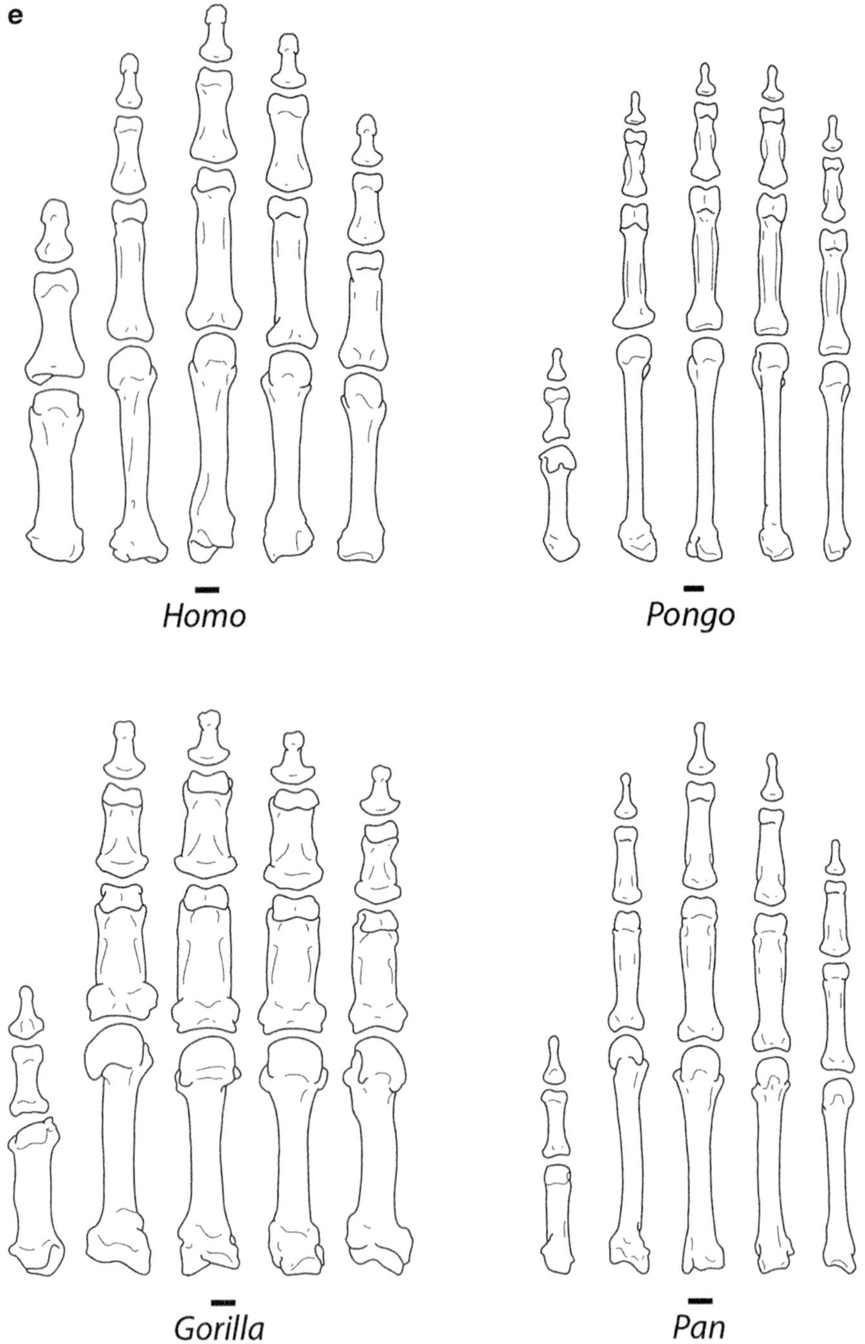

Fig. 4.1 (continued)

2 Evolutionary History of Hand Rays

Primates, like the earliest mammals from the Late Jurassic/Early Cretaceous are *pentadactyl*, meaning that they have five rays (Ji et al. 1999, 2002; Chen and Luo 2013). In fact, the pentadactyl limb arose early in tetrapod evolution; it was present in the amniote ancestor of synapsids (mammals and close relatives) and sauropsids (reptiles and their relatives) (Romer 1956). In this primitive amniote *Bauplan*, each digit is comprised of a metacarpal and several phalanges. The number of phalanges from ray 1 to 5 was 2–3–4–5–3, respectively. The pollex lost a phalanx likely before amniotes first evolved; but it was not until later, at some point or points during the therapsid lineage (a group of synapsids that includes mammals and their ancestors), that ray 3 lost one phalanx and ray 4 lost two phalanges (Hopson 1995). Thus, the first mammals had two phalanges in the pollex and three in each ulnar ray. The digital rays of primates (and other euarchontan mammals) follow this primitive mammalian pattern: the four ulnar rays comprise a metacarpal and three phalanges (proximal, intermediate/middle, and distal), and the pollical ray comprises a metacarpal and two phalanges (proximal and distal). Therefore, there are 19 bones in total across all five rays. However, it should be noted that there are exceptions to this pattern with several primate species exhibiting extreme reduction of a portion or an entire digit.

3 Lengths and Proportions

The most well-documented morphological attributes of primate rays are the absolute and relative lengths of the bones that make up each ray. Length measures have been emphasized in the literature because they are easy variables to quantify and, more importantly, because they vary along meaningful phylogenetic and functional lines (e.g., strepsirrhine vs. haplorhine; arboreal vs. terrestrial; palmigrade vs. digitigrade vs. knuckle-walking quadruped; Fig. 4.1). In fact, numerous authorities have used a series of biometric ratios describing relative lengths of different rays and bones within rays (i.e., intrinsic proportions) to infer function and behavior (Mivart 1867; Schultz 1930; Napier and Davis 1959; Napier and Napier 1967; Etter 1973, 1974; Jouffroy and Lessertisseur 1979; Jouffroy et al. 1993; Lemelin 1999; Watkins 2003; Lemelin and Schmitt 2007; Kirk et al. 2008; Patel et al. 2009; Patel and Wunderlich 2010; Boyer et al. 2013; Venkataraman et al. 2013). These include, but are not limited to, the phalangeal index, thumb length index, opposability index, or some variation of these (Tables 4.2 and 4.3; see also Napier and Napier 1967; Jouffroy et al. 1993). These indices have demonstrated several exceptional differences among primates. For example, the two major radiations of living primates, strepsirrhines and haplorhines, can be distinguished by which ray of the hand is the longest, fourth or third, respectively (Jouffroy et al. 1993; Boyer et al. 2013). There are, however, exceptions to this dichotomy, such as *Ateles*, which tends to have a longer fourth ray (Jouffroy et al. 1993). Relatively long pollices (compared to the length of other rays) evolved convergently in humans and *Theropithecus* (and possibly *Cebus*), an adaptation that is thought to be associated with enhanced precision grasping in both groups (Etter 1973; Susman 2004). Also, *Ateles*, *Brachyteles*, and African colobines (*Colobus* and *Procolobus*) all have reduced pollices to varying degrees (and often considered rudimentary), possibly as adaptations for increased forelimb suspension, to enhance hook-like grips, and/or to aid in forelimb-first landing after leaps (Straus 1942; Stern and Oxnard 1973; Morbeck 1976; Rosenberger and Strier 1989; but see Tuttle 1972). Lorises have reduced second rays, with the digit portion of some species amounting only to a small bump—this morphology mimics that of their feet, which are well adapted for powerful grasping during locomotion on fine branches (Jouffroy and Lessertisseur 1979; Jouffroy 1993; Lemelin 1996; Lemelin and Jungers 2007; Gyambibi and Lemelin 2013; Kingston et al., 2010). Both *Tarsius* and *Daubentonia* have very long phalanges that have been functionally linked to their specialized diets on small animals or grubs (Niemitz 1979; Erickson 1994; Lemelin and Jungers 2007; Kirk et al. 2008). Finally, habitually terrestrial monkeys (the papionins, *Erythrocebus*) and African apes (*Pan*, *Gorilla*) are distinguished from committed arborealists in having relatively short proximal and intermediate phalanges, an adaptation that likely attenuates bending moments on the fingers during ground-dwelling locomotion (Etter 1973; Preuschoft 1973a; Nieschalk and Demes 1993; Patel and Wunderlich 2010). Some of these patterns are illustrated in Fig. 4.1 and quantified in Tables 4.2 and 4.3.

Table 4.2 Intrinsic proportions for manual ray 3 in primates[a]

Taxon	*N*	%Mc3	%PP3	%IP3	Phalangeal index	PP index	IP index	IPPI index
Allenopithecus nigroviridis[b]	2	46.87	32.58	20.56	113.5	69.5	44.0	63.4
Alouatta spp.[c]	11	37.66	37.07	25.27	165.6	98.5	67.1	68.2
Aotus trivirgatus	5	39.29	35.78	24.92	154.6	91.1	63.5	69.7
Arctocebus calabarensis	11	43.18	36.89	19.94	131.7	85.5	46.2	54.2
Ateles belzebuth[c]	4	43.10	33.11	23.80	132.3	77.0	55.3	72.2
Ateles fusciceps[c]	2	42.22	34.34	23.44	136.9	81.4	55.5	68.3
Ateles geoffroyi[c]	1	41.17	35.37	23.47	142.9	85.9	57.0	66.4
Ateles paniscus[c]	2	42.12	34.39	23.49	137.5	81.7	55.8	68.3
Avahi laniger	10	42.71	35.30	21.99	134.2	82.7	51.5	62.3
Brachyteles arachnoides[c]	1	42.36	34.91	22.73	136.1	82.4	53.7	65.1
Callicebus moloch	6	37.87	37.13	25.00	164.4	98.2	66.2	67.4
Callithrix jacchus	8	42.67	34.59	22.74	134.6	81.1	53.5	65.9
Cebus albifrons[c]	1	42.08	33.96	23.97	137.7	80.7	57.0	70.6
Cebus apella[c]	7	41.27	34.60	24.12	142.4	83.9	58.5	69.7
Cebus capucinus[c]	2	41.48	34.16	24.37	141.1	82.4	58.8	71.3
Cercocebus agilis[b]	2	45.63	31.87	22.50	119.4	70.0	49.4	70.6
Cercocebus galeritus[c]	1	47.85	30.99	21.16	109.0	64.8	44.2	68.3

(continued)

Table 4.2 (continued)

Taxon	*N*	%Mc3	%PP3	%IP3	Phalangeal index	PP index	IP index	IPPI index
Cercocebus torquatus[b]	5	46.08	31.88	22.03	117.1	69.2	47.8	69.1
Cercopithecus ascanius[b]	2	44.79	32.27	22.95	123.3	72.1	51.2	71.1
Cercopithecus campbelli[b]	2	44.64	32.18	23.19	124.0	72.1	51.9	72.1
Cercopithecus cephus[b]	1	43.85	32.57	23.58	128.1	74.3	53.8	72.4
Cercopithecus diana[b]	1	45.88	32.21	21.91	118.0	70.2	47.8	68.0
Cercopithecus mitis[b]	3	43.26	33.58	23.16	131.3	77.7	53.6	69.0
Cercopithecus mona[b]	5	43.49	33.38	23.13	130.1	76.8	53.2	69.3
Cercopithecus nictitans[b]	1	42.49	34.63	22.88	135.4	81.5	53.9	66.1
Cheirogaleus major	8	40.77	36.46	22.78	145.7	89.7	56.0	62.6
Cheirogaleus medius	6	40.25	36.01	23.74	148.5	89.5	59.0	65.9
Chlorocebus aethiops[b]	6	45.52	32.63	21.85	119.9	71.8	48.1	66.9
Colobus angolensis[c]	1	45.21	32.17	22.61	121.2	71.2	50.0	70.3
Colobus guereza[b]	8	42.69	33.36	23.96	134.4	78.2	56.2	71.8
Colobus polykomos[b]	2	41.49	33.46	25.06	141.1	80.7	60.4	74.9
Daubentonia madagascariensis	9	37.10	43.54	19.35	170.4	117.9	52.5	44.5
Erythrocebus patas[b]	1	56.06	25.59	18.35	78.4	45.6	32.7	71.7
Eulemur coronatus	2	40.40	35.74	23.88	147.6	88.5	59.1	66.8
Eulemur fulvus fulvus	13	41.12	35.72	23.16	143.2	86.9	56.4	64.9
Eulemur macaco macaco	7	40.69	35.75	23.55	145.8	87.9	57.9	65.9
Eulemur mongoz	7	40.68	35.40	23.92	145.9	87.1	58.8	67.6
Eulemur rubriventer	2	40.27	36.41	23.33	148.3	90.4	57.9	64.1
Euoticus elegantulus	11	36.34	39.71	23.94	175.2	109.3	65.9	60.3
Galago alleni	5	36.84	38.36	24.80	171.5	104.1	67.4	64.7
Galago moholi	7	37.29	36.93	25.78	168.3	99.1	69.2	69.8
Galago senegalensis	14	36.72	37.78	25.51	172.4	102.9	69.5	67.6
Galago zanzibaricus	1	37.50	36.37	26.13	166.7	97.0	69.7	71.9
Galagoides demidoff	11	35.70	38.21	26.10	180.4	107.2	73.2	68.4
Gorilla spp.[c]	24	47.50	30.82	21.69	110.7	65.0	45.7	70.4
Hapalemur griseus	12	39.93	36.57	23.50	150.6	91.7	58.9	64.3
Hapalemur simus	2	39.92	36.19	23.90	150.6	90.7	59.9	66.0
Homo sapiens[c]	10	46.66	31.80	21.50	114.4	68.3	46.1	67.6
Hylobates spp.[c]	20	42.74	33.06	24.20	134.1	77.4	56.7	73.2
Indri indri	8	42.81	35.63	21.56	133.7	83.3	50.4	60.5
Lagothrix lagothricha[c]	13	38.70	36.98	24.33	158.5	95.6	62.9	65.8

(continued)

Table 4.2 (continued)

Taxon	*N*	%Mc3	%PP3	%IP3	Phalangeal index	PP index	IP index	IPPI index
Lemur catta	11	41.56	35.09	23.36	140.8	84.5	56.3	66.6
Leontopithecus rosalia	7	46.28	32.27	21.45	116.1	69.7	46.4	66.5
Lepilemur leucopus	7	38.31	37.69	24.00	161.1	98.4	62.7	63.8
Lepilemur mustelinus	5	39.76	37.46	22.77	151.5	94.2	57.3	60.8
Lepilemur ruficaudatus	1	37.17	39.50	23.32	169.0	106.3	62.7	59.0
Lophocebus albigena[b]	3	45.68	32.11	22.20	118.9	70.3	48.6	69.2
Lophocebus aterrimus[c]	2	43.14	33.78	23.09	132.1	78.4	53.7	68.3
Loris tardigradus	11	36.78	39.54	23.68	172.1	107.6	64.5	60.1
Macaca arctoides[b]	2	43.86	34.11	22.04	128.0	77.8	50.2	64.7
Macaca assamensis[b]	1	41.75	33.33	24.92	139.5	79.8	59.7	74.8
Macaca fascicularis[b]	14	45.65	32.37	21.98	119.4	71.2	48.3	68.1
Macaca fuscata[b]	4	46.75	31.26	21.99	114	66.9	47.0	70.4
Macaca mulatta[b]	1	45.40	31.93	22.67	120.2	70.3	49.9	71.0
Macaca nemestrina	15	45.55	32.20	22.25	120.0	70.9	49.0	69.2
Macaca nigra[b]	2	45.00	32.28	22.73	122.3	71.8	50.5	70.5
Macaca ochreata[c]	1	44.85	32.46	22.69	123.0	72.4	50.6	69.9
Macaca radiata[c]	1	46.76	30.89	22.35	113.9	66.1	47.8	72.4
Macaca sylvanus[b]	2	45.85	32.11	22.05	118.8	70.6	48.2	69.2
Macaca thibetana[b]	1	43.14	33.95	22.90	131.8	78.7	53.1	67.5
Macaca tonkeana[b]	6	45.32	32.38	22.30	120.7	71.5	49.2	68.9
Mandrillus leucophaeus[b]	6	48.93	30.74	20.33	104.7	63.0	41.7	66.1
Mandrillus sphinx[b]	8	49.05	30.26	20.69	104.1	61.8	42.3	68.5
Microcebus murinus	17	38.32	35.22	26.46	161.4	92.1	69.3	75.2
Miopithecus talapoin[b]	2	40.82	34.69	24.49	145.0	85.0	60.0	70.8
Mirza coquereli	1	40.35	37.46	22.18	147.8	92.8	55.0	59.2
Nasalis larvatus[b]	10	41.68	34.39	23.93	140.0	82.6	57.5	69.6
Nycticebus coucang	17	37.65	40.13	22.22	166.2	106.8	59.4	55.4
Nycticebus pygmaeus	5	35.47	40.43	24.10	182.0	114.0	68.0	59.6
Otolemur crassicaudatus	9	37.53	38.20	24.26	166.5	101.8	64.7	63.6
Otolemur garnettii	7	36.91	39.30	23.78	170.9	106.5	64.4	60.5
Pan paniscus[c]	11	48.50	30.15	21.34	106.2	62.2	44.0	70.8
Pan troglodytes[c]	29	46.35	31.28	22.36	115.8	67.5	48.3	71.5
Papio anubis[b]	7	52.34	28.88	18.78	91.1	55.2	35.9	65.1
Papio cynocephalus[b]	6	51.99	28.88	19.14	93.4	56.1	37.3	66.1
Papio hamadryas[b]	3	51.22	29.74	19.04	95.4	58.2	37.2	64.2

(continued)

Table 4.2 (continued)

Taxon	*N*	%Mc3	%PP3	%IP3	Phalangeal index	PP index	IP index	IPPI index
Papio papio[b]	6	52.02	28.99	19.00	92.30	55.8	36.6	65.6
Papio ursinus[b]	9	53.31	28.20	18.49	87.70	53.0	34.7	65.6
Perodicticus potto	12	38.25	40.64	21.11	161.7	106.4	55.3	52.0
Piliocolobus badius[b]	4	41.04	34.42	24.55	143.8	83.9	59.9	71.4
Pongo spp.[c]	21	44.81	33.52	21.67	123.3	74.9	48.4	64.7
Presbytis comata[b]	3	41.60	33.40	24.90	140.4	80.3	60.0	74.7
Presbytis frontata[b]	3	42.30	32.80	24.90	136.4	77.4	58.9	76.2
Presbytis melalophos[b]	3	41.62	33.25	25.13	140.3	79.9	60.4	75.6
Propithecus diadema	7	41.86	35.64	22.49	138.9	85.2	53.7	63.1
Propithecus verreauxi	12	41.79	35.87	22.35	139.4	85.9	53.5	62.3
Pygathrix nemaeus[c]	2	40.70	34.40	24.90	146.4	84.8	61.6	72.5
Pygathrix nigripes[c]	1	40.80	34.20	25.00	144.9	83.7	61.2	73.1
Rhinopithecus roxellana[b]	1	41.99	34.17	23.84	138.2	81.4	56.8	69.8
Saguinus midas	8	43.51	34.72	21.78	129.9	79.8	50.1	62.7
Saguinus oedipus	13	43.13	35.76	21.11	132.0	83.0	49.0	59.1
Saimiri sciureus	5	40.80	33.79	25.41	145.2	82.9	62.3	75.3
Semnopithecus entellus[b]	7	47.16	30.67	22.17	112.6	65.4	47.2	72.5
Symphalangus syndactylus[c]	9	43.60	35.50	22.90	129.4	76.9	52.5	68.3
Tarsius bancanus	11	32.37	39.38	28.25	209.2	121.8	87.4	72.0
Tarsius pumilus[c]	1	32.18	38.64	29.18	210.7	120.0	90.6	75.5
Tarsius syrichta	12	34.93	38.18	26.89	186.3	109.3	77.0	70.5
Theropithecus gelada[b]	4	57.00	26.34	16.66	75.5	46.2	29.2	63.3
Trachypithecus auratus[c]	1	43.33	32.85	23.82	130.8	75.8	55.0	72.5
Trachypithecus cristatus[b]	7	43.10	33.27	23.63	132.1	77.2	54.9	71.0
Trachypithecus obscurus[b]	6	41.80	34.56	23.64	139.3	82.7	56.6	68.4
Trachypithecus phayrei[b]	4	42.06	33.38	24.57	137.9	79.4	58.5	73.6
Varecia variegata	10	40.71	36.45	22.85	145.9	89.7	56.2	62.7

[a]Data from Kirk et al. (2008) except where noted with a [b]for data from Patel and Wunderlich (2010) and [c]which is new data. **%Mc3**: (metacarpal 3 length × 100)/total ray length. **%PP3**: (proximal phalanx 3 length × 100)/total ray length. **%IP3**: (intermediate phalanx 3 length × 100)/total ray length. [Ray length: metacarpal 3 length + proximal phalanx 3 length + intermediate phalanx 3 length]. **Phalangeal index**: (proximal phalanx 3 length + intermediate phalanx 3 length × 100)/metacarpal 3 length. **Proximal phalanx (PP) Index**: (proximal phalanx 3 length × 100)/metacarpal 3 length. **Intermediate phalanx (IP) index**: (intermediate phalanx 3 length × 100)/metacarpal 3 length. **Interphalangeal (IPPI) index**: (intermediate phalanx 3 length × 100)/proximal phalanx 3 length

Table 4.3 Relative lengths of metacarpals (1, 2, and 4) and digital rays (1, 2, and 4)[a]

Genus	Mc1/Mc3			Mc2/Mc3			Mc4/Mc3			Ray 1/ray 3			Ray 2/ray 3			Ray 4/ray 3		
	N	Mean	SD	*N*	Mean	SD	*N*	Mean	SD	*N*	Mean	SD	*N*	Mean	SD	*N*	Mean	SD
Alouatta	8	62.3	2.6	8	88.3	3.6	8	96.7	7.1	8	53.8	1.6	8	86.4	1.8	8	97.7	5.7
Aotus	8	65.5	4.1	8	90.2	5.8	8	97.2	5.5	8	55.6	1.4	8	89.6	3.9	8	99.1	3.0
Arctocebus	14	103.5	12.5	14	63.8	9.4	14	103.9	7.4	18	93.4	5.7	18	53.1	3.8	18	127.2	5.0
Ateles	9	42.3	4.5	9	91.1	2.9	9	98.0	3.5	9	19.7	2.4	9	88.9	1.7	9	100.0	1.9
Avahi	7	46.0	3.4	7	82.7	1.8	7	99.3	7.0	7	42.1	2.7	7	80.3	4.1	7	106.4	6.3
Brachyteles	1	32.0	–	1	95.2	–	1	100.8	–	1	25.4	–	1	90.2	–	1	98.5	–
Callicebus	3	67.2	2.3	3	89.8	3.6	3	98.0	1.8	3	53.2	2.9	3	87.8	1.4	3	100.1	0.6
Callithrix	6	66.8	1.6	6	92.5	6.2	6	96.3	5.4	6	61.9	2.4	6	92.9	4.5	6	98.2	4.6
Cebus	11	73.5	5.2	11	100.2	10.4	11	97.0	3.1	11	62.5	4.2	11	95.3	4.1	11	97.4	2.1
Cercopithecus	11	63.6	2.5	11	95.1	4.0	11	95.7	3.1	11	49.3	1.8	11	86.5	4.6	11	95.5	4.6
Cheirogaleus	10	69.9	5.5	10	95.1	5.8	10	89.7	6.9	9	60.1	2.7	9	90.3	2.2	9	99.4	3.8
Colobus	10	44.9	5.0	10	90.6	7.2	10	99.0	1.9	10	22.3	5.8	10	84.6	5.0	10	100.7	5.8
Daubentonia	9	40.4	3.5	9	52.7	2.4	9	68.4	3.8	10	45.2	2.0	10	76.7	2.2	10	116.3	5.0
Euoticus	20	78.3	5.6	20	93.3	4.2	20	98.5	9.1	20	60.3	2.5	20	82.9	2.2	20	107.4	3.7
Galago[b]	59	80.3	8.4	59	87.2	7.0	59	94.3	6.7	53	59.6	5.8	53	78.4	6.4	53	107.0	5.8
Gorilla	6	56.6	6.8	6	103.6	7.2	6	95.9	3.0	6	45.9	3.8	6	93.3	4.3	6	96.9	2.0
Hapalemur	9	65.7	5.6	9	87.4	4.3	9	98.1	4.3	8	56.6	1.8	8	79.2	3.0	8	103.5	2.5
Homo	13	76.3	7.9	13	111.8	13.0	13	88.9	10.0	3	65.6	3.2	3	97.8	3.4	3	93.3	2.2
Hylobates	8	61.1	5.0	8	109.1	4.4	8	92.0	2.1	8	45.0	1.9	8	96.8	3.4	8	93.0	1.9
Indri	7	61.2	3.7	7	89.4	4.5	7	103.1	2.7	7	55.6	2.7	7	82.6	2.9	7	106.3	1.5
Lagothrix	6	66.6	5.7	6	94.1	3.7	6	99.1	2.6	6	55.5	3.2	6	88.8	1.5	6	100.1	1.2
Lemur[c]	40	59.1	4.1	40	91.3	4.7	40	95.8	3.9	39	53.4	2.7	39	87.0	2.7	39	102.7	3.4
Leontopithecus	7	67.4	9.8	7	86.9	4.2	7	95.9	3.8	7	54.8	5.6	7	83.4	5.0	7	92.0	3.8
Lepilemur	12	59.8	3.3	12	84.6	4.8	12	101.1	3.4	12	50.4	2.5	12	80.9	3.2	12	107.6	1.9

Loris	9	90.8	6.6	9	79.7	9.7	9	96.2	6.5	9	66.3	2.2	9	73.0	4.3	9	104.9	3.0
Macaca	12	65.7	5.7	12	98.0	5.4	12	98.3	5.6	12	50.5	4.0	12	88.5	5.3	12	98.8	3.6
Microcebus	19	70.1	7.6	19	93.2	6.9	19	95.4	6.0	17	57.9	5.9	17	87.8	4.0	17	104.0	4.8
Miopithecus	7	61.2	6.5	7	92.2	5.8	7	93.8	3.8	7	45.8	2.9	7	85.2	2.4	7	93.8	2.4
Nycticebus	20	76.0	6.2	20	79.6	5.6	20	95.8	5.7	22	60.1	9.1	22	66.9	4.7	22	106.0	4.1
Pan	14	46.8	3.4	14	100.8	1.8	14	93.5	2.7	14	41.0	2.4	14	89.4	2.9	14	93.5	2.6
Papio	10	65.5	3.5	10	98.7	3.8	10	96.8	2.4	10	54.6	4.0	10	91.6	4.1	10	99.3	2.9
Perodicticus	6	78.2	7.9	6	66.2	8.9	6	95.2	4.7	8	66.0	4.0	8	46.1	4.1	8	110.0	5.7
Phaner	3	69.5	0.5	3	91.7	0.5	3	94.7	1.7	2	57.5	1.4	2	86.0	0.7	2	102.0	0.7
Pongo	9	47.4	3.6	9	101.5	2.5	9	95.7	2.7	9	36.6	2.7	9	94.4	3.0	9	95.2	4.2
Presbytis	10	51.6	4.0	10	94.9	1.3	10	96.4	1.8	10	37.3	3.6	10	85.7	1.9	10	98.7	0.8
Propithecus	9	66.3	7.2	9	78.3	9.5	9	100.4	5.5	6	54.7	2.5	6	78.7	1.3	6	107.8	3.8
Rhinopithecus	2	44.8	4.3	2	98.9	3.8	2	101.0	2.0	2	34.3	1.0	2	85.6	0.3	2	100.1	1.1
Saimiri	8	69.7	4.7	8	93.0	4.8	8	89.7	5.0	8	58.7	5.9	8	89.3	3.4	8	96.4	2.8
Semnopithecus	5	52.0	4.1	5	96.6	9.1	5	100.9	4.0	5	37.0	6.2	5	91.9	5.8	5	99.7	5.4
Symphalangus	3	58.5	2.7	3	106.1	0.6	3	92.3	2.4	3	42.3	1.0	3	94.1	1.2	3	92.9	1.1
Tarsius	12	75.2	6.6	12	91.4	4.7	12	87.8	6.4	11	50.9	3.8	11	86.9	4.6	11	89.2	5.8
Trachypithecus	2	44.1	0.8	2	95.7	2.5	2	101.2	3.0	2	33.4	0.1	2	85.6	2.8	2	100.8	0.4
Varecia	8	60.4	5.4	8	90.7	4.1	8	95.2	1.6	9	53.6	2.8	9	88.7	2.9	9	100.3	12.1

[a]Data from Jouffroy et al. (1993). **Mc1/Mc3**: metacarpal 1 length × 100/metacarpal 3 length. **Mc2/Mc3**: metacarpal 2 length × 100/metacarpal 3 length. **Mc4/Mc3**: metacarpal 4 length × 100/metacarpal 3 length. **Ray1/ray3**: (metacarpal 1 length + proximal phalanx 1 length + distal phalanx 1 length × 100)/(metacarpal 3 length + proximal phalanx 3 length + intermediate phalanx 3 length + distal phalanx 3 length). **Ray2/ray3**: (metacarpal 2 length + proximal phalanx 2 length + intermediate phalanx 2 length + distal phalanx 2 length × 100)/(metacarpal 3 length + proximal phalanx 3 length + intermediate phalanx 3 length + distal phalanx 3 length). **Ray4/ray3**: (metacarpal 4 length + proximal phalanx 4 length + intermediate phalanx 4 length + distal phalanx 4 length × 100)/(metacarpal 3 length + proximal phalanx 3 length + intermediate phalanx 3 length + distal phalanx 3 length)

[b]These data likely include species of both *Galago*, *Galagoides*, and *Otolemur*

[c]These data likely include species of both *Lemur* and *Eulemur*

Based, in part, on these different intrinsic hand proportions, several authorities (e.g., Jouffroy et al. 1993; Preuschoft et al. 1993) have identified three basic morphotypes of digital ray patterning among primates: *ectaxonic*, *paraxonic*, and *mesaxonic*. An ectaxonic hand is found in most extant strepsirrhines (e.g., *Lepilemur*, *Propithecus*, *Arctocebus*) and is characterized by a fourth ray that is absolutely the longest, relatively long digits (>56 % of total hand length), an absolutely long pollex, and relatively short metacarpals (<29 % of total hand length) (Jouffroy and Lessertisseur 1979; Jouffroy et al. 1993). A paraxonic hand is similar to an ectaxonic hand, except that the third and fourth rays are similar in length. Some strepsirrhines (e.g., *Otolemur*, *Varecia*, and *Cheirogaleus*), platyrrhines (e.g., *Ateles*, *Lagothrix*), and catarrhines (e.g., *Pongo*) can show this morphology (Jouffroy et al. 1993). In a mesaxonic hand, which is typical of most catarrhines and platyrrhines, the third ray is always the longest, the digits are relatively short (<57 % of total hand length), the pollex is short, and the metacarpals are relatively long (>27 % of total hand length) (Jouffroy and Lessertisseur 1979; Jouffroy et al. 1993). Most pentadactyl mammals, including tree shrews, display a mesaxonic hand shape. This, along with the fact that nearly all haplorhines have mesaxonic patterning, suggests that this is likely the primitive condition for primates and that the ectaxonic pattern of strepsirrhines is probably derived for this group (Jouffroy et al. 1993; Boyer et al. 2013). Although original biomechanical models proposed that different types of hand configurations are related to hand positioning on arboreal substrates (i.e., an ectaxonic hand should be more ulnarly deviated and a mesaxonic hand should have a more neutral position when walking quadrupedally; Preuschoft et al. 1993), actual observations of primate locomotion have shown that hand positioning is related more to substrate size rather than just hand anatomy (Lemelin and Schmitt 1998).

4 Metacarpals

4.1 General Morphology

Each **metacarpal** (Mc) has a **base** (proximal epiphysis), a **shaft**, and a **head** (distal epiphysis). These key anatomical features are illustrated on the human Mc3 in Fig. 4.2. The base and head have similar radioulnar width and dorsopalmar height dimensions. The base can have a single (as in Mc1) or multiple (as in Mc2–Mc5) articular facets that create joints with only one carpal or with multiple carpals and other adjacent metacarpal(s), respectively. Metacarpals 1–5 strongly differ in the morphology of their bases and accompanying articular facets reflecting differences in the joints they share with adjacent bones (see below). The proximal articular surface of Mc1 is the most unusual; not only does it articulate with only one carpal (the trapezium), it is also saddle shaped (i.e., articular surface convex in the radioulnar plane and concave in the dorsopalmar plane), thus allowing for a greater range of motion in multiple planes (Lewis 1989; Marzke et al. 2010). The metacarpal shaft is the radioulnarly narrowest at mid-length, and its cross-sectional geometry

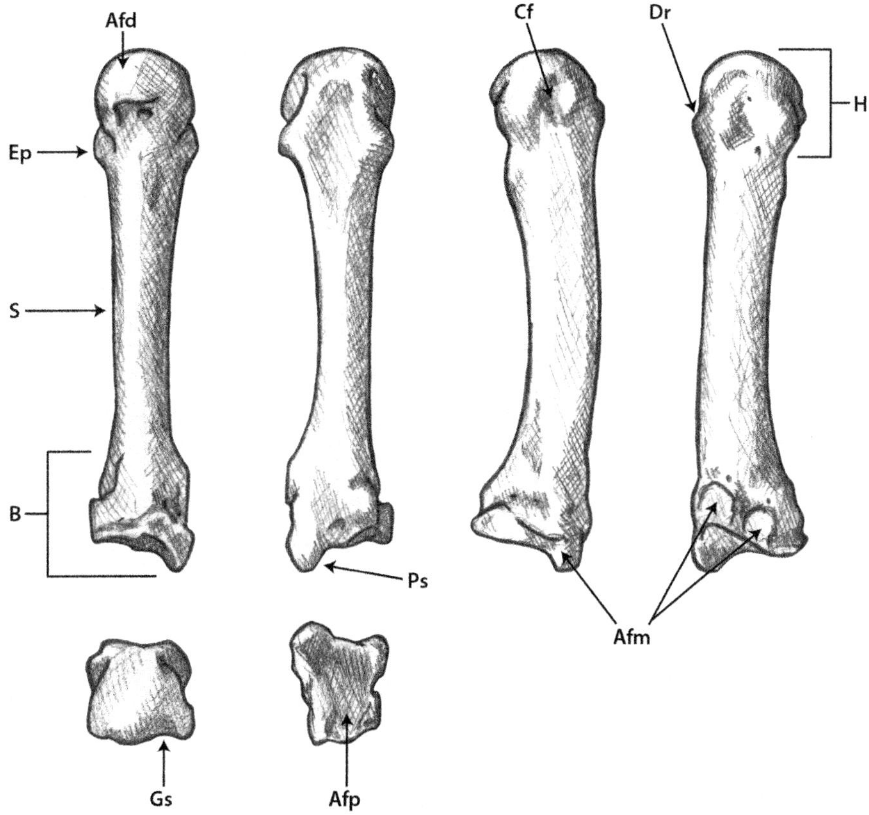

Fig. 4.2 Right third metacarpal (Mc3) of *Homo sapiens* in (from *left* to *right*) palmar, dorsal, radial, ulnar (*top* row), distal, and proximal (*bottom* row) views. Abbreviations: *Afd* distal articular facet, *Afm* metacarpal articular facets, *Afp* proximal articular facet, *B* base, *Cf* collateral fossa, *Dr* dorsal ridge, *Ep* epicondyle, *Gs* sesamoid gutter, *H* head, *Ps* styloid process, *S* shaft

and shape can be quite variable across the rays (Lazenby 1998; Marchi 2005). Its dorsal surface is flat and relatively smooth, while its palmar surface generally shows some degree of concavity (most pronounced in Mc1 and Mc2) and scarring for interosseous muscle attachments. The distal epiphyses (or heads) are variably shaped across the different rays, with the distal articular surfaces tending to be compressed more dorsally than palmarly in Mc2 and Mc3 (Susman 1979). These head shape differences across the rays are related to the subtle but functionally important differences by which the metacarpophalangeal (McP) joints flex and extend (see Lewis 1977, 1989). At the most proximal extent of the articular surface of the Mc2–Mc5 heads lies a slight transverse (dorsal) ridge. On both sides of the head are epicondyles and collateral fossae that serve as attachment sites for ligaments that make up the joint capsule complex for the McP joints; the Mc1 has the least developed epicondyles and can often be asymmetrical in development. The metacarpal length pattern in humans is always 2 > 3 > 4 > 5 > 1 (Susman 1979).

4.2 Morphological Diversity

Compared to primates, other euarchontans (i.e., tree shrews and flying lemurs) have metacarpals that differ in morphology. For example, the metacarpal heads of *Tupaia* show strong midline keels that extend palmarly, resulting in two prominent gutters on either side of the keel to accommodate sesamoid bones; *Cynocephalus* has a double keel on the palmar side of the head (see Fig. 19 in Boyer et al. 2013). The metacarpal heads of non-primate euarchontans can also extend dorsally as if to create a "domed" shape, similar to the metatarsal heads of bipedal human feet. This suggests a greater range for dorsiflexion (and hyperextension) at the McP joints in non-primates (cf. Latimer and Lovejoy 1990).

In general, the metacarpals of strepsirrhine primates scale with positive allometry, meaning that larger animals have relatively longer metacarpals. This pattern may correlate with adaptations in volar skin morphology (e.g., flatter and coalesced volar surface; see Chap. 8) that facilitate the use of larger-diameter and/or vertical substrates by larger animals (Lemelin and Jungers 2007). Although Mc3 and Mc4 are typically the longest metacarpals across strepsirrhines, their metacarpal length pattern is variable and can be clade specific (Jouffroy et al. 1993; Boyer et al. 2013; Table 4.3). Among strepsirrhines, lorisids and *Daubentonia* demonstrate the most extreme metacarpal modifications. As noted above, lorisids (*Nycticebus*, *Loris*, *Arctocebus*, and *Perodicticus*) have reduced second rays, including a shorter Mc2 length (Straus 1942; Jouffroy et al. 1993; Tague 2002; Fig. 4.1; Table 4.3). This reduction results in an unusual pincer-like appearance of the hand that closely resembles that of the foot (Jouffroy and Lessertisseur 1979; Jouffroy 1993; Lemelin 1996; Lemelin and Jungers 2007; Gyambibi and Lemelin 2013). As a whole, the distal ends of the metacarpals of both lorisids and galagids appear more robust and radioulnarly broader than in other nonhuman primates. This may be reflective, in part, of the radioulnarly compressed Mc4 and Mc5 bases in these taxa. The Mc5 head of all lorisiforms, as well as *Daubentonia* and the haplorhine *Tarsius*, is especially hyperinflated. It is unclear as to if and what functional significance such a morphology has, but it is hypothesized to be related to some mode of grasping, possibly emphasizing large forces acting on the fifth ray when landing after a leap (Napier and Walker 1967). *Daubentonia* is the only primate to have an extremely elongated and narrowed Mc3. In fact, it is so elongated that the McP joint of this ray is located distally beyond the limits of the palm (Soligo 2005). This specialized morphology reflects the use of this narrowed ray as a percussive and extractive probe in foraging behaviors (Erickson 1994).

Within haplorhines, anthropoids have a relatively consistent metacarpal length pattern with 3 > 2 > 4 > 5 > 1. The exceptions to this pattern are found in living hominoids and *Cebus*, which have a longer Mc2 than Mc3 (Table 4.3). Second metacarpal elongation may be related to enhanced manipulative capabilities in hominoids and capuchin monkeys. In contrast, the metacarpal length formula of tarsiers is more variable because the Mc1 can often be the same length as, or even exceed, the length of the Mc5. Among cercopithecoids, species like *Papio*, *Theropithecus*, *Mandrillus*, and *Erythrocebus* tend to have metacarpals that are relatively short overall (when scaled to body mass and/or humerus length) and have relatively more narrow dorsopalmar diameters at midshaft. Both are possible adaptations for manual digitigrady during terrestrial quadrupedal locomotion (Patel 2010a).

One of the most intriguing morphological adaptations of anthropoid metacarpals is the reduction in both length and robusticity of the Mc1 in *Ateles*, *Brachyteles*, and the African colobines (*Colobus*, *Procolobus*, and *Piliocolobus*; Straus 1942; Tague 2002). In African colobines, the Mc1 is reduced yet still retains its overall shape and can still resemble the Mc1 of sister taxa (see Fig. 1 in Tague 2002). In contrast, the Mc1 of *Ateles* and *Brachyteles* is almost rudimentary, with the distal end often lacking a distinctive head (Fig. 4.1). This morphology often coincides with the complete lack of proximal and distal pollical phalanges in *Ateles* and the frequent absence of pollical phalanges in African colobines (Straus 1942; pers. obs.). It is often thought that thumb reduction is an adaptation for increased forelimb suspension, to enhance hook-like grips, and to aid in forelimb-first landing after leaps (e.g., Straus 1942; Stern and Oxnard 1973; Morbeck 1976; Rosenberger and Strier 1989; but see Tuttle 1972). But these explanations are unsatisfactory because *Hylobates* has a long thumb (relative to body size; Lorenz 1974) and is the most forelimb suspensory primate, and leaping strepsirrhines also have relative long thumbs (see Table 4.3). Thus, there could be other reasons for its independent evolution in these two groups, with no clear consensus as to why African colobines and atelines have reduced thumbs.

Differences in robusticity across anthropoid metacarpals (e.g., the shaft and head of Mc3 is always larger than Mc5), either in terms of external dimensions or cross-sectional geometry, may be related to different forces and pressure experienced by different digits during locomotion. For example, Mc2 and Mc3 have more cortical area and larger polar moments of area than Mc5 in *Pan troglodytes* (Marchi 2005; Wallace and Patel 2013), which may be associated with higher peak palmar pressure on the radial side of the hand (Wunderlich and Jungers 2009; but see Matarazzo 2013). Similarly, the Mc3 and Mc4 in *Gorilla* have strong muscle scarring on their palmar surfaces that give them a more triangular appearance in cross section (Susman 1979), which may serve to further act as a buttresses against axial compression during quadrupedal weight support in this large-bodied primate. Interestingly, Wallace and Patel (2013) have found that the chimpanzee Mc1 can actually have relatively more cortical area and a larger polar moment of area than all other metacarpals, thus suggesting that relative metacarpal size may not be simply related to the functional demands of only quadrupedal locomotion.

Hominoid metacarpals are the best studied because of the unique locomotor and manipulative behaviors used by African apes and humans, respectively. The predominant metacarpal length formula in hominoids is 2>3>4>5>1, but there can be tremendous variation (e.g., Mc4 can be longer than Mc3 in *Pongo*; Susman 1979). Coupled with a long Mc2, the heads of both Mc2 and Mc3 in humans are more axially twisted in a radial (pronated) direction compared with other hominoids, which likely improves precision grasping and dexterous manipulation (Susman 1979; Matsuura et al. 2010). The human Mc3 is also unique among primates because the base has a long styloid process on its dorsoradial aspect that fits tightly with the capitate (Ward et al. 2014). This morphology is thought to help to stabilize the hand against forces directed toward the palmar side of the Mc3 head, preventing hyperextension of the metacarpal and palmar displacement of its base

(Marzke and Marzke 1987). In other anthropoid primates, one study (Orr 2010, unpublished data) showed that individual movements between the Mc3 and capitate in quadrupedal monkeys and *Pongo* are greater during extension (mean range of 8–13°) than in knuckle-walkers (mean of 4°; Table 4.4). More details of these carpometacarpal joints are also discussed in Chaps. 3 and 9 of this volume.

In all nonhuman hominoids, the base of Mc5 is quadrangular in shape and the proximal articular surface lacks a significant distal extension onto the dorsal surface. Also, the palmar side of the proximal articular surfaces of Mc4 and Mc5 articulates tightly with the hamulus of the hamate in nonhuman hominoids. Together, these traits likely limit the ability to extend between the Mc4/Mc5 and the hamate (Whitehead 1993; Orr personal communication). This is also a fundamental difference that distinguishes all nonhuman hominoids from other anthropoid primates, which have a more triangular-shaped Mc5 base with a distally expanded articular surface on its dorsal side. The hominoid morphology may be related to limiting ability to extend the ulnar side of the wrist and providing a more stable joint complex for weight support, whereas the monkey-like condition may be related to more pliable hand postures and greater range of extension at this carpometacarpal joint (Lewis 1977; Whitehead 1993; Patel 2009, 2010b; Orr 2010; Patel and Polk 2010). Whether this morphology has been independently derived in each of the living hominoids or that humans have reverted back to a more primitive condition is still debated (Lovejoy et al. 2009).

Unlike all other primates, African apes have Mc3 and Mc4 heads with radioulnarly wider articular surfaces on the dorsal side than on the palmar side. This feature has been suggested to be related to knuckle-walking and a reduced ability to extend at the McP joints in a close-packed position (Susman 1979). Available data on maximum range of extension and hyperextension in these joints support this hypothesis (Tuttle 1969b; Table 4.4). Although it is often presumed that digitigrade terrestrial monkeys also have a limited range of extension at their McP joints compared with palmigrade arboreal monkeys (Tuttle 1969b), the radioulnar width of their metacarpal heads is always greater on the palmar side (the opposite of the African ape pattern) (Patel 2010a). This is also the pattern found in hylobatids and *Pongo* (Susman 1979). In addition to flexing and extending, the McP joints 2–5 also allow for abduction/adduction (Lewis 1977), and among hominoids, *Pongo* may have the greatest ability to do so (Rose 1988).

Of all the presumed knuckle-walking adaptations of the metacarpals of *Pan* and *Gorilla*, one of the most hotly contested relates to the functional significance of the transverse dorsal ridge on the articular surface of the metacarpal head. Some authors have suggested that these ridges act as bony stops to facilitate close-packed position of the McP joints during knuckle-walking and prevent further hyperextension of the joints (e.g., Tuttle 1967; see also Richmond et al. 2001). Others, however, have suggested that these ridges are not related to knuckle-walking, or specific hand postures per se, as dorsal ridges are variably developed on the heads of Mc2 and Mc5, are also found on metatarsals, and may be related to changes in body size throughout ontogeny (e.g., largest in adult male gorillas) (Lewis 1977; Susman 1979; Inouye and Shea 2004). Additionally, larger males of some of the largest cercopithecoid taxa (e.g., *Papio*, *Nasalis*) and the subfossil lemur *Archaeolemur* have fairly well-developed transverse dorsal ridges on the heads of Mc3 and Mc4 (Richmond et al. 2001; Jungers et al. 2005; Allen 2008; Patel et al. 2009).

Table 4.4 Range of motion for several joints of the manual rays in nonhuman primates

	DIP extension[a]				McP hyperextension[a]			Capitate-Mc3 extension[b]			
Taxon	*N*	Mean	Min	Max	Mean	Min	Max	*N*	Mean	Min	Max
Ateles geoffroyi								1	8.0		
Cercocebus atys	10	54.0	74.6	93.4	140.0	107.7	172.3				
Cercopithecus cephus	1	70.0			110.0						
Cercopithecus diana	1	60.0			110.0						
Chlorocebus aethiops	10	90.0			156.0	138.3	173.7				
Colobus guereza								1	15.0		
Erythrocebus patas	10	84.0	74.6	93.4	126.0	106.3	145.7				
Gorilla gorilla	10	0.0			54.0	41.2	66.8				
Hylobates lar	10	26.0	−9.8	61.8	0.0						
Macaca arctoides	10	85.0	75.4	94.6	146.0	128.3	163.7				
Macaca fascicularis	10	86.0	76.6	95.4	116.0	96.3	135.7				
Macaca maura	4	82.5	70.8	94.2	117.5	87.9	147.1				
Macaca mulatta	10	87.0	78.2	95.8	137.0	114.1	159.9	2	11.0	9.0	12.0
Macaca nemestrina	10	81.0	67.5	94.5	112.0	89.5	134.5				
Macaca niger	10	79.0	68.6	89.4	99.0	78.9	119.1				
Macaca radiata	10	90.0			136.0	116.3	155.7				
Pan troglodytes	21	0.0			44.0	20.9	67.1	5	4.0	1.0	7.0
Papio anubis	3							3	12.0	10.0	13.0
Papio cynocephalus	20	79.5	73.2	84.8	103.0	81.8	122.4				
Papio papio	4	80.0			107.5	85.0	130.0				
Pongo pygmaeus	27	3.0	−5.8	11.8	19.0	5.5	32.5	5	13.0	5.0	20.0
Semnopithecus entellus	10	81.0	75.2	86.8	151.0	135.0	167.0				
Theropithecus gelada	10	78.0	61.2	94.8	108.0	87.2	128.8				
Trachypithecus cristata	6	81.7	73.4	89.9	125.0	103.8	146.2				

[a]*DIP* distal interphalangeal joint, *McP* metacarpophalangeal joint; data from Tuttle (1969b)
[b]*Mc3* third metacarpal; data from Orr (2010), Orr (unpublished data)

5 Proximal and Intermediate Phalanges

5.1 General Morphology

Each **proximal phalanx** (PP) has a **base** (proximal epiphysis), a **shaft**, and a **head** (distal epiphysis) (Fig. 4.3). In humans, the base is the radioulnarly widest and dorsopalmarly tallest part of the bone, which articulates with the head of the corresponding metacarpal (to form the McP joint) via a single, wider than tall articular

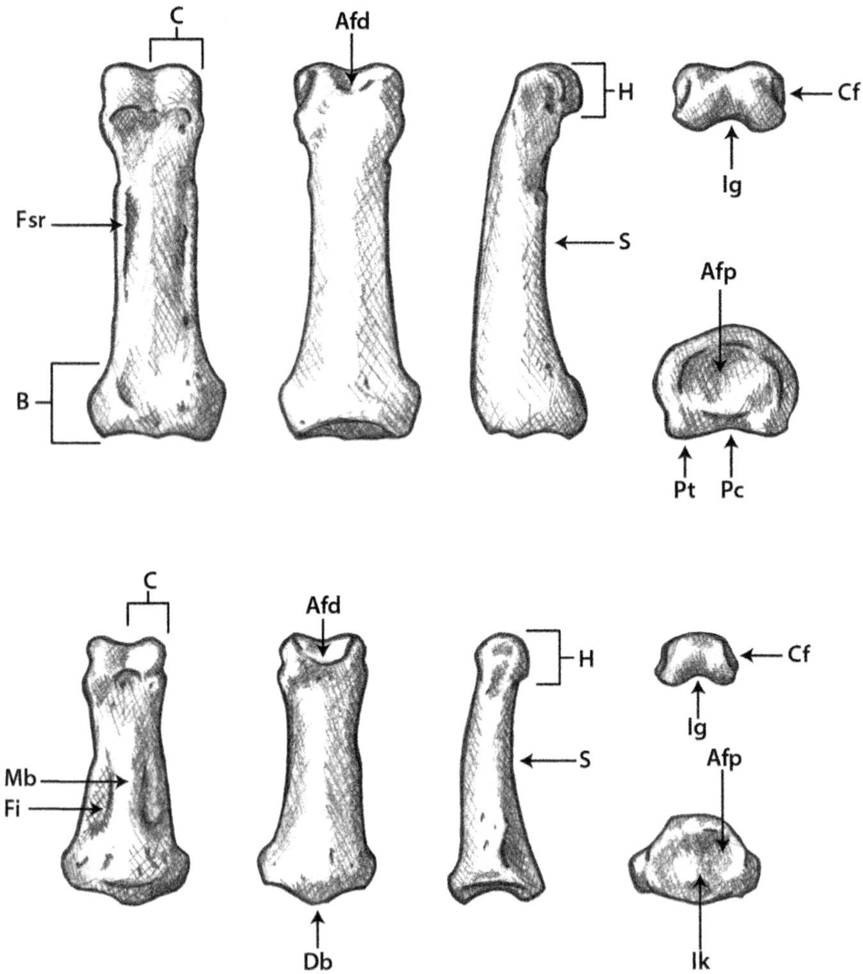

Fig. 4.3 Right third proximal phalanx (PP3; *top* row) and intermediate phalanx (IP3; *bottom* row) of *Homo sapiens*. For both rows, palmar, dorsal, lateral, distal (*above*), and proximal (*below*) views are illustrated from *left* to *right*. Abbreviations: *Afd* distal articular facet, *Afp* proximal articular facet, *B* base, *C* trochlear condyle, *Cf* collateral fossa, *Db* dorsal beak, *Fsr* flexor sheath ridge, *Fi* insertion of flexor tendon, *H* head, *Ik* interarticular keel, *Ig* intercondylar groove, *Mb* median bar, *Pc* palmar channel, *Pt* palmar tubercle, *S* shaft

facet that is vertically oriented. There are two tubercles on the palmar side of the base that, in part, serve as attachment sites for strong ligaments of the McP joint capsule and to channel or guide the digital flexor tendons toward their more distal insertions. The shaft extends distally from the base where it is radioulnarly wide but narrows distally before reaching the head. The dorsal surface of the shaft is relatively smooth, but its palmar surface generally shows some degree of concavity and clear muscle markings. Along the radial and ulnar sides of the palmar surface of the shaft are variably developed ridges that serve as the attachment sites for the flexor sheaths, often referred to as the flexor sheath ridges (Susman 1979). These attachments serve to hold the digital flexor tendons in place along the palmar aspect of the digit. Flexor sheath ridges are often most well developed in PP3 and PP4. The distal epiphysis (or head) is shaped liked a trochlea with asymmetrical radial and ulnar halves; one side tends to be more bulbous. The head is moderately tall and does not typically extend dorsally above the level of the shaft in humans. The distal articular surface does extend proximally on the dorsal aspect of the shaft and accommodates the dorsal beak found in the intermediate phalanx (see below). Finally, there are radial and ulnar pits that serve as attachment sites for collateral ligaments that stabilize the distal interphalangeal (DIP) joint. The proximal phalangeal length pattern in humans is usually 3>4>2>5>1 (Susman 1979), although the difference in length between digits 2 and 4 can vary depending on developmental conditions such as in utero hormone levels (Nelson et al. 2011; see Chap. 5).

The pollical proximal phalanx (PP1) is typically the shortest and most robust (in primates that possess one). While a pulley system that serves to hold the long pollical flexor tendon in place does attach to certain portions along the palmar margins of this phalanx, it does not typically leave well-developed flexor sheath ridges as in the PPs of the fingers. In general, the dorsal surface of the shaft is not very proximodistally curved, though it can appear curved in lateral view because it has dorsopalmarly tall distal condyles. The distal end of the proximal pollical phalanx articulates with its distal phalanx, which can be highly specialized, especially in humans (see below).

The **intermediate** (or middle) **phalanges** (IP) are the least well studied of all the primate hand bones. All rays of the hand have an intermediate phalanx, except for the pollex. Each intermediate phalanx has a **base** (proximal epiphysis), a **shaft**, and a **head** (distal epiphysis) (Fig. 4.3). In humans, the base is the radioulnarly widest and dorsopalmarly tallest part of the bone, and its proximal articular surface articulates with the proximal phalanx via two elliptical (taller than wide) articular facets separated by a moderately developed middle vertical keel (i.e., bipartite). Each proximal facet receives the two condyles of the trochlea-shaped head of the proximal phalanx. On its dorsal side, there is a proximally extending beak that helps lock it into place with the distal end of the proximal phalanx during full extension (see below). The shaft extends distally from the base where it is radioulnarly wide but narrows distally before reaching the head. Again, the dorsal surface of the shaft is flat and relatively smooth since it lacks any strong inserting components of the digital extensor expansion complex. Where the shaft meets the head, there is almost no dorsal concavity (i.e., the head does not extend dorsally above the height of shaft). The palmar surface of all intermediate phalanges is concave, but this curvature never exceeds that seen in a corresponding proximal phalanx (Matarazzo 2008). Typically, an intermediate phalanx has a middle keel (or palmar median bar) running along its length, with deep to shallow gutters (i.e., lateral fossae) bounded by lateral ridges on either side that can serve as the insertion site for the bifurcated tendons of flexor digitorum superficialis (Susman 2004; Marzke et al. 2007). The palmar median bar is the narrowest at the location of where the lateral fossae are the widest. The development of the median bar, lateral fossae, and lateral ridges varies in shape and size among primates and may not serve to adequately represent the size of the muscle or its tendons (Marzke et al. 2007). Like in the proximal phalanges, the intermediate phalanges have a head shaped like a trochlea, although the radial and ulnar halves are not as bulbous. The middle groove of the trochlea accommodates the intercondylar crest of the base of the distal phalanx. The palmar aspect of the distal articular surface tends to sweep proximally on both the radial and ulnar sides, thereby leaving a distinct V-shaped morphology (see palmar view in Fig. 4.3b). Fossae on the sides of the head are the attachment sites for collateral ligaments that stabilize the distal interphalangeal joint, but these pits are generally not as well developed in the intermediate phalanges as they are in the proximal phalanges. The phalangeal length pattern for intermediate phalanges in humans is typically 3>4>2>5, but variations are common (Susman 1979).

The articulation between the head of the proximal phalanx and the base of the intermediate phalanx of digits 2–5 makes up the proximal interphalangeal (PIP) joints. In these same digits, the head of the intermediate phalanx articulates with the base of the distal phalanx to comprise the distal interphalangeal (DIP) joints. In humans, all four PIP and DIP joints are restricted to one degree of freedom in the flexion-extension plane. Extension is limited in the PIP and DIP joints by the proximally projecting dorsal beaks of the more distal bony element (see below). However, in many terrestrial cercopithecoids that apply considerable forces on their fingertips when walking (Patel and Wunderlich 2010), the DIP joints are capable of up to 95° of hyperextension (Tuttle 1969b; Table 4.4).

5.2 *Morphological Diversity*

There exists some subtle diversity in primate proximal and intermediate phalangeal morphology, and this again is likely related to both functional and phylogenetic causes. Other than the quantification of relative lengths (as reviewed above) and curvature (see below) of these elements, only a few studies have quantified the finer details of phalangeal morphology and shape. In those morphological studies, researchers focused their attention on humans and other hominoids (e.g., Susman 1979), with the occasional comparison to cercopithecoids and New World monkeys such as *Alouatta*, *Ateles*, and *Cebus* (e.g., Meldrum and Yuerong 1988; Begun 1993; Hamrick et al. 1995; Nakatsukasa et al. 2003; Marzke et al. 2007). These elements are rarely discussed in much detail for strepsirrhines and most platyrrhines (Hamrick et al. 1995, 1999; Boyer et al. 2013). Even fewer studies have compared the morphology of primate proximal and intermediate phalanges to closely related non-primate mammals (Hamrick et al. 1999).

In general, the proximal and intermediate phalanges of strepsirrhines are relatively long and scale isometrically with body mass (Lemelin and Jungers 2007; Kirk et al. 2008). Typically, the proximal and intermediate phalanges of the fourth ray are absolutely the longest (and widest), but sometimes they are the same length as the third proximal phalanx (e.g., *Otolemur*; Fig. 4.1). Thus, the strepsirrhine proximal (non-pollical) and intermediate phalangeal length pattern would be 4>3>2>5. Typically, the non-pollical metacarpals are longer than their corresponding proximal phalanges; however, in some species the pollical proximal phalanx is longer than its corresponding first metacarpal (e.g., *Daubentonia*, *Eoticus*, *Galago*, *Galagoides*, *Otolemur*, *Nycticebus*, *Loris*, *Perodicticus*) (Kirk et al. 2008; Fig. 4.1). The trochlea-shaped head tends to be quite tall and dorsally flexed in non-anthropoids, often extended well above the level of the shaft (Hamrick et al. 1999). The distal articular surface can also extend proximally on the dorsal side, most evident in lorises. This morphology likely permits lorises to "kink" or "bow up" their phalanges (i.e., flex their PIP joints while hyperextending their DIP and McP joints) to functionally shorten their relatively long digits to reduce phalangeal bending moments during locomotion (Nieschalk and Demes 1993). Other strepsirrhines, tarsiers, and quadrupedal anthropoids adopt similar digital ray postures for the same reasons, and thus, this is not a loris-specific biomechanical phenomenon (Lemelin 1996; Patel and Wunderlich 2010). Although proximal phalanx curvature has only been reported for a few strepsirrhine species, the variation observed is quite dramatic (Jungers et al. 2002; Boyer et al. 2013; Table 4.5). For example, lorisiforms have relatively flat proximal phalanges, while taxa like *Indri* and *Propithecus* are intermediate, and *Varecia* and *Daubentonia* have relatively marked curvatures. The most extreme proximal phalanx curvatures are seen in the subfossil sloth lemurs from Madagascar (Jungers et al. 1997, 2002, 2005; see Chap. 15).

In the proximal phalanx, dorsal canting (i.e., orientation) of the proximal articular surface appears minimal to be absent in lorisids, vertical clinging and leaping lemurs (e.g., *Propithecus*, *Lepilemur*), and even among many arboreal palmigrade quadrupeds such as *Daubentonia*. Nevertheless, this feature is present in more committed quadrupeds that are habitually palmigrade such as *Varecia* and *Lemur*. This morphology is thought to allow for increased range of hyperextension at the McP joints in committed quadrupedal species (Table 4.4; see below). Close relatives such as *Cynocephalus* and *Tupaia* also have dorsally canted non-pollical proximal phalanges (Boyer et al. 2013), which not only reflects their more generalized quadrupedal hand posture like many primates but also the possibility that this configuration of the McP joints may be primitive for primates.

Some taxa such as *Eulemur* have robust palmar tubercles at the base of the proximal phalanx, another trait possibly related to palmigrady according to some authors (Rose 1986; Almécija et al. 2009). Well-developed palmar tubercles are also found in *Tupaia* and *Cynocephalus* (Boyer et al. 2013). However, other primates that are not habitually quadrupedal, such as *Tarsius* (Fig. 4.4a), also have relatively (and absolutely) large palmar tubercles at the base of their proximal phalanges, raising doubt that this feature may have something to do with quadrupedalism.

Table 4.5 Proximal phalanx curvature—included angle (IA) in degrees (°)

	Stern et al. (1995)			Jungers et al. (2002)			Matarazzo (2008)[b]			Rein (2011)[b]			Patel (unpublished)[b,c]		
Taxon[a]	*N*	Mean	SD	*N*	Mean	SD	*N*	Mean	SD	*N*	Mean	SD	*N*	Mean	SD
Alouatta seniculus										15	36.9	6.3			
Alouatta spp.													20	38.4	5.0
Ateles geoffroyi										7	50.8	8.3			
Ateles spp.							14	51.0	5.9				4	52.1	8.7
Brachyteles arachnoides													1	57.2	
Cebus apella							15	36.0	3.9						
Cebus olivaceus													7	13.9	5.4
Cercopithecus mitis										21	31.0	7.6			
Chlorocebus aethiops										20	24.2	7.7	12	24.8	4.1
Colobus guereza										20	30.1	5.7			
Colobus spp.													16	32.1	5.4
Erythrocebus patas													18	21.5	4.4
Gorilla g. beringei							11	55.0	4.9						
Gorilla g. gorilla							23	54.0	5.6						
Gorilla spp.	88	37.2	4.2							20	41.1	5.3	19	38.3	5.0
Homo sapiens	146	25.1	4.1							20	29.1	4.2	31	25.3	3.7
Hylobates lar							97	52.0	4.8	20	45.5	4.8	12	45.2	2.4
Hylobates spp.	68	47.7	5.5												
Indri indri				66	35.0	6.0									
Lagothrix lagothricha										13	53.3	5.0	3	54.9	8.2
Lagothrix spp.				12	62.4	6.0									
Lophocebus albigena										6	29.5	10.1			
Macaca fascicularis							20	45.0	6.6	20	16.2	6.6	11	22.0	7.0

Macaca mulatta													16	22.0	3.7
Macaca nemestrina							16	39.0	4.9				12	23.7	4.9
Nasalis larvatus				14	35.3	3.8							18	24.2	3.7
Pan paniscus	38	44.8	4.2												
Pan troglodytes	63	42.4	4.8				25	55.0	4.4	21	38.4	5.3	20	37.0	5.9
Papio anubis										19	14.7	5.0			
Papio spp.													9	17.5	7.8
Pongo pygmaeus[d]	68	63.1	7.2				19	65.0	5.8	19	64.2	13.2	27	62.8	10.4
Presbytis melalophos										11	25.4	4.8			
Procolobus badius										14	46.1	9.1			
Propithecus diadema				20	35.8	4.4									
Symphalangus syndactylus	20	55.1	5.6										1	50.9	
Theropithecus gelada													1	14.2	
Trachypithecus cristata													12	35.1	3.9
Trachypithecus obscurus										20	33.3	7.3			
Varecia variegata				12	51.6	5.4									

[a]Boyer et al. (2013) have also reported manual proximal phalanx IA values for *Tarsius* sp. ($n=16$, mean$=28.3°$, SD$=5.2°$)
[b]Values for the third digit (PP3) only
[c]Unpublished data with IA values measured and calculated using 3D landmarks following Rein (2011)
[d]Data from multiple species or subspecies likely

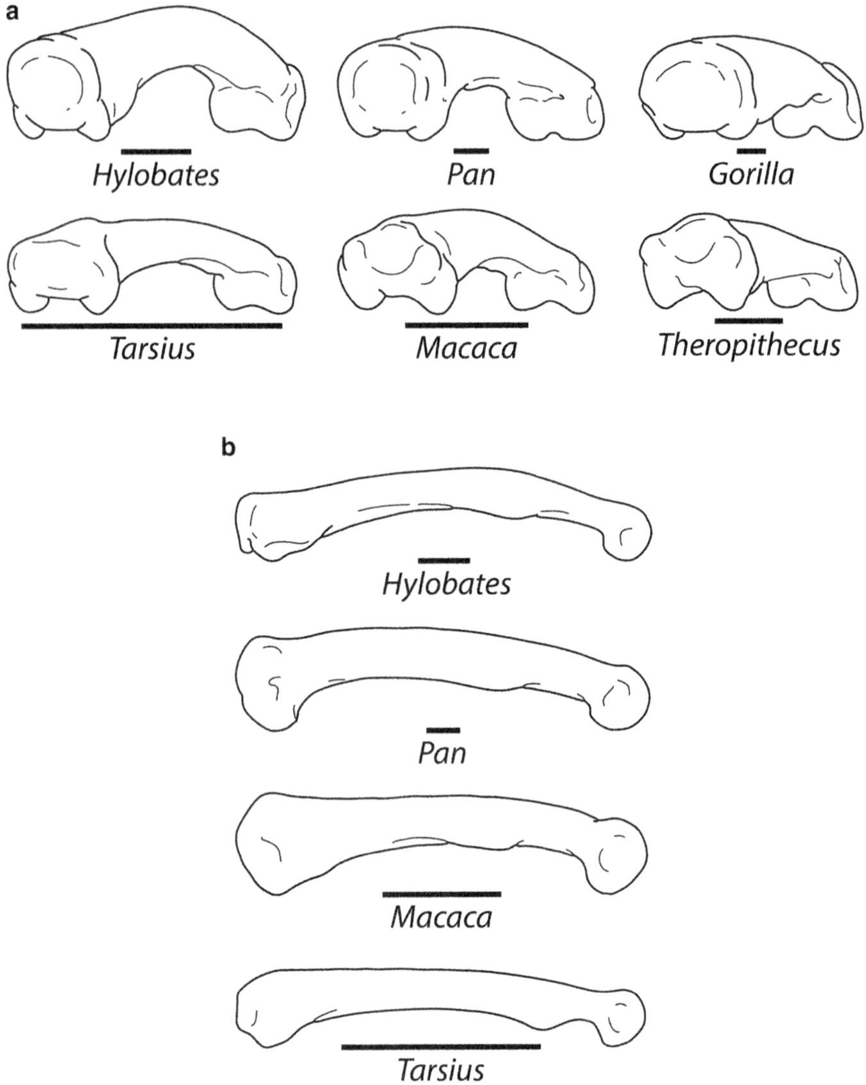

Fig. 4.4 Proximal phalanges of nonhuman primates in proximo-oblique (**a**) and lateral (**b**) views. Note the excavated and dorsally canted proximal articular surface in quadrupedal monkeys, the lack of well-developed palmar tubercles in African apes, and differences in dorsoventral curvature. Scale bars are 1 cm

Among primates, the third of ray of *Daubentonia* contains the relatively longest but also narrowest bones (Fig. 4.1; Soligo 2005). The extremely narrow phalanges of the third digit are still capable of maintaining their structural integrity, most likely because the digits experience relatively lower pressures compared to the rest of the hand during quadrupedal walking (Kivell et al. 2010). Tarsiers also have relatively long proximal and intermediate phalanges (Fig. 4.1); the proximal phalanges are typically longer than their corresponding metacarpals (Hill 1955; Kirk et al. 2008). Pollical length, however, is variable among tarsiers with *Tarsius pumilus* having a relative longer PP1 than the other species of *Tarsius*. Because they are not habitually quadrupedal or suspensory, tarsier proximal phalanges are straight (Boyer et al. 2013; Table 4.5) and do not have any noticeable dorsal canting of the proximal articular facets (Fig. 4.4).

On the whole, platyrrhines and most cercopithecoid monkeys are similar in their respective proximal and intermediate phalangeal morphologies. As noted above, these primarily arboreal or terrestrial quadrupedal primates tend to have proximal phalanges with dorsally excavated and canted proximal articular facets (Fig. 4.4a). Lewis (1977), Rose (1986), Hamrick et al. (1995), and Hayama et al. (1994) all suggested that this morphology is only found in cercopithecoids. However, Rein and McCarty (2012) have shown that dorsal canting of the phalanges in *Cebus* is similar to that of many Old World monkeys. We also observed dorsal canting in *Callicebus*, as well as *Varecia* and *Lemur* as noted above. The suspensory *Ateles*, *Brachyteles*, and *Lagothrix*, however, are more similar to suspensory hominoids with little to no dorsal canting (Hayama et al. 1994; Rein and McCarty 2012). These suspensory New World monkeys also have substantially more curved proximal and intermediate phalanges (Matarazzo 2008; Rein 2011; Fig. 4.5; Table 4.5). For the most part, platyrrhines and most cercopithecoid monkeys have proximal (non-pollical) and intermediate phalangeal length patterns that are 3>4>2>5, but *Ateles* appears unique among anthropoids because some individuals have their longest proximal and intermediate phalanges along ray 4 (see also Jouffroy et al. 1993; Fig. 4.1). Callitrichines are distinct among platyrrhines because they have very short intermediate phalanges relative to their proximal phalanges, a possible adaptation for vertical feeding postures (Kirk et al. 2008; Fig. 4.1). The non-pollical proximal and intermediate phalanges are also relatively short when compared with metacarpal length in terrestrial cercopithecoids such as *Theropithecus*, *Papio*, *Mandrillus*, and *Erythrocebus* (Napier and Napier 1967; Etter 1973; Jungers et al. 2005; Patel and Wunderlich 2010; Venkataraman et al. 2013; Fig. 4.1; Tables 4.2 and 4.3). This latter adaptation is related to minimizing potentially high bending moments that could arise during ground dwelling when substrate reaction forces tend to be high (Schmitt 1994).

As mentioned previously, the reduced pollex of African colobines, *Ateles*, and *Brachyteles* may include reduced or completely absent pollical phalanges (Straus 1942; Tague 2002). In contrast, *Theropithecus gelada* is unique because it possesses a relatively long thumb (Napier and Napier 1967). Apparent pollical elongation in *Theropithecus* is actually related to absolutely shortened proximal and intermediate phalanges of the index finger (ray 2) rather than an elongated pollex (see also Etter 1973; Susman 2004; Jungers et al. 2005; Fig. 4.1; Table 4.3).

The comparative anatomy of the proximal and intermediate phalanges has been best studied in hominoids (Susman 1979). African apes have short intermediate phalanges relative to the length of the proximal phalanges compared with *Pongo* and the hylobatids, a condition that has been linked to knuckle-walking hand postures. The PP1 (and pollex as a whole) is relatively short in all great apes and appears the most

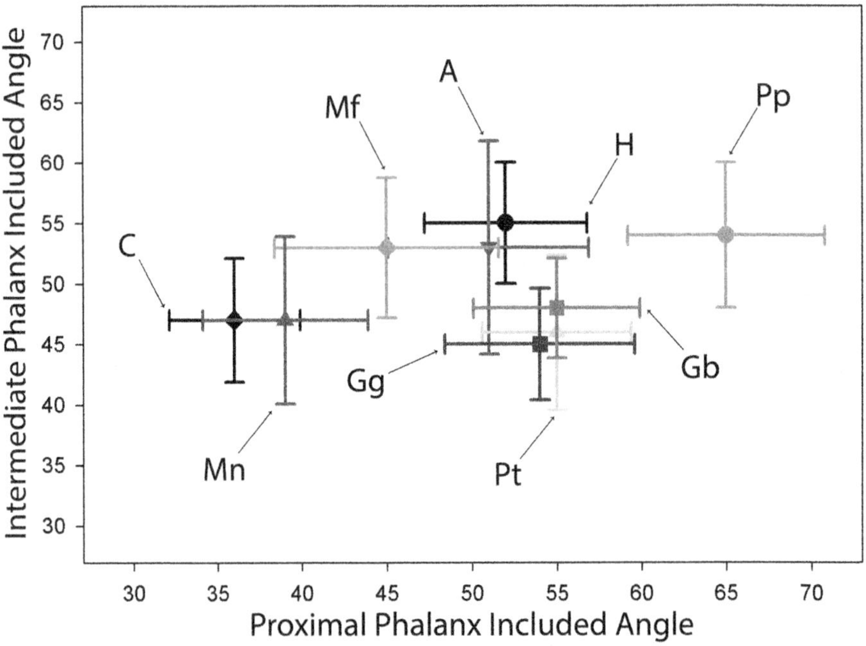

Fig. 4.5 Included angle (IA) values in degrees (°) for proximal vs. middle phalanges of digit three (data from Matarazzo 2008). Note that suspensory taxa have the most curved phalanges and that proximal phalanges are always more curved than middle phalanges in all taxa. See text for further details. *A Ateles* spp., *C Cebus apella*, *Gb Gorilla beringei*, *Gg Gorilla gorilla*, *H Hylobates lar*, *Mf Macaca fascicularis*, *Mn Macaca nemestrina*, *Pt Pan troglodytes*, *Pp Pongo pygmaeus*

diminutive in *Pongo* because of its substantially longer fingers (Straus 1942). Relative pollical length reduction in *Pongo*, although originally hypothesized to be related to reduced use of this digit during manipulative tasks (Tuttle 1969a), may not necessarily be true according to observations of animals in the wild (McClure et al. 2012). Overall, the great ape and hylobatid pollex are never as reduced compared to *Ateles*, *Brachyteles*, or African colobine monkeys. The proximal (non-pollical) and intermediate phalangeal length patterns are consistent across humans, African apes, and hylobatids (3>4>2>5) but are much more variable in *Pongo* (Susman 1979). *Pongo* and hylobatids also have the most curved proximal phalanges of all extant primates (Figs. 4.4b and 4.5); the latter also have curved intermediate phalanges (Matarazzo 2008; Fig. 4.5; Table 4.5). The proximal articular surfaces of the proximal phalanges are not dorsally canted or excavated in any of the extant nonhuman hominoids (Hayama et al. 1994; Rein and McCarty 2012; Fig. 4.4a), which correlates well with the reduced ability to hyperextend their McP joints (Tuttle 1969b; Table 4.4). Our own observations, however, reveal that humans have more dorsally canted proximal phalanges than other hominoids, and we propose that this may be related to increased range of passive dorsiflexion during tool use or other manipulative behaviors that experience higher than normal forces in a palmar to dorsal direction.

The intermediate phalanges of *Gorilla* and *Pan* are radioulnarly broad relative to their dorsopalmar height, possibly providing a much larger platform to support body weight during knuckle-walking (Susman 1979). Also, their medullary cavities are very narrow because of increased cortical thickness, and they often contain trabecular struts running in a dorsopalmar direction (Susman 1979; Nakatsukasa et al. 2003; Wallace and Patel 2013). The cortex of proximal phalanges is also relatively thicker in African apes compared with humans, *Pongo*, and hylobatids, with the distal ends being thicker than the midshafts (Susman 1979). Finally, in *Gorilla*, and to a lesser extent in *Pan*, the flexor sheath ridges of the proximal phalanges are extremely well developed (Fig. 4.1). It is unlikely, however, that these are specific knuckle-walking adaptations since digital flexor muscles show minimal activity during quadrupedal locomotion compared with other behaviors such as forelimb suspension and grasping of food objects (at least in *Pan*) (Susman and Stern 1979). But this, too, cannot be the explanation since *Pongo* and hylobatids do not have very large flexor sheath ridges on the palmar surfaces of their proximal phalanges.

5.3 Phalangeal Curvature

Phalangeal curvature is a predominant topic in the comparative primate literature, and discussions surrounding its functional meaning have played a major role in the interpretation of hand function during locomotion in extant and fossil primates (Susman 1979; Stern and Susman 1983; Susman et al. 1984; Stern et al. 1995; Jungers et al. 1997; Richmond 1998; Deane et al. 2005; Deane and Begun 2008; Matarazzo 2008; Rein 2011). When compared to most mammals, including euarchontans like tree shrews, nonhuman primates have relatively curved phalanges. Much of this curvature is owed to the fact that nearly all nonhuman primates grasp curved arboreal supports of varying diameters during locomotion and posture. When grasping a curved surface, the proximal phalanges are subjected to substrate reaction forces (SRF) from the branch, muscle forces arising from the contraction of superficial and deep digital flexors muscles, and joint reaction forces at the PIP and McP joints (Richmond 1998, 2007). Coupled together, these forces can cause relatively high levels of stresses and strains on hand bones (Preuschoft 1973b). Some of these forces, such as the SRF, can effectively be lowered in an arboreal setting by adopting a compliant gait (Schmitt 1999). Others, however, such as muscle forces, may be accentuated when on top or below a branch (Susman and Stern 1979; Patel et al. 2012a). Biomechanical theories have postulated that when gripping a branch with flexed PIP and McP joints, dorsopalmarly curved proximal phalanges should experience lower strains compared to straight phalanges (Preuschoft 1973b; Oxnard 1973). Recent biomechanical experiments using validated finite-element (FE) models of curved and straight proximal phalanges have generally supported these theoretical claims (Richmond 1998, 2007; Jungers et al. 2002; Nguyen et al. 2014; see Chap. 9). For example, these FE models have predicted that curved proximal phalanges experience roughly half the compressive and tensile strains (on the dorsal and palmar surfaces, respectively) than those of mathematically modeled straight phalanges. Moreover, ontogenetic studies have shown that some degree of phalangeal curvature is plastic and thus sensitive to mechanical loading throughout an animal's life (Richmond 1998; Jungers et al. 2002; Congdon 2012). An alternative hypothesis, although not mutually exclusive, for greater phalangeal curvature in arboreal and suspensory primates proposes that curvature: (1) increases the surface area over which the palmar skin can contact the substrate, thereby reducing strains in deeper soft tissues, and (2) allows the digits to circumduct larger diameter supports (Hunt 1991). Regardless of which of these hypotheses are correct, it is clear that curved fingers are an adaptation for arboreal and especially suspensory behaviors.

There are several methods used to estimate phalangeal curvature quantitatively. The most common approach considers the bone's included angle (IA), which assumes that the bone's curvature represents an arc length on the perimeter of a circle (Susman et al. 1984; Stern et al. 1995). An alternative method uses high-resolution polynomial curve fitting (HR-PCF), which assumes that a second-order polynomial or a parabolic shape provides a better representation of an anatomical curve (Deane et al. 2005; Deane and Begun 2008). Despite theoretical differences and assumptions, both methods show that increased curvature of phalanges is a functional adaptation for arboreality and is most accentuated in suspensory species (Figs. 4.4b and 4.5; Table 4.5). However, it is important to note that there can be substantial variation and overlap in phalangeal curvature across species (see, e.g., Fig. 3 in Deane and Begun 2008). Still, the most terrestrial quadrupedal species have both flat proximal and middle phalanges. African apes, which are both forelimb suspensory and terrestrial quadrupeds (i.e., knuckle-walkers), have curved proximal phalanges, but flat middle phalanges (Fig. 4.5). When compared to other euarchontans, strepsirrhines and other primates tend to have more curved intermediate phalanges (see Fig. 21 in Boyer et al. 2013). Moreover, with the exception of suspensory Asian hominoids and atelines, some strepsirrhines (e.g., *Mirza*) have more curved intermediate phalanges than similar body-sized platyrrhines, *Tarsius*, and cercopithecoids (Boyer et al. 2013). It is important to note that the methods used to measure IA, or any other variable trying to estimate curvature, can be highly susceptible to measurement error and thus can lead to discrepancies in published values. For example, IA estimates in *Pan troglodytes* differ between authors who apply the method differently [mean values of 42.4° in Stern et al. (1995); 55.0° in Matarazzo (2008); and 38.4° in Rein (2011); see Table 4.5 for other discrepancies between researchers].

6 Distal Phalanges

6.1 General Morphology

Like proximal and middle phalanges, the **distal phalanx** can also be divided into three distinct components: the **base** (proximal epiphysis), the **shaft**, and the **apical tuft** (Mittra et al. 2007; Fig. 4.6). In humans, as well as in most other primates, the

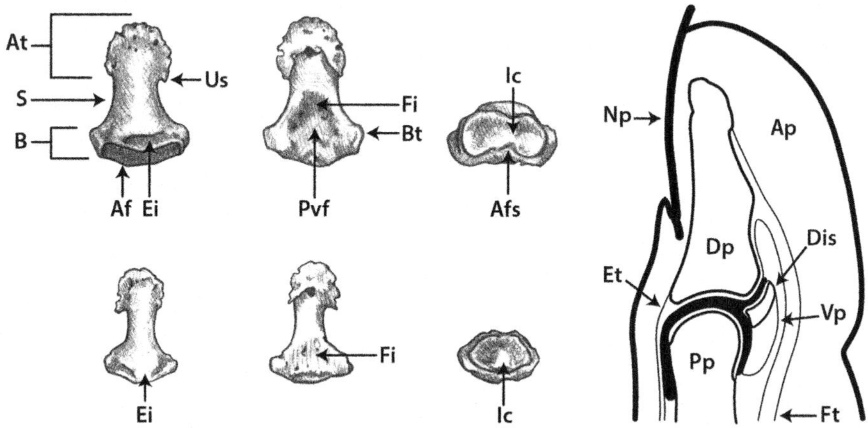

Fig. 4.6 Distal phalanges of *Homo sapiens*. *Top* row (*left* to *right*) shows dorsal, volar, and proximal views of a right pollical distal phalanx. *Bottom* row (*left* to *right*) shows dorsal, volar, and proximal views of a left distal phalanx of the third ray. *Right* drawing shows schematic median sagittal section through the pollex demonstrating tendon insertions and relationship of phalanx to soft tissue structures. Abbreviations: *Ap* apical pad, *Af* proximal articular facet, *Afs* articular facet for distal interphalangeal sesamoid, *At* apical tuft, *B* base, *Bt* basal tubercle, *Dis* distal interphalangeal sesamoid, *Dp* distal phalanx, *Ei* insertion of extensor tendon, *Et* extensor tendon, *Fi* insertion of flexor tendon, *Ft* flexor tendon and sheath, *Ic* intercondylar crest, *Np* nail plate, *Pp* proximal phalanx, *Pvf* proximal volar fossa, *S* shaft, *Us* ungual spine, *Vp* volar plate

base is most often the radioulnarly widest and dorsopalmarly tallest part of the phalanx, flaring beyond the radial and ulnar margins of the shaft. It articulates with the intermediate (or proximal in the case of the pollex) phalanx via the proximal articular facet. This facet accommodates the two condyles of the trochlea-shaped head of the articulating phalanx, with a blunt keel or intercondylar crest usually dividing the facet into two reciprocal concavities (Shrewsbury et al. 2003). The shaft extends distally from the base, which may be canted palmarly or dorsally. Compared to claw-bearing species (see below), the shaft is radioulnarly wide and dorsopalmarly shallow (Clark 1936; Hershkovitz 1977; Spearman 1985; Hamrick 1998; Soligo and Müller 1999; Maiolino et al. 2011) and tends to taper distally in both dimensions. An expanded bony flange surrounds the distal extremity of the shaft. This structure has been referred to variously as the apical tuft (Susman 1979; Aiello and Dean 1990; Jungers et al. 2005; Mittra et al. 2007; Maiolino et al. 2012), distal phalangeal tuberosity (Walker et al. 2011), ungual tuft (Marzke 1997), ungual tuberosity (Shrewsbury and Johnson 1975; Nakatsukasa et al. 2003), or *tuberositas unguicularis* (Wilkinson 1951; Day and Napier 1966). Different terms are often used to designate the same structure depending on the species; for example, some authors reserve the term "tuft" solely for the extremely expanded and rugous structure of humans, while "tuberosity" refers to the condition observed in nonhuman primates (Shrewsbury and Sonek 1986). Others have argued that since the term "tuft" does not adequately describe primate morphology, the term "shield" should be used instead (Koenigswald et al. 2012). Regardless of the preferred nomenclature, this structure is highly variable in shape and shows a great deal of diversity within and across primate taxa (Bruhns 1910; Shrewsbury et al. 2003; Mittra et al. 2007; Koenigswald et al. 2012). It may be rounded or pointed, radioulnarly wide, or narrow and can vary in the degree to which it surrounds the shaft and in the degree of rugosity along its margins. The tuft is closely associated with the nail plate as the distal portion of the nail bed is firmly attached to its dorsal surface (Shrewsbury et al. 2003). The apical pad may be situated against the palmar surface of the tuft (Clark 1936; Shrewsbury and Johnson 1975), but in many primates, especially strepsirrhines, a large portion of the tuft appears to project beyond the apical pad where it is surrounded only by the nail and associated tissues (Bruhns 1910; Maiolino personal observations).

In humans, the tendons of the flexor digitorum profundus (FDP) and the flexor pollicis longus (FPL) muscles insert on the non-pollical and pollical distal phalanges, respectively. Each tendon has its primary insertion on the palmar aspect of the shaft, which may be marked by a roughened tuberosity (Shrewsbury and Johnson 1975; Susman 1998; Shrewsbury et al. 2003). This attachment site may be positioned proximally or even as far distal as the proximal margin of the apical tuft (Shrewsbury et al. 2003). The insertion of an extensor tendon (i.e., extensor pollicis longus (EPL) for the distal pollical phalanx and terminal tendon of the extensor assembly of extensor digitorum (ED) for all other distal phalanges) is on the dorsal surface of the distal phalanx, near or on its base, and typically does not leave a well-distinguished extensor tubercle (Clark 1936).

Non-pollical distal phalanges usually cannot be distinguished from one another on the basis of shape alone. However, those of the third and fourth digits do tend to be absolutely longer than those of the second and fifth digits (Susman 1979; Ricklan 1988; Case and Heilman 2006).

The pollical distal phalanx is clearly distinguishable from other distal phalanges. The base and articular facet are kidney bean in shape when viewed proximally (Susman 1979), while those of the other digits tend to be more oval. The pollical distal phalanx is often distinguished by a large proximal palmar fossa that is often coupled with a V-shaped insertion of the FPL tendon on its distal margin (Shrewsbury et al. 2003). In humans, this fossa has been suggested to accommodate a sesamoid embedded within the palmar plate of the pollical interphalangeal joint during flexion (Marzke et al. 1998). Other primate species have been observed to have such a sesamoid (Shrewsbury et al. 2003). When reviewing earlier literature, the reader should be aware that this fossa was once assumed to be the insertion site for the FPL tendon, a viewpoint that has been rescinded in light of dissection-based evidence (Marzke et al. 1998; Shrewsbury et al. 2003). The portion of the articular facet that contacts the radial condyle of the proximal phalanx has a larger, more concave surface area than that for the ulnar condyle, which is thought to facilitate conjunct pronation and flexion at the interphalangeal joint (Shrewsbury et al. 2003).

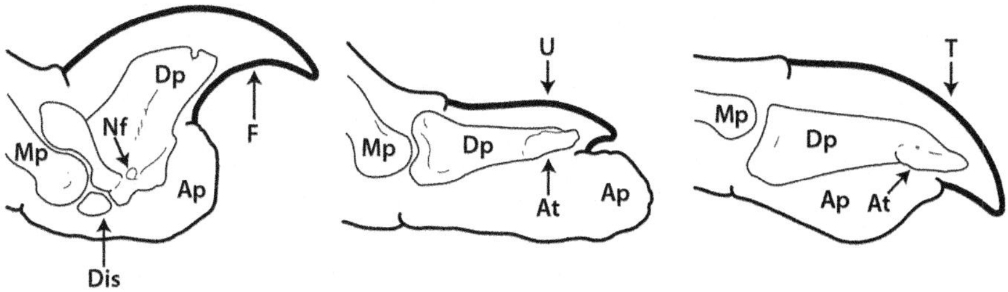

Fig. 4.7 Key anatomical differences among falcular (*Ptilocercus lowii*: *left*), ungular (*Galago senegalensis*: middle), and tegular (*Saguinus fuscicollis*: *right*) distal phalanges of the manual third ray. Integument and other soft tissue are indicated by outlines surrounding the distal phalanges. Abbreviations: *Ap* apical pad, *At* apical tuft, *Dis* distal interphalangeal sesamoid, *Dp* distal phalanx, *F* falcula, *Mp* middle (intermediate) phalanx, *Nf* nutrient foramen, *T* tegula, *U* ungula

6.2 Tegula-Bearing Distal Phalanges

Callitrichine platyrrhines and *Daubentonia* possess specialized claw-like structures on their manual digits called tegulae (Weber 1904; Clark 1936; Hershkovitz 1977; Spearman 1985; Soligo and Müller 1999; see Chap. 8). This specialization allows these primates to cling and climb on relatively large tree trunks (Cartmill 1974; Garber 1980; Hamrick 1998). Tegula-bearing (tegular) distal phalanges differ in several ways from those bearing nails (called ungulae) and those bearing claws (called falculae) typical of non-primate mammals. Falcular distal phalanges are relatively narrow and tall, lack apical tufts, usually have associated interphalangeal sesamoids, possess a nutrient foramen located on the palmar aspect of each side of the shaft near its junction with the base, and usually have well-developed tubercles for insertions of the long extensor and flexor tendons (Clark 1936). The flexor tubercle is located on a distinctive, proximally positioned palmar process (sometimes referred to as the volar process) that is associated with a proximally restricted apical pad (Maiolino et al. 2011). Tegular phalanges are narrow and tall like falcular phalanges, but resemble ungular phalanges in most other ways. Like ungular phalanges, they possess apical tufts and typically lack interphalangeal sesamoids, well-defined nutrient foramina, and well-developed extensor tubercles (Clark 1936; Thorndike 1968; Garber 1980; Maiolino et al. 2011). Furthermore, the apical pads of tegular digits are more extensive, expanding further distally along the palmar surface of the distal phalanx compared to falcular digits (Rosenberger 1977; Garber 1980). These differences are illustrated in Fig. 4.7.

6.3 Morphological Diversity

Although they are the smallest component of the rays, distal phalanges have received a great deal of attention in anatomical and comparative studies. The relative size and shape of the apical tuft has been compared in primates exhibiting different locomotor behaviors. For example, suspensory taxa have been shown to have the radioulnarly narrowest tufts, while terrestrial species have the widest (Mittra et al. 2007; Table 4.6). The manual distal phalanges of strepsirrhines are highly variable among species but on average possess relatively large apical tufts. Many lemurs (e.g., *Lemur*, *Eulemur*, *Hapalemur*) have distinctively pointed nails that follow the contours of a distal phalanx that is triangular in cross section with a shallow dorsal apex and a mediolaterally wide base (Bruhns 1910). Pointed nails are also present on the digits of *Euoticus* and *Phaner*, which are suggested to assist these small-bodied species in clinging on large trunks (Charles-Dominique 1977). Other taxa have more rounded apical tufts, though the degree to which they are radioulnarly expanded is highly variable (e.g., *Galago* and *Microcebus*; Fig. 4.8). Among strepsirrhines, *Daubentonia* is unique in possessing radioulnarly narrow tegulae on all of its digits (with the exception of the hallux and the second pedal digit). In lorisids, the tendency for reduction of the second manual digit (see above) often leads to phalangeal loss and in its most extreme form results in a nailless stump comprised of two minuscule phalanges (e.g., *Arctocebus* and *Perodicticus*; Straus 1942).

Like strepsirrhines, the distal phalanges of tarsiers are also quite variable. Most notably, those of *Tarsius pumilus* are somewhat mediolaterally narrow and have well-developed, pointed apical tufts that support pointed, keeled nails. This is generally thought to assist the animal in clinging on the moss-covered substrates that are prevalent in their high-altitude habitat (Musser and Dagosto 1987). However, some have contended that these small claw-like nails may not be capable of penetrating the tough, slippery moss (Grow and Gursky-Doyen 2010). On the other end of the spectrum, *Tarsius* (*Cephalopachus*) *bancanus* has particularly long, narrow distal phalanges with small, rounded apical tufts that support diminutive nails (Clark 1936; Cartmill 1974). *T. bancanus* has been noted to have longer phalanges than other tarsiers (Niemitz 1979), and its distal phalanges certainly conform to this pattern. These long digits may assist *T. bancanus* in catching insects or be reflective of its choice of substrate (Niemitz 1979; Musser and Dagosto 1987). Other tarsiers fall into the range of distal phalanx shapes between *T. pumilus* and *T. bancanus*, with *T. spectrum* more similar to *T. pumilus*, and *T. syricta* more similar to *T. bancanus*.

Among anthropoids, platyrrhine nails tend to be strongly curved or keeled along the sagittal plane (Hershkovitz 1977). In contrast, the underlying distal phalanges (excluding those of callitrichines) tend to be straight, cylindrical in cross section and have small apical tufts (e.g., *Pithecia* and *Cebus*; Mittra et al. 2007; Fig. 4.8). Variation in distal phalangeal morphology seen between ungulae-bearing platyrrhines, such as *Saimiri*, and tegulae-bearing callitrichines has been associated with substrate preferences; species that prefer terminal branches have wider and shallower distal phalanges than those that prefer large diameter vertical substrates (Hamrick 1998). Platyrrhine genera typically lacking an external pollex (i.e., *Ateles* and *Brachyteles*) have highly variable pollical distal phalanges that may be lost, reduced, or fused to the proximal phalanx (Straus 1942; Tague 2002).

Cercopithecoid distal phalanges vary in shape from narrow and slender to wide and robust (Fig. 4.8). Apical tuft morphology is also highly variable and is most well developed in *Theropithecus* and *Papio* (Mittra et al. 2007; Table 4.6). Among

Table 4.6 Distal phalanx apical tuft shape[a]

		Expansion index[b]		Robusticity index[c]	
Taxon	*N*	Mean	SD	Mean	SD
Alouatta seniculus	2	52.29	10.50	26.63	9.55
Arctocebus calabarensis	8	54.34	8.76	26.18	4.53
Ateles spp.	7	56.94	6.23	23.71	2.44
Avahi laniger	10	64.33	7.75	39.22	7.91
Brachyteles arachnoides	1	53.00	5.71	20.79	2.31
Cebus apella	2	42.45	6.47	17.43	1.93
Cebus olivaceus	1	43.73	6.86	17.60	3.14
Cheirogaleus major	7	61.55	5.55	49.41	5.37
Cheirogaleus medius	3	72.46	5.30	59.61	8.25
Chlorocebus aethiops	4	63.75	11.97	40.17	8.25
Eulemur fulvus	12	69.46	4.81	56.59	5.43
Eulemur macaco	7	71.44	5.82	59.63	5.25
Eulemur mongoz	8	71.10	5.80	56.74	5.27
Euoticus elegantulus	9	76.28	6.16	55.70	6.45
Galago moholi	6	57.03	7.91	42.87	6.30
Galago senegalensis	12	55.27	5.93	40.96	5.55
Galagoides alleni	2	54.59	3.94	32.71	2.36
Galagoides demidoff	8	67.97	11.59	43.14	5.48
Gorilla gorilla	13	67.56	13.40	36.81	4.52
Hapalemur griseus	10	74.59	5.68	48.37	4.50
Homo sapiens	48	66.77	5.63	42.92	5.74
Hylobates lar	11	45.51	8.36	20.38	4.43
Indri indri	8	68.54	7.54	40.13	9.23
Lagothrix lagothricha	2	41.18	6.54	16.12	1.48
Lemur catta	11	74.29	5.47	59.16	6.30
Lepilemur leucopus	6	71.25	6.84	51.21	5.53
Lepilemur mustelinus	5	66.69	7.77	45.14	6.23
Loris tardigradus	8	55.68	6.72	38.73	8.38
Microcebus murinus	9	56.69	8.32	50.56	7.51
Nycticebus coucang	16	51.02	6.71	42.69	6.50
Otolemur crassicaudatus	6	49.58	8.44	35.10	3.90
Otolemur garnettii	2	51.93	7.02	38.32	6.40
Pan paniscus	1	67.06	5.08	30.83	8.16
Pan troglodytes	11	58.74	6.11	28.88	5.71
Perodicticus potto	14	44.22	6.81	32.40	6.34
Pongo abelii	11	44.05	5.04	22.14	3.24
Pongo pygmaeus	8	46.49	7.06	23.94	4.36
Propithecus diadema	6	65.03	6.19	37.34	6.22
Propithecus verreauxi	8	66.41	9.04	37.75	7.25
Symphalangus syndactylus	5	40.91	8.27	19.07	4.56

(continued)

Table 4.6 (continued)

Taxon	*N*	Expansion index[b]		Robusticity index[c]	
		Mean	SD	Mean	SD
Tarsius bancanus	5	72.51	5.29	44.24	6.01
Tarsius syrichta	8	82.13	7.28	48.00	4.07
Theropithecus gelada	2	74.91	6.48	42.11	5.78
Varecia variegata	8	68.24	7.87	52.80	6.90

[a]Data from Mittra et al. (2007)
[b]Expansion index: apical tuft width divided by base width (×100) from rays 2 to 5
[c]Robusticity index: apical tuft width divided by length (×100) from rays 2 to 5

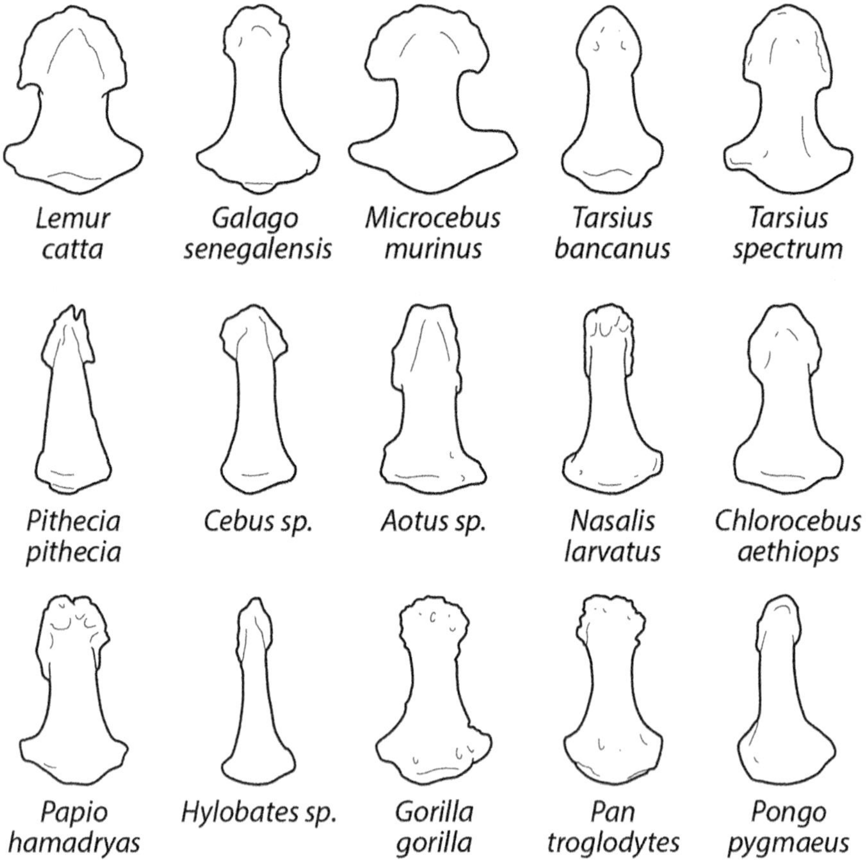

Fig. 4.8 Dorsal view of distal phalanges of the manual third ray of various nonhuman primates (scaled to the same length). Note differences in width of base, shaft, and apical tuft

cercopithecoids, the pollical distal phalanx of *Papio* has been described as often possessing a pair of small, proximally directed spines (termed ungual spines) on the sides of the apical tuft (Susman 1998; Shrewsbury et al. 2003). Like *Ateles* and *Brachyteles*, the distal pollical phalanx of the reduced pollex of *Colobus* may be present, absent, or fused to the proximal phalanx (Tague 2002).

Hominoid distal phalanges vary in length, robusticity, and apical tuft expansion (Fig. 4.8). Hylobatids and *Pongo* have relatively narrow apical tufts, while those of *Gorilla* and *Pan* are wider, and those of *Homo* are the widest (Susman 1979; Mittra et al. 2007). *Gorilla* has the most robust distal phalanges of all the hominoids (measured as the robusticity index; Table 4.6), which stand in sharp contrast to the relatively long and slender phalanges observed in hylobatids and other great apes (Susman 1979; Susman and Creel 1979; Mittra et al. 2007). Finally, human distal phalanges are unique in combining relatively large apical tufts with well-developed ungual spines and dorsopalmarly shallow shafts (Susman and Creel 1979). Many of the features observed in humans have been discussed as functionally linked to precision gripping (e.g., Shrewsbury and Sonek 1986; Marzke 1997; Susman 1998; Shrewsbury et al. 2003; Almécija et al. 2010), though they can be found individually or in different combinations in other primates (e.g., relatively large apical tufts in strepsirrhines and well-developed ungual spines in *Papio*; Susman 1998; Mittra et al. 2007).

7 Future Directions

Despite over a century of research on the comparative anatomy of primate hand bones (e.g., Mivart 1867), additional qualitative observations and quantitative metric data are needed across the entire Primate order. Current data sampling is biased toward hominoids; undeniably, other nonhuman primates need more attention. Comparative details (both qualitative and quantitative) across all anthropoids and "prosimians" are needed about phalangeal and metacarpal shaft robustness (e.g., cortical bone data does not exist for non-hominoids) and curvatures (e.g., no data exists on metacarpal curvature), along with proximal and distal articular surface shapes. As recovery of new fossil material typically dictates new research directions, we can only envision that the discovery of new non-hominoid fossil hand bone material from the Paleocene to the Holocene will stimulate the study of more comprehensive and detailed extant primate hand bone morphology (e.g., Hamrick et al. 1995; Jablonski et al. 2002; Jungers et al. 2005; Boyer 2009; Franzen et al. 2009; Rossie et al. 2012; Boyer et al. 2013; DeSilva et al. 2013; Halenar and Rosenberger 2013).

Primate hand bones are small and irregular in shape. As such, quantification of their morphology has primarily been limited to linear measurements, which are easy to acquire with calipers (e.g., Begun 1993; Almécija et al. 2009; Patel 2010a) or two-dimensional measurements from photographs and radiographs (e.g., Susman 1979; Richmond 1998; Rolian 2009). More recently, landmarking techniques with a point digitizer have been used to capture three-dimensional angle data and for the purpose of geometric morphometric analyses; however, reliable data with acceptable accuracy and precision are only obtainable on larger hand bones like metacarpals (Rein 2011; Rein and McCarty 2012). In order to overcome size-limiting factors, the use of digital imaging techniques, including laser scanning and computed tomography (both medical-grade and high-resolution micro-CT), should be employed to visualize and quantify morphology of smaller hand bones in three dimensions (see Chap. 9).

The use of 3D digital imaging techniques should allow more comprehensive data to be gathered to describe the overall shape and the details of hand bones (e.g., Almécija et al. 2015), as well as to gather more functional/biomechanical variables such as lengths, angles, curves, areas, and volumes (e.g., Boyer et al. 2013; Carnation et al. 2013; Patel et al. 2014, Patel et al. 2015). Moreover, the use of medical and micro-CT scanning should also allow for visualization and quantification of other aspects of internal skeletal form such as cortical bone cross-sectional geometry (e.g., Marchi 2005; Wallace and Patel 2013) and trabecular bone architecture (e.g., Tsegai et al. 2013) that can further inform us about hand bone functional morphology. Furthermore, the 3D images acquired from laser/CT scanning should also facilitate new observations on small hand bones and enhance meticulous character-state lists that can be used for analyses of phylogenetic relationships (e.g., Patel et al. 2012b). Finally, when these functional and phylogenetic studies are coupled with our understanding of the primate hand fossil record (Boyer et al. 2013; see also Chaps. 14–19), reconstruction of evolutionary trends and rates in the primate hand will become more feasible (e.g., Patel et al. 2012b; Kivell et al. 2013). Thus, future studies should continue to use these technologies and analytical techniques in order to further our understanding of the primate digital rays and hand.

Acknowledgments We would like to offer our thanks to the editors for inviting us to contribute to this volume. Specifically, we would like to acknowledge the comments and helpful suggestions offered by Pierre Lemelin and Tracy Kivell during the writing process. Both Sergio Almécija and Campbell Rolian provided much of the raw data needed to create Fig. 4.1. Finally, we thank William Jungers, Caley Orr, and Doug Boyer for providing original unpublished metric and 3D scan data and Eileen Westwig for the access to museum specimens housed in the American Museum of Natural History. Funding for this work was provided in part from The Leakey Foundation to each author and the National Science Foundation (BCS 1317047 to BAP; BCS 1341075 to SAM).

References

Aiello L, Dean C (1990) An introduction to human evolutionary anatomy. Academic Press, New York

Allen KL (2008) The hand of *Kenyapithecus africanus* from Maboko Island, Kenya. M.A. thesis, New Mexico State University

Almécija S, Alba DM, Moyà-Solà S (2009) *Pierolapithecus* and the functional morphology of Miocene ape hand phalanges: paleobiological and evolutionary implications. J Hum Evol 57:284–297

Almécija S, Moyà-Solà S, Alba DM (2010) Early origin for human-like precision grasping: a comparative study of pollical distal phalanges in fossil hominins. PLoS One 5:e11727

Almécija S, Orr CM, Tocheri MW, Patel BA, Jungers WL (2015) Exploring phylogenetic and functional signals in complex morphologies: the hamate of extant anthropoids as a test-case study. Anat Rec 298:212–229

Begun DR (1993) New catarrhine phalanges from Rudabánya (Northeastern Hungary) and the problem of parallelism and convergence in hominoid postcranial morphology. J Hum Evol 24:373–402

Boyer DM (2009) New cranial and postcranial remains of late Paleocene Plesiadapidae ("Plesiadapiformes," Mammalia) from North America and Europe: description and evolutionary implications. Ph.D. dissertation, Stony Brook University

Boyer DM, Yapuncich GS, Chester SGB, Bloch JI, Godinot M (2013) Hands of early primates. Yearb Phys Anthropol 57:33–78

Bruhns F (1910) Der Nagel der Halbaffen und Affen, ein Beitrag zur Phylogenie des menschichen Nagels. Morphol Jahr 40:501–609

Carnation SL, Orr CM, Patel BA (2013) The relationship between thumb reduction and carpal volume in African colobines. Am J Phys Anthropol 150(Suppl 56):95 (abstract)

Cartmill M (1974) Pads and claws in arboreal locomotion. In: Jenkins FA Jr (ed) Primate locomotion. Academic Press, New York, pp 45–83

Case DT, Heilman J (2006) New siding techniques for the manual phalanges: a blind test. Int J Osteoarchaeol 16:338–346

Charles-Dominique P (1977) Ecology and behavior of nocturnal primates. Columbia University Press, New York

Chen M, Luo Z (2013) Postcranial skeleton of the Cretaceous mammal *Akidolestes cifellii* and its locomotor adaptations. J Mammal Evol 20:159–189

Le Gros Clark WE (1936) The problem of the claw in primates. Proc Zool Soc Lond 106:1–24

Congdon KA (2012) Interspecific and ontogenetic variation in proximal pedal phalangeal curvature of great apes (*Gorilla gorilla*, *Pan troglodytes*, and *Pongo pygmaeus*). Int J Primatol 33:418–427

Day MH, Napier JR (1966) A hominid toe bone from Bed I Olduvai Gorge Tanzania. Nature 211:929–930

Deane AS, Begun DR (2008) Broken fingers: retesting locomotor hypotheses for fossil hominoids using fragmentary proximal phalanges and high-resolution polynomial curve fitting (HR-PCF). J Hum Evol 55:691–701

Deane AS, Kremer EP, Begun DR (2005) New approach to quantifying anatomical curvatures using high-resolution polynomial curve fitting (HR-PCF). Am J Phys Anthropol 128:630–638

DeSilva JM, Steininger CM, Patel BA (2013) Cercopithecoid primate postcranial fossils from Cooper's D, South Africa. GeoBios 46:381–394

Erickson CJ (1994) Tap-scanning and extractive foraging in aye-ayes, *Daubentonia madagascariensis*. Folia Primatol 62:125–135

Etter HF (1973) Terrestrial adaptations in the hands of cercopithecinae. Folia Primatol 20:331–350

Etter HF (1974) Morphologisch- und metrisch-vergleichende untersuchung am handskelet rezenter Primaten. Gegenbaurs Morphol Jahrb 120:1–484

Franzen JL, Gingerich PD, Habersetzer J, Hurum JH, von Koenigswald W, Smith BH (2009) Complete primate skeleton from the middle Eocene of Messel in Germany: morphology and paleobiology. PLoS One 4:e5723

Garber PA (1980) Locomotor behavior and feeding ecology of the Panamanian tamarin (*Saguinus oedipus geoffroyi*, Callitrichidae, Primates). Int J Primatol 1:185–201

Grow N, Gursky-Doyen S (2010) Preliminary data on the behavior, ecology, and morphology of pygmy tarsiers (*Tarsius pumilus*). Int J Primatol 31:1174–1191

Gyambibi A, Lemelin P (2013) Comparative and quantitative myology of the forearm and hand of prosimian primates. Anat Rec 296:1196–1203

Halenar LB, Rosenberger AL (2013) A closer look at the "*Protopithecus*" fossil assemblages: new genus and species from Bahia. Brazil J Hum Evol 65:374–390
Hamrick MW (1998) Functional and adaptive significance of primate pads and claws: evidence from New World anthropoids. Am J Phys Anthropol 106:113–127
Hamrick MW, Meldrum DJ, Simons EL (1995) Anthropoid phalanges from the Oligocene of Egypt. J Hum Evol 28:121–145
Hamrick MW, Rosenman BA, Brush JA (1999) Phalangeal morphology of the Paromomyidae (?Primates, Plesiadapiformes): the evidence for gliding behavior reconsidered. Am J Phys Anthropol 109:397–413
Hayama S, Chatani K, Nakatsukasa M (1994) The digitigrade hand and terrestrial adaptation in Japanese macaques. Anthropol Sci 102(Suppl):115–125
Hershkovitz P (1977) Living new world monkeys (Platyrrhini), vol 1. University of Chicago Press, Chicago
Hill WCO (1955) Primates: comparative anatomy and taxonomy. II Haplorhini: Tarsioidea. University Press, Edinburgh
Hopson JA (1995) Patterns of evolution in the manus and pes of non-mammalian therapsids. J Vert Paleontol 15:615–639
Hunt KD (1991) Mechanical implications of chimpanzee positional behavior. Am J Phys Anthropol 86:521–536
Inouye SE, Shea BT (2004) The implications of variation in knuckle-walking features for models of African hominoid locomotor evolution. J Anthropol Sci 82:67–88
Jablonski NG, Leakey MG, Antón M (2002) A new skeleton of *Theropithecus brumpti* (Primates: Cercopithecidae) from Lomekwi, West Turkana, Kenya. J Hum Evol 43:887–923
Ji Q, Luo Z, Ji S (1999) A Chinese triconodont mammal and mosaic evolution of the mammalian skeleton. Nature 398:326–330
Ji Q, Luo Z-X, Yuan C-X, Wible JR, Zhang J-P, Georgi JA (2002) The earliest known eutherian mammal. Nature 416:816–822
Jouffroy FK (1993) Primate hands and the human hand: the tool of tools. In: Berthelet A, Chavaillon J (eds) The use of tools by human and non-human primates. Clarendon Press, Oxford, pp 6–33
Jouffroy FK, Lessertisseur J (1979) Relationships between limb morphology and locomotor adaptations among prosimians: an osteometric study. In: Morbeck ME, Preuschoft H, Gomberg N (eds) Environment, behavior, and morphology: dynamic interactions in primates. Gustav Fischer, New York, pp 143–181
Jouffroy FK, Godinot M, Nakano Y (1993) Biometrical characteristics of primate hands. In: Preuschoft H, Chivers DJ (eds) Hands of primates. Springer-Verlag, Vienna, pp 133–171
Jungers WL, Godfrey LR, Simons EL, Chatrath PS (1997) Phalangeal curvature and positional behavior in extinct sloth lemurs (Primates, Palaeopropithecidae). Proc Natl Acad Sci U S A 94:11998–12001
Jungers WL, Godfrey LR, Simons EL, Wunderlich RE, Richmond BG, Chatrath PS (2002) Ecomorphology and behavior of giant extinct lemurs from Madagascar. In: Plavcan JM, Kay RF, Jungers WL, van Schaik CP (eds) Reconstructing behavior in the primate fossil record. Kluwer Academic, New York, pp 371–411
Jungers W, Lemelin P, Godfrey L, Wunderlich R, Burney D, Simons E, Chatrath P, James H, Randria G (2005) The hands and feet of *Archaeolemur*: metrical affinities and their functional significance. J Hum Evol 49:36–55
Kingston AK, Boyer DM, Patel BA, Larson SG, Stern JT (2010) Hallucal grasping in Nycticebus coucang: further implications for the functional significance of a large peroneal process. J Hum Evol 58:33–42
Kirk EC, Lemelin P, Hamrick MW, Boyer DM, Bloch JI (2008) Intrinsic hand proportions of euarchontans and other mammals: implications for the locomotor behavior of plesiadapiforms. J Hum Evol 55:278–299
Kivell TL, Schmitt D, Wunderlich RE (2010) Hand and foot pressures in the aye-aye (*Daubentonia madagascariensis*) reveal novel biomechanical trade-offs required for walking on gracile digits. J Exp Biol 213:1549–1557

Kivell TL, Barros AP, Smaers JB (2013) Different evolutionary pathways underlie the morphology of wrist bones in hominoids. BMC Evol Biol 13:229

von Koenigswald W, Habersetzer J, Gingerich PD (2012) Pedal distal phalanges of the Eocene adapoids *Europolemur* and *Darwinius* compared to the phalanges of *Notharctus* and other primates. Palaeobio Palaeoenv 92:539–565

Latimer B, Lovejoy CO (1990) Metatarsophalangeal joints of *Australopithecus afarensis*. Am J Phys Anthropol 83:13–23

Lazenby RA (1998) Second metacarpal cross-sectional geometry: rehabilitating a circular argument. Am J Hum Biol 10:747–756

Lemelin P (1996) The evolution of manual prehensility in primates. A comparative study of prosimian primates and didelphid marsupials. Ph.D. dissertation, State University of New York at Stony Brook

Lemelin P (1999) Morphological correlates of substrate use in didelphid marsupials: implications for primate origins. J Zool 247:165–175

Lemelin P, Jungers WL (2007) Body size and scaling of the hands and feet of prosimian primates. Am J Phys Anthropol 133:828–840

Lemelin P, Schmitt D (1998) The relation between hand morphology and quadrupedalism in primates. Am J Phys Anthropol 105:185–197

Lemelin P, Schmitt D (2007) Origins of grasping and locomotor adaptations in primates: comparative and experimental approaches using an opossum model. In: Ravosa MJ, Dagosto M (eds) Primate origins: adaptations and evolution. Springer, New York, pp 329–380

Lewis OJ (1977) Joint remodeling and the evolution of the human hand. J Anat 123:157–201

Lewis OJ (1989) Functional morphology of the evolving hand and foot. Clarendon Press, Oxford

Lorenz R (1974) On the thumb of the Hylobatidae. In: Rumbaugh DM (ed) Gibbon and Siamang, vol 3. Karger, Basel, pp 157–175

Lovejoy CO, Simpson SW, White TD, Asfaw B, Suwa G (2009) Careful climbing in the Miocene: the forelimbs of *Ardipithecus ramidus* and humans are primitive. Science 326:70e1–70e8

Maiolino S, Boyer DM, Rosenberger A (2011) Morphological correlates of the grooming claw in distal phalanges of platyrrhines and other primates: a preliminary study. Anat Rec 294:1975–1990

Maiolino S, Boyer DM, Bloch JI, Gilbert CC, Groenke J (2012) Evidence for a grooming claw in a North American adapiform primate: implications for anthropoid origins. PLoS One 7:e29135

Marchi D (2005) The cross-sectional geometry of the hand and foot bones of the Hominoidea and its relationship to locomotor behavior. J Hum Evol 49:743–761

Marzke MW (1997) Precision grips, hand morphology, and tools. Am J Phys Anthropol 102:91–110

Marzke MW, Marzke RF (1987) The third metacarpal styloid process in humans: origin and functions. Am J Phys Anthropol 73:415–431

Marzke MW, Toth N, Schick K, Reece S, Steinberg B, Hunt K, Linscheid RL, An KN (1998) EMG study of hand muscle recruitment during hard hammer percussion manufacture of Oldowan tools. Am J Phys Anthropol 105:315–332

Marzke MW, Shrewsbury MM, Horner KE (2007) Middle phalanx skeletal morphology in the hand: can it predict flexor tendon size and attachments? Am J Phys Anthropol 134:141–151

Marzke MW, Tocheri MW, Steinberg B, Femiani JD, Reece SP, Linscheid RL, Orr CM, Marzke RF (2010) Comparative 3D quantitative analyses of trapeziometacarpal joint surface curvatures among living catarrhines and fossil hominins. Am J Phys Anthropol 141:38–51

Matarazzo S (2008) Knuckle walking signal in the manual digits of *Pan* and *Gorilla*. Am J Phys Anthropol 135:27–33

Matarazzo S (2013) Manual pressure distribution patterns of knuckle-walking apes. Am J Phys Anthropol 152:44–50

Matsuura Y, Ogihara N, Nakatsukasa M (2010) A method for quantifying articular surface morphology of metacarpals using quadric surface approximation. Int J Primatol 31:263–274

McClure NK, Phillips AC, Vogel ER, Tocheri MW (2012) Unexpected pollex and hallux use in wild *Pongo pygmaeus wurmbii*. Am J Phys Anthropol 147(Suppl 54):208 (abstract)

Meldrum D, Yuerong P (1988) Manual proximal phalanx of *Laccopithecus robustus* from the latest Miocene site of Lufeng. J Hum Evol 17:719–731

Mittra ES, Smith HF, Lemelin P, Jungers WL (2007) Comparative morphometrics of the primate apical tuft. Am J Phys Anthropol 134:449–459

Mivart SG (1867) On the appendicular skeleton of the primates. Philos Trans R Soc Lond 157:299–429

Morbeck ME (1976) Leaping, bounding and bipedalism in *Colobus guereza*: a spectrum of positional behavior. Yearb Phys Anthropol 20:408–420

Musser GG, Dagosto M (1987) The identity of *Tarsius pumilus*, a pygmy species endemic to the montane mossy forests of Central Sulawesi. Am Mus Novit 2867:1–53

Nakatsukasa M, Kunimatsu Y, Nakano Y, Takano T, Ishida H (2003) Comparative and functional anatomy of phalanges in *Nacholapithecus kerioi*, a Middle Miocene hominoid from northern Kenya. Primates 44:371–412

Napier J (1980) Hands. Pantheon Books, New York

Napier JR, Davis PR (1959) The fore-limb skeleton and associated remains of *Proconsul africanus*. Fossil Mammals of Africa, No. 16. British Museum (Natural History), London

Napier JR, Napier PH (1967) A handbook of living primates. Academic Press, New York

Napier JR, Walker AC (1967) Vertical clinging and leaping—a newly recognized category of locomotor behavior of primates. Folia Primat 6:204–219

Nelson E, Rolian C, Cashmore L, Shultz S (2011) Digit ratios predict polygyny in early apes, *Ardipithecus*, Neanderthals and early modern humans but not in *Australopithecus*. Proc Biol Sci 278:1556–1563

Nguyen NH, Pahr DH, Gross T, Skinner MM, Kivell TL (2014) Micro-finite element (μFE) modeling of the siamang (*Symphalangus syndactylus*) third proximal phalanx: the functional role of curvature and the flexor sheath ridge. J Hum Evol 67:60–75

Niemitz C (1979) Relationships among anatomy, ecology, and behavior: a model developed in the genus *Tarsius*, with thoughts about phylogenetic mechanisms and adaptive interactions. In: Morbeck ME, Preuschoft H, Gomberg N (eds) Environment, behavior and morphology: dynamic interactions in primates. Gustav Fischer, New York, pp 119–137

Nieschalk U, Demes B (1993) Biomechanical determinants of reduction of the second ray in Lorisinae. In: Preuschoft H, Chivers DJ (eds) Hands of primates. Springer-Verlag, Vienna, pp 225–234

Orr CM (2010) Adaptations to knuckle-walking and digitigrady: a three-dimensional kinematic and morphometric analysis of the anthropoid wrist. Ph.D. dissertation, Arizona State University

Oxnard CE (1973) Form and pattern in human evolution. University of Chicago Press, Chicago

Patel BA (2009) Not so fast: speed effects on forelimb kinematics in cercopithecine monkeys and implications for digitigrade postures in primates. Am J Phys Anthropol 140:92–112

Patel BA (2010a) Functional morphology of cercopithecoid primate metacarpals. J Hum Evol 58:320–337

Patel BA (2010b) The interplay between speed, kinetics and hand postures during primate terrestrial locomotion. Am J Phys Anthropol 141:222–234

Patel BA, Polk JD (2010) Distal forelimb kinematics in *Erythrocebus patas* and *Papio anubis* during walking and galloping. Int J Primatol 31:191–207

Patel BA, Wunderlich RE (2010) Dynamic pressure patterns in the hands of olive baboons (*Papio anubis*) during terrestrial locomotion: implications for cercopithecoid primate hand morphology. Anat Rec 293:710–718

Patel BA, Susman RL, Rossie JB, Hill A (2009) Terrestrial adaptations in the hands of *Equatorius africanus* revisited. J Hum Evol 57:763–772

Patel BA, Larson SG, Stern JT Jr (2012a) Electromyography of wrist and finger flexor muscles in olive baboons (*Papio anubis*). J Exp Biol 215:115–123

Patel BA, Seiffert ER, Boyer DM, Jacobs RL, St. Clair EM, Simons EL (2012b) New primate first metatarsals from the Paleogene of Egypt and the origin of the anthropoid big toe. J Hum Evol 63:99–120

Patel BA, Krishnan A, Perez AC, Lewis AL, Orr CM (2014) A new automated method to quantify primate phalangeal curvature from 3D virtual models. Presentation for the International Primatological Society XXV Congress, Hanoi, Vietnam, 16 Aug 2014 (abstract)

Patel BA, Varghese B, Perez AC, Lewis AL, Jashashvili T, Tocheri M, Almécija S, Jungers WL, Orr CM (2015) A new (semi-) automated method to quantify primate phalangeal curvature from 3D virtual models: new estimates for fossil hominin phalanges. Presented at the 2015 Annual Meeting of the Paleoanthropology Society, San Francisco, CA. April 14–15, 2015 (abstract)

Preuschoft H (1973a) Body posture and locomotion in some East African Miocene Dryopithecinae. In: Day M (ed) Human evolution. Taylor and Francis, London, pp 13–46

Preuschoft H (1973b) Functional anatomy of the upper extremity. In: Bourne GH (ed) The chimpanzee, vol 6. Karger, Basel, pp 34–120

Preuschoft H, Godinot M, Beard C, Nieschalk U, Jouffroy FK (1993) Biomechanical considerations to explain important morphological characters of primate hands. In: Preuschoft H, Chivers DJ (eds) Hands of primates. Springer-Verlag, Vienna, pp 245–246

Rein TR (2011) The correspondence between proximal phalanx morphology and locomotion: implications for inferring the locomotor behavior of fossil catarrhines. Am J Phys Anthropol 146:435–445

Rein TR, McCarty LA (2012) Metacarpophalangeal joint orientation in anthropoid manual phalanges. Anat Rec 295:2057–2068

Richmond BG (1998) Ontogeny and biomechanics of phalangeal form in primates. Ph.D. dissertation, State University of New York at Stony Brook

Richmond BG (2007) Biomechanics of phalangeal curvature. J Hum Evol 53:678–690

Richmond BG, Begun DR, Strait DS (2001) Origin of human bipedalism: the knuckle-walking hypothesis revisited. Yearb Phys Anthropol 44:70–105

Ricklan DE (1988) A functional and morphological study of the hand bones of early and recent South African hominids. Ph.D. dissertation, University of Witwatersrand

Rolian C (2009) Integration and evolvability in primate hands and feet. Evol Biol 36:100–117

Romer AS (1956) Osteology of the reptiles. The University of Chicago Press, Chicago

Rose MD (1986) Further hominoid postcranial specimens from the late Miocene Nagri Formation of Pakistan. J Hum Evol 15:333–367

Rose MD (1988) Functional anatomy of the cheiridia. In: Schwartz JH (ed) Orangutan biology. Oxford University Press, New York, pp 299–310

Rosenberger AL (1977) *Xenothrix* and ceboid phylogeny. J Hum Evol 6:461–481

Rosenberger AL, Strier KB (1989) Adaptive radiation of the ateline primates. J Hum Evol 18:717–750

Rossie JB, Guitierrez M, Goble E (2012) Fossil forelimbs of *Simiolus* from Morourot, Kenya. Am J Phys Anthropol 147(Suppl 54):252 (abstract)

Schmitt D (1994) Forelimb mechanics as a function of substrate type during quadrupedalism in two anthropoid primates. J Hum Evol 26:441–457

Schmitt D (1999) Compliant walking in primates. J Zool 248:149–160

Schultz AH (1930) The skeleton of the trunk and limbs of higher primates. Hum Biol 2:303–438

Shrewsbury M, Johnson RK (1975) The fascia of the distal phalanx. J Bone Joint Surg Am 57:784–788

Shrewsbury MM, Sonek A (1986) Precision holding in humans, non-human primates, and Plio-Pleistocene hominids. Hum Evol 1:233–242

Shrewsbury MM, Marzke MW, Linscheid RL, Reece SP (2003) Comparative morphology of the pollical distal phalanx. Am J Phys Anthropol 121:30–47

Soligo C (2005) Anatomy of the hand and arm in *Daubentonia madagascariensis*: a functional and phylogenetic outlook. Folia Primatol 76:262–300

Soligo C, Müller AE (1999) Nails and claws in primate evolution. J Hum Evol 36:97–114

Spearman RIC (1985) Phylogeny of the nail. J Hum Evol 14:57–61

Stern JT Jr, Oxnard CE (1973) Primate locomotion: some links with evolution and morphology. Primatologia 4:1–93
Stern JT Jr, Susman RL (1983) The locomotor anatomy of *Australopithecus afarensis*. Am J Phys Anthropol 60:279–317
Stern JT Jr, Jungers WL, Susman RL (1995) Quantifying phalangeal curvature: an empirical comparison of alternative methods. Am J Phys Anthropol 97:1–10
Straus WL Jr (1942) Rudimentary digits in primates. Q Rev Biol 17:228–243
Susman RL (1979) Comparative and functional morphology of hominoid fingers. Am J Phys Anthropol 50:215–236
Susman RL (1998) Hand function and tool behavior in early hominids. J Hum Evol 35:23–46
Susman RL (2004) *Oreopithecus bambolii*: an unlikely case of hominidlike grip capability in a Miocene ape. J Hum Evol 46:105–117
Susman RL, Creel N (1979) Functional and morphological affinities of the subadult hand (O.H. 7) from Olduvai Gorge. Am J Phys Anthropol 51:311–331
Susman RL, Stern JT Jr (1979) Telemetered electromyography of flexor digitorum profundus and flexor digitorum superficialis in *Pan troglodytes* and implications for interpretation of the O.H. 7 hand. Am J Phys Anthropol 50:565–574
Susman RL, Stern JT Jr, Jungers WL (1984) Arboreality and bipedality in the Hadar hominids. Folia Primatol 43:113–156
Tague RG (2002) Variability of metapodials in primates with rudimentary digits: *Ateles geoffroyi*, *Colobus guereza*, and *Perodicticus potto*. Am J Phys Anthropol 117:195–208
Thorndike EE (1968) A microscopic study of the marmoset claw and nail. Am J Phys Anthropol 28:247–268
Tsegai ZJ, Kivell TL, Gross T, Nguyen NH, Pahr DH, Smaers JB, Skinner MM (2013) Trabecular bone structure correlates with hand posture and use in hominoids. PLoS One 8:e78781
Tuttle RH (1967) Knuckle-walking and the evolution of hominoid hands. Am J Phys Anthropol 26:171–206
Tuttle RH (1969a) Quantitative and functional studies on the hands of the Anthropoidea. I The Hominoidea. J Morphol 128:309–363
Tuttle RH (1969b) Terrestrial trends in the hands of the Anthropoidea. In: Hofer H (ed) Proceedings 2nd international congress of primatology, vol 2. Karger, Basel, pp 192–200
Tuttle RH (1972) Functional and evolutionary biology of hylobatid hands and feet. In: Rumbaugh DM (ed) Gibbon and Siamang, vol 1. Karger, Basel, pp 137–206
Venkataraman VV, Rolian C, Gordon AD, Patel BA (2013) A resampling approach and implications for estimating the phalangeal index from unassociated hand bones in fossil primates. Am J Phys Anthropol 151:280–289
Walker MJ, Ortega J, Lopez MV, Parmova K, Trinkaus E (2011) Neandertal postcranial remains from the Sima de las Palomas del Cabezo Gordo, Murcia, Southeastern Spain. Am J Phys Anthropol 144:505–515
Wallace IJ, Patel BA (2013) Cross-sectional geometry of chimpanzee finger bones. Am J Phys Anthropol 150(Suppl 56):283 (abstract)
Ward CV, Tocheri MW, Plavcan JM, Brown FH, Manthi FK (2014) Early Pleistocene third metacarpal from Kenya and the evolution of modern human-like hand morphology. Proc Natl Acad Sci U S A 111:121–124
Watkins BT (2003) Hand bone ratios and their utility in predicting general substrate use in primates. Cour Forsch-Inst Senckenberg 243:47–59
Weber M (1904) Die Säugetiere. Einführung in die Anatomie und Systematik der recenten und fossilen Mammalia. Gustav Fisher, Jena
Whitehead PF (1993) Aspects of the anthropoid wrist and hand. In: Gebo DL (ed) Postcranial adaptation in nonhuman primates. Northern Illinois University Press, DeKalb, pp 96–120
Wilkinson JL (1951) The anatomy of an oblique proximal septum of the pulp space. Br J Surg 38:454–459
Wunderlich RE, Jungers WL (2009) Manual digital pressures during knuckle-walking in chimpanzees (*Pan troglodytes*). Am J Phys Anthropol 139:394–403

Chapter 5
The Role of Genes and Development in the Evolution of the Primate Hand

Campbell Rolian

1 Introduction

Compared with the hands of other mammals, the primate hand seems remarkably unspecialized. During the Cenozoic, other mammal lineages were evolving highly specialized hand morphologies, which involved not only the loss of digital rays (artiodactyls, perissodactyls) but also extreme modifications in the size and shape of the hand in response to the functional requirements of flight (chiropterans) or swimming (cetaceans). In contrast, primates retain the basic pentadactyl manus present in some of the earliest tetrapods (Carroll 1988), with none of the dramatic morphological modifications seen in horses, bats, and whales. This apparent conservatism does not mean, however, that there is no diversity in the morphology of primate hands. On the contrary, it does not take a trained eye to identify significant anatomical differences in the hands of lorises, baboons, or gibbons (Fig. 5.1). These subtle anatomical differences are functionally related to habitual locomotor and manipulative behaviors. For example, in lorisids, the unique pincerlike hands with short second digits, diametrically divergent thumbs, and derived flexor muscles allow them to grasp branches securely, whether above or below the substrate (Lemelin and Jungers 2007). The shorter, straighter digits of baboons are related to their digitigrade postures during terrestrial locomotion, which increase stride length and reduce contact time and propulsive effort (Patel 2010). In contrast, the longer and more curved digits of gibbons and other hominoids appear to be functionally related to suspensory postures and locomotion, reducing loads on the phalanges while contributing to pendular length during brachiation (Fleagle 1974; Richmond 2007).

C. Rolian (✉)
Department of Comparative Biology and Experimental Medicine, Faculty of Veterinary Medicine, University of Calgary, 3330 Hospital Dr NW, Calgary, T2N 0L8, AB, Canada
e-mail: cprolian@ucalgary.ca

T.L. Kivell et al. (eds.), *The Evolution of the Primate Hand*, Developments in Primatology: Progress and Prospects, DOI 10.1007/978-1-4939-3646-5_5

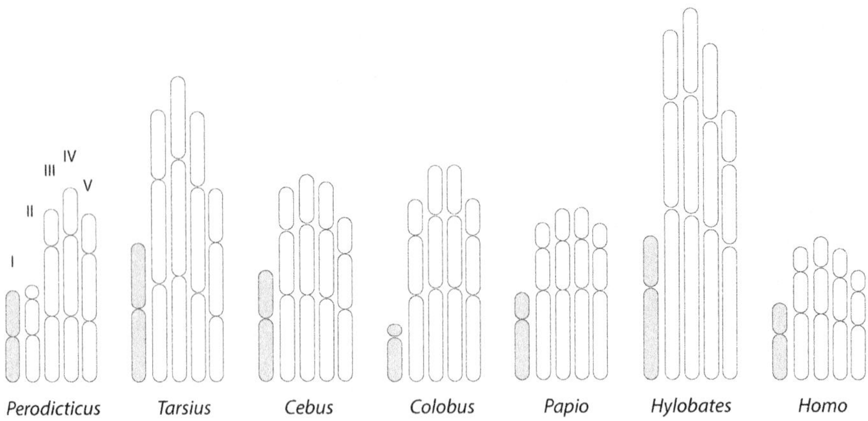

Fig. 5.1 Variation in manual skeletal proportions among primates. *Gray* elements represent lengths of the first metacarpal and proximal phalanx (I), and *white* elements represent lengths of the metacarpal, proximal, and middle phalanges of the second to fifth digits (II–V). Based on skeletal data collected by the author (Rolian 2009) or kindly provided by Pierre Lemelin. All lengths are scaled to the cube root of body mass, based on Smith and Jungers (1997), and thus illustrate variation in manual proportions within the hand, as well as between the hand and body size

The observation that anatomical differences among primate hands are related to distinct locomotor and manipulative functions suggests that such anatomical differences have an adaptive origin—that is, they evolved by means of natural selection acting on phenotypic variation in hand anatomy and morphology. Specifically, evolutionary theory predicts that particular phenotypic variants in hand hard and soft anatomy were selected in specific functional and ecological contexts, because they conferred a reproductive advantage on individuals that possessed them (i.e., increased their evolutionary fitness). But while evolutionary theory explains *why* certain hand morphologies ultimately evolve, it does not tell us *how* hand morphology evolves through proximate mechanisms (Mayr 1961). To answer this question, we must understand how phenotypic variation itself is produced in the first place.

Heritable phenotypic variation is the product of underlying genetic variation, environmental effects, and normal processes of organismal development. Development plays two fundamental roles in evolutionary biology. First, development translates the many "instructions" encoded in the genome into a functional and integrated organism. Organismal development links genotype to phenotype, thus providing something for selection to act upon. Second, it translates the *variation* included in these genomic "instructions," as well as modulating phenotypic variation itself through genetic and epigenetic phenomena (Jamniczky et al. 2010). Organismal development not only makes phenotypic variation possible; it also determines its parameters, such as magnitude, direction, and the correlation between traits. As a complex morphological structure, the primate hand is no exception: hand morphological diversity within primates can be viewed as the large-scale, long-term result of evolutionary processes acting on subtle phenotypic differences among individuals, produced and patterned by organismal development. In other words, the evolution of primate hands is intimately tied to genetic, cell- and tissue-level processes that govern vertebrate limb development.

This chapter provides a review of limb development in tetrapod vertebrates, with a focus on soft and hard tissues of the forelimb and hand. The chapter has two objectives: the first is to provide a detailed account of recent progress in deciphering the genetic and developmental processes that direct the formation of the vertebrate hand. The second objective is to discuss, as far as is known, how these genetic and developmental mechanisms contribute to selectable variation in the different tissue systems in the vertebrate hand. To address these objectives, I will use an evolutionary developmental biology (evo-devo) framework, emphasizing not only broad differences and similarities between developmental genetic programs in model organisms but also how the same developmental programs may mediate evolutionary change at lower taxonomic levels, such as within populations (Hendrikse et al. 2007). In this chapter, standard developmental orientation and terminology are used. These differ from "adult" terminology used throughout the remainder of the book because development of the limbs is organized and described in basic tetrapod anatomical orientation. For example, anteroposterior orientation in the hand (thumb to fifth ray) is equivalent to radioulnar in the adult hand. In the developmental literature, the terms **preaxial** and **postaxial** are often used to define the radial and ulnar borders of the hand, especially with reference to the dermatomes (i.e., areas of the hand innervated by a single ventral ramus of a mixed spinal nerve) (see Sadler and Langman 2009; Schoenwolf et al. 2009).

2 Overview of Vertebrate Development

Organismal development in amniotes is divided into three phases: pre-embryonic, embryonic, and fetal/postnatal. The **pre-embryonic phase** is the shortest and in humans lasts from 0 to 2 weeks gestation. Two important events must unfold at this time to set the stage for the subsequent development of a fully formed forelimb. The first is gastrulation, a process whereby the bilaminar germ disc becomes trilaminar, establishing three germ layers: **endoderm**, **mesoderm**, and **ectoderm**. The latter two layers will give rise to all the tissues of the developing limbs. The second event, occurring in parallel with gastrulation, is the establishment of the notochord. This rod-shaped mesoderm derivative extends the length of the future axial skeleton and plays a major inductive role in embryonic development (Sadler and Langman 2009; Schoenwolf et al. 2009). The **embryonic phase** is the period of organogenesis and morphogenesis and lasts from 2 to 8 weeks in humans. During this phase, the three germ layers give rise to all major tissues, organ systems, and structures of a developing organism. By the end of morphogenesis, major features of the internal and external organism are recognizable, including the limbs and digits. During the embryonic phase, the germ disc also undergoes several folding events (e.g., neurulation), taking on the characteristic three-dimensional shape of the vertebrate body

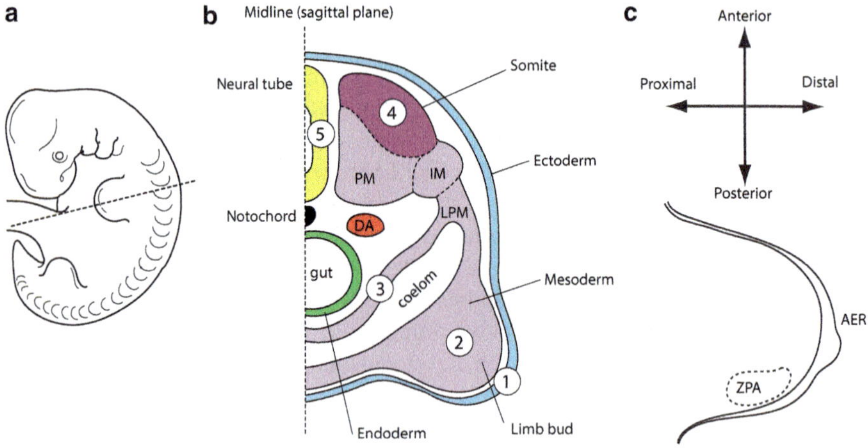

Fig. 5.2 (**a**) Appearance of a human embryo around 30 days gestation. (**b**) Schematic transverse section through the embryo at the level of the forelimb (*dashed line* in (**a**)), showing the layout of the germ layers and embryonic structures. For clarity, only the *left* side of the bilaterally symmetrical embryo is shown. Abbreviations: *PM* paraxial mesoderm, *IM* intermediate mesoderm, *LPM* lateral plate mesoderm, *DA* dorsal aorta. Circled numbers show embryonic origins of limb tissues: (1) integument (ectoderm), (2) skeletal/cartilaginous/connective tissues (somatic layer of LPM), (3) vascular tissue (visceral layer of LPM), (4) muscle (dermomyotome), (5) peripheral nerves (neural tube). (**c**) Schematic layout of the embryonic limb bud showing the relationship of the apical ectodermal ridge (AER) and zone of polarizing activity (ZPA) in relation to the proximodistal and anteroposterior axes of limb development and patterning

plan. During this time, the mesoderm divides into three distinct but continuous regions on either side of the midline: paraxial mesoderm (adjacent to the neural tube and notochord), intermediate mesoderm, and lateral plate mesoderm (closest to the body wall) (Fig. 5.2).

Paraxial mesoderm becomes further segmented along the cephalocaudal axis of the embryo into somitomeres (in the head and neck regions only) and somites (throughout the remainder of the body). The ventromedial region of the somites (the sclerotome) will give rise to mesenchyme, a loose connective tissue whose cells will contribute to the axial skeleton (vertebral column and ribs). The dorsal region of the somites (the dermomyotome) will give rise to progenitor cells for all skeletal muscle in the axial, appendicular, and body wall regions. **Intermediate mesoderm** will give rise to parts of the urogenital and reproductive system. **Lateral plate mesoderm** (LPM) can be further divided into an inner visceral layer and an outer somatic layer (Fig. 5.2). The visceral layer contributes to the circulatory system and to the smooth muscle surrounding the gut, while the outer somatic layer is the source of all bone- and cartilage-forming mesenchyme in the limbs.

The **fetal period** is the third and longest phase of development in amniotes. In humans, this period begins in the third month of gestation. The fetal period is marked by continuing maturation of the systems and structures established during the embryonic phase and by rapid growth of the body (Sadler and Langman 2009; Schoenwolf et al. 2009). With respect to the limbs, all patterning occurs before the end of the embryonic period, and hence most subsequent changes during the fetal period are concerned with changes in the size and shape of the different limb components. Although the fetal period ends at birth, postnatal changes in the structure, size, and/or shape of the limbs employ the same basic growth mechanisms as in utero and hence can be considered part of the same developmental phase that establishes the fully mature forelimb and hand. Of the three main phases of organismal development, genetic and developmental events during the embryonic and fetal/postnatal phases are the most likely sources of phenotypic variation in forelimb and hand morphology (Atchley and Hall 1991; Sanger et al. 2011).

3 Embryonic Patterning of the Limb Skeleton

3.1 *Setting the Stage*

3.1.1 Establishment of Limb Fields

The limbs develop as outgrowths of the body wall at fixed positions along the flank of the embryo known as limb fields. Limb fields are initially composed of undifferentiated cells from the LPM that express specific combinations of *Hox* transcription factors (i.e., proteins that bind specific regions of DNA and regulate the timing and/or magnitude of their transcription into mRNA) in response to axial cues at specific somite levels in the trunk (Noro et al. 2011). Mouse data indicate that the expression territories for different members of the *HoxC* cluster correspond to the presumptive forelimb field (*HoxC4, HoxC5*), interlimb flank (*HoxC6, HoxC8*), and presumptive hind limb field (*HoxC9-11*) (Duboc and Logan 2011b).

3.1.2 Limb Identity

Determining limb identity is an important step in early limb development, given the substantial morphological differences between the fore- and hind limb in tetrapods. Recent evidence suggests that a highly conserved family of transcription factors called T-box genes are involved in specifying whether a particular limb field will develop into a fore- or hind limb (reviewed in King et al. 2006). In several vertebrate models, two of these transcription factors, *Tbx4* and *Tbx5*, show restrictive expression in the hind- and forelimb, respectively (Agarwal et al. 2003; Naiche and Papaioannou 2003). *Tbx4* and *Tbx5* may play an important role in determining early structural differentiation between paired appendages; however, other studies report that they are not sufficient by themselves to determine limb-specific morphologies (Minguillon et al. 2005). The key role of these transcription factors may thus be to determine limb-type *identity* rather than *morphology* (King et al. 2006; Duboc and Logan 2011b). So far, a single transcription factor, *Pitx1*, has been unambiguously implicated in determining hind limb-type morphology (DeLaurier et al. 2006; Duboc and Logan 2011a).

3.1.3 Initiation of Limb Bud Outgrowth

Without initiation of limb outgrowth, there can be no limb development. The mechanism of limb bud outgrowth is based on a fibroblast growth factor (FGF) feedback loop (Ohuchi et al. 1997). During this feedback loop, mesenchymal cells in the LPM secrete FGF10, a diffusible growth factor which induces the overlying ectoderm to produce FGF8. This induction simultaneously causes the distal epithelium to form the **apical ectodermal ridge** (AER), a key structure of embryonic limb patterning whose integrity is necessary for continued outgrowth (see below). FGF8 production from the AER, in turn, maintains high expression of FGF10 in the mesenchyme through a positive feedback loop. The high expression of FGF10 in the LPM mesenchyme promotes cell proliferation and limb bud expansion.

3.2 Proximodistal Patterning of the Limb

Once limb bud outgrowth is initiated, the developing forelimb must be properly patterned along three axes: proximodistal (from humerus to digits), anteroposterior (from thumb to fifth ray), and dorsoventral (extensor and flexor surfaces, respectively). Initial development along the proximodistal axis produces three limb segments: the **stylopod** (the humerus and femur), **zeugopod** (the radius and ulna or tibia and fibula), and **autopod** (the hand and foot). Proximodistal patterning is directed by the AER, a thickened ridge of ectodermal cells that lies along the entire length of the anteroposterior axis of the limb, at the interface between its dorsal and ventral sides (see Capdevila and Izpisua Belmonte 2001 for a review of its induction). As mentioned above, a key protein secreted by the AER is FGF8, which maintains underlying mesenchymal cells in a proliferative and undifferentiated state. This pool of proliferative cells adjacent to the AER has long been recognized as a functionally important part of proximodistal limb patterning, known as the progress zone.

During limb outgrowth, proximal structures are formed first, followed sequentially (spatially and temporally) by more distal structures. Several models have been proposed on how this process of limb outgrowth occurs. One model, known as the progress zone model of proximodistal patterning (Summerbell 1977), posits that the cells in the progress zone, under the influence of the AER, acquire information about their position by measuring the amount of time they spend in the progress zone. As the limb bud grows out from the flank and cells exit the progress zone proximally at successively later stages, they are fated to contribute to more distal limb skeletal elements (Zeller et al. 2009). An alternative model proposes that the limb segments are "prespecified" at a very early stage (Dudley et al. 2002; Sun et al. 2002). In this model of proximodistal patterning, the AER functions in preventing cell death and maintaining cell proliferation, rather than controlling cell fate (Duboule 2002; see also Tickle and Wolpert 2002; Tabin and Wolpert 2007).

Recently, a third model of proximodistal limb patterning has been proposed in an attempt to reconcile conflicting data from the progress zone and prespecification models (Tabin and Wolpert 2007; Cooper et al. 2011; Rosello-Diez et al. 2011). This model, known as the two-signal model, proposes that the fate of cells that exit the undifferentiated pool adjacent to the AER is determined by a dynamic balance of two signals: a proximalizing signal derived from the embryonic flank and a distalizing signal produced by the AER. The model proposes that undifferentiated cells are continuously proliferating and leaving the undifferentiated zone (equivalent to the progress zone) as the limb grows out. When a given cell exits this zone through the "differentiation front" (Tabin and Wolpert 2007), the regional balance of the two signals will determine to which limb element this cell will contribute. Cells that leave early, when the bud is close to the flank, will still be under the influence of the proximalizing signal and will contribute to the most proximal limb element, the stylopod. Cells that leave the progress zone later are beyond the range of the proximal signal and become committed to more distal limb elements in response to distal AER signals (zeugopod and autopod). The two-signal model also suggests that differences in the relative lengths of the limb segments (e.g., arm vs. forearm vs. hand length) can be influenced in a dose-dependent manner by the balance of the two proximodistal signals. For example, if the dose (level of expression) of the distalizing signal is greater, then this may result in a longer distal-to-proximal diffusible range for that signal. Accordingly, cells that exit the progress zone will remain within this signaling range longer and establish a relatively larger distal segment.

The proposed signals for this model are retinoic acid (RA) in the flank region and FGFs in the AER (in particular FGF8). The FGFs appear to serve two functions: (1) maintaining the undifferentiated zone cells in a proliferative state and (2) initiating the deployment of *Hox* transcription factors along the proximodistal axis of the developing limb. Specifically, AER-FGF signaling establishes a morphogenetic gradient (Hashimoto et al. 1999; Vargesson et al. 2001), which sets up nested expression domains for members of the *HoxA* and *HoxD* gene clusters. For example, *HoxA9 and HoxD9* are expressed throughout the limb bud, while *HoxA10-13* and *HoxD10-13* are progressively restricted toward more distal parts of the developing limb (Zakany and Duboule 1999, 2007). Moreover, several *HoxA* domains become localized so that they coincide precisely with the boundaries of the future distal limb elements, including *HoxA11* in the zeugopod and *HoxA13* in the autopod (Yokouchi et al. 1991).

3.3 *Proximodistal Patterning in the Autopod*

Subsequent to proximodistal patterning of the growing limb into three segments, the autopod (i.e., the hand and foot) must also be patterned along the proximodistal axis. In the forelimb, the hand comprises a conserved sequence of skeletal elements from proximal to distal, beginning with carpals, followed by metacarpals and ending with phalanges. Proximodistal patterning mechanisms within the autopod are relatively poorly documented. The autopod begins as an undifferentiated plate of mesenchymal progenitor cells. As in the whole limb, the distal elements of the hand (i.e., digits) are established after the more proximal elements of the carpus. The digital rays are established as rodlike condensations and subsequently elongate while periodically segmenting into a specific number of phalanges (Stricker and Mundlos 2011).

Phalangeal segmenting mechanisms are also poorly understood. In the chick autopod, there appears to be a temporal aspect to the cyclical segmentation of the digits mediated by a segmentation "clock" gene (*hairy2*, mouse homolog, *HES1*) (Pascoal et al. 2007). Alternatively, or perhaps in concert with such a temporal signal, inhibitory effects from both the AER and cavitating interphalangeal joints ensure that subsequent (and more distal) joints are induced at a minimum distance from more proximal joints. Thus, in addition to promoting growth and maintaining cells in an undifferentiated state, the AER also appears to inhibit the formation of joints immediately subjacent to it within the developing autopod mesenchyme (Stricker and Mundlos 2011).

Distal phalanges are partially under autonomous genetic and developmental mechanisms distinct from those acting more proximally (Hamrick 2012). These mechanisms include the expression of unique transcription factors such as *Msx1* and *Msx2*, *HoxC13* in the digit tip ectoderm, as well as *BAMBI*, a negative feedback regulator of bone morphogenetic protein signaling (Grotewold et al. 2001). Interestingly, in the chick *BAMBI* is only expressed in the tips of digits that form claws, suggesting it could play a role in the formation of claws vs. nails.

The nested pattern of proximodistal expression of posterior *HoxA* and *HoxD* genes introduced above also plays a role in ensuring that the proper sequence, type, and size of each element develop within the autopod (Zakany et al. 1997; Zakany and Duboule 2007). Sequential disruptions of *Hox* group genes 11–13 produce progressively more distal abnormal limb phenotypes, generally involving the loss or length reduction of specific elements. Thus, *Hox11* disruption affects the distal zeugopod and proximal carpal row, *Hox12* disruption affects the distal carpal row and digits, and disruption of *Hox13* not only affects the digits but in double mutants can lead to the entire absence of the autopod (Davis et al. 1995; Delpretti et al. 2012). *Hox* gene function in autopod proximodistal patterning may be mediated via downstream targets that control the production of cortical bone and secondary ossification centers in tubular bones. Disruption of *HoxD13* activity in mouse metacarpals leads to abnormal perichondrium formation, lack of cortical bone, irregular or absent joints, and bony elements that are covered in cartilage. In other words, the metacarpals in these mice undergo homeotic transformations and take on carpal morphologies (Villavicencio-Lorini et al. 2010).

3.4 Anteroposterior Patterning in the Autopod

Anteroposterior patterning mechanisms in the developing limb bud ensure that the correct number and identity of skeletal elements develop in the zeugopod (e.g., radius and ulna) and especially in the autopod (e.g., carpals, metacarpals, and digits). Like proximodistal patterning, anteroposterior patterning is under the control of an organizing center, known as the zone of polarizing activity (ZPA). The ZPA is established early on in limb bud outgrowth and becomes spatially restricted to the distal posterior margin of the limb bud, adjacent to the AER (Fig. 5.2). Its induction and localization are reviewed in Zeller et al. (2009).

ZPA patterning activity is mediated via a secreted factor known as sonic hedgehog (Shh). Shh is a morphogen, a diffusible molecule that provides positional information to cells based on its concentration in their vicinity (Zeller et al. 2009). Interestingly, the identity of the posterior digits (2–5) is determined by both autocrine and paracrine signaling (reviewed in Harfe et al. 2004). In contrast, digit 1 development (pollex/hallux) does not depend on Shh signaling (see below).

Shh signaling also induces downstream targets to establish a morphogenetic gradient that similarly dictates the development of the autopod. One such target is the *HoxD* clusters, which are redeployed in a second "wave" of expression along the anteroposterior (Zakany and Duboule 1999, 2007). Much like the first domains along the PD axis, this second wave confers segment identity to the digit condensations in combination with the local concentration of Shh. Evidence from the mouse further suggests that the number and length of the digit primordia are affected in a dose-dependent manner by loss or gain of function in specific *HoxD* genes (Zakany et al. 1997; Zakany and Duboule 1999).

Woltering and Duboule (2010) suggested that the two waves of *HoxD* deployment in proximodistal and anteroposterior patterning may be linked to developmental patterning of the carpals. Specifically, they identify a region at the boundary between the two expression domains with no or very low posterior *HoxD* expression (see Reno et al. 2008), coinciding with the location of the presumptive carpals. This "no-*HoxD*" territory may induce the development of non-tubular, irregular-shaped bones that lack true growth plates (i.e., carpals) sandwiched between two sets of tubular bones with both primary and secondary ossification centers (radius/ulna vs. digital rays) (Reno et al. 2008; Villavicencio-Lorini et al. 2010).

3.4.1 Thumbness

The thumb skeleton is anatomically distinct from the ulnar digits because it has only two phalanges (no intermediate phalanx) and the metacarpal growth plate located proximally instead of distally. Increasing evidence suggests that there are two main developmental processes distinct from those that produce the ulnar digits that are responsible for determining the unique morphology of the thumb (Wagner and Vargas 2008). First, only *HoxD13* is expressed, at low doses, in the presumptive thumb (and big toe) domain (Montavon et al. 2008; Reno et al. 2008). Second, and undoubtedly related, is the relative absence of Shh signaling in the anterior autopod domain (Kraus et al. 2001). In essence, the unique morphology of the thumb and big toe may be developmental by-products of the asymmetric expression patterns of *Hox* genes along the anteroposterior or pre-postaxial (thumb to fifth digit) axis (Wagner and Vargas 2008).

3.5 Dorsoventral Patterning of the Autopod

From a skeletal point of view, patterning along the dorsoventral axis may not seem as important as anteroposterior or proximodistal patterning; yet the marked differences in soft tissue structures between dorsal and ventral compartments of vertebrate appendages show that correct patterning along this axis is an integral process of limb development. The AER is induced at the margin between dorsal and ventral aspects of the ectoderm in the developing limb bud. Once the AER is established, *Wnt7a*, a secreted glycoprotein, is expressed only in dorsal ectoderm, and *En-1*, a transcription factor, is expressed only in ventral ectoderm (reviewed in Capdevila and Izpisua Belmonte 2001). The cells in these two compartments give rise to independent cell lineages that remain restricted to their respective compartments throughout development, separated distally by the AER (Kimmel et al. 2000).

In the dorsal compartment, the secretion of *Wnt7a* product from the ectoderm into the underlying mesenchyme activates the expression of *Lmx-1b* in these cells. The mechanism by which the expression of *Lmx-1b* in the dorsal mesenchyme determines dorsal structures is unclear. However, mutant mice homozygous for *Lmx-1b* display double-ventral phenotypes, which suggest that *Lmx-1b* is involved in modifying a default ventral pattern. Once the dorsoventral fate is determined in the ectoderm, mesenchymal cells are restricted to dorsal and ventral compartments (Arques et al. 2007). They can disperse along the anteroposterior and proximodistal axes, as well as within the dorsal *or* ventral compartment, but they do not disperse across a sharp planar dorsoventral boundary. This boundary coincides precisely with dorsal cells that express *Lmx1b* and ventral cells that do not. The condensations of future skeletal elements also occur at this boundary, but with recruitment of mesenchymal cells from both the dorsal and ventral compartments (Arques et al. 2007; Li et al. 2010).

4 Postembryonic Growth and Development in the Limb Skeleton

During the embryonic period described above, all elements of the future limb skeleton are laid down as distinct populations of cartilage cells known as condensations (or ***Anlagen***). *Anlagen* arise through differentiation, when mesenchymal cells in the limb bud are assigned to one of two fates: to aggregate and become cartilage cells (chondrocytes) and associated connective tissues, or to undergo apoptosis (programmed cell death) and form interdigital spaces (Kavanagh et al. 2002; Zuzarte-Luis and Hurle 2005). In this context, the primary role of the genes, pathways, and organizing centers described above is to ensure that cartilage condensations are properly patterned with respect to their spatial relationships, joint locations, and relative sizes.

The postembryonic period is concerned with growth and maturation of cartilaginous templates established during the embryonic period. Most limb skeletal elements follow a stereotypical sequence of events, starting from *Anlagen* with relatively simple morphologies and ending as recognizable osseocartilaginous units capable of directed growth and programmed changes in size and shape (Fig. 5.3). First, beginning near the end of the embryonic period, chondrocytes in the center of a condensation undergo hypertrophy and produce extracellular matrix (ECM) rich in collagen. At the same time, cells in the perichondrium begin to express *Runx2*, the "master switch" that instructs these cells to differentiate into bone-forming osteoblasts (Ducy et al. 1997). These cells form the bone collar, a structure responsible for the formation of cortical bone.

Next, death of the hypertrophic chondrocytes leaves behind an ECM scaffold that promotes mineralization and vascularization through the bone collar. The invasion of blood vessels brings osteoblasts into the center of the condensation, where they will deposit bone on the ECM scaffold (Ballock and O'Keefe 2003; Karsenty et al. 2009). This area of bone formation, known as the primary ossification center, gives rise to trabecular bone within mature long bones. Ossification progresses away from the initial center, as chondrocytes proliferate, hypertrophy, and eventually die and are replaced with invading osteoblasts that deposit bone on the ECM scaffold left behind. The chondrocytes become organized into columns that are parallel to the direction of longitudinal growth (Fig. 5.3). Eventually, these columns give rise to growth plates (physes), structures that are primarily responsible for *elongation* of the bones. Meanwhile, in the bone collar, osteoblasts in the perichondrium (at this stage known as periosteum) continue to deposit bone in sheets, contributing to *appositional growth* of the long bone. This process, where bone replaces a cartilage model, is known as **endochondral bone growth**. All hand bones grow via this process.

In some postcranial bones, including carpals, a single ossification center contributes to centrifugal growth of the element via endochondral bone growth. In many long bones, including metacarpals and phalanges, secondary centers of ossification form on the side of each physis farthest from the bone center. These centers, which will become epiphyses, are formed in the same way as the primary center of ossification, and together they "sandwich" the growth plate, forming a narrow line of cartilage between two bone fronts (Fig. 5.3). Most long bones have a proximal and distal secondary ossification centers, and often the proximal end develops several secondary ossification centers, allowing the growth of unique morphologies in multiple directions. In contrast, metacarpals and phalanges develop a single secondary ossification center and associated growth plate. In all phalanges, the growth plate is located at the proximal end. In the ulnar metacarpals, the growth plate is located distally, but is at the proximal end in the first metacarpal of the thumb (and metatarsal of the hallux).

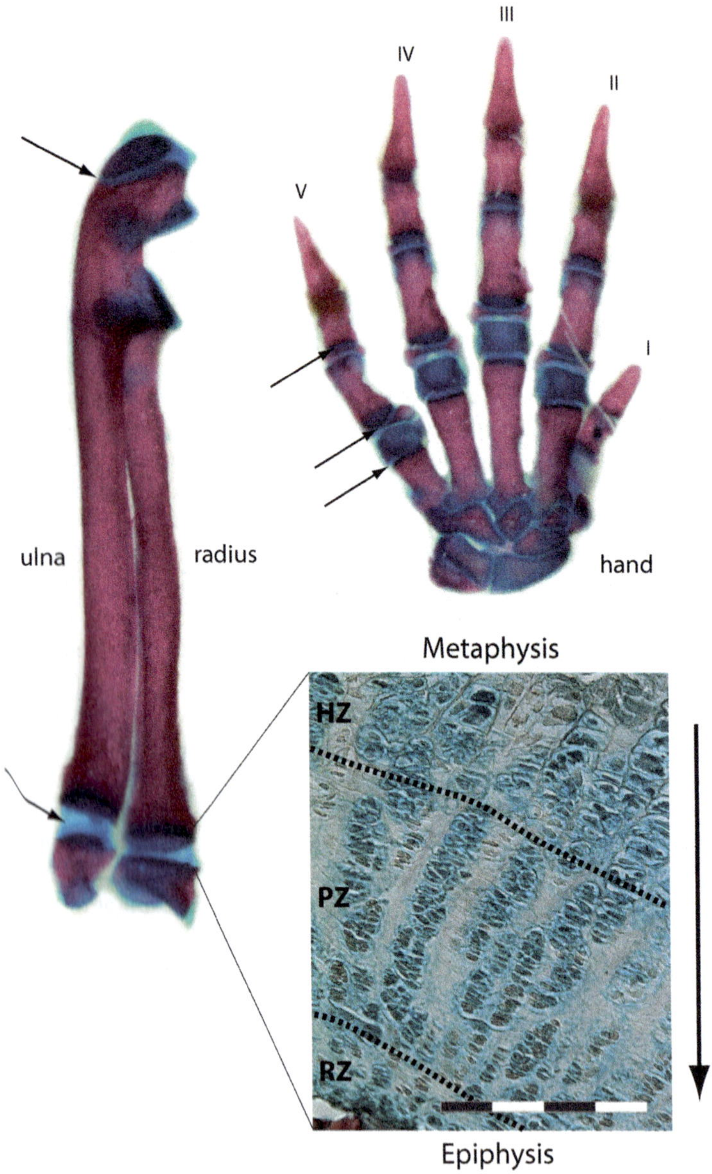

Fig. 5.3 Cleared and stained forearm and hand bones of a 3-week-old gerbil (*Meriones unguiculatus*) showing the location of several growth plates (*arrows*). The specimen is double stained with alizarin red (bone) and Alcian blue (cartilage), the latter revealing the physes of the long bones and articular cartilage around the epiphyses of the long bones and carpals. The *inset* shows a thin section through the distal radius growth plate from the contralateral side of the same individual. Physeal chondrocytes are organized into columns oriented parallel to the direction of longitudinal growth (*black arrow*). Initially dormant chondrocytes in the resting zone (RZ), adjacent to the epiphysis, eventually undergo a highly orchestrated life cycle of cell proliferation (proliferating zone, PZ), hypertrophy (hypertrophic zone, HZ), and apoptosis near the metaphysis. Thin section scale bar is 200 μm

Genetic and developmental regulation of the growth plate is well documented (reviewed in Ballock and O'Keefe 2003), and there are several potential postembryonic mechanisms to modify the relative lengths of metapodials and phalanges. Within a growth plate, chondrocytes undergo a highly orchestrated life cycle of proliferation, hypertrophy, and cell death (Fig. 5.3). Local differences in these mechanisms, for example, in the number of cells dividing, their rates of division, or the size to which they hypertrophy before dying, can modulate the growth rate and final size of the long bones (Kirkwood and Kember 1993; Wilsman et al. 1996). What is less clear, however, is the relative contribution of each of these mechanisms to the final size, as well as the relative importance of systemic, limb-specific, and local regulation of these mechanisms in generating selectable differences in relative bone length (van der Eerden et al. 2003).

In primates, as in most mammals, the growth plates of the long bones eventually senesce and fuse, marking the end of skeletal growth. Different growth plates fuse at slightly different times during postembryonic development (e.g., Bolter and Zihlman 2012). Thus, the final size and shape of a hand bone is the outcome of two main phases: (1) an early embryonic patterning phase in which differences in condensation size may set up differences in relevant downstream morphological parameters such chondrocyte pool size and (2) a later, fetal, and postnatal phase in which these differences are further established through changes in the growth parameters of primary (all hand bones) and secondary ossification centers (metacarpals and phalanges).

5 Soft Tissue Patterning in the Vertebrate Limb

Soft tissues of the hand—namely, muscles, tendons/ligaments, fascia and other connective tissues, vasculature, nerves, and integumentary structures—can vary among primate taxa just as the morphology of the hand skeleton. In the skeleton, interspecific variation relates primarily to the size, shape, and proportion of individual elements (Fig. 5.1). In contrast, soft tissue variation is more qualitative, concerning, for example, the presence/absence of individual muscles or tendons (Aversi-Ferreira et al. 2010; Diogo and Wood 2011), the branching patterns of specific vessels and nerves, or the presence of specific integumentary structures (e.g., claws, Hamrick 2003). The patterns of occurrence of these synapomorphies among different primate taxa indicate that they have a genetic and developmental basis.

Genetic and molecular pathways that control the development of the limb's soft tissues have received significantly less attention than those of the limb's skeletal tissues. Two factors may explain this disparity. First, until the advent of specific molecular markers and novel 3D imaging techniques, the analysis of developmental phenotypes was more practical in hard tissues, such as cartilage or bone. Second, the emphasis on hard tissues may have been unwittingly motivated by a desire to apply what is learned from vertebrate developmental biology to the fossil record (in which mostly mineralized tissues of bones and teeth are recovered), to understand better the evolutionary diversification of vertebrates.

5.1 Limb Muscle Patterning

Limb myogenic precursor cells for developing muscle arise from the somites at the level of each limb bud, specifically from the region known as the hypaxial dermomyotome (Chevallier et al. 1977). From this location, the formation of muscles in the embryonic limb involves the following steps (Duprez 2002): (1) migration of the precursor cells into the growing limb bud, (2) activation of the myogenic cell pathway, (3) proliferation, (4) formation of dorsal and ventral muscle masses, (5) differentiation into muscle tissue, and (6) splitting into anatomically recognizable muscles.

In the earliest steps, the transcription factor *Pax3* has been identified as a "master switch" gene of the muscle precursor cells in the dermomyotome and is involved in mediating the migration of these cells into the limb buds. Inactivating *Pax3* leads to a limb phenotype with no muscles present (Buckingham et al. 2003). Once in the limb bud, subsequent steps in muscle development are largely under the influence of the limb bud mesoderm environment and the organizing centers discussed in the previous sections (reviewed in Duprez 2002).

In the limb bud, precursor cells express myogenic fate genes (e.g., *MyoD*), likely in response to positional cues from the dorsal ectoderm and ZPA (Buckingham et al. 2003) and from *Pax3* itself. The next step involves maintaining these myogenic cells (myoblasts) in a proliferative state, in order to achieve critical muscle mass. Here again, the same AER signals (e.g., FGFs), which maintain the limb mesenchyme in an undifferentiated, proliferative state, likely play a similar role in the growing muscle mass. Following a period of proliferation, distinct dorsal and ventral premuscular masses are formed, and the cells undergo further differentiation into physiologically active muscle tissue (e.g., production of actin and myosin). The establishment of dorsal and ventral compartments is almost certainly related to the patterning mechanism that sets up a sharp dorsoventral boundary in the skeletal system discussed above (Li et al. 2010).

The final step in the embryonic patterning of limb muscles—cleaving of premuscular masses into anatomically distinct muscles—is the most evolutionarily and functionally relevant for understanding muscle diversity across primate hands. Unfortunately, this final step is also the least well understood. The myogenic precursor cells are naïve, that is, not predetermined to form specific muscles before entering the limb (Kardon et al. 2002). The dorsal and ventral premuscular masses are distributed throughout the proximodistal and anteroposterior extents of the embryonic limb. Thus, much like the bony elements, dorsal and ventral premuscular masses must be further separated and patterned along the anteroposterior (e.g., radial vs. ulnar carpal flexors) and proximodistal axes (e.g., forearm muscles vs. intrinsic hand muscles). No genes underlying individual muscles have been identified so far, although disruption of specific genes, such as *Lbx1* (Brohmann et al. 2000) and *Mox2* (Mankoo et al. 1999), has been shown to differentially affect groups of muscles (e.g., flexor vs. extensor muscle groups, hind limb vs. forelimb muscles). These genes affect muscle development by reducing the volume of, or entirely eliminating, the premuscular mass of a given muscle group.

The lack of predetermination in myogenic precursors, the absence of genes "for" individual muscles, and the similarity of bone and muscle patterning events along the three cardinal axes all suggest that muscle patterning is under the extrinsic control of the limb environment and/or its organizing centers. This has prompted developmental biologists to search for extrinsic factor(s) or tissue(s) responsible for muscle patterning. Tozer et al. (2007) have recently proposed that vascular development plays a key role in splitting the muscle masses and thus could be responsible for muscle variability. Their hypothesis does not, however, properly address the question of how variation in the presence/absence of individual muscles arises during development, but merely shifts the focus of analysis to developmental variation in the vascular system as the source of variability in muscle formation. Ironically, a recent study shows that the forming limb skeleton itself plays a central role in patterning the developing limb vasculature (Eshkar-Oren et al. 2009). For the time being, the limb mesenchyme and molecular instructions issued by the AER and ZPA remain the most likely candidates as extrinsic organizers and providers of positional identity for soft tissues such as musculature and vasculature.

5.2 *Limb Tendon and Ligament Development*

Tendons, ligaments, and aponeuroses are related tissues with similar histological and gene expression profiles (Hasson 2011). Unlike muscle, however, these soft tissues (hereafter "tendons") are derived from the lateral plate mesoderm (LPM). One of the earliest molecular markers of tendon progenitor cells is scleraxis (*scx*). Scx-positive cells can be detected alongside myogenic precursors in mixed dorsal and ventral regions beneath the ectoderm and appear to be induced by the ectoderm itself (reviewed in Schweitzer et al. 2010). This early mixing of the two progenitor cells suggests that the development of muscle and tendon is linked, which would be expected given their intimate anatomical relationships. However, *Pax3* muscle-less limbs show correct *scx* activation and tendon initiation, suggesting that *Pax3* plays a bigger role in muscle than in tendon differentiation and/or maintenance (Kardon 1998; Schweitzer et al. 2010). Still, both cell types are likely influenced by the mesenchymal environment, particularly interactions between the ectoderm and condensing cartilage elements. For example, in the autopod, late removal of interdigital ectoderm induces the condensation of an ectopic cartilage element, followed by the induction of a tendon along its length (Hurle et al. 1990).

5.3 *Limb Vasculature Development*

Forelimb and hand vasculature can be quite variable within and between primate taxa (Ikeda et al. 1988). Vasculature emerges early in embryogenesis, to ensure proper nutrient supply and adequate gas exchange for growing tissues. Vascular tissue is derived from the visceral layer of the LPM (Schoenwolf et al. 2009). Limb vascularization entails two steps: (1) vasculogenesis, in which a primary vascular plexus is formed within the limb mesenchymal core as an outgrowth of the dorsal aorta, and (2) angiogenesis, during which the plexus is expanded and extensively remodeled through growth, migration, branching, and regression of vessels, leading to a characteristic vascular tree (Sato et al. 2002). Several early studies showed that angiogenesis and skeletogenesis are tightly coordinated, although the directions of the regulatory interactions were unclear. In a recent study, Eshkar-Oren et al. (2009) argue that the forming skeleton, in particular condensing mesenchymal cores, acts as an essential signaling center in angiogenesis and patterning of the limb vascular system.

5.4 Limb Nerve Development

The peripheral nervous system in the limb is a derivative of the embryonic neural tube. Peripheral nerves (both motor and sensory) consist of bundles of neurons whose cell bodies are in the central nervous system, but whose fibers (axons) run uninterrupted to and from their targets in the limb tissues (e.g., the muscles, skin, etc.). Thus, unlike other soft tissues, which grow via migration and cell proliferation, nerves grow from the spine all the way into the hand by adding length to the fiber of each neuron. To ensure a connection with appropriate synaptic targets within the peripheral limb, axonal growth must be guided with high fidelity and precision throughout the growing limb. Empirical evidence since the 1980s has confirmed that the correct growth and branching of nerve fibers to their peripheral targets depend on local "guidance cues" within the limb (Landmesser 2001). In other words, much like the other soft tissues, limb nerves take their cues from the immediate limb environment.

Experiments based on limb muscle ablation or disruption of myogenic precursor genes have revealed that muscles do not provide positional or guidance cues to the growing neurons, as the nerve network is faithfully reproduced in the absence of muscle (Landmesser 2001). More recently, the development of the peripheral nervous system has been shown to depend on interactions with the developing vascular network (reviewed in Eichmann et al. 2005). Vascular precursor cells and axons both produce a cell surface receptor (neuropilin-1 or NRP1) with affinity for a class of axonal "pathfinding" molecules known as semaphorins. NRP1 also shows affinity for VEGF (important in vascular development), and it is thought that competition between these two molecules for their common receptor can indirectly influence the path and branching of one or the other (Mackenzie and Ruhrberg 2012).

5.5 Limb Integumentary Structures

Integumentary structures such as nails and papillary ridges are well developed in primates and have been important to the success of the primate radiation (Hamrick 2003). Nails and claws develop as epithelial thickenings on the dorsal surface of the digit tip. Like other aspects of dorsoventral patterning, their induction is dependent on *Wnt7a* expression in the dorsal ectoderm (Parr and McMahon 1995). Homeobox genes *Msx1* and *Msx2* are also involved in claw/nail development: the former maintains epithelial cells in an undifferentiated proliferative state, while the latter accelerates their differentiation into cells competent to form claw/nail appendages. The molecular basis of variation in the presence of claws vs. nails, both within the autopods of some species (e.g., nails and grooming claws in strepsirrhines), but also among related primate species (e.g., secondary derived claws in callitrichids vs. nails in all other platyrrhines), is not known. Given that nail/claw morphology closely matches the underlying size and shape of distal phalanges (Maiolino et al. 2011; see Chaps. 4, 8), it is likely that the morphology of the developing distal phalanx influences whether a nail or claw develops.

Papillary ridges (fingerprints) are integrated neural-dermal structures containing specialized mechanoreceptors involved in tactile discrimination. Two such receptors, Pacinian and Meissner's corpuscles, are found at high densities in the glabrous skin of primate hands and feet, especially in those of humans (Hoffmann et al. 2004; Chaps. 6, 8). They develop earlier in primates than in mice, reflecting the increased importance of tactile discrimination in the former (Renehan and Munger 1990; Saxod 1996). Although much is known about histological aspects of their ontogeny in mammals, less is known regarding the molecular basis of their development, particularly the interactions between the growing afferent sensory axons and the dermal/epidermal structures they invest (Vega et al. 2009). On the neuronal side, the formation of sensory corpuscles involves a subpopulation of afferent nerves that depend on growth factors known as neurotrophins. Interestingly, changes in the expression levels of either neurotrophins or their receptors in the skin affect the size and density of Meissner's corpuscles, providing evidence of a potential dose-dependent mechanism for generating variation in tactile sensory acuity in the vertebrate hand (reviewed in Vega et al. 2009).

6 Discussion

The preceding overview of forelimb and, specifically, hand (i.e., autopod) development, although thorough, is still somewhat superficial. While I have attempted to highlight the most important genes, molecules, and developmental events involved in limb patterning and growth, countless additional factors—known and unknown—come into play to produce a complete and functional hand. But even the few key genes, organizing centers, and developmental processes outlined above reveal that limb development is a highly complex and integrated process, involving precise interactions among different cell types over developmental time and space. How, if at all, can we use the growing wealth of information on developmental mechanics—gathered mostly in model vertebrates—to understand the evolution of hand morphological and anatomical diversity in primates?

The difficulty in answering this question is that, despite advances in limb developmental biology, we are still only beginning to understand large-scale patterning mechanisms involved in creating a limb across divergent organisms such as a mouse or chick. Selection, however, does not act at such a macroscopic level or even on such vastly different limb morphologies. It acts on small-scale variation in primarily continuous phenotypic traits (so long as variation in such traits differentially affects fitness across individuals). From an evo-devo perspective, understanding how primate hand diversity has evolved means understanding how individual-level differences in the developmental processes discussed above lead to structured quantitative phenotypic variation in the hand. The nature of vertebrate limb patterning and growth mechanisms discovered to date highlights two important aspects of phenotypic variation in the vertebrate hand, one concerning its origins and the other its structure.

6.1 The Developmental Origin(s) of Limb Phenotypic Variation

One of the most important discoveries to come out of developmental biology is the fact that the same genes, molecules, and developmental pathways are used at different times and places throughout organismal development. This is true within an individual, at the population level, and across vastly divergent species (Carroll 2005). Thus, despite the tremendous morphological differences that exist between fish, birds, and mammals, we all share conserved genetic "toolkits" that trace their origins in the distant past, when we last shared common ancestors (Shubin et al. 2009). More than simply being present in each species, the coding sequences themselves are also highly conserved, to the extent that they are often functionally interchangeable between species like a mouse and chick (Martin et al. 2012).

If all individuals within a population, or indeed almost all vertebrates, share the same genes and genetic toolkits for producing a hand, then where does heritable anatomical variation come from? The answer to this question lies in the *regulation* of these genes (Wray 2007; Carroll 2008). Variation among individuals, populations, and even species is produced primarily by changes in how, when, and what quantity genes are expressed (i.e., their regulation), rather than by changes to the coding sequence of the genes or their products (i.e., their structure) (Kiefer 2010; Sholtis and Noonan 2010). For example, *HoxD* genes are almost certainly involved in promoting digit development in all primates. However, it is most likely differences in how, where, when, and for how long these transcription factors are expressed, rather than changes to their protein sequences, that explain the relative differences in digit size and shape between species in Fig. 5.1 (Zakany et al. 1997).

The idea that morphology evolves primarily through changes (mutations) in gene regulatory mechanisms is known as the "*cis*-regulatory hypothesis" (Carroll et al. 2008; Stern and Orgogozo 2008), in reference to *cis*-regulatory regions. *Cis*-regulatory regions are noncoding DNA regions, distributed over many kilobases of DNA, but located on the same DNA strand as the gene(s) they regulate. They include different types of genetic elements (e.g., promoters, enhancers, silencers) that typically act as binding sites for transcription factors, thereby modulating the spatiotemporal expression patterns of their target gene(s) in a context-specific manner (Stern and Orgogozo 2008). *Cis*-regulatory elements (CREs) enable the same genes to be deployed at different times and in different contexts, promoting gene recycling and tinkering without undermining the structural integrity of a gene or its product.

Several recent studies highlight the potential of CREs for driving limb development and evolution. Cretekos et al. (2008) compared the structure and function of *Prx1* and its CRE in mice and bats. *Prx1* is a transcription factor that promotes bone lengthening and is expressed in limb mesoderm. Their analysis revealed similar expression patterns during embryogenesis in both mammals, along with >99 % homology in the protein's structure. They sequenced two known *Prx1* enhancer regions in the mouse and homologous regions in the bat, revealing lower overall homology between the two (~93.5 % sequence similarity). Cretekos et al. (2008) used these CREs to develop transgenic mice with the endogenous *Prx1* gene but with the bat enhancers. The forelimbs in these transgenic mice were patterned normally, but were ~6 % longer than the controls. This innovative experiment shows that CREs in one species can substitute functionally for another's and, more importantly, produce phenotypic changes that correspond to their roles in each. The most relevant example of CRE-driven morphological evolution in primates concerns a gain-of-function mutation (i.e., a mutation leading to a new function, or new expression domain, for the target of the CRE) in a human-specific CRE (Prabhakar et al. 2006, 2008). Prabhakar and colleagues have identified conserved noncoding sequences (CNS) that underwent rapid evolution in humans relative to other primates and vertebrates, including one with 16 human-specific substitutions within a short region that is otherwise highly conserved in other primates and vertebrates (*HACNS1*). In order to visualize what, if anything, this CNS does, Prabhakar et al. (2008) created transgenic mice in which the *HACNS1* homologs in humans, chimpanzees, and macaques drive the expression of a reporter gene (i.e., a gene whose product is easily visualized/quantified, such as β-galactosidase). Their analysis revealed that *HACNS1* is an enhancer whose role is most strongly associated with **preaxial** (radial) limb development in humans. Transgenic mice with the chimpanzee and macaque orthologs showed virtually no β-gal staining in the limbs. In contrast, the mouse with the human *HACNS1* showed strong but transient β-gal staining in the **preaxial** side of the limbs, including the putative domain of the pollex. The gene target of this enhancer is still unknown, but Prabhakar et al. (2008) speculate that it underlies derived manual proportions of humans, specifically our relatively longer and more robust thumbs.

These examples illustrate how changes in regulatory regions can lead to different morphological phenotypes among individuals (i.e., variation). The relative importance of CREs, mutations in coding sequences (structural mutations), and other genetic phenomena such as gene duplication, in driving morphological evolution, is still a matter of debate (Hoekstra and Coyne 2007). Of course, all these genetic mechanisms have the potential for driving evolutionary divergence in morphological traits over macroevolutionary scales. Given that most genetic toolkits and developmental pathways are highly conserved across metazoans, however, mutations in CREs remain the most likely candidate for enabling microevolutionary change, that is, producing continuous variation in quantitative traits, such as hand bone size/shape, at the level of the population.

6.2 The Structure of Limb Phenotypic Covariation

The second important aspect of the relationship between organismal development and phenotypic variation concerns the role of the former in patterning the latter. More than just "reading out" instructions contained in the genotype, development also imparts structure to phenotypic variation. The variation of any phenotypic trait is not random: it has magnitude and directionality, and some of it even depends on variation in other traits. The result is that genes and development introduce a bias in the structure of selectable variation (Arthur 2004), such that some phenotypic traits are more variable than others (e.g., phalangeal length *vs.* number of phalanges), while others tend to covary (e.g., phalangeal length between the fingers and toes, see below) (Hallgrimsson et al. 2009). The tendency of complex phenotypes to be organized into groups of traits that covary more strongly and independently of other such groups is referred to as modularity, with semiautonomous groups of phenotypic traits known as modules (Wagner 1996; Wagner et al. 2007).

The notion that development influences the magnitude and direction of phenotypic variation within and among traits is not new. Charles Darwin discussed this phenomenon under the term *correlations of growth* in the *Origin of Species*: "I mean by this expression that the whole organization is so tied together during its growth and development, that when slight variations in one part occur, and are accumulated through natural selection, other parts become modified" (Darwin 1859: 147). Without the benefit of twenty-first-century developmental biology tools or even a working theory of genetics, Darwin recognized that interactions between phenotypic traits during development could bias the production of variation within and among traits, with consequences for the evolutionary potential of complex organisms (Rolian and Willmore 2009; Rolian 2014). The advent of developmental genetics and genomics tools simply led to the empirical confirmation that the mechanistic basis of this bias lies primarily in the extensive "recycling" of the same genetic toolkits during the development and growth of complex structures (i.e., in their pleiotropic effects on multiple phenotypic traits [Cheverud 1996; Pavlicev et al. 2011]). Below are three examples of the relationship between development, phenotypic covariation, and evolvability as applied to the forelimb and hand.

6.3 Covariation with Body Size

At a basic level, all morphological structures in the primate hand covary with body size. Simply put, larger animals tend to have absolutely larger hands, whether one is comparing different primate taxa (e.g., mouse lemurs vs. gorillas), different individuals within a species (e.g., males and females of sexually dimorphic species), or even over the ontogeny of an individual (e.g., juveniles and adults) (Jungers 1984; Inouye 1992; Lemelin and Jungers 2007). From a functional point of view, the scaling of hand size over ontogeny and across animals of different sizes may ensure that there is no mismatch between the size of an individual at any stage of its development and the functional demands placed on the hands in the context of locomotion or manipulation. Mechanistically, ontogenetic and phylogenetic covariation between hand and body size is likely mediated by genes promoting body size growth and variation, mostly circulating growth factors such as insulin-like growth factors (IGFs), growth hormone, and/or estrogens (Bernstein et al. 2007; Sutter et al. 2007).

6.4 Covariation within the Forelimb and Hand

The genetic and developmental mechanisms that pattern the embryonic limb along the three cardinal axes represent the second source of phenotypic covariation among different structures of the forelimb and hand. Here again, pleiotropy appears to be a major cause underlying this covariation, via two related mechanisms. First, phenotypic covariation within the limb may be due to the deployment of the same developmental pathway in multiple developmental contexts. For example, the dual function of posterior *HoxD* genes, first in proximodistal and then in anteroposterior patterning, may lead to increased covariation in size and shape between the distal forelimb and posterior hand (Reno et al. 2008). Second, an individual developmental pathway may affect multiple phenotypic traits simultaneously. For example, the secondary anteroposterior deployment of *HoxD* genes and their downstream effects on Shh function set up two anatomically distinct regions: an anterior Shh-free domain that gives rise to the first digit and a posterior Shh-dependent domain that includes all remaining digits. In this example, the two regions can be considered modules: the relative autonomy of the first digit from its posterior neighbors indicates that it is more independently variable from these digits, while conversely shared patterning mechanisms in the posterior digits suggest a greater degree of covariation between these digits.

Hamrick (2012) recently proposed the existence of three variational domains (i.e., modules) within the autopod rays: (1) a **postaxial** digit domain for rays 2–5, (2) a **preaxial** domain affecting the first ray, and (3) a digit tip domain affecting the terminal phalanges across all five rays (Fig. 5.4). Each of these domains is relatively

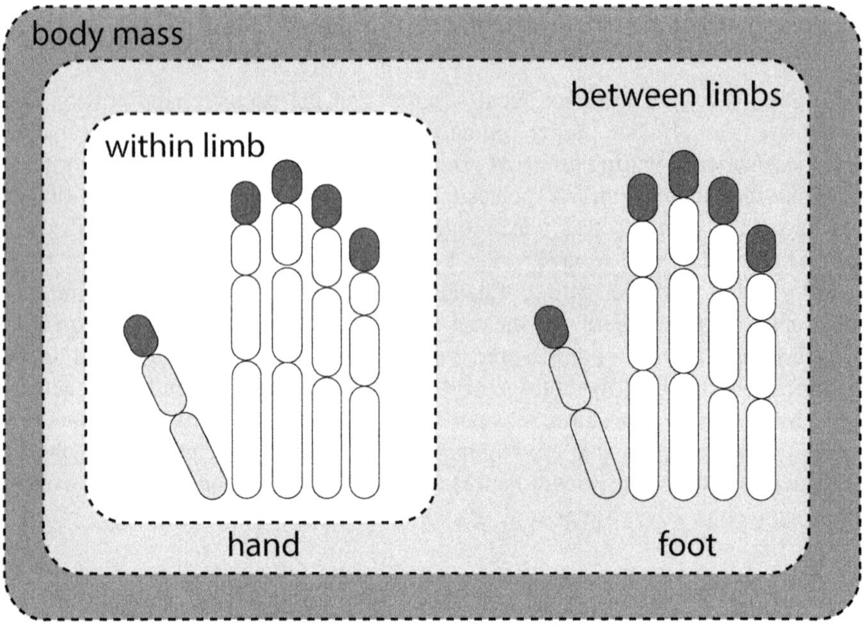

Fig. 5.4 Diagram showing the modular nature of variation in the hand skeleton, as well as the major sources of covariation in size and shape (*dashed boxes*). Within the hand, the skeleton can be divided into a number of semiautonomous modules that include the posterior digits (*white*), thumb (*light gray*), and distal phalanges (*dark gray*). The size and shape in these elements in turn covary with other skeletal traits within the forelimb (*dashed white box*), with serially homologous elements in the hind limb and foot (*light gray dashed box*) and with overall body size (*dark gray dashed box*). See text for details

autonomous from the others with respect to variation in morphology, setting up independent foci of evolutionary change in the primate (and presumably vertebrate) autopod and enabling mosaic evolution of its constituent parts. Hamrick (2012) focused on the digital rays, but there are undoubtedly additional variational modules in the hard (e.g., carpals) and soft tissues of the hand (e.g., thenar vs. hypothenar muscles) that facilitate its mosaic evolution.

6.5 *Covariation Between Fore- and Hind Limb*

The third potential source of bias in phenotypic variation of the hand is serial homology between the fore- and hind limb. Serial homology occurs when a structure's underlying genetic program is duplicated, in part or in whole, and expressed at a new time or place during development (Hall 1995). By definition, serially homologous structures share much of their genetic architecture and should thus be prone to the effects of pleiotropy. All the patterning mechanisms described above, and the vast majority of underlying genes and genetic pathways, regulate embryonic and postembryonic development in *both* limbs, from stylopod to zeugopod and from anterior to posterior digits. In fact, there are more genes that are expressed in both developing limbs than genes that are limb specific (Margulies et al. 2001; Shou et al. 2005; Taher et al. 2011).

The consequence of this extensive pleiotropy between the forelimb and hind limb is that phenotypic covariation between serially homologous structures in the hand and foot is expected to be relatively strong (Young and Hallgrimsson 2005). For example, the posterior (ulnar) fingers form a semiautonomous module *within* the hand, but the posterior fingers and toes may also form a module, representing an additional and superimposed source of covariation in the size and shape of the metacarpals and phalanges within the hand (Fig. 5.4). Increased phenotypic covariation between hand and foot structures implies that selective pressures on one may lead to correlated evolutionary changes in the other. I have argued previously that such correlated responses to selection can be advantageous, particularly in primate species where the hand and foot share similar functions in locomotion (e.g., arboreal or terrestrial quadrupeds, Rolian 2009). In such contexts, strong covariation between the hand and foot promotes coordinated, and presumably more rapid, evolutionary changes in digital size, shape, or proportions (Rolian 2014).

Conversely, however, strong covariation between serially homologous structures may constrain their ability to evolve independently in response to selection. In some situations, exploiting a new ecological niche might favor functional and morphological divergence between the hand and foot, and covariation would be expected to hinder this independent morphological evolution. In other situations, strong homology-based covariation has likely created new evolutionary opportunities. We have shown that selection acting on hominin foot proportions in the context of the evolution of bipedal locomotion likely caused parallel evolutionary changes in hominin manual proportions that facilitated the evolution of precision grasping and emergence of lithic technology (Rolian et al. 2010). We hypothesized that the "emancipation" of the hominin hand from functional constraints related to arboreal locomotion was accompanied by a relaxation of selective pressures on hand morphology. Manual digital morphology then evolved via a correlated response to strong selective pressures that caused the dramatic reorganization of the foot phalanges during the transition from a partly arboreal apelike creature to a fully committed terrestrial biped.

At first glance, our conclusions contradict the hypothesis by Prabhakar et al. (2008) that *HACNS1* evolved as a thumb-specific enhancer. However, in their experiments, *HACNS1* drove β-gal expression in the anterior autopod of *both* fore- and hind limb (see Fig. 3 in Sholtis and Noonan 2010). Thus, it is entirely plausible that *HACNS1* is actually an evolved hallux-specific enhancer in humans whose function is also present in the pollex. Moreover, pleiotropic effects due to serial homology may not be limited to the skeleton (Straus 1930). For example, the flexor hallucis longus (foot) and flexor pollicis longus (hand) are homologous muscles that show remarkably convergent derived characteristics in humans compared with other hominoids: they are both hypertrophic, have become separated along their entire course from the neighboring deep digital flexors to the posterior digits, and are heavily tendonized as they pass into the autopod to insert on the distal phalanx of their respective first digits (Straus 1942). One might speculate that parallel changes in the deep flexor of the hallux and pollex in humans are also the result of serial homology, specifically the gain-of-function mutations in *HACNS1*.

Based on the preceding discussion, one could conclude that it is virtually impossible for hand features to evolve independently of each other and/or of other phenotypic traits such as body size or other fore- and hind limb traits. This is not the case, of course, and many primate species are exceptions that appear to "break the rules" regarding the bias imposed by covariation on hand morphological evolution (Fig. 5.1). For example, posterior (ulnar) manual digits can be highly modified independently of each other (e.g., the second digits of *Perodicticus*), and homologous hand and foot bones can be morphologically disparate (e.g., vestigial thumbs but well-developed halluces in *Colobus*). Such exceptions serve as a reminder that most morphological evolution is likely driven by genetic changes in context-specific regulatory regions that control shared developmental pathways.

7 Future Directions

Developmental biologists have made tremendous inroads into the patterning mechanisms that govern vertebrate limb development. Empirical evidence from classical and emerging model organisms has revealed many important molecular, cellular, and tissue-level processes that turn an undifferentiated outgrowth of the embryonic body wall into a functionally integrated limb. At the same time, ongoing advances in comparative genomics and genetic engineering provide greater insights into how such processes are tweaked over developmental and evolutionary timescales to create diversity in tetrapod hand morphology. We are still only scratching the proverbial surface, however, and much remains to be learned about limb development and evolution. One area of emphasis for future research will be the soft tissues of the limb, which have received comparatively less attention than the skeletal system. Of particular interest will be the study of inductive interactions between developing systems, not only between soft tissue systems (e.g., limb vasculature and nervous system, Vieira et al. 2007) but also between soft tissues and the skeletal system, which appears to act as a master regulator of soft tissue patterning in the developing limb (e.g., Eshkar-Oren et al. 2009).

More broadly, one of the great challenges facing evo-devo is the study of the developmental and genetic basis of phenotypic variation at finer levels of phylogenetic analysis, such as among sister taxa or even within populations (where selection operates). To fully appreciate how and why organismal development matters to evolution, we need to understand how development produces and structures phenotypic variation, the "raw material" for evolution by natural selection. This will require broadening the current focus on deciphering patterning mechanisms and developmental pathways, to also include the study of how these pathways vary among individuals and how this variation translates into heritable phenotypic variation at the population level. We now know enough about key patterning mechanisms and have the right transgenic and genomic tools, to begin asking how differences among individuals in developmental mechanisms lead to continuous phenotypic variation in quantitative traits.

There are at least two potentially useful and complementary approaches to tackling this challenge. The first focuses on *cis*-regulatory elements (CREs), discussed above. Comparative genomics increasingly allows us to identify genetic differences (polymorphisms) in regulatory regions among individuals (Mahr et al. 2006) and among closely related species (Jones et al. 2012; Prabhakar et al. 2006). As this knowledge increases, so too does our ability to manipulate CREs experimentally, in the same way developmental biologists have been manipulating coding regions for years. For example, we might switch CREs between species or between individuals, make them induce or repress their targets ectopically in time or space, or delete them in whole or in part.

The second approach makes use of the actual targets of CREs. This approach may solve a persistent problem in genomics: we do not always know the regulatory region for a coding sequence nor conversely the precise genomic target(s) of putative CREs (e.g., Prabhakar et al. 2008). The solution is to manipulate empirically the transcription of known patterning genes and molecules in a manner that emulates the effects of their regulatory gene networks, even if these are unknown. For example, incrementally changing the expression of a known morphogen (e.g., BMPs, Shh) might have a quantifiable dose-dependent effect on a quantitative trait. This approach is already common in overexpression studies, some of which have revealed dose-dependent effects of patterning molecules on the size and shape of limb skeletal elements (Buxton et al. 2001; Duprez et al. 1996; Holmberg et al. 2008).

Phenotypic variation remains a fundamental link between organismal development and evolution by selection and by extrapolation between micro- and macroevolutionary processes. Evo-devo studies that focus specifically on the origin(s), structure, and/or evolutionary consequences of phenotypic *variation* will have the greatest impact on our understanding of the mechanistic basis of morphological and anatomical diversity in the hand.

Acknowledgments Thank you to the editors for inviting me to participate in this volume, especially to Pierre Lemelin for sharing his strepsirrhine hand data. Heather Jamniczky and Benedikt Hallgrimsson provided comments on a draft of this chapter, and their input is gratefully acknowledged.

References

Agarwal P, Wylie JN, Galceran J, Arkhitko O, Li CL, Deng CX, Grosschedl R, Bruneau BG (2003) *Tbx5* is essential for forelimb bud initiation following patterning of the limb field in the mouse embryo. Development 130:623–633

Arques CG, Doohan R, Sharpe J, Torres M (2007) Cell tracing reveals a dorsoventral lineage restriction plane in the mouse limb bud mesenchyme. Development 134:3713–3722

Arthur W (2004) The effect of development on the direction of evolution: toward a twenty-first century consensus. Evol Dev 6:282–288

Atchley W, Hall B (1991) A model for development and evolution of complex morphological structures. Biol Rev Camb Philos Soc 66:101–157

Aversi-Ferreira TA, Diogo R, Potau JM, Bello G, Pastor JF, Aziz MA (2010) Comparative anatomical study of the forearm extensor muscles of *Cebus libidinosus* (Rylands et al., 2000; Primates, Cebidae), modern humans, and other primates, with comments on primate evolution, phylogeny, and manipulatory behavior. Anat Rec 293:2056–2070

Ballock RT, O'Keefe RJ (2003) The biology of the growth plate. J Bone Joint Surg Am 85:715–726

Bernstein RM, Leigh SR, Donovan SM, Monaco MH (2007) Hormones and body size evolution in papionin primates. Am J Phys Anthropol 132:247–260

Bolter DR, Zihlman AL (2012) Skeletal development in *Pan paniscus* with comparisons to *Pan troglodytes*. Am J Phys Anthropol 147:629–636

Brohmann H, Jagla K, Birchmeier C (2000) The role of *Lbx1* in migration of muscle precursor cells. Development 127:437–445

Buckingham M, Bajard L, Chang T, Daubas P, Hadchouel J, Meilhac S, Montarras D, Rocancourt D, Relaix F (2003) The formation of skeletal muscle: from somite to limb. J Anat 202:59–68

Buxton P, Edwards C, Archer C, Francis-West P (2001) Growth/differentiation factor-5 (GDF-5) and skeletal development. J Bone Joint Surg Am 83(Suppl 1):s23–s30

Capdevila J, Izpisua Belmonte J (2001) Patterning mechanisms controlling vertebrate limb development. Annu Rev Cell Dev Biol 17:87–132

Carroll RL (1988) Vertebrate paleontology and evolution. W.H. Freeman, New York

Carroll SB (2005) Endless forms most beautiful: the new science of Evo Devo and the making of the animal kingdom. Norton, New York

Carroll SB (2008) Evo-devo and an expanding evolutionary synthesis: a genetic theory of morphological evolution. Cell 134:25–36

Carroll SB, Prud'homme B, Gompel N (2008) Regulating evolution. Sci Am 298:60–67

Chevallier A, Kieny M, Mauger A (1977) Limb-somite relationship: origin of limb musculature. J Embryol Exp Morphol 41:245–258

Cheverud JM (1996) Developmental integration and the evolution of pleiotropy. Am Zool 36:44–50

Cooper KL, Hu JKH, ten Berge D, Fernandez-Teran M, Ros MA, Tabin CJ (2011) Initiation of proximal-distal patterning in the vertebrate limb by signals and growth. Science 332:1083–1086

Cretekos CJ, Wang Y, Green ED, Martin JF, NISC Comparative Sequencing Program, Rasweiler JJ, Behringer RR (2008) Regulatory divergence modifies limb length between mammals. Genes Dev 22:141–151

Darwin C (1859) On the origin of species by means of natural selection. J. Murray, London

Davis AP, Witte DP, Hsiehli HM, Potter SS, Capecchi MR (1995) Absence of radius and ulna in mice lacking *Hoxa-11* and *Hoxd-11*. Nature 375:791–795

DeLaurier A, Schweitzer R, Logan M (2006) *Pitx1* determines the morphology of muscle, tendon, and bones of the hindlimb. Dev Biol 299:22–34

Delpretti S, Zakany J, Duboule D (2012) A function for all posterior *Hoxd* genes during digit development? Dev Dyn 241:792–802

Diogo R, Wood B (2011) Soft-tissue anatomy of the primates: phylogenetic analyses based on the muscles of the head, neck, pectoral region and upper limb, with notes on the evolution of these muscles. J Anat 219:273–359

Duboc V, Logan MPO (2011a) Pitx1 is necessary for normal initiation of hindlimb outgrowth through regulation of Tbx4 expression and shapes hindlimb morphologies via targeted growth control. Development 138:5301–5309

Duboc V, Logan MPO (2011b) Regulation of limb bud initiation and limb-type morphology. Dev Dyn 240:1017–1027

Duboule D (2002) Developmental biology: making progress with limb models. Nature 418:492–493

Ducy P, Zhang R, Geoffroy V, Ridall AL, Karsenty G (1997) Osf2/Cbfa1: a transcriptional activator of osteoblast differentiation. Cell 89:747–754

Dudley AT, Ros MA, Tabin CJ (2002) A re-examination of proximodistal patterning during vertebrate limb development. Nature 418:539–544

Duprez D (2002) Signals regulating muscle formation in the limb during embryonic development. Int J Dev Biol 46:915–925

Duprez D, Bell EJD, Richardson MK, Archer CW, Wolpert L, Brickell PM, Francis-West PH (1996) Overexpression of BMP-2 and BMP-4 alters the size and shape of developing skeletal elements in the chick limb. Mech Dev 57:145–157

Eichmann A, Makinen T, Alitalo K (2005) Neural guidance molecules regulate vascular remodeling and vessel navigation. Genes Dev 19:1013–1021

Eshkar-Oren I, Viukov SV, Salameh S, Krief S, Oh CD, Akiyama H, Gerber HP, Ferrara N, Zelzer E (2009) The forming limb skeleton serves as a signaling center for limb vasculature patterning via regulation of Vegf. Development 136:1263–1272

Fleagle JG (1974) Dynamics of a brachiating siamang [*Hylobates* (*Symphalangus*) *syndactylus*]. Nature 248:259–260

Grotewold L, Plum M, Dildrop R, Peters T, Ruther U (2001) *Bambi* is coexpressed with *Bmp-4* during mouse embryogenesis. Mech Dev 100:327–330

Hall BK (1995) Homology and embryonic development. Evol Biol 28:1–37

Hallgrimsson B, Jamniczky H, Young NM, Rolian C, Parsons TE, Boughner JC, Marcucio RS (2009) Deciphering the palimpsest: studying the relationship between morphological integration and phenotypic covariation. Evol Biol 36:355–376

Hamrick MW (2003) Evolution and development of mammalian limb integumentary structures. J Exp Zool B 298B:152–163

Hamrick MW (2012) The developmental origins of mosaic evolution in the primate limb skeleton. Evol Biol 39:447–455

Harfe BD, Scherz PJ, Nissim S, Tian F, McMahon AP, Tabin CJ (2004) Evidence for an expansion-based temporal Shh gradient in specifying vertebrate digit identities. Cell 118:517–528

Hashimoto K, Yokouchi Y, Yamamoto M, Kuroiwa A (1999) Distinct signaling molecules control *Hoxa-11* and *Hoxa-13* expression in the muscle precursor and mesenchyme of the chick limb bud. Development 126:2771–2783

Hasson P (2011) "Soft" tissue patterning: muscles and tendons of the limb take their form. Dev Dyn 240:1100–1107

Hendrikse JL, Parsons TE, Hallgrimsson B (2007) Evolvability as the proper focus of evolutionary developmental biology. Evol Dev 9:393–401

Hoekstra HE, Coyne JA (2007) The locus of evolution: evo devo and the genetics of adaptation. Evolution 61:995–1016

Hoffmann JN, Montag AG, Dominy NJ (2004) Meissner corpuscles and somatosensory acuity: the prehensile appendages of primates and elephants. Anat Rec A 281A:1138–1147

Holmberg J, Ingner G, Johansson C, Leander P, Hjalt TA (2008) PITX2 gain-of-function induced defects in mouse forelimb development. BMC Dev Biol 8:25

Hurle JM, Ros MA, Ganan Y, Macias D, Critchlow M, Hinchliffe JR (1990) Experimental analysis of the role of ECM in the patterning of the distal tendons of the developing limb bud. Cell Differ Dev 30:97–108

Ikeda A, Matsumoto S, Tanaka H (1988) Comparative anatomical study of arteriographs of the hand of primates. Deep palmar arterial arches and their correlating arteries in Cercopithecidae, Pongidae and Hominidae. Acta Anat 133:30–34

Inouye SE (1992) Ontogeny and allometry of African ape manual rays. J Hum Evol 23:107–138

Jamniczky HA, Boughner JC, Rolian C, Gonzalez PN, Powell CD, Schmidt EJ, Parsons TE, Bookstein FL, Hallgrimsson B (2010) Rediscovering Waddington in the post-genomic age. Bioessays 32:553–558

Jones FC, Grabherr MG, Chan YF, Russell P, Mauceli E, Johnson J, Swofford R, Pirun M, Zody MC, White S, Birney E, Searle S, Schmutz J, Grimwood J, Dickson MC, Myers RM, Miller CT, Summers BR, Knecht AK, Brady SD, Zhang HL, Pollen AA, Howes T, Amemiya C et al (2012) The genomic basis of adaptive evolution in threespine sticklebacks. Nature 484:55–61

Jungers WL (1984) Aspects of size and scaling in primate biology with special reference to the locomotor skeleton. Yearb Phys Anthropol 27:73–97

Kardon G (1998) Muscle and tendon morphogenesis in the avian hind limb. Development 125:4019–4032

Kardon G, Campbell JK, Tabin CJ (2002) Local extrinsic signals determine muscle and endothelial cell fate and patterning in the vertebrate limb. Dev Cell 3:533–545

Karsenty G, Kronenberg HM, Settembre C (2009) Genetic control of bone formation. Annu Rev Cell Dev Biol 25:629–648
Kavanagh E, Abiri M, Bland YS, Ashhurst DE (2002) Division and death of cells in developing synovial joints and long bones. Cell Biol Int 26:679–688
Kiefer JC (2010) Primer and interviews: molecular mechanisms of morphological evolution. Dev Dyn 239:3497–3505
Kimmel RA, Turnbull DH, Blanquet V, Wurst W, Loomis C, Joyner AL (2000) Two lineage boundaries coordinate vertebrate apical ectodermal ridge formation. Genes Dev 14: 1377–1389
King M, Arnold JS, Shanske A, Morrow BE (2006) T-genes and limb bud development. Am J Med Genet A 140A:1407–1413
Kirkwood JK, Kember NF (1993) Comparative quantitative histology of mammalian growth plates. J Zool 231:543–562
Kraus P, Fraidenraich D, Loomis CA (2001) Some distal limb structures develop in mice lacking Sonic hedgehog signaling. Mech Dev 100:45–58
Landmesser LT (2001) The acquisition of motoneuron subtype identity and motor circuit formation. Int J Dev Neurosci 19:175–182
Lemelin P, Jungers WL (2007) Body size and scaling of the hands and feet of prosimian primates. Am J Phys Anthropol 133:828–840
Li Y, Qiu Q, Watson SS, Schweitzer R, Johnson RL (2010) Uncoupling skeletal and connective tissue patterning: conditional deletion in cartilage progenitors reveals cell- autonomous requirements for *Lmx1b* in dorsal-ventral limb patterning. Development 137:1181–1188
Mackenzie F, Ruhrberg C (2012) Diverse roles for VEGF-A in the nervous system. Development 139:1371–1380
Mahr S, Burmester GR, Hilke D, Göbel U, Grützkau A, Häupl T, Hauschild M, Koczan D, Krenn V, Neidel J, Perka C, Radbruch A, Thiesen HJ, Müller B (2006) *Cis*- and *trans*-acting gene regulation is associated with osteoarthritis. Am J Hum Genet 78:793–803
Maiolino S, Boyer DM, Rosenberger A (2011) Morphological correlates of the grooming claw in distal phalanges of platyrrhines and other primates: a preliminary study. Anat Rec 294: 1975–1990
Mankoo BS, Collins NS, Ashby P, Grigorieva E, Pevny LH, Candia A, Wright CVE, Rigby PWJ, Pachnis V (1999) *Mox2* is a component of the genetic hierarchy controlling limb muscle development. Nature 400:69–73
Margulies EH, Kardia SLR, Innis JW (2001) A comparative molecular analysis of developing mouse forelimbs and hindlimbs using Serial Analysis of Gene Expression (SAGE). Genome Res 11:1686–1698
Martin A, Maher S, Summerhurst K, Davidson D, Murphy P (2012) Differential deployment of paralogous Wnt genes in the mouse and chick embryo during development. Evol Dev 14:178–195
Mayr E (1961) Cause and effect in biology. Science 134:1501–1506
Minguillon C, Del Buono J, Logan MP (2005) *Tbx5* and *Tbx4* are not sufficient to determine limb-specific morphologies but have common roles in initiating limb outgrowth. Dev Cell 8:75–84
Montavon T, Le Garrec JF, Kerszberg M, Duboule D (2008) Modeling *Hox* gene regulation in digits: reverse collinearity and the molecular origin of thumbness. Genes Dev 22:346–359
Naiche LA, Papaioannou VE (2003) Loss of *Tbx4* blocks hindlimb development and affects vascularization and fusion of the allantois. Development 130:2681–2693
Noro M, Yuguchi H, Sato T, Tsuihiji T, Yonei-Tamura S, Yokoyama H, Wakamatsu Y, Tamura K (2011) Role of paraxial mesoderm in limb/flank regionalization of the trunk lateral plate. Dev Dyn 240:1639–1649
Ohuchi H, Nakagawa T, Yamamoto A, Araga A, Ohata T, Ishimaru Y, Yoshioka H, Kuwana T, Nohno T, Yamasaki M, Itoh N, Noji S (1997) The mesenchymal factor, FGF10, initiates and maintains the outgrowth of the chick limb bud through interaction with FGF8, an apical ectodermal factor. Development 124:2235–2244
Parr BA, McMahon AP (1995) Dorsalizing signal *Wnt-7a* required for normal polarity of D-V and A-P axes of mouse limb. Nature 374:350–353

Pascoal S, Carvalho CR, Rodriguez-León J, Delfini MC, Duprez D, Thorsteindóttir S, Palmeirim I (2007) A molecular clock operates during chick autopod proximal-distal outgrowth. J Mol Biol 368:303–309

Patel BA (2010) The interplay between speed, kinetics, and hand postures during primate terrestrial locomotion. Am J Phys Anthropol 141:222–234

Pavlicev M, Norgard EA, Fawcett GL, Cheverud JM (2011) Evolution of pleiotropy: epistatic interaction pattern supports a mechanistic model underlying variation in genotype-phenotype map. J Exp Zool B 316B:371–385

Prabhakar S, Noonan JP, Paabo S, Rubin EM (2006) Accelerated evolution of conserved noncoding sequences in humans. Science 314:786

Prabhakar S, Visel A, Akiyama JA, Shoukry M, Lewis KD, Holt A, Plajzer-Frick I, Morrison H, FitzPatrick DR, Afzal V, Pennacchio LA, Rubin EM, Noonan JP (2008) Human-specific gain of function in a developmental enhancer. Science 321:1346–1350

Renehan WE, Munger BL (1990) The development of Meissner corpuscles in primate digital skin. Brain Res Dev Brain Res 51:35–44

Reno PL, McCollum MA, Cohn MJ, Meindl RS, Hamrick M, Lovejoy CO (2008) Patterns of correlation and covariation of anthropoid distal forelimb segments correspond to Hoxd expression territories. J Exp Zool B 310B:240–258

Richmond BG (2007) Biomechanics of phalangeal curvature. J Hum Evol 53:678–690

Rolian C (2009) Integration and evolvability in primate hands and feet. Evol Biol 36:100–117

Rolian C (2014) Genes, development, and evolvability in primate evolution. Evol Anthropol 23:93–104

Rolian C, Lieberman DE, Hallgrimsson B (2010) The coevolution of human hands and feet. Evolution 64:1558–1568

Rolian C, Willmore KE (2009) Morphological integration at 50: patterns and processes of integration in biological anthropology. Evol Biol 36:1–4

Rosello-Diez A, Ros MA, Torres M (2011) Diffusible signals, not autonomous mechanisms, determine the main proximodistal limb subdivision. Science 332:1086–1088

Sadler TW, Langman J (2009) Langman's medical embryology, 11th edn. Lippincott William & Wilkins, Baltimore

Sanger TJ, Norgard EA, Pletscher LS, Bevilacqua M, Brooks VR, Sandell LJ, Cheverud JM (2011) Developmental and genetic origins of murine long bone length variation. J Exp Zool B 316B:146–161

Sato TN, Loughna S, Davis EC, Visconti RP, Richardson CD (2002) Selective functions of angiopoietins and vascular endothelial growth factor on blood vessels: the concept of "vascular domain". Cold Spring Harb Symp Quant Biol 67:171–180

Saxod R (1996) Ontogeny of the cutaneous sensory organs. Microsc Res Tech 34:313–333

Schoenwolf GC, Bleyl SB, Brauer PR, Francis-West PH (2009) Larsen's Human Embryology, 4th edn. Churchill Livingstone/Elsevier, Philadelphia

Schweitzer R, Zelzer E, Volk T (2010) Connecting muscles to tendons: tendons and musculoskeletal development in flies and vertebrates. Development 137:2807–2817

Sholtis SJ, Noonan JP (2010) Gene regulation and the origins of human biological uniqueness. Trends Genet 26:110–118

Shou SM, Scott V, Reed C, Hitzemann R, Stadler HS (2005) Transcriptome analysis of the murine forelimb and hindlimb autopod. Dev Dyn 234:74–89

Shubin N, Tabin C, Carroll S (2009) Deep homology and the origins of evolutionary novelty. Nature 457:818–823

Smith RJ, Jungers WL (1997) Body mass in comparative primatology. J Hum Evol 32:523–559

Stern DL, Orgogozo V (2008) The loci of evolution: how predictable is genetic evolution? Evolution 62:2155–2177

Straus WL (1930) Parallelism in the interosseous muscles of primate hand and foot. Anat Rec 45:279 (abstract)

Straus WL (1942) Rudimentary digits in primates. Q Rev Biol 17:228–243

Stricker S, Mundlos S (2011) Mechanisms of digit formation: human malformation syndromes tell the story. Dev Dyn 240:990–1004

Summerbell D (1977) Reduction of rate of outgrowth, cell density, and cell-division following removal of apical ectodermal ridge of chick limb-bud. J Embryol Exp Morphol 40:1–21

Sun X, Mariani FV, Martin GR (2002) Functions of FGF signalling from the apical ectodermal ridge in limb development. Nature 418:501–508

Sutter NB, Bustamante CD, Chase K, Gray MM, Zhao K, Zhu L, Padhukasahasram B, Karlins E, Davis S, Jones PG, Quignon P, Johnson GS, Parker HG, Fretwell N, Mosher DS, Lawler DF, Satyaraj E, Nordborg M, Lark KG, Wayne RK, Ostrander EA (2007) A single *IGF1* allele is a major determinant of small size in dogs. Science 316:112–115

Tabin C, Wolpert L (2007) Rethinking the proximodistal axis of the vertebrate limb in the molecular era. Genes Dev 21:1433–1442

Taher L, Collette NM, Murugesh D, Maxwell E, Ovcharenko I, Loots GG (2011) Global gene expression analysis of murine limb development. PLoS One 6:e28358

Tickle C, Wolpert L (2002) The progress zone—alive or dead? Nat Cell Biol 4:E216–E217

Tozer S, Bonnin MA, Relaix F, Di Savino S, Garcia-Villalba P, Coumailleau P, Duprez D (2007) Involvement of vessels and PDGFB in muscle splitting during chick limb development. Development 134:2579–2591

van der Eerden B, Karperien M, Wit J (2003) Systemic and local regulation of the growth plate. Endocr Rev 24:782–801

Vargesson N, Kostakopoulou K, Drossopoulou G, Papageorgiou S, Tickle C (2001) Characterisation of *Hoxa* gene expression in the chick limb bud in response to FGF. Dev Dyn 220:87–90

Vega JA, Garcia-Suarez O, Montano JA, Pardo B, Cobo JM (2009) The Meissner and Pacinian sensory corpuscles revisited new data from the last decade. Microsc Res Tech 72:299–309

Vieira JM, Schwarz Q, Ruhrberg C (2007) Selective requirements for nrp1 ligands during neurovascular patterning. Development 134:1833–1843

Villavicencio-Lorini P, Kuss P, Friedrich J, Haupt J, Farooq M, Turkmen S, Duboule D, Hecht J, Mundlos S (2010) Homeobox genes d11-d13 and a13 control mouse autopod cortical bone and joint formation. J Clin Invest 120:1994–2004

Wagner GP (1996) Homologues, natural kinds and the evolution of modularity. Am Zool 36:36–43

Wagner GP, Pavlicev M, Cheverud JM (2007) The road to modularity. Nat Rev Genet 8:921–931

Wagner GP, Vargas AO (2008) On the nature of thumbs. Genome Biol 9:213

Wilsman N, Farnum C, Leiferman E, Fry M, Barreto C (1996) Differential growth by growth plates as a function of multiple parameters of chondrocytic kinetics. J Orthop Res 14:927–936

Woltering JM, Duboule D (2010) The origin of digits: expression patterns versus regulatory mechanisms. Dev Cell 18:526–532

Wray GA (2007) The evolutionary significance of *cis*-regulatory mutations. Nat Rev Genet 8:206–216

Yokouchi Y, Sasaki H, Kuroiwa A (1991) Homeobox gene-expression correlated with the bifurcation process of limb cartilage development. Nature 353:443–445

Young NM, Hallgrimsson B (2005) Serial homology and the evolution of mammalian limb covariation structure. Evolution 59:2691–2704

Zakany J, Duboule D (1999) *Hox* genes in digit development and evolution. Cell Tissue Res 296:19–25

Zakany J, Duboule D (2007) The role of *Hox* genes during vertebrate limb development. Curr Opin Genet Dev 17:359–366

Zakany J, Fromental-Ramain C, Warot X, Duboule D (1997) Regulation of number and size of digits by posterior *Hox* genes: a dose-dependent mechanism with potential evolutionary implications. Proc Natl Acad Sci U S A 94:13695–13700

Zeller R, Lopez-Rios J, Zuniga A (2009) Vertebrate limb bud development: moving towards integrative analysis of organogenesis. Nat Rev Genet 10:845–858

Zuzarte-Luis V, Hurle JM (2005) Programmed cell death in the embryonic vertebrate limb. Semin Cell Dev Biol 16:261–269

Chapter 6
Organization and Evolution of the Neural Control of the Hand in Primates: Motor Systems, Sensory Feedback, and Laterality

Andrey Verendeev, Chet C. Sherwood, and William D. Hopkins

1 Introduction

Hands play an important role in the lives of primates (Connolly 1998; Wilson 1999). They use their hands to explore their surroundings, obtain food, and interact with social partners. This chapter presents an overview of the organization and evolution of the neural control of the hand in primates. We first provide a general survey of the motor system and then focus on the cortical areas most relevant to the movement of the hand. There is no single brain area that is solely responsible for motor control of the hand, but rather a number of neural structures that are strongly interconnected, acting together to coordinate arm and hand movements and contribute to manual ability. Moreover, since there is no singular "primate brain," we review variation among primates in the neural organization of control of the hand and how it might relate to dexterity. In this context, we briefly consider tool use since the evolution of this behavior was made possible by neurobiological reorganization that allowed more sophisticated motor control of the arm and the hand. Finally, we discuss lateralization of hand preference in relation to brain anatomy, with the left hemisphere dominance of hand use being one of the best examples of neuroanatomical specializations in modern humans.

A. Verendeev (✉) • C.C. Sherwood
Department of Anthropology and Center for the Advanced Study of Human Paleobiology, The George Washington University, 800 22nd St NW, Suite 6000, Washington, DC 20052, USA
e-mail: andrey.verendeev@gmail.com

W.D. Hopkins
Neuroscience Institute and Language Research Center, Georgia State University, Atlanta, GA 30302, USA

Division of Developmental and Cognitive Neuroscience, Yerkes National Primate Research Center, Atlanta, GA 30332, USA

T.L. Kivell et al. (eds.), *The Evolution of the Primate Hand*, Developments in Primatology: Progress and Prospects, DOI 10.1007/978-1-4939-3646-5_6

2 Manual Skill of Primates

Primates are a diverse group of placental mammals, represented by over 200 extant species today (Martin 1990; Purvis 1995). The capacity for dexterous control of the hands is one feature that distinguishes primates from most other mammals. As a clade, primates are characterized by a number of adaptations that allowed their ancestors to exploit fine-branch niche environments. Although many modern primates are predominantly terrestrial in their daily activities, the ability to climb trees and employ visually guided foraging set early primates apart from other mammals (Cartmill 1992). Among the defining characteristics of primates are large forward-facing eyes, grasping hands and feet, flattened nails instead of claws on most digits, a reduced olfactory apparatus, enlargement of visual structures, and overall increase in brain size (Preuss 2009; see also Martin 1990; Ravosa and Dagosto 2007). Presumably, these characteristics—stereopsis from forward-facing eyes and increased body stability from grasping extremities with flattened nails—allowed early primates to safely navigate and forage in the challenging three-dimensional environment of the trees, particularly slender and unstable branches.

Predation might also have been an important part of the adaptive complex that propelled early primate evolution. Cartmill (1972, 1974) proposed that forward-facing eyes were an adaptation for visually guided predation, something that other avian and mammal predators share. Moreover, he argued that ocular convergence was not required for successful arboreal lifestyle (e.g., some arboreal species do not have stereopsis). In this view, stereoscopic vision enabled by forward-facing eyes evolved as an adaptation for visually guided grasping of prey, such as insects and small vertebrates. Others (Sussman and Raven 1978; Sussman et al. 2013) argued instead that ocular convergence and grasping ability coevolved to allow early primates to forage on fruit and/or flowers of angiosperms (i.e., flowering plants). Regardless of which hypothesis is correct (and the two are not necessarily mutually exclusive), it is nevertheless important to note the remarkable manual skill of primates and the extent to which they depend on vision to guide the use of their forelimbs in the handling of food items and other objects. In this chapter, we explore the neural aspects of this extraordinary ability of primates.

3 Organization of Motor Control Areas in the Central Nervous System and Their Contribution to Hand Movements

The control of motor function in the primate central nervous system is hierarchically organized (see Fig. 6.1; reviewed in Purves et al. 2008). At the base of this hierarchy are the neurons, known as **motor neurons**, which innervate the skeletal musculature and ultimately generate bodily movements, including hand movements. The motor neurons located in the brainstem are mostly involved in the

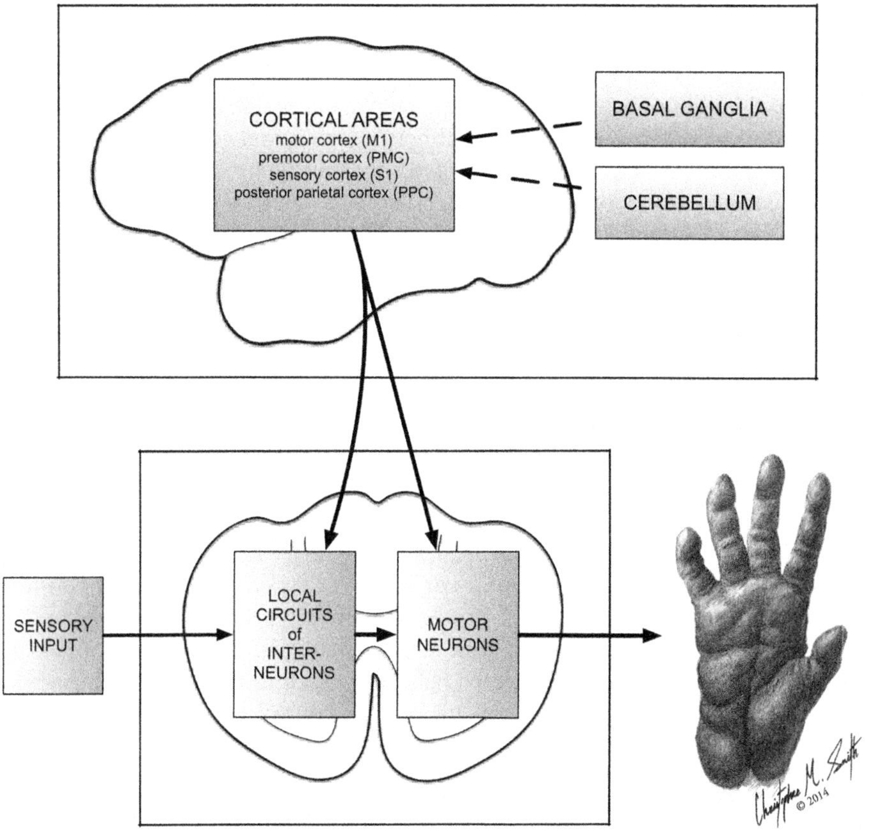

Fig. 6.1 Hierarchical organization of the neural control of the hand in primates. *Above* Cortical and subcortical (i.e., basal ganglia and cerebellum) areas involved in hand control. *Below* Organization of the motor control at the level of the spinal cord. Image after Purves et al. (2008)

control of muscles of the head and neck, whereas those located in the spinal cord innervate trunk and limb musculature. The activity of motor neurons within the spinal cord and brain stem is organized by local circuits of interneurons. The motor neurons and the local circuits are in turn controlled by neurons that send descending axonal inputs from the primary motor cortex and other cortical areas, as well as higher brainstem centers. Additionally, they receive sensory feedback from muscle spindle receptors that provide information on the position of limbs and other body parts and help regulate motor programs. The corticospinal neurons (i.e., neurons that originate in the cerebral cortex and terminate in the spinal cord) coordinate the activity of motor neurons and organize pathways for voluntary, goal-directed movement. Finally, these systems receive further input from the basal ganglia and the cerebellum. These subcortical structures indirectly influence the activity of motor neurons by regulatory activation of the cortical areas and descending corticospinal neurons. The basal ganglia and the cerebellum (both subcortical areas of the brain, see Fig. 6.1 and below) are responsible, respectively, for proper initiation and regulation of movement, as well as coordination of ongoing behavior, and have been implicated in a number of motor disorders (Purves et al. 2008).

The cell bodies of motor neurons that innervate the trunk and limb muscles are located in the gray matter of the ventral horn of the spinal cord (see below). The motor neurons are organized in an orderly manner along the longitudinal axis (rostral-caudal dimension), as well as across the coronal axis (medial-lateral dimension) of the spinal cord. Along the rostral-caudal dimension, motor neurons innervating the forelimbs (including the hand) are located along the cervical segment of the spinal cord, whereas the motor neurons innervating the muscles of the hind limbs originate from the lumbar segment of the cord. Across the spinal cord sections, the neurons are organized in a somatotopic fashion (i.e., corresponding to different areas of the body), with the more medial neurons innervating more proximal muscles (e.g., shoulder and elbow) and the more lateral neurons innervating more distal muscles (e.g., muscles of the hand and individual digits). This organization essentially provides a spatial map of the somatic musculature of the body (Purves et al. 2008).

The activity of motor neurons is controlled by the local circuits of interneurons (known as **central pattern generators** or CPGs) of the spinal cord and brainstem that organize muscle movements in an ethologically and behaviorally relevant manner. One demonstration of this is that direct stimulation of the local circuits in experimental animals elicits organized behaviors (such as those resembling walking), even in the absence of input from the cerebral cortex (i.e., when the spinal cord has been severed from the brain; Grillner and Zangger 1979; Whelan 1996; see also Dimitrijevic et al. 1998 for similar results in humans suffering from spinal injury). Additionally, these spinal circuits receive sensory information, and their intermediate position allows for another set of behaviors: by connecting sensory input and motor output they mediate the familiar sensory-motor reflexes (such as the knee jerk reflex).

As mentioned above, in addition to local circuits, the activity of motor neurons is organized by descending inputs from the neurons with cell bodies located in the motor cortex and other cortical areas (e.g., premotor cortex), as well as several brainstem centers. These systems form direct (i.e., monosynaptic) connections that synchronize the complex spatiotemporal sequences of hand movements and appear especially important for the fine motor ability of primates (Lemon 2008; discussed below). The most relevant of these for the present discussion are the descending projections from the primary motor cortex and the surrounding premotor cortex. We discuss these structures next.

The **primary motor cortex** (or M1) occupies the anterior wall of the central sulcus and adjacent precentral gyrus on the lateral surface of the cerebral cortex and, medially, the paracentral lobule (Fig. 6.2). It was designated area 4 by Brodmann (1909), which is distinct in its cytoarchitecture (i.e., cellular composition) from area 6 (premotor cortex) and area 3 (somatosensory cortex), located rostrally and caudally, respectively. M1 is primarily recognized by the lack of layer IV and the presence of giant pyramidal neurons in layer V known as Betz cells

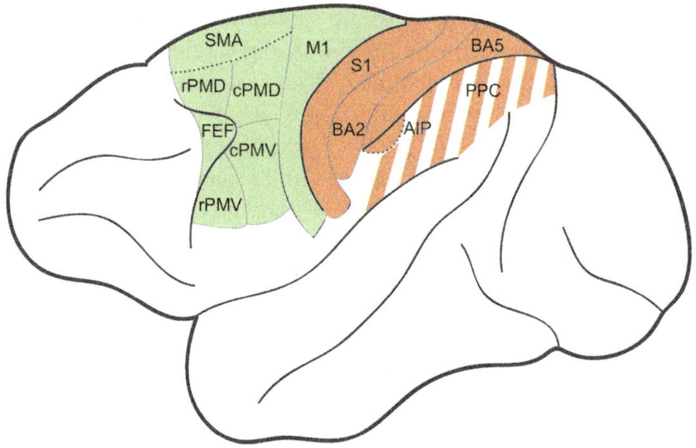

Fig. 6.2 Cortical areas of the frontal and parietal lobes involved in movement overlaid on a template Old World monkey brain. Motor and premotor cortex areas are in green; sensory and parietal cortex areas in orange. *Dotted lines* represent cortical area on the medial surface of the premotor cortex and anterior portion of the lateral wall of intraparietal sulcus. *M1* motor cortex, *PM* premotor, *PMD* dorsal premotor, *PMV* ventral premotor, *SMA* supplementary motor area, *FEF* frontal eye field, *S1* sensory cortex, *BA5* area 5 (after Brodmann 1909), *BA2* area 2 (after Brodmann 1909), *AIP* anterior intraparietal area, *PPC* posterior parietal cortex, *r* rostral, *c* caudal. Image after Rizzolatti et al. (1998)

(Brodmann 1909; Rivara et al. 2003). These and other pyramidal neurons in deep cortical layers give rise to corticobulbar and corticospinal projections that penetrate the brainstem and the spinal cord, respectively, and organize voluntary, goal-directed behavior. It is important to note, however, that M1 is not the single origin of these descending corticospinal projections; additional sources of input can be found to originate from cell bodies located in premotor and somatosensory cortices as well (Dum and Strick 1991; discussed below).

Physiologically, M1 is characterized by the low threshold of electrical current necessary to elicit movement, which confirms a direct pathway to the motor neurons of the brainstem and the spinal cord (Sessle and Wiesendanger 1982). In its spatial arrangement, M1 is musculotopically organized with a clear demarcation between the upper and lower body parts, wherein the head, arms, forearms, and hands are represented more ventrally, and the thighs, legs, and feet are represented more dorsally (Levine et al. 2012).

Just rostral to M1 lies the **premotor cortex** (Brodmann's area 6). Unlike the narrow band of M1, the premotor cortex is expansive and has been described as a "complex mosaic of interconnected frontal lobe areas" that contribute to motor function (Purves et al. 2008: 444; see also Rizzolatti et al. 1998; Rizzolatti and Luppino 2001; Kaas 2004).

In primates, the premotor cortex can be subdivided into several specialized subregions, such as **supplementary motor area** (SMA), **frontal eye field** (FEF), **dorsal** and **ventral premotor areas** (PMD and PMV, respectively), and others

(Kaas 2004; see Fig. 6.2). These areas are specialized to the extent that they can be functionally differentiated using brain imaging studies, single-neuron recordings, intracortical microstimulation, as well as differential behavioral effects in patients with frontal lobe injuries (Rizzolatti et al. 1998; Matelli et al. 1991; Purves et al. 2008). These premotor cortex areas have representations of various body parts, including the face and eyes, the limbs, and the trunk. Together, these areas make up the "motor association cortex" that integrates information from the sensory cortex and other areas of the parietal cortex and more anterior regions of the prefrontal cortex and organizes behavior in a purposeful manner, relevant for each body part. The best-studied primate species in the organization of the premotor cortex are macaques and humans. In these species, premotor areas may be even further partitioned into dorsal and medial SMA, **supplementary eye field** (SEF), rostral and caudal portions of PMD and PMV, as well as other areas (for more detail, see Rizzolatti et al. 1998; Kaas 2004; see Fig. 6.2).

The various divisions of the premotor cortex receive extensive input from the parietal lobes as well as more rostral divisions of the prefrontal cortex and play an essential role in coordination of visually guided goal-directed motor actions (Burnod et al. 1999). Together with M1, the premotor cortex gives rise to corticospinal and corticobulbar projections that directly control the activity of local circuits and motor neurons in the spinal cord and the brainstem. This has been confirmed in studies using retrograde neural tracers (Keizer and Kuypers 1989; He et al. 1993, 1995). The premotor cortex also influences motor behavior indirectly through extensive connections with M1. Together, the premotor cortex and M1 comprise highly interconnected areas of the posterior frontal lobe that contain significant populations of descending corticospinal neurons and sit atop the hierarchical organization of motor function (see Fig. 6.1; Purves et al. 2008).

Another set of cortical areas that is intimately involved in motor function in primates is the **posterior parietal cortex** (PPC) (see Fig. 6.2). This is revealed by the extensive connections between the PPC and the motor and premotor areas of the frontal cortex (Petrides and Pandya 1984; for review, see Rizzolatti et al. 1998). Indeed, networks of connections between these cortical areas (i.e., PPC and motor and premotor areas) are essential for visually guided movements of the forelimb, including the hand (Caminiti et al. 1996; Rizzolatti et al. 1997; Burnod et al. 1999). Specifically, PPC receives both somatosensory and visual information and seems to be especially well positioned to integrate this input and relay it to the motor areas. For example, PPC contains a number of representation areas for both upper and lower extremities, as well as the face. It is notable that in primates the forelimb is represented extensively within this cortical area. This has been confirmed in studies that reported complex forelimb movements (e.g., reaching, hand-to-mouth movements, defensive arm postures, etc.) evoked by electrical microstimulation of this area that are organized in manner that resembles behaviorally significant patterns in galagos (Stepniewska et al. 2005).

The connections between the PPC and the motor areas of the frontal lobe, although extensive, are not random. It is possible, for example, to differentiate a number of separate parieto-frontal circuits (Rizzolatti et al. 1997), each of which represents a functional unit of the motor system and is characterized by a specific sensory input and motor output (reviewed in Rizzolatti et al. 1998). For instance, **superior parietal lobule** (SPL), architectonically identified as area 5 by Brodmann (1909) and area PE by von Economo (1929), contains circuits that encode the location of the forelimb in relation to the trunk (Lacquaniti et al. 1995). The SPL has extensive connections with M1 (Petrides and Pandya 1984), and these connections are important in providing the motor cortex with the information necessary for the control of limbs relative to the body (Rizzolatti et al. 1998). This circuit may be especially important in the evolution of primates in the context of adaptations to the fine-branch niche of arboreal environments (see below).

Another parieto-frontal circuit relevant to organization of hand control in primates is the network of connections between the anterior portion of the intraparietal sulcus (**anterior intraparietal area** or AIP; Gallese et al. 1994) and the **rostral part of PMV** (rPMV, also called F5ab). It has been observed that AIP is active during hand and digit movements related to grasping activity (Taira et al. 1990; Sakata et al. 1995). This area has extensive connections with rPMV, which has been previously shown to mediate the movements of the hand, including grasping, holding, and manipulating movements. Together, these connections seem to be responsible for regulating hand movements appropriate for the object to be manipulated, such as adjusting the hand and the fingers for the precision or whole-hand grip depending on the size and other properties of the object (Jeannerod et al. 1995; Gallese et al. 1996; Rizzolatti et al. 1998; Fogassi et al. 2001; reviewed in Castiello 2005).

Given the extensive connections of the PPC to the motor cortex and other cortical areas involved in motor function and its role in regulating the movements of the forelimbs, it is not surprising that PPC has been implicated in tool use. Support for this comes from lesion and functional neuroimaging studies that collectively show the importance of parietal cortex areas and parieto-frontal circuits in tool manipulation and use (Heilman et al. 1982; Goldenberg and Haggmann 1998; Binkofski et al. 1999; Haaland et al. 2000; Johnson-Frey et al. 2005). Additionally, these studies confirm the distinction between the cognitive aspect of tool use (i.e., knowledge of tools) and its motor component (i.e., motor skill necessary for tool handling and use) and provide evidence that the PPC is involved in both of these processes (Johnson-Frey 2004; Ramayya et al. 2010).

Non-cortical areas are also essential for controlling complex movement of the hand and other body parts. The basal ganglia and the cerebellum both influence movement indirectly through the organization of the activity of cortical structures. The basal ganglia are a large group of subcortical nuclei (striatum, substantia nigra, nucleus accumbens, and globus pallidus) that have broad connections to the cortex, thalamus, and the brainstem. Together, this system ensures a smooth course of voluntary action with proper initiation, regulation, and termination of movement. Damage to this system has been implicated in a number of motor disorders, such as Parkinson's disease, characterized by the inability to properly organize and initiate movement. The cerebellum, like the basal ganglia, does not have direct connections to the motor neurons. Instead, it regulates the activity of corticospinal neurons and coordinates ongoing behavior by detecting and correcting "motor error." Recently, the cerebellum has been suggested to be involved in cognitive functions as well (see Purves et al. 2008 for a more detailed discussion on the role of basal ganglia and cerebellum in movement in general).

4 Variation Among Primates in the Neural Organization of Hand Representation

Evidence from primate comparative analyses demonstrates phylogenetic variation in the motor systems that control the hand and forelimb. This neural variation might be related to the evolution of manual skill of the hand among primates. Unfortunately, detailed analyses of the neuroanatomy of the motor control of the hand are absent in most primate species. However, data are available for several New World and Old World monkeys, as well as apes. Together, these data allow for some confidence in the reconstruction of the evolution of primate manual skill and its underlying neurobiology. We begin by considering early primates in comparison to other mammalian species.

Most placental mammals have anatomically differentiated motor and somatosensory cortices (Woolsey 1958; Kaas 2004). This is not true, however, among monotremes and marsupials. Electrophysiological studies revealed that marsupial opossums possess a primary somatosensory cortex, but with no separate motor area (Lende 1963; Beck et al. 1996). This undifferentiated cortical area with both sensory and motor properties has been referred to as a "sensorimotor amalgam" (Lende 1963) and may be primitive for mammals. It is interesting to note in this context that the somatosensory cortex of primates also has some motor properties and is responsive to electrical stimulation (albeit at higher threshold values than the motor cortex proper), even when the motor cortex has been experimentally removed. Nevertheless, comparative studies suggest that a separate motor cortex distinct from its somatosensory neighbor is present in most placental mammals, including all primates (reviewed in Kaas 2009; see also Kaas 2008).

Galagos are small nocturnal primates that resemble early primates in brain size and shape (Radinsky 1967; Martin 1990) and are often used as a model in the study of early primate brain evolution. Galagos exhibit a primary motor cortex and, based on microstimulation and cytoarchitectonic studies, show evidence for additional premotor areas as well (reviewed in Kaas 2004). The M1 of galagos is well differentiated with a representation of the trunk, the limbs, and the face. Although forelimb movements have a large representation in M1, there is little control of individual digits. This conclusion is based on the observation that intracortical stimulation elicits some hand movements but almost no single digit movements (Wu et al. 2000). In addition to the primary motor cortex, galagos demonstrate many secondary areas of the premotor cortex as well, including PMD and PMV, SMA, as well as FEF, all of which are absent in other mammals (Kaas 2009).

Compared to strepsirrhine primates, the primary motor cortex of anthropoid primates (monkeys, apes, and humans) has been shown to have more cortical space dedicated to movement of individual digits (Gould et al. 1986). Moreover, the premotor cortex underwent further modifications such that PMD and PMV could be further subdivided into their rostral and caudal portions, SMA added dorsal and medial areas, and other premotor areas have been augmented as well (Rizzolatti et al. 1998; Kaas 2004; see discussion above). In addition to reorganization of the premotor cortex, anthropoid primates (especially Old World monkeys and apes) demonstrate significant expansion of the posterior parietal cortex that presumably assumed greater involvement in visually guided reaching and manipulation (Marconi et al. 2001; Hill et al. 2010; Chaplin et al. 2013) through development of wide connections to M1 and different areas of the premotor cortex.

As described above, the **central sulcus** divides the primary motor and sensory cortices that occupy the anterior and posterior walls of the sulcus, respectively. An interesting morphological characteristic of the central sulcus is the presence of the "hand knob" area, so called because it corresponds to the area of the primary motor cortex mediating the hand and individual finger movements (Yousry et al. 1997). Recently, Hopkins et al. (2014) examined the morphology of the central sulcus using MRI in several Old World monkeys and apes, with a special focus on the motor hand area of the precentral gyrus. They found that great apes have a relatively large central sulcus compared to Old World monkeys. Moreover, a well-formed motor hand area (i.e., "hand knob") was only evident in humans and other apes, further corroborating previous findings of this area in great apes (Hopkins and Pilcher 2001; see below for more discussion on the "hand knob" in relation to lateralization of hand use).

It has been argued (and is generally accepted) that direct monosynaptic connections from cortical regions to spinal motor neurons play an important role in the control of digital dexterity, making possible the superb manual skill that characterizes some primate species, including humans (Lemon 2008; see also Iwaniuk et al. 1999). Comparative analyses reveal, for example, that corticospinal connections are dense in primates, but are more sparse or absent in other mammals (Illert et al. 1976; Yang and Lemon 2003; Alstermark and Ogawa 2004; Alstermark et al. 2004). Within primates, there is considerable interspecies variability in the density and anatomical terminations of these projections, while numerous in some Old World and New World monkeys (e.g., macaques and capuchin monkeys), they are sparser in others (e.g., squirrel monkeys) (Bortoff and Strick 1993). Corticospinal projections appear to be more extensive in great apes, including humans (Kuypers 1981). Importantly, these differences appear to correlate well with the dexterity of these species (Heffner and Masterton 1975, 1983; Courtine et al. 2007; see also Iwaniuk et al. 1999 and discussion below).

It is important to note that corticospinal projections consist of more than one tract in primates, and these travel in any of the three **funiculi** (dorsal, lateral, and ventral) on either the ipsilateral or contralateral side of the spinal cord (relative to their cortical origins), making up six possible tracts that traverse the spinal cord (Nudo and Frost 2009; see Fig. 6.3). In primates, the majority of the corticospinal fibers travel through the contralateral tract of the spinal cord carrying approximately 90 % of all the fibers. This is not the case for all mammalian species, however, and the analysis of the largest corticospinal tracts in placental mammals shows variation among different orders, but little variation within orders (Noback and Shriver 1966;

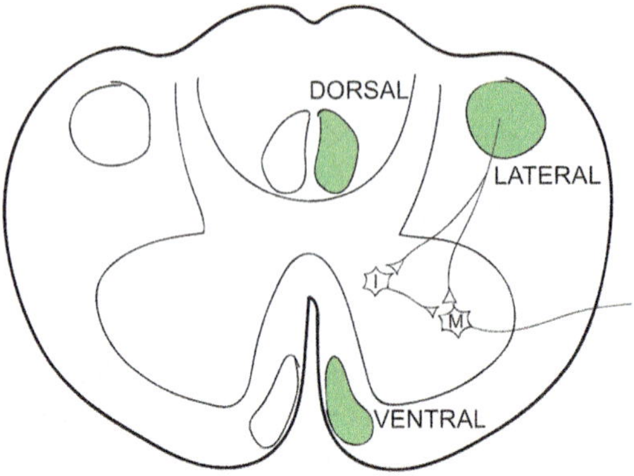

Fig. 6.3 Corticospinal neuron tracts in the spinal cord of primates travel in three funiculi (dorsal, lateral, and ventral) on either the ipsilateral or contralateral side of the spinal column (relative to their cortical origins). The principal (i.e., largest) corticospinal tract in primates is contralateral, containing approximately 90 % of all corticospinal fibers. *M* motor neuron, *I* interneuron. Image after Nudo and Frost (2009)

Nudo and Frost 2009). Thus, whereas in primates the largest tract (carrying more and larger fibers) is contralateral, it is contradorsal in tree shrews, among the closest living relatives of primates. This is one of the many ways in which tree shrews are different from primates in the structure of their nervous system (see Campbell 1980; Kaas and Preuss 1993), and it further justifies the separation of tree shrews from the order of primates in the phyletic relationship (Martin 1990).

Whereas the spinal route of the corticospinal tracts is generally consistent across primate species, the termination patterns of these fibers vary. It was suggested by some investigators that this variation might be related to manual skill. For example, Heffner and Masterton (1975, 1983) proposed a relationship between digital dexterity and the length and depth of penetration of corticospinal tracts within the spinal cord. Using the patterns of distribution of corticospinal fibers in 69 mammalian species and a measure of digital dexterity, they concluded that species that are more dexterous in their ability to handle objects had tracts that penetrated more caudally within the spinal cord and deeper within the ventral horn (which is the location of motor neuron cell bodies) of the spinal gray matter. These results, however, have been questioned. Using modern comparative statistics (i.e., taking into account the phylogenetic relationships among the species examined), Iwaniuk et al. (1999) found that only the length of corticospinal tract fibers was related to the variation in digital dexterity. Moreover, they found that the length of corticospinal projections showed a significant correlation with hand-eye coordination, but did not correlate with either arboreality or diet, suggesting a specific relationship to visually guided reaching and manipulation that is not dependent on the locomotor or dietary habits of the animal. Without much additional data on the characteristics of the neurons and axons involved in the corticospinal projections, it is difficult to speculate why the length, but not the depth of penetration, of corticospinal tracts within the spinal cord results in greater manual ability. One possibility is that the length of corticospinal tracts may be associated with an increased diameter of these axon fibers, resulting in a greater capacity for rapid signals to be synchronized as they are sent to cervical segments of the spinal cord controlling the forelimbs, including the hand.

One of the most exciting discoveries in recent decades has been that of the mirror neurons in the ventral premotor cortex of macaques (Di Pellegrino et al. 1992; Rizzolatti et al. 1996; Rizzolatti and Luppino 2001). These neurons were reported to fire when the experimental subject preformed an action *and* when the same action was being observed in another monkey subject or human experimenter. Mirror neurons have now been documented in humans as well (Fabbri-Destro and Rizzolatti 2008; Mukamel et al. 2010). Several attempts have been made to delineate their function but without general agreement (Casile 2013). Relevant to the present discussion is the finding that mirror neurons are most active during the actions performed by the hand, such as grasping, holding, and manipulation of objects (Rizzolatti and Luppino 2001). Recent findings suggest that mirror neurons may be a part of a larger circuit that connects frontal, parietal, and temporal areas (Rizzolatti et al. 2001; Rizzolatti and Sinigaglia 2010). Although mirror neurons have been described in species recognized for their manual dexterity (i.e., macaques and humans), it remains to be seen whether mirror neurons are essential for the fine motor ability in these and other species.

Taken together, these comparative analyses suggest that over time primates evolved more complex cortical sensorimotor systems that presumably were helpful in the fine-branch niche of arboreal environments. This was achieved through expansion of both the primary motor and sensory cortices, the addition of new premotor and posterior parietal areas, and development of extensive reciprocal connections between these structures. Moreover, evolution of more caudal extension of corticospinal fibers within the spinal cord led to increased digital dexterity that allowed fine motor skill in reaching, grasping, and manipulation of food items and other objects.

5 Sensory Aspects of the Neural Organization of the Hand

The neurobiological modifications associated with the control of the hand in primates allowed for better manipulation skills that surely facilitated foraging and predation abilities. In addition to descending cortical control of motor function, however, sensory system has played an important role in the evolution of the primate hand and its characteristic dexterity. The sensory system conveying tactile information has coevolved in parallel with the motor system and contributed to the manipulation abilities of primates. As mentioned above, the expansion of parietal cortical areas that receive, process, and integrate sensory information has contributed to primate dexterity. In this section we briefly describe the mechanoreceptors responsible for gathering tactile information, the ascending fibers directly involved in transmitting this information to the primary sensory cortex, and how the evolution of this system facilitated the manual ability of primates.

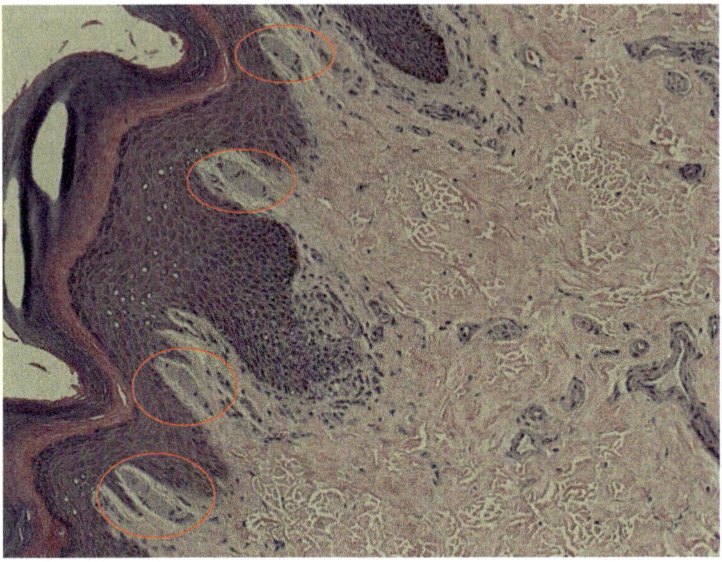

Fig. 6.4 Meissner's corpuscles (MCs) in a rhesus macaque (*Macaca mulatta*) fingertip. MCs are outlined in *red circles*

The skin of the hand is innervated with afferent nerve fibers that collect and transmit tactile information regarding touch, temperature, vibration, and pain to the somatosensory cortex and surrounding parietal areas. The glabrous (i.e., hairless) skin of the digits is characterized by the presence of four types of tactile mechanoreceptors: Merkel cells, Meissner's and Pacinian corpuscles, and Ruffini endings. **Meissner's corpuscles** (MCs) make up the majority of the tactile mechanoreceptors in the human hand (Purves et al. 2008). MCs are localized in the dermal papillae and are the most superficial of the tactile mechanoreceptors (see Fig. 6.4; Chap. 8). They are therefore very sensitive to skin deformations and may be essential for the control of grip.

The tactile information from cutaneous mechanoreceptors is transmitted to the somatosensory cortex through a series of ascending relays. The first in this chain are the afferent fibers that gather the information from the skin's surface and transmit it via the dorsal columns of the spinal cord, where they synapse onto the cell bodies of the cuneate nucleus located in the caudal medulla. These neurons then convey the information via the medial lemniscus tract to the ventral posterior lateral nucleus of the thalamus. The thalamus is known as the sensory relay center in the brain that gathers sensory information and transfers it to the relevant cortical areas. The tactile information is relayed to the **primary somatosensory cortex** (S1) and more posterior parietal areas (e.g., areas 2 and 5), where it is processed and integrated into the appropriate motor response.

The parietal areas 2 and 5 (after Brodmann 1909; see Fig. 6.2), located in the rostral parietal lobe, have been implicated in the evolution of greater manual skill among certain primate species. Area 2 comprises the most posterior part of the primary sensory cortex and lies parallel to somatosensory area 3 (3a and 3b in anthropoid primates) and area 1. It gathers sensory information from both the cutaneous and muscle spindle receptors. Immediately posterior to it is area 5, which has been shown to be broadly involved in sensorimotor processing, specifically in relation to the control of the forelimbs (Lacquaniti et al. 1995). As noted above, it has extensive connections with the primary motor cortex and conveys information about the position of the forelimbs relative to the body. Together, these areas have been shown to be involved in reaching, grasping, and manipulation of objects in primates.

It has been noted that parietal areas 2 and 5 are easily differentiated in dexterous species such as macaques (Pons et al. 1985), but they are poorly developed or altogether absent in less dexterous species of New World monkeys (Merzenich et al. 1978; Sur et al. 1982; Coq et al. 2004; see also Padberg et al. 2005, 2007). Recently, Padberg et al. (2007) examined these areas in capuchin monkeys, which are known to demonstrate exceptionally complex manual skills, including precision grips (Christel and Fragaszy 2000; Spinozzi et al. 2004). In the wild, capuchin monkeys employ fine motor skill in manipulating stone hammers to crack nuts and encased seeds against anvils, as well as to use sticks to extract prey from inside rock crevices (Fragaszy et al. 2004; Ottoni and Izar 2008; Spagnoletti et al. 2011). Using electrophysiological recordings, Padberg et al. (2007) mapped the functional organization of the anterior parietal cortex in capuchin monkeys with specific focus on areas 2 and 5. They found that both areas were clearly defined in the capuchin monkey, while absent in a close relative—the squirrel monkey. In capuchin monkeys, a complete representation of the contralateral body, including the trunk and the limbs, was described in area 2, with a large portion of it dedicated to the hand and face. Area 5 mostly served the forelimb and the hand, with little cortical space dedicated to other body parts (Padberg et al. 2007). These results resembled those for the macaque, another primate species known for its dexterity (see Iwaniuk et al. 1999), in both the location and organization of parietal areas 2 and 5. Since capuchin monkeys (platyrrhine) and macaques (catarrhine) are more distantly related, these results suggest independent evolution of similar reorganization of the parietal cortex related to skilled hand use.

6 Lateralization in Hand Preference and the Brain

We now turn to the issue of lateralization in primate hand use and examine its relationship to brain anatomy, both at the macrostructural and microstructural levels. Lateralization of hand preference is relevant to the present discussion because this asymmetry in behavior—specifically, the predominance of right-hand preference that is observed in humans—is characterized by asymmetry in brain anatomy and is one of the best examples of neuroanatomical specializations (i.e., left hemisphere specialization for motor skill). Examining this issue in humans, apes, and monkeys then might help us better understand the neural organization of fine motor ability of primates.

Lateralization in hand use is especially evident in humans, with the majority of people preferentially using their right hand in everyday activities. In humans, the right versus left-hand preference ratio is at least 8:1 (and possibly as high as 9:1) (Marchant and McGrew 1994; Marchant et al. 1995; Raymond and Pontier 2004; Faurie et al. 2005) and has been documented in many different cultures worldwide (Porac and Coren 1981; Perelle and Ehrman 1994). This right-hand bias has long been assumed to be a distinctively human characteristic related to our superb manual skills. The close association between the two is revealed by the term *dexterity*, which is derived from Latin for "right hand" and has been used in English to communicate fine motor ability (see Corballis 2012). However, studies in both captive and wild primates show that individual hand preference may not be restricted to humans (McGrew and Marchant 1997; Rogers and Andrew 2002; Papademetriou et al. 2005; Hopkins 2007; Meguerditchian et al. 2011). It has been debated whether hand preference seen in nonhuman primates is equivalent to human handedness, with some researchers arguing that the difference is simply a matter of quantitative degree (Hopkins 2013a, b), while others arguing that there is a more significant discontinuity between human species-wide handedness versus nonhuman hand preference (Cashmore et al. 2008: Marchant and McGrew 2013).

Initially, the discussion of handedness in nonhuman primates was supported by little data (e.g., Finch 1941; Walker 1980; Warren 1980). However, over the last 25 years or so, there has been a proliferation of studies to help us better understand the expression of handedness in other primate species. In this section, we provide a brief overview of the published data (see Hopkins et al. 2007; Hopkins 2013b; McGrew and Marchant 1997 for in-depth reviews), highlighting some of the key issues and caveats in the current research on nonhuman primate handedness.

First, it is important to note that there is no a priori reason to assume that lateralization in hand use is a feature unique to humans. Although our species exhibits exceptional behavioral and cognitive abilities, the Darwinian approach would suggest that perhaps *continuity* of form and function should be the null hypothesis in our attempt to understand ourselves until shown otherwise. Indeed, early speculations on handedness considered the right-hand bias (together with language) in humans to be a part of the overall left hemisphere dominance restricted to ourselves (Broca 1861). However, it is now well established that asymmetry at the level of both neuroanatomy and function is much more ancient and widespread than previously thought. This has been especially well documented in songbirds and other avian species, but also in different species of fish, amphibians, and other vertebrates (Rogers and Andrew 2002). As mentioned above, there has been increased interest in nonhuman handedness over the last couple of decades (MacNeilage et al. 1987), and the available data suggest that lateralization in hand use may be evident in other primates.

Another significant issue in handedness research has been the one of valid assessment (Hopkins 2013b). What makes a good measure of hand preference and which behavioral tasks are better at eliciting preference? Hand use in nonhuman primates has been studied using a number of different measures, ranging from simple reaching to much more challenging tasks that require coordinated manipulation by both hands (e.g., tube task) (Hopkins et al. 2007). Interestingly, some of these measures show a pronounced hand preference, whereas others fail to do so. Consequently, it has been argued by some researchers that nonhuman primates do not show true "handedness," defined as consistent hand use across different tasks and in most individuals in a population (McGrew and Marchant 1997). However, as has been pointed out by Hopkins (2013b), this argument assumes that all measures of hand preference are equally good at eliciting a bias. This is not the case and different tasks can vary dramatically in their ability to elicit consistent hand preference even within *the same individual*. The more demanding coordinated bimanual tasks such as the "tube task" (i.e., probing for a food item inside a tube with one hand while the other hand holds the tube in place) have been argued to be a better measure of handedness than simple tasks such as reaching because they are better at eliciting consistent hand preferences, and they remove the potential role of situational and postural factors on hand use (see Hopkins et al. 2007 for data on chimpanzees; see Fragaszy and Mitchell 1990; Westergaard and Suomi 1996; Spinozzi and Truppa 1999 for data on other primate species). Moreover, individual variation in the tube task, but not simple reaching, has been linked to asymmetries in the morphology of the hand area of the primary motor cortex (see above) in chimpanzees and capuchin monkeys (Hopkins and Cantalupo 2004; Phillips and Sherwood 2005; Dadda et al. 2006; see below). In assessments of human handedness, tasks that do not elicit consistent hand preference are often removed from the analyses as being poor measures (i.e., having little construct validity; see Hopkins 2013b). Accordingly, it has been suggested that for a meaningful comparison to human handedness, behavioral measures that are consistent in showing a hand preference be used instead (Hopkins 2013b).

Finally, it is important to note that despite the abundance of data, the interpretation of findings is not always straightforward. One specific point of disagreement has been whether the available data suggest that population-level handedness in nonhuman primates is similar to that of the species-level handedness in humans. As mentioned above, the right-hand bias in humans is particularly strong and stable across different tasks. In chimpanzees, however, the distribution of right-handed versus left-handed individuals dramatically deviates from that observed in human subjects, with some studies showing strong to moderate right-hand preference in about half of the individuals studied (Hopkins 2013b) and other studies reporting left-handedness in a majority of subjects (Bogart et al. 2012) or no significant population-level handedness (see McGrew and Marchant 1997 for a review). A number of attempts have been made to explain these inconsistencies, focusing on methodological, statistical, and subject variables that differ among studies (McGrew and Marchant 1997; Hopkins 2013b; Marchant and McGrew 2013). Despite an apparent lack of consensus on population-level handedness in nonhuman primates (which might be task specific), it nevertheless has been noted that hand preference in individual subjects can be consistent over time and across different tasks (Meguerditchian et al. 2011). It is therefore possible to examine the neuroanatomical correlates of preferential hand use in individual subjects. A number of studies have examined this question at both the macrostructural and microstructural levels, as well as in terms of functional asymmetry, in both humans and nonhuman primates.

In humans, several studies have examined the morphology of the central sulcus in relation to handedness (see Hammond 2002 for a review). Using MRI-based morphometry, Amunts et al. (1996) showed that the central sulcus is deeper in the dominant than the nondominant hemisphere in males of either handedness (see also Amunts et al. 2000). This finding has been partially replicated in both sexes of known handedness: right-handed subjects had deeper central sulcus in the opposite hemisphere, but not left-handed subjects (Foundas et al. 1998). In subjects of unknown handedness, the central sulcus was found to be deeper in the left hemisphere compared with the right (White et al. 1994; although see White et al. 1997a, b for inconsistent results). Overall, these studies suggest sulcal asymmetry in relation to handedness with the dominant hemisphere (usually left) being greater in depth than the nondominant hemisphere.

In nonhuman primates, a number of studies have examined the relation of hand preference and neuroanatomical asymmetries, including the morphology of the central sulcus. In chimpanzees, for example, Dadda et al. (2006) examined asymmetry in the depth of the central sulcus along the dorsal-ventral axis using MRI. They found that right-handed chimpanzees differed from left-handed chimpanzees in the more dorsal portion of the central sulcus. Further analysis revealed that this difference was primarily due to asymmetry in the "hand knob" area of the precentral gyrus (see also Hopkins and Pilcher 2001; Hopkins and Cantalupo 2004; Sherwood et al. 2007). These results have been further confirmed in male capuchin monkeys in that the central sulcus was deeper in the right hemisphere of the left-handed monkeys, but was more symmetrical in right-handed subjects (Phillips and Sherwood 2005). Taken together, these results parallel the human data and suggest a deeper (i.e., more asymmetrical) central sulcus in the dominant hemisphere compared with nondominant hemisphere in nonhuman primates.

In addition to morphological asymmetry, it is also possible to examine cerebral asymmetry at a microstructural level in terms of cellular architecture. Amunts et al. (1997; see also Amunts et al. 1996) examined postmortem human brains of unknown handedness in postnatal development of interhemispheric asymmetry in the cytoarchitecture of the primary motor cortex. They found greater neuropil (interneuronal space containing axons, dendrites, synapses, as well as processes of glial cells) volume in the left hemisphere than the right (i.e., lower volume of cell bodies). This evidence suggests greater interconnectivity among neurons in the presumably dominant left hemisphere compared with the right. Importantly, the pattern of postnatal maturation of cytoarchitectonic asymmetry implies a possible connection with the functional development of motor skill and hand preference (Amunts et al. 1997).

In chimpanzees, some studies have also found interhemispheric differences at the microstructural level. For example, Sherwood et al. (2007) showed cytoarchitectonic asymmetry in the region of hand representation of the primary motor cortex between the two hemispheres. Specifically, they found leftward bias for higher neuron density in layer II/III, known for its role in interhemispheric corticocortical connections. Furthermore, the distribution of synaptophysin (a presynaptic vesicle-associated protein) immunoreactivity was shown to correlate with handedness as measured using the bimanual tube task in chimpanzees, such that non-right-handed (i.e., left-handed and ambidextrous) individuals showed a rightward bias compared with right-handed individuals that showed no overall asymmetry (Sherwood et al. 2010).

7 Conclusion

In this chapter, we presented an overview of the organization and evolution of the neural control of the hand in primates. Motor control of the hand is hierarchically organized, with several structures of the central nervous system contributing to different aspects of planning and execution of movement sequences. Although primates as a clade are very diverse, the capacity for dexterous control of the hands is one of the features that distinguish them from other mammals. Nonetheless, within primates there is significant variation in manual ability, with some species being more dexterous than others. For example, chimpanzees and macaques demonstrate highly dexterous precision grips, whereas galagos show almost no single digit control. These behavioral differences are most certainly related to variation in the complexity of the sensorimotor systems of these species, and neuroanatomical comparisons confirm this view. Lateralization of hand use is another manifestation of motor specialization in primates. Although very pronounced in humans, species-level handedness might not be restricted to us and may be evident in other species as well, such as chimpanzees (for which the behavioral data are most abundant). Both human and nonhuman primate data show that asymmetry in hand use is related to both the macrostructural and microstructural differences in the organization of the motor cortex. These differences may be important for neurobiological and cognitive developments that made possible tool use and manufacture in certain primate species, including ourselves.

8 Future Directions

As described above, much is known about the organization of the neural control of the hand in primates. Most of our knowledge comes from the experimental work in two primate species—humans and macaques. Studies using electrical microstimulation pioneered in the early twentieth century allowed for extensive mapping of the motor cortex and other cortical areas involved in movement. Lesions, either experimentally induced (in macaques) or a result of brain injury (in humans), corroborated physiological studies in outlining the organization of motor areas. Recent advances in functional imaging resulted in our better understanding of the interconnected nature (i.e., the extensive networks of frontal and parietal areas) of motor function. Comparative analyses, on the other hand, have allowed us to infer the likely scenario of the evolution of motor control of the hand in primates.

Despite these advances, however, certain areas of research require further inquiry. For example, the ontogenetic development of motor function (especially in relation to the control of the hand and individual digits) across a range of primate species remains relatively unknown. Additionally, as we mentioned above, the manipulation of objects is facilitated by cutaneous sensory feedback collected from the glabrous surface of the hands. Although Meissner's corpuscles have been examined in several primate species in relation to digital dexterity (Verendeev et al. 2015), with inconclusive results, it is of special interest whether their distribution patterns relate to the complexity of manipulative abilities in different primate species. Finally, it is important to investigate how different hand actions are used in different foraging and social contexts. This insight may further our understanding of the neurobiology and evolution of manual ability in primates.

Acknowledgments We would like to thank Christopher Smith for the drawing of the hand used in this chapter.

References

Alstermark B, Ogawa J (2004) In vivo recordings of bulbospinal excitation in adult mouse forelimb motoneurons. J Neurophysiol 92:1958–1962

Alstermark B, Ogawa J, Isa T (2004) Lack of monosynaptic corticomotoneuronal EPSPs in rats: disynaptic EPSPs mediated via reticulospinal neurones and polysynaptic EPSPs via segmental interneurones. J Neurophysiol 91:1832–1839

Amunts K, Schlaug G, Schleicher A, Steinmetz H, Dabringhaus A, Roland PE, Zilles K (1996) Asymmetry in the human motor cortex and handedness. Neuroimage 4:216–222

Amunts K, Schmidt-Passos F, Schleicher A, Zilles K (1997) Postnatal development of interhemispheric asymmetry in the cytoarchitecture of human area 4. Anat Embryol 196:393–402

Amunts K, Jäncke L, Mohlberg H, Steinmetz H, Zilles K (2000) Interhemispheric asymmetry of the human motor cortex related to handedness and gender. Neuropsychologia 38:304–312

Beck PD, Pospichal MW, Kaas JH (1996) Topography, architecture, and connections of somatosensory cortex in opossums: evidence for five somatosensory areas. J Comp Neurol 366:109–133

Binkofski F, Buccino G, Posse S, Seitz RJ, Rizzolatti G, Freund H (1999) A fronto-parietal circuit for object manipulation in man: evidence from an fMRI study. Eur J Neurosci 11:3276–3286

Bogart SL, Pruetz JD, Ormiston LK, Russell JL, Meguerditchian A, Hopkins WD (2012) Termite fishing laterality in the Fongoli savanna chimpanzees (*Pan troglodytes verus*): further evidence of a left hand preference. Am J Phys Anthropol 149:591–598

Bortoff GA, Strick PL (1993) Corticospinal terminations in two New World primates: further evidence that corticomotoneuronal connections provide part of the neural substrate for manual dexterity. J Neurosci 13:5105–5118

Broca P (1861) Remarques sur le siège de la faculté du langage articulé, suivies d'une observation d'aphémie (perte de la parole). Bull Soc Anthropol 6:330–357

Brodmann K (1909) Vergleichende Lokalisationslehre der Großhirnrinde: in ihren Prinzipien dargestellt auf Grund des Zellenbaues. Barth, Leipzig

Burnod Y, Baraduc P, Battaglia-Mayer A, Guigon E, Koechlin E, Ferraina S, Lacquaniti F, Caminiti R (1999) Parieto-frontal coding of reaching: an integrated framework. Exp Brain Res 129:325–346

Caminiti R, Ferraina S, Johnson PB (1996) The sources of visual information to the primate frontal lobe: a novel role for the superior parietal lobule. Cereb Cortex 6:319–328

Campbell CBG (1980) The nervous system of the Tupaiidae: its bearing on phyletic relationships. In: Luckett WP (ed) Comparative biology and evolutionary relationships of tree shrews. Plenum Press, New York, pp 219–242

Cartmill M (1972) Arboreal adaptations and the origin of the order Primates. In: Tuttle R (ed) The functional and evolutionary biology of primates. Aldine-Atherton, Chicago, pp 97–122

Cartmill M (1974) Rethinking primate origins. Science 184:436–443
Cartmill M (1992) New views on primate origins. Evol Anthropol 1:105–111
Cashmore L, Uomini N, Chapelain A (2008) The evolution of handedness in humans and great apes: a review and current issues. J Anthropol Sci 86:7–35
Casile A (2013) Editorial. Neurosci Lett 540:1–2
Castiello U (2005) The neuroscience of grasping. Nat Rev Neurosci 6:726–736
Chaplin TA, Hsin-Hao Y, Soares JGM, Gattass R, Rosa MGP (2013) A conserved pattern of differential expression of cortical areas in simian primates. J Neurosci 33:15120–15125
Christel MI, Fragaszy DM (2000) Manual function in *Cebus apella*. Digital mobility, preshaping, and endurance in repetitive grasping. Int J Primatol 21:697–719
Connolly KJ (1998) The psychobiology of the hand. Lavenham, London
Corballis MC (2012) Lateralization of the human brain. In: Hofman MA, Falk D (eds) Progress in brain research. Elsevier, Amsterdam, pp 103–121
Courtine G, Bunge MB, Fawcett JW, Grossman RG, Kass JH et al (2007) Can experiments in nonhuman primates expedite the translation of treatments for spinal cord injury in humans? Nat Med 13:561–566
Coq JO, Qi H, Collins CE, Kaas JH (2004) Anatomical and functional organization of somatosensory areas of the lateral fissure of the New World titi monkey (*Callicebus moloch*). J Comp Neurol 476:363–387
Dadda M, Cantalupo C, Hopkins WD (2006) Further evidence of an association between handedness and neuroanatomical asymmetries in the primary motor cortex of chimpanzees (*Pan troglodytes*). Neuropsychologia 44:2582–2586
Dimitrijevic MR, Gerasimenko Y, Pinter MM (1998) Evidence for a spinal central pattern generator in humans. Ann N Y Acad Sci 860:360–376
Di Pellegrino G, Fadiga L, Fogassi L, Gallese V, Rizzolatti G (1992) Understanding motor events: a neurophysiological study. Exp Brain Res 91:176–180
Dum RP, Strick PL (1991) The origin of corticospinal projections from the premotor areas in the frontal lobe. J Neurosci 11:667–689
von Economo CF (1929) The cytoarchitectonics of the human cerebral cortex. Oxford University Press, London
Fabbri-Destro M, Rizzolatti G (2008) Mirror neurons and mirror systems in monkeys and humans. Physiology 23:171–179
Faurie C, Schiefenhövel W, Le Bomin S, Billiard S, Raymond M (2005) Variation in the frequency of left-handedness in traditional societies. Curr Anthropol 46:142–147
Finch G (1941) Chimpanzee handedness. Science 94:117–118
Fogassi L, Gallese V, Buccino G, Craighero L, Fadiga L, Rizzolatti G (2001) Cortical mechanism for the visual guidance of hand grasping movements in the monkey: a reversible inactivation study. Brain 124:571–586
Foundas AL, Hong K, Leonard CM, Heilman KM (1998) Hand preference and magnetic resonance imaging asymmetries of the central sulcus. Cogn Behav Neurol 11:65–71
Fragaszy DM, Izar P, Visalberghi E, Ottoni EB, de Oliveira MG (2004) Wild capuchin monkeys (*Cebus libidinosus*) use anvils and stone pounding tools. Am J Primatol 64:359–366
Fragaszy DM, Mitchell SR (1990) Hand preference and performance on unimanual and bimanual tasks in capuchin monkeys (*Cebus apella*). J Comp Psychol 104:272–282
Gallese V, Murata A, Kaseda M, Niki N, Sakata H (1994) Deficit of hand preshaping after muscimol injection in monkey parietal cortex. Neuroreport 5:1525–1529
Gallese V, Fadiga L, Fogassi L, Rizzolatti G (1996) Action recognition in the premotor cortex. Brain 119:593–609
Goldenberg G, Haggmann S (1998) Tool use and mechanical problem solving in apraxia. Neuropsychologia 36:581–589
Gould HJ, Cusick CG, Pons TP, Kaas JH (1986) The relationship of corpus callosum connections to electrical stimulation maps of motor, supplementary motor, and the frontal eye fields in owl monkeys. J Comp Neurol 247:297–325

Grillner S, Zangger P (1979) On the central generation of locomotion in the low spinal cat. Exp Brain Res 34:241–261

Haaland KY, Harrington DL, Knight RT (2000) Neural representations of skilled movement. Brain 123:2306–2313

Hammond G (2002) Correlates of human handedness in primary motor cortex: a review and hypothesis. Neurosci Biobehav Rev 26:285–292

He SQ, Dum RP, Strick PL (1993) Topographic organization of corticospinal projections from the frontal lobe: motor areas on the lateral surface of the hemisphere. J Neurosci 13:952–980

He SQ, Dum RP, Strick PL (1995) Topographic organization of corticospinal projections from the frontal lobe: motor areas on the medial surface of the hemisphere. J Neurosci 15:3284–3306

Heffner R, Masterton B (1975) Variation in form of the pyramidal tract and its relationship to digital dexterity. Brain Behav Evol 12:188–200

Heffner RS, Masterton RB (1983) The role of the corticospinal tract in the evolution of human digital dexterity. Brain Behav Evol 23:165–183

Heilman KM, Rothi LJ, Valenstein E (1982) Two forms of ideomotor apraxia. Neurology 32:342–346

Hill J, Inder T, Neil J, Dierker DL, Harwell J, Van Essen D (2010) Similar patterns of cortical expansion during human development and evolution. Cereb Cortex 8:13135–13140

Hopkins WD (ed) (2007) Evolution of hemispheric specialization in primates. Elsevier, Oxford

Hopkins WD (2013a) Neuroanatomical asymmetries and handedness in chimpanzees (*Pan troglodytes*): a case for continuity in the evolution of hemispheric specialization. Ann N Y Acad Sci 1288:17–35

Hopkins WD (2013b) Comparing human and nonhuman primate handedness: challenges and a modest proposal for consensus. Dev Psychobiol 55:621–636

Hopkins WD, Cantalupo C (2004) Handedness in chimpanzees (*Pan troglodytes*) is associated with asymmetries of the primary motor cortex but not with homologous language areas. Behav Neurosci 118:1176–1183

Hopkins WD, Pilcher DL (2001) Neuroanatomical localization of the motor hand area with magnetic resonance imaging: the left hemisphere is larger in great apes. Behav Neurosci 115:1159–1164

Hopkins WD, Russell JL, Lambeth S, Schapiro SJ (2007) Handedness and neuroanatomical asymmetries in captive chimpanzees: a summary of 15 years of research. In: Hopkins WD (ed) Evolution of hemispheric specialization in primates. Elsevier, Oxford, pp 146–181

Hopkins WD, Meguerditchian A, Coulon O, Bogart S, Mangin J-F, Sherwood CC, Grabowski MW, Bennett AJ, Pierre PJ, Fears S, Woods R, Hof PR, Vauclair J (2014) Evolution of the central sulcus morphology in primates. Brain Behav Evol 84:19–30

Illert M, Lundberg A, Tanaka R (1976) Integration in descending motor pathways controlling the forelimb in the cat. Exp Brain Res 26:509–519

Iwaniuk AN, Pellis SM, Whishaw IQ (1999) Is digital dexterity really related to corticospinal projections?: a re-analysis of the Heffner and Masterton data set using modern comparative statistics. Behav Brain Res 101:173–187

Jeannerod M, Arbib MA, Rizzolatti G, Sakata H (1995) Grasping objects: the cortical mechanisms of visuomotor transformation. Trends Neurosci 18:314–320

Johnson-Frey SH (2004) The neural bases of complex tool use in humans. Trends Cogn Sci 8:71–78

Johnson-Frey SH, Newman-Norlund R, Grafton ST (2005) A distributed left hemisphere network active during planning of everyday tool use skills. Cereb Cortex 15:681–695

Kaas JH (2004) Evolution of somatosensory and motor cortex in primates. Anat Rec A Discov Mol Cell Evol Biol 281:1148–1156

Kaas JH (2008) The evolution of the complex sensory and motor systems of the human brain. Brain Res Bull 75:384–390

Kaas JH (2009) The evolution of sensory and motor systems in primates. In: Kaas JH (ed) Evolutionary neuroscience. Academic Press, Oxford, pp 847–870

Kaas JH, Preuss TM (1993) Archontan affinities as reflected in the visual system. In: Szalay FS, Novacek MJ, McKenna MC (eds) Mammal phylogeny. Springer-Verlag, New York, pp 115–128

Keizer K, Kuypers HG (1989) Distribution of corticospinal neurons with collaterals to the lower brain stem reticular formation in monkey (*Macaca fascicularis*). Exp Brain Res 74:311–318

Kuypers HGJM (1981) Anatomy of the descending pathways. In: Brooks VB (ed) The nervous system. Handbook of physiology. Williams and Wilkins, Baltimore, pp 597–666

Lacquaniti F, Guigon E, Bianchi L, Ferraina S, Caminiti R (1995) Representing spatial information for limb movement: role of area 5 in the monkey. Cereb Cortex 5:391–409

Lemon RN (2008) Descending pathways in motor control. Annu Rev Neurosci 31:195–218

Lende RA (1963) Cerebral cortex: a sensorimotor amalgam in the marsupial. Science 141:730–732

Levine AJ, Lewallen KA, Pfaff SL (2012) Spatial organization of cortical and spinal neurons controlling motor behavior. Curr Opin Neurobiol 22:812–821

MacNeilage PF, Studdert-Kennedy MG, Lindblom B (1987) Primate handedness reconsidered. Behav Brain Sci 10:247–303

Marchant LF, McGrew WC (1994) Human handedness: an ethological perspective. Hum Evol 13:221–228

Marchant LF, McGrew WC (2013) Handedness is more than laterality: lessons from chimpanzees. Ann N Y Acad Sci 1288:1–8

Marchant LF, McGrew WC, Eibl-Eibesfeldt I (1995) Is human handedness universal? Ethological analyses from three traditional cultures. Ethology 101:239–258

Marconi B, Genovesio A, Battaglia-Mayer A, Ferraina S, Squatrito S, Molinari M, Lacquaniti F, Caminiti R (2001) Eye–hand coordination during reaching I: anatomical relationships between parietal and frontal cortex. Cereb Cortex 11:513–527

Martin RD (1990) Primate origins and evolution. A phylogenetic reconstruction. Princeton University Press, Princeton

Matelli M, Luppino G, Rizzolatti G (1991) Architecture of superior and mesial area 6 and of the adjacent cingulate cortex. J Comp Neurol 311:445–462

McGrew WC, Merchant LF (1997) On the other hand: current issues in and meta-analysis of the behavioral laterality of hand function in non-human primates. Yrbk Phys Anthropol 40:201–232

Meguerditchian A, Molesti S, Vauclair J (2011) Right-handedness in 162 baboons (*Papio anubis*) for gestural communication: consistency across time and groups. Behav Neurosci 125:653–660

Merzenich MM, Kaas JH, Sur M, Lin CS (1978) Double representation of the body surface within cytoarchitectonic areas 3b and 1 in "SI" in the owl monkey (*Aotus trivirgatus*). J Comp Neurol 181:41–74

Mukamel R, Ekstrom AD, Kaplan J, Iacoboni M, Fried I (2010) Single-neuron responses in humans during execution and observation of actions. Curr Biol 20:750–756

Noback CR, Shriver JE (1966) The phylogeny of the corticospinal tract. Anat Rec 154:479

Nudo RJ, Frost SB (2009) The evolution of motor cortex and motor systems. In: Kaas JH (ed) Evolutionary neuroscience. Academic Press, Oxford, pp 727–756

Ottoni EB, Izar P (2008) Capuchin monkey tool use: overview and implications. Evol Anthropol 17:171–178

Padberg J, Disbrow E, Krubitzer L (2005) The organization and connections of anterior and posterior parietal cortex in titi monkeys: do New World monkeys have an area 2? Cereb Cortex 15:1938–1963

Padberg J, Franca JG, Cooke DF, Soares JG, Rosa MG, Fiorani M, Gattass R, Krubitzer L (2007) Parallel evolution of cortical areas involved in skilled hand use. J Neurosci 27:10106–10115

Papademetriou E, Sheu CF, Michel GF (2005) A meta-analysis of primate hand preferences, particularly for reaching. J Comp Psychol 119:33–48

Perelle IB, Ehrman L (1994) An international study of human handedness: the data. Behav Genet 24:217–227

Petrides M, Pandya DN (1984) Projections to the frontal cortex from the posterior parietal region in the rhesus monkey. J Comp Neurol 228:105–116

Phillips KA, Sherwood CC (2005) Primary motor cortex asymmetry is correlated with handedness in capuchin monkeys (*Cebus apella*). Behav Neurosci 119:1701–1704

Pons TP, Garraghty PE, Cusick CG, Kaas JH (1985) The somatotopic organization of area 2 in macaque monkeys. J Comp Neurol 241:445–466

Porac C, Coren S (1981) Lateral preferences and human behavior. Springer, New York

Preuss TM (2009) Primate brain evolution. In: Kaas JH (ed) Evolutionary neuroscience. Academic Press, Oxford, pp 793–825

Purves D, Augustine GJ, Fitzpatrick D, Hall WC, LaMantia AS, McNamara JO, White LE (2008) Neuroscience, 4th edn. Sinauer, Sunderland

Purvis A (1995) A composite estimate of primate phylogeny. Philos Trans R Soc Lond B 348:405–421

Radinsky LB (1967) The oldest primate endocast. Am J Phys Anthropol 27:385–388

Ramayya AG, Glasser MF, Rilling JK (2010) A DTI investigation of neural substrates supporting tool use. Cereb Cortex 20:507–516

Ravosa MJ, Dagosto M (eds) (2007) Primate origins: adaptations and evolution. Springer, New York

Raymond M, Pontier D (2004) Is there geographical variation in human handedness? Laterality 9:35–51

Rivara CB, Sherwood CC, Bouras C, Hof PR (2003) Stereologic characterization and spatial distribution patterns of Betz cells in the human primary motor cortex. Anat Rec A Discov Mol Cell Evol Biol 270:137–151

Rizzolatti G, Luppino M (2001) The cortical motor system. Neuron 31:889–901

Rizzolatti G, Sinigaglia C (2010) The functional role of the parieto-frontal mirror circuit: interpretations and misinterpretations. Nat Rev Neurosci 4:264–274

Rizzolatti G, Fadiga G, Gallese V, Fogassi L (1996) Premotor cortex and recognition of motor actions. Cogn Brain Res 3:131–141

Rizzolatti G, Fogassi L, Gallese V (1997) Parietal cortex: from sight to action. Curr Opin Neurobiol 7:562–567

Rizzolatti G, Luppino M, Matelli M (1998) The organization of the cortical motor system: new concepts. Electroencephalogr Clin Neurophysiol 106:283–296

Rizzolatti G, Fogassi L, Gallese V (2001) Neurophysiological mechanisms underlying the understanding and imitation of action. Nat Rev 2:661–670

Rogers LJ, Andrew JR (2002) Comparative vertebrate lateralization. Cambridge University Press, Cambridge

Sakata H, Taira M, Murata A, Mine S (1995) Neural mechanisms of visual guidance of hand action in the parietal cortex of the monkey. Cereb Cortex 5:429–438

Sessle BJ, Wiesendanger M (1982) Structural and functional definition of the motor cortex in the monkey (*Macaca fascicularis*). J Physiol 323:245–265

Sherwood CC, Wahl E, Erwin JM, Hof PR, Hopkins WD (2007) Histological asymmetries of primary motor cortex predict handedness in chimpanzees (*Pan troglodytes*). J Comp Neurol 503:525–537

Sherwood CC, Duka T, Stimpson CD, Schenker NM, Garrison AR, Schapiro SJ, Baze WB, McArthur MJ, Erwin JM, Hof PR, Hopkins WD (2010) Neocortical synaptophysin asymmetry and behavioral lateralization in chimpanzees (*Pan troglodytes*). Eur J Neurosci 31:1456–1464

Spagnoletti N, Visalberghi E, Ottoni E, Izar P, Fragaszy D (2011) Stone tool use by adult wild bearded capuchin monkeys (*Cebus libidinosus*): frequency, efficiency and tool selectivity. J Hum Evol 61:97–107

Spinozzi G, Truppa V (1999) Hand preferences in different tasks by tufted capuchins (*Cebus apella*). Int J Primatol 20:827–849

Spinozzi G, Truppa V, Lagana T (2004) Grasping behavior in tufted capuchin monkeys (*Cebus apella*): grip types and manual laterality for picking up a small food item. Am J Phys Anthropol 125:30–41

Stepniewska I, Fang P-C, Kaas JH (2005) Microstimulation reveals specialized subregions for different complex movements in posterior parietal cortex of prosimian galagos. Proc Natl Acad Sci U S A 102:4878–4883

Sur M, Nelson RJ, Kaas JH (1982) Representations of the body surface in cortical areas 3b and 1 of squirrel monkeys: comparisons with other primates. J Comp Neurol 211:177–192

Sussman RW, Raven PH (1978) Pollination by lemurs and marsupials: an archaic coevolutionary system. Science 200:731–736

Sussman RW, Rasmussen DT, Raven PH (2013) Rethinking primate origins again. Am J Primatol 75:95–106

Taira M, Mine S, Georgopoulos AP, Murata A, Sakata H (1990) Parietal cortex neurons of the monkey related to the visual guidance of hand movement. Exp Brain Res 83:29–36

Verendeev A, Thomas C, McFarlin SC, Hopkins WD, Phillips KA, Sherwood CC (2015) Comparative analysis of Meissner's corpuscles in the fingertips of primates. J Anat 227:72–80

Walker SF (1980) Lateralization of functions in the vertebrate brain: a review. Br J Psychol 71:329–367

Warren JM (1980) Handedness and laterality in humans and other animals. Physiol Psychol 8:351–359

Westergaard GC, Suomi SJ (1996) Hand preference for a bimanual task in tufted capuchins (*Cebus apella*) and rhesus macaques (*Macaca mulatta*). J Comp Psychol 110:406–411

Whelan PJ (1996) Control of locomotion in the decerebrate cat. Prog Neurobiol 49:481–515

White LE, Lucasm G, Richards A, Purves D (1994) Cerebral asymmetry and handedness. Nature 368:197–198

White LE, Andrews TJ, Hulette C, Richards A, Groelle M, Paydarfar J, Purves D (1997a) Structure of the human sensorimotor system. I: Morphology and cytoarchitecture of the central sulcus. Cereb Cortex 7:18–30

White LE, Andrews TJ, Hulette C, Richards A, Groelle M, Paydarfar J, Purves D (1997b) Structure of the human sensorimotor system. II: Lateral symmetry. Cereb Cortex 7:31–47

Wilson FR (1999) The hand: how its use shapes the brain, language, and human culture. Vintage Books, New York

Woolsey CN (1958) Organization of somatic sensory and motor areas of the cerebral cortex. In: Harlow HF, Woolsey CN (eds) Biological and biochemical bases of behavior. The University of Wisconsin Press, Madison, pp 63–82

Wu CW-H, Bichot NP, Kaas JH (2000) Converging evidence from microstimulation, architecture, and connections for multiple motor areas in the frontal and cingulate cortex of prosimian primates. J Comp Neurol 423:140–177

Yang H-W, Lemon RN (2003) An electron microscopic examination of the corticospinal projection to the cervical spinal cord in the rat: lack of evidence for cortico-motoneuronal synapses. Exp Brain Res 149:458–469

Yousry TA, Schmid UD, Alkadhi H, Schmidt D, Peraud A, Buettner A, Winkler P (1997) Localization of the motor hand area to a knob on the precentral gyrus. A new landmark. Brain 120:141–157

Chapter 7
Anatomy, Function, and Evolution of the Primate Hand Musculature

Pierre Lemelin and Rui Diogo

1 Introduction

The musculature responsible for movement of the hand has been a topic of choice among anatomists, biological anthropologists, clinicians, and other students of the human body. Throughout the twentieth century, comparative anatomists were preoccupied in establishing hand muscle homologies and documenting differences between taxa in order to interpret morphological change across vertebrates, often in terms of phylogenetic "progression" (amphibian, reptile, marsupial, prosimian, monkey, ape, human) (e.g., Bardeleben 1894; McMurrich 1903a, b; Howell 1936; Campbell 1939; Haines 1939, 1950; Straus 1941a, b, 1942b; Lewis 1989). At the same time, clinical anatomists aimed at uncovering how muscles produce complex movements of the hand such as the action of the interossei and lumbricals across multiple joints during finger extension (e.g., Duchenne 1867; Braithwaite et al. 1948; Landsmeer 1949, 1955; Eyler and Markee 1954; Stack 1962; Valentin 1981). The refinement and wider use of electromyography (EMG) provided data on the timing and intensity of muscle activity that dramatically improved our knowledge of hand muscle function in humans (Backhouse and Catton 1954; Long and Brown 1964; Landsmeer and Long 1965; Long et al. 1970; Long 1981) and nonhuman primates (Tuttle and Basmajian 1974; Susman and Stern 1979, 1980; Susman et al. 1982; Patel et al. 2012). EMG methods have also been employed to test relationships between muscle anatomy and specific grips in order to better understand the evolution of the hand and toolmaking behavior in humans (Hamrick et al. 1998; Marzke et al. 1998).

P. Lemelin (✉)
Division of Anatomy, Department of Surgery, Faculty of Medicine and Dentistry, University of Alberta, 5-05A Medical Sciences Building, Edmonton, AB, Canada T6G 2H7
e-mail: plemelin@ualberta.ca

R. Diogo
Department of Anatomy, College of Medicine, Howard University, Washington, DC 20059, USA

T.L. Kivell et al. (eds.), *The Evolution of the Primate Hand*, Developments in Primatology: Progress and Prospects, DOI 10.1007/978-1-4939-3646-5_7

This multidisciplinary perspective using comparative, functional, and even clinical approaches is recurrent in the writings of Wood Jones and Napier on the hand musculature. In his treatise on the hand, Wood Jones (1942: 275) warned readers of the inherent complexity of the intrinsic muscles of the hand, pointing out that "numerous difficulties are presented to the investigator who attempts to unravel the story of these little muscles." At the same time, Wood Jones offered clear and enduring observations on the anatomy and function of hand muscles, including those intricate and short muscles found in the palm of the hand. As a matter of fact, his description of the thenar muscles (the short muscles at the base of the thumb or pollex) and their compound action in producing opposition of the thumb are both accurate and insightful, especially considering the limited experimental techniques available in the early twentieth century. In the same vein, Napier (1952) combined precise dissection observations with physiological data available at the time to assess impairment of the abductor pollicis brevis (one of the thenar muscles) during opposition in patients. This early study, which evinces Napier's background as a physician, paved the way to later interests on the comparative anatomy of thenar muscles in primates (Day and Napier 1963).

In this chapter, we want to preserve this multidisciplinary theme espoused by Wood Jones, Napier, and many other scholars when presenting contributions that have advanced our knowledge of the musculature of the hand of humans and other primates. Our goal is to provide a succinct yet comprehensive background that will pique interest and stimulate further research on the comparative and functional myology of the hand of primates. The chapter has three main sections: (1) summary of the basic organization and homologies of the musculature of the human hand, (2) comparison of major differences in hand musculature between primates and other pentadactyl mammals and presentation of these differences in a phylogenetic context, and (3) suggestions for areas of future research on the hand musculature of primates. The functional roles of selected hand muscles, primarily from anatomical and clinical studies of modern humans and from experimental studies of nonhuman primates, are briefly presented throughout the chapter.

2 Basic Organization and Homologies of the Human Hand Musculature

An understanding of how the musculature of the hand is organized is paramount before any comparative or functional considerations. To achieve this goal, we present basic organization of the myology of the hand in what is probably the best-documented and most accessible primate species for anatomical study using dissection: humans (Figs. 7.1 and 7.2). Countless textbooks, atlases, and guides describe, discuss, and illustrate the muscular anatomy of the human hand (e.g., Henle 1855; Wood Jones 1942; Bunnell 1944; Kaplan 1953; Landsmeer 1976; Tubiana 1981; Lewis 1989; Napier 1993; Williams 1995; Schmidt and Lanz 2004; Netter 2006; Schuenke et al. 2006), and not enough pages are contained in the present chapter to cite them all.

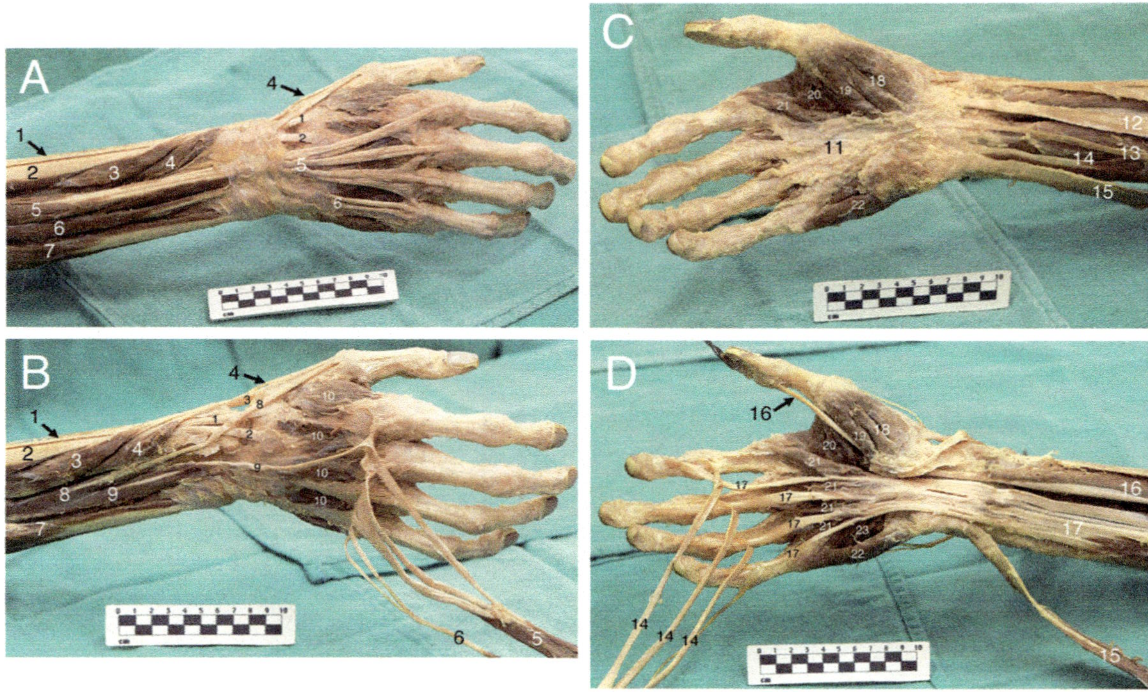

Fig. 7.1 Extrinsic and intrinsic muscles of the human hand (right side). Extrinsic muscles of the posterior compartment of forearm (superficial layer, **a**; deep layer, **b**) and anterior compartment of forearm (superficial layer. **c**; deep layer, **d**) are shown, along with intrinsic muscles of anterior compartment of the hand (**b**–**d**). **1**, extensor carpi radialis longus; **2**, extensor carpi radialis brevis; **3**, abductor pollicis longus, **4**, extensor pollicis brevis; **5**, extensor digitorum; **6**, extensor digiti minimi; **7**, extensor carpi ulnaris; **8**, extensor pollicis longus; **9**, extensor indicis; **10**, dorsal interossei; **11**, palmar aponeurosis; **12**, flexor carpi radialis; **13**, tendon of palmaris longus; **14**, flexor digitorum superficialis; **15**, flexor carpi ulnaris; **16**, flexor pollicis longus; **17**, flexor digitorum profundus; **18**, abductor pollicis brevis; **19**, flexor pollicis brevis (superficial head); **20**, adductor pollicis; **21**, lumbricals; **22**, abductor digiti minimi; **23**, flexor digiti minimi. Scale bars are 10 cm

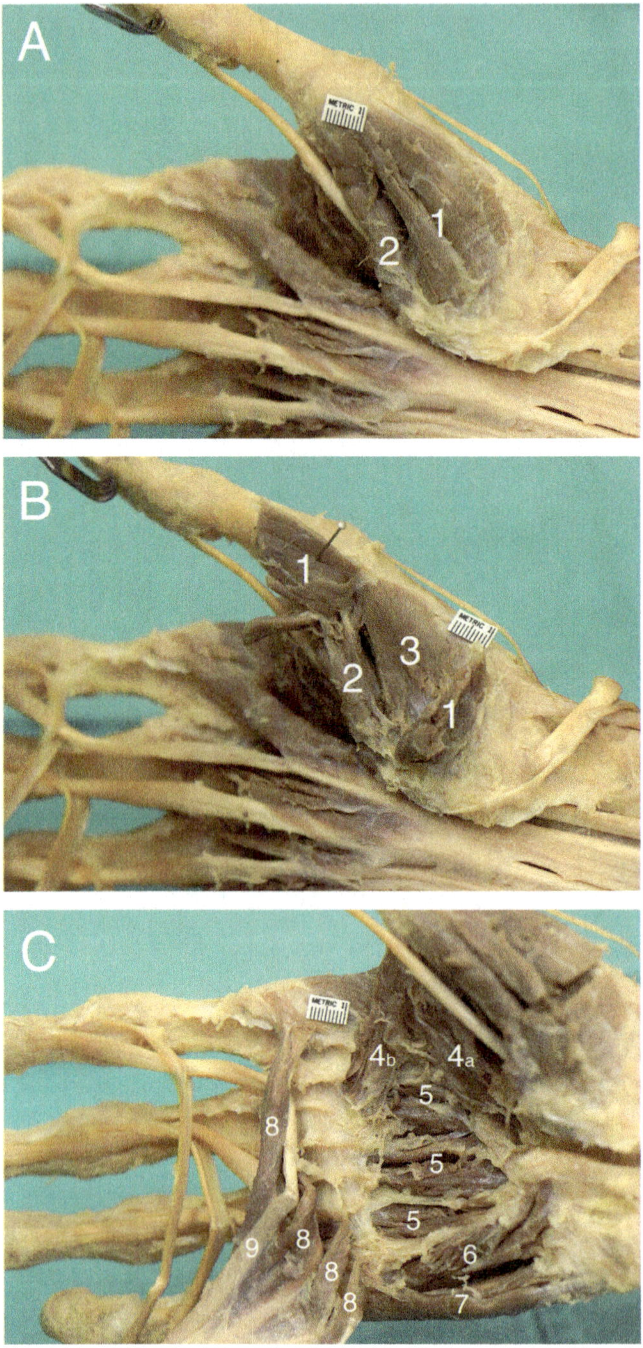

Fig. 7.2 Intrinsic muscles of the human hand (right side). Muscles of the thenar eminence (**a**, **b**), hypothenar eminence (**c**), and different muscle layers of the palm (**c**) are shown. **1**, abductor pollicis brevis; **2**, flexor pollicis brevis (superficial head); **3**, opponens pollicis; **4**, adductor pollicis, oblique head (*a*) and transverse head (*b*); **5**, palmar interossei (flexores breves profundi); **6**, flexor digiti minimi; **7**, abductor digiti minimi; **8**, lumbricals; **9**, tendon of flexor digitorum profundus. Scale bars are 1 cm

The musculature of the upper limb (or forelimb) of humans, including the hand, is derived from cervical (C5 to C8) and upper thoracic (T1) somites, specifically the portion of the somites referred to as the hypaxial dermomyotomes or hypomeres (Schoenwolf et al. 2009; see Chap. 5). As development proceeds, C5-T1 hypaxial dermomyotomes invade the forelimb bud on its dorsal and ventral aspects, creating specific muscle compartments. In turn, these muscle compartments are innervated by C5-T1 anterior (or ventral) rami, which are branches of the spinal nerves originating from the spinal cord. C5-T1 anterior rami contribute to the formation of what is known as the brachial plexus, a complex neural network that sends nerves to specific muscle compartments, including those responsible for hand movements.

There are three muscle compartments and associated nerves that are relevant to the basic organization of the human hand: (1) **posterior (dorsal) compartment of the forearm** with extrinsic hand muscles (i.e., longer muscles with proximal attachments in the forearm and distal attachments in the hand) all innervated by the radial nerve (C5-T1), (2) **anterior (ventral) compartment of the forearm** with extrinsic hand muscles innervated mostly by the median nerve (C5 or 6-T1) and some by the ulnar nerve (C8-T1), and (3) **anterior (ventral) compartment of the hand** with intrinsic muscles (i.e., shorter muscles with both attachments within the hand) innervated mostly by the ulnar nerve and some by the median nerve—the reverse of the condition observed in the anterior compartment of the forearm. Note that there is no posterior muscle compartment in the hand (as in other tetrapods; see below).

2.1 Posterior (Dorsal) Compartment of the Forearm

Superficial and deep layers are found in this forearm compartment with most muscles involved primarily in extension of the hand and digits, hence the term *extensor* compartment commonly used by anatomists (Fig. 7.1a, b; Tables 7.1 and 7.2). The superficial layer comprises five extrinsic hand muscles, all of which have a common attachment site on the lateral (radial) epicondyle of the humerus. The **extensor carpi radialis longus, extensor carpi radialis brevis**, and **extensor carpi ulnaris** all cross the radiocarpal joint to attach onto the base of metacarpals (Mc) 2, 3, and 5, respectively (Fig. 7.1a, b). In addition to extending the hand, these muscles can also move it in radial deviation (abduction; both muscles on the radial side) or ulnar deviation (adduction; the muscle on the ulnar side). **Extensor digitorum** and **extensor digiti minimi** have individualized tendons destined to the middle and distal phalanges via a complex network of intertwining fibers known as the **extensor expansion**, extensor assembly, or dorsal aponeurosis (Landsmeer 1949; Fig. 7.1a, b). Both muscles are involved in extension of the fingers (digits 2–5 by the extensor digitorum and digit 5 by the extensor digiti minimi).

The deep layer of the posterior compartment has four extrinsic muscles with attachments onto the radius, ulna, and interosseous membrane (a fibrous band connecting the two bones of the forearm). Three of these muscles have long tendons going onto the pollical metacarpal and phalanges and produce different actions across various joints of the thumb: the **abductor pollicis longus** (pollical abduction and

Table 7.1 Putative homologies of the forearm and hand muscles in representative tetrapod taxa (adapted from Diogo et al. 2009; Diogo and Abdala 2010; Diogo and Tanaka 2012; Diogo and Molnar 2014)

	Timon (lizard)	*Ornithorhynchus* (platypus)	*Rattus* (rat)	*Cynocephalus* (colugo)	*Tupaia* (tree shrew)	*Homo* (human)
Dorsal Compartment of Forearm	Extensor antebrachii et carpi radialis	Extensor carpi radialis	Extensor carpi radialis longus	Extensor carpi radialis longus	Extensor carpi radialis longus	Extensor carpi radialis longus
	–	–	Extensor carpi radialis brevis	Extensor carpi radialis brevis	Extensor carpi radialis brevis	Extensor carpi radialis brevis
	–	Brachioradialis	–	Brachioradialis	Brachioradialis	Brachioradialis
	–	Supinator	Supinator	Supinator	Supinator	Supinator
	Extensor antebrachii et carpi ulnaris	Extensor carpi ulnaris	Extensor carpi ulnaris	Extensor carpi ulnaris	Extensor carpi ulnaris	Extensor carpi ulnaris
	–	Anconeus	Anconeus	Anconeus	Anconeus	Anconeus
	Extensor digitorum	Extensor digitorum	Extensor digitorum	Extensor digitorum	Extensor digitorum	Extensor digitorum
	Extensores digitorum breves	–	–	–	–	–
	–	Extensor digit minimi	Extensor digiti minimi	Extensor digiti minimi	Extensor digiti minimi	Extensor digiti minimi
	–	–	Extensor digiti quarti	–	–	–
	–	Extensor digiti III proprius	–	–	–	–
	–	Extensor indicis	Extensor indicis	Extensor indicis	Extensor indicis	Extensor indicis
	–	Extensor pollicis longus	Extensor pollicis longus	Extensor pollicis longus	Extensor pollicis longus	Extensor pollicis longus
	Abductor pollicis longus	Abductor pollicis longus	Abductor pollicis longus	Abductor pollicis longus	Abductor pollicis longus	Abductor pollicis longus
	–	–	–	–	–	Extensor pollicis brevis
	Dorsometacarpales	–	–	–	–	–
Ventral Compartment of Forearm and Hand	Flexor digitorum longus	Flexor digitorum longus + tendons of Flexor breves superficiales	Flexor digitorum profundus	Flexor digitorum profundus	Flexor digitorum profundus	Flexor digitorum profundus
	–	–	–	–	–	Flexor pollicis longus
	–	–	Flexor digitorum superficialis	Flexor digitorum superficialis	Flexor digitorum superficialis	Flexor digitorum superficialis
	–	–	Palmaris longus	Palmaris longus	Palmaris longus	Palmaris longus
	Flexor carpi ulnaris	Flexor carpi ulnaris	Flexor carpi ulnaris	Flexor carpi ulnaris	Flexor carpi ulnaris	Flexor carpi ulnaris
	Flexor carpi radialis	Flexor carpi radialis	Flexor carpi radialis	Flexor carpi radialis	Flexor carpi radialis	Flexor carpi radialis
	Pronator teres	Pronator teres	Pronator teres	Pronator teres	Pronator teres	Pronator teres
	Pronator quadratus	–	Pronator quadratus	–	Pronator quadratus	Pronator quadratus
	Pronator accessorius	–	–	–	–	–
	Epitrochleoanconeus	Epitrochleoanconeus	Epitrochleoanconeus	Epitrochleoanconeus	Epitrochleoanconeus	–
	Palmaris profundus 1	–	–	–	–	–
	Flexores breves superficiales	–	–	–	–	–
	–	–	Palmaris brevis	Palmaris brevis	Palmaris brevis	Palmaris brevis
	–	–	–	Flexor digitorum brevis manus	Flexor digitorum brevis manus	–
	Lumbricales	Lumbricales	Lumbricales	Lumbricales	Lumbricales	Lumbricales
	Contrahentes digitorum	–	Contrahentes digitorum	Contrahentes digitorum	Contrahentes digitorum	–
	–	Adductor pollicis	Adductor pollicis	Adductor pollicis	Adductor pollicis	Adductor pollicis
	–	–	–	–	–	Adductor pollicis accessorius[1]
	Flexores breves profundi	Flexores breves profundi	Flexores breves profundi	–	–	Flexor brevis profundus 2[2]
	–	–	–	Interossei[3]	Interossei	Interossei palmares[4]
	–	–	–	–	–	Interossei dorsales[5]
	Intermetacarpales	–	Intermetacarpales	–	–	–
	–	–	Flexor pollicis brevis	Flexor pollicis brevis	Flexor pollicis brevis	Flexor pollicis brevis
	–	–	Flexor digiti minimi	Flexor digiti minimi	Flexor digiti minimi	Flexor digiti minimi
	–	–	–	–	–	Opponens pollicis[6]
	–	–	Opponens digiti minimi[6]	–	–	Opponens digiti minimi
	Abductor pollicis brevis	Abductor pollicis brevis	Abductor pollicis brevis	Abductor pollicis brevis	Abductor pollicis brevis	Abductor pollicis brevis
	Abductor digiti minimi	Abductor digiti minimi	Abductor digiti minimi	Abductor digiti minimi	Abductor digiti minimi	Abductor digiti minimi

From left to right, the black arrows indicate hypothetical correspondence (homology) of the forearm and hand muscles between taxa and do not imply direct evolution from one muscle to the next. For instance, the extensor digiti quarti and minimi of the rat correspond (are homologous) to the extensor digiti minimi of the colugo; however, fusion of two muscles to produce a single one is not implied

[1]Corresponds to the first palmar interosseous or interosseous volaris primus of Henle

[2]The flexor brevis profundus 2 corresponds to the deep head of the flexor pollicis brevis of humans and is found in many pentadactyl mammals, including *Rattus* and most primates. It is completely fused with the flexor pollicis brevis or lost in *Tupaia* and *Cynocephalus*

[3]All seven flexor breves profundi (3–9) and four intermetacarpales (1–4) are deeply mixed and form a uniform muscle layer that cannot be subdivided into palmar and dorsal interossei

[4]Interossei palmares 1–3 correspond to flexores breves profundi 4, 7, and 9

[5]Interossei dorsales 1–4 correspond to flexores breves profundi 3, 5, 6, and 8 fused with intermetacarpales 1, 2, 3, and 4

[6]Opponens pollicis and opponens digiti minimi are most likely derived from the flexor pollicis brevis and flexor digiti minimi, respectively (Čihák 1972)

Table 7.2 Putative homologies of the forearm and hand muscles in representative primate taxa (adapted from Diogo and Wood 2011, 2012a)

	Lemur (lemur)	*Tarsius* (tarsier)	*Aotus* (owl monkey)	*Macaca* (macaque)	*Hylobates* (gibbon)	*Gorilla* (gorilla)	*Pan* (chimpanzee)	*Homo* (human)
Dorsal Compartment of Forearm	Ext. carpi radialis longus	Ext. carpi radialis longus	Ext. carpi radialis longus	Ext. carpi radialis longus	Ext. carpi radialis longus	Ext. carpi radialis longus	Ext. carpi radialis longus	Ext. carpi radialis longus
	Ext. carpi radialis brevis	Ext. carpi radialis brevis	Ext. carpi radialis brevis	Ext. carpi radialis brevis	Ext. carpi radialis brevis	Ext. carpi radialis brevis	Ext. carpi radialis brevis	Ext. carpi radialis brevis
	Brachioradialis	Brachioradialis	Brachioradialis	Brachioradialis	Brachioradialis	Brachioradialis	Brachioradialis	Brachioradialis
	Supinator	Supinator	Supinator	Supinator	Supinator	Supinator	Supinator	Supinator
	Extensor carpi ulnaris	Extensor carpi ulnaris	Extensor carpi ulnaris	Extensor carpi ulnaris	Extensor carpi ulnaris	Extensor carpi ulnaris	Extensor carpi ulnaris	Extensor carpi ulnaris
	Anconeus	Anconeus	Anconeus	Anconeus	–	Anconeus	Anconeus	Anconeus
	Extensor digitorum	Extensor digitorum	Extensor digitorum	Extensor digitorum	Extensor digitorum	Extensor digitorum	Extensor digitorum	Extensor digitorum
	Extensor indicis	Extensor indicis	Extensor indicis	Extensor indicis	Extensor indicis	Extensor indicis	Extensor indicis	Extensor indicis
	Extensor digiti minimi	Extensor digit minimi	Extensor digit minimi	Extensor digit minimi	Extensor digiti minimi	Extensor digiti minimi	Extensor digiti minimi	Extensor digiti minimi
	Extensor pollicis longus	Extensor pollicis longus	Extensor pollicis longus	Extensor pollicis longus	Extensor pollicis longus	Extensor pollicis longus	Extensor pollicis longus	Extensor pollicis longus
	Abductor pollicis longus	Abductor pollicis longus	Abductor pollicis longus	Abductor pollicis longus	Abductor pollicis longus	Abductor pollicis longus	Abductor pollicis longus	Abductor pollicis longus
	–	–	–	–	Extensor pollicis brevis[1]	–	–	Extensor pollicis brevis[1]
Ventral Compartment of Forearm and Hand	Flexor digit. profundus	Flexor digit. profundus	Flexor digit. profundus	Flexor digit. profundus	Flexor digit. profundus	Flexor digit. profundus	Flexor digit. profundus	Flexor digit. profundus
	–	–	–	–	Flexor pollicis longus[1]	–	–	Flexor pollicis longus[1]
	Flexor digit. superficialis	Flexor digit. superficialis	Flexor digit. superficialis	Flexor digit. superficialis	Flexor digit. superficialis	Flexor digit. superficialis	Flexor digit. superficialis	Flexor digit. superficialis
	Palmaris longus	Palmaris longus	Palmaris longus	Palmaris longus	Palmaris longus	Palmaris longus	Palmaris longus	Palmaris longus
	Flexor carpi ulnaris	Flexor carpi ulnaris	Flexor carpi ulnaris	Flexor carpi ulnaris	Flexor carpi ulnaris	Flexor carpi ulnaris	Flexor carpi ulnaris	Flexor carpi ulnaris
	Flexor carpi radialis	Flexor carpi radialis	Flexor carpi radialis	Flexor carpi radialis	Flexor carpi radialis	Flexor carpi radialis	Flexor carpi radialis	Flexor carpi radialis
	Pronator teres	Pronator teres	Pronator teres	Pronator teres	Pronator teres	Pronator teres	Pronator teres	Pronator teres
	Pronator quadratus	Pronator quadratus	Pronator quadratus	Pronator quadratus	Pronator quadratus	Pronator quadratus	Pronator quadratus	Pronator quadratus
	Epitrochleanconeus	Epitrochleanconeus	Epitrochleanconeus	Epitrochleanconeus	–	–	Epitrochleanconeus	–
	Palmaris brevis	Palmaris brevis	Palmaris brevis	Palmaris brevis	Palmaris brevis	Palmaris brevis	Palmaris brevis	Palmaris brevis
	Lumbricales	Lumbricales	Lumbricales	Lumbricales	Lumbricales	Lumbricales	Lumbricales	Lumbricales
	Adductor pollicis[2]	Adductor pollicis	Adductor pollicis	Adductor pollicis	Adductor pollicis	Adductor pollicis	Adductor pollicis	Adductor pollicis
	–	–	–	–	–	–	–	Add. pollicis accessorius[3]
	Contrahentes digit. (2)[4]	Contrahentes digit. (8)	Contrahentes digit. (3)	Contrahentes digit. (3)	Contrahentes digit. (3)	–	Contrahentes digit. (2)	–
	Interossei accessorii	–	–	–	Interossei accessorii	–	–	–
	Flex. brevis profundus 2[5]	Flex. brevis profundus 2	–	Flex. brevis profundus 2	Flex. brevis profundus 2	Flex. brevis profundus 2	Flex. brevis profundus 2	Flex. brevis profundus 2
	Flex. breves profundi[6]	Flex. breves profundi	Interossei palmares[8]	Flex. breves profundi	Interossei palmares	Interossei palmares	Flex. breves profundi[10]	Interossei palmares
	–	–	Interossei dorsales[9,10]	–	Interossei dorsales	Interossei dorsales	–	Interossei dorsales
	Intermetacarpales[7]	Intermetacarpales	–	Intermetacarpales	–	–	Intermetacarpales[10]	–
	Flexor pollicis brevis	Flexor pollicis brevis	Flexor pollicis brevis[11]	Flexor pollicis brevis	Flexor pollicis brevis	Flexor pollicis brevis	Flexor pollicis brevis	Flexor pollicis brevis
	Flexor digiti minimi	Flexor digiti minimi	Flexor digiti minimi	Flexor digiti minimi	Flexor digiti minimi	Flexor digiti minimi	Flexor digiti minimi	Flexor digiti minimi
	Opponens pollicis	Opponens pollicis	Opponens pollicis	Opponens pollicis	Opponens pollicis	Opponens pollicis	Opponens pollicis	Opponens pollicis
	Opponens digiti minimi	Opponens digiti minimi	Opponens digiti minimi	Opponens digiti minimi	Opponens digiti minimi	Opponens digiti minimi	Opponens digiti minimi	Opponens digiti minimi
	Abductor pollicis brevis	Abductor pollicis brevis	Abductor pollicis brevis	Abductor pollicis brevis	Abductor pollicis brevis	Abductor pollicis brevis	Abductor pollicis brevis	Abductor pollicis brevis
	Abductor digiti minimi	Abductor digiti minimi	Abductor digiti minimi	Abductor digiti minimi	Abductor digiti minimi	Abductor digiti minimi	Abductor digiti minimi	Abductor digiti minimi

From left to right, the black arrows indicate hypothetical correspondence (homology) of the forearm and hand muscles between taxa and do not imply direct evolution from one muscle to the next. For instance, the abductor pollicis longus and extensor pollicis brevis of the gibbon correspond (are homologous) to the abductor pollicis longus of the gorilla; however, fusion of two muscles to produce a single one is not implied

[1]Extensor pollicis brevis and flexor pollicis longus evolved independently in humans and gibbons

[2]Corresponds to the first contrahens muscle

[3]Corresponds to the first palmar interosseous or interosseous volaris primus of Henle

[4]Number in parentheses indicate number of contrahentes muscles (excluding adductor pollicis)

[5]Corresponds to the deep head of the flexor pollicis brevis

[6]Include seven flexor breves profundi (3–9)

[7]Include four intermetacarpales (1–4)

[8]Interossei palmares 1–3 correspond to flexores breves profundi 4, 7, and 9

[9]Interossei dorsales 1–4 correspond to flexores breves profundi 3, 5, 6, and 8 fused with intermetacapales 1, 2, 3, and 4

[10]Interossei dorsales either (1) evolved independently in New World monkeys and in hominoids, with a reversion to the plesiomorphic/ancestral condition (i.e., distinct intermetacarpales and flexor breves profundi) in chimpanzees or (2) evolved in anthropoids, with a reversion to the plesiomorphic/ancestral condition in Old World monkeys and chimpanzees. Both options are equally parsimonious and involve three evolutionary steps

[11]Flexor brevis profundus 2 fuses with the flexor pollicis brevis in New World monkeys to form a single muscle

extension), **extensor pollicis longus** (pollical extension including the distal phalanx), and **extensor pollicis brevis** (pollical extension excluding the distal phalanx) (Fig. 7.1a, b). The fourth muscle, the **extensor indicis**, sends a tendon that blends with the extensor expansion of the index finger (Fig. 7.1b). This additional tendon provides the index finger with greater independence during extension compared to digits 3 and 4, which can only be extended by the extensor digitorum.

Three additional muscles with no attachment on the hand are also found in the superficial and deep layers of the posterior compartment of the forearm: the anconeus, brachioradialis, and supinator (Tables 7.1 and 7.2). The latter muscle assists the biceps brachii muscle (a powerful muscle of the anterior compartment of the arm) in moving the forearm and hand into supination.

2.2 *Anterior (Ventral) Compartment of Forearm*

Most muscles found in this compartment are involved in flexion of the hand and digits, hence the term *flexor* compartment (Fig. 7.1c, d; Tables 7.1 and 7.2). Like the posterior compartment, superficial and deep muscle layers are found on the anterior aspect of the forearm. Four extrinsic muscles moving the hand and digits at various joints are found superficially: the **flexor carpi radialis**, **flexor carpi ulnaris**, **palmaris longus**, and **flexor digitorum superficialis**. All these muscles have a common attachment tendon onto the medial (ulnar) epicondyle of the humerus. As such, anterior and posterior compartment muscles of the forearm can easily be distinguished on the basis of which humeral epicondyle they originate from. Both flexor carpi radialis and flexor carpi ulnaris cross the radiocarpal joint before attaching onto the base of Mc2-3 and Mc5 (with the latter muscle having other attachments onto carpal bones as well). Not only can these muscles flex the hand, but they can also move it in abduction (flexor carpi radialis) or adduction (flexor carpi ulnaris). Palmaris longus is a thin muscle with a long tendon blending with the **palmar aponeurosis**, a triangular fibrous sheet just deep to the skin of the palm (Fig. 7.1c). The muscle is sometimes absent (between 5 and 20 % of humans depending on study; see Gibbs 1999), and its contraction can cause weak hand flexion at the radiocarpal joint. Flexor digitorum superficialis (FDS) is the deepest and bulkiest of all superficial forearm flexors. FDS has two distinctive heads: a humeroulnar head with attachment onto the proximal ulna (in addition to the medial epicondyle of the humerus) and a radial head with attachment onto the radial shaft. Each head gives off a pair of long tendons going to specific digits. All four tendons travel through the carpal tunnel—an enclosed space made of carpal bones and covered by a thick fibrous band known as the **flexor retinaculum**—and then split before attaching on the palmar (anterior) surface of each middle phalanx of digits 2–5 (Fig. 7.1d). The splitting of the tendon allows the tendon of the flexor digitorum profundus to pass through (see below).

The deep muscle layer has two extrinsic muscles with attachments onto the shaft of the radius or ulna and the interosseous membrane: the **flexor pollicis longus** (FPL) and **flexor digitorum profundus** (FDP). Like FDS, the tendons of FPL and

FDP enter the carpal tunnel before insertion onto the palmar aspect of the distal phalanges. Each tendon of FDP passes through the split tendon of FDS before reaching a distal phalanx. In this way, each finger receives two long flexor tendons (FDS and FDP), whereas the thumb receives only one (FPL). By virtue of their path along the palmar side of the rays, the long flexor tendons have the ability to flex several joints, including carpometacarpal (CM), metacapophalangeal (McP), and interphalangeal (IP) joints (proximal and distal in the case of FDP).

Two additional muscles with no attachment on the hand are also found in the superficial and deep layers of the anterior compartment of the forearm: the pronator teres and pronator quadratus. As their names indicate, these muscles are involved in moving the forearm and hand into pronation.

2.3 Anterior (Ventral) Compartment of the Hand

Many intrinsic muscles moving the digits are found in the palm of the hand and are arranged in more superficial and deeper layers (Figs. 7.1c, d and 7.2; Tables 7.1 and 7.2). Major intrinsic hand muscles include the thenar, hypothenar, palmaris brevis, lumbrical, adductor pollicis, and interosseous muscles.

Muscles forming the **thenar eminence**, the prominent bulge at the base of the thumb, comprise the **abductor pollicis brevis**, **flexor pollicis brevis**, and **opponens pollicis** (Fig. 7.2a, b). Abductor pollicis brevis is the more superficial of the thenar muscles with attachments from the flexor retinaculum and carpal bones to the proximal pollical phalanx and dorsal aponeurosis of the thumb. The flexor pollicis brevis muscle is more ulnar to the abductor pollicis brevis and has two heads: superficial and deep heads separated at their origin by the tendon of FPL. Usually, the superficial head is innervated by the median and the deep head by the ulnar nerve, although dual innervation to each head is not uncommon (Belson et al. 1976). The deep head of the flexor pollicis brevis is most likely derived from a deeper palmar layer known as flexores breves profundi (Čihák 1972; Diogo and Wood 2011, 2012a; Tables 7.1 and 7.2; see Sects. 2.4 and 3.3 for more details). Both heads attach from carpal bones (with the superficial head having attachment onto the flexor retinaculum) and proceed toward the radial side of the McP joint of the thumb to attach on the sesamoid (the pollical McP joint has two notable radial and ulnar sesamoid bones) and base of the proximal phalanx (PP). The opponens pollicis is deepest and requires reflection of the abductor pollicis brevis to have a full view of the muscle (Fig. 7.2b). Muscle fibers cross the trapeziometacarpal (CM1) joint only, from the carpal bones and flexor retinaculum to the radial side of the pollical metacarpal shaft. Each muscle can produce independent action on the thumb (abduction, flexion, and medial rotation); acting together or synergistically, the compound action of thenar muscles produces opposition (Long et al. 1970), a vital movement for achieving precision grips and pulp-to-pulp contact between the thumb and the fingers (Napier 1955, 1956). EMG data have shown steady increase in activity of all three thenar muscles as the tip of the thumb opposes the tip of the index finger, middle finger, and so on (Basmajian 1979). This is something you can easily test on yourself: press the thenar eminence of the right hand with the left index finger and then gently oppose the tip of your thumb with the tip of each finger of your right hand. You will notice a progressive tightening of the muscles of the thenar eminence as opposition takes place between the thumb and ulnarmost fingers. Clever cooks use this trick to assess meat doneness by feel when grilling.

Three muscles form the less prominent **hypothenar eminence** at the base of the little finger: the **abductor digiti minimi**, **flexor digiti minimi**, and **opponens digiti minimi** (Figs. 7.1c, d and 7.2c). A thin muscle just deep to the skin of the palm on the ulnar side—palmaris brevis—can also be observed during careful dissection of the hand. Attachment patterns and topographical relationships of the hypothenar muscles are generally similar to those of their thenar counterparts. Differences also exist, with flexor digiti minimi having a single head, no sesamoid bone attachment (as no sesamoid is present in the McP5 joint), and may be absent or fused to the abductor digiti minimi (Fahrer 1981). Hypothenar muscles move the little finger in planes similar to those of the thumb, however, with more limited ranges of motion as the shape of the CM5 joint, although mildly saddle shaped, is much different than the saddle-shaped trapeziometacarpal joint (see Chap. 3). Like their thenar counterparts, hypothenar muscles show a progressive increase in activity as the thumb opposes the more ulnar fingers, with the opponens digiti minimi showing disproportionally higher activity as the tip of the thumb touches the side and tip of the little finger (Basmajian 1979).

The **lumbricals** are found deeper into the palm and consist of four muscle slips (that are wormlike in appearance as the name implies) (Figs. 7.1d and 7.2c). Lumbricals are unusual among hand muscles as they link two tendons. From its attachment from a tendon (or adjacent tendons) of FDP on the palmar side, each lumbrical proceeds dorsally and blends on the radial side of the extensor expansion of a corresponding finger. The triangular area where the lumbrical attaches on the extensor expansion is sometimes referred to as the wing tendon (Landsmeer 1955; Stack 1962). Because of their attachments and path, lumbricals have the ability to extend the IP joints of a finger whether the McP joint is flexed or extended (Backhouse and Catton 1954; Landsmeer and Long 1965). EMG studies have clarified the functional role of the lumbricals in transferring viscoelastic force from an FDP tendon palmarly to the extensor expansion dorsally while the finger is extending at McP and IP joints via contraction of the extensor digitorum (Long and Brown 1964; Landsmeer and Long 1965; Long et al. 1970; see Long 1981 and Valentin 1981 for comprehensive reviews of those findings).

Deep to the FDP tendons and lumbricals, the **adductor pollicis** muscle can be observed at the base of the thumb (Fig. 7.2c). It has a smaller oblique head (with attachment onto the capitate and adjacent Mc bases) and a larger triangular transverse head (with attachment onto the shaft of the Mc3) (Fig. 7.2c). In some people, a small bundle originating from the deep head of the flexor pollicis brevis can be observed joining the oblique head of the adductor pollicis (Day and Napier 1961). Both heads merge as they reach the ulnar sesamoid of the pollical McP joint and ulnar side of the PP base. Because of this pollical attachment, some anatomists consider it to be a thenar muscle (e.g., Wood Jones 1942). The adductor pollicis can flex the thumb at the CM1 and McP1 joints as it brings it from an abducted to adducted position (toward the palm).

The deepest intrinsic muscles of the palm are the **palmar** and **dorsal interossei** muscles found on various surfaces of metacarpal shafts (Figs. 7.1b and 7.2c). A palmar interosseous muscle can be found for each digit, except the middle finger. The first palmar interosseous often runs from the pollical metacarpal base to the ulnar sesamoid of the McP1 joint, PP base, and extensor expansion. This highly variable muscle first described by Henle (1855) is often overlooked in anatomy textbooks, although the muscle is present in most humans (Susman et al. 1999; Bello-Hellegouarch et al. 2013). It is sometimes termed "interosseous volaris primus of Henle" (e.g., Susman et al. 1999), although the newer term "adductor pollicis accessorius" (Diogo et al. 2012; Bello-Hellegouarch et al. 2013) may be more appropriate because of its derivation from the adductor pollicis muscle, innervation pattern, and comparative evidence in other primates (Diogo and Wood 2011, 2012a; see Sect. 3.3). The remaining three palmar interossei attach on the wing tendon of the extensor expansion of the index finger (ulnar side), ring finger (radial side), and little finger (radial side).

Dorsal intersossei are the deepest muscles of the palm and are better seen from the posterior side of the hand in between the metacarpals, despite being part of the anterior (ventral) musculature (Fig. 7.1b). A total of four dorsal interossei are present: first for the index finger (radial side), second and third on either side of the middle finger, and the fourth for the ring finger (ulnar side). They are bipennate muscles with attachments on adjacent metacarpal shafts (e.g., first dorsal interosseous muscle attaches on Mc1 and Mc2) with two different components: dorsal (deep) component with attachment onto a PP base of a finger and superficial (palmar or volar) component with attachment onto the wing tendon of the extensor expansion. Typically, some dorsal interossei have both components attaching onto the PP base only (e.g., first dorsal interosseus) or extensor expansion only (e.g., third dorsal interosseous) (Valentin 1981). The primary action of the dorsal interossei is described as abducting the index, middle, and ring fingers (away from the central axis of the hand that passes through the third ray), whereas the palmar interossei adduct the index, ring, and little fingers toward the middle finger. The absence of a palmar interosseous muscle for the middle finger explains the lack of adduction for this digit. EMG data show that both sets of interossei also enable finger flexion at McP joints, extension at IP joints, and some rotation at McP joints (Landsmeer and Long 1965; Long et al. 1970; Long 1981). Although subtle, the latter action is vital in conforming the palmar surface of the fingers to an object (e.g., a ball) during power grips (Long et al. 1970; Long 1981; Valentin 1981).

2.4 *Evolution and Homologies of the Human Hand Musculature*

Our knowledge of the basic organization of the hand musculature can be extended by comparing humans to other primates, pentadactyl mammals, and even other tetrapods such as reptiles and amphibians. Comparisons of homologous hand muscles (i.e., muscles acquired through common ancestry) between animals of different phylogenetic affinities are essential to provide insights into the evolution of the primate and human hand.

For most muscles, homologies can be established on the basis of topographical and other anatomical criteria readily accessible following skillful dissection. For example, the fact that the flexor carpi radialis attaching onto the medial epicondyle of the humerus is innervated by a branch of the median nerve, has the pronator teres and palmaris longus as neighboring muscles, and has a tendinous insertion onto the palmar aspect of the base of Mc2-3 are several of the many criteria that can used to establish homology of the muscle between a human and a rhesus macaque [see Howell and Straus (1933) for a detailed description of the hand musculature of this nonhuman primate].

For some hand muscles, however, establishing homologies can be challenging. A classic example is the interossei muscles, which "… have presented generations of morphologists with a perplexing puzzle" (Lewis 1989: 145). There are three main hypotheses regarding the homologies of the interossei muscles of humans and other vertebrates [see Lewis (1989) for a thorough review of various theories here summarized]. In many nonhuman primates and other tetrapods, two distinctive muscle layers can be found in the deepest aspect of the palm: flexores breves profundi and intermetacarpales (Cunningham 1882; Windle 1883; McMurrich 1903b; Forster 1917; Campbell 1939; Haines 1950; Lewis 1989; Diogo et al. 2009; Diogo and Wood 2011). Some comparative anatomists proposed that in humans, the palmar interossei and dorsal interossei are homologous, respectively, to specific flexores breves profundi (with remainder flexors disappearing) and intermetacarpales (Cunningham 1882). Others adopted a different viewpoint, suggesting that only flexores breves profundi are homologous to both palmar and dorsal interossei and that the intermetacarpales have no representation in the human hand (Windle 1883; Campbell 1939; Haines 1950). A third view—originally proposed by McMurrich (1903b) from observations of human embryonic hands and supported by extensive comparative evidence compiled by Forster (1917)—emphasizes the compound nature of the dorsal interossei. Each dorsal interosseous muscle comprises an individual slip of flexor brevis profundus muscle that fuses with an intermetacarpal muscle; some individual slips of flexores breves profundi that persist give rise to the palmar interossei (with other slips contributing to the thenar and hypothenar muscles; see below). This view is consistent with the bipennate configuration and dual attachment that typifies human dorsal interossei (see above) and is confirmed by detailed developmental data of human embryonic hands (Čihák 1972), as well as recent comparative studies of major tetrapod groups (Diogo et al. 2009; Diogo and Abdala 2010) and studies of human variations and birth defects (Diogo and Wood 2012b).

Developmental data represent a powerful source of information to sort out hand muscle homologies and should be considered, whenever available, with other criteria listed above. In this regard, the ontogenetic study of Čihák (1972) of human embryonic hand series remains an invaluable reference in spite of being over 40 years old. In addition to clarifying the homology of the interossei muscles, this study showed the transient nature of the contrahentes (i.e., muscle layer typically found between the FDP tendons and interossei in most nonhuman primates and many pentadactyl mammals) during human development. The adductor pollicis (first contrahens) represents the only remaining contrahentes in the human adult hand (Forster 1917; Napier 1961; Čihák 1972). This developmental study also provided evidence that the thenar muscles are derived from two different muscle primordia: (1) a superficial layer for the abductor pollicis brevis and (2) a deeper layer consisting of flexores breves profundi 1 (for the superficial head of the flexor pollicis brevis and opponens pollicis) and 2 (for the deep head of the flexor pollicis brevis) (Čihák 1972; see following section for further details).

Diogo and colleagues carried out detailed dissections, as well as regenerative and developmental studies, of the forelimb on large sample sizes of tetrapods (including many nonhuman primates and humans) to clarify the origin and evolution of the human hand musculature (Diogo et al. 2009, 2014a, b; Diogo and Abdala 2010; Diogo and Wood 2011, 2012a; Diogo and Tanaka 2012, 2014; Diogo and Molnar 2014; Diogo and Ziermann 2014). Homologies of the forearm and hand musculature between a lizard, platypus, rat, colugo, tree shrew, and human are provided in Table 7.1 and those of representative primate taxa in Table 7.2. These homologies form the framework of the next section, which focuses on morphological differences and evolution of the hand musculature in primates and selected pentadactyl mammals, with an emphasis on humans. Readers should keep in mind that homologies presented in these tables are hypotheses that require further testing in future studies. Moreover, correspondence of muscles between adjacent taxa in the tables does not imply direct evolution and ancestor-descendant relationships from one taxon to the other.

3 Major Differences in the Hand Musculature Between Humans, Nonhuman Primates, and Other Pentadactyl Mammals

From the homology data compiled in Tables 7.1 and 7.2, readers can easily verify one major characteristic of the primate and mammalian hand musculature: it is conservative. Most hand muscles observed in humans can be found and homologized to those of other primates, euarchontans (tree shrews and colugos), and pentadactyl mammals (e.g., rats and even monotremes). The notion of "*scalae naturae*" and "progress" toward a higher complexity of the human hand promoted by early comparative anatomists, which has persisted in some contemporary anatomy textbooks, is not borne out by the evidence (Diogo et al. 2015). For instance, humans usually have 21 intrinsic hand muscles, whereas tetrapods such as lizards may have more than 40 muscles (Howell 1936; Straus 1941a, b, 1942b; Haines 1950; Jouffroy 1971; Diogo et al. 2009; Diogo and Abdala 2010; Table 7.1) and phylogenetically more basal primate taxa such as *Tarsius* may have up to 36 muscles (Howell and Straus 1933; Jouffroy 1962, 1975; Schultz 1984; Diogo and Wood 2011, 2012a; Table 7.2). That being said, humans, like many other primates, have hand muscle characteristics that are apomorphic (i.e., characteristic that is derived or specialized relative to a plesiomorphic or ancestral condition) or even autapomorphic (i.e., characteristic that is derived and exclusive to a clade). The following section surveys major differences by muscle compartment for humans, as well as other non-human primates, tree shrews (*Tupaia*), colugos (*Cynocephalus*), and other selected pentadactyl mammals (e.g., *Rattus*). Tables 7.3, 7.4, and 7.5 provide a summary of some of the major differences for all these taxa, including muscle character polarity (**0**, plesiomorphic; **1**, apomorphic; **1–3**, apomorphic for a multistate character).

3.1 Dorsal Compartment of the Forearm

The dorsal (extensor) compartment of the forearm of primates and other pentadactyl mammals contains muscles that primarily extend the digits. In reptiles, such as the lizard *Timon*, the extensor compartment muscles are divided into two major layers: the superficial layer (including the radial sector or antebrachii et carpi radialis, ulnar sector or antebrachii et carpi ulnaris, and intermediate segment or extensor digitorum) and deep layer (including the extensor digitorum brevis and abductor pollicis longus) (Brooks 1889; Howell 1936; Haines 1939; Straus 1941a, b; Lewis 1989; Diogo et al. 2009; Diogo and Abdala 2010; Table 7.1). The extensor carpi radialis longus, extensor carpi radialis brevis, brachioradialis, and supinator of primates and other pentadactyl mammals correspond to the extensor antebrachii et carpi radialis found in reptiles (Lewis 1989; Diogo et al. 2009; Diogo and Abdala 2010; Table 7.1). Similarly, the extensor carpi ulnaris and anconeus correspond to the extensor antebrachii et carpi ulnaris, whereas the extensor pollicis longus, extensor indicis, and extensor digiti minimi correspond to the extensores digitorum breves (Lewis 1989; Diogo et al. 2009; Diogo and Abdala 2010; Table 7.1). Both the extensor digitorum and abductor pollicis longus are distinct and independent muscles (from the superficial and deep layers, respectively) that can be homologized across tetrapods (Lewis 1989; Diogo et al. 2009; Diogo and Abdala 2010; Tables 7.1 and 7.2).

Much of the variation in the musculature of the dorsal compartment of the forearm in primates, other euarchontans, and *Rattus* relates to the number of tendons originating from some of the muscles of the deep layer (Table 7.3). These muscles are sometimes designated as bundles of "extensor digitorum proprius," which are said to constitute the deep layer of the dorsal forearm musculature, together with the abductor pollicis longus (Brooks 1889; Howell and Straus 1933; Howell 1936; Haines 1939; Straus 1941a, b; Lewis 1989). We prefer to use the more familiar terms extensor pollicis longus, extensor digiti minimi, and extensor indicis as these muscles can be homologized across pentadactyl mammals and are consistent with the formal nomenclature of human anatomy.

Many primates display the plesiomorphic condition with the extensor pollicis longus sending a single tendon to the pollex and extensor digiti minimi sending tendons to digits 4 and 5 (Diogo and Wood 2011, 2012a; Fig. 7.3; Table 7.3). In New World monkeys and *Colobus*, however, the extensor pollicis longus is deeply blended with the extensor indicis and contributes a tendon to the pollex and one to

Table 7.3 Summary of some of the morphological differences for selected muscles of the dorsal (extensor) compartment of the forearm in primates and other pentadactyl mammals (from Diogo and Wood 2011)

Muscle	Comparative remarks[a]
Extensor carpi ulnaris	**0**, Attachment onto the lateral epicondyle of the humerus and ulna (many pentadactyl mammals, including *Cynocephalus*, *Tupaia*, and most primates)
	1, Ulnar attachment absent (New World and Old World monkeys such as *Aotus*, *Pithecia*, *Saimiri*, *Cercopithecus*, *Colobus*, *Macaca*, and *Papio*)
Extensor digiti minimi	**0**, Separate extensor digiti quarti going to digit 4 and "extensor digiti quinti proprius" going to digit 5 (many pentadactyl mammals such as *Rattus*)
	1, Fusion of both muscles into an extensor digiti minimi (*Cynocephalus*, *Tupaia*, and primates)
	0: Tendon insertion onto digits 4, 5 (*Tupaia* and many primates, including *Lemur*, *Propithecus*, *Tarsius*, *Aotus*, *Pithecia*, *Saimiri*, *Cercopithecus*, *Colobus*, *Macaca*, *Papio*, and *Pongo*)
	1, Tendon insertion onto digit 5 only (*Loris*, *Nycticebus*, *Potto*, *Hylobates*, *Gorilla*, *Pan*, and *Homo*)
	2, Tendon insertion onto digits 3, 4, 5 (*Cynocephalus*)
Extensor indicis	**0**, Tendon insertion onto digits 2, 3 (*Rattus*, *Tupaia*, and many primates such as *Loris*, *Nycticebus*, *Tarsius*, *Cercopithecus*, *Colobus*, *Macaca*, *Papio*, and *Pongo*)
	1, Tendon insertion onto digits 1, 2, 3 (*Cynocephalus*)
	2, Tendon insertion onto digits 2, 3, 4 (*Hylobates* and New World monkeys such as *Aotus*, *Callithrix*, *Pithecia*, and *Saimiri*)
	3, Tendon insertion onto digit 2 only (*Gorilla*, *Pan*, and *Homo*)
Extensor pollicis longus	**0**, Separate muscle from extensor indicis (*Rattus*, *Cynocephalus*, and most primates)
	1, Deeply blended with extensor indicis (*Tupaia*, *Colobus*, and New World monkeys such as *Aotus*, *Callithrix*, *Pithecia*, and *Saimiri*)
	0, Only one tendon goes to digit 2, either from the extensor pollicis longus or extensor indicis (most primates)
	1, Two tendons go to digit 2, one from the extensor pollicis longus and the other from the extensor indicis (*Colobus* and New World monkeys such as *Aotus*, *Callithrix*, *Pithecia*, and *Saimiri*)
Abductor pollicis longus	**0**, Attachment onto the carpal region and/or metacarpal 1 (pentadactyl mammals, including *Tupaia* and most primates)
	1, Same attachment with distal extension onto the proximal pollical phalanx (*Gorilla* and *Homo*)
Extensor pollicis brevis	**0**, Absent (pentadactyl mammals, including most primates)
	1, Distinct muscle (*Hylobates* and *Homo*)

0, plesiomorphic (primitive) muscle characteristic; **1**, apomorphic (derived) muscle characteristic; **1–3**, apomorphic (derived) for multistate muscle characteristic

[a]Details of data sources (bibliographical and list of specimens dissected) can be found in Diogo and Wood (2011)

Table 7.4 Summary of some of the morphological differences for selected muscles of the ventral (flexor) compartment of the forearm in primates and other pentadactyl mammals (from Diogo and Wood 2011)

Muscle	Comparative remarks[a]
Flexor carpi ulnaris	**0**: Origin of the muscle from the medial epicondyle of the humerus and ulna (pentadactyl mammals, including *Rattus*, *Tupaia*, and most primates)
	1: Origin of the muscle from the humerus is absent (*Cynocephalus*)
Flexor carpi radialis	**0**: Origin of the muscle from the humerus only (pentadactyl mammals, including *Rattus*, *Tupaia*, *Cynocephalus*, and most primates)
	1: Origin of the muscle from the humerus and radius (most hominoids, including *Pongo*, *Gorilla*, *Pan*, and *Homo*)
	0: Tendon insertion onto metacarpal 2 or 3 (*Rattus* and some primates, including *Lemur*, *Propithecus*, *Loris*, and *Nycticebus*)
	1: Tendon insertion onto both metacarpals 2 and 3 (*Tupaia* and most primates)
	2: Tendon insertion onto trapezium and/or trapezoid (*Cynocephalus*)
Flexor digitorum superficialis	**0**: Origin of the muscle from the medial epicondyle of the humerus (including the common flexor tendon) (pentadactyl mammals, including *Rattus*, *Tupaia*, *Cynocephalus*, and most primates)
	1: Origin of the muscle from the medial epicondyle of the humerus (including the common flexor tendon), radius, and ulna (hominoids, including *Hylobates*, *Pongo*, *Gorilla*, *Pan*, and *Homo*)
	0: Tendon insertion onto digits 2, 3, and 4 only (pentadactyl mammals such *Rattus* and *Tupaia*)
	1: Tendon insertion onto digits 2, 3, 4, and 5 (*Cynocephalus* and primates)
	0: Tendon insertion onto digit 2 (pentadactyl mammals such as *Rattus*, *Tupaia*, *Cynocephalus*, and most primates)
	1: Tendon insertion onto digit 2 absent (*Loris* and *Perodicticus*)
Flexor digitorum profundus	**0**: Origin of the muscle from medial epicondyle of the humerus (and/or the common flexor tendon), radius, ulna, and/or interosseous membrane (pentadactyl mammals, including *Rattus*, *Tupaia*, *Cynocephalus*, and most primates)
	1: Origin of the muscle from the radius and/or ulna and often from the interosseous membrane (*Macaca*, *Pongo*, *Gorilla*, and *Homo*)
	0: Presence of the pollical tendon (or flexor pollicis longus tendon) (pentadactyl mammals such as *Rattus*, *Tupaia*, *Cynocephalus*, and most primates)
	1: Vestigial or absent pollical tendon (*Colobus*, *Pongo*, *Gorilla*, and *Pan*)
	0: Innervation by median and ulnar nerves (pentadactyl mammals, including *Rattus*, *Cynocephalus*, *Tupaia*, and most primates)
	1: Innervation by the median nerve only (*Colobus*, *Macaca*, and *Papio*)
Flexor pollicis longus	**0**: Integral part of the flexor digitorum profundus (pentadactyl mammals such as *Rattus*, *Cynocephalus*, *Tupaia*, and most primates)
	1: Distinct muscle from the flexor digitorum profundus (*Hylobates* and *Homo*)

0, plesiomorphic (primitive) muscle characteristic; **1**, apomorphic (derived) muscle characteristic; **1–2**, apomorphic (derived) for multistate muscle characteristic

[a]Details of data sources (bibliographical and list of specimens dissected) can be found in Diogo and Wood (2011)

Table 7.5 Summary of some of the morphological differences for selected muscles of the ventral compartment of the hand in primates and other pentadactyl mammals (from Diogo and Wood 2011)

Muscle	Comparative remarks[a]
Lumbricales	**0**: Total of four muscles (pentadactyl mammals, including *Rattus*, *Tupaia*, and most primates)
	1: Total of three muscles, with lumbrical to digit 5 missing (*Hylobates*)
	2: Total of seven muscles, with two for digits 2, 3, and 4 and one for digit 5 (*Cynocephalus*)
Contrahentes digitorum	**0**: Total of 3, 4, or 5 muscles, including the adductor pollicis (pentadactyl mammals, including *Rattus*, *Tupaia*, *Cynocephalus*, and most primates)
	1: Only one muscle—adductor pollicis—is present (*Pongo*, *Gorilla*, and *Homo*)
Adductor pollicis	**0**: Insertion of muscle onto the pollical proximal phalanx, MP joint capsule, and/or associated ulnar sesamoid (pentadactyl mammals, including *Rattus*, *Tupaia*, *Cynocephalus*, and most primates)
	1: Insertion of the muscle onto the pollical metacarpal (*Hylobates*)
	0: No distinct (blended) oblique and transverse heads (pentadactyl mammals such as *Rattus*, *Tupaia*, *Cynocephalus*, and strepsirrhine primates, including *Lemur*, *Propithecus*, *Loris*, and *Nycticebus*)
	1: Partly differentiated oblique and transverse heads, but blended at insertion onto the pollex (*Tarsius* and New World monkeys, including *Aotus*, *Pithecia*, and *Saimiri*)
	2: Completely differentiated oblique and transverse heads (all hominoids and Old World monkeys, including *Colobus*, *Cercopithecus*, *Macaca*, and *Papio*)
Flexor brevis profundis 2[b]	**0**: Distinct and separate muscle (pentadactyl mammals such as *Rattus* and most primates, including *Lemur*, *Propithecus*, *Loris*, *Nycticebus*, *Tarsius*, *Cercopithecus*, *Macaca*, *Papio*, *Colobus*, and all hominoids)
	1: Absent or completely fused to the flexor pollicis brevis muscle (*Tupaia*, *Cynocephalus*, and New World monkeys, including *Aotus*, *Callithrix*, *Pithecia*, and *Saimiri*)
Dorsal interossei	**0**: Flexores breves profundi are not fused with the intermetacarpales (pentadactyl mammals such as *Rattus* and many primates, including *Lemur*, *Propithecus*, *Loris*, *Nycticebus*, *Tarsius*, *Colobus*, *Cercopithecus*, *Macaca*, *Papio*, and *Pan*)
	1: Flexores breves profundi 3, 5, 6, 8 are fused with all four intermetacarpales to form the dorsal interossei (*Tupaia*, *Cynocephalus*, *Aotus*, *Callithrix*, *Pithecia*, *Saimiri*, *Hylobates*, *Pongo*, *Gorilla*, and *Homo*)
	0: Muscles attach on both sides of third ray, which represents the functional axis (pentadactyl mammals such as *Rattus*, *Tupaia*, *Cynocephalus*, and haplorhine primates)
	1: Muscles attach on both sides of fourth ray, which represents the functional axis (strepsirrhine primates such as *Lemur*, *Propithecus*, *Loris*, and *Nycticebus*)

(continued)

Table 7.5 (continued)

Muscle	Comparative remarks[a]
Opponens pollicis	**0**: Not present as a distinct muscle (most pentadactyl mammals, including *Rattus*, *Tupaia*, *Cynocephalus*, and *Callithrix*)
	1: Present as a distinct muscle (all primates, except *Callithrix*)
	0: Insertion of muscle onto the pollical metacarpal shaft, ½ to ¾ proximal (*Lemur*, *Propithecus*, *Tarsius*, *Aotus*, *Pithecia*, and *Saimiri*)
	1: Insertion of muscle extends to the distal portion of the pollical metacarpal (*Colobus*, *Cercopithecus*, *Macaca*, *Papio*, *Pongo*, *Gorilla*, *Pan*, and *Homo*), sometimes involving attachment onto the pollical phalanges (*Hylobates*) or exclusively onto the distal portion of the pollical metacarpal (*Loris* and *Nycticebus*)
Flexor digiti minimi	**0**: Origin of the muscle does not include attachment onto the pisiform (pentadactyl mammals such as *Rattus*, *Tupaia*, and most primates)
	1: Origin of the muscle often includes partial attachment onto the pisiform (*Cynocephalus* and *Hylobates*)
	0: Insertion of the muscle onto the MP joint, base of proximal phalanx, and extensor expansion of digit 5 (pentadactyl mammals, including most primates)
	1: Insertion of the muscle extends distally onto the middle and/or distal phalanx of digit 5 (*Nycticebus* and *Hylobates*)
Opponens digiti minimi	**0**: Not present as a distinct muscle (pentadactyl mammals such as *Tupaia* and *Cynocephalus*)
	1: Separate muscle present (pentadactyl mammals such as *Rattus* and all primates)
	0, When present, the muscle is undivided (*Rattus*, *Tarsius*, strepsirrhine primates, including *Lemur*, *Propithecus*, *Loris* and *Nycticebus*, and New World monkeys, including *Aotus*, *Callithrix*, *Pithecia*, and *Saimiri*)
	1: When present, the muscle is slightly divided into superficial and deep bundles (hominoids)
	2: When present, the muscle is deeply divided into separate superficial and deep heads (Old World monkeys, including *Colobus*, *Cercopithecus*, *Macaca*, and *Papio*)
	0: When present, the muscle inserts onto the distal metacarpal and/or proximal phalanx of ray 5 (*Rattus*)
	1: When present, the muscle inserts along the entire proximodistal length of metacarpal 5 (all primates)
Abductor digiti minimi	**0**: The muscle is not or slightly divided at its origin (pentadactyl mammals, including *Rattus*, *Tupaia*, *Cynocephalus*, and most primates)
	1: The muscle is well differentiated into two heads connected by a common tendon of insertion (*Macaca* and *Papio*)

0, plesiomorphic (primitive) muscle characteristic; **1**, apomorphic (derived) muscle characteristic; **1–2**, apomorphic (derived) for multistate muscle characteristic

[a]Details of data sources (bibliographical and list of specimens dissected) can be found in Diogo and Wood (2011)

[b]Corresponds to the deep head of the flexor pollicis brevis muscle

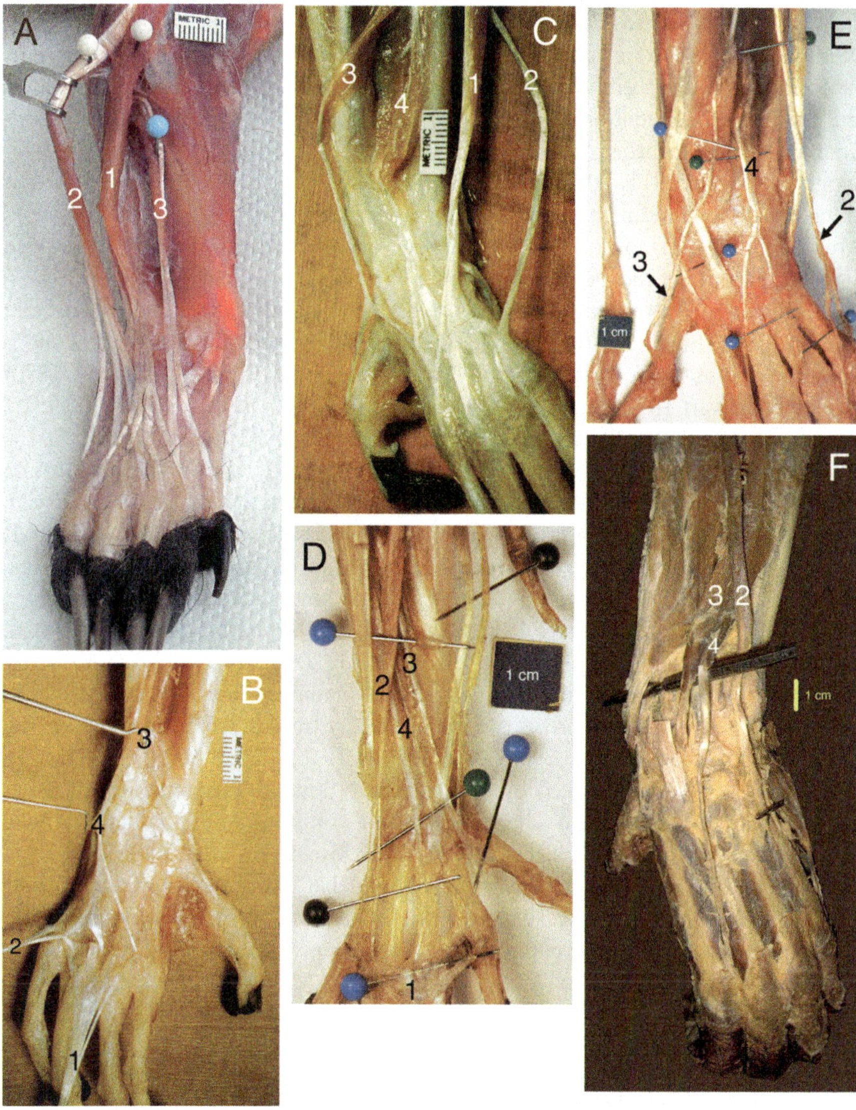

Fig. 7.3 Comparative morphology of the extensor digitorum (**1**) and other muscles of the dorsal compartment of the forearm in nonhuman primates and a pentadactyl mammal (coati). Note the contribution of the extensor digiti minimi (**2**) to digits 3–5 in the coati (**a**, *Nasua*; right side) compared to contribution to digits 4–5 in ruffed lemur (**b**, *Varecia variegata*; right side), sifaka (**c**, *Propithecus verreauxi*; left side), saki monkey (**d**, *Pithecia pithecia*; right side), baboon (**e**, *Papio anubis*; left side), and chimpanzee (**f**, *Pan troglodytes*; left side). Extensor pollicis longus (**3**) sends tendons to digits 1–2 in the coati, sifaka, and saki monkey, whereas a single tendon goes to the pollex in the ruffed lemur, baboon, and chimpanzee. The extensor indicis (**4**) sends tendons to digits 2 and 3 in the sifaka, saki monkey, and baboon, whereas tendons to digits 2–4 are found in the ruffed lemur. A single tendon to the index finger typifies the chimpanzee. Scale bars are 1 cm. Photographs in **d**–**f** are adapted from Diogo and Wood (2012a).

the second digit (Jouffroy and Lessertisseur 1960; Jouffroy 1962; Kaneff 1980a, b; Dunlap et al. 1985; Aziz and Dunlap 1986; Diogo and Wood 2011, 2012a; Fig. 7.3; Table 7.3). In lorises, gibbons, African apes, and humans, the extensor digiti minimi sends a single tendon to digit 5 (Bischoff 1870; Murie and Mivart 1872; Deniker 1885; Kohlbrügge 1890-1892; Hepburn 1892; Straus 1941a; Miller 1943; Diogo and Wood 2011, 2012a; Fig. 7.3). The same muscle in colugos is different in having three tendons inserting onto digits 3, 4, and 5 (Leche 1886; Diogo and Wood 2011, 2012a), a condition also observed in the coati (a South American mammal part of the raccoon family; Fig. 7.3a).

In most pentadactyl mammals, the extensor indicis provides a tendon to more than one digit (Fig. 7.3). For example, the muscle sends tendons to digits 2 and 3 (sometimes 4 as well) in *Rattus*, tree shrews, and most primates (Bischoff 1870; Murie and Mivart 1872; Deniker 1885; Kohlbrügge 1890-1892; Hepburn 1892; Le Gros Clark 1924; Woollard 1925; Howell and Straus 1933; Greene 1935; Straus 1941a; Miller 1943; Jouffroy 1962, 1971; George 1977; Kaneff 1980b; Schultz 1984; Dunlap et al. 1985; Aziz and Dunlap 1986; Diogo and Wood 2011, 2012a; Fig. 7.3; Table 7.3). Colugos are unique in that this muscle sends tendons to digits 1, 2, and 3 (Leche 1886; Diogo and Wood 2011, 2012a). In African apes and humans, the muscle lives up to its name by having a single tendon attaching onto the index finger (Wilder 1862; Macalister 1873; Deniker 1885; Hepburn 1892; Straus 1941a, b; Tuttle 1969; Lewis 1989; Diogo and Wood 2011, 2012a; Fig. 7.3f; Table 7.3).

Compared to other primates and pentadactyl mammals, humans are unusual in having an extensor pollicis brevis, a muscle distinct from the abductor pollicis longus that appears late during development (Lewis 1910; Kaneff 1959; Fig. 7.1a, b). Gibbons are also characterized by such extrinsic pollical muscle, although theirs attaches to the Mc1 base and adjacent carpals, as opposed to the pollical PP as in humans (Bischoff 1870; Kohlbrügge 1890-1892; Michilsens et al. 2009; Diogo and Wood 2011, 2012a). The extensor pollicis brevis that has been reported in gorillas (e.g., Straus 1941a, b; Raven 1950) is actually an abductor pollicis longus with a bipartite tendon that has one part attaching onto the proximal PP (Diogo and Wood 2011, 2012a).

3.2 Ventral Compartment of the Forearm

The ventral (flexor) compartment of the forearm of primates, like that of most pentadactyl mammals, contains muscles that primarily flex the pollex and fingers, which is critical for grasping. Three layers can be homologized among tetrapods (deep, intermediate, and superficial), with the middle and superficial layers leading to most of the ventral musculature of the forearm observed in mammals (Windle 1889; McMurrich 1903a; Straus 1942b; Haines 1950; Lewis 1989). The flexor digitorum longus of reptiles such as *Timon* has five distinctive parts and sectors derived from two different layers mentioned above: (1) ulnar and radial parts from the intermediate layer and (2) condyloradialis, condyloulnaris, and centralis sectors from the superficial layer (Windle 1889; McMurrich 1903a; Lewis 1989). Together, the flexor digitorum longus and flexores breves superficiales of reptiles correspond to the flexor digitorum profundus, flexor digitorum superficialis (FDS; also termed flexor digitorum sublimis by comparative anatomists), and palmaris longus of most mammals, including primates (Diogo et al. 2009; Diogo and Abdala 2010; Table 7.1). Recent developmental evidence in mice shows that the portion of FDS corresponding to the flexores breves superficiales of reptiles first develops in the palm as intrinsic muscles and then relocates as myofibers in the forearm in its final position (Huang et al. 2013). Other muscles of the ventral compartment can be homologized across tetrapods as similar entities (e.g., flexor carpi ulnaris) (Diogo et al. 2009; Diogo and Abdala 2010; Tables 7.1 and 7.2).

Within the ventral compartment of the forearm, the arrangement and morphology of the FDP, FDS, and palmaris longus are most variable among primates and other pentadactyl mammals. Because of their intimate relationships in the forearm and pairs of tendons given to the fingers, major differences in FDS and FDP can be reviewed together. Lewis (1989) pointed out the "progressive trend" from more primitive mammals to higher primates in having the FDS receiving a greater contribution from the common flexor mass, thus becoming larger relative to FDP. For example, opossums, rats, coatis, tree shrews, strepsirrhines, and tarsiers all have a relatively poorly developed FDS compared to FDP (Murie and Mivart 1872; Le Gros Clark 1924; Greene 1935; Jouffroy 1962; George 1977; Stein 1981; Schultz 1984; Lewis 1989; Fig. 7.4a, e). In contrast, the FDS of anthropoids, and particularly hominoids, is bulkier relative to the FDP (Howell and Straus 1933; Jouffroy and Lessertisseur 1960; Tuttle 1969; Lewis 1989). Underlying this trend is the greater contribution of the condyloradialis, condyloulnaris, and centralis sectors of the superficial muscle layer to the formation of the FDS (Lewis 1989). The functional reasons behind this trend are unclear and probably not linked to locomotor specializations in anthropoid primates. To that effect, EMG data in chimpanzees and baboons show that locomotor behaviors such as knuckle walking or digitigrade walking elicit less activity of the FDS and FDP compared to hook grips used during suspensory postures, when holding food objects, or even scratching (Susman and Stern 1979; Patel et al. 2012; see Chap. 10). This experimental evidence underscores the fundamental role played by the extrinsic digital flexors for basic grasping functions of the hand.

There is also variation in the number of tendons sent by the FDS and FDP muscles, although not to the extent observed for the extensor muscles. In *Rattus* and *Tupaia*, there is no FDS tendon for digit 5, unlike in *Cynocephalus* and primates (Leche 1886; Le Gros Clark 1924; Straus 1942b; Jouffroy 1971; George 1977; Diogo and Abdala 2010; Table 7.4). Reduced length of particular digits in primates is often associated with absence of an FDS or FDP tendon. For example, the FDS tendon for the diminutive second digit of *Loris* and *Perodicticus* is missing (Murie and Mivart 1872; Forster 1934; Straus 1942a; Miller 1943; Diogo and Wood 2011; Table 7.4), and the FDP tendon for the reduced pollex of *Colobus* and great apes is either vestigial or absent (Polak 1908; Straus 1942b; Jouffroy and Lessertisseur

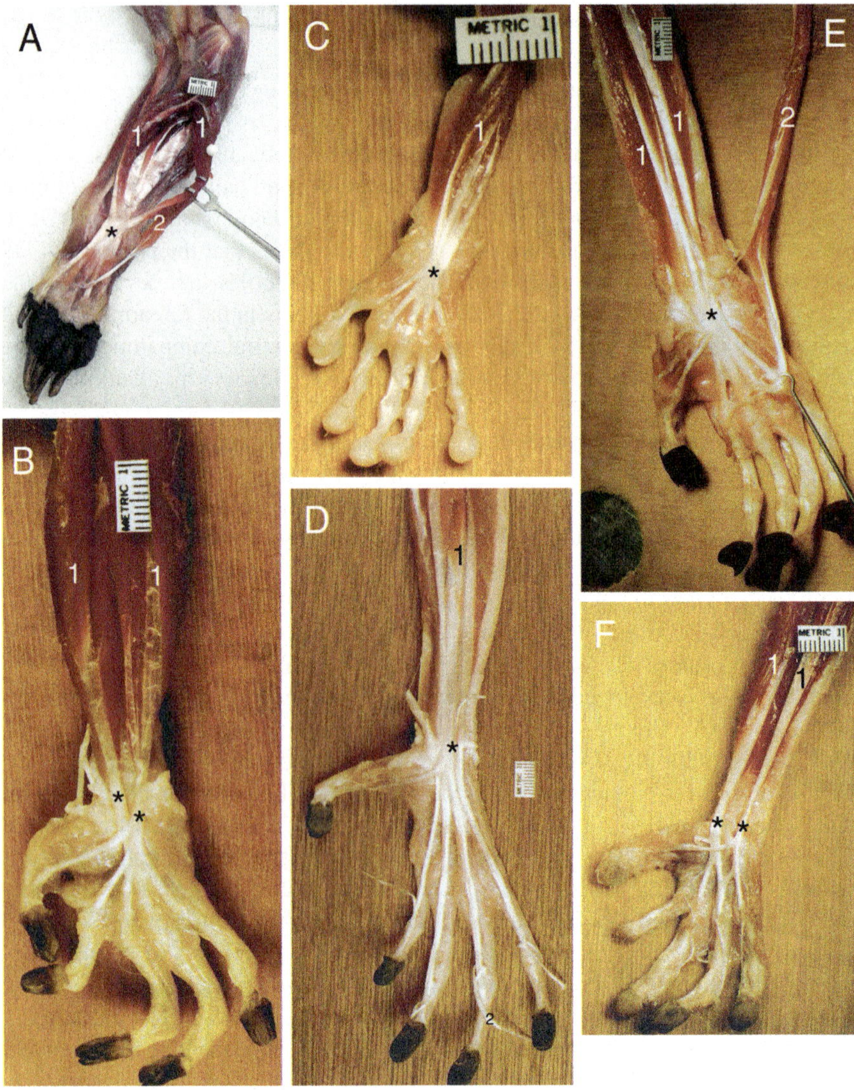

Fig. 7.4 Comparative morphology of the flexor digitorum profundus (FDP) (**1**) and flexor digitorum superficialis (**2**) in nonhuman primates and a pentadactyl mammal (coati). Note the varying degree of fusion of the tendinous lamina of the FDP (indicated by *stars*) in the coati (**a**, *Nasua*; right side), greater bushbaby (**b**, *Otolemur garnettii*; right side), fat-tailed dwarf lemur (**c**, *Cheirogaleus medius*; right side), sifaka (**d**, *Propithecus verreauxi*; right side), ruffed lemur (**e**, *Varecia variegata*; right side), slow loris (**f**, *Nycticebus coucang*, right side), Western tarsier (**g**, *Tarsius* [or *Cephalopachus*] *bancanus*; left side), baboon (**h**, *Papio anubis*; left side), squirrel monkey (**i**, *Saimiri sciureus*; left side), gibbon (**j**, *Hylobates gabriellae*; left side), and orangutan (**k**, *Pongo pygmaeus*; right side). Compared to most other taxa, the FDP of the galago, and especially the slow loris, has well-separated muscle heads giving rise to individual tendons inserting onto digits. The gibbon is the only nonhuman primate with a distinct flexor pollicis longus (**3**), with an occasional connection to the FDP tendon for the index finger (*arrow* in **j**). Some intrinsic hand muscles are labeled for the baboon [**4**, abductor pollicis brevis; **5**, flexor pollicis brevis, superficial head (*a*), and deep head (*b*); **6**, adductor pollicis, oblique head (*a*), and transverse head (*b*); **7**, abductor digiti minimi (with two heads); **8**, flexor digiti minimi; **9**, lumbricals]. Scale bars are 1 cm. Photographs in **d** and **f** are adapted from Gyambibi and Lemelin (2013) and in **h**–**k** from Diogo and Wood (2012a)

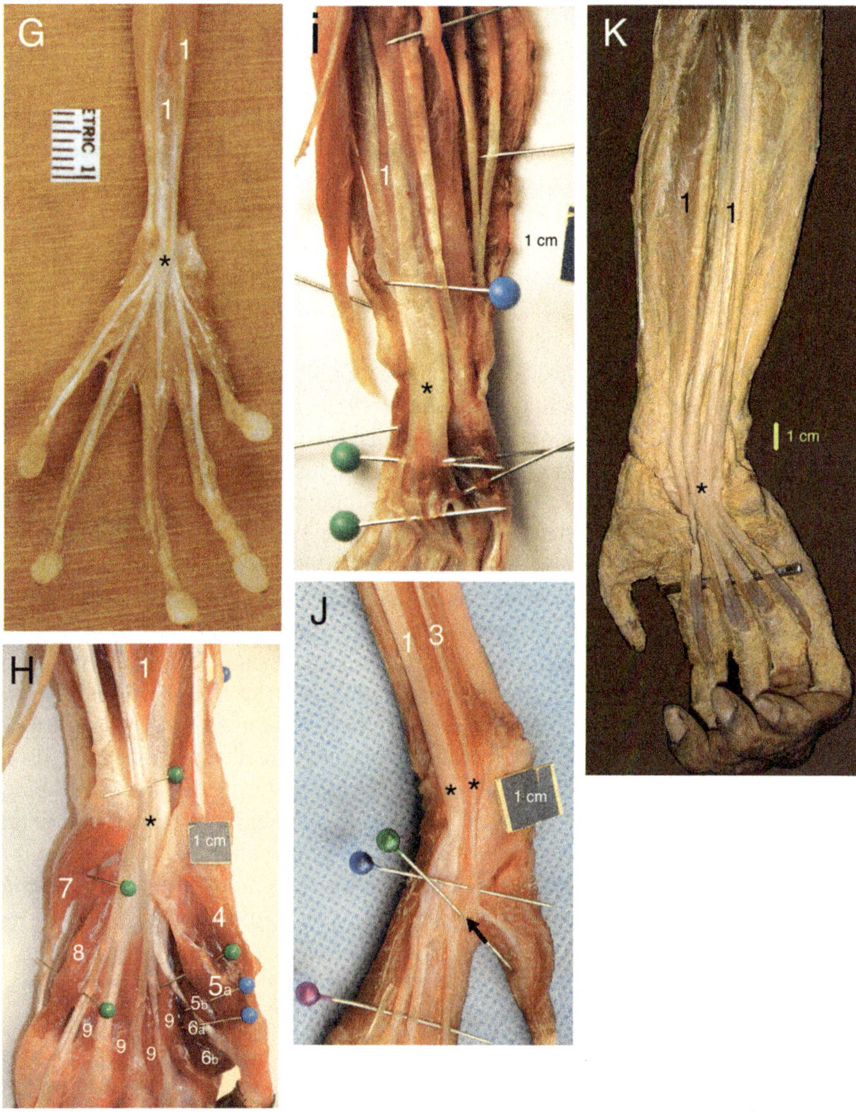

Fig. 7.4 (continued)

1960; Tuttle 1969, 1970; Schultz 1986; Susman 1994; Diogo and Wood 2011; Table 7.4).

On its own, the FDP muscle shows some important variation among primates. In most primates and other pentadactyl mammals, the muscle bellies of FDP are intimately connected, and associated tendons are fused and form a single flexor tendon plate (lamina) at the level of the carpal tunnel (Murie and Mivart 1872; Le Gros Clark 1924; Howell and Straus 1933; Greene 1935; Straus 1942a; Haines

1950; Jouffroy 1962, 1971; George 1977; Stein 1981; Schultz 1986; Lewis 1989; Gyambibi and Lemelin 2013; Fig. 7.4). In lorises, the FDP tendons are separated at the carpal tunnel and originate from two well-segregated muscle bellies in the forearm (Murie and Mivart 1872; Forster 1934; Straus 1942a; Gyambibi and Lemelin 2013; Fig. 7.4f). This condition correlates with the unusual "pincerlike" appearance of the hand, with the longest rays of that pincer (pollex and digit 4) receiving contributions from both muscle bellies (Gyambibi and Lemelin 2013; Fig. 7.4f). In great apes (*Pongo*, *Gorilla*, and *Pan*), the FDP muscle shows some degree of separation with two chief tendons: one for the index finger (and pollex when tendon present) and one for the remaining digits (Hepburn 1892; Sonntag 1923, 1924; Lewis 1989; Fig. 7.4k). Finally, gibbons and humans have a separate belly and tendon of FDP that is exclusive to the pollex (Deniker 1885; Kohlbrügge 1890-1892; Hepburn 1892; Jouffroy and Lessertisseur 1960; Tuttle 1969, 1972a; Susman 1998; Diogo and Wood 2011, 2012a; Diogo et al. 2012; Figs. 7.1c, d and 7.4j). More has been written (and debated) about the flexor pollicis longus (FPL) than any other hand muscle, from its uniqueness to its functional significance for toolmaking during human evolution (e.g., Susman 1994, 1998; Marzke 1997; Tocheri et al. 2008; see Chaps. 10, 11, and 18). Whether or not the FPL of gibbons is truly separate from the FDP [see Susman (1998) for arguments in favor of the FPL as a muscle unique to humans], the fact remains that a more independent muscle belly characterizes the two hominoids that have the longest thumbs. EMG data show FPL activity during forceful toolmaking behaviors (Hamrick et al. 1998; Marzke et al. 1998; see Chaps. 10 and 11); highest recruitment levels of the muscle were observed with increased pressure on the pollical apical pad whether power or precision grips were used (Hamrick et al. 1998) or when using cylindrical objects during digging or clubbing with a power squeeze grip (Marzke et al. 1998). On the basis of this experimental evidence, one could easily argue in favor of convergent evolution of an FPL in gibbons and humans as a mean to increase force at the tip of a long pollex, whether a branch, club, or hammerstone is being gripped.

3.3 Ventral Compartment of the Hand

The ventral compartment of the hand contains a myriad of muscles that are vital for movements of the digits in different planes and used for different grips. As exemplified above with the case of the interossei muscles, tracing the homologies of various muscle layers of the hand across tetrapods can be both challenging and controversial. Different hand layers in the palm of reptiles and mammals have been homologized (from superficial to deep): flexores breves superficiales, abductors, lumbricals, contrahentes, flexores breves profundi, and intermetacarpales (Cunningham 1882; Windle 1883; McMurrich 1903b; Forster 1917; Campbell 1939; Haines 1950; Lewis 1989; Diogo et al. 2009; Diogo and Abdala 2010; Table 7.1).

In many pentadactyl mammals, the flexores breves superficiales are represented by the palmaris brevis (found in most primates, except *Hylobates* and *Pongo*) and flexor digitorum brevis manus (Diogo et al. 2009; Diogo and Abdala 2010; Diogo and Wood 2011, 2012a; see also references therein). The latter muscle—found in tree shrews and colugos—is distinct from the palmaris brevis and has a perforating tendon attaching onto the middle phalanx of digit 5 (like the FDS tendon), thus supporting the notion that it is derived from an ancestral flexor brevis superficialis layer (Leche 1886; Le Gros Clark 1924; Haines 1955; George 1977; Diogo et al. 2009; Diogo and Abdala 2010). A superficial palmar layer has also been described in lemurs, lorises, and galagos, with muscle slips arising from the palmar aponeurosis and inserting onto the extensor hood of the central digits (Murie and Mivart 1872; Forster 1917; Jouffroy 1962; Kanagasuntheram and Jayawardene 1957; Lemelin 1992). However, on the basis of innervation by the deep branch of the ulnar nerve (Kanagasuntheram and Jayawardene 1957), these slips found in strepsirrhines appear more likely to be components of a deeper muscle layer (contrahentes or flexores breves profundi) that migrated superficially. As such, this muscle layer has been identified as interossei accessorii (see Diogo and Wood 2011 and references therein; Fig. 7.5; Table 7.2). Interossei accessorii have been observed in gibbons as well (Kohlbrügge 1890-1892; Fitzwilliams 1910; Forster 1917; Jouffroy and Lessertisseur 1960; Tuttle 1969, 1972a), although superficial attachment onto the palmar aponeurosis appears to be lacking. The interosseous accessorius for the index finger shows considerable EMG activity during pinch grasping of a small object between the elongated thumb and side of the index finger (Susman et al. 1982; see Chap. 10).

Two marginal abductors (abductor pollicis brevis and abductor digiti minimi) and four lumbricals are consistent across tetrapods (Diogo et al. 2009; Diogo and Abdala 2010; Fig. 7.4h; Tables 7.1 and 7.2). Gibbons sometimes lack a lumbrical for digit 5 (Tuttle 1969; Diogo and Wood 2011, 2012a; Table 7.5), while colugos are unusual in having a total of seven lumbricals attaching onto both sides of digits 2–4 and the radial side of digit 5 (Leche 1886; Diogo and Wood 2011, 2012a; Table 7.5).

The contrahentes can be observed just deep to the lumbricals and FDP tendons. In primates, the largest contrahens is represented by the adductor pollicis, which shows varying degrees of separation between transverse and oblique heads (Murie and Mivart 1872; Haines 1958; Jouffroy and Lessertisseur 1959; Napier 1961; Day and Napier 1963; Jouffroy 1971, 1975; Dunlap et al. 1985; Diogo and Wood 2011, 2012a; Figs. 7.4h and 7.6; Table 7.5). Since most pentadactyl mammals (including tarsiers and anthropoids) possess a mesaxonic hand (i.e., ray 3 is longest and forms the central axis of the hand), the contrahens muscle layer arises from a seamlike line (known as a raphe) along the Mc3 (Forster 1917; Haines 1958; Jouffroy and Lessertisseur 1959; Napier 1961; Jouffroy 1962, 1971, 1975; Dunlap et al. 1985; Lewis 1989). Strepsirrhine primates have an ectaxonic hand (i.e., ray 4 is longest); as such, the contrahentes tend to be organized around the Mc4 (Jouffroy and Lessertisseur 1959; Jouffroy 1962, 1975). The number of contrahentes varies between three and five among most primates and pentadactyl mammals, with only the adductor pollicis remaining in orangutans, gorillas, and humans (in contrast, chimpanzees appear to retain several contrahentes depending on the specimen) (see Diogo and Wood 2011, 2012a for in-depth review of the literature on the subject; Table 7.5). Tarsiers are highly unusual in having five contrahentes separated into

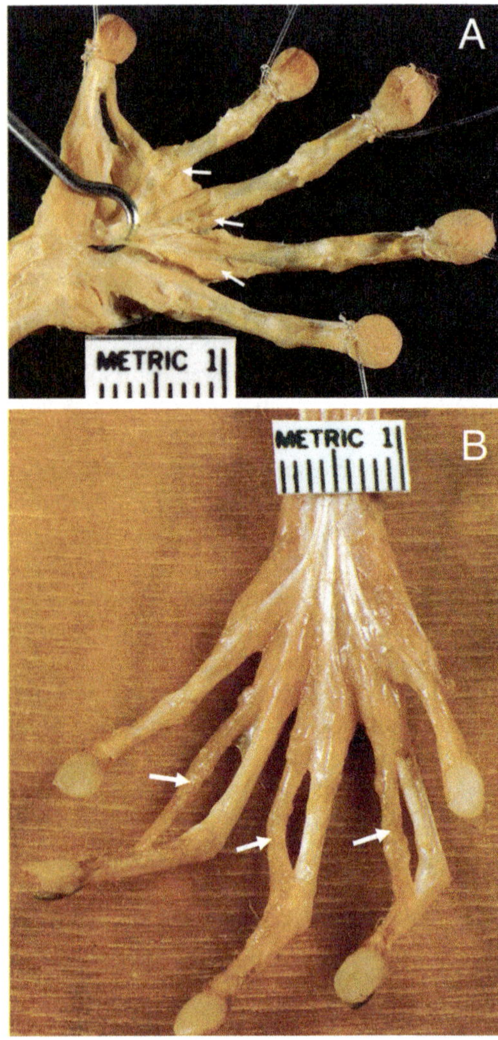

Fig. 7.5 Palmar musculature in two small-bodied primates. *Arrows* point to slips of interossei accessorii in the lesser bushbaby (**a**, *Galago senegalensis*; left side) and long slips of contrahentes in the Western tarsier (**b**, *Tarsius* [or *Cephalopachus*] *bancanus*; left side). The dissecting hook is lifting up the palmar aponeurosis onto which interossei accessorii are attaching. Note also the additional slip of the adductor pollicis muscle on the thumb of the lesser bushbaby. Scale bars are 1 cm

superficial and deep layers, with fibers of the superficial layer attaching onto the distal phalanges of the central digits before splitting around the middle phalanges (Day and Iliffe 1975; Schultz 1984; Fig. 7.5b). In some ways, the anatomy of these superficial contrahentes resembles the surperficial palmar layer found in basal tetrapods (Diogo and Wood 2011). Long slips of interossei accessorii in strepsirrhines and contrahentes in tarsiers have been functionally linked to "bowed-up" finger postures or "finger adhesion" (i.e., hyperextension of McP joint and flexion of proximal IP joint) (Day and Iliffe 1975; Lemelin 1992).

The flexores breves profundi and intermetacarpales contribute to thenar, hypothenar, and interossei muscles, all of which are vital for digital movements in various planes. Plesiomorphically, a total of ten flexores breves profundi going to the

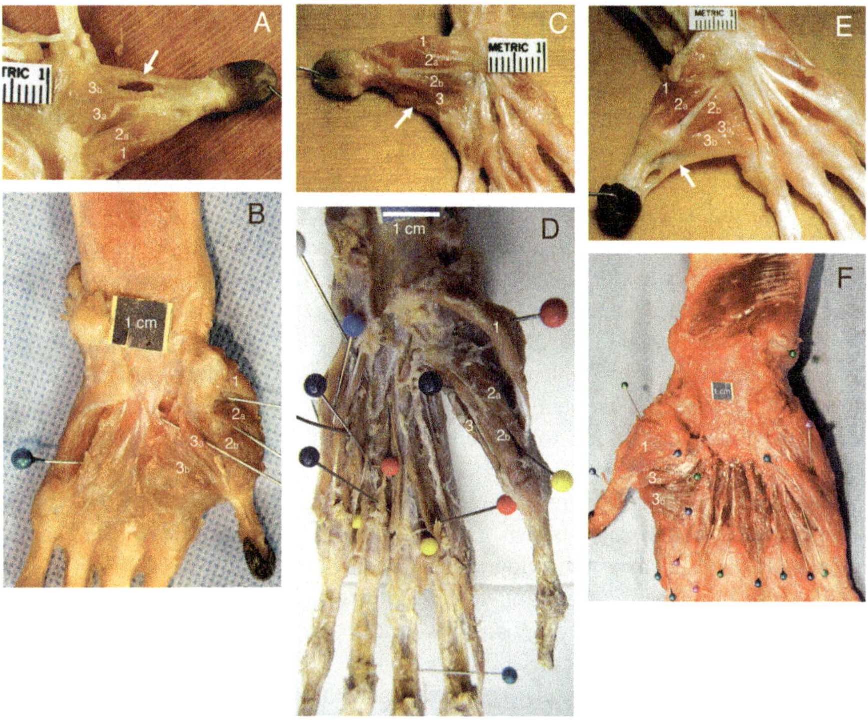

Fig. 7.6 Comparative morphology of pollical muscles in nonhuman primates. Thenar muscles [**1**, abductor pollicis brevis; **2**, flexor pollicis brevis, superficial head (*a*), and deep head (*b*)] and adductor pollicis [**3**, oblique head (*a*) and transverse head (*b*)] are shown in the greater bushbaby (**a**, *Otolemur garnettii*; right side), crab-eating macaque (**b**, *Macaca fascicularis*; left side), slow loris (**c**, *Nycticebus coucang*; right side), gibbon (**d**, *Hylobates lar*; left side), ruffed lemur (**e**, *Varecia variegata*; right side), and gorilla (**f**, *Gorilla gorilla*; right side). The *arrow* in **a**, **c**, and **e** points to an additional slip of the adductor pollicis commonly found in strepsirrhine primates. Scale bars are 1 cm. Photographs in **b**, **d**, and **f** adapted from Diogo and Wood (2012a)

radial and ulnar sides of all digits can be found in the hand of pentadactyl mammals such as rats (Lewis 1989; Diogo et al. 2009; Diogo and Abdala 2010; Diogo and Wood 2011, 2012a). The first flexor brevis profundus can be homologized with the flexor pollicis brevis (superficial head) and opponens pollicis, the second one to the flexor pollicis brevis (deep head), and the last one to the flexor digiti minimi and opponens digiti minimi (Čihák 1972; Lewis 1989; Diogo et al. 2009; Diogo and Abdala 2010; Diogo and Wood 2011, 2012a; Diogo et al. 2012; Tables 7.1 and 7.2). All other flexores breves profundi either fuse with the intermetacarpales to form dorsal interossei or remain individualized to become palmar interossei (Čihák 1972; Lewis 1989; Diogo et al. 2009; Diogo and Abdala 2010; Diogo and Wood 2011, 2012a; Diogo et al. 2012; Tables 7.1 and 7.2).

The opponens pollicis is present in all primates (except *Callithrix*), with varying attachment patterns onto the pollex, but is absent in other pentadactyl mammals

(Day and Napier 1963; Jouffroy 1971; Dunlap et al. 1985; Lewis 1989; Diogo and Wood 2011, 2012a; Table 7.5). In turn, the deep head of the flexor pollicis brevis found in most Old World monkeys and humans (but absent in apes, except *Pongo*) is thought to reflect enhanced abilities for pollical opposability (e.g., Day and Napier 1963; Susman 1994). However, more recent comparative evidence has shown the prevalence of this muscle in most primates and rats (Diogo et al. 2009; Diogo and Abdala 2010; Diogo and Wood 2011, 2012a; Diogo et al. 2009; Fig. 7.6; Table 7.5). As a matter of fact, absence of the deep head of the flexor pollicis brevis in tree shrews, colugos, and New World monkeys has been interpreted as apomorphic relative to other pentadactyl mammals (Diogo et al. 2009; Diogo and Abdala 2010; Diogo and Wood 2011, 2012a; Table 7.5). By the same token, the so-called first volar interosseous of Henle found in humans (and occasionally in African apes) has been reinterpreted as being a small slip derived from the adductor pollicis muscle and renamed "adductor pollicis accessorius" (Diogo et al. 2012; Bello-Hellegouarch et al. 2013).

Plesiomorphically, flexores breves profundi and intermetacarpales remain unfused as evinced by rats, strepsirrhines, tarsiers, Old World monkeys, and chimpanzees (Diogo et al. 2009; Diogo and Abdala 2010; Diogo and Wood 2011, 2012a; Fig. 7.7; Table 7.5). Varying degrees of fusion between these muscle layers are found in tree shrews, colugos, New World monkeys, and most hominoids, including humans (Diogo et al. 2009; Diogo and Abdala 2010; Diogo and Wood 2011, 2012a; Tables 7.2 and 7.5). The functional reasons why such fusion occurred in various anthropoid lineages and its evolutionary sequence remain unclear. It is however interesting to note that in spite of being unfused, chimpanzees recruit the interossei muscles in the same way humans do, that is, during McP joint flexion and rapid extension of the IP joints (Susman and Stern 1980; see Chap. 10). Like the contrahentes, the interossei tend to be organized around the central axis of the hand by attaching on both sides of digit 3 (most pentadactyl mammals) or digit 4 (strepsirrhines) (Jouffroy and Lessertisseur 1959; Jouffroy 1962, 1975; Table 7.5).

3.4 Final Remarks on the Primate and Human Hand Musculature

Several key points can be reviewed from this comparative review of the hand musculature. First and foremost, the primate hand musculature is conservative vis-à-vis that of other euarchontans and other pentadactyl mammals. Opponens pollicis appears to be the only hand muscle specific to all primates (except *Callithrix*). In spite of this overall conservatism, there is considerable homoplasy (i.e., resemblance in morphology not due to common descent) in the hand musculature, including reversions, convergence, and parallelism (Fig. 7.8). For example, the plesiomorphic condition of having distinct flexores breves profundi and intermetacarpales muscles (instead of interossei muscles) is found in taxa as diverse as *Rattus*, *Lemur*, *Tarsius*, *Macaca*, and *Pan* (Fig. 7.8; Tables 7.1, 7.2 and 7.5). Similarly, gibbons and humans

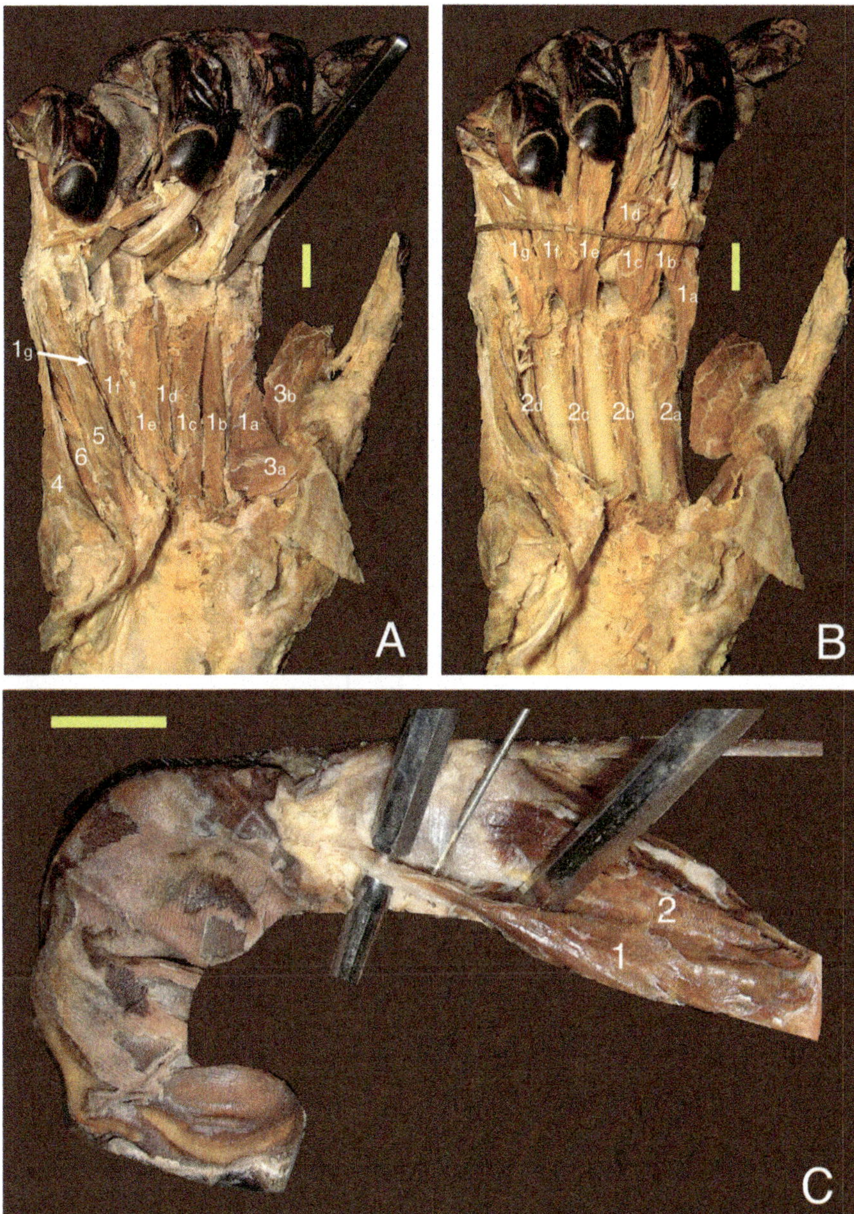

Fig. 7.7 Palmar musculature in the chimpanzee (*Pan troglodytes*, right side). In **a**, long digital flexor muscle tendons have been reflected to show flexores breves profundi muscles (**1**) for various digits (*a* and *b*, digit 2; *c* and *d*, digit 3; *e* and *f*, digit 4; *g*, digit 5). Adductor pollicis (**3***a*, oblique head; **3***b*, transverse head) and hypothenar muscles (**4**, abductor digiti minimi; **5**, flexor digiti minimi; **6**, opponens digiti minimi) are also labeled. In **b**, flexores breves profundi muscles have been reflected to show intermetacarpal muscles (**2**) to various digits (*a*, digit 2; *b* and *c*, digit 3; *d*, digit 4). Note that the flexor brevis profundus and intermetacarpal muscles are distinct and unfused in the chimpanzee (**c**). Scale bars are 1 cm. Photographs adapted from Diogo and Wood (2012a)

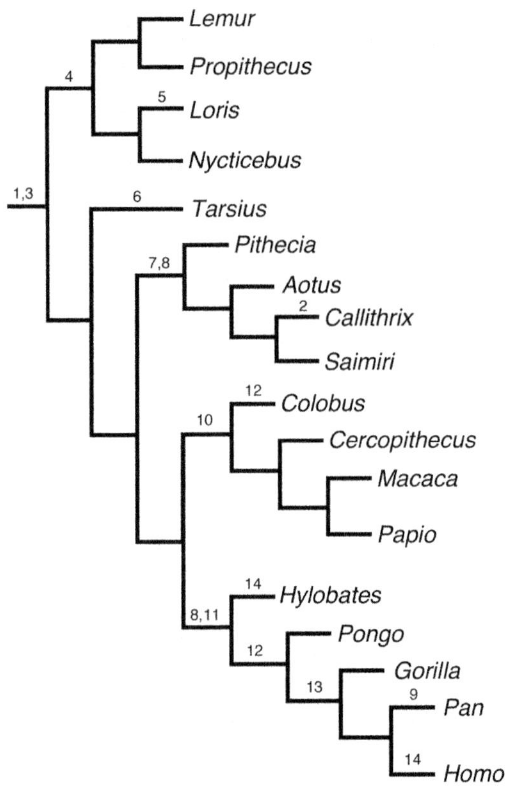

Fig. 7.8 Selected hand muscle characters mapped onto a simplified phylogeny of primates [from muscle data and tree published in Diogo and Wood (2011)]. **1**, presence of opponens pollicis; **2**, secondary undifferentiation of opponens pollicis; **3**, presence of opponens digiti minimi; **4**, intermetacarpales organized around digit 4; **5**, no tendon of flexor digitorum superficialis on digit 2; **6**, highly specialized slips of contrahentes; **7**, flexor brevis profundus 2 not a distinctive muscle; **8**, fusion of flexores breves profundi with intermetacarpales to form dorsal interossei; **9**, reversion to unfused flexores breves profundi with intermetacarpales; **10**, well-differentiated heads of adductor pollicis and opponens digiti minimi; **11**, origin of flexor digitorum superficialis from both the radius and ulna; **12**, vestigial or absent pollical tendon of flexor digitorum profundus; **13**, extensor indicis has a single tendon attaching onto digit 2; **14**, presence of flexor pollicis longus and extensor pollicis brevis. Note that some characters are derived and exclusive to a clade or autapomorphic (e.g., character 5 in *Loris*; character 6 in tarsiers; character 10 in Old World monkeys), while others are convergent or homoplastic (e.g., character 12 in *Colobus* and great apes; character 14 in gibbons and humans)

share hand muscle specializations such as an extensor pollicis brevis and separate FPL muscle, while *Colobus* and great apes (distantly related primate taxa with reduced thumbs) possess a vestigial pollical tendon of FDP, albeit with considerable variation in some taxa (Fig. 7.8). Finally, contrary to what is often suggested in the literature, very few hand muscles are autapomorphic (i.e., unique and exclusive) to humans. In many ways, what makes the human hand musculature derived compared to that of nonhuman primates and other pentadactyl mammals is the reduced overall number of muscles attaching onto the fingers combined with an emphasis on muscles attaching and moving the thumb, especially two extrinsic muscles: the extensor pollicis brevis and FPL. Indeed, more muscles attach onto the pollex of humans more than in almost all other primates, reinforcing the hypothesis that focal thumb movements probably played an important role in human evolution.

4 Future Directions

Given the breath of the literature reviewed in this chapter, readers may be tempted to conclude that very little remains to be investigated with regard to the primate hand musculature. We contend that this is not the case and here suggest several potential areas for future research. For well over a century, a vast body of comparative data has been accumulated on the hand musculature of primates and other pentadactyl mammals. However, comparative observations reported in the literature are sometimes contradictory and underscore partial sampling of the natural variation occurring within a species. This is not surprising as sample sizes in comparative studies can be low because of limited access to rare and endangered species. Therefore, additional observations on the primate hand musculature are warranted for most taxa, especially for species with ambiguous data (see appendices in Diogo and Wood 2011) or representatives of primate families that have received less attention (e.g., cheirogaleids and lepilemurids). Such new data would be valuable when used in a phylogenetic context to clarify the primitive or derived states and levels of variation of certain muscles, in addition to providing a more precise picture of hand myology evolution (see Diogo and Wood 2013). In collecting new muscle data, special attention should be given in gathering quantitative characteristics such as muscle mass, fiber length, tendon length, and the like. Such data have been useful to compare different functional muscle groups across primate and nonprimate taxa (e.g., Tuttle 1969, 1970, 1972a, b; Alexander 1993; Ker 1993; Rauwerdink 1993) and uncover scaling relationships of some of these parameters with body mass (e.g., Alexander et al. 1981; Myatt et al. 2012; Gyambibi and Lemelin 2013). More recently, other quantitative aspects such as physiological cross-sectional area (PCSA) have been documented in orangutans and African apes (Marzke et al. 1999; Ogihara et al. 2005; Oishi et al. 2009; Myatt et al. 2012), and collecting such data in other primates would prove invaluable.

As noted above, ontogenetic data from Čihák (1972) on the human fetal hand and foot material remain widely cited despite being over 40 years old. It is unfortunate—but not surprising considering the painstaking work involved and extensive sampling required—that little such data exist for nonhuman primates. Granted that access to primate fetal material would be possible, ontogenetic series of the hand musculature for representative nonhuman primate taxa would represent a lasting contribution to the field of primate comparative biology. Similarly, recent advances in molecular techniques should allow a wealth of new data to be collected on the genetic bases underlying hand muscle development and patterning, as well as covariation between the hand and foot musculature due to pleiotropic effects and integration (see Chap. 5).

One last area of future investigation involves collection of EMG data on hand muscles of primates, including humans. Much remains to be discovered about hand muscle function as only a handful of experimental studies have been performed on nonhuman primates in the past 40 years. For example, activity of thenar muscles has never been documented for any nonhuman primates, including those with enhanced pollical abilities such as baboons and other Old World monkeys. Since living nonhuman primates can be difficult to access for experimental research, reliance on human subjects should also be considered as well. Future EMG research could focus on nonhuman primate behaviors not typically practiced in daily activities by humans such as climbing. Much could be learned on the evolution of the hand musculature of humans by gathering data for a wide repertoire of gripping behaviors others than tool manufacture and object manipulation.

Acknowledgments P.L. is grateful to the curators and staff of the Duke Lemur Center and Edmonton Valley Zoo for facilitating access to some of the nonhuman primate and nonprimate specimens used for this chapter, as well as the Department of Surgery (Division of Anatomy) at the University of Alberta for providing access to human specimens and laboratory facilities. R.D.'s work was funded by a Howard University College of Medicine start-up package. Tracy Kivell, Brian Richmond, and Daniel Schmitt read an earlier version of this chapter and provided very insightful comments. The groundbreaking work of Frederic Wood Jones and John Napier on the primate hand greatly inspired our chapter.

References

Alexander RMN (1993) Joints and muscles of hands and paws. In: Preuschoft H, Chivers DJ (eds) Hands of primates. Springer, Vienna, pp 199–205

Alexander RMN, Jayes AS, Maloiy GMO, Wathuta EM (1981) Allometry of the leg muscles in mammals. J Zool 194:539–552

Aziz MA, Dunlap SS (1986) The human extensor digitorum profundus muscle with comments on the evolution of the human hand. Primates 27:293–319

Backhouse KM, Catton WT (1954) An experimental study of the functions of the lumbrical muscles in the human hand. J Anat 88:133–141

Bardeleben K (1894) On the bones and muscles of the mammalian hand and foot. Proc Zool Soc Lond 1894:354–376

Basmajian JV (1979) Muscles alive. Their functions revealed by electromyography, 4th edn. Williams and Wilkins, Baltimore

Bello-Hellegouarch G, Aziz MA, Ferrero EM, Kern M, Francis N, Diogo R (2013) "Pollical palmar interosseous muscle" (*Musculus adductor pollicis accessorius*): attachments, innervation, variations, phylogeny, and implications for human evolution and medicine. J Morphol 274:275–293

Bischoff TLW (1870) Beiträge zur Anatomie des *Hylobates leuciscus* und zu einer vergleichenden Anatomie der Muskeln der Affen und des Menschen. Abhandlungen der Bayerischen Akademie der Wissenschaften, Mathematisch-Physikalische Klasse 10:197–297

Belson P, Smith LK, Puentes J (1976) Motor innervation of the flexor pollicis brevis: an electromyographic study. Am J Phys Med 55:122–138

Braithwaite F, Channell GD, Moore FT, Whillis J (1948) The applied anatomy of the lumbrical and interosseous muscles of the hand. Guys Hosp Rep 97:185–197

Brooks HSJ (1889) On the morphology of the muscles of the extensor aspect of the middle and distal segments of the limbs: with an account of the various paths which are adopted by the nerve-trunks in these segments, 1. Studies from the Museum of Zoology in University College, Dundee, pp 1–17

Bunnell S (1944) Surgery of the hand. J.B. Lippincott, Philadelphia

Campbell B (1939) The comparative anatomy of the dorsal interosseous muscles. Anat Rec 73:115–125

Čihák R (1972) Ontogenesis of the skeleton and intrinsic muscles of the human hand and foot. Ergeb Anat Entwicklungsgesch 46:1–194

Cunningham DJ (1882) Report on some points of the anatomy of the thylacine (*Thylacinus cynocephalus*), cuscus (*Phalangista maculata*) and phascogale (*Phascogale calura*) collected during the voyage of H.M.S. Challenger in the years 1873–1876; with an account of the comparative anatomy of the intrinsic muscles and nerves of the mammalian pes. Challenger Zool Rep 5:1–192

Day MH, Napier JR (1961) The two heads of flexor pollicis brevis. J Anat 95:123–130

Day MH, Napier J (1963) The functional significance of the deep head of flexor pollicis brevis in primates. Folia Primatol 1:122–134

Day MH, Iliffe SR (1975) The contrahens muscle layer in *Tarsius*. Folia Primatol 24:241–249

Deniker J (1885) Recherches anatomiques et embryologiques sur les singes anthropoïdes, foetus de gorille et de gibbon. Arch Zool Exp Gén 3:1–265

Diogo R, Abdala V (2010) Muscles of vertebrates. Comparative anatomy, evolution, homologies, and development. CRC Press (Taylor and Francis Group), Boca Raton

Diogo R, Molnar JL (2014) Comparative anatomy, evolution and homologies of the tetrapod hindlimb muscles, comparisons with forelimb muscles, and deconstruction of the forelimb-hindlimb serial homology hypothesis. Anat Rec 297:1047–1075

Diogo R, Tanaka EM (2012) Anatomy of the pectoral and forelimb muscles of wildtype and GFP-transgenic axolotls and comparison with other tetrapods including humans: a basis for regenerative, evolutionary and developmental studies. J Anat 221:622–635

Diogo R, Tanaka EM (2014) Development of fore- and hindlimb muscles in GFP-transgenic axolotls: morphogenesis, the tetrapod Bauplan, and new insights on the forelimb-hindlimb enigma. J Exp Zool B 322:106–127

Diogo R, Wood BA (2011) Soft-tissue anatomy of the primates: phylogenetic analyses based on the muscles of the head, neck, pectoral region and upper limb, with notes on the evolution of these muscles. J Anat 219:273–359

Diogo R, Wood B (2012a) Comparative anatomy and phylogeny of primate muscles and human evolution. Science Publishers, St. Helier

Diogo R, Wood B (2012b) Violation of Dollo's law: evidence of muscle reversions in primate phylogeny and their implications for the understanding of the ontogeny, evolution, and anatomical variations of modern humans. Evolution 66:3267–3276

Diogo R, Wood B (2013) The broader lessons to be learned from a comparative and phylogenetic analysis of primate muscle morphology. Biol Rev 88:988–1001

Diogo R, Ziermann JM (2014) Development of fore- and hindlimb muscles in frogs: morphogenesis, homeotic transformations, digit reduction, and the forelimb-hindlimb enigma. J Exp Zool B 322:86–105

Diogo R, Abdala V, Aziz MA, Lonergan N, Wood BA (2009) From fish to modern humans—comparative anatomy, homologies and evolution of the pectoral and forelimb musculature. J Anat 214:694–716

Diogo R, Richmond BG, Wood B (2012) Evolution and homologies of primate and modern human hand and forearm muscles, with notes on thumb movements and tool use. J Hum Evol 63:64–78

Diogo R, Nacu E, Tanaka EM (2014a) Is salamander limb regeneration really perfect? Anatomical and morphogenetic analysis of forelimb muscle regeneration in GFP-transgenic axolotls as a basis for regenerative, developmental and evolutionary studies. Anat Rec 297:1076–1089

Diogo R, Molnar JL, Smith T (2014b) The anatomy and ontogeny of the head, neck, pectoral and upper limb muscles of *Lemur catta* and *Propithecus verreauxi* (Primates): discussion on the parallelism between ontogeny and phylogeny and implications for evolutionary and developmental biology. Anat Rec 297:1435–1453

Diogo R, Ziermann JM, Linde-Medina M (2015) Is evolutionary biology becoming too politically correct? A reflection on the *scala naturae*, phylogenetically basal clades, anatomically plesiomorphic taxa, and 'lower' animals. Biol Rev 90:502–521

Duchenne GB (1867) Physiologie des movements. J.B. Ballière et Fils, Paris

Dunlap SS, Thorington RW, Aziz MA (1985) Forelimb anatomy of the New World monkeys: myology and the interpretation of primitive anthropoid models. Am J Phys Anthropol 68:489–517

Eyler DL, Markee JE (1954) The anatomy and function of the intrinsic musculature of the fingers. J Bone Joint Surg Am 36A:1–9

Fahrer M (1981) The hypothenar eminence. In: Tubiana R (ed) The hand, vol 1. W.B. Saunders, Philadelphia, pp 259–263

Fitzwilliams DCL (1910) The short muscles of the hand of the agile gibbon (*Hylobates agilis*), with comments on the morphological position and function of the short muscles of the hand of man. Proc R Soc Edinb 30:202–218

Forster A (1917) Die Mm. contrahentes und interossei manus in der Säugeteirreihe und beim Menschen. Arch Anat Physiol 1916:101–378

Forster A (1934) La pince palmaire et la pince plantaire du *Perodicticus potto*. Arch d'Anat d'Histol d'Embryol 17:181–246

George RM (1977) The limb musculature of the Tupaiidae. Primates 18:1–34

Gibbs S (1999) Comparative soft tissue morphology of the extant Hominoidea, including man. Ph.D. dissertation, University of Liverpool

Greene EC (1935) Anatomy of the rat. Hafner, New York

Gyambibi A, Lemelin P (2013) Comparative and quantitative myology of the forearm and hand of prosimian primates. Anat Rec 296:1196–1206

Haines RW (1939) A revision of the extensor muscles of the forearm in tetrapods. J Anat 73:211–233

Haines RW (1950) The flexor muscles of the forearm and hand in lizards and mammals. J Anat 84:13–29

Haines RW (1955) The hand anatomy of certain insectivores. Proc Zool Soc Lond 125:761–777

Haines RW (1958) Arboreal or terrestrial ancestry of placental mammals. Q Rev Biol 33:1–23

Hamrick MW, Churchill SE, Schmitt D, Hylander WL (1998) EMG of the human flexor pollicis longus muscle: implications for the evolution of hominid tool use. J Hum Evol 34:123–136

Henle J (1855) Handbuch der Systematischen Anatomie des Menschen. Friedrich Vieweg und Sohn, Braunschweig

Hepburn D (1892) The comparative anatomy of the muscles and nerves of the superior and inferior extremities of the anthropoid apes. Part I. J Anat Physiol 26:149–186

Howell AB (1936) Phylogeny of the distal musculature of the pectoral appendage. J Morphol 60:287–315

Howell AB, Straus WL Jr (1933) The muscular system. In: Hartman GG, Straus WL Jr (eds) The anatomy of the rhesus monkey (*Macaca mulatta*). Hafner, New York, pp 89–175

Huang AH, Riordan TJ, Wang L, Eyal S, Zelzer E, Brigande JV, Schweitzer R (2013) Repositioning forelimb superficialis muscles: tendon attachment and muscle activity enable active of functional myofibers. Dev Cell 26:544–551

Jouffroy FK (1962) La musculature des membres chez les lémuriens de Madagascar: étude descriptive et comparative. Mammalia 26:1–326

Jouffroy FK (1971) Musculature des membres. In: Grassé PP (ed) Traité de Zoologie. Anatomie, Systématique, Biologie. Tome XVI (Mammifères). Masson, Paris, pp 1–475

Jouffroy FK (1975) Osteology and myology of the lemuriform postcranial skeleton. In: Tattersall I, Sussman RW (eds) Lemur biology. Plenum Press, New York, pp 149–192

Jouffroy FK, Lessertisseur J (1959) Réflexions sur les muscles contracteurs des doigts et desorteils (contrahentes digitorum) chez les primates. Ann Sci Nat Zool (Sér 12) 1:211–235

Jouffroy FK, Lessertisseur J (1960) Les spécialisations anatomiques de la main chez les singes à progression suspendue. Mammalia 24:91–151

Kanagasuntheram R, Jayawardene FLW (1957) The intrinsic muscles of the hand in the slender loris. Proc Zool Soc Lond 128:301–312

Kaneff A (1959) Über die evolution des m. abductor pollicis longus und m. extensor pollicis brevis. Mitteil Morphol Inst Bulg Akad Wiss 3:175–196

Kaneff A (1980a) Évolution morphologique des musculi extensores digitorum et abductor pollicis longus chez l'homme. II. Évolution morphologique des m. extensor digiti minimi, abductor pollicis longus, extensor pollicis brevis et extensor pollicis longus chez l'homme. Gegenbaurs Morphol Jahrb 126:594–630

Kaneff A (1980b) Évolution morphologique des musculi extensores digitorum et abductor pollicis longus chez l'homme. III. Évolution morphologique du m. extensor indicis chez l'homme, conclusion générale sur l'évolution morphologique des musculi extensores digitorum et abductor pollicis longus chez l'homme. Gegenbaurs Morphol Jahrb 126:774–815

Kaplan EB (1953) Functional and surgical anatomy of the hand. J.B. Lippincott, Philadelphia

Ker RF (1993) Elasticity of hand and forefoot tendons. In: Preuschoft H, Chivers DJ (eds) Hands of primates. Springer, Vienna, pp 257–269

Kohlbrügge JHF (1890-1892) Versuch einer Anatomie des Genus *Hylobates*. In: Weber M (ed) Zoologische Ergebnisse Einer Reise in Niederländisch Ost-Indien, vol 1. Verlag von EJ Brill, Leiden. pp 211–354. vol 2, pp 138–208

Landsmeer JMF (1949) The anatomy of the dorsal aponeurosis of the human finger and its functional significance. Anat Rec 104:31–44

Landsmeer JMF (1955) Anatomical and functional investigations on the articulations of the human fingers. Acta Anat Suppl 24:1–69

Landsmeer JMF (1976) Atlas of anatomy of the hand. Churchill Livingstone, New York

Landsmeer JMF, Long C II (1965) The mechanism of finger control, based on electromyograms and location analysis. Acta Anat 60:330–347

Le Gros Clark WE (1924) The myology of the tree-shrew (*Tupaia minor*). Proc Zool Soc Lond 1924:461–497

Leche W (1886) Uber die säugethiergattung Galeopithecus: eine morphologische untersuchung. Kongl Svenska vetenskap-akad handlingar 21:3–92

Lemelin P (1992) Anatomy of the superficial palmar musculature of *Galago senegalensis*. Am J Phys Anthropol 35(Suppl 14):108 (abstract)

Lewis WH (1910) The development of the muscular system. In: Keibel F, Mall FP (eds) Manual of human embryology, vol 1. Lippincott, Philadelphia, pp 454–522

Lewis OJ (1989) Functional morphology of the evolving hand and foot. Clarendon Press, Oxford

Long C II (1981) Electromyographic studies of hand function. In: Tubiana R (ed) The hand, vol 1. W.B. Saunders, Philadelphia, pp 427–440

Long C II, Brown ME (1964) Electromyographic kinesiology of the hand muscles moving the long finger. J Bone Joint Surg Am 46A:1683–1706

Long C II, Conrad PW, Hall EA, Furler SL (1970) Intrinsic-extrinsic muscle control of hand in power grip and precision handling. An electromyographic study. J Bone Joint Surg [Am] 52A:853–867

Macalister A (1873) The muscular anatomy of the gorilla. Proc R Irish Acad Sci 1:501–506

Marzke MW (1997) Precision grips, hand morphology, and tools. Am J Phys Anthropol 102:91–110

Marzke MW, Toth N, Schick K, Reece S, Steinberg B, Hunt K, Linscheid RL, An KN (1998) EMG study of hand muscle recruitment during hard hammer percussion manufacture of Oldowan tools. Am J Phys Anthropol 105:315–332

Marzke MW, Marzke RF, Linscheid RL, Smutz P, Steinberg B, Reece S, An KN (1999) Chimpanzee thumb muscle cross sections, moment arms and potential torques, and comparisons with humans. Am J Phys Anthropol 110:163–178

McMurrich JP (1903a) The phylogeny of the forearm flexors. Am J Anat 2:177–209
McMurrich JP (1903b) The phylogeny of the palmar musculature. Am J Anat 2:463–500
Michilsens F, Vereecke EE, D'Août K, Aerts P (2009) Functional anatomy of the gibbon forelimb: adaptations to a brachiating lifestyle. J Anat 215:335–354
Miller RA (1943) Functional and morphological adaptations in the forelimbs of the slow lemurs. Am J Anat 73:153–183
Murie J, Mivart SG (1872) On the anatomy of the Lemuroidea. Trans Zool Soc Lond 7:1–113
Myatt JP, Crompton RH, Payne-Davis RC, Vereecke EE, Isler K, Savage R, D'Août K, Günther MM, Thorpe SKS (2012) Functional adaptations in the forelimb muscles of non-human great apes. J Anat 220:13–28
Napier JR (1952) The attachments and function of the abductor pollicis brevis. J Anat 86:335–341
Napier JR (1955) The form and function of the carpo-metacarpal joint of the thumb. J Anat 89:362–369
Napier JR (1956) The prehensile movements of the human hand. J Bone Joint Surg [Br] 38B:902–913
Napier JR (1961) Prehensility and opposability in the hands of primates. Symp Zool Soc Lond 5:115–132
Napier J (1993) Hands. Revised edition by Russell H. Tuttle. Princeton University Press, Princeton
Netter FH (2006) Atlas of human anatomy, 4th edn. Saunders Elsevier, Philadelphia
Ogihara N, Kunai T, Nakatsukasa M (2005) Muscle dimensions in the chimpanzee hand. Primates 46:275–280
Oishi M, Ogihara N, Endo H, Ichihara N, Asari M (2009) Dimensions of forelimb muscles in orangutans and chimpanzees. Anat Rec 215:373–382
Patel BA, Larson SG, Stern JT Jr (2012) Electromyography of wrist and finger flexor muscles in olive baboons (*Papio anubis*). J Exp Biol 215:115–123
von Polak C (1908) Die Anatomie des Genus *Colobus*. Verh Koninklijke Akad Wetenschappen Amsterdam Afdeeling Natuurkunde 14:1–247
Rauwerdink GP (1993) Muscle fibre and tendon lengths in primate extremities. In: Preuschoft H, Chivers DJ (eds) Hands of primates. Springer, Vienna, pp 207–223
Raven HC (1950) Regional anatomy of the gorilla. In: Gregory WK (ed) The Henry Cushier Raven Memorial Volume: the anatomy of the gorilla. Columbia University Press, New York, pp 15–188
Schmidt H-M, Lanz U (2004) Surgical anatomy of the hand. Georg Thieme Verlag, Stuttgart
Schoenwolf GC, Bleyl SB, Brauer PR, Francis-West PH (2009) Larsen's human embryology, 4th edn. Churchill Livingstone/Elsevier, Philadelphia
Schuenke M, Schulte E, Schumacher U (2006) Thieme atlas of anatomy. General anatomy and musculoskeletal system. Georg Thieme Verlag, Stuttgart
Schultz M (1984) Osteology and myology of the upper extremity of *Tarsius*. In: Niemitz C (ed) Biology of tarsiers. Gustav Fischer Verlag, Stuttgart, pp 143–165
Schultz M (1986) The forelimb of the Colobinae. In: Swindler DR, Erwin J (eds) Comparative primate biology: systematics, evolution and anatomy. Alan R. Liss, New York, pp 559–661
Sonntag CF (1923) On the anatomy, physiology, and pathology of the chimpanzee. Proc Zool Soc Lond 93:323–429
Sonntag CF (1924) On the anatomy, physiology, and pathology of the orangutan. Proc Zool Soc Lond 94:323–429
Stack HG (1962) Muscle function in the fingers. J Bone Joint Surg [Br] 44B:899–909
Stein BR (1981) Comparative limb myology of two opossums, *Didelphis* and *Chironectes*. J Morphol 169:113–140
Straus WL Jr (1941a) The phylogeny of the human forearm extensors. Hum Biol 13:23–50
Straus WL Jr (1941b) The phylogeny of the human forearm extensors. Hum Biol 13:203–238
Straus WL Jr (1942a) Rudimentary digits in primates. Q Rev Biol 17:228–243
Straus WL Jr (1942b) The homologies of the forearm flexors: urodeles, lizards, mammals. Am J Anat 70:281–317
Susman RL (1994) Fossil evidence for early hominid tool use. Science 265:1570–1573

Susman RL (1998) Hand function and tool behavior in early hominids. J Hum Evol 35:23–46
Susman RL, Stern JT Jr (1979) Telemetered electromyography of flexor digitorum profundus and flexor digitorum superficialis in *Pan troglodytes* and implications for interpretation of the O.H.7 hand. Am J Phys Anthropol 50:565–574
Susman RL, Stern JT Jr (1980) EMG of the interosseous and lumbrical muscles in the chimpanzee (*Pan troglodytes*) hand during locomotion. Am J Anat 157:389–397
Susman RL, Jungers WL, Stern JT Jr (1982) The functional morphology of the accessory interosseous muscle in the gibbon hand: determination of locomotor and manipulatory compromises. J Anat 134:111–120
Susman RL, Nyati L, Jassal MS (1999) Observation on the pollical palmar interosseous muscle (of Henle). Anat Rec 254:159–165
Tocheri MW, Orr CM, Jacofsky MC, Marzke MW (2008) The evolutionary history of the hominin hand since the last common ancestor of *Pan* and *Homo*. J Anat 212:544–562
Tubiana R (ed) (1981) The hand, vol 1. W.B. Saunders, Philadelphia
Tuttle RH (1969) Quantitative and functional studies on the hands of the Anthropoidea, I, the Hominoidea. J Morphol 128:309–363
Tuttle RH (1970) Postural, propulsive and prehensile capabilities in the cheiridia of chimpanzees and other great apes. In: Bourne GH (ed) The chimpanzee, vol 2. Karger, Basel, pp 167–263
Tuttle RH (1972a) Functional and evolutionary biology of hylobatid hands and feet. In: Rumbaugh DM (ed) Gibbon and Siamang, vol 1. Karger, Basel, pp 136–206
Tuttle R (1972b) Relative mass of cheiridial muscles in catarrhine primates. In: Tuttle R (ed) The functional and evolutionary biology of primates. Aldine, Chicago, pp 262–291
Tuttle RH, Basmajian JV (1974) Electromyography of forearm musculature in *Gorilla* and problems related to knuckle-walking. In: Jenkins FA Jr (ed) Primate locomotion. Academic Press, New York, pp 293–347
Valentin P (1981) The interossei and the lumbricals. In: Tubiana R (ed) The hand, vol 1. W.B. Saunders, Philadelphia, pp 244–254
Wilder B (1862) Contributions to the comparative myology of the chimpanzee. Boston J Nat Hist 6:353–384
Williams PL (ed) (1995) Gray's anatomy, 38th edn. Churchill Livingstone, New York
Windle BCA (1883) On the embryology of the mammalian muscular system. No. 1—the short muscles of the human hand. Trans R Irish Acad 28:211–240
Windle BCA (1889) The flexors of the digits of the hand. I. The muscular masses of the forearm. J Anat 24:72–84
Woollard HH (1925) The anatomy of *Tarsius spectrum*. Proc Zool Soc Lond 1925:1071–1184
Wood Jones F (1942) The principles of anatomy as seen in the hand, 2nd edn. Ballière, Tindall, and Cox, London

Chapter 8
Comparative and Functional Morphology of the Primate Hand Integument

Stephanie A. Maiolino, Amanda K. Kingston, and Pierre Lemelin

1 Introduction

The primate hand is covered by a thick layer of hairless skin on its palmar surface, while the dorsal surfaces of the fingertips terminate in flattened nails (with a few exceptions; see Sect. 4.2). The volar (i.e., palmar) skin and nails are part of what is defined in human anatomy as the **integumentary system**—the organ system that covers and protects the body (Montagna and Parakkal 1974; Williams 1995). Primates use the integument of their hands to interact with the outside world, and as such, it plays a vital role in sensation, manipulation, and locomotion (e.g., climbing on arboreal supports). Therefore, the integument is well adapted to provide a durable, secure gripping surface while still managing to provide one of the most sensitive tactile surfaces of the body. The volar integument of human beings, like other primates, is also critical for daily activity. It plays a role in myriads of activities such as social communication (e.g., judging the grip of a hand shake), food selection (e.g., testing the ripeness of a fruit at a grocery store), personal grooming, and, of

S.A. Maiolino
Department of Pathology and Anatomical Sciences, University of Missouri School of Medicine, Columbia, MO 65211, USA

Interdepartmental Doctoral Program in Anthropological Sciences, Stony Brook University, Stony Brook, NY 11794-4364, USA

A.K. Kingston
Interdepartmental Doctoral Program in Anthropological Sciences, Stony Brook University, Stony Brook, NY 11794-4364, USA

P. Lemelin (✉)
Division of Anatomy, Department of Surgery, Faculty of Medicine and Dentistry, University of Alberta, 5-05A Medical Sciences Building, Edmonton, AB, Canada, T6G 2H7
e-mail: plemelin@ualberta.ca

T.L. Kivell et al. (eds.), *The Evolution of the Primate Hand*, Developments in Primatology: Progress and Prospects, DOI 10.1007/978-1-4939-3646-5_8

course, tool use (e.g., writing with a pen). Our hand integument is so important for us that we take meticulous, and often expensive, care of it (e.g., hand washing and moisturizing, manicures).

The integument of the hand has been featured prominently in comparative studies of primate anatomy and evolution. Since the late nineteenth century, primates have been recognized as the only mammalian order in which every living member possesses a nail, as opposed to a hoof or a claw, on at least one of its digits (Mivart 1873; Cartmill 1972, 1974a, b; Martin 1986; Wible and Covert 1987). This is not to say that primates are the only mammals that possess nails (a few rodents and marsupials also have them; Musser 1972; Russell 1986), but that nails are relatively uncommon in other clades. The volar skin of the hand has been emphasized as well, particularly in the writings of Wood Jones and Napier. For example, the tactile nature of the hand (through neural receptors found in the volar skin) is a key element of Wood Jones' (1916) theory of "emancipation" of the forelimb in primates. In his usual witty prose, Napier (1993: 29) summarized the function of the skin as the organ that "keeps the blood in, and the rain out."

In this chapter, we aim to provide a comprehensive review of the integument of the hand in primates. We begin with a brief description of the gross and microscopic anatomy of the volar skin and nail using humans as a model and from which comparisons with nonhuman primates and other mammals are made. Some of the functional roles of the hand integument are then briefly presented. We end the chapter by providing several avenues for future research on this relatively understudied aspect of the primate hand.

2 Basic Structure of the Primate Volar Skin

The integument of primates is not a homogeneous organ. The hair-bearing skin that covers much of the limbs and torso is the most familiar, but smaller regions of the integument—the lips, sexual skin, palms of the hands, and soles of the feet—are highly specialized and may only superficially resemble one another. While all regions of the integument share major roles in protecting deep tissues from ultraviolet damage, preventing dehydration, regulating body temperature, and housing sensory organs, different regions are additionally specialized to carry out specific functions. The lips, for example, are packed full of touch and heat receptors which inform the brain of the texture and temperature of food about to enter the mouth and thereby help to prevent ingestion of items that are dangerous (too sharp, too hot, etc.) to the digestive tract. Here, we are interested in the primate volar skin or the thick, ridged skin that covers the palms and digits of our hands.

2.1 *Microstructure and Histology of the Volar Skin*

Before discussing the gross morphology of the human volar skin and comparing it to other primates and mammals, it is necessary to have a basic understanding of its anatomy at the microscopic level. Regardless of where it is found on the body, the

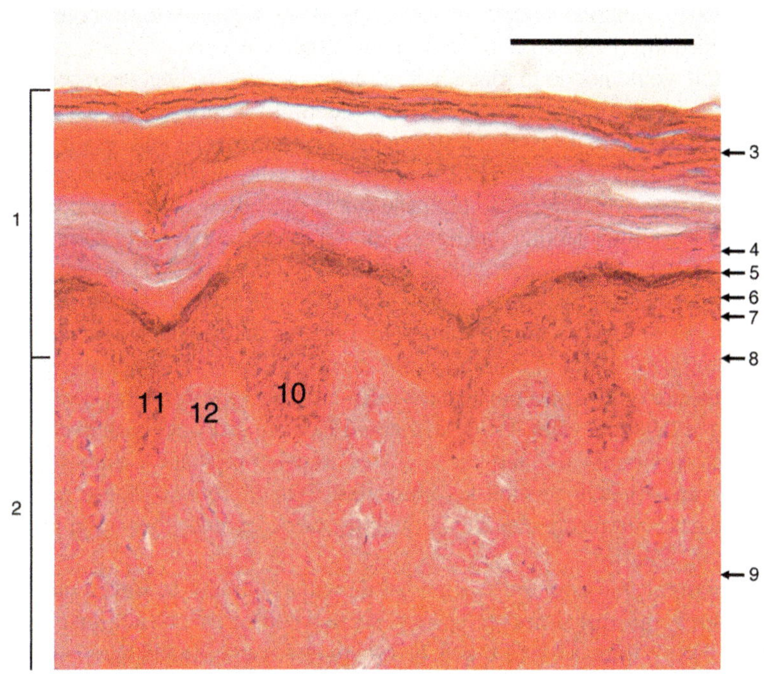

Fig. 8.1 Photomicrograph of histological section through the apical pad of a human neonate finger. The epidermis (**1**) can be differentiated from the dermis (**2**). From superficial (*top*) to deep (*bottom*), the following layers of the volar skin can be identified: stratum corneum (**3**), stratum lucidum (**4**), stratum granulosum (**5**), stratum spinosum (**6**), stratum basale (or germinativum) (**7**), papillary layer (**8**), and reticular layer (**9**). At the epidermal-dermal junction (at the level of the stratum basale), intermediate ridges (**10**) and limiting ridges (**11**) (fingerlike projections of the epidermal layer going downward) can be observed alternating. The dermal papilla (**12**) is therefore the resulting space in the dermis between these two epidermal ridges. Microscopic slide provided courtesy of Professor Norman Taslitz (University of New Mexico). Scale bar is approximately 100 μm

skin is divided anatomically into two layers: the deep-lying **dermis** and the superficial **epidermis** (Montagna and Parakkal 1974; Leeson et al. 1985; Fawcett 1994; Williams 1995; Fig. 8.1). It is not only the location of a cell within the integument that determines whether it is considered part of the dermis or epidermis; the *types* of cells that make up each layer vary as well. The cells comprising the dermis and epidermis are unique to their respective layers and do not travel between them (Montagna and Parakkal 1974; Williams 1995).

The epidermis is a layer in flux, shedding and generating cells at a consistent rate, with cells arising in deeper layers and being pushed superficially until they are sloughed off and replaced by younger cells (Montagna and Parakkal 1974). Because the epidermis comprises the outermost covering of the body, the cells here are subjected to a constant barrage of trauma from ultraviolet light, wind, water, chemicals, and friction. Cells that have reached the outer surface of the epidermis are shed very quickly—usually within 2 weeks of arrival—due in large part to these hostile conditions (Montagna and Parakkal 1974; Williams 1995). Because of this constant shifting and shedding, the types of cells comprising the epidermis are remarkably consistent; it would not do to root a hair or tuck a blood vessel between cells that are constantly being forced outward. In fact, the epidermis is entirely avascular (it receives nourishment from blood vessels located in the dermis below it), and only a very small number of nerve fibers pass into it. The majority of cells found here are keratinocytes, but melanocytes (which produce pigment that is passed into the keratinocytes) and Merkel cells (a slow-adapting touch receptor discussed later in the chapter) are common here as well (Montagna and Parakkal 1974; Williams 1995; Halata et al. 2003).

The epidermis can be further broken down into five layers, or strata, that keratinocytes move through before they are shed. Like the dermis and epidermis at large, the types of tissue found in each layer are slightly different. The deepest stratum—**stratum basale** or **stratum germinativum**—consists of a single layer of live cells (Leeson et al. 1985; Fawcett 1994; Williams 1995; Fig. 8.1). Here, stem cells divide to produce keratinocytes, and melanocytes transfer pigment granules (melanosomes) to these newly formed keratinocytes. Tucked between these are Merkel cells, which communicate chemically with the nerve endings housed in the dermis (Halata et al. 2003).

The keratinocytes from the stratum basale will eventually be forced upward into the **stratum spinosum** as new cells are generated beneath them (Leeson et al. 1985; Fawcett 1994; Williams 1995; Fig. 8.1). Melanocytes and Merkel nerve endings are common in this stratum as well.

In the **stratum granulosum**, keratinocytes begin to solidify and fill with keratin (Leeson et al. 1985; Fawcett 1994; Williams 1995; Fig. 8.1). As they move through the stratum (usually three to five cell layers), the keratinocytes become thinner and flatter as their membranes become thicker and less permeable. Eventually, oxygen and nutrients will no longer be able to enter the cell, the nucleus and organelles will starve and disintegrate, and the cell will die. Though no longer living cells, these flattened keratinocytes—now called corneocytes—are not finished with their journey and will continue to move through additional strata (Montagna and Parakkal 1974).

From the stratum granulosum, the majority of corneocytes will enter directly into the **stratum corneum**, the outermost layer of the epidermis (Leeson et al. 1985; Fawcett 1994; Williams 1995; Fig. 8.1). This most superficial stratum is also the thickest: it is composed of anywhere from 20 to 30 layers of corneocytes that interlock to form a tight outer barrier. Upon entering the stratum corneum, the corneocytes have a life span averaging 2 weeks before they are shed (Fawcett 1994; Williams 1995).

While the majority of the integument is formed of just these four strata, a fifth is found in the volar skin. The **stratum lucidum**, set between the strata granulosum and corneum, is formed of half-keratinized keratinocytes exiting the stratum granulosum slightly earlier than their counterparts in hairy skin. These cells pause on their journey to the skin's surface to form a translucent fifth stratum of cells, two to three cell layers thick, before completing the process of keratinization and continuing to the stratum corneum.

In stark contrast to the constant turnover of the epidermis, the cells in the dermis are relatively stable. After death, they are resorbed by the body rather than forced outward, creating a steady bedrock in which accessory structures of the integument root themselves. Because of this, the tissues present in the dermis are far more varied than those in the epidermis. A meshwork of collagen and elastin fibers holds together an amalgamation of nervous tissue, vascular tissue, glands, muscle, and hair follicles with only the barest hint of logic in their arrangement (Leeson et al. 1985; Fawcett 1994; Williams 1995).

The dermis can be divided into two layers. The most superficial is the **papillary layer**, which is named for the many **dermal papillae** (or "nipple-like" structures) that project upward and interlock with the **intermediate ridges** of the stratum basale of the epidermis (Cauna 1956; Leeson et al. 1985; Fawcett 1994; Williams 1995; Fig. 8.1). In between each intermediate ridge, a shallower **limiting ridge** (also from the stratum basale of the epidermis) can be observed as well; the dermal space created between an intermediate ridge and limiting ridge is therefore the dermal papilla (Cauna 1956; Fig. 8.1). Dermal papillae maximize the contact area between the dermis and epidermis, which is especially important as all blood and nutrients reach the avascular epidermis by flowing through the limited spaces between cells at this boundary. This enhanced contact area may also be important to keep the integrity of the dermal-epidermal junction as shearing forces are applied onto the volar surface (MacKenzie and Iberall 1994). Because of this, blood vessels within the dermis will often form loops within papillae to further utilize the expanded area. Meissner's corpuscles (rapidly adapting mechanoreceptors and encapsulated nerve endings; see below and Chap. 6) are common here and usually sit directly within a papilla (Cauna 1956; Winkelmann 1963; Halata 1975; Williams 1995; Purves et al. 2008).

The **reticular layer** sits deep to the papillary layer and has the most diverse composition of any layer of the skin (Leeson et al. 1985; Fawcett 1994; Williams 1995; Fig. 8.1). Blood vessels traverse vertically across this layer, connecting the tiny capillaries of the papillary layer to the larger vessels sitting in the subcutaneous tissues. These same vessels also supply a number of structures within the reticular layer itself. Hair follicles are rooted here and are invariably surrounded by sebaceous oil glands that secrete their products directly onto the hair shafts. The hair follicles are further connected to surrounding tissues by arrector pili muscles: bands of smooth muscle that pull the hair follicle toward the surface of the skin (commonly referred to as "hair standing on end" or "goose bumps"). Interspersed around these complexes are the bulbs of sweat glands and the very large, onion-shaped Pacinian corpuscles (rapidly adapting mechanoreceptors and encapsulated nerve endings similar to Meissner's corpuscles but working at a lower response threshold and thus at higher frequencies; Williams 1995; Purves et al. 2008). Deep to these lies the hypodermis or subcutaneous tissue, which will not be discussed here.

2.2 Gross Morphology of the Volar Skin and Dermatoglyphics

Turning from microstructure to visible external anatomy, the palmar surface of the hand can be examined with the naked eye. This exercise reveals some interesting anatomical facts: the palmar surface of the hand can be divided into discrete areas (**volar pads**) separated by creases or lines (Fig. 8.2). On the palm itself, a thenar,

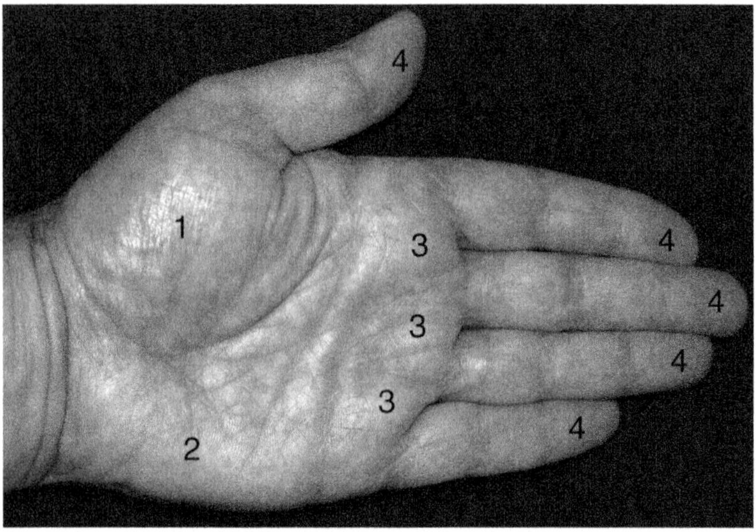

Fig. 8.2 Volar pads of the human hand. Thenar (**1**), hypothenar (**2**), interdigital (**3**), and apical (**4**) pads can be distinguished on the palmar surface of the hand

hypothenar, and three interdigital pads can be discerned, as well as an apical pad surmounting the tip of each digit. These pads are separated by flexure lines, which are organized with the underlying joints of the hand (see Wood Jones 1942). The surface of the volar skin, especially the apical pads, is characterized by **papillary ridges** that are organized into intricate patterns of arches, loops, and whorls known as **dermatoglyphics** (Midlo and Cummings 1942). Because of their heritable nature, dermatoglyphics (*derma*=skin, *glyph*=carving) have been of interest to anthropologists and criminologists alike ever since Sir William Herschel first noticed that the collection of palm prints he accepted in lieu of signatures could be used to identify their owners. Papillary ridges are not simple foldings on the surface of the epidermis. They extend deeper into the epidermal layer (all the way to the stratum basale) where they transition into intermediate ridges (see above; Fig. 8.1). Napier (1993) used the iceberg analogy to describe the depth of the papillary ridges and underlying intermediate ridges. The valleys that border each papillary ridge are known as **papillary grooves**, which transition into limiting ridges in the deeper epidermal layer (see above; Fig. 8.1). As discussed in the previous section, the epidermal-dermal junction is not smooth but punctuated by an alternating sequence of dermal papillae and intermediate and limiting ridges (see Fig. 8.1). Interestingly, the pattern of papillary ridges found on the surface of the volar skin is preserved in the dermal layer as a mirror image and forms the **dermal ridges** (Leeson et al. 1985; Okajima and Asai 1985). Papillary ridges are punctured by small funnel-shaped pores, which drain fluid from underlying sweat glands (Fawcett 1994; Williams 1995). In this way, papillary ridges provide a frictional surface with varying degrees of moisture when the hand contacts an object (see Sect. 5.1).

3 Basic Structure of the Primate Nail

The shape and form of nails (called ungulae, sing. ungula) vary considerably among primate groups, but all share the same basic anatomy (Bruhns 1910). This anatomy is described using humans as a model [following Johnson and Cohen (1975), Beaven and Brooks (1984), Stenn and Fleckman (2000), Stone et al. (2000) and Zook (2003), with the exception of the eponychium (see below)].

3.1 Microscopic and Gross Morphology of the Nail

The **ungula** (or nail plate; Figs. 8.3a and 8.4a) is derived from the epidermis, is homologous with the stratum corneum of the skin, and consists of corneocytes (i.e., hardened, dead cells containing a protein known as keratin; Fawcett 1994; Williams 1995; see Sect. 2.1). The nail plate drapes over the dorsal surface of the distal phalanx and is surrounded on three sides by folds of the skin (see Chap. 4 for more information on the distal phalanx and illustrations of its relationship to the ungula). The two lateral folds are called the **paronychia** (sing. paronychium; Fig. 8.3c), while the proximal fold is referred to as the **proximal nail fold** (Figs. 8.3b and 8.4b, c). A strip of paronychium is what pulls away from the rest of the digit in what is colloquially referred to as a "hang nail." The proximal nail fold is a portion of the skin that folds under itself such that it has a dorsal layer (the visible skin just proximal to the nail; Figs. 8.3c and 8.4b) and a volar portion (Fig. 8.4c) that lies in contact with the nail plate; it covers roughly one quarter of the plate. The **cuticle** (Figs. 8.3d and 8.4d) is a thin layer that extends from the volar surface of the fold onto a short portion of the dorsal surface of the nail plate. It is a cornified layer, meaning that the cells within it are keratinized, like those of the outermost layer of the rest of the skin and the nail plate. Various portions of the proximal nail fold have been called the eponychium, but this can be a confusing term because different authorities use it in different ways. It is sometimes used interchangeably with proximal nail fold (e.g., Le Gros Clark 1936), or it can refer to the band of tissue at the distal apex of the fold (e.g., Stone et al. 2000), the cuticle (e.g., Homberger et al. 2009), or the cuticle-generating volar portion of the fold (e.g., Perrin et al. 2004). Here, we will avoid using the term eponychium, preferring proximal nail fold instead (which includes all relevant portions and outgrowths). The nail plate is attached to the **nail bed**, the epidermal layers that lie deep to the nail. The portion of the nail bed that is located under the proximal region of the nail plate is called the **matrix**, and it is responsible for producing the plate. As the nail is homologous to the stratum corneum, the matrix is homologous to deeper layers of the epidermis. In some references, especially from the nonhuman primate literature, the matrix is referred to as the germinal matrix (Fig. 8.4e), while the remainder of the nail bed is referred to as the sterile matrix (Le Gros Clark 1936; Fig. 8.4f). Here, we simplify this terminology by using the term matrix to refer to the germinal (nail-producing)

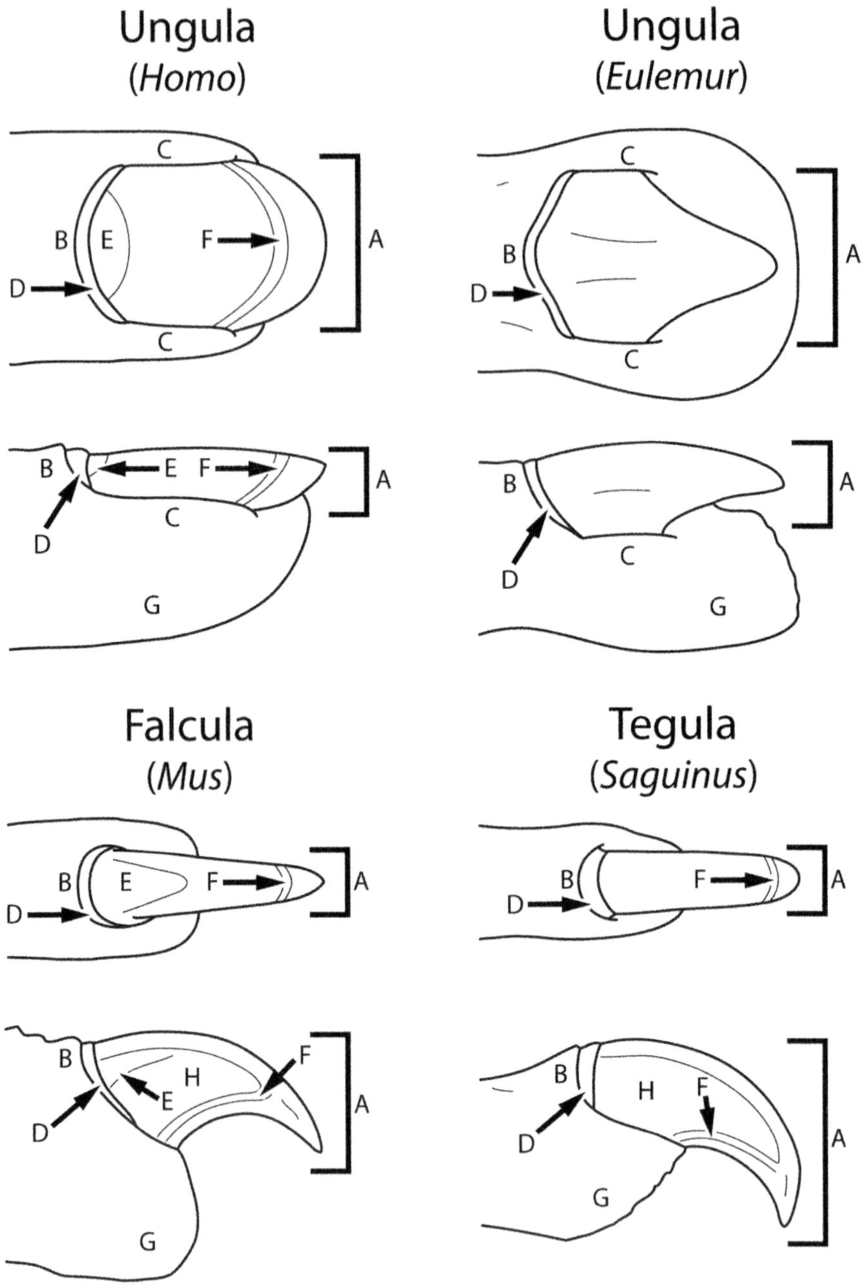

Fig. 8.3 Ungual morphology in three primates (*Homo*, *Eulemur*, and *Saguinus*) and mouse (*Mus*). Pigmentation (not illustrated) of the unguis differs among species: humans and mice have ungues that are largely translucent, lemurs have heavily pigmented ungues, and tamarins are intermediate. Therefore, structures that lie deep to ungues are only illustrated for species in which the unguis is translucent enough to show them. Humans and lemurs have flattened nails (ungulae) compared to the claw-like nails (tegulae) of tamarins and claws (falculae) of mice. In spite of these differences

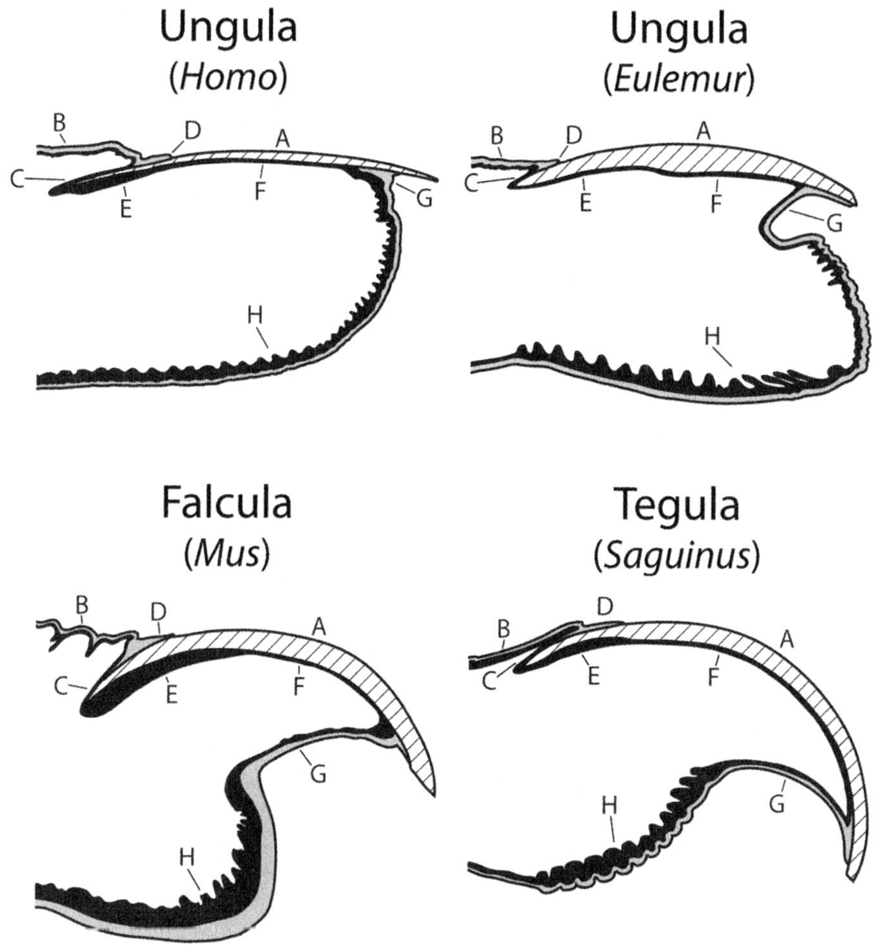

Fig. 8.4 Epidermal structures shown in sagittal sections through the fingertips of three primates (*Homo*, *Eulemur*, and *Saguinus*) and mouse (*Mus*). Stratum corneum and cornified tissues are *lightly shaded*, while deeper layers of the epidermis are shown in *black*. Humans and lemurs have flattened nails (ungulae) compared to the claw-like nails (tegulae) of tamarins and claws (falculae) of mice. In spite of these differences in ungual morphology, homologous structures and regions can be identified: unguis (nail/claw plate) (**A**), dorsal portion of proximal nail/claw fold (**B**), volar portion of proximal nail/claw fold (**C**), cuticle (**D**), matrix (germinal portion of nail/claw bed) (**E**), sterile portion of nail/claw bed (**F**), hyponychium (*black* layer) and subunguis (*gray* layer) (**G**), and epidermis of apical pad (**H**). Specimens not on same scale

Fig. 8.3 (continued) in gross ungual morphology, homologous structures and regions can be identified: unguis (nail/claw plate) (**A**), proximal nail/claw fold (**B**), paronychia (**C**), cuticle (**D**), lunula (**E**), distal edge of nail/claw bed seen through a translucent unguis (often called the onychodermal band) (**F**), apical pad (**G**), and dermis of claw bed seen through translucent overlying structures (often called the quick in non-primate mammals) (**H**). Specimens not on same scale portion of the nail bed. The **hyponychium** (Fig. 8.4g, black layer) is a portion of the epidermis that lies between the nail bed and the epidermis of the apical pad (Figs. 8.3g and 8.4h); it contains an extremely high proliferation of white blood cells (Pardo-Castello 1960). The hyponychium has a cornified outer layer called the subunguis or solehorn (Fig. 8.4g, gray layer), which attaches to the volar surface of the nail plate. Together, the hyponychium and subunguis form a plug that prevents foreign matter from penetrating the nail bed. Unlike many other primates, the human ungula is generally translucent, allowing for the observation of features that cannot be seen in species with heavily pigmented nails: the distal end of the matrix may be visible through the nail plate as a light-colored crescent called the **lunula** (Fig. 8.3e), and a thin band that marks the junction between the hyponychium and nail bed is often seen as the **onychodermal band** (also called the yellow line; Fig. 8.3f).

4 Comparative Aspects of the Hand Integument

The integument of the hand presents considerable variation among primates: some species have discrete and bulbous volar pads with digits that bear claw-like nails, whereas others have flatter and coalesced volar surfaces with digits bearing flattened nails (Whipple 1904; Biegert 1959, 1961, 1963; Cartmill 1974a, 1979; Hamrick 1998). Variation in integument morphology is even more dramatic when examining some of its microstructure and when close relatives of primates (i.e., tree shrews and dermopterans) and other pentadactyl mammals are considered. The following section reviews some of that diversity in volar skin and ungual morphology.

4.1 Variation in Volar Surface Morphology in Primates and Other Mammals

When comparing the histological section of an apical pad of a human specimen (Fig. 8.1) to that of other primates (Fig. 8.5), the resemblances are quite obvious. Representatives of strepsirrhines, tarsiers, and anthropoids show the same alternating pattern of dermal papillae flanked by a deeper intermediate ridge on one side and the shallower limiting ridge on the other (see Fig. 8.5h). The stratum basale lies directly above this and can be thought of as a transition zone between the deep and superficial ridging patterns; its deep edge adheres closely to the contours of the dermal papillae and dermal ridges; however, its superficial edge grades into a much simpler undulation of ridges and grooves roughly of equal amplitudes (Fig. 8.5). The stratum granulosum, stratum lucidum, and stratum corneum follow the same ridging pattern; the external surface of the stratum corneum is folded into a series of equally deep papillary grooves and papillary ridges (Fig. 8.5). In spite of these microstructural similarities, the depth, shape, and relative size of the papillary ridges/grooves in the superficial epidermis and intermediate/limiting ridges in the deep epidermis appear to vary considerably across primate species (Hamrick 1998, 2003; Lemelin 2000; Fig. 8.5). However, the full range of variation remains undocumented at present.

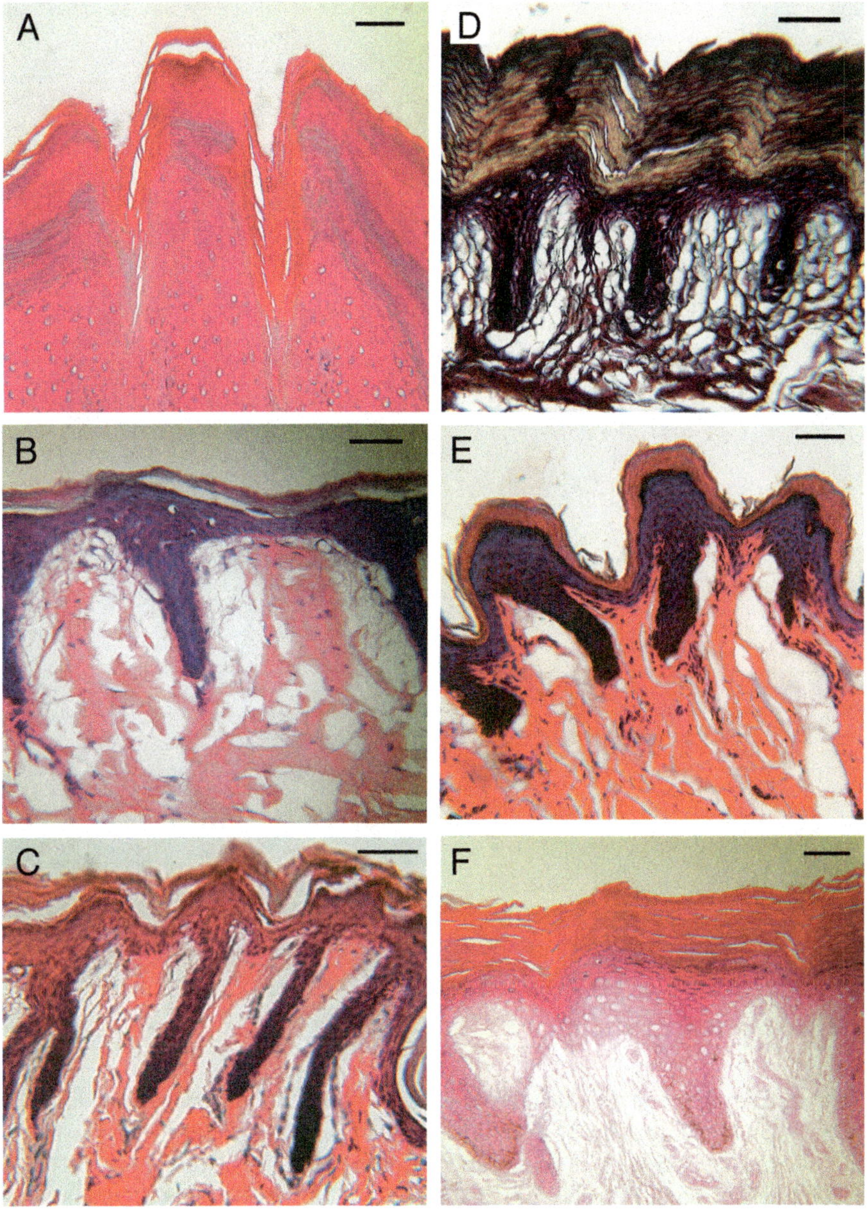

Fig. 8.5 Photomicrographs of histological sections through the apical pad of primates and other mammals. Note the differences in size and shape of papillary ridges (**1**) and grooves (**2**) (superficial epidermis of each specimen at *top* of each panel, labeled in panel **h**) and intermediate ridges (**3**) and limiting ridges (**4**) (deep epidermis at *bottom* of each panel, labeled in panel **h**) between the Virginia opossum (**a**, *Didelphis virginiana*), greater bushbaby (**b**, *Otolemur crassicaudatus*), western tarsier (**c**, *Tarsius* (or *Cephalopachus*) *bancanus*), tree shrew (**d**, *Tupaia glis*), sifaka(**e**, *Propithecus verreauxi*), common marmoset (**f**, *Callithrix jacchus*), Goeldi's monkey (**g**, *Callimico goeldii*), capuchin monkey (**h**, *Cebus* (or *Sapajus*) *apella*), squirrel monkey (**i**, *Saimiri sciureus*), and rhesus macaque (**j**, *Macaca mulatta*). *White arrows* in panels **h** and **j** point to Meissner's corpuscles. Specimen information can be found in Lemelin (2000). Scale bars are approximately 100 μm

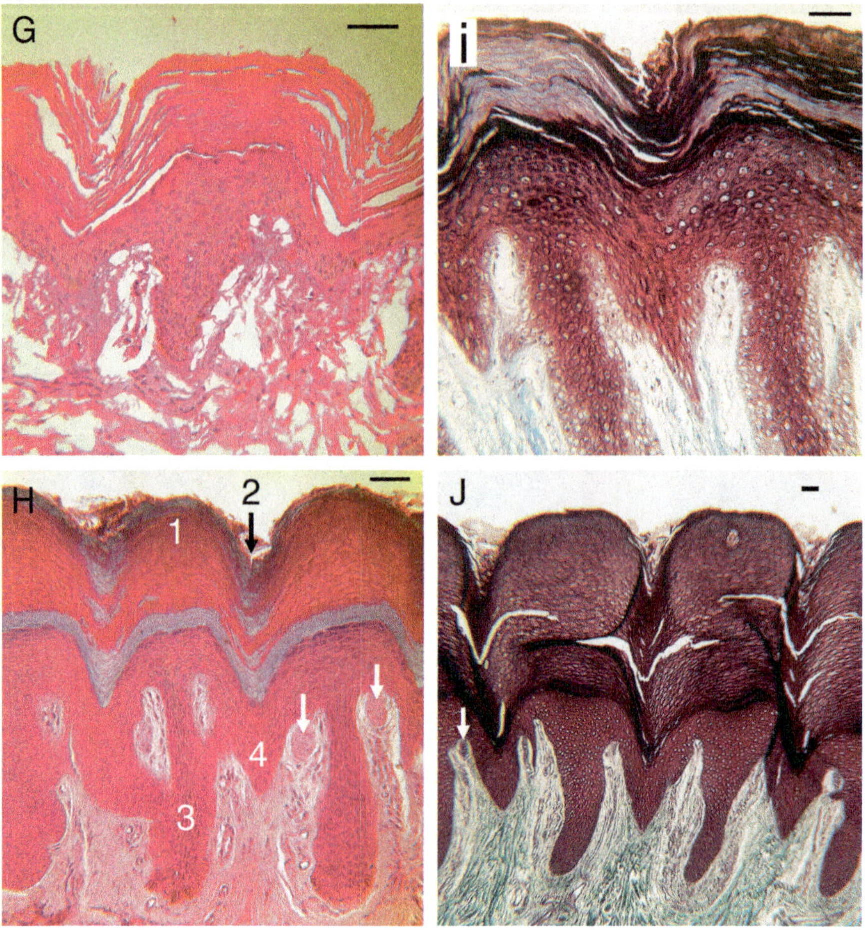

Fig. 8.5 (continued)

Outside of the primate order, the volar skin of some non-primate mammals such as marsupials and tree shrews also shows distinct papillary ridges with corresponding intermediate ridges in the deepest layer of the epidermis (Rosenberg and Rose 1999; Lemelin 2000; Hamrick 2001b, 2003; Fig. 8.5a, d). The same alternating pattern of respectively deeper and shallower intermediate and limiting ridges on the dermal surface can be observed, especially in tree shrews (Lemelin 2000; Fig. 8.5d). Other mammals such as raccoons have papillary ridges on the volar surface of their paws. However, the epidermal-dermal junction is characterized by a pattern of deep limiting ridges and shallow intermediate ridges, the opposite pattern observed in primates, tree shrews, and didelphids (see Fig. 8 in Munger and Pubols 1972). The rat presents an interesting combination of traits: in spite of having relatively smooth-surfaced volar pads, the epidermal-dermal junction has a characteristic groove-and-rampart appearance with deep fingerlike epidermal projections surrounding the dermal papillae (Okajima and Asai 1985; Okajima 1991).

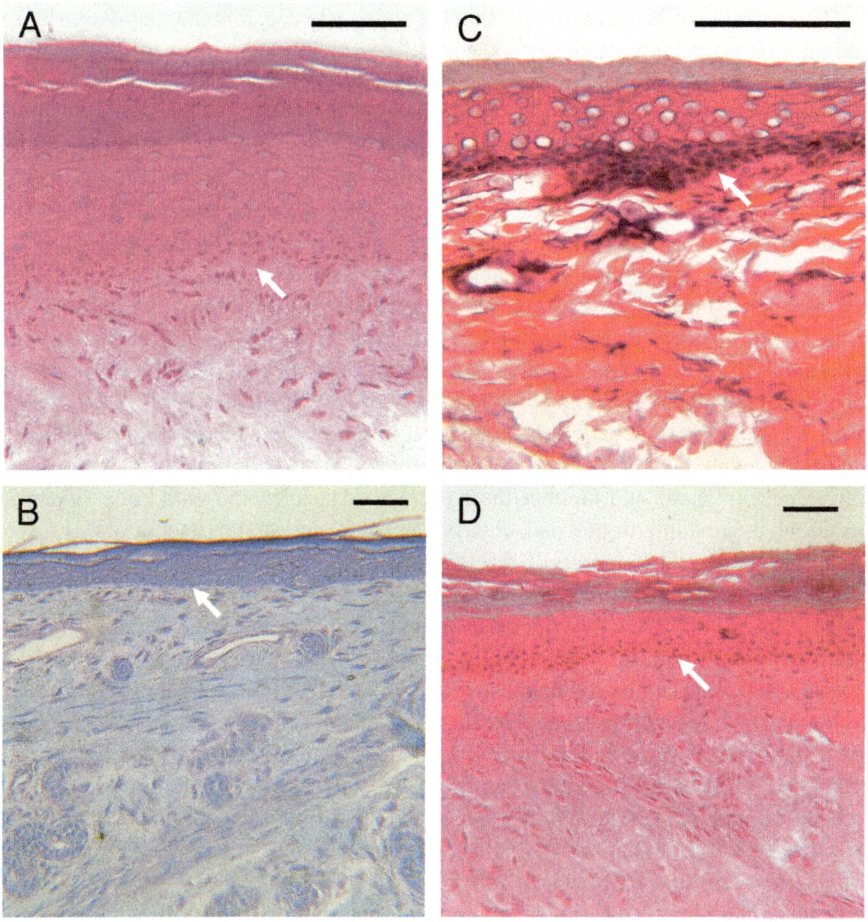

Fig. 8.6 Photomicrographs of histological sections through the thenar or apical pads of bats, a dermopteran, and an insectivore. Note similarities in the smooth epidermis (superficial epidermis is at *top* of panels) and linear epidermal-dermal junction with no or very small ridges (*white arrows*) between the short-tailed shrew (**a**, *Blarina* sp.), colugo (**b**, *Cynocephalus* sp.), leaf-nose bat (**c**, *Phyllostomus* sp.), and fruit bat (**d**, *Macroglossus* sp.). Specimen information can be found in Lemelin (2000). Scale bars are approximately 100 μm

Other pentadactyl mammals have volar skin with shallow or no papillary ridges. Unlike primates, the surface of the thumb pads of some bat species and the apical pads of dermopterans or colugos (*Cynocephalus*) and some insectivores (the short-tailed shrew *Blarina*) is smooth, and the corresponding epidermal-dermal junction is linear with no or very few small ridges (Thewissen and Etnier 1995; Lemelin 2000; Hamrick 2003; Fig. 8.6). However, it should be pointed out that the volar surface of the thumb pad of some bat species (e.g., *Thyroptera* and *Pipistrellus*) has distinctive folds (Wimsatt and Villa-R 1970; Thewissen and Etnier 1995; Hamrick 2003).

There are two distinct morphs within the superorder Euarchonta: papillary ridges are found in primates and tree shrews, while dermopterans maintain a much smoother volar surface (Lemelin 2000). Since bats are no longer believed to be close relatives of euarchontans on the basis of molecular evidence (Springer et al. 2004), similarities in volar skin between some bats and dermopterans are most likely convergent. The primitive condition for Euarchonta is still unclear, although the highly derived nature of the hand and postcranial anatomy of dermopterans do suggest that smooth epidermis could also be autapomorphic (i.e., uniquely derived) for this clade.

The number and shape of volar pads vary among mammalian species. The arrangement, most likely primitive for mammals, consists of six volar pads on the palm of the hand: a thenar pad placed proximal to the pollex, a hypothenar pad proximal to the fifth digit, and four interdigital pads located near the bases of the proximal phalanges (Whipple 1904; Haines 1955, 1958; Biegert 1959, 1961, 1963). Broadly speaking, this pattern has persisted in many pentadactyl mammal groups, and primates by and large retain this primitive volar pad configuration, with minor differences in shape and number that vary across species and with body size. For example, the first interdigital pad, clearly identifiable in galagos and lorises, is fused with the thenar pad in lemurs (Biegert 1959, 1961, 1963; Bishop 1964; Fig. 8.7). These variations have been cataloged (Whipple 1904; Haines 1955, 1958; Midlo 1934; Biegert 1959, 1961, 1963) and can be easily visualized within the sample of strepsirrhine primates of varying body size shown in Fig. 8.7. Small primates like galagos, lorises, and cheirogaleids have more discrete, well-separated, and bulbous volar pads, while larger primates like humans and gorillas have flatter and coalesced pads that form a single and uniform traction surface (Biegert 1961, 1963; Cartmill 1974a, 1979). These differences can be easily verified even within a more closely related primate clade with a narrower range of body size variation such as lemurs (Cartmill 1979; Lemelin and Jungers 2007; compare the volar surface of the hand of *Cheirogaleus* to that of the larger *Propithecus* in Fig. 8.7). Cartmill (1974a, 1979) attributed these differences to improving frictional force and maintaining functional equivalence for climbing in larger primates (see Sect. 5.1 below). There are exceptions to this rule [e.g., the smaller *Arctocebus* has more coalesced volar pads compared to the larger *Perodicticus*; see Biegert (1959, 1961) and Cartmill (1979)], and it is unlikely that body size alone is responsible for the full range of variation. For example, Hamrick (1998) reported important differences in the size and histology of the apical pads of three New World monkeys, independent of body mass differences. He observed the largest apical pads with the most prominent papillary ridges in *Saimiri*, which spends more time moving and foraging on small branches compared to *Callithrix* and *Saguinus*.

As pointed out above, dermatoglyphics are a hallmark of the volar skin of the human hand. Perhaps because of their ubiquitousness on our own volar surfaces, it may come as a surprise that it is not a universal configuration among primates. In fact, for many strepsirrhine primates, dermatoglyphics are limited to specific portions of the volar surface—primarily the pads—and may be interrupted by patches of smooth skin or punctuated with "wart-like" nodules (Whipple 1904; Biegert 1959, 1961, 1963). Moreover, the surface of the apical pads of strepsirrhines, tarsi-

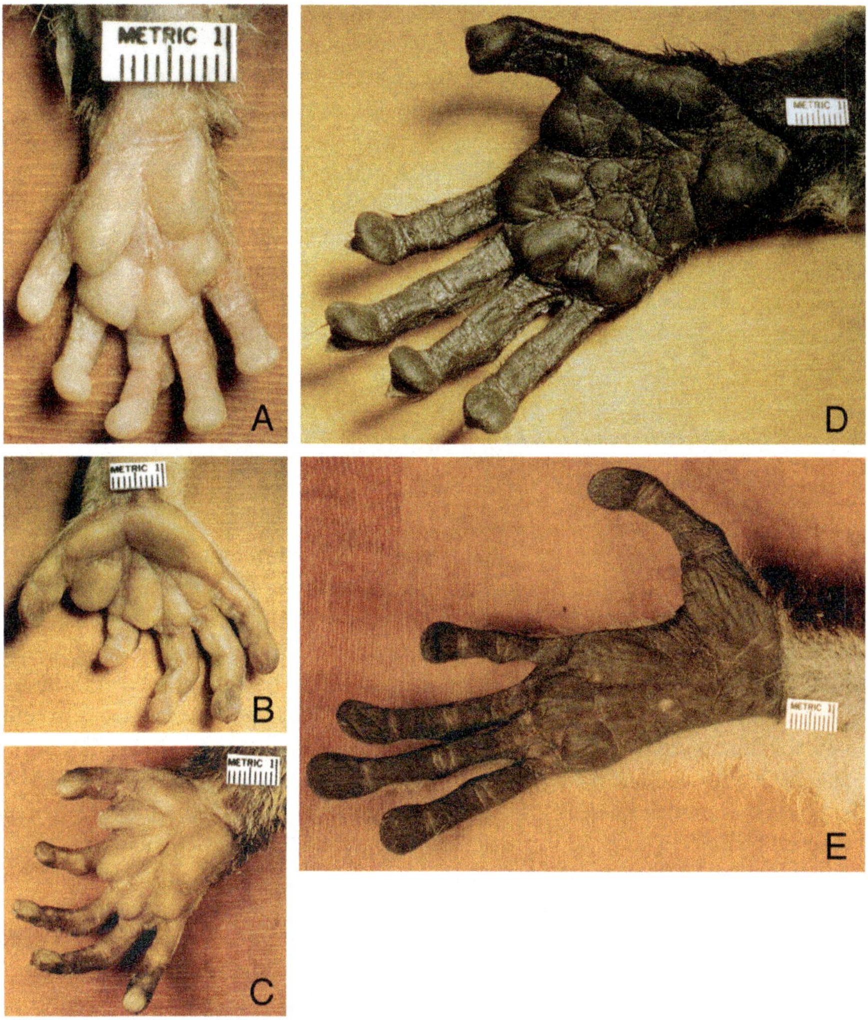

Fig. 8.7 Body size and morphological diversity of the volar pads of the hand (right side) in strepsirrhine primates. Note the more discrete, well-separated, and bulbous pads of the fat-tailed dwarf lemur (**a**, *Cheirogaleus medius*), slow loris (**b**, *Nycticebus coucang*), greater bushbaby (**c**, *Otolemur garnettii*) compared to the flatter and more coalesced pads of the ruffed lemur (**d**, *Varecia variegata*), and sifaka (**e**, *Propithecus verreauxi*, right side). The first interdigital pad (between thumb and index finger) is distinctive in galagos and lorises (**b** and **c**) but fused with the thenar pad in lemurs (**a**, **d**, and **e**). Scale bars are 1 cm. Photographs in **a** and **e** are adapted from Lemelin and Jungers (2007)

ers, and many arboreal marsupials is covered by a series of papillary ridges oriented parallel to the long axis of the hand, unlike the more complex patterns seen in humans and most other anthropoids (Whipple 1904; Biegert 1959, 1961, 1963; Cartmill 1974a). In the tree shrew *Tupaia* and many other pentadactyl mammals, the papillary ridges have a similar arrangement, but are oriented perpendicular to the long axis of the hand (Haines 1955, 1958; Biegert 1959, 1961, 1963; Cartmill 1974a). These differences in papillary ridge orientation have been interpreted as reflecting differences in orientation of the hand and foot relative to the substrate during locomotion [see Cartmill (1974a) for a full discussion]. Niemitz (1990) suggested that concentric and round dermatoglyphics (i.e., loops and whorls) are found in areas of the palm and sole for which high compressive forces are frequent and V-shaped papillary ridges of the volar surface of the palm, sole, and even prehensile tails of atelines are oriented along the shearing force vector. Experimental evidence of human dermatoglyphics indicates that greatest frictional resistance is achieved when force acts perpendicular to the orientation of the papillary ridges (Buck and Bär 1993). Additional in vivo data are needed to test these functional hypotheses.

4.2 Variation in Ungual Morphology in Primates and Other Mammals

Although homologous, primate ungulae appear quite different from the claws of non-primate mammals (called **falculae**, sing. falcula; Figs. 8.3 and 8.4). Note that the generalized term for a keratinized digital structure, regardless of form (hoof, nail, claw, etc.), is **unguis** (pl. ungues). Falculae are mediolaterally compressed ungues that project well beyond (above and distal to) the boundaries of a proximally restricted apical pad. Falculae are surrounded by the same tissues that surround ungulae, although they differ in size and shape (Homberger et al. 2009; Fleckman et al. 2013; Figs. 8.3 and 8.4). A typical falcula (or claw plate) appears as a sheet of keratin that is draped over a claw bed surrounding the dorsum and lateral sides of the distal phalanx. Therefore, there is usually a gap (at least partially, but this varies among mammals) between the two volar margins of the falcula. It is here that the hyponychium (Fig. 8.4g, black layer) and its cornified layer, the subunguis (Fig. 8.4g, gray layer), can be found (Le Gros Clark 1936; Homberger et al. 2009; Fleckman et al. 2013). The dermis of the claw bed is highly vascularized, so it bleeds profusely when damaged. This is what is called the quick by pet groomers. If the claw plate is not strongly pigmented, it can be seen as a pink region deep to the plate (Fleckman et al. 2013; Fig. 8.3h). Like the nail bed of ungulae, the claw bed is divided into a matrix (that generates the claw and lies partially under a proximal claw fold) and sterile portion (Le Gros Clark 1936; Hamrick 2001a; Homberger er al., 2009). The distal portion of the matrix can be seen through translucent falculae as a lunula (Fig. 8.3e). A cuticle also grows from the volar layer of the claw fold (Fleckman et al. 2013; Fig. 8.3d). In many mammal species, especially carnivorans, the base of the claw plate (and matrix) actually lies deep to an enlarged bony flange called an ungual crest or unguicular hood (Bryant et al. 1996; Homberger et al. 2009). If the apical pad is extensive, as in some marsupials, it may overlap with the claw plate beyond the proximal fold creating a "paronychium" of sorts. However, the term paronychium is not generally used in discussion of falcular anatomy. It should be noted that falculae and ungulae are not necessarily discrete character states; some mammals have ungula-like falculae, and others have falcula-like ungulae creating a morphological continuum (Hamrick 1998, 2001a, 2003; see below). Further, it is now widely recognized that primate ungulae are derived from falculae (Le Gros Clark 1936; Cartmill 1974a; Szalay and Dagosto 1980; Spearman 1985; Godinot and Beard 1991), although historically, it was thought that the ancestral mammalian condition was to bear ungulae (Wood Jones 1916; Panzer 1932).

Interestingly, not all primates possess ungulae on all digits. The aye-aye (*Daubentonia*) and callitrichine monkeys (marmosets and tamarins) have falcula-like ungues called **tegulae** (Le Gros Clark 1936; Hershkovitz 1977; Rosenberger 1977; Soligo 2005; Figs. 8.3 and 8.4). Some authors have used the term tegulae to refer to the ungulae of non-callitrichine platyrrhines as they may be relatively narrow compared to those of other primates (e.g., Hershkovitz 1977). However, they are not nearly so narrow as the extreme cases seen in callitrichines and *Daubentonia*, which are referred to as claws by the same authors. To avoid confusion and to emphasize anatomical differences, we will refer solely to the structures of callitrichines and *Daubentonia* as tegulae and reserve the term "falcula" for those of non-primates. Tegulae are found on all digits of callitrichines save the halluces (which bear ungulae) and all digits of the aye-aye except the halluces (which also bear ungulae) and the second pedal digits (which bear specialized ungues used for grooming—the "grooming claw" as in all other strepsirrhines). They are similar to falculae as they are associated with a more proximally restricted apical pad and appear as a keratinized sheet draped over the distal phalanx that is open inferiorly (Le Gros Clark 1936; Thorndike 1968; Rosenberger 1977; Garber 1980). The gap between the two sides of the tegula is filled with subunguis (Fig. 8.4g, gray layer) that is generated by a hyponychium (Le Gros Clark 1936; Fig. 8.4g, black layer). In *Daubentonia*, the two sides of the unguis closely approximate one another in the proximal region near the apical pad, such that the gap can only be observed distally, but this does not occur in callitrichines (Le Gros Clark 1936). The same tissues (e.g., proximal fold, cuticle, paronychium, matrix, and sterile portion of unguis bed) that surround falculae and ungulae can also be observed in tegula-bearing digits (Le Gros Clark 1936; Thorndike 1968; Figs. 8.3 and 8.4).

The origin and homologies of tegulae have been the subject of much debate. Anatomically, they appear different from both ungulae and falculae but are more falcula-like than the ungulae of other primates. Some have contended that tegulae are actually retained, but slightly modified, falculae from a falcula-bearing primate ancestor (Le Gros Clark 1936; Cartmill 1974a; Hershkovitz 1977; Spearman 1985). Others suggest that tegulae are derived from primate ungulae and are convergent in morphology to the falculae of non-primates (Pocock 1917; Rosenberger 1977; Ford 1980; Garber 1980; Martin 1992; Hamrick 1998; Soligo and Müller 1999). Today, most researchers agree that the latter is the case, as a number of detailed studies lend support to this hypothesis (e.g., Hamrick 1998; Soligo and Müller 1999).

Early studies on the affiliations among falculae, ungulae, and tegulae focused on the histology of the keratinized structures themselves. As stated earlier, the matrix is a portion of the unguis bed that is responsible for producing the unguis. The matrix in most falcula-bearing mammals can be divided into two parts: the basal matrix containing flat cells (squamous) and the terminal matrix containing taller cells (columnar; Le Gros Clark 1936). This leads to the formation of two layers (or strata) of the falcula itself: the basal matrix forms a superficial stratum, while the terminal matrix forms a deep stratum. The two layers can be distinguished from one another based on the orientation of lamellae (thin plates of unguis substance that are sequentially laid down by the matrix). The tegulae of *Daubentonia* and callitrichines also have two layers, although the deep stratum is thinner and less pronounced than what is seen in non-primates (Le Gros Clark 1936; Thorndike 1968; Soligo and Müller 1999). The story with primate ungulae is a bit more convoluted. Originally, it was thought that all ungulae lacked a deep stratum (Le Gros Clark 1936). This presumed difference between falculae and ungulae seemed to ally tegulae with falculae. However, once a broader sample of ungulae was examined, it became clear that a one-layered ungula is not ubiquitous among primates, as the presence of a deep layer is quite variable (Thorndike 1968; Soligo and Müller 1999). Further, more recent studies have shown that falcula and ungula plates can be divided into additional, histologically defined strata and regions beyond the simplistic superficial and deep dichotomy (Garson et al. 2000; Homberger et al. 2009). Regardless, it is clear that the presence and absence of a deep stratum as defined in early studies have not elucidated the evolutionary origins of primate tegulae.

Perhaps more informative regarding tegula evolution are the distinctive anatomical features of the distal digital segment. These structures demonstrate dissimilarities between falcula-bearing distal phalanges and primate distal phalanges (regardless of whether they bear tegulae or ungulae). This morphology is explained and illustrated in more detail in Fig. 4.6, so it is not repeated here. Overall, tegular digital segments are more similar to ungular segments than to falcular segments in a number of ways (Rosenberger 1977; Garber 1980) and, in fact, have been shown to form a morphological continuum with those that bear ungulae (Hamrick 1998).

While diverse, strepsirrhines and tarsiers share some similarities in ungual morphology. Perhaps the most striking are the nails of lemuriform strepsirrhines. Almost all species have wide, pointed nails that can be described as possessing a median keel or ridge running down the center (Bruhns 1910; Montagna and Yun 1963; see *Lepilemur* and the less pointed *Avahi* in Fig. 8.8). However, a few exceptions exist. For example, the keel is less prominent in *Cheirogaleus* and nearly absent in *Microcebus*, which both possess flattened nails that terminate in more rounded distal edges. The nails of lorisiform strepsirrhines (apart from *Euoticus*) are relatively wide and flat with rounded distal edges (e.g., *Galagoides* and *Otolemur*; Fig. 8.8). *Euoticus*, in contrast, possesses a median keel like those described for most lemuriforms (Biegert 1959; Charles-Dominique 1977; Fig. 8.8). As mentioned earlier, *Daubentonia* is unique among strepsirrhines in possessing tegulae rather than nails on its digits (Fig. 8.8). The nail morphology of tarsiers shows similarity to that of strepsirrhines. Most tarsier species have nails that are small relative to their expanded apical pads [*Cephalopachus* (or *Tarsius*) *bancanus*] but, like many lemuriforms, end in points and possess a median keel (Le Gros Clark 1936; Cartmill 1974a; Fig. 8.8). *Tarsius pumilus* differs from other tarsiers as it has relatively narrow nails with very pronounced median keels that resemble tegulae more than nails (Musser and Dagosto 1987; Fig. 8.8).

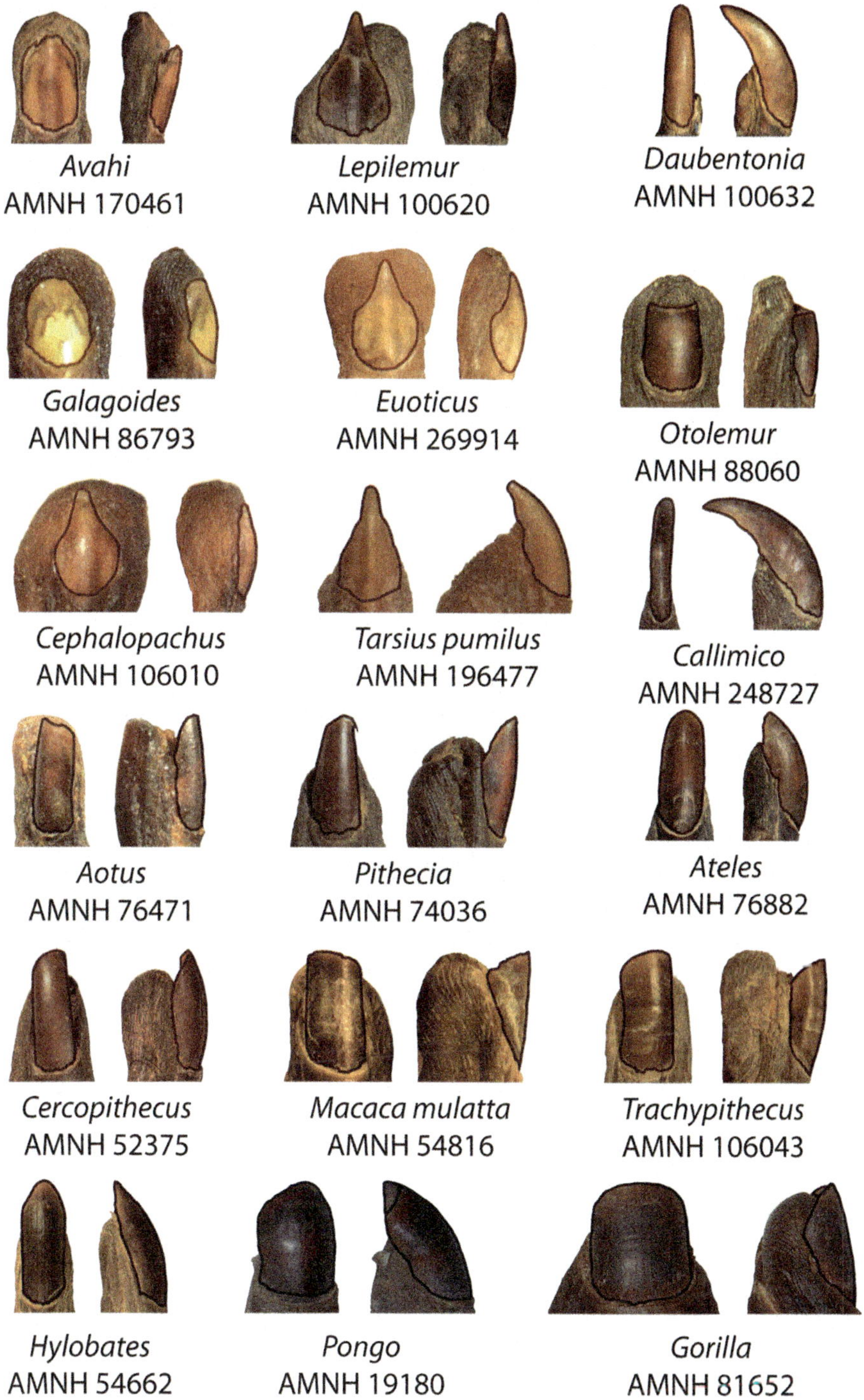

Fig. 8.8 Diversity in ungual morphology in nonhuman primates. For each primate genus, the margins of the ungula (or nail plate) are outlined in *black* in both dorsal (*left*) and lateral (*right*) views. Specimens not on same scale. *AMNH,* American Museum of Natural History

Anthropoid primates have nails that lack a median keel and end in a rounded distal tip (Bruhns 1910). Platyrrhines have mediolaterally compressed nails that are strongly keeled (e.g., *Ateles*, but less keeled in *Aotus* and *Pithecia*; Hershkovitz 1977; Fig. 8.8). Note that in platyrrhines (especially *Ateles*), the entire nail is keeled rather than just a median ridge as in many strepsirrhines and tarsiers (Fig. 8.8). Callitrichines, as discussed earlier, are particularly notable in having extremely mediolaterally compressed tegulae and somewhat proximally restricted apical pads (e.g., *Callimico*; Fig. 8.8). Cercopithecoids (e.g., *Cercopithecus*, *Macaca*, and *Trachypithecus*) have nails that tend to be less keeled than platyrrhines and can range from relatively narrow (e.g., *Colobus* and *Cercopithecus*) to relatively broad (e.g., *Papio* and *Macaca mulatta*; Fig. 8.8). Gibbons (e.g., *Hylobates*) possess nails that are relatively narrow and similar to those of many cercopithecoids, while great apes (e.g., *Pongo* and *Gorilla*) and humans are notable in having particularly wide, flat nails (Fig. 8.8).

In addition to gross morphological variation, primate ungulae also show a high degree of histological variation. In terms of the superficial and deep strata dichotomy, the strepsirrhine *Lemur*, the platyrrhines *Cebus* and *Lagothrix*, and the cercopithecoids *Presbytis*, *Cercopithecus*, and *Macaca* have two layers, while the strepsirrhines *Microcebus*, *Nycticebus*, and *Galago*, the tarsiers *Tarsius* and *Cephalopachus*, the cercopithecoids *Chlorocebus* and *Papio*, and the hominoids *Hylobates* and *Homo* all have one (Thorndike 1968; Soligo and Müller 1999). Additionally, the nails of the hominoids *Pan* and *Pongo* have also been shown to have two layers [referred to as ventral and dorsal nail in Sprankel (1969a, b)]. The distribution of two-layered nails among living primates led Soligo and Müller (1999) to suggest that the last common ancestor was large bodied and possessed nails with two layers, the deep layer subsequently being lost in several lineages due to dwarfing and/or terrestriality. However, Hamrick (1999) pointed out that the loss of a deep layer is likely related to shortening of the distal phalanges rather than small body size per se. Shortened distal phalanges also have shortened regions of matrix whose proximo-distal length is directly related to the dorso-volar thickness of the nail plate (Hamrick 1999, 2001a). Regardless, the possible link between ungula thickness (and/or shortened distal phalanges) and its role during grasping requires more data to ascertain.

5 Functional Aspects of the Hand Integument

The integument plays vital functional roles when the hand grasps or feels an object. For example, when the surface of the volar skin contacts the surface of an object, a resistance between the two surfaces is created and results in a frictional force. At the same time, when pressure is applied on the surface of the volar skin, specialized nerve cells send tactile information back to the brain. The following section briefly discusses some of these functional roles.

5.1 Functional Roles of the Volar Skin

The volar skin plays a vital role in providing a friction surface, and all primates rely on friction when grasping an object with the hand. Friction is defined as the resistance experienced between two objects when one slides over the other and is dependent on two factors: the normal force (or force acting perpendicular between the two objects) and the frictional qualities of the two surfaces (expressed as the coefficient of static friction; Cochran 1982; Tomlinson et al. 2007). The relationship between these two factors is known as Amonton's first law and states that frictional force is proportional to both load (normal force) and the coefficient of static friction. Load is influenced by body size—a larger animal will exert greater normal force—while the coefficient of friction is largely influenced by the surface morphology of the volar skin. Asperities, such as dermatoglyphics or the wart-like patches found in strepsirrhines, can interlock with irregularities on substrates and produce a greater coefficient of friction (Cartmill 1974a, 1979, 1985; Buck and Bär 1993; Hamrick 1998; Rosenberg and Rose 1999; Lemelin 2000; Derler et al. 2009; however, see Warman and Ennos 2009 for an opposing viewpoint).

Structures that enhance friction through surface interlocking should not be confused with structures that allow for surface adhesion. Unfortunately, the differences between the two are not always clear. For instance, the feathertail glider (*Acrobates pygmaeus*) has papillary ridges that are visually indistinguishable from those found in primates and which might be used to increase surface interlocking friction. However, closer inspection of clinging behavior in feathertail gliders shows something very different indeed. The ducts of underlying eccrine sweat glands pierce the papillary ridges to release their secretions, and the gliders are small enough in mass (10–14 g) that the surface tension generated by these excretions is enough to adhere an individual to a smooth surface (Turner and McKay 1989; Rosenberg and Rose 1999). Adhesive adaptations (dry and wet) are quite common among arthropods, amphibians, and reptiles but much less so among mammals (Cartmill 1985; Fig. 8.9). Even so, wet adhesive adaptations can be found in several groups, most notably among some bat species (Thewissen and Etnier 1995; Riskin and Racey 2010; Schliemann and Goodman 2011).

Because surface adhesion requires an individual's mass to be smaller than the force generated by surface tension, animals relying on it are necessarily small in size (of the three genera mentioned previously, the feathertail glider is the largest at 10–14 g). The smallest primate, believed to be Madame Berthe's mouse lemur (*Microcebus berthae*), weighs roughly 30 g or nearly triple the mass of known adhering mammals. However, while primates' greater mass prevents them from adhering to surfaces, it does not preclude other adaptations to their volar glands that may enhance their grasping ability. Hashimoto and colleagues (1986) noted that eccrine glands in rodents are not the major heat dissipating organs that they are in primates and draw upon empirical findings of moist skin's frictional properties to suggest that they may provide moisture to enhance friction and keep the feet from slipping (Naylor 1955). Anatomical evidence seems to support this hypothesis: terrestrial mice have a relatively smaller number of eccrine glands with excretory ducts

Fig. 8.9 Volar surface of the hand of a gecko (*Gekko* sp.) clinging on a glass surface. Note the apical ridges on the palmar surface of the digits allowing for dry adhesion on the substrate. Photograph courtesy of Karen Schultz

oriented caudocranially in the volar skin, such that the gland is compressed first at the heel during a step cycle and its excretions carried cranially as the body's weight is rolled across the pad; in contrast, climbing mice have a larger number of glands with ducts that open in a multitude of directions (Haffner 1998). These arrangements ensure that excretion is not interrupted during locomotion on discontinuous substrates when the pads are compressed in different orientations.

The anatomy of and variation in primate volar glands are still poorly understood, but what evidence has been published supports the hypothesis that their excretions may enhance grip as well. A comparison of the closely related *Galago senegalensis* and *Loris tardigradus* species finds several key differences that appear to be linked to their different modes of locomotion (Nieschalk and Klauer 1989). *G. senegalensis* moves about its environment largely through vertical clinging and leaping, and its volar eccrine glands are relatively small with tightly coiled ducts that sit deep within the fat body of its volar pads. The fat bodies provide a means of diffusing reaction forces applied to the volar surface such that only relatively large reaction forces (i.e., those generated during landing after a leap) will travel through them to compress the glands (Nieschalk and Klauer 1989). *L. tardigradus*, by contrast, moves by deliberate, quadrumanous climbing and does not leap between supports. Its glands are located in the superficial layer of the volar fat body, and their ducts are covered with sheets of collagen that prevent them from collapsing. This arrangement likely does not diffuse reaction forces to the extent found in *Galago*; instead, it allows compression of the glands by a relatively small, constant force, and the sheaths of collagen around the ducts allow them to remain open under the same conditions (Nieschalk and Klauer 1989). Further study should reveal whether this pattern is common among primates.

Mammals that apply all or part of the volar surfaces directly to a substrate during locomotion often rely upon a cushion of sorts to diffuse reaction forces applied to their limbs and protect the underlying bones and vessels. For most mammals, this cushion takes the form of fat pads which sit deep to the volar skin. The fatty tissues that comprise these pads are surrounded by septa of collagen and elastin fibers. This configuration results in volar pads that behave mechanically as a semisolid, viscoelastic material and can be described by Fung's quasi-linear viscoelastic tissue model (Bennett and Ker 1990; Ker 1990; Aerts et al. 1995, 1996; Pawluk and Howe 1999; Jindrich et al. 2003). In simple terms, this means that the pads are able to deform with applied force and return to their original shape once the stimulus is removed. The force required to deform the pads is converted to heat and released and, thereby, does not pass on to the underlying structures. This deformability also has consequences for the frictional properties of the volar surfaces. Cartmill (1979) showed that the volar surfaces of primates do not follow Amonton's law directly, but rather that frictional force is proportional to load raised to an exponent reflecting the surface area and deformability of the volar surface (see also Comaish and Bottoms 1971). In this way, the discreet, protuberant pads of smaller primates allow them to take advantage of the enhanced friction provided by pliability, while the flatter and more coalesced pads of larger primates allow for greater surface contact with substrates. Readers are encouraged to read Cartmill (1974a, 1979, 1985) for more details on this difficult but very important topic.

While most hypotheses concerning the anatomy of the volar skin revolve around its role as a friction surface during locomotion, other factors likely shape its anatomy as well. The foremost of these is the volar skin's role in receiving sensation from the outside world. Cauna (1954) noted that despite being 4–20 times thicker than hairy skin, volar skin still manages to act as a highly efficient sensory conduit in humans. In fact, volar skin is more sensitive than any area of hairy skin. Because human volar skin is characterized by papillary ridges, it is reasonable to hypothesize that these may play a role in transmitting tactile sensation. As discussed previously, Merkel cells are found in the epidermis, but other tactile end organs are located deeper in the dermis, or even the hypodermis, and could hypothetically benefit from assisted tactile transmission. Of these remaining end organs, Meissner's corpuscles in particular are well positioned within the dermal papillae to be directly impacted by the epidermal ridges. In this position, each corpuscle is flanked by an intermediate ridge on one side and a limiting ridge on the other (Fig. 8.5h, j). The limiting ridges are quite short and so likely do not influence the shape of the dermal papillae, but the intermediate ridges are long enough that their movement against the dermal papillae may act as a lever to deform the papilla and stimulate the Meissner's corpuscles (Cauna 1954).

Meissner's corpuscles have garnered attention because of their vital role in tactile acuity and their preponderance in the volar skin of primates (Winkelmann 1962, 1963, 1964, 1965; Hoffmann et al. 2004; Organ et al. 2011; Verendeev et al. 2015). These end organs populate the apical pads of the digits of primates in different densities and sizes depending on the species and diet. For example, the apical pads of the thumb and index finger of more frugivorous primates appear to have a greater density of Meissner's corpuscles compared to less frugivorous species (Hoffmann et al. 2004), although other factors may also be involved. More recently, Verendeev and colleagues (2015) found no such relationship between Meissner's corpuscle density and diet or manual dexterity, only with differences in body size (Verendeev et al. 2015). Human studies have shown variability in the density of Meissner's corpuscles throughout the hand (with the highest densities at the tip of the digits; Johansson and Vallbo 1979) and decrease in density with age or occupation (e.g., manual labor; Cauna 1956).

5.2 Functional Roles of Claws and Nails

Claws fulfill a myriad of functions in mammals. For example, the large and fast-growing claws of fossorial mammals such as moles and ground squirrels are used to dig holes and tunnels (Hildebrand 1985), whereas the retractable claws of felids act as formidable weapons to take down prey (Gonyea and Ashworth 1975). Claws (both falculae and tegulae) are also very useful in arboreal locomotion because they allow mammals to climb up and down vertical substrates that are too large for them to grip in any other manner (Cartmill 1974a, 1985; Garber 1992). The tips of claws are interlocked with the substrate (tree bark) to assist in supporting the animal's weight. In some arboreal mammals such as tree kangaroos and sloths, claws evolved as large hooks that facilitate grasping of branches or food items (Mendel 1981; Iwaniuk et al. 1998).

Although of considerable interest, functional arguments to explain the prevalence of nails (as opposed to claws) in primates have been unconvincing. The lack of claws is usually thought to be associated with an increased reliance on grasping; primarily, they are either thought to be actively selected against because they impede grasping (e.g., Le Gros Clark 1959; Napier 1993) or to be a passive result of a broadened apical pad associated with grasping (e.g., Cartmill 1974a; Hamrick 1998; Godinot 2007). There are a number of counterexamples to the former in which mammals that possess claws (or claw-like nails) are not hindered while grasping (e.g., Cartmill 1974a; Rasmussen 1990; Lemelin and Grafton 1998), and it is not clear that the latter is necessarily the case as some grasping mammals have broadened their apical pads, but have lost any vestige of an unguis (i.e., the grasping halluces of many diprotodont marsupials have no keratinized structure at all; Szalay 1994). Therefore, the functional roles that nails themselves can play should also be taken into account when considering claw loss in primates.

The clinical literature abounds with arguments about the functional roles achieved by human nails (ungulae). These include protection and support of the fingertip, a role in sensory perception, assistance in the handling and manipulation of small objects, and, surprisingly, for the use as a defensive weapon (e.g., Johnson and Cohen 1975; Beaven and Brooks 1984; Zook 2003). In vivo data in humans show greater strains (i.e., deformation) of the nail in specific planes during compression of the fingertip (Sakai and Shimawaki 2007). This suggests that, like bone, nails have anisotropic properties (i.e., varying material properties depending on orientation) and may be functionally adapted to resist mechanical stresses in certain planes (see Farren et al. 2004). Pertaining to sensory perception, the human paronychium is populated by specialized sensory mechanoreceptors that respond to movements of the nail plate against the skin folds (Birznieks et al. 2009). This likely helps to determine the direction of forces placed upon the fingertip. Nonhuman primate nails are also suggested to serve similar functions: to prevent damage or distortion of the apical pad (Le Gros Clark 1936; Napier 1993), to redistribute forces placed on the fingertip (Preuschoft 1970, 1973), to scratch one's self, and, for primates with increased manual dexterity, to assist in manipulation of objects (Spearman 1985; Napier 1993). However, the high degree of variation in nail form among primates (Fig. 8.8) suggests that nails may serve different roles in different species. For example, the nails of some tarsiers are too diminutive when compared to their apical pads to prevent damage or distortion (Cartmill 1974a; Spearman 1985), and the median keel in some strepsirrhines and tarsiers is suspected to play a role in vertical clinging and climbing (Charles-Dominique 1977; Tilden 1990). Clearly, far more research will be necessary to explain the functional significance of shape variation in nail morphology across primates.

6 Future Directions

In many respects, the study of the hand integument of primates remains in its infancy. Many authors have made in-roads in the form of descriptive tomes or limited-sample contrasts, but as yet there exists no comprehensive, functional evaluation of the morphology of the volar skin and pads. Quantitative anatomical data are sparse and only now beginning to trickle into the published literature. Further study in this area will benefit from the continued quantification of both micro- and macro-anatomical traits across a broad spectrum of primates and evaluation of these data within phylogenetic, functional, and ecological contexts. Similarly, the majority of research on primate ungues has focused on elucidation of the origins of claw-like nails (i.e., tegulae) via histological analysis, while the functional significance of nail morphology remains poorly understood. Future work should address the relationship between morphology and function by utilizing behavioral, kinematic, and observational data. With the exception of claw climbing in callitrichines, it is not entirely clear if, when, and how different primates use their nails in locomotor or other activities. Experimental techniques used in a comparative context could also prove invaluable to shed more light on the functional morphology of primate volar skin and nails.

Acknowledgments We are very grateful to David Begg, Tracy Kivell, Brian Richmond, and Daniel Schmitt for providing insightful comments on a draft of this chapter. S.M. acknowledges Neil Duncan and especially Eileen Westwig for access to the mammalogy collections at the American Museum of Natural History and Drs. John Fleagle and Susan Larson for access to specimens held at Stony Brook University. Her work was funded by grants from the National Science Foundation (BCS-1341075) and the Leakey Foundation. A.K. acknowledges Drs. Susan Larson

(University at Stony Brook), Magdalena Muchlinski (University of Kentucky), and Nathan Kley (University at Stony Brook) and the Duke Lemur Center for guidance, access to study specimens, and access to histology equipment. Her work was funded by a grant from the National Science Foundation (BCS-1097438). P.L. acknowledges Drs. Matt Cartmill (Boston University), John Fleagle (University at Stony Brook), Norman Taslitz (University of New Mexico), and Hans Thewissen (Northeast Ohio Medical University), the Duke Lemur Center, and the Carnegie Museum of Natural History (Sue MacLaren) for access to primate and non-primate volar skin specimens, Jeannette Killius and Janice Walas (Northeast Ohio Medical University) for their expertise and help during preparation of the histological slides, Dr. Daniel Livy (University of Alberta) for access to photomicroscopic equipment, and Elizabeth Hodges for her invaluable help with bibliographical research.

References

Aerts P, Ker RF, De Clercq D, Ilsley DW, Alexander RM (1995) The mechanical properties of the human heel pad: a paradox resolved. J Biomech 28:1299–1308

Aerts P, Ker RF, De Clercq D, Ilsley DW (1996) The effects of isolation on the mechanics of the human heel pad. J Anat 188:417–423

Beaven DW, Brooks SE (1984) Color Atlas of the nail in clinical diagnosis. Year Book Medical Publishers, Inc., Chicago

Bennett MB, Ker RF (1990) The mechanical properties of the human subcalcaneal fat pad in compression. J Anat 171:131–138

Biegert J (1959) Die Ballen, Leisten, furchen und Nägel von Hand und Fuß der Halbaffen. Z Morphol Anthropol 49:316–409

Biegert J (1961) Volarhaut der Hände und Füsse. In: Hofer T, Schultz AH, Stark D (eds) Primatologia, vol 2. Karger, Basel, pp 1–326

Biegert J (1963) The evaluation of characteristics of the skull, hands, and feet for primate taxonomy. In: Washburn SL (ed) Classification and human evolution. Aldine, Chicago, pp 116–145

Birznieks I, Macefield VG, Westling G, Johansson RS (2009) Slowly adapting mechanoreceptors in the borders of the human fingernail encode fingertip forces. J Neurosci 29:9370–9379

Bishop A (1964) Use of the hand in lower primates. In: Buettner-Janusch J (ed) Evolutionary and genetic biology of primates. Academic Press, New York, pp 133–223

Bruhns F (1910) Der Nagel der Halbaffen und Affen: ein Beitrag zur Phylogenie des menschlichen Nagels. Gegenbaurs Morphol Jahrb 40:501–609

Bryant HN, Russell AP, Laroiya R, Powell GL (1996) Claw retraction and protraction in the Carnivora: skeletal microvariation in the phalanges of the Felidae. J Morphol 229:289–308

Buck C, Bär H (1993) Investigations on the biomechanical significance of dermatoglyphic ridges. In: Preuschoft H, Chivers DJ (eds) Hands of primates. Springer, Vienna, pp 285–306

Cartmill M (1972) Arboreal adaptations and the origin of the order Primates. In: Tuttle R (ed) The functional and evolutionary biology of primates. Aldine-Atherton, Chicago, pp 97–122

Cartmill M (1974a) Pads and claws in arboreal locomotion. In: Jenkins FA Jr (ed) Primate locomotion. Academic Press, New York, pp 45–83

Cartmill M (1974b) Rethinking primate origins. Science 184:436–443

Cartmill M (1979) The volar skin of primates: its frictional characteristics and their functional significance. Am J Phys Anthropol 50:497–509

Cartmill M (1985) Climbing. In: Hildebrand M, Bramble DM, Liem KF, Wake DB (eds) Functional vertebrate morphology. Belknap Press, Cambridge, MA, pp 73–88

Cauna N (1954) Nature and functions of the papillary ridges of the digital skin. Anat Rec 119:449–468

Cauna N (1956) Nerve supply and nerve ending in Meissner's corpuscles. Am J Anat 99:315–350

Charles-Dominique P (1977) Ecology and behaviour of nocturnal primates. Columbia University Press, New York

Cochran GVB (1982) A primer of orthopaedic biomechanics. Churchill Livingstone, New York

Comaish S, Bottoms E (1971) The skin and friction: deviations from Amonton's laws, and the effects of hydration and lubrication. Br J Derm 84:37–43

Derler S, Huber R, Feuz H-P, Hadad M (2009) Influence of surface microstructure on the sliding friction of plantar skin against hard substrates. Wear 267:1281–1288

Farren L, Shayler S, Ennos AR (2004) The fracture properties and mechanical design of human fingernails. J Exp Biol 207:735–741

Fawcett DW (1994) Bloom and Fawcett: a textbook of histology, 12th edn. Chapman and Hall, New York

Fleckman P, Jaeger K, Silva KA, Sundberg JP (2013) Comparative anatomy of mouse and human nail units. Anat Rec 296:521–532

Ford SM (1980) Callitrichids as phyletic dwarfs, and the place of the Callitrichidae in Platyrrhini. Primates 21:31–43

Garber PA (1980) Locomotor behavior and feeding ecology of the Panamanian tamarin (*Saguinus oedipus geoffroyi*, Callitrichidae, Primates). Int J Primatol 1:185–201

Garber PA (1992) Vertical clinging, small body size, and the evolution of feeding adaptations in the Callitrichinae. Am J Phys Anthropol 88:469–482

Garson JC, Baltenneck F, Leroy F, Riekel C, Müller M (2000) Histological structure of human nail as studied by synchrotron X-ray microdiffraction. Cell Mol Biol (Noisy-le-Grand) 46:1025–1034

Godinot M (2007) Primate origins: reappraisal of historical data favoring tupaiid affinities. In: Ravosa MJ, Dagosto M (eds) Primate origins: adaptations and evolution. Springer, New York, pp 403–435

Godinot M, Beard KC (1991) Fossil primate hands: a review and an evolutionary inquiry emphasizing early forms. Hum Evol 6:307–354

Gonyea W, Ashworth R (1975) The form and function of retractile claws in the Felidae and other representatives carnivorans. J Morphol 145:229–238

Haffner M (1998) A comparison of the gross morphology and micro-anatomy of the foot pads in two fossorial and two climbing rodents (Mammalia). J Zool 244:287–294

Haines RW (1955) The anatomy of the hand of certain insectivores. Proc Zool Soc Lond 125:761–777

Haines RW (1958) Arboreal or terrestrial ancestry of placental mammals. Q Rev Biol 33:1–23

Halata Z (1975) The mechanoreceptors of the mammalian skin. Adv Anat Embryol Cell Biol 50:1–77

Halata Z, Grim M, Bauman KI (2003) Friedrich Sigmund Merkel and his "Merkel Cell", morphology, development, and physiology: review and new results. Anat Rec 271A:225–239

Hamrick MW (1998) Functional and adaptive significance of primate pads and claws: evidence from New World anthropoids. Am J Phys Anthropol 106:113–127

Hamrick MW (1999) Pattern and process in the evolution of primate nails and claws. J Hum Evol 37:293–297

Hamrick MW (2001a) Development and evolution of the mammalian limb: adaptive diversification of nails, hooves, and claws. Evol Dev 3:355–363

Hamrick MW (2001b) Morphological diversity in digital skin microstructure of didelphid marsupials. J Anat 198:683–688

Hamrick MW (2003) Evolution and development of mammalian limb integumentary structures. J Exp Zool B 298B:152–163

Hashimoto K, Hori K, Aso M (1986) Sweat glands. In: Bereiter-Hahn J, Matoltsy AG, Richards KS (eds) Biology of the integument. Springer-Verlag, Berlin, pp 339–356

Hershkovitz P (1977) Living New World monkeys (Platyrrhini), vol 1. University of Chicago Press, Chicago

Hildebrand M (1985) Digging of quadrupeds. In: Hildebrand M, Bramble DM, Liem KF, Wake DB (eds) Functional vertebrate morphology. Belknap Press, Cambridge, MA, pp 89–109

Hoffmann JN, Montag AG, Dominy NJ (2004) Meissner corpuscles and somatosensory acuity: the prehensile appendages of primates and elephants. Anat Rec 281A:1138–1147

Homberger DG, Ham K, Ogunbakin T, Bonin JA, Hopkins BA, Osborn ML, Hossain I, Barnett HA, Matthews KL II, Butler LG, Bragulla HH (2009) The structure of the cornified claw sheath in the domesticated cat (*Felis catus*): implications for the claw-shedding mechanism and the evolution of cornified digital end organs. J Anat 214:620–643

Iwaniuk AN, Nelson JE, Ivanco TL, Pellis SM, Whishaw IQ (1998) Reaching, grasping, and manipulation of food objects by two tree kangaroo species, *Dendrolagus lumholtzi* and *Dendrolagus matschiei*. Aust J Zool 46:235–248

Jindrich DL, Zhou Y, Becker T, Dennerlein JT (2003) Non-linear viscoelastic models predict fingertip pulp force-displacement characteristics during voluntary tapping. J Biomech 36: 497–503

Johansson RS, Vallbo ÅB (1979) Tactile sensibility in the human hand: relative and absolute densities of four types of mechanoreceptive units in glabrous skin. J Physiol 286:282–300

Johnson MK, Cohen MJ (1975) The hand atlas. Charles C. Thomas, Springfield

Ker RF (1990) The time-dependent mechanical properties of the human heel pad in the context of locomotion. J Exp Biol 199:1501–1508

Leeson CR, Leeson TS, Paparo AA (1985) Textbook of histology. W.B. Saunders, Philadelphia

Le Gros Clark WE (1936) The problem of the claw in primates. Proc Zool Soc Lond 106:1–24

Le Gros Clark WE (1959) The antecedents of man. Edinburgh University Press, Edinburgh

Lemelin P (2000) Micro-anatomy of the volar skin and interordinal relationships of primates. J Hum Evol 38:257–267

Lemelin P, Grafton BW (1998) Grasping performance in *Saguinus midas* and the evolution of hand prehensility in primates. In: Strasser E, Fleagle J, Rosenberger A, McHenry H (eds) Primate locomotion: recent advances. Plenum Press, New York, pp 131–144

Lemelin P, Jungers WL (2007) Body size and scaling of the hands and feet of prosimian primates. Am J Phys Anthropol 133:828–840

MacKenzie CL, Iberall T (1994) The grasping hand. Elsevier, Amsterdam

Martin RD (1986) Primates: a definition. In: Wood B, Martin L, Andrews P (eds) Major topics in primate and human evolution. Cambridge University Press, Cambridge, pp 1–31

Martin RD (1992) Goeldi and the dwarfs: the evolutionary biology of the small New World monkeys. J Hum Evol 22:367–393

Mendel FC (1981) The hand of two-toed sloths (*Choloepus*): its anatomy and potential uses relative to size of support. J Morphol 169:1–19

Mivart SG (1873) On *Lepilemur* and *Cheirogaleus*, and on the zoological rank of the Lemuroidea. Proc Zool Soc Lond 1873:485–510

Montagna W, Parakkal PF (1974) The structure and function of skin, 3rd edn. Academic Press, New York

Montagna W, Yun JS (1963) The skin of Primates. XVI The skin of *Lemur mongoz*. Am J Phys Anthropol 21:371–381

Midlo C (1934) Form of the hand and foot in primates. Am J Phys Anthropol 19:337–389

Midlo C, Cummings H (1942) Palmar and plantar dermatoglyphics in the Primates. Am Anat Mem 20:1–198

Munger BL, Pubols LM (1972) The sensorineural organization of the digital skin of the raccoon. Brain Behav Evol 5:367–393

Musser GG (1972) The species of *Hapalomys* (Rodentia, Muridae). Am Mus Novit 2503:1–27

Musser GG, Dagosto M (1987) The identity of *Tarsius pumilus*, a pygmy species endemic to the montane mossy forests of Central Sulawesi. Am Mus Novit 2867:1–53

Napier J (1993) Hands. Revised edition by Russell H. Tuttle. Princeton University Press, Princeton

Naylor PFD (1955) The skin surface and friction. Br J Dermatol 67:239–248

Niemitz C (1990) The evolution of primate skin structures in relation to gravity and locomotor patterns. In: Jouffroy FK, Stack MH, Niemitz C (eds) Gravity, posture and locomotion in primates. Il Sedicesimo, Florence, pp 129–156

Nieschalk U, Klauer GJ (1989) Functional morphology of the palmar pads in *Loris tardigradus* and *Galago senegalensis*. In: Splechtna H, Hilgers H (eds) Trends in vertebrate morphology. Gustav Fischer Verlag, Stuttgart, pp 605–607
Okajima M (1991) Dermal ridge development on the volar pads of the rat (*Rattus norvegicus*) and comparative study of pattern formation using inbred strains. Am J Anat 191:23–34
Okajima M, Asai Y (1985) Anatomical and microscopic study of the volar dermal ridges of the rat (*Rattus norvegicus*). Am J Phys Anthropol 67:81–88
Organ JM, Muchlinski MN, Deane AS (2011) Mechanoreceptivity of prehensile tail skin varies between ateline and cebine primates. Anat Rec 294:2064–2072
Panzer W (1932) Beiträge zur biologischen Anatomie des Baumkletterns der Säugetiere. I Das Nagel-Kralle-Problem. Z Anat Entwicklungs 98:147–198
Pardo-Castello V (1960) Disease of the nail, 3rd edn. Charles C. Thomas, Springfield
Pawluk DTV, Howe RD (1999) Dynamic lumped element response of the human fingerpad. J Biomech Eng 121:178–183
Perrin C, Langbein L, Schweizer J (2004) Expression of hair keratins in the adult nail unit: an immunohistochemical analysis of the onychogenesis in the proximal nail fold, matrix and nail bed. Br J Dermatol 151:362–371
Pocock RI (1917) The genera of the Hapalidae. Ann Mag Nat Hist 20:247–258
Preuschoft H (1970) Functional anatomy of the lower extremity. In: Bourne GH (ed) The chimpanzee, vol 3. Karger, Basel, pp 221–294
Preuschoft H (1973) Functional anatomy of the upper extremity. In: Bourne GH (ed) The chimpanzee, vol 6. University Park Press, Baltimore, pp 34–120
Purves D, Augustine GJ, Fitzpatrick D, Hall WC, LaMantia A-S, McNamara JO, White LE (eds) (2008) Neuroscience, 4th edn. Sinauer, Sunderland
Rasmussen DT (1990) Primate origins: lessons from a neotropical marsupial. Am J Primatol 22:263–277
Riskin DK, Racey PA (2010) How do sucker-footed bats hold on, and why do they roost head-up? Biol J Linn Soc 99:233–240
Rosenberg HI, Rose R (1999) Volar adhesive pads of the feathertail glider, *Acrobates pygmaeus* (Marsupialia; Acrobatidae). Can J Zool 77:233–248
Rosenberger AL (1977) *Xenothrix* and ceboid phylogeny. J Hum Evol 6:461–481
Russell EM (1986) Observations on the behaviour of the honey possum, *Tarsipes rostratus* (Marsupialia: Tarsipedidae) in captivity. Aust J Zool Suppl Ser 34:1–63
Sakai N, Shimawaki S (2007) Strain in the nail at fingertip compression. Skin Res Technol 13:449–453
Schliemann H, Goodman SM (2011) A new study on the structure and function of the adhesive organs of the Old World sucker-footed bat (Myzopoda: Myzopodidae) of Madagascar. Verh Naturwiss Ver Hamburg 46:313–330
Soligo C (2005) Anatomy of the hand and arm in *Daubentonia madagascariensis*: a functional and phylogenetic outlook. Folia Primatol 76:262–300
Soligo C, Müller AE (1999) Nails and claws in primate evolution. J Hum Evol 36:97–114
Spearman RIC (1985) Phylogeny of the nail. J Hum Evol 14:57–61
Sprankel H (1969a) Comparative microscopic studies of nail plate and surrounding soft tissue of some Hominoidea. In: Hofer HO (ed) Proceedings of the second international congress of primatology, Atlanta 1968, vol 2, Recent advances in primatology. S. Karger, Basel, pp 82–86
Sprankel H (1969b) Observation on growth structure of the nail in Hominoidea. Folia Primatol 10:161–171
Springer MS, Stanhope MJ, Madsen O, de Jong WW (2004) Molecules consolidate the placental mammal tree. Trends Ecol Evol 19:430–438
Stenn K, Fleckman P (2000) Hair and nail physiology. In: Hordinsky MK, Sawaya ME, Scher RK (eds) Atlas of hair and nails. Churchill Livingstone, Philadelphia, pp 3–8

Stone M, Styles AR, Cockerell CJ (2000) Histology of the normal nail unit. In: Hordinsky MK, Sawaya ME, Scher RK (eds) Atlas of hair and nails. Churchill Livingstone, Philadelphia, pp 18–26

Szalay FS (1994) Evolutionary history of the marsupials and an analysis of osteological characters. Cambridge University Press, Cambridge

Szalay FS, Dagosto M (1980) Locomotor adaptations as reflected on the humerus of Paleogene primates. Folia Primatol 34:1–45

Thewissen JGM, Etnier SA (1995) Adhesive devices on the thumb of vespertilionoid bats (Chiroptera). J Mammal 76:925–936

Thorndike E (1968) A microscopic study of marmoset claw and nail. Am J Phys Anthropol 28:247–268

Tilden CD (1990) A study of locomotor behavior in a captive colony of red-bellied lemurs (*Eulemur rubriventer*). Am J Primatol 22:87–100

Tomlinson SE, Lewis R, Carré MJ (2007) Review of the frictional properties of finger-object when grasping. Proc Inst Mech J Eng J Eng Tribol 221:841–850

Turner V, McKay GM (1989) Burramyidae. In: Walton DW, Richardson BJ (eds) Fauna of Australia, vol 1B, Mammalia. Australian Government Publishing Services, Canberra, pp 652–664

Verendeev A, Thomas C, McFarlin SC, Hopkins WD, Phillips KA, Sherwood CC (2015) Comparative analysis of Meissner's corpuscles in the fingertips of primates. J Anat 227:72–80

Warman PH, Ennos R (2009) Fingerprints are unlikely to increase the friction of primate fingerpads. J Exp Biol 212:2016–2022

Whipple IL (1904) The ventral surface of the mammalian chiridium. Z Morphol Anthropol 7:261–368

Wible JR, Covert HH (1987) Primates: cladistic diagnosis and relationships. J Hum Evol 16:1–22

Williams PL (ed) (1995) Gray's anatomy, 38th edn. Churchill Livingstone, New York

Wimsatt WA, Villa-R B (1970) Locomotor adaptations in disc-winged bat *Thyroptera tricolor*. 1. Functional organization of adhesive discs. Am J Anat 129:89–119

Winkelmann RK (1962) Cutaneous sensory end organs of some anthropoid apes. Science 136:384–386

Winkelmann RK (1963) Nerve ending in the skin of primates. In: Buettner-Janusch J (ed) Evolutionary and genetic biology of Primates, vol 1. Academic Press, New York, pp 229–259

Winkelmann RK (1964) Nerve endings of the North American opossum (*Didelphis virginiana*): a comparison with nerve endings of primates. Am J Phys Anthropol 22:253–258

Winkelmann RK (1965) Innervation of the skin: notes on a comparison of primate and marsupial nerve endings. In: Lyne AG, Short BF (eds) Biology of the skin and hair growth. American Elsevier, New York, pp 171–182

Wood Jones F (1916) Arboreal man. Edward Arnold, London

Wood Jones F (1942) The principles of anatomy as seen in the hand, 2nd edn. Ballière Tindall, and Cox, London

Zook EH (2003) Anatomy and physiology of the perionychium. Clin Anat 16:1–8

Part II
Biomechanical, Experimental and Behavioral Evidence

Chapter 9
Functional Morphology of the Primate Hand: Recent Approaches Using Biomedical Imaging, Computer Modeling, and Engineering Methods

Caley M. Orr

1 Introduction

A thorough understanding of the form and function of the hand and wrist is germane to many questions relating to primate evolution. These research areas range from the earliest origins of the primate order (with its broad emphasis on manual grasping capabilities) and the relative "emancipation" of the forelimb compared with other animals (Wood Jones 1916). Detailed information on manual mechanics and morphology can tell us how the hands of extant primates are adapted to their particular lifestyles, as well as shed light on the mechanical constraints acting on the forelimb depending on specific locomotor or manipulative behaviors. In turn, such information is essential for reconstructing hand function in extinct species and for fully understanding the adaptive significance of evolutionary changes throughout the primate clade. For example, the alleviation of locomotor demands on the forelimb may have had a particularly important influence on the morphological adaptation of the hand in bipedal hominins as the hand shifted from a presumed primitive condition in which it bore loads during locomotion, to a derived organ of precision manipulation (Napier 1956, 1960). The evolution of such functional "trade-offs" might be brought to light by better understanding how morphology reflects aspects of hand function, such as joint movement and load transmission.

Previous chapters have surveyed the morphological diversity of the primate hand, and chapters to follow cover behavioral and biomechanical aspects of hand function as understood from in vivo experiments and naturalistic studies. The important and logistically challenging in vivo studies provide key insights into what primates do with their hands and in some cases have quantified the gross

C.M. Orr (✉)
Department of Cell and Developmental Biology, University of Colorado School of Medicine, Anschutz Medical Campus, Aurora, CO 80045, USA
e-mail: caley.orr@ucdenver.edu

T.L. Kivell et al. (eds.), *The Evolution of the Primate Hand*, Developments in Primatology: Progress and Prospects, DOI 10.1007/978-1-4939-3646-5_9

movements (kinematics) and external forces (kinetics) involved. However, the necessary observational scale of most in vivo studies (and limits on how invasive a researcher can be) can make it difficult to assess the connection between what an animal does with its hands and how the detailed anatomy itself (i.e., the complex system of small joints, muscles, and ligaments that composes the hand) is adapted to such activities. This chapter provides examples of recent methodological approaches inspired by technological advances in biomedical imaging, computer modeling, and engineering that bridge observations on pure form with in vivo observations of function, thus providing a middle-range biomechanical link that helps connect the two sets of data.

2 Three-Dimensional (3D) Imaging and Surface Modeling

In the postcranial skeleton, only the foot rivals the hand in terms of articular complexity. Although the hinge and condylar joints of the fingers are relatively simple, the joints of the palm and wrist interlock in a highly intricate arrangement. Quantification of that geometry is important for comparing movement potential and possible patterns of load transmission. Until recently, it was primarily linear measurements that were used as shape descriptors in the hand (e.g., McHenry and Corruccini 1975; Corruccini 1978; McHenry 1983; Rose 1984; Kivell and Begun 2009; Kivell and Schmitt 2009; Begun and Kivell 2011; Kivell et al. 2011a, 2013). However, given the unusual shapes of the bones, three-dimensional (3D) approaches can quantify aspects of form that are not possible using more traditional methods (e.g., Tocheri et al. 2003, 2005, 2007; Tocheri 2007, 2009; Marzke et al. 2010, 2012; Matsuura et al. 2010; Orr 2010; Orr et al. 2013).

Modern imaging techniques are important tools in the study of comparative morphology and permit relatively rapid acquisition of 3D data that can be analyzed in a variety of ways (Zollikofer and Ponce de León 2005). These methods typically capture a dense cloud of XYZ coordinate points, either from the surface of an object (e.g., laser scanning) or by reconstructing a similar dataset from radiographic techniques such as computed tomography (CT) and micro-computed tomography (μCT) scans (Zollikofer and Ponce de León 2005). From that cloud of points, a surface is modeled digitally by connecting the points with a triangular polygon mesh (Fig. 9.1). Surface laser scanning has become an increasingly common form of data capture due to its portability (especially important when research involves travel to museums) and reductions in the cost of equipment and software. Wood et al. (1998) demonstrated the potential of laser scanning and surface modeling for studying primate morphology, and later applications by Tocheri, Marzke, Orr, and colleagues (Tocheri et al. 2003, 2005, 2007; Tocheri 2007; Marzke et al. 2010, 2012; Orr 2010; Orr et al. 2013) have focused on detailed analyses of the hand and wrist in particular. CT-based data acquisition can be more rapid in the initial data collection phases than surface scanning (at least for standard medical CT scanners), but requires more time in post-processing as the necessary thresholding protocols (e.g., Coleman and

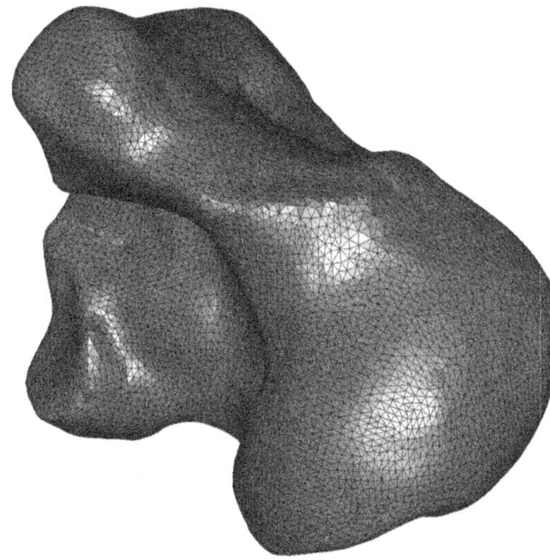

Fig. 9.1 Three-dimensional polygon model of a chimpanzee capitate derived from laser scanning. XYZ coordinates forming a point cloud are "nodes" connected by a triangular polygon mesh. Polygon models can also be constructed from computed tomography (CT) data

Colbert 2007) are applied and the polygon models are generated. These techniques also require a CT scanner to be on site or for specimens to be taken on loan (not always feasible). μCT is especially useful for accurately modeling smaller specimens such as the tiny carpals of small-bodied primates (e.g., smaller than *Cercopithecus*), but scanning times are much longer. If only external surfaces are required, high-end laser scanners are also a viable option for small structures. When done carefully, laser scanning and CT-based models can often be pooled for many analyses (Tocheri et al. 2011). Once generated, the polygon models are useful for quantifying various aspects of bone shape once features of interest are identified. Such features (Fig. 9.2) may include whole-bone surface areas, surface areas of more specific regions of interest, cross sections, joint surface angles, or individual landmarks on the bone (e.g., tubercles or junctions between articular surfaces) that can be used to take traditional linear measurements (e.g., Maiolino et al. 2011, 2012) or to conduct geometric morphometric analyses (see Zelditch et al. (2012) for an overview of geometric morphometrics).

Surface areas and angles between joint surfaces are among the best-studied aspects of carpal form using 3D digital polygon models and provide good examples of such applications in a comparative context. Surface areas can be quantified from the polygon models by summing the surface areas of the triangles that compose the bone model or the region of interest (e.g., a joint surface). Articular angles can be quantified easily by mathematically fitting a plane to joint surfaces and measuring the angle between the normal vectors of the planes (i.e., vectors perpendicular to the planes) (Tocheri et al. 2003). For example, Tocheri et al. (2003, 2005) and Tocheri (2007) used these methods to demonstrate that the radial side of the wrist in later hominins (modern humans, Neandertals, and their immediate ancestors) is highly distinctive in its arrangement of the bones at the base of the thumb relative to that of the apes and more primitive hominins (including

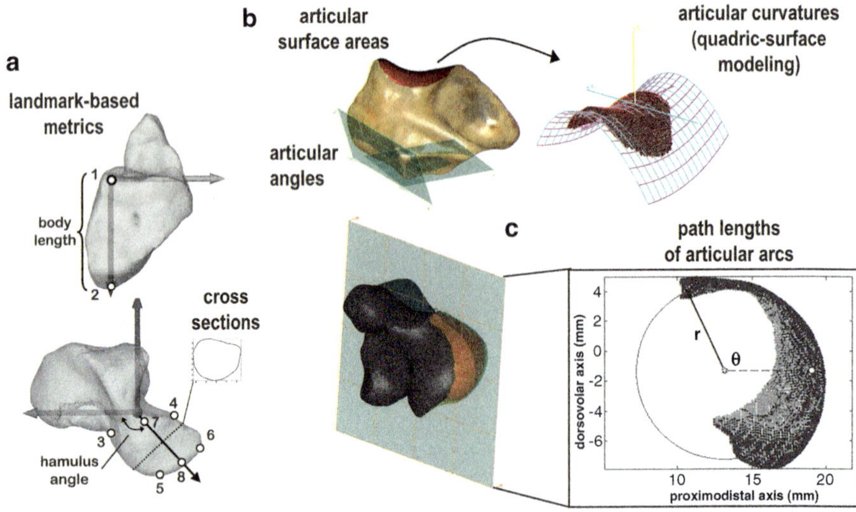

Fig. 9.2 Examples of how three-dimensional polygon models and various morphometric techniques can be used to quantify articular geometry and other aspects of bone shape in the hand. Various modeling approaches can be applied to capture joint shape. (**a**) Hamate with single landmarks (which can be used for traditional linear metrics or for use in geometric morphometric analyses; see Zelditch et al. 2012) and a cross section demonstrated on a hamate; (**b**) trapezium demonstrating joint surface areas, angles, and curvatures; and (**c**) capitate with the head segmented to quantify the joint's dorsal articular arc

Homo floresiensis: Tocheri et al. 2007; Orr et al. 2013). Analysis of the surface areas and joint angles of the trapezium, trapezoid, capitate, scaphoid, and first and second metacarpals shows clearly that the derived condition shared by humans and Neandertals is a blocky trapezoid that articulates palmarly with the capitate along with corresponding changes in the surrounding bones. This is in contrast to the primitive wedge-shaped trapezoid that partially defines a deeper carpal arch in the apes. Based on a small number of specimens, Lewis (1989) qualitatively described the differences in arrangement of the carpal bones in humans versus other primates. Accurate quantification using 3D methods permitted rigorous verification of the taxonomic distribution. The functional hypothesis proposed by Lewis (1989) is that the humanlike configuration facilitates load transmission in an oblique, radial-to-ulnar direction during powerful pollical grasping by providing a stable arrangement of the bones at the base of the thumb. The 3D quantification of joint sizes and orientations is consistent with that functional hypothesis; however, corroborating evidence using direct and indirect methods of studying load distribution is necessary for confirmation.

More sophisticated mathematical modeling can quantify other aspects of joint shape using 3D polygon models (e.g., Marzke et al. 2010; Matsuura et al. 2010). For example, Marzke et al. (2010) modeled the pollical carpometacarpal joint as a quadric paraboloid surface (i.e., a mathematically defined surface that has parabolic curvatures in two directions; see Fig. 9.2b) for comparisons among hominins and other primates. This technique confirmed that the apes and other catarrhines also possess saddle-shaped joints, suggesting a basic similarity of the joint kinematics among these taxa. Specifically, the saddle shape of the articulation permits opposition (i.e., a conjunct rotation involving a combination of abduction and flexion) at the pollical carpometacarpal joint (Napier 1952, 1956). Qualitative observations by Napier (1960) and Rose (1992), humans and other catarrhines are quite similar with regard to the morphology and motion potential at this joint. As such, the limited abilities of apes in opposing the thumb to the fingers during precision grasping (Napier 1960) may have less to do with differences in the shape of the trapeziometacarpal joint than it does with the shortness of the thumb relative to the length of the palm and fingers. However, humans are derived relative to apes and *Australopithecus afarensis* in that the trapeziometacarpal saddle surface is flatter in both the dorsopalmar and radioulnar directions (Marzke et al. 2010). As with the trapezoid-capitate geometry, these differences in trapeziometacarpal curvature probably have important consequences for the loading at the interface between the thumb and the wrist. The less pronounced curvature of humans probably results in a somewhat less stable articulation than the deeply curved joints of apes and *A. afarensis* in terms of resisting joint dislocation; nevertheless, the flatter surface might also facilitate forceful precision grips between the thumb and the other digits by offering a greater surface area to directly resist axial loads produced by relatively powerful thumb musculature (Marzke et al. 1998), thus representing a morphological trade-off. However, understanding the exact functional consequences of documented differences in joint curvature and orientation requires further comparative and experimental work focused on joint kinematics and kinetics during actual hand use (e.g., during grasping or hand posture during locomotion) as well as load transmission. This highlights the need to integrate morphometric data with experimental techniques to better understand function.

3 Analysis of Hand and Wrist Joint Kinematics Using Computed Tomography Imaging

Robust hypothesis testing in functional morphology cannot rely solely on the structural and morphometric analysis of bone shape. This is especially true when attempting to draw functional inferences regarding complex structures such as the hand and wrist. In such cases, the biomechanics should be studied in extant taxa, using experimental approaches when possible. Understanding the mechanical basis of mobility and stability in the primate wrist is especially important for reconstructing the evolution of locomotor and manipulative capabilities in primates (e.g., Schön and Ziemer 1973; O'Connor 1975; Richmond and Strait 2000; Richmond et al. 2001; Lovejoy et al. 2009; Williams et al. 2010). The kinematics of the wrist are particularly complex, and the carpals are difficult to study without the aid of internal visualization, because alternative methods (e.g., dissection or the implantation of pins to track the movements) require the disruption of the joint complex.

A number of biomedically oriented studies have employed advanced visualization techniques to study 3D carpal kinematics to determine how individual bones contribute to the overall movement at the wrist (e.g., Kobayashi et al. 1997a, b; Wolfe et al. 1997, 2000, 2006; Patterson et al. 1998, 2007; Crisco et al. 1999, 2001, 2003, 2005; Moojen et al. 2001, 2002a, b, 2003; Neu et al. 2001; Kaufmann et al. 2005, 2006; Moritomo et al. 2003, 2004, 2006; Camus et al. 2004; Werner et al. 2004; Goto et al. 2005; Calfee et al. 2008; Leventhal et al. 2008, 2010; Rainbow et al. 2008, 2013). However, there have been few such studies of the wrist in nonhuman primates (Orr et al. 2010; Daver et al. 2012). Orr and colleagues (2010) adapted a CT-based method designed for tracking 3D bone kinematics in humans (Wolfe et al. 1997; Crisco et al. 1999, 2001, 2003; Neu et al. 2000, 2001; Coburn et al. 2007; Moore et al. 2007) for general application in comparative analyses. The raw data acquisition for this approach involves serial CT scanning of cadaveric specimens using a custom-designed jig to position the hand at regular intervals throughout the wrist's range of motion (Fig. 9.3). Once CT scans have been acquired, the individual bones from each image are segmented into separate objects using a commercial software package such as Mimics® or Amira.® Using the shapes of the bones (based on the calculated inertial properties of each object; Crisco et al. 1999), the corresponding elements from each scan are then registered (matched up). The rotation and translation required to match corresponding elements can then be used as a description and quantification of the motion from one wrist position to the next following the standard kinematic calculations (e.g., Bottema and Roth 1979; Beggs 1983; McCarthy 1990; Zatsiorsky 1998).

The quantitative description of 3D carpal kinematics can take a variety of forms, but the most straightforward is the use of helical axes of motions. Helical axes of motion quantify the motion based on a single, unique axis in space around which a bone rotates and along which it translates (see Orr et al. 2010 for a summary). Mathematically, any bone in the model can be held constant, so that motion of a bone of interest can be quantified relative to any other. A series of coordinate axis systems have been defined for general application in comparative analyses across primate species (Orr et al. 2010). Following biomedical convention, the "global" position of the hand is tracked as the motion of the third metacarpal relative to the radius. Importantly, the method allows for excellent visualization of the movements involved, providing an intuitive method of study and a "reality check" on the numbers. Given enough scans, motion from one position to the next can be animated with minimal interpolation.

The analysis of wrist extension (or dorsiflexion) in primates is a good example of the application of the 3D kinematics technique, because of its relevance for understanding modifications of the primate wrist to particular hand postures such as knuckle-walking, digitigrady, and palmigrady (Tuttle 1967, 1969a, b; Schön and Ziemer 1973; Jenkins and Fleagle 1975; Tuttle and Watts 1985; Richmond and Strait 2000; Richmond et al. 2001; Orr 2005, 2010). Wrist dorsiflexion is also a critical component of the "wrist snap" involved in toolmaking and used in hominins (Williams et al. 2010; see Chap. 11). Whether a high degree of possible wrist extension is a

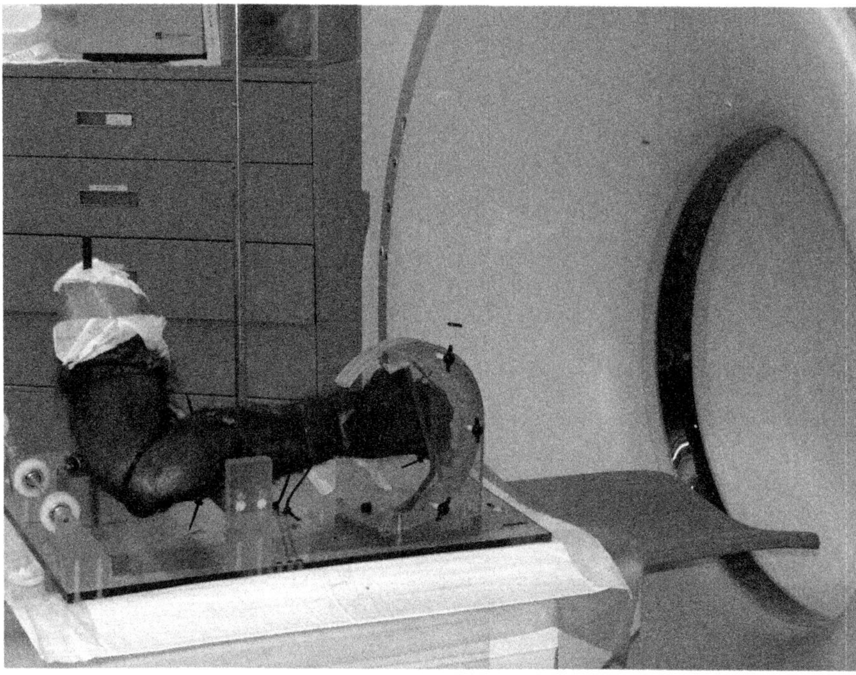

Fig. 9.3 Custom polycarbonate jig designed to place the hand of primate cadavers through a range of motion at the wrist joints (chimpanzee specimen shown here). The apparatus is outfitted with goniometers to measure precise wrist joint positions at regular intervals, which are then scanned using computed tomography. Modified from Orr et al. (2010)

derived adaptation in the hominin clade related to intensified tool behaviors or a primitive condition retained from a mobile-wristed ancestor is debated (Richmond and Strait 2000; Lovejoy et al. 2001, 2009; Richmond et al. 2001; Begun 2004; Orr 2005; Kivell and Schmitt 2009). Resolving this issue requires data on how the mechanics of wrist extension differ among extant primates, as well as determining morphological correlates of wrist mobility that might ultimately be recognized in the fossil record.

Wrist extension in humans is well known to occur as a combination of motions between the bones that compose the more proximal radiocarpal and more distal midcarpal complexes (Levangie and Norkin 2011). Three-dimensional analysis of anthropoid wrist dorsiflexion demonstrates that chimpanzees and orangutans are differentiated in wrist extension from palmigrade Old and New World monkeys in exhibiting relatively greater mobility between the scaphoid and lunate as wrist extension proceeds (Fig. 9.4a). That more pliable proximal carpal row might allow effective transfer of loads through the wrist in a variety of positions. Such an arrangement might be beneficial for the diversity of hand positions required during climbing. However, as a consequence, to allow for more stability during quadrupedal hand posturing and propulsion (i.e., knuckle-walking for the African apes), the lunate must be dynamically stabilized in chimpanzees. This occurs via a

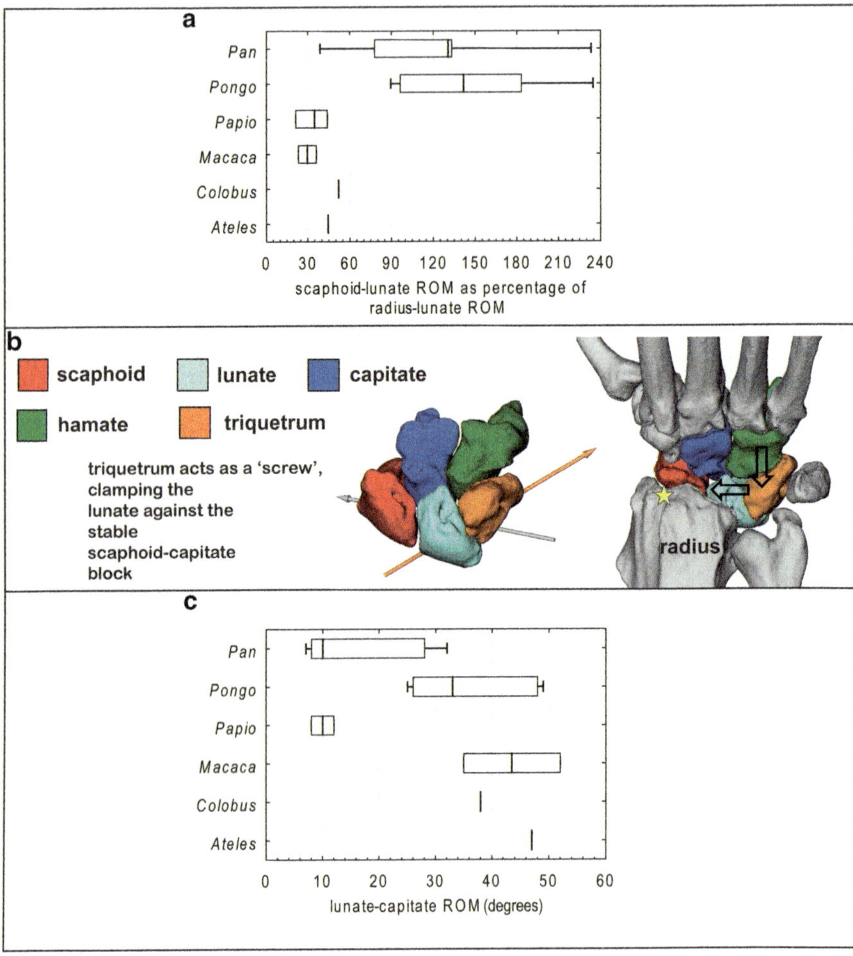

Fig. 9.4 Three-dimensional computed tomography-based analysis of the kinematics of the proximal carpal and midcarpal joint complexes in *Pan troglodytes* ($n=5$), *Pongo spp.* ($n=5$), *Papio anubis* ($n=3$), *Macaca mulatta* ($n=2$), *Colobus guereza* ($n=1$), and *Ateles geoffroyi* ($n=1$). (**a**) The range of motion (ROM) of the intercarpal scaphoid-lunate joint is determined as the maximum rotation around a helical axis of motion from the neutral position to maximum extension of the wrist. The ROM is expressed as a percentage of the lunate's rotation on the radius. These data show that the apes have relatively greater mobility within the proximal carpal row during extension, whereas in monkeys, the proximal carpal row acts as a more rigid unit. (**b**) Screw-clamp action of the midcarpal joint in a chimpanzee, in which the lunate is dynamically stabilized between the scaphoid/capitate and the triquetrum. (**c**) In conjunction with the screw-clamp action stabilizing the proximal carpal row, chimpanzees exhibit lower ranges of motion between the lunate and capitate (i.e., the lunatocapitate joint) than in orangutans (which reflects the overall mobility in extension of the midcarpus). In **a** and **c**, the vertical ticks are the medians (or individual values), the *rectangles* represent the interquartile range, and the *whiskers* represent the full range. The *yellow star* in **b** represents the dorsal ridge on the distal radius

"screw-clamp" mechanism in which the lunate is pinned between the triquetrum and scaphoid as wrist extension proceeds (Orr 2010; Orr et al. 2009, 2010; Fig. 9.4b). An overall more rapid engagement of the midcarpal complex thus occurs in chimpanzees than is seen in orangutans, resulting in lower mobility in extension between the lunate and capitate (which reflects overall midcarpal mobility, Fig. 9.4c). This rapid engagement in *Pan* is facilitated in part by the fusion of the centrale to the scaphoid, which allows the scaphoid to act as a more stable unit as the lunate is pressed to it during extension. The human midcarpus also functions by way of this "screw-clamp" mechanism (MacConaill 1941; Wolfe et al. 2000).

Integrating such experimental kinematics work with the morphometrics of the bone structure (as discussed in the previous section) is then the critical next step for demonstrating the link between form and function. The use of polygon models of bones is one such way of integrating morphology and kinematics. For example, a polygon model of the capitate permits measurement of the "lunatocapitate arc" (i.e., the articular space available for lunate movement on the capitate head). This can be done by means of finding a circle that "best fits" the points on the head using a mathematical technique (Figs. 9.4c and 9.5). The correlation of lunatocapitate arc with the experimentally determined range of motion of the lunate on the capitate during wrist extension (Fig. 9.5) provides a decent anatomical indication of midcarpal mobility across primates. Because primate cadavers are difficult to acquire, full investigation of the variation in the lunatocapitate arc can proceed using larger sample sizes of polygon models collected from dry-bone specimens housed in museums, and these datasets can be applied to interpreting fossils. Early hominin wrists are most similar

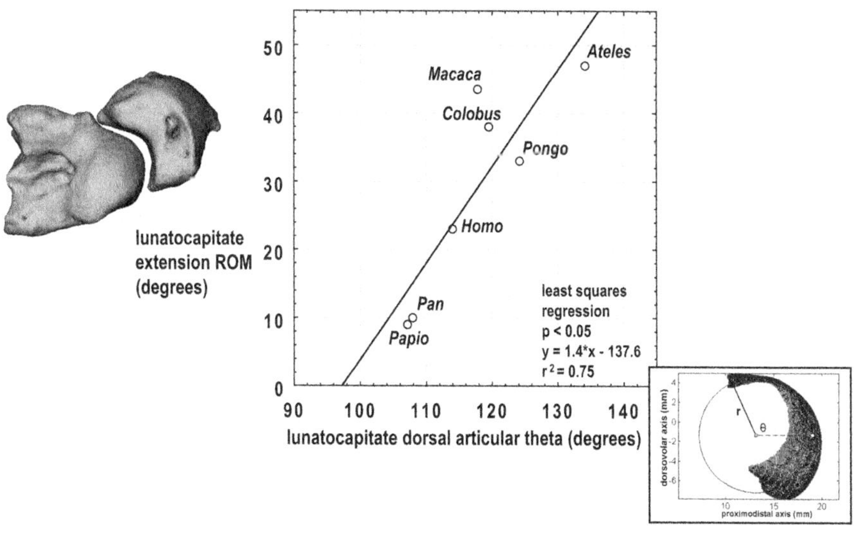

Fig. 9.5 Combination of carpal kinematic data and morphometric data. The median values for the articular arc path length of the dorsal aspect of the capitate (expressed as the "theta" angle) are correlated with the median lunatocapitate range of motion (ROM). Articular path length thus provides a decent approximation of midcarpal mobility in extension, which can be measured in fossils to estimate range of motion

to the African apes in having abbreviated lunatocapitate articular arcs and an os centrale that is fused to the scaphoid (Orr 2013). Given these findings and that humans share scaphoid-centrale fusion with chimpanzees and gorillas (see Kivell and Begun 2007), it suggests that hominins and chimpanzees (and probably gorillas) share a derived pattern of midcarpal kinematics relative to other anthropoids.

In addition to the mechanical limits on midcarpal mobility, all extant apes exhibit constraints on mobility at the more proximal radiocarpal joint complex (i.e., the joints between the proximal row of carpals and the radius) when compared with palmigrade monkeys and modern humans. This limited mobility is reflected by the palmar declination of the distal radius and a consequent development of a dorsal ridge of the bone (Fig. 9.4b) that limits the dorsal rotation of the scaphoid (or at least marks the limit thereof) during wrist extension (Orr 2012). In contrast, palmigrade monkeys (which fully extend the wrist to place the palm flat on the substrate) and modern humans do not develop such a ridge and display high mobility of the scaphoid relative to the radius during extension. Because early hominins are more ape-like in this regard (Richmond and Strait 2000; Richmond et al. 2001; Orr 2005, 2010, 2012, 2013; Tallman 2012), modern humans appear to have lost the dorsal ridge of the radius secondarily, reflecting a more mobile radiocarpal joint than that of extant apes.

The integrated study of 3D joint kinematics and joint shape demonstrates that humans are uniquely derived in their combination of a midcarpal screw-clamp (a wrist mechanism shared with chimpanzees and possibly gorillas) with a secondarily derived radiocarpal joint complex. This unique combination of features allows for an overall higher degree of extension of the hand at the wrist in humans than in the extant apes, but in a way that is biomechanically distinctive from how palmigrade monkeys produce such mobility. The arrangement observed in humans may have resulted from relaxation on selection for climbing proficiency (the presumed primitive condition for hominins) in favor of a wrist that functions effectively for the manipulative tasks. In turn, this shift in the biomechanics of the wrist may have facilitated the advent of such hominin-specific behaviors as flint knapping, shown through in vivo experiments to require a high degree of overall dorsiflexion for effective performance (Williams et al. 2010; see Chap. 11). The combined experimental and structural approach to morphology outlined here therefore allows for a better understanding of how a complex anatomical structure such as the wrist evolved in a mosaic fashion within the hominin clade from a primitive condition (in this case, a rigid wrist) to produce a derived function (high overall range of motion in extension) that appears to facilitate novel behaviors.

4 Understanding Load Transmission Through the Primate Hand

The analysis of load transmission through joints is a critical component to understanding the functional morphology of primate limbs. In theory, habitual loading regimes should vary according to locomotor mode, limb posture, and substrate use (terrestrial vs. arboreal). In the hand, the habitual loading regime of bipedal hominins is expected to be especially distinct with the forelimb having become freed from locomotor duty and possibly adapted to the intensified reliance on various manipulative behaviors such as tool making and use. Establishing the correlation between loading history and specific aspects of bone biology accessible from fossils has the potential to be the ultimate tool for reconstructing specific behaviors in extinct species. The bone is expected to respond adaptively to its mechanical environment, but the exact mechanisms and extent to which the bone is modified to habitual activities within an individual's lifetime are not yet fully understood (Pearson and Lieberman 2004; Ruff et al 2006). However, it is generally accepted that some aspects of internal skeletal structure are at least a partial reflection of its loading history (e.g., Carter and Beaupré 2001; Frost 2001; Currey 2002; Ruff et al. 2006). A number of techniques have been employed in an effort to infer information about the loading history of the hand that might inform us about its use and adaptation to particular behaviors. These include the analysis of trabecular architecture, subchondral bone density, and modeling approaches such as finite element analysis.

4.1 Trabecular Architecture

It is generally thought that the internal trabeculae of the skeleton are particularly subject to variations in the mechanical environment that occurs under different habitual loading regimes. That is, the orientation, size, and density of the bony struts filling the inside of the bone should be modified to effectively resist compression (e.g., Odgaard 1997; Pontzer et al. 2006; Barak et al. 2011). Typical variables used to quantify these aspects of trabecular form are summarized in Table 9.1. Quantification relies on high-resolution μCT and sophisticated image-processing techniques to visualize in 3D the fine filaments that form the spider-web-like architecture of the trabecular bone.

Trabecular bone architecture has received considerable attention in a number of anatomical regions other than the hand (Fajardo and Müller 2001; Fajardo et al. 2002, 2007; MacLatchy and Müller 2002; Robson-Brown et al. 2002; Ryan and Ketcham 2002a, b, 2005; Ryan and van Rietbergen 2005; Maga et al. 2006; Ryan and Krovitz 2006; Ryan and Walker 2010; Ryan and Shaw 2012). There have been few comparative studies of the trabecular architecture of the primate hand, although important initial forays have been made by several researchers (Lazenby et al. 2008a, b, 2010, 2011; Kivell et al. 2011b; Tsegai et al. 2013; Schilling et al. 2014). These careful studies have been informative with regard to the prospects and limitations of analyzing the trabecular structure of the bones of the hand.

How the hand's trabecular architecture responds to different functional demands is not fully understood. One approach to this problem is to analyze different joints within the hands of single individuals based on theoretical expectations about how the joints should behave. For example, Lazenby and colleagues (2008a, b) have shown that within individual humans, some aspects of trabecular structure vary in

Table 9.1 Common morphometric variables used to quantify trabecular architecture

Variable (and symbol)	Definition[a]
Bone volume fraction (BV/TV)	The volume of trabeculae relative to the total VOI and is one of the best determinants of overall bone strength (Odgaard 1997, 2001)
Bone surface density (BS/TV)	Bone surface area within the VOI relative to total volume (Odgaard 1997, 2001)
Intersection surface (i.S)	Total area of trabecular bone that is "cut off" by the surface of the VOI. As the number, volume, and thickness of the trabeculae increases, the intersection will also increase (Odgaard 2001)
Trabecular thickness (Tb.Th)	The mean minimum thickness of trabeculae within the VOI measured using a series of spheres fit within the trabecular struts (Hildebrand and Rüegsegger 1997a)
Trabecular number (Tb.N)	An estimate of the number of trabeculae calculated as the ratio of the bone volume fraction to trabecular thickness
Trabecular separation (Tb.Sp)	The average width between trabeculae within the VOI
Trabecular bone pattern factor (Tb.Pf)	An indirect measure of how connected trabeculae are within the VOI. Higher values indicate less connectivity with isolated struts, while a lower value reflects greater overall connectivity (Hahn et al. 1992)
Structure model index (SMI)	The proportion of platelike (longitudinally oriented) versus rodlike (transversely oriented) trabecular struts within the VOI, with values ranging from 0 (all plates) to 3 (all rods) (Hildebrand and Rüegsegger 1997b). Platelike trabeculae are more closely correlated with the elastic mechanical properties of the bone (Liu et al. 2008)
Degree of anisotropy (DA)	A measure of the overall "preference" of the trabeculae to be oriented in a particular direction or not. This can be done using a variety of methods (Odgaard 1997, 2001). A common method has been the mean intercept length (Harrigan and Mann 1984; Odgaard 2001), which returns a range of values running from 0 (complete isotropy or no preferred direction) to 1 (complete anisotropy in which trabeculae are all oriented the same). Greater anisotropy presumably reflects more stereotyped loading, and the primary orientation of the trabeculae can be determined using the orientation of the first eigenvector of the struts relative to an anatomically based coordinate system (e.g., Kivell et al. 2011b)

[a]Note VOI = volume of interest, which is the standard volume (typically a sphere or cube) used to sample an anatomical region for study of the trabeculae

ways expected based on theoretical loading regimes at different joints and according to handedness. The head of the second metacarpal (Mc2) should exhibit a varied and intense loading regime given its contribution to the mobile metacarpophalangeal (McP) joint and the regular use of the index finger in a variety of human tasks. In contrast, the proximal end of the bone with its stable, and nearly immobile, articulation with the wrist should display signs of a less demanding mechanical environment. Similarly, because right handedness is predominant in humans, it is expected that at a population level, bones from the right side should, on average, show more robust internal architecture than those from the left side.

Indeed, for both the between-joint and between-side comparisons, these expectations bear out with regard to bone volume fraction and trabecular number (Lazenby et al. 2008a, b; Table 9.1). However, other structural variables, including those capturing thickness, shape (platelike or rodlike), and connectivity of the trabecular struts, do not follow the same pattern, suggesting that these measures may not be tightly linked to history of loading.

Anisotropy (i.e., the degree to which trabecular struts vary in their orientation) also differs between the Mc2 head and base; however, Lazenby et al. (2008a) found a pattern contrary to theoretical predictions, as the head (expected to be loaded in a more diverse manner) is significantly more anisotropic (i.e., struts are more aligned in a particular orientation) than the base. These findings may indicate that the bone resists loads more effectively at certain positions of the McP joint, while the proximal base is comparably resistant in different loading directions. Although this is contrary to the initial predictions, these findings may not be surprising. The proximal base of the Mc2 is likely to receive loads via the thumb and index finger and, to a lesser degree, by recruitment of muscles on the ulnar side of the hand (e.g., during a strong five-jaw chuck or power squeeze grips in which the other fingers are used to stabilize an object in the hand, such that the palm becomes radioulnarly compressed). Consequently, load transfer at the base of the Mc2 probably occurs in both proximodistal and radioulnar directions. If loading of the McP joint occurs at stereotyped finger postures (e.g., at a regular degree of flexion), then trabecular buttressing of the joint in a particular direction might be expected, thereby resulting in greater anisotropy. This alternative hypothesis highlights the need to combine the fine-scale structural and biomechanical analyses exemplified by Lazenby et al. (2008a, b, 2010, 2011) with the in vivo experiments that quantify hand kinematics and kinetics. Interspecific comparisons among primates that habitually load their hands in different ways would also provide an indirect test. Nevertheless, the intra-individual approach to trabecular architecture demonstrates that analysis of internal structure may be a promising area of inquiry for understanding loading patterns in the primate hand. However, a major challenge is to identify functional signals correlated with specific behaviors, which requires extensive interspecific comparative work (Kivell et al. 2011b; Tsegai et al. 2013; Schilling et al. 2014).

In comparison with other primates, it is expected that humans should exhibit a trabecular architecture of the hand (e.g., low bone volume fraction and other variables that reflect the strength of the structure; see Table 9.1) that reflects our derived adaptation to bipedality and the drastically reduced load-bearing role of the forelimbs. In contrast, quadrupedal primates (whether palmigrade, digitigrade, or knuckle-walking) are predicted to display hand bones with a different trabecular architecture that reflects their ability to transmit loads effectively. Suspensory primates such as orangutans, gibbons, and spider monkeys might be expected to display a somewhat intermediate state due to the presumably reduced compressive loads that occur when the forelimb is used primarily to suspend the body below a support (in which case, loading is mostly a function of muscle forces acting across various bones and joints of the hand). If such a locomotor signal could be identified, then trabecular architecture might provide a powerful means of tracing the evolution of a more "emancipated" hand (Wood Jones 1916), as primates of various lineages (especially hominins) adapted their forelimbs to activities other than strict quadrupedal weight bearing.

Somewhat unexpectedly, consistent differences relating to hand function across primates have been rather elusive or, at least, complicated. For example, although humans exhibit low bone volume fraction in the scaphoid, lunate, and capitate compared to nonhuman primates (as predicted), other trabecular architecture variables do not distinguish us from other primate taxa (Schilling et al. 2014). Moreover, other structural variables fail to differentiate among primate locomotor modes (bipedal, quadrupedal, and suspensory) altogether (Schilling et al. 2014). Schilling and colleagues (2014) suggest that the inability to detect a clear locomotor signal may be due in part to the intricate geometry of the wrist and associated complexity of load transmission through that region of the hand, but also to problems arising from standard methodology. Specifically, analyses of trabecular architecture rely on the use of a "volume of interest" (VOI) in each specimen (usually a sphere standardized to some aspect of the specimen size). The architectural variables are then analyzed within that VOI, which (hopefully) provides a means of comparing the same (i.e., homologous or at least biomechanically analogous) region of the bone in each specimen and taxon under study. Unfortunately, many of these architectural variables are highly sensitive to the size and placement of the VOI when used to analyze small bones of the hand (Kivell et al. 2011b), and the error that accrues as a result might be obscuring functional signals to some degree.

Going beyond the VOI to assess the trabecular architecture of whole bones, or at least whole epiphyses (the articular ends of bones), may hold more promise for assessing trabecular architecture in the skeletal elements of the hand (Tsegai et al. 2013; Gross et al. 2014). Indeed, a "whole-epiphysis" analysis of the trabecular architecture of the third metacarpal (Mc3) head among hominoids demonstrates more clearly the predicted loading patterning and locomotor groupings (Tsegai et al. 2013). The Mc3 heads of knuckle-walking African apes generally exhibit the highest relative bone volumes and more anisotropic patterning of the trabecular struts (consistent with both dorsal and palmar loading of the metacarpophalangeal joint during knuckle-walking and flexed-finger positions while climbing). In contrast, humans exhibit both relatively low bone volume fractions (as shown for the carpals by Schilling et al. 2014) and lower degrees of anisotropy (i.e., a clear orientation to the struts consistent with an emphasis on the palmarly oriented loads expected to occur during grasping). As predicted by their emphasis on powerful grasping with flexed fingers, the suspensory Asian apes (orangutans, gibbons, and siamangs) fall somewhat between the knuckle-walkers and bipeds for these two key variables (Tsegai et al. 2013). Imaging the trabecular structure for qualitative assessment further clarifies and reinforces the results of the quantitative analysis (Fig. 9.6). As such, trabecular architecture of the Mc3 head appears to track with reasonable fidelity the habitual hand postures among the hominoids. If results similar to those found for the Mc3 distal epiphysis can be demonstrated to hold for other components of the hand skeleton, then trabecular architecture may indeed prove to be an invaluable tool for evaluating primate hand function and its evolution.

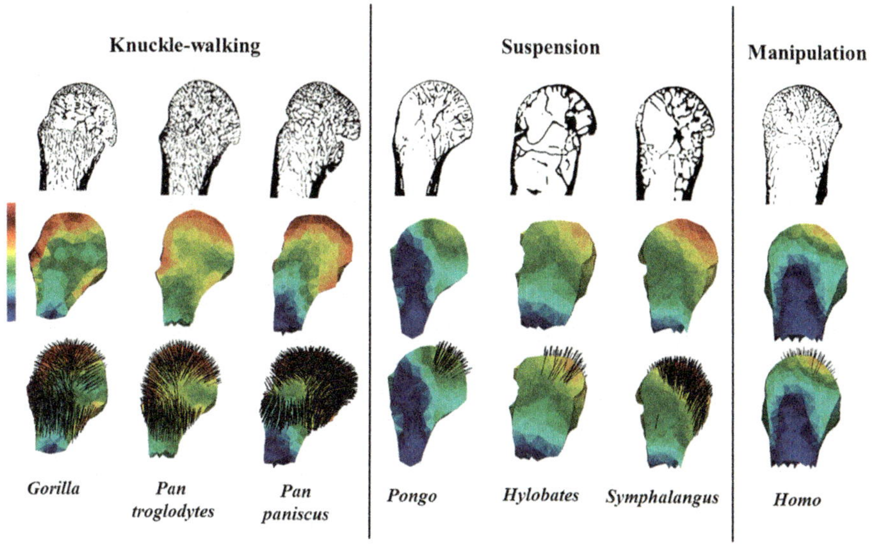

Fig. 9.6 Analysis of the trabecular architecture of the third metacarpal head in hominoids. The *top* row shows the actual trabecular struts in a sagittal cross section from micro-CT scans of the bone. The *middle* row displays color maps demonstrating differences in the distribution of bone volume fraction (BV/TV; see Table 9.1) among the specimens (*cool colors* indicate low BV/TV, while *warm colors* indicate high BV/TV). The *bottom* row displays vectors indicating the directions of greatest stiffness (greater than elastic modulus of 1000 Pa) throughout the head. The stiffness vectors indicate that African apes have greater stiffness in their trabecular structure throughout the metacarpal head, compared with Asian apes and humans, where overall stiffness is lower and is concentrated to the palmar region of the head. Image provided by Tracy Kivell, as originally published in Tsegai et al. (2013)

4.2 *Subchondral Bone Density*

Another means of evaluating loading patterns in animal joints is through the analysis of the cortical bone directly deep to the articular surface. The mineral density of the bone is a primary determinant of its compressive strength (Currey 1984, 2002; Keller 1994). The idea behind examining subchondral bone is that the response to loading of the joints via transarticular muscle activity and weight support will result in varying levels of bone density depending on the magnitude of the habitual compression. Apparent density of the subchondral plate can be measured using CT osteoabsorptiometry, which measures density based on the degree to which X-rays passing through the bone are absorbed (Müller-Gerbl et al. 1989, 1992; Eckstein et al. 1995). Pixel brightness of the CT image serves as a proxy for bone density, with brighter pixels indicating greater absorption and therefore greater apparent density. In especially congruent joints (which are primarily subject to compression), patterns of subchondral density can provide insight into the loading history.

The efficacy of osteoabsorptiometric methods to evaluate patterns of peak habitual loading within a joint based on subchondral bone density is supported by comparative anatomical evidence. These data demonstrate that morphological regions expected to be the most heavily loaded typically show the regions of highest bone density (Müller-Gerbl et al. 1989, 1992; Eckstein et al. 1995; Hoogbergen et al. 2002; Lewis et al. 2005; Carlson and Patel 2006; Patel and Carlson 2007; Dickomeit et al. 2011; Carlson et al. 2013). Furthermore, in vivo data in sheep have shown that alterations to the positioning of the region of maximum subchondral apparent density track changes in the habitual posture of the hind limb as the grade of the treadmill varies (Polk et al. 2008). Although these findings hold potential for reconstructing habitual joint postures in extinct forms, its direct application is probably limited to "subfossil" (unfossilized or partially fossilized) remains (e.g., Polk et al. 2010), as bone mineralization resulting from the fossilization process appears to obscure evidence of bone density. Nevertheless, the approach can provide valuable insights into the basic comparative biomechanics critical for informing our understanding of the evolution of joint form and function in primates and other taxa (e.g., Nowak et al. 2010; Carlson et al. 2013).

Studies of the subchondral bone density relating to the functional morphology of the hand and wrist in primates have thus far been restricted to the radiocarpal articulation. Hoogbergen and colleagues (2002) investigated the region of highest apparent bone density of the distal radius and ulna in ten cadaveric humans (and therefore the inferred region of highest joint stress in the antebrachiocarpal joint complex). Despite the small sample, they found that the region of highest apparent density was shifted toward the radial side in individuals with pathological conditions of the wrist region (e.g., Colles' fracture of the radius, scaphoid fracture, and scapholunate instability). These observations suggest that even within a single individual's lifetime, the loading history of the wrist joint complex is reflected in the subchondral bone density patterns.

To date, Carlson and Patel are the only researchers who have conducted comparative analyses of subchondral bone density in the wrist across primates (Carlson and Patel 2006; Patel and Carlson 2007). Their work has shown that patterns of apparent density reflect habitual locomotor modes used by primates. These differences were predicted based on theoretical loading regimes derived from habitual hand postures. In bipedal humans and suspensory primates, lower compressive loading of the forelimb is manifested by a more diffuse patterning of the densest areas of subchondral bone across the entire carpal articular surface of the distal radius (Fig. 9.7). Knuckle-walking African apes show clear foci of higher apparent bone density on the palmar aspect of the articular surface, fitting expectations of limited dorsiflexion during the forelimb stance phase. In contrast to knuckle-walking, suspensory, and bipedal primates, digitigrade and palmigrade primates exhibit higher density patches of subchondral bone within the scaphoid notch along the dorsal edge of the radius. Palmigrade taxa tend to be differentiated from digitigrade taxa in displaying a greater frequency of high-density regions on the palmar side of the joint surface as well. This last observation does not fit the theoretical expectations, which predicted greater bone density on the dorsal side of the joint in

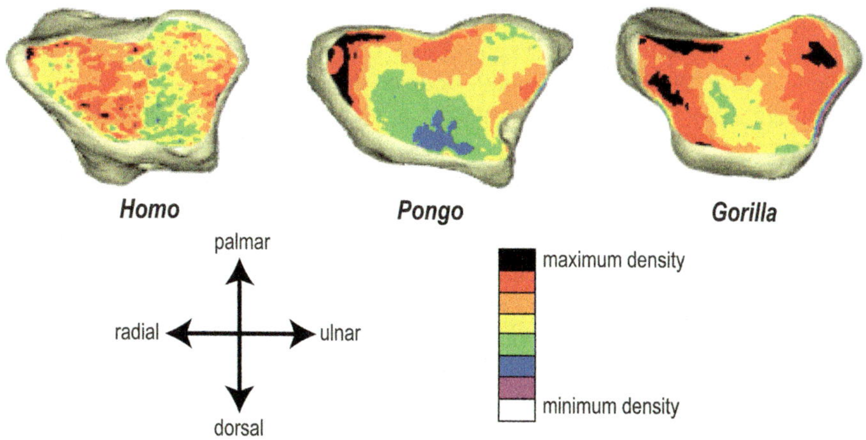

Fig. 9.7 Apparent subchondral bone density in the distal radius of a human, orangutan, and gorilla. Note the greater overall density in the quadrupedal knuckle-walking gorilla, which is expected to have much greater habitual compressive loads acting on the wrist during locomotion and posture. Figure provided by Biren Patel (Modified from Carlson and Patel 2006)

the digitigrade hand compared to the palmigrade hand. Nevertheless, the patterning of apparent bone density using CT osteoabsorptiometry clearly provides insights into load transmission through the radius across the broader categories of primate locomotor behavior. These findings are corroborated by comparative analyses of xenarthrans (New World sloths, anteaters, and armadillos) with quadrupedal taxa (anteaters) and suspensory taxa (sloths) exhibiting patterns of subchondral bone density in the distal radius that closely match primates with similar locomotor behaviors (Patel and Carlson 2008).

4.3 Finite Element Modeling and Analysis

Morphological structures are often too complex to predict or estimate how loads are transmitted and what the resulting pattern of stress and strain in the bones might be. Formally, stress (σ) is an estimate of the internal forces acting within a structure and is defined as force per unit area, with the units usually given in *pascals* (=Newtons/m^2). Strain ($\boldsymbol{\varepsilon}$) is a dimensionless expression of the deformation of a structure defined as a change in length relative to the original length, such that $\varepsilon = \Delta L/L$. As with other complex biomechanical systems, such as the jaws and teeth (Strait et al. 2005, 2007, 2009, 2010; Chalk et al. 2011), simple models for determining the response of the highly intricate hand skeleton to external loads are inadequate. Finite element modeling (FEM), which is commonly used in engineering fields, is a method of approximating strains at various locations on a complex structure under a variety loading conditions. Detailing the mathematical underpinnings of FEM is outside the scope of this chapter, but good introductions are available elsewhere (e.g., Beaupré and Carter 1992; Liu and Quek 2013; Richmond et al. 2005; Johnson 2009; Cook et al. 2011).

In brief, FEM allows the approximation of stresses and strains by modeling the geometry of a structure as a finite series of smaller and simpler geometric elements and quantifying the "deformation" of the mesh that occurs when a load is applied. The elements can be a variety of 2D or 3D shapes, and these are connected at their vertices (the coordinate points of which are referred to as "nodes") to compose a mesh. The mesh itself can be derived from any number of raw data acquisition methods from which a polygon model can be produced (see Fig. 9.1). However, morphologists must usually take into account internal structure, so CT data are preferred. The stresses and strains can then be solved as a summation of deformations about the nodes given a set of parameters defining kinematic constraints, material properties, and loading conditions.

As with all models, FEMs are only as good as the data used to construct the model and define its parameters. Of critical importance are the material properties of the structure and the boundary conditions providing the kinematic constraints that prevent rigid-body displacement of the model (i.e., movement of the FEM when loads are applied). The kinematic constraints are necessary to enable a unique solution (Richmond et al. 2005). The most relevant material properties describe the elasticity of the structures involved (e.g., bones, ligament, cartilage, etc.) and include the elastic modulus (or Young's modulus) and the Poisson ratio (see Currey 2002 for full discussion). Young's modulus (E) describes the stiffness of the structure as the slope of the materials' stress-strain curve (Currey 2002) or, in other words, the ratio of the stress (σ) to the resulting strain (ε): $E=\sigma/\varepsilon$. Thus, E describes how much the structure will be either stretched or compressed at a given stress. The Poisson ratio (ν) is similar, but describes how much the sides of the structure will expand outward (or "bulge") at a given stress, and is therefore calculated as the ratio of the stress (σ) to the transverse strain ($\varepsilon_{\text{trans}}$), which is the change in the transverse dimension or "width" ($\nu=\sigma/\varepsilon_{\text{trans}}$). These two elastic properties are then applied in the model to determine how much the FEM mesh (or a defined subregion of the mesh) responds to the stresses generated by a particular loading regime. Material properties typically vary within a particular tissue type (e.g., within the bone) as well as among different types (e.g., bone versus cartilage) (e.g., Peterson and Dechow 2003; Wang et al. 2006). It is best to incorporate such regional variation into the model and to use experimentally determined values for the structure of interest when assigning elastic properties to the FEM (Strait et al. 2005).

Selection of the boundary conditions is also of critical importance for producing an accurate FEM. The kinematic constraints should not affect the calculation of strains but still pin down the model to allow for a solution by restricting the model from rigid translation in all dimensions. For example, if the interest is in modeling strains on the metacarpus when a primate's hand is in quadrupedal locomotor posture, constrained nodes might be selected on digits contacting the ground and at the distal radius. Once kinematically constrained, a loading regime must be applied such that it captures as realistically as possible the load orientations and magnitudes as they occur in vivo. For example, ground reaction force data might be incorporated from in vivo locomotor experiments (e.g., Patel 2010), or experiments designed to determine the kinetics of stone toolmaking might be used to guide models investigating strains in the hand that arise from manipulative tasks (e.g., Rolian et al. 2011). Muscle forces can be approximated using in vitro measures of physiological cross-sectional area (which is proportional to muscle force potential, e.g., Marzke et al. 1999), while in vivo electromyographic experiments can provide further data on muscle activation and timing (Tuttle et al. 1972; Susman and Stern 1979, 1980; Marzke et al. 1998; Ross et al. 2005).

After the model is built and the appropriate parameters established, the model can be solved. The resulting deformation of the structure can be visualized on the model by exaggerating the deformation mathematically and/or by color mapping the values of various strain indices. The color maps are particularly helpful in identifying areas of high stress or strain during loading (Fig. 9.8), and the results should be validated against experimental observations of the biomechanical activity of interest. For example, researchers working on FEM of primate masticatory mechanics have validated their models using in vivo data from strain gages attached to the cranium and mandible during feeding in various primates (Ross et al. 2005, 2011). Although much experimental work has been done on the skull to estimate the sensitivity of the FEM to variation in the input parameters (e.g., Ross et al. 2005; Strait et al. 2005), only limited attempts have yet been made to validate FEM models of the primate hand (e.g., Richmond 2007; Varga et al. 2013).

With regard to the comparative functional morphology of the hand in primates, the most detailed and well-validated FEMs thus far have been constructed for the third ray proximal phalanx (PP3) (Richmond 1998, 2007; Nguyen et al. 2014). These models were constructed to test biomechanical hypotheses proposed to explain why phalangeal curvature correlates with the frequency of suspensory behavior in primates (Preuschoft 1973; Susman 1979; Stern and Susman 1983; Stern et al. 1995; Jungers et al. 1997; Richmond 1998; Deane and Begun 2008;

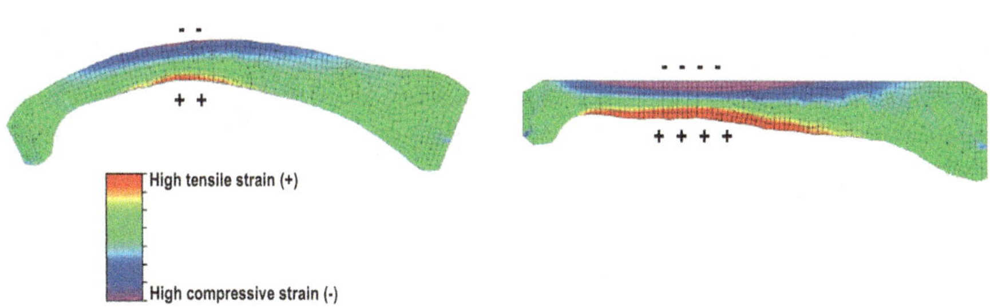

Fig. 9.8 A 2D finite element model of a third proximal phalanx in a siamang (*Symphalangus syndactylus*) during unimanual suspension (i.e., hanging by one hand). The curved model (*left*) exhibits overall lower strains compared to the "straightened" model (*right*). These findings support the hypothesis that bone curvature decreases strains on the fingers during suspension by aligning the metacarpophalangeal joint with the joint reaction forces arising from finger flexion and grasping. Figure provided by Brian Richmond (Modified from Richmond 2007)

Matarrazzo 2008; Congdon 2012)—an observation that has played an important role in debates concerning the locomotor behavior of early hominins (Stern and Susman 1983; Stern 2000; Ward 2002). Material properties were assigned based on typical values for mammalian cortical bone (from Currey 2002), and a variety of boundary conditions (kinematic constraints and loading regimes) were applied with similar overall results (Richmond 1998, 2007; Nguyen et al. 2014). Muscle, joint reaction, and substrate reaction forces were approximated using data on muscle geometry and activation patterns, and these were applied to a FEM constructed using a still image from video footage of a gibbon in unimanual suspension (Richmond 1998, 2007). In turn, the FEM model was validated using a strain gage data captured from the third proximal phalanx of a cadaveric gibbon forelimb, which was posed in a suspensory posture with weights hung to simulate body weight and digital flexor muscle activation.

Models of the phalangeal shaft that are mathematically "straightened" provide a comparative baseline for understanding the consequences of a curved shaft on the strain patterns (Richmond 1998, 2007; Nguyen et al. 2014; Fig. 9.8). In both the straight and the curved models, compressive strains are concentrated dorsally, while tensile strains are concentrated palmarly, but overall strain magnitudes are lower in the curved phalanx. These results support Preuschoft's (1970, 1973) model, which holds that curved phalanges reduce overall bone strain in flexed postures by acting to align the shaft with the metacarpophalangeal joint reaction force during manual suspension. In this way, the FEM bolsters the purely morphological studies by providing a solid biomechanical basis underlying the correlation between suspensory behavior and phalangeal curvature. Together, these results suggest that the curved phalanges of hominins such as *Orrorin tugenensis* (Senut et al. 2001; Richmond and Jungers 2008) or *Australopithecus afarensis* (Stern and Susman 1983; Stern 2000) indicate that the hand maintained its primitive role as a locomotor structure (i.e., for climbing/suspensory locomotion) for a considerable duration following the evolution of bipedality. Current evidence indicates that this primitive role fulfilled by curved phalanges persisted for at least three to four million years following the evolution of bipedality if *O. tugenensis* (Senut et al. 2001) or *Sahelanthropus tchadensis* (Brunet et al. 2002) truly represent the earliest bipeds.

Finite element modeling of the load transmission through the complex geometry of the wrist is likely to be a more difficult endeavor. However, with careful validation and attention to the model, FEM has the potential to contribute to a better understanding of the functional significance of derived morphological changes of the wrist in the human lineage. Macho and colleagues (2011) have provided some preliminary efforts in this regard. These authors built FEMs of the capitate from a human, chimpanzee, gorilla, orangutan, *Australopithecus* cf. *afarensis* (KNM-WT 22944), and *Australopithecus anamensis* (KNM-KP 31724). These were loaded on the proximal end and constrained distally using both models assuming homogenous elastic properties throughout and a more detailed model incorporating the elastic properties of the trabeculae. For the most part, the results are similar for the two models, and the data suggest that in the nonhuman sample, a greater portion of the total stress (55 %) is transmitted to the radial side of capitate in humans. In contrast, less than 40 % of the total stress was transmitted to the radial side in the capitate models from the nonhuman taxa. The human pattern might reflect the reorganization of the radial side of the wrist in which the Mc2 facet is more obliquely aligned with the long axis of the capitate, such that a portion of the proximodistal loading can be transferred along the radial aspect of the bone (Macho et al. 2011). In contrast, the greater ulnar-side transmission of stress in the nonhuman sample may reflect the use of adducted hand postures and ulnarly oriented substrate reaction forces during arboreal behaviors.

The *A. anamensis* capitate exhibits the nonhuman pattern of load transmission (consistent with its primitive, sagittally oriented Mc2 facet; Leakey et al. 1998), and *A. afarensis* exhibits somewhat greater radial-side transfer of stress (although closer to the gorilla specimen than the human) consistent with its more obliquely oriented Mc2 facet (Macho et al. 2011). These findings might reflect slightly different hand use in these australopiths. Although intriguing, higher-resolution models, more work on intrataxon variability, and assumption testing of the FEM will be needed for verification. However, the preliminary findings are consistent with the hypothesis that modern human wrist geometry is derived and specially adapted to transmitting loads obliquely and radially by strong contraction of thenar muscles during manipulative tasks (Lewis 1989; Tocheri 2007; Tocheri et al. 2007, 2008).

5 Conclusions

Although much work remains to be done, the data on joint morphology, kinematics, and load transmission obtained thus far from the biomedical imaging, and engineering techniques reviewed here have already provided important insights into the functional and evolutionary morphology of the primate hand—especially within the hominin clade. For example, the radial side of the wrist in humans appears to be reorganized to withstand transversely oriented loads (as might occur during manipulative tasks) when compared with nonhuman primates. In addition, humans and chimpanzees share a derived pattern of carpal kinematics during wrist extension, suggesting that humans evolved increased wrist mobility from a more rigid-wristed ancestor. Indicators of load transmission through the hand (e.g., trabecular architecture and subchondral bone density) reflect the reduction in locomotor use of the forelimb in humans. Meanwhile, the interpretation of traditional morphological indicators of arboreal behavior (e.g., phalangeal curvature) has been bolstered through careful application of FEM. As these techniques continue to be improved and applied to the fossil record, they should greatly facilitate our understanding of how and when the full commitment to terrestrial bipedality lifted locomotor constraints on the hominin forelimb and the influence that this behavioral shift had on subsequent adaptation of the hand to novel tasks such as tool making and use.

6 Future Directions

Methodology should be driven by theory and questions of interest rather than vice versa. Nevertheless, new technologies and techniques can provide novel insights in functional morphology. Each approach reviewed here has seen limited application to questions in comparative anatomy of the hand, and there is ample room for future work in terms of expanding the taxonomic diversity of the samples and examining other regions of the hand. The compilation of large, taxonomically broad datasets of 3D polygon models of primate hand bones (e.g., Tocheri et al. 2003, 2005, 2007; Boyer et al. 2013; Orr et al. 2013) will provide a rich dataset for comparative analyses of articular structure across extant primates and for interpreting the fossil record. Many aspects of hand morphology have not been quantified, and the polygon models provide incredible flexibility for researchers to design their own morphometrics for comparing features relevant to their particular projects.

Expanding and improving studies of load transmission through the hand (via trabecular architectural analysis, subchondral bone density, and finite element modeling) are particularly important for further testing the biomechanical basis for apparent correlations between particular morphological features (e.g., quantified using 3D morphometrics) and specific behaviors. For example, the studies of the radiocarpal complex (e.g., Carlson and Patel 2006; Patel and Carlson 2007, 2008) could be extended to the ulna and proximal carpals to help determine whether or not the load distribution in the forelimb has shifted radially in the apes, which are derived in having lost an ulnocarpal contact through retraction of the ulnar styloid process (Sarmiento 1985, 1988; Lewis 1989; see Chap. 3). If apparent bone density is similar across the proximal radiocarpus in monkeys, but concentrated radially in the apes, it would indicate reduced loading through the ulnar side in extant hominoids. This in turn would help test the hypothesis that the predominantly knuckle-walking apes evolved a derived scaphoid (i.e., large and robust with a fused os centrale) as an accommodation to increased radial-side load transmission. Similar studies on the radial side of the wrist complex (i.e., trapezium, trapezoid, scaphoid, capitate, and Mc2) might help to test the hypothesis that the derived carpal geometry of modern humans and Neandertals more effectively transmits loads radioulnarly (as during strong pollical grasping associated with tool behaviors) compared to the primitive primate condition configured for proximodistal loading during locomotion (Lewis 1989; Tocheri 2007; Tocheri et al. 2007, 2008).

More comprehensive modeling efforts would also be beneficial to the study of primate hand function. With regard to the load transmission, FEM of the whole hand and carpus will be challenging given the numerous articulations, ligaments, cartilage, and muscles exerting loads on the structures; however, efforts have been made for humans (Ulrich et al. 1999; Carrigan et al. 2003; Anderson et al. 2005; Gislason et al. 2009, 2010, 2012). Additional complications will arise when attempting to model the stresses incurred during locomotion as substrate reaction forces must also be considered. Basic data on material properties of various hand structures and carefully collected in vivo kinematic and kinetic data will be required to provide input variables and to validate the models.

In addition to FEMs to study stress distribution in the hand, 3D musculoskeletal simulations could help shed light on the morphological basis of grip capabilities among primates. For example, forward dynamic modeling, which has been used to model locomotor gaits of early hominins (e.g., Nagano et al. 2005), could be applied to the hand. Niewoehner et al. (2003) used 3D modeling to demonstrate precision grasping in Neandertals based on the geometry of the digits and simple range of motion estimates, which may be adequate as these hominins are quite similar to modern humans in overall hand morphology (Tocheri et al. 2008). However, given debates over manipulative capabilities in more primitive taxa (e.g., Marzke 1997; Susman 1998), 3D musculoskeletal simulation of hand movements incorporating data on muscle architecture (Marzke et al. 1999), inertial properties of hand segments, and joint torque estimates may help assess grip performance. Such simulation studies coupled with experimental validation (preferably in vivo) in living taxa would be especially useful for testing classic hypotheses about grasping performance among primates. Ultimately, this may allow us to better explain the morphological basis of and evolution of humans' apparently unique ability (among living primates) to produce true precision grips (Napier 1960), providing the manipulative capabilities to fully exploit the tool-using niche.

Acknowledgments Thanks to the editors for inviting this contribution and for their great service in compiling a much needed update on the state of our knowledge about primate hand evolution. Tracy Kivell, Biren Patel, and Brian Richmond generously provided figures illustrating the results of their work. Orr's research has been funded by the National Science Foundation (BCS-622515 and BCS-1317029) and the Wenner-Gren Foundation. Many of the ideas developed here were inspired by work conducted as a part of Arizona State University's Interdisciplinary Graduate Research Training Program in Neural and Musculoskeletal Adaptation in Form and Function and the pioneering efforts of Mary Marzke.

References

Anderson DD, Deshpande BR, Daniel TE, Baratz ME (2005) A three-dimensional finite element model of the radiocarpal joint: distal radius fracture step-off and stress transfer. Iowa Orthop J 25:108–117

Barak MM, Lieberman DE, Hublin JJ (2011) A Wolff in sheep's clothing: trabecular bone adaptation in response to changes in joint loading orientation. Bone 49:1141–1151

Beaupré GS, Carter DS (1992) Finite element analysis in biomechanics. In: Biewener AA (ed) Biomechanics—structures and systems: a practical approach. IRL Press at Oxford University Press, New York, pp 149–174

Beggs JS (1983) Kinematics. Hemisphere Publishing Corporation, Washington

Begun DR (2004) Knuckle-walking and the origin of human bipedalism. In: Meldrum DJ, Hilton CE (eds) From biped to strider: the emergence of modern human walking, running, and resource transport. Kluwer Academic/Plenum Publishers, New York, pp 9–33

Begun DR, Kivell TL (2011) Knuckle-walking in *Sivapithecus*: the combined effects of homology and homoplasy and implications for the origin of human bipedalism. J Hum Evol 60:158–170

Bottema O, Roth B (1979) Theoretical kinematics. North-Holland Publishing Co., Amsterdam

Boyer DM, Yapuncich GS, Chester SGB, Bloch JI, Godinot M (2013) Hands of early primates. Yearb Phys Anthropol 57:33–78

Brunet M, Guy F, Pilbeam D, Mackaye HT, Likius A, Ahounta D, Beauvilain A, Blondel C, Bocherens H, Boisserie JR, De Bonis L, Coppens Y, Dejax J, Denys C, Duringer P, Eisenmann V, Fanone G, Fronty P, Geraads D, Lehmann T, Lihoreau F, Louchart A, Mahamat A, Merceron G, Mouchelin G, Otero O, Pelaez Campomanes P, Ponce De León M, Rage JC, Sapanet M, Schuster M, Sudre J, Tassy P, Valentin X, Vignaud P, Viriot L, Zazzo A, Zollikofer C (2002) A new hominid from the Upper Miocene of Chad, Central Africa. Nature 418:145–151

Calfee RP, Leventhal EL, Wilkerson J, Moore DC, Akelman E, Crisco JJ (2008) Simulated radioscapholunate fusion alters carpal kinematics while preserving dart-thrower's motion. J Hand Surg [Am] 33:503–510

Camus EJ, Millot F, Lariviere J, Raoult S, Rtaimate M (2004) Kinematics of the wrist using 2D and 3D analysis: biomechanical and clinical deductions. Surg Radiol Anat 26:399–410

Carlson KJ, Patel BA (2006) Habitual use of the primate forelimb is reflected in the material properties of subchondral bone in the distal radius. J Anat 208:659–670

Carlson KJ, Jashashvili T, Houghton K, Westaway MC, Patel BA (2013) Joint loads in marsupial ankles reflect habitual bipedalism versus quadrupedalism. PLoS One 8:e58811

Carrigan SD, Whiteside RA, Pichora DR, Small CF (2003) Development of a three-dimensional finite element model for carpal load transmission in a static neutral posture. Ann Biomed Eng 31:718–725

Carter DR, Beaupré GS (2001) Skeletal function and form: mechanobiology of skeletal development, aging, and regeneration. Cambridge University Press, New York

Chalk J, Richmond BG, Ross CF, Strait DS, Wright BW, Spencer MA, Wang Q, Dechow PC (2011) A finite element analysis of masticatory stress hypotheses. Am J Phys Anthropol 145:1–10

Coburn JC, Upal MA, Crisco JJ (2007) Coordinate systems for the carpal bones of the wrist. J Biomech 40:203–209

Coleman MN, Colbert MW (2007) Technical note: CT thresholding protocols for taking measurements on three-dimensional models. Am J Phys Anthropol 133:723–725

Congdon KA (2012) Interspecific and ontogenetic variation in proximal pedal phalangeal curvature of great apes (*Gorilla gorilla*, *Pan troglodytes*, and *Pongo pygmaeus*). Int J Primatol 33:418–427

Cook RD, Malkus DS, Plesha ME, Witt RJ (2011) Concepts and applications of finite element analysis, 4th edn. Wiley, Hoboken

Corruccini RS (1978) Comparative osteometrics of the hominoid wrist joint, with special reference to knuckle-walking. J Hum Evol 7:307–321

Crisco JJ, McGovern RD, Wolfe SW (1999) Noninvasive technique for measuring *in vivo* three-dimensional carpal bone kinematics. J Orthop Res 17:96–100

Crisco JJ, Wolfe SW, Neu CP, Pike S (2001) Advances in the *in vivo* measurement of normal and abnormal carpal kinematics. Orthop Clin North Am 32(2):219–231, vii

Crisco JJ, Pike S, Hulsizer-Galvin DL, Akelman E, Weiss AP, Wolfe SW (2003) Carpal bone postures and motions are abnormal in both wrists of patients with unilateral scapholunate interosseous ligament tears. J Hand Surg [Am] 28:926–937

Crisco JJ, Coburn JC, Moore DC, Akelman E, Weiss APC, Wolfe SW (2005) *In vivo* radiocarpal kinematics and the dart thrower's motion. J Bone Joint Surg Am 87:2729–2740

Currey J (1984) Comparative mechanical properties and histology of bone. Am Zool 24:5–12

Currey J (2002) Bones: structure and mechanics. Princeton University Press, Princeton

Daver G, Berillon G, Grimaud-Hervé D (2012) Carpal kinematics in quadrupedal monkeys: towards a better understanding of wrist morphology and function. J Anat 220:42–56

Deane AS, Begun DR (2008) Broken fingers: retesting locomotor hypotheses for fossil hominoids using fragmentary proximal phalanges and high-resolution polynomial curve fitting (HR-PCF). J Hum Evol 55:691–701

Dickomeit MJ, Bottcher P, Hecht S, Liebich HG, Maierl J (2011) Topographic and age-dependent distribution of subchondral bone density in the elbow joints of clinically normal dogs. Am J Vet Res 72:491–499

Eckstein F, Muller-Gerbl M, Steinlechner M, Kierse R, Putz R (1995) Subchondral bone density in the human elbow assessed by computed tomography osteoabsorptiometry: a reflection of the loading history of the joint surfaces. J Orthop Res 13:268–278

Fajardo RJ, Müller R (2001) Three-dimensional analysis of nonhuman primate trabecular architecture using micro-computed tomography. Am J Phys Anthropol 115:327–336

Fajardo RJ, Ryan TM, Kappelman J (2002) Assessing the accuracy of high-resolution X-ray computed tomography of primate trabecular bone by comparisons with histological sections. Am J Phys Anthropol 118:1–10

Fajardo RJ, Müller R, Ketcham RA, Colbert M (2007) Nonhuman anthropoid primate femoral neck trabecular architecture and its relationship to locomotor mode. Anat Rec 290:422–436

Frost HM (2001) From Wolff's law to the Utah paradigm: insights about bone physiology and its clinical applications. Anat Rec 262:398–419

Gislason MK, Nash DH, Nicol A, Kanellopoulos A, Bransby-Zachary M, Hems T, Condon B, Stansfield B (2009) A three-dimensional finite element model of maximal grip loading in the human wrist. Proc Inst Mech Eng H 223:849–861

Gislason MK, Stansfield B, Nash DH (2010) Finite element model creation and stability considerations of complex biological articulation: the human wrist joint. Med Eng Phys 32:523–531

Gislason MK, Stansfield B, Bransby-Zachary M, Hems T, Nash DH (2012) Load transfer through the radiocarpal joint and the effects of partial wrist arthrodesis on carpal bone behaviour: a finite element study. J Hand Surg [Eur] 37:871–878

Goto A, Moritomo H, Murase T, Oka K, Sugamoto K, Arimura T, Masumoto J, Tamura S, Yoshikawa H, Ochi T (2005) *In vivo* three-dimensional wrist motion analysis using magnetic resonance imaging and volume-based registration. J Orthop Res 23:750–756

Gross T, Kivell TL, Skinner MM, Nguyen NH, Pahr DH (2014) A CT-image-based framework for the holistic analysis of cortical and trabecular bone morphology. Palaeontol Electron 17:33A

Hahn M, Vogel M, Pompesius-Kempa M, Delling G (1992) Trabecular bone pattern factor—a new parameter for simple quantification of bone microarchitecture. Bone 13:327–330

Harrigan TP, Mann RW (1984) Characterization of microstructural anisotropy in orthotropic materials using a second rank tensor. J Mater Sci 19:761–767

Hildebrand T, Rüegsegger P (1997a) A new method for the model-independent assessment of thickness in three-dimensional images. J Microsc 185:67–75

Hildebrand T, Rüegsegger P (1997b) Quantification of bone microarchitecture with the structure model index. Comput Methods Biomech Biomed Engin 1:15–23

Hoogbergen MM, Niessen WJ, Schuurman AH, Spauwen PH, Kauer JM (2002) Subchondral bone mineral density patterns representing the loading history of the wrist joint. J Hand Surg [Br] 27:150–154

Jenkins FA Jr, Fleagle JG (1975) Knuckle-walking and the functional anatomy of the wrists in living apes. In: Tuttle RH (ed) Primate functional morphology and evolution. Mouton, The Hague, pp 213–227

Johnson C (2009) Numerical solution of partial differential equations by the finite element method. Dover Publications, Mineola

Jungers WL, Godfrey LR, Simons EL, Chatrath PS (1997) Phalangeal curvature and positional behavior in extinct sloth lemurs (Primates, Palaeopropithecidae). Proc Natl Acad Sci U S A 94:11998–12001

Kaufmann R, Pfaeffle J, Blankenhorn B, Stabile K, Robertson D, Goitz R (2005) Kinematics of the midcarpal and radiocarpal joints in radioulnar deviation: an in vitro study. J Hand Surg [Am] 30:937–942

Kaufmann RA, Pfaeffle HJ, Blankenhorn BD, Stabile K, Robertson D, Goitz R (2006) Kinematics of the midcarpal and radiocarpal joint in flexion and extension: an in vitro study. J Hand Surg [Am] 31:1142–1148

Keller TS (1994) Predicting the compressive mechanical behavior of bone. J Biomech 27:1159–1168

Kivell TL, Begun DR (2007) Frequency and timing of scaphoid-centrale fusion in hominoids. J Hum Evol 52:321–340

Kivell TL, Begun DR (2009) New primate carpal bones from Rudabanya (late Miocene, Hungary): taxonomic and functional implications. J Hum Evol 57:697–709

Kivell TL, Schmitt D (2009) Independent evolution of knuckle-walking in African apes shows that humans did not evolve from a knuckle-walking ancestor. Proc Natl Acad Sci U S A 106:14241–14246

Kivell TL, Kibii JM, Churchill SE, Schmid P, Berger LR (2011a) *Australopithecus sediba* hand demonstrates mosaic evolution of locomotor and manipulative abilities. Science 333:1411–1417

Kivell TL, Skinner MM, Lazenby R, Hublin JJ (2011b) Methodological considerations for analyzing trabecular architecture: an example from the primate hand. J Anat 218:209–225

Kivell TL, Guimont I, Wall CE (2013) Sex-related shape dimorphism in the human radiocarpal and midcarpal joints. Anat Rec 296:19–30

Kobayashi M, Berger RA, Linscheid RL, An KN (1997a) Intercarpal kinematics during wrist motion. Hand Clin 13:143–149

Kobayashi M, Berger RA, Nagy L, Linscheid RL, Uchiyama S, Ritt MJ (1997b) Normal kinematics of carpal bones: a three-dimensional analysis of carpal bone motion relative to the radius. J Biomech 30:787–793

Lazenby RA, Angus S, Cooper DM, Hallgrimsson B (2008a) A three-dimensional microcomputed tomographic study of site-specific variation in trabecular microarchitecture in the human second metacarpal. J Anat 213:698–705

Lazenby RA, Cooper DM, Angus S, Hallgrimsson B (2008b) Articular constraint, handedness, and directional asymmetry in the human second metacarpal. J Hum Evol 54:875–885

Lazenby RA, Skinner MM, Kivell TL, Hublin JJ (2010) Scaling VOI size in 3D muCT studies of trabecular bone: a test of the over-sampling hypothesis. Am J Phys Anthropol 144:196–203

Lazenby RA, Skinner MM, Hublin JJ, Boesch C (2011) Metacarpal trabecular architecture variation in the chimpanzee (*Pan troglodytes*): Evidence for locomotion and tool-use? Am J Phys Anthropol 144:215–225

Leakey MG, Feibel CS, McDougall I, Ward C, Walker A (1998) New specimens and confirmation of an early age for *Australopithecus anamensis*. Nature 393:62–66

Levangie PK, Norkin CC (2011) Joint structure and function: a comprehensive analysis, 5th edn. F.A. Davis Co., Philadelphia

Leventhal EL, Wolfe SW, Moore DC, Akelman E, Weiss AP, Crisco JJ (2008) Interfragmentary motion in patients with scaphoid nonunion. J Hand Surg [Am] 33:1108–1115

Leventhal EL, Moore DC, Akelman E, Wolfe SW, Crisco JJ (2010) Carpal and forearm kinematics during a simulated hammering task. J Hand Surg [Am] 35:1097–1104

Lewis CW, Williamson AK, Chen AC, Bae WC, Temple MM, Wong WV, Nugent GE, James SP, Wheeler DL, Sah RL, Kawcak CE (2005) Evaluation of subchondral bone mineral density associated with articular cartilage structure and integrity in healthy equine joints with different functional demands. Am J Vet Res 66:1823–1829

Lewis OJ (1989) Functional morphology of the evolving hand and foot. Clarendon Press, Oxford

Liu GR, Quek SS (2013) The finite element method: a practical course. Butterworth-Heinemann, Waltham

Liu XS, Sajda P, Saha PK, Wehrli FW, Bevill G, Keaveny TM, Guo XE (2008) Complete volumetric decomposition of individual trabecular plates and rods and its morphological correlations with anisotropic elastic moduli in human trabecular bone. J Bone Miner Res 23:223–235

Lovejoy CO, Heiple KG, Meindl RS (2001) Palaeoanthropology: did our ancestors knuckle-walk? Nature 410:325–326

Lovejoy CO, Simpson SW, White TD, Asfaw B, Suwa G (2009) Careful climbing in the Miocene: the forelimbs of *Ardipithecus ramidus* and humans are primitive. Science 326(70):70e170e8

MacConaill MA (1941) The mechanical anatomy of the carpus and its bearing on some surgical problems. J Anat 75:166–175

Macho GA, Spears IR, Leakey MG, McColl DJ, Jiang Y, Abel R, Nakatsukasa M, Kunimatsu Y (2011) An exploratory study on the combined effects of external and internal morphology on load dissipation in primate capitates: its potential for an understanding of the positional and locomotor repertoire of early hominins. Folia Primatol 81:292–304

MacLatchy L, Müller R (2002) A comparison of the femoral head and neck trabecular architecture of *Galago* and *Perodicticus* using micro-computed tomography (μCT). J Hum Evol 43:89–105

Maga M, Kappelman J, Ryan TM, Ketcham RA (2006) Preliminary observations on the calcaneal trabecular microarchitecture of extant large-bodied hominoids. Am J Phys Anthropol 129:410–417

Maiolino S, Boyer DM, Rosenberger A (2011) Morphological correlates of the grooming claw in distal phalanges of platyrrhines and other primates: a preliminary study. Anat Rec 294:1975–1990

Maiolino S, Boyer DM, Bloch JI, Gilbert CC, Groenke J (2012) Evidence for a grooming claw in a North American adapiform primate: implications for anthropoid origins. PLoS One 7, e29135

Marzke MW (1997) Precision grips, hand morphology and tools. Am J Phys Anthropol 102:91–110

Marzke MW, Toth N, Schick K, Reece S, Steinberg B, Hunt K, Linscheid RL, An KN (1998) EMG study of hand muscle recruitment during hard hammer percussion manufacture of Oldowan tools. Am J Phys Anthropol 105:315–332

Marzke MW, Marzke RF, Linscheid RL, Smutz P, Steinberg B, Reece S, An KN (1999) Chimpanzee thumb muscle cross sections, moment arms and potential torques, and comparisons with humans. Am J Phys Anthropol 110:163–178

Marzke MW, Tocheri MW, Steinburg B, Femiani JD, Linscheid RL, Orr CM, Marzke RF (2010) Comparative 3D quantitative analyses of trapeziometacarpal joint surface curvatures among living catarrhines and fossil hominins. Am J Phys Anthropol 141:38–51

Marzke MW, Tocheri MW, Marzke RF, Femiani JD (2012) Three-dimensional quantitative comparative analysis of trapezial-metacarpal joint surface curvatures in human populations. J Hand Surg [Am] 37:72–76

Matarazzo S (2008) Knuckle walking signal in the manual digits of *Pan* and *Gorilla*. Am J Phys Anthropol 135:27–33

Matsuura Y, Ogihara N, Nakatsukasa M (2010) A method for quantifying articular surface morphology of metacarpals using quadric surface approximation. Int J Primatol 31:263–274

McCarthy JM (1990) An introduction to theoretical kinematics. The MIT Press, Cambridge

McHenry HM (1983) The capitate of *Australopithecus afarensis* and *A. africanus*. Am J Phys Anthropol 62:187–192

McHenry HM, Corruccini RS (1975) The wrist of *Proconsul africanus* and the origin of hominoid postcranial adaptations. In: Ciochon RL, Corruccini RS (eds) New interpretations of ape and human ancestry. Plenum Press, New York, pp 353–367

Moojen TM, Snel JG, Ritt MJ, Venema HW, den Heeten GJ, Bos KE (2001) Pisiform kinematics *in vivo*. J Hand Surg [Am] 26:901–907

Moojen TM, Snel JG, Ritt MJ, Kauer JM, Venema HW, Bos KE (2002a) Three-dimensional carpal kinematics in vivo. Clin Biomech 17:506–514

Moojen TM, Snel JG, Ritt MJ, Venema HW, Kauer JM, Bos KE (2002b) Scaphoid kinematics *in vivo*. J Hand Surg [Am] 27:1003–1010

Moojen TM, Snel JG, Ritt MJ, Venema HW, Kauer JM, Bos KE (2003) *In vivo* analysis of carpal kinematics and comparative review of the literature. J Hand Surg [Am] 28:81–87

Moore DC, Crisco JJ, Trafton TG, Leventhal EL (2007) A digital database of wrist bone anatomy and carpal kinematics. J Biomech 40:2537–2542

Moritomo H, Goto A, Sato Y, Sugamoto K, Murase T, Yoshikawa H (2003) The triquetrum-hamate joint: an anatomic and in vivo three-dimensional kinematic study. J Hand Surg [Am] 28:797–805

Moritomo H, Murase T, Goto A, Oka K, Sugamoto K, Yoshikawa H (2004) Capitate-based kinematics of the midcarpal joint during wrist radioulnar deviation: an *in vivo* three-dimensional motion analysis. J Hand Surg [Am] 29:668–675

Moritomo H, Murase T, Goto A, Oka K, Sugamoto K, Yoshikawa H (2006) *In vivo* three-dimensional kinematics of the midcarpal joint of the wrist. J Bone Joint Surg Am 88:611–621

Müller-Gerbl M, Putz R, Hodapp N, Schulte E, Wimmer B (1989) Computed tomographyosteoabsorptiometry for assessing the density distribution of subchondral bone as a measure of long-term mechanical adaptation in individual joints. Skeletal Radiol 18:507–512

Müller-Gerbl M, Putz R, Kenn R (1992) Demonstration of subchondral bone density patterns by three-dimensional CT osteoabsorptiometry as a noninvasive method for in vivo assessment of individual long-term stresses in joints. J Bone Miner Res 7(Suppl 2):411–418

Napier JR (1952) The form and function of the carpo-metacarpal joint of the thumb. J Anat 89:362–370

Napier JR (1956) The prehensile movements of the human hand. J Bone Joint Surg [Br] 38:902–913
Napier JR (1960) Studies of the hands of living primates. Proc Zool Soc Lond 134:647–657
Nagano A, Umberger BR, Marzke MW, Gerritsen KG (2005) Neuromusculoskeletal computer modeling and simulation of upright, straight-legged, bipedal locomotion of *Australopithecus afarensis* (A.L. 288-1). Am J Phys Anthropol 126:2–13
Neu CP, McGovern RD, Crisco JJ (2000) Kinematic accuracy of three surface registration methods in a three-dimensional wrist bone study. J Biomech Eng 122:528–533
Neu CP, Crisco JJ, Wolfe SW (2001) *In vivo* kinematic behavior of the radio-capitate joint during wrist flexion-extension and radio-ulnar deviation. J Biomech 34:1429–1438
Nguyen NH, Pahr DH, Gross T, Skinner MM, Kivell TL (2014) Micro-finite element (μFE) modeling of the siamang (*Symphalangus syndactylus*) third proximal phalanx: the functional role of curvature and the flexor sheath ridge. J Hum Evol 67:60–75
Niewoehner WA, Bergstrom A, Eichele D, Zuroff M, Clark JT (2003) Digital analysis: manual dexterity in Neanderthals. Nature 422:395
Nowak MG, Carlson KJ, Patel BA (2010) Apparent density of the primate calcaneocuboid joint and its association with locomotor mode, foot posture, and the "midtarsal break". Am J Phys Anthropol 142:180–193
O'Connor BL (1975) The functional morphology of the cercopithecoid wrist and inferior radioulnar joints, and their bearing on some problems in the evolution of the Hominoidea. Am J Phys Anthropol 43:113–121
Odgaard A (1997) Three-dimensional methods for quantification of cancellous bone architecture. Bone 20:315–328
Odgaard A (2001) Quantification of cancellous bone architecture. In: Cowin SC (ed) Bone mechanics handbook, 2nd edn. CRC, New York, pp 14–19
Orr CM (2005) The knuckle-walking anteater: a convergence test of adaptation for purported knuckle-walking features of the African Hominidae. Am J Phys Anthropol 128:639–658
Orr CM (2010) Adaptations to knuckle-walking and digitigrady: a three-dimensional kinematic and morphometric analysis of the anthropoid wrist. Ph.D. dissertation, Arizona State University
Orr CM (2012) Kinematics and morphometrics of the radiocarpus in anthropoids with implications for reconstructing the evolution of hominin wrist mechanics. Am J Phys Anthropol 147(Suppl 54):229–230
Orr CM (2013) The earliest hominins were rigid wristed. Am J Phys Anthropol 150(Suppl 56):211
Orr CM, Leventhal EL, Chivers FS, Menzies SL, Larson S, Crisco JJ (2009) Three-dimensional midcarpal kinematics during wrist extension in five anthropoid species. Am J Phys Anthropol 138(Suppl 48):203
Orr CM, Leventhal EL, Chivers SF, Marzke MW, Wolfe SW, Crisco JJ (2010) Studying primate carpal kinematics in three-dimensions using a computed-tomography-based markerless registration method. Anat Rec 293:692–709
Orr CM, Tocheri MW, Burnett SE, Awe RD, Saptomo EW, Sutikna T, Jatmiko, Wasisto S, Morwood MJ, Jungers WL (2013) New wrist bones of *Homo floresiensis* from Liang Bua (Flores, Indonesia). J Hum Evol 64:109–129
Patel BA (2010) The interplay between speed, kinetics, and hand postures during primate terrestrial locomotion. Am J Phys Anthropol 141:222–234
Patel BA, Carlson KJ (2007) Bone density spatial patterns in the distal radius reflect habitual hand postures adopted by quadrupedal primates. J Hum Evol 52:130–141
Patel BA, Carlson KJ (2008) Apparent density patterns in subchondral bone of the sloth and anteater forelimb. Biol Lett 4:486–489
Patterson RM, Nicodemus CL, Viegas SF, Elder KW, Rosenblatt J (1998) High-speed, three-dimensional kinematic analysis of the normal wrist. J Hand Surg [Am] 23:446–453
Patterson RM, Williams L, Andersen CR, Koh S, Viegas SF (2007) Carpal kinematics during simulated active and passive motion of the wrist. J Hand Surg [Am] 32:1013–1019
Pearson OM, Lieberman DE (2004) The aging of Wolff's "law": ontogeny and responses to mechanical loading in cortical bone. Yearb Phys Anthropol 39:63–99

Peterson JV, Dechow PC (2003) Material properties of the human cranial vault and zygoma. Anat Rec 274A:785–797

Polk JD, Blumenfeld J, Ahluwalia K (2008) Knee posture predicted from subchondral apparent density in the distal femur: an experimental validation. Anat Rec 291:293–302

Polk JD, Godfrey LR, Peterson JV, Roseman CC, Williams SA (2010) Subchondral bone apparent density and locomotor behavior in extant primates and subfossil lemurs *Hadropithecus* and *Pachylemur*. Int J Primatol 31:275–299

Pontzer H, Lieberman DE, Momin E, Devlin MJ, Hallgrimsson B, Cooper DM (2006) Trabecular bone in the bird knee responds with high sensitivity to changes in load orientation. J Exp Biol 209:57–65

Preuschoft H (1970) Functional anatomy of the lower extremity. In: Bourne GH (ed) The chimpanzee, vol 3. Karger, Basel, pp 221–294

Preuschoft H (1973) Functional anatomy of the upper extremity. In: Bourne GH (ed) The chimpanzee, vol 6. Karger, Basel, pp 34–120

Rainbow MJ, Crisco JJ, Moore DC, Wolfe SW (2008) Gender differences in capitate kinematics are eliminated after accounting for variation in carpal size. J Biomech Eng 130:041003

Rainbow MJ, Kamal RN, Leventhal E, Akelman E, Moore DC, Wolfe SW, Crisco JJ (2013) *In vivo* kinematics of the scaphoid, lunate, capitate, and third metacarpal in extreme wrist flexion and extension. J Hand Surg [Am] 38:278–288

Richmond BG (1998) Ontogeny and biomechanics of phalangeal form in primates. Ph.D. dissertation, State University of New York at Stony Brook

Richmond BG (2007) Biomechanics of phalangeal curvature. J Hum Evol 53:678–690

Richmond BG, Jungers WL (2008) *Orrorin tugenensis* femoral morphology and the evolution of hominin bipedalism. Science 319:1662–1665

Richmond BG, Strait DS (2000) Evidence that humans evolved from a knuckle-walking ancestor. Nature 404:382–385

Richmond BG, Begun DR, Strait DS (2001) Origin of human bipedalism: the knuckle-walking hypothesis revisited. Yearb Phys Anthropol 44:70–105

Richmond BG, Wright BW, Grosse I, Dechow PC, Ross CF, Spencer MA, Strait DS (2005) Finite element analysis in functional morphology. Anat Rec A Discov Mol Cell Evol Biol 283:259–274

Robson-Brown KA, Davies EN, McNally DS (2002) The angular distribution of vertebral trabeculae in modern humans, chimpanzees and the Kebara 2 Neanderthal. J Hum Evol 43:189–205

Rolian CP, Lieberman DE, Zermano JP (2011) Hand biomechanics during simulated stone tool use. J Hum Evol 61:26–41

Rose MD (1984) Hominoid postcranial specimens from the middle Miocene Chinji Formation. Pakistan J Hum Evol 13:503–516

Rose ME (1992) Kinematics of the trapezium-1st metacarpal joint in extant anthropoids and Miocene hominoids. J Hum Evol 22:255–266

Ross CF, Patel BA, Slice DE, Strait DS, Dechow PC, Richmond BG, Spencer MA (2005) Modeling masticatory muscle force in finite element analysis: sensitivity analysis using principal coordinates analysis. Anat Rec 283:288–299

Ross CF, Berthaume MA, Dechow PC, Iriarte-Diaz J, Porro LB, Richmond BG, Spencer M, Strait D (2011) *In vivo* bone strain and finite-element modeling of the craniofacial haft in catarrhine primates. J Anat 218:112–141

Ruff C, Holt B, Trinkaus E (2006) Who's afraid of the big bad Wolff?: "Wolff's law" and bone functional adaptation. Am J Phys Anthropol 129:484–498

Ryan TM, Ketcham RA (2002a) Femoral head trabecular bone structure in two omomyid primates. J Hum Evol 43:241–263

Ryan TM, Ketcham RA (2002b) The three-dimensional structure of trabecular bone in the femoral head of strepsirrhine primates. J Hum Evol 43:1–26

Ryan TM, Ketcham RA (2005) Angular orientation of trabecular bone in the femoral head and its relationship to hip joint loads in leaping primates. J Morphol 265:249–263

Ryan TM, Krovitz GE (2006) Trabecular bone ontogeny in the human proximal femur. J Hum Evol 51:591–602

Ryan TM, Shaw CN (2012) Unique suites of trabecular bone features characterize locomotor behavior in human and non-human anthropoid primates. PLoS One 7:e41037

Ryan TM, van Rietbergen B (2005) Mechanical significance of femoral head trabecular bone structure in *Loris* and *Galago* evaluated using micromechanical finite element models. Am J Phys Anthropol 126:82–96

Ryan TM, Walker A (2010) Trabecular bone structure in the humeral and femoral heads of anthropoid primates. Anat Rec 293:719–729

Sarmiento EE (1985) Functional differences in the skeleton of wild and captive orangutans and their adaptive significance. Ph.D. dissertation, New York University

Sarmiento EE (1988) Anatomy of the hominoid wrist joint: its evolutionary and functional implications. Int J Primatol 9:281–345

Schön MA, Ziemer LK (1973) Wrist mechanism and locomotor behavior of *Dryopithecus* (*Proconsul*) *africanus*. Folia Primatol 20:1–11

Schilling AM, Tofanelli S, Hublin JJ, Kivell TL (2014) Trabecular bone structure in the primate wrist. J Morphol 275:572–585

Senut B, Pickford M, Gommery D, Mein P, Cheboi K, Coppens Y (2001) First hominid from the Miocene (Lukeino Formation, Kenya). C R Acad Sci 332:137–144

Strait DS, Wang Q, Dechow PC, Ross CF, Richmond BG, Spencer MA, Patel BA (2005) Modeling elastic properties in finite-element analysis: how much precision is needed to produce an accurate model? Anat Rec 283A:275–287

Strait DS, Richmond BG, Spencer MA, Ross CF, Dechow PC, Wood BA (2007) Masticatory biomechanics and its relevance to early hominid phylogeny: an examination of palatal thickness using finite-element analysis. J Hum Evol 52:585–599

Strait DS, Weber GW, Neubauer S, Chalk J, Richmond BG, Lucas PW, Spencer MA, Schrein C, Dechow PC, Ross CF, Grosse IR, Wright BW, Constantino P, Wood BA, Lawn B, Hylander WL, Wang Q, Byron C, Slice DE, Smith AL (2009) The feeding biomechanics and dietary ecology of *Australopithecus africanus*. Proc Natl Acad Sci U S A 106:2124–2129

Strait DS, Grosse IR, Dechow PC, Smith AL, Wang Q, Weber GW, Neubauer S, Slice DE, Chalk J, Richmond BG, Lucas PW, Spencer MA, Schrein C, Wright BW, Byron C, Ross CF (2010) The structural rigidity of the cranium of *Australopithecus africanus*: implications for diet, dietary adaptations, and the allometry of feeding biomechanics. Anat Rec 293:583–593

Stern JT Jr (2000) Climbing to the top: a personal memoir of *Australopithecus afarensis*. Evol Anthropol 9:113–133

Stern JT Jr, Susman RL (1983) The locomotor anatomy of *Australopithecus afarensis*. Am J Phys Anthropol 60:279–317

Stern JT Jr, Jungers WL, Susman RL (1995) Quantifying phalangeal curvature: an empirical comparison of alternative methods. Am J Phys Anthropol 97:1–10

Susman RL (1979) Comparative and functional morphology of hominoid fingers. Am J Phys Anthropol 50:215–236

Susman RL (1998) Hand function and tool behavior in early hominids. J Hum Evol 35:23–46

Susman RL, Stern JT Jr (1979) Telemetered electromyography of flexor digitorum profundus and flexor digitorum superficialis in *Pan troglodytes* and implications for interpretation of the O.H. 7 hand. Am J Phys Anthropol 50:565–574

Susman RL, Stern JT Jr (1980) EMG of the interosseous and lumbrical muscles in the chimpanzee (*Pan troglodytes*) hand during locomotion. Am J Anat 157:389–397

Tallman M (2012) Morphology of the distal radius in extant hominoids and fossil hominins: implications for the evolution of bipedalism. Anat Rec 295:454–464

Tocheri MW (2007) Three-dimensional riddles of the radial wrist: derived carpal and carpometacarpal joint morphology in the genus Homo and the implications for understanding the evolution of stone-tool related behaviors in hominins. PhD dissertation, Arizona State University

Tocheri MW (2009) Laser scanning: 3D analysis of biological surfaces. In: Sensen CW, Hallgrimsson B (eds) Advanced imaging in biology and medicine. Springer, Berlin, pp 85–101

Tocheri MW, Marzke MW, Liu D, Bae M, Jones GP, Williams RC, Razdan A (2003) Functional capabilities of modern and fossil hominid hands: three-dimensional analysis of trapezia. Am J Phys Anthropol 122:101–112

Tocheri MW, Razdan A, Williams RC, Marzke MW (2005) A 3D quantitative comparison of trapezium and trapezoid relative articular and nonarticular surface areas in modern humans and great apes. J Hum Evol 49:570–586

Tocheri MW, Orr CM, Larson SG, Sutikna T, Jatmiko, Jatmiko EW, Due RA, Djubiantono T, Morwood MJ, Jungers WL (2007) The primitive wrist of *Homo floresiensis* and its implications for hominin evolution. Science 317:1743–1745

Tocheri MW, Orr CM, Jacofsky MC, Marzke MW (2008) The evolutionary history of the hominin hand since the last common ancestor of *Pan* and *Homo*. J Anat 212:544–562

Tocheri MW, Solhan CR, Orr CM, Femiani J, Frohlich B, Groves CP, Harcourt-Smith WE, Richmond BG, Shoelson B, Jungers WL (2011) Ecological divergence and medial cuneiform morphology in gorillas. J Hum Evol 60:171–184

Tsegai ZJ, Kivell TL, Gross T, Nguyen NH, Pahr DH, Smaers JB, Skinner MM (2013) Trabecular bone structure correlates with hand posture and use in hominoids. PLoS One 8:e78781

Tuttle RH (1967) Knuckle-walking and the evolution of hominoid hands. Am J Phys Anthropol 26:171–206

Tuttle RH (1969a) Knuckle-walking and the problem of human origins. Science 166:953–961

Tuttle RH (1969b) Terrestrial trends in the hands of the Anthropoidea. Proc 2nd Int Congr Primatol 2:192–200

Tuttle RH, Watts DP (1985) The positional behavior and adaptive complexes of *Pan gorilla*. In: Kondo S (ed) Primate morphophysiology, locomotor analyses, and human bipedalism. University of Tokyo Press, Tokyo, pp 261–288

Tuttle R, Basmajian JV, Regenos E, Shine G (1972) Electromyography of knuckle-walking: results of four experiments on the forearm of *Pan gorilla*. Am J Phys Anthropol 37:255–265

Ulrich D, van Rietbergen B, Laib A, Ruegsegger P (1999) Load transfer analysis of the distal radius from in-vivo high-resolution CT-imaging. J Biomech 32:821–828

Varga P, Schefzig P, Unger E, Mayr W, Zysset PK, Erhart J (2013) Finite element based estimation of contact areas and pressures with the human scaphoid in various functional positions of the hand. J Biomech 46:984–990

Wang Q, Strait DS, Dechow PC (2006) A comparison of cortical elastic properties in the craniofacial skeletons of three primate species and its relevance to the study of human evolution. J Hum Evol 51:375–382

Ward CV (2002) Interpreting the posture and locomotion of *Australopithecus afarensis*: where do we stand? Yearb Phys Anthropol 45:185–215

Werner FW, Green JK, Short WH, Masaoka S (2004) Scaphoid and lunate motion during a wrist dart throw motion. J Hand Surg [Am] 29:418–422

Williams EM, Gordon AD, Richmond BG (2010) Upper limb kinematics and the role of the wrist during stone tool production. Am J Phys Anthropol 143:134–145

Wolfe SW, Crisco JJ, Katz LD (1997) A non-invasive method for studying *in vivo* carpal kinematics. J Hand Surg [Br] 22:147–152

Wolfe SW, Neu C, Crisco JJ (2000) *In vivo* scaphoid, lunate, and capitate kinematics in flexion and in extension. J Hand Surg [Am] 25:860–869

Wolfe SW, Crisco JJ, Orr CM, Marzke MW (2006) The dart-throwing motion of the wrist: is it unique to humans? J Hand Surg [Am] 31:1429–1437

Wood B, Aiello L, Wood C, Key C (1998) A technique for establishing the identity of 'isolated' fossil hominin limb bones. J Anat 193:61–72

Wood Jones F (1916) Arboreal man. Edward Arnold, London

Zatsiorsky VM (1998) Kinematics of human motion. Human Kinetics, Urbana-Champaign

Zelditch ML, Swiderski DL, Sheets HD (2012) Geometric morphometrics for biologists: a primer, 2nd edn. Academic Press, Boston

Zollikofer CPE, Ponce de León MS (2005) Virtual reconstruction: a primer in computer-assisted paleontology and biomedicine. Wiley, Hoboken

Chapter 10
Experimental Research on Hand Use and Function in Primates

Evie E. Vereecke and Roshna E. Wunderlich

1 Introduction

The primate hand has long intrigued researchers of different disciplines. The extensive work of Napier (1955, 1956, 1960, 1961, 1980) remains elegant, with careful observations about the anatomy and function of the primate hand. While a lot can be learned and inferred by observation to achieve a more complete insight into the functioning of a complex organ such as the hand, experimental work is needed to illuminate patterns of joint movement, muscle activity and loads. In the last decades, researchers have collected a wealth of information about hand morphology and function by setting up and conducting laboratory-based experiments. In this chapter, we aim to give a comprehensive overview of the experimental work that has been done since Napier's publications in the 1950s. The main focus will be on the nonhuman primate hand, though comparative studies that include the human hand will also be reviewed. Studies on the human hand are much more numerous than on the nonhuman primate hand and often have a clinical framework or application. A comprehensive review of these studies is beyond the scope of this chapter. The experimental studies discussed in this chapter are grouped by the primary methodology (electromyography, force, etc.) and/or research question, rather than chronological order. We hope that readers find this, as we did, a logical and useful way of reviewing the available literature. For those who prefer a chronological overview, we refer to Table 10.1.

E.E. Vereecke (✉)
Jan Palfijn Anatomy Lab, Department Development & Regeneration,
University of Leuven, Leuven, Belgium
e-mail: Evie.Vereecke@kuleuven.be

R.E. Wunderlich
Department of Biology, James Madison University, Harrisonburg, VA 22801, USA

T.L. Kivell et al. (eds.), *The Evolution of the Primate Hand*, Developments in Primatology: Progress and Prospects, DOI 10.1007/978-1-4939-3646-5_10

Table 10.1 Overview of studies on primate hands in chronological order

Reference	Technique	Species	Topic	Remarks
Napier (1955, 1956, 1960, 1961, 1980)	Behavior	Primates		Prehension, grip types, hand use
Jolly (1970)	Behavior	*Theropithecus gelada*		Locomotion, position
Tuttle et al. (1972), Tuttle and Basmajian (1974)	EMG	*Gorilla gorilla*	Forearm muscles	Knuckle walking
Jenkins and Fleagle (1975)	Cineradiography	*Pan troglodytes*	Carpal motion	Knuckle walking, wrist function
Susman and Tuttle (1976)	Behavior	*Pongo pygmaeus*		Knuckle walking, locomotion
Rose (1977)	Behavior	*Papio anubis*		Positional behavior
Jenkins et al. (1978)	Cineradiography	*Ateles geoffroyi*, *A. paniscus*	Shoulder motion	Brachiation
Susman and Stern (1979)	EMG	*Pan troglodytes*	Flexor digitorum superficialis and profundus	Knuckle walking
Susman and Stern (1980)	EMG	*Pan troglodytes*	Interossei, lumbricals	Locomotion
Stern et al. (1980)	EMG	*Hylobates lar*, *Ateles* spp., *Lagothrix lagotricha*, *Alouatta seniculus*	Forelimb muscles	Brachiation, forelimb muscles
Jenkins (1981)	Cineradiography	*Ateles paniscus*, *Lagothrix lagotricha*	Carpal motion	Brachiation, wrist function
Jungers and Stern (1981)	EMG	*Ateles* spp., *Hylobates* spp.	Forelimb muscle activation	Brachiation, shoulder function
Susman et al. (1982)	EMG	*Hylobates lar*	Accessory interosseus muscle	Hand function, grasping
Tuttle et al. (1983, 1992), Tuttle and Basmajian (1978a, b)	EMG	*Pan troglodytes*, *Pongo pygmaeus*, *Gorilla gorilla*	Forelimb muscle activation	Terrestrial and suspensory locomotion and posture
Marzke and Shackley (1986)	Video	*Homo sapiens*		Hand function during tool making and use
Marzke and Marzke (1987)	Video	*Homo sapiens*	MC3 styloid process	Hand function during hammering

Costello and Fragaszy (1988)	Behavior	*Cebus apella*, *Saimiri sciureus*		Grasping
Castiello et al. (1992)	Kinematics, IR cameras	*Homo sapiens*	3D kinematics	Reaching, grasping
Marzke et al. (1992)	Behavior	*Homo sapiens*, *Pan troglodytes*		Squeeze and power grip
Toth et al. (1993)	Behavior	*Pan paniscus*		Grip types during toolmaking
Jude (1993)	Behavior	*Papio hamadryas*		Manipulation
Christel (1993)	Behavior	Hominoid primates (22 spp.)		Grasping
Inouye (1994)	Behavior	African apes	Hand posture	Knuckle walking
Schmitt (2003b, c)	Kinematics and kinetics	*Macaca mulatta*, *M. fascicularis*, *Erythrocebus patas*, *Chlorocebus aethiops*, *Papio anubis*, *Ateles geoffroyi*	Arboreal and terrestrial substrates, mediolateral forces	Forelimb kinematics
Marzke and Wullstein (1996)	Behavior	*Pan troglodytes*	Grip types	Locomotion, manipulation
Marzke (1997), Marzke and Marzke (2000)	Review paper	*Homo sapiens*	Grip types, tool making and use	Evolution of human hand
Marzke et al. (1998)	EMG	*Homo sapiens*	Hand muscles	Stone tool making
Hamrick et al. (1998)	EMG	*Homo sapiens*	Flexor pollicis longus muscle	Tool making and use
Christel et al. (1998)	Video	*Pan paniscus*, *Homo sapiens*		Reaching and grasping, spatiotemporal, performance
Schmitt (1999)	Kinematics, kinetics	*Macaca fascicularis*, *M. mulatta*, *Papio anubis*, *Chlorocebus aethiops*, *Erythrocebus patas*	Arboreal and terrestrial substrates, gait compliance	Forelimb kinematics, quadrupedalism
Crisco et al. (1999)	Medical imaging, semi-dynamic CT	*Homo sapiens*	Carpal motion	3D wrist kinematics
Kaufman et al. (1999)	EMG	*Homo sapiens*	Thumb muscles	Muscle function during thumb motion
Schick et al. (1999)	Behavior	*Pan paniscus*		Stone tool use and making
Schmidt and Fischer (2000)	Cineradiography	*Eulemur fulvus*	Forelimb kinematics	Quadrupedalism

(continued)

Table 10.1 (continued)

Reference	Technique	Species	Topic	Remarks
Christel and Fragaszy (2000)	Behavior, spatiotemporal	*Cebus apella*	Hand	Grasping, preference, performance
Wolfe et al. (2000)	Medical imaging	*Homo sapiens*		Carpal kinematics
Chang et al. (2000)	Kinematics, kinetics	*Hylobates lar*	3D kinematics, torque	Brachiation
Jouffroy and Medina (2002)	X-ray	Hominoids, cercopithecoids, platyrrhines, strepsirrhines		Wrist radioulnar deviation
Christel and Billard (2002)	Kinematics	*Macaca mulatta*, *Homo sapiens*	Grasping behavior	Prehension, motion control
Moojen et al. (2002)	Medical imaging, Semi-dynamic CT	*Homo sapiens*		Extension-flexion, radioulnar deviation
Schmitt (2003a)	Kinematics, kinetics	*Callithrix jacchus*	Instrumented pole	Forelimb kinematics, walking on fine branches
Carelsen et al. (2005)	Medical imaging, 4D CT	*Homo sapiens*	Carpal kinematics	Wrist joint motion
Courtine et al. (2005)	Kinematics, EMG	*Macaca mulatta*	3D kinematics	Quadrupedal locomotion, control
Chan and Moran (2006)	Musculoskeletal modeling	*Macaca* spp.		Reaching kinematics
Klein Breteler et al. (2007)	EMG	*Homo sapiens*		Neuromotor control of hand movement
Banks et al. (2007)	Pinch force	*Macaca fascicularis*		Pinch strength
Vigouroux et al. (2007)	EMG	*Homo sapiens*		Biomechanical modeling
Moore et al. (2007)	Medical imaging	*Homo sapiens*		Database of carpal kinematics, wrist motion
Spinozzi et al. (2007, 2004)	Behavior	*Cebus apella*	Grasping	Hand use, performance
Bury et al. (2009)	Grip forces	*Saimiri* spp.		Power grip force
Pouydebat et al. (2009, 2006a, b)	Behavior, spatiotemporal	*Papio papio*, *Gorilla gorilla*, *Pan troglodytes*, *Macaca fuscata*, *Cebus apella*, *Pongo pygmaeus*, *Homo sapiens*	Hand, thumb, fingers	Grip type, grasping

Patel (2009, 2010), Patel and Polk (2010)	Kinematics, kinetics	*Papio anubis*, *Erythrocebus patas*, *Macaca mulatta*	Forelimb	Hand posture, digitigrade
Wunderlich and Jungers (2009)	Digital pressures	*Pan troglodytes*	Terrestrial and arboreal substrates	Knuckle walking
Macfarlane and Graziano (2009)	Behavior	*Macaca mulatta*		Grasping, grip types
Foumani et al. (2009)	medical imaging 4D CT	*Homo sapiens*	3D kinematics	Wrist motion
Patel and Wunderlich (2010)	Palmar pressure	*Papio anubis*	Terrestrial substrates	Quadrupedal locomotion
Kivell et al. (2010)	Palmar pressure	*Daubentonia madagascariensis*	Arboreal substrates, ascent/descent	Quadrupedal locomotion
Williams et al. (2010)	Kinematics, IR cameras	*Homo sapiens*		Tool making
Higurashi et al. (2010)	Palmar pressure	*Macaca fuscata*	Arboreal substrates	Quadrupedal locomotion
Johnston et al. (2010)	EMG	*Homo sapiens*	Activation of intrinsics and extrinsics	Neuromotor control of precision grip
Orr et al. (2010)	Medical imaging	*Gorilla*, *Pan*, *Pongo*, *Papio*, *Colobus*, *Macaca*, *Ateles*	3D carpal kinematics	Mechanics of wrist motion in *Pan* and *Pongo*
Rolian et al. (2011)	Kinematics, kinetics	*Homo sapiens*		Tool making and use
Daver et al. (2012)	Kinematics, X-ray	*Papio*, *Macaca*, *Cercopithecus*, *Chiropotes*, *Aotus*, *Saguinus*	Carpal range of motion	Flexion-extension, radioulnar deviation of wrist
Patel et al. (2012)	EMG	*Papio anubis*	EMG of extrinsic hand muscles	Digitigrade hand posture
Williams et al. (2012)	Palmar pressure	*Homo sapiens*		Stone tool production
Reghem et al. (2013)	Kinematics	*Sapajus*, *Lemur*, *Gorilla*, *Pan*, *Homo*	Prehension	Hand and forelimb motion, behavior
Rainbow et al. (2008, 2013)	Medical imaging	*Homo sapiens*		Wrist flexion-extension
Matarazzo (2013)	Digital pressures	*Pan troglodytes*, *Gorilla gorilla*	Hand position, pressure distribution	Knuckle walking
Pontzer et al. (2014)	Kinematics, kinetics	*Pan troglodytes*		Quadrupedal and bipedal locomotion

2 Behavioral Studies

Primates are particularly adept at manual grasping, both for locomotion on arboreal substrates and for manipulation. Manipulation in primates is used primarily for feeding and grooming and is characterized by a diversity of grip types (Fig. 10.1). Enhancing our understanding of hand function in primates starts with observation. Napier's work (1956, 1960, 1961) is a nice illustration of this, with plenty of careful observations resulting in a classification in two basic grip types: the precision grip and the power grip. Since then, many studies have investigated grip behavior in nonhuman primates and categorized grasping into many subtypes (see Chap. 12 for a review). While many early studies focused primarily on classification of grip types, subsequent studies have attempted to use the behavioral observation of grasping to (1) interpret the neuromotor complexities of primate grasping and changes within the order or (2) develop biomechanical correlates for grasping behaviors that can be used to interpret the fossil record. It is in these two areas that experimental approaches become critical.

Studies focusing on the behavioral and spatiotemporal aspects of reaching and grasping typically use an experimental setup consisting of a maze or box in which objects are placed and manipulated by the subject, and one or two video cameras to record the behavior. Often, these studies are oriented toward neuromotor control and laterality rather than to the biomechanics of prehension movement because dexterity is not only determined by the functional morphology of the hand but also, importantly, by refined neuromotor control (e.g., Lemon 1999; Christel and Billard 2002; Pizzimenti et al. 2007; see Chap. 6). This was demonstrated by Costello and Fragaszy (1988) who studied grasping behavior in capuchins and squirrel monkeys. They showed that capuchins have a pseudo-opposable thumb [sensu Napier's (1961) definition; see Chap. 12] and are able to use a precision grip, while other New World

Fig. 10.1 *Lemur catta* grasping food items using a full-hand grasp typical of smaller primates (Reghem et al. 2013). Although numerous studies have examined hand posture and grasping in catarrhine primates, research to date is still quite taxonomically restricted, and few studies have examined strepsirrhines or platyrrhines (Reghem et al. 2013)

monkeys, such as squirrel monkeys, lack this ability. Later on, Christel and Fragaszy (2000) demonstrated that capuchins use lateral opposability grips (tip to side), but that precise finger coordination in capuchins is more effortful than in catarrhine primates. Christel and colleagues (Christel 1993; Christel et al. 1998) studied grasping and hand preferences in all hominoid primates and documented different subtypes for precision grasping.

Other researchers have extended the research of Fragaszy and Christel and provided insight both into further details of hand movements and into the universality and variation of hand use across primates. For example, Spinozzi and colleagues (2004, 2007) studied manual dexterity and laterality in tufted capuchins using an experimental setup consisting of a narrow, transparent tube with food items. The animals were videotaped using two laterally placed cameras while retrieving food items from the tube. Capuchins predominantly used their index or index and middle fingers to extract food from the tube (86 % of manipulations). Precision grips between the index and thumb were also observed, but were only occasionally used (6 % of manipulations).

Moving to Old World monkeys, Macfarlane and Graziano (2009) made a detailed analysis of basic grip behavior in macaques using video footage of a large colony of semi-free-ranging macaques. During food manipulation, they could distinguish 15 different grip types according to the hand surfaces that were used to contact the gripped object. These observations complement earlier observations on manipulative behavior in hamadryas and olive baboons (Napier 1961; Jolly 1970; Rose 1977; Jude 1993). Pouydebat and colleagues (2006a, b, 2009) also recorded grasping behavior in a large sample of semi-free-ranging primates (including great apes, cercopithecines, capuchins) using a high-speed camera during ad libitum sampling. They focused on interspecific differences in grasping techniques to pick up objects of different sizes. One of the conclusions of the study is that precision grasping is used preferentially to pick up small objects by all primates.

Behavioral studies focusing specifically on understanding the evolution of the human hand have been more limited since much of the experimental work in this area has been based on electromyography (EMG; see Sect. 5 below). Marzke and colleagues (Marzke and Shackley 1986; Marzke and Marzke 1987; Marzke et al. 1992; Marzke and Wullstein 1996) presented substantial work on grasping types in humans and nonhuman primates with a particular focus on hand function during tool making and use. This work was aimed at understanding tool use in early hominins and the evolution of the human hand. They filmed subjects while performing different manual tasks and used the recorded images to make a detailed description of grip types, hand position, and use. In some studies, they also recorded EMG data (see Sect. 5) in combination with their behavioral observations (Marzke et al. 1998). Their studies proposed a new classification of precision grips (Marzke and Shackley 1986; Marzke and Wullstein 1996) and used observational studies of modern humans crafting stone tools to hypothesize specific associations between anatomical features and stone tool making in early hominins (Marzke 1997; Marzke and Marzke 2000). Toth and Schick (1993) recorded human subjects during contemporary stone tool making behavior in native populations and complemented this study with observations of stone tool making in a trained bonobo (Toth et al. 1993; Schick et al. 1999). This comparison showed that stone tools can be made by nonhuman primates, using different techniques and grip types than observed in humans (see Chap. 11).

3 Kinematics and Kinetics

Although the behavioral studies detailed above have taught us a lot about hand use in primates, they do not allow a detailed appraisal of kinematics of specific joints in the hand. In contrast to standard video recording, which is often used to study and describe locomotor behavior and prehension movement in primates, motion capture techniques allow a quantification of the motion of interest. The simplest setup consists of one video camera recording at regular speed of 50 or 60 Hz (though sometimes analyzed as low as 30 Hz or at speeds higher than 100 Hz) positioned perpendicular to the direction of motion, allowing analysis of speed of movement, patterns of contact, and 2D joint and segment angles in each frame. The method is quite straightforward and in its simplest form can also be applied in the field to collect 2D kinematics of free-ranging primates during daily activities (Stevens et al. 2006). When two or more cameras are used and synchronized, a 3D analysis is possible (Bishop 2007; Hedrick 2008; Channon et al. 2010, 2012). With untrained primates, one often has to resort to manual digitization of the recorded images, which is time consuming and difficult, especially in animals with dark skin and long fur. Shaving the skin around the joint areas and the use of nontoxic white paint can, however, simplify the segmentation process and augment accuracy. In gait labs, reflective markers or a custom-made suit with markers can be used in combination with infrared cameras to automate the digitization process (Fig. 10.2b). Much work is currently under way to develop automatic video acquisition for primates with minimal markers (Schwartz et al. 2007) using Kinect or Wii technology (Metcalf et al. 2013; Choppin et al. 2014) and completely markerless systems with multiple cameras based on photogrammetric 3D reconstruction (Sellers and Hirasaki 2014).

Fig. 10.2 Examples of hand postures during locomotion in (**a**) *Lemur catta*, (***b***) *Papio*, and (**c**) *Pongo*. Lemurs regularly use palmigrade postures. Monkeys use palmigrade and digitigrade postures depending on speed. Some orangutans also use a palmigrade posture, but most use a fist walking posture (c), unlike the knuckle walking hand posture of the African apes

The latter technique, however, requires excellent light conditions and high-quality images, making application to both field and lab work on primate prehension and locomotion difficult.

The kinematics of prehensile movement in primates have only been studied by a few research groups, and the main focus often lies on neural control of reaching. Castiello and colleagues (1992) used a motion analysis system to collect the kinematics of reaching and grasping in humans. Specifically, they investigated the differences between grasping of small and large objects. Small objects were grasped with a precision grip, large objects with a whole-hand grip. Christel and Billard (2002) used two synchronized cameras to capture forelimb kinematics during grasping in macaques and humans. The study revealed clear differences in prehension kinematics between both species, with a distinct elbow and shoulder posture. In addition, the study also demonstrated that reaching is characterized by a bell-shaped velocity profile and typical timing of joint activity, a finding that has been confirmed by other studies (e.g., Castiello et al. 1992). Reghem and colleagues (2013) have also focused on the 3D kinematics of prehension movement across different primates (*Homo*, *Gorilla*, *Pan*, *Sapajus*, and *Lemur*). Using five synchronized cameras, they recorded the forelimb and hand motion of primates while grasping static food items. They found that grasp types varied among species, with a preference for whole-hand grips in monkeys and for precision grips in African apes and humans. During grasping, the shoulder kinematics of humans were more similar to monkeys than to the apes (Reghem et al. 2013).

Three-dimensional kinematics of the human hand during stone tool production have been studied recently by Williams et al. (2010) using a motion capture system with multiple infrared cameras and reflective markers attached to the hand, wrist, and forearm (Fig. 10.3b; see Chap. 11). This is an accurate way of recording forelimb kinematics, but is only applicable in humans or trained nonhuman primates. Williams et al. (2010) showed that the knapping strategy is associated with strong wrist extension followed by rapid wrist flexion, suggesting that the human wrist joint is adapted to increase efficiency for this specific motion pattern.

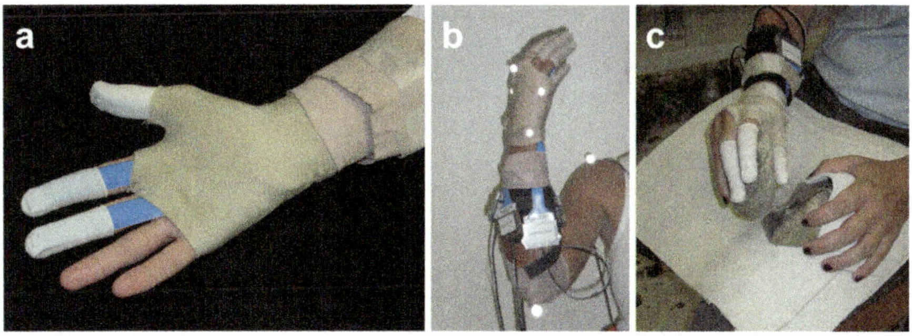

Fig. 10.3 Strips of pressure distribution measurement material can be used to measure loads on the fingers during stone tool making (photos courtesy of Erin Marie Williams-Hatala). In (**a**) the pressure strips can be seen covering the palmar surfaces of the first three digits, (**b**) kinematic markers are placed on each segment of the upper limb, and (**c**) the subject can make or use stone tools naturally, while load distribution is quantified on the digits. For more information on this study, see Williams et al. (2012) and Chap. 11

Most studies of the primate hand in an explicitly experimental context have, however, focused on forelimb kinematics (including the wrist and hand) during primate locomotion, with some studies that commented directly on hand mechanics during locomotion. For example, Patel and colleagues (Patel 2009; Patel and Polk 2010) compared 3D forelimb kinematics in patas monkeys, macaques, and baboons using four synchronized cameras to better understand the effect of speed on forelimb kinematics and hand posture. Contrary to expectations, subjects of all three species demonstrated a more palmigrade hand posture with increasing speed. This finding was later confirmed using dynamic pressure measurements (Patel and Wunderlich 2010).

Depending on the setting, kinematics can be recorded simultaneously with forces (kinetics), which can provide important information on loading of the hand, and other elements up the kinetic chain from the hand in different activities (e.g., manipulation, locomotion). Ground reaction forces can be recorded using force plates that are usually built in a walkway (Fig. 10.4a). Alternatively, dedicated transducers can be integrated in a pole to measure forces on arboreal substrates (Fig. 10.4b). There is a range of force platforms and transducers available on the market that operates with slightly different technologies. Most commonly used in biomechanical studies are platforms based on strain gauges or piezoelectric sensors allowing measurement of forces and torques in three directions. These devices can be synchronized with the video cameras, so that kinematics and kinetics can be collected together. Several researchers have used an integrated setup to study arboreal forelimb kinematics and kinetics in nonhuman primates (e.g., Kimura et al. 1979; Reynolds 1985). Schmitt (1999, 2003a, b) examined forelimb and hind limb kinematics in different quadrupedal catarrhine primates using setups with integrated force plates (e.g., raised horizontal pole mounted on top of a force plate or runway with built-in force plate) and surrounded by video cameras. Chang and colleagues (2000) used an instrumented "brachiation runway" to study the dynamics of gibbon brachiation. Their setup

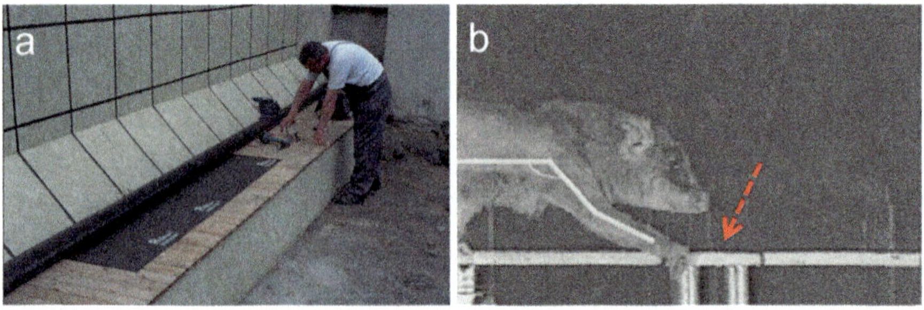

Fig. 10.4 Examples of two different arrangements in which kinematics are integrated with force data collection. In (**a**) force platforms and pressure mats are integrated into a platform runway, and cameras filming the animal can take advantage of regularly spaced grid markings to measure speed. (**b**) Experimental setup modified from Schmitt and Lemelin (2002) in which a segment of the pole (*dashed arrow*) is isolated from the rest of the pole and mounted on a force transducer. Video allows quantification of kinematic variables (arm protraction is illustrated)

(equipped with built-in force transducers) allowed for simultaneous recording of forces and kinematics using a single camera, so they could focus on the load of the hand. Michilsens and colleagues (2012) studied siamang brachiation using a similar runway with separate handholds, allowing for recording of kinetics.

Force transducers are not only used to record ground reaction forces; they can also be applied to measure grip forces, either as principal components of hand dynamometers or as part of custom-built measuring devices. Most grip force recordings of nonhuman primates are reported in clinically oriented studies, often focusing on the impact of neurological disorders on hand function. For example, Banks et al. (2007) recorded pinch forces in macaques, while Bury et al. (2009) reported power grip forces in squirrel monkeys before and after a cortical injury. Both studies used custom-made devices to record forces, comparable to hand dynamometers used in the assessment of human subjects. So while force data for different primates might exist, there is, to our knowledge, no comparative study of grip forces across primate species. Linking force production during power or pinch grips to specific anatomical traits of the hand could offer interesting insights in primate hand function.

4 Dynamic Palmar Pressure

The way the hand is used during locomotion differs substantially from primate to primate. For example, patas monkeys, macaques, and baboons are digitigrade; gorillas, bonobos, and chimpanzees are knuckle walkers, while orangutans are so-called fist walkers (Fig. 10.2; see Chap. 13). But hand posture can also vary within a species, according to substrate type, inclination, or speed. While force plates can provide insight into whole limb forces or torques (see Sect. 3), understanding of the distribution of force under specific manual (or pedal) elements requires the use of pressure-sensitive pads or platforms. These devices are very versatile, as they can be used to measure pressure during both terrestrial and arboreal locomotion underneath any structure of interest (e.g., hand, foot, fingers). Pressure platforms and mats provide valuable information about hand posture and load distribution during primate locomotion, as they allow recording of average and peak pressures underneath the hand/fingers (Fig. 10.5). Because pressures are recorded dynamically, information is also gained about roll-off patterns and changes in contact area. Furthermore, palmar pressure distribution is the only mechanism by which we can currently obtain information about forces on the individual elements of the hand, thereby allowing hypotheses about specific hand bones to be tested.

The first studies that investigated pressure were conducted on the foot rather than the hand. Elftman (1934) and Elftman and Manter (1934) developed the first apparatus designed to continuously record foot pressures to resolve a debate regarding the functional axis of the foot using what we now call the center of pressure of the foot. In 1935, the same authors published footprints of a human and a chimpanzee generated by their pressure-sensing apparatus in which they used a rubber mat with pyramidal projections that was placed on a glass surface with

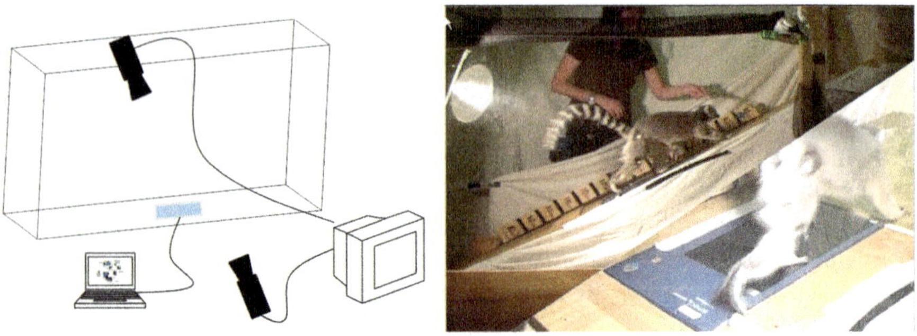

Fig. 10.5 Typical experimental setup for collection of 3D kinematic data integrated with other data. Here the two cameras are synchronized by way of a special effect generator, so multiple images from any time frame can be viewed or analyzed

some intermediate reflective fluid. They then filmed the displacement of the pyramids to give a dynamic view of pressure distribution, but never quantified pressure distribution from the images (Elftman 1934; Elftman and Manter 1934, 1935). Systems designed to *quantify* plantar pressure in different regions of the foot were developed starting in the 1930s (Schwartz and Heath 1937) and proliferated in the 1960s and 1970s when strain gauges, capacitors, resistors, piezoceramic transducers, and pressure-sensitive foils were incorporated into the devices. In 1976, Nicol and Hennig (Nicol and Hennig 1976; Hennig and Nicol 1978) made one of the greatest recent advancements in pressure-measuring techniques when they used multiplexing to collect pressure information from a matrix of capacitive pressure transducers in a flexible mat.

The first use of pressure distribution in nonhuman primates was not until early in the twenty-first century and again focused on foot pressures. The first studies to publish *quantitative* data on pressure distributions during locomotion in nonhuman primates were conducted on chimpanzees and monkeys (Wunderlich 1999) and bonobos and gibbons (Vereecke et al. 2003, 2005). The first manual pressure data were published later in a study on knuckle-walking chimpanzees (Wunderlich and Jungers 2009). In this study, a pressure platform and a flexible pressure-sensitive pad that could be wrapped around a pole were used to allow comparisons of arboreal ("pole") versus terrestrial locomotion ("platform") in chimpanzees. Differences in pressure distribution underneath the digits were found between substrates, but also pressure differences related to age and hand posture. More recently, Matarazzo (2013) used a pressure mat in a larger and ontogenetically broader sample of both chimpanzees and gorillas to compare hand position, contact patterns, and pressure distribution across the manual digits of chimpanzees and gorillas knuckle walking on the ground. Her results resembled those of Wunderlich and Jungers (2009), as well as kinematic results of Tuttle (1969), and Inouye (1994). Furthermore, Matarazzo's (2013) large sample highlighted the high levels of variability of hand position in chimpanzees and the importance of hand position to pressure distribution in both African ape taxa.

While these studies examined the relationship between structure and function at specific manual digits, other studies have used palmar pressure distribution to ask questions about force distribution across the hand or the influence of particular kinematic factors on the manual interface with the substrate. For example, Patel and Wunderlich (2010) collected data from baboons running over pressure-sensitive mats at varying speeds. This study showed that baboons gradually shift from digitigrade to palmigrade hand postures with increasing speed (Fig. 10.6). The palmigrade postures at higher speeds may be a way to distribute higher forces over a larger contact area. While the hands become increasingly palmigrade, the feet become increasingly digitigrade at high speeds, indicating a possible differentiation of hind limb and forelimb mechanisms (Wunderlich and Patel 2008) and supporting evidence from various species that primates actively moderate forelimb relative to hind limb loads [e.g., Schmitt and Hanna (2004)].

One of the most distinctive aspects of primate locomotion is the diversity of arboreal and terrestrial supports used by primates and the extent to which primates exhibit morphological and/or behavioral specializations, especially in the cheiridia, for particular substrate environments. A number of studies have taken advantage of pressure distribution technology to examine the interaction between hands and specific substrate demands. For example, Kivell et al. (2010) studied aye-aye (*Daubentonia*) hand posture and pressure during horizontal locomotion, ascent, and descent using a pressure platform. Together with videography, palmar pressure data clearly showed that the aye-aye consistently curls its fingers during locomotion on all slopes. Furthermore, using hand and foot pressures from the same stride, it was demonstrated that aye-ayes actively shift their weight posteriorly during descent in order to avoid unwanted load increases on their forelimbs, particularly their delicate third finger specialized for foraging (Kivell et al. 2010).

Higurashi and colleagues (2010) approached the challenges of substrate orientation in a different way. They collected palmar pressures during locomotion of a macaque on arboreal substrates with different orientations. They used flexible pressure pads that were wrapped around arboreal substrates, a horizontal pole (parallel to craniocaudal axis of animal), and a horizontal ladder (rungs perpendicular to craniocaudal axis of animal). Gait characteristics were recorded simultaneously using four cameras. Their study showed considerable differences in hand use

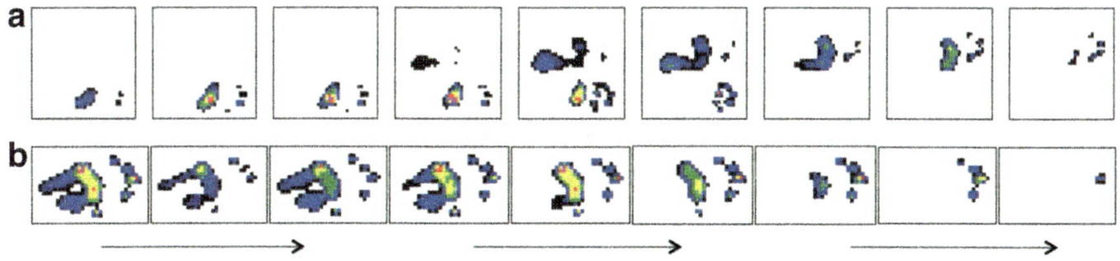

Fig. 10.6 Output from plantar and palmar pressure measurement. The animal is traveling from *left* to *right*; *top* frames (**a**) show left hand and foot traveling slowly. The *bottom* frames (**b**) illustrate pressures under the left hand at higher speeds. Note the digitigrade loading of the hand in (**a**) and the more palmigrade loading in (**b**).

between the two arboreal substrate orientations, suggesting a horizontal pole might be a more challenging substrate than a horizontal ladder.

Measurement of pressure distribution during human grasping has been mostly limited to the areas of ergonomics (Hall 1997; Aldien et al. 2005) and robotics (Yousef et al. 2011). Recently, however, Williams and colleagues (Williams and Richmond 2012; Williams et al. 2012; see Chap. 11) developed an elegant experiment to test hypotheses about the relationship of human hand structure to stone tool use and manufacture in early hominins. They attached strips of flexible pressure mats to the thumb and fingers of experienced and inexperienced stone tool knappers (Fig. 10.3) and found that, contrary to expectations, the long robust human thumb is probably not associated with resistance of high forces generated during stone tool production (Williams et al. 2012). Rather, they were able to demonstrate that high forces were present on the thumb during grasping of flakes with a two-jaw pad-to-side grip, a grasping posture used during the use of flakes for cutting or scraping (Williams and Richmond 2012).

Rolian et al. (2011), however, found conflicting results using a different method. They instrumented a simulated stone tool and stone flake with a force transducer to examine the forces imparted on the tool by the digits during stone tool production and use in novice knappers. They found high tool reaction forces on the thumb during stone tool making and precision-grip use of stone tool flakes. While the results of these studies both supported the argument that improved performance of grasping during flake use was likely the greater selective pressure on human thumb morphology, the variation in the results of these studies illustrates some of the advantages and disadvantages of the technology used in each and highlight the need for integrative experimental research. In the studies by Williams and colleagues (Williams and Richmond 2012; Williams et al. 2012), palmar pressure distribution affords the opportunity to measure individuals performing in their natural environment, making real stone tools in this case, and simply measuring the pressure on each finger during this activity. However, it should be reiterated that these pressures only reflect the normal (perpendicular to the pressure sensor) load on the finger. Alternatively, an artificial tool can be created to measure forces imparted on the tool itself (Rolian et al. 2011; see Chap. 11). While these forces may or may not represent the actual "natural" forces produced during stone tool production, they are three-dimensional forces that, when combined with three-dimensional kinematic analysis, can be used to calculate joint moments. Combining these methods would provide tremendous insight into the biomechanics of digit use during stone tool production and use.

Pressure pads and platforms are valuable tools to study primate hand function, yet they vary considerably in quality, flexibility, and affordability. The less cost-prohibitive devices require simultaneous data collection on a force plate for calibration. This makes the setup less portable and adds the cost of a force plate to the experimental setup. These setups have often been limited to use in animal gait labs, with an unavoidably limited number of subjects [e.g., two chimpanzees (Wunderlich and Jungers 2009); two baboons (Patel and Wunderlich 2010); one macaque (Higurashi et al. 2010)]. However, as zoos and sanctuaries have become increasingly open to noninvasive research, more portable systems can be used in these settings, affording the opportunity to measure a more diverse and larger sample [e.g., five aye-ayes (Kivell et al. 2010), eight chimpanzees, and seven gorillas (Matarazzo 2013)]. Pressure distribution can also be used to study semi-free-ranging animals, as shown by Vereecke et al. (2003) on ten bonobos and Vereecke et al. (2005) on four gibbons. However, in this setting, data collection of single-handed pressures is hampered if too many animals traverse the pressure platform simultaneously. Also, the mats should be well secured and protected, to avoid being damaged by all too curious primates. Pressure mats also differ from force plates in that they only measure pressure/force normal to the substrate. Therefore, any attempt to understand completely the substrate reaction forces at individual digits or individual joint moments requires the combination of pressure technology with a device that measures substrate reaction forces in three directions (i.e., a force plate). There are numerous models for modifications of standard force plates for measurement of substrate reaction forces on arboreal supports [e.g., Schmitt (1994, 1998, 1999); Schmitt and Lemelin (2002); Lammers (2007)], for measuring substrate torque [e.g., Lammers and Gauntner (2008)] and for measuring substrate reaction forces on asymmetrically shaped objects [e.g., artificial stone tools (Rolian et al. 2011)]. It is imperative that these technologies are used together if we are to understand fully the interface between hands (and feet) and the substrates or objects they grasp.

5 Electromyography

If you want to learn about the myology of the hand, you perform a dissection (see Chap. 7), but if you want to understand how the muscles work, you need to use electromyography. Electromyography (EMG) has played a prominent role in experimental analyses of hand use in nonhuman primates and comprises some of the first true manipulative experiments (other than film analysis) performed to understand the function of the primate hand. There are two types of EMG: surface EMG and fine-wire EMG. The first is commonly used in human subjects, as it is noninvasive and easy to apply. The small size and deep or overlapping position of many of the extrinsic and intrinsic hand muscles makes application of surface electrodes for registration of muscle activation patterns difficult and mostly inaccurate (e.g., cross talk; Solomonow et al. 1994). Therefore, fine-wire EMG is the preferred technique in nonhuman primate subjects. This method is more invasive than other techniques described here, and it requires administration of anesthesia to the animal prior to positioning of the fine-wire electrodes. With good training and good knowledge of the anatomy, correct placement of the electrodes is possible, but not straightforward. Tools that can increase accuracy include back stimulation of the electrodes, to verify correct placement of the electrode, and ultrasound-guided needle placement (Corneil et al. 2012). For this latter technique, a standard ultrasound device with a linear probe is used to visualize the muscles. Most needles are radiopaque and will be visible on the ultrasound image too. In this way, feedback about electrode placement is acquired during the procedure, allowing correct placement of the needle in the selected muscle (Burgar et al. 1997; Rudroff 2008). Correct needle placement is important and requires quite some time, which together with its invasiveness, makes it difficult to study large samples.

Tuttle and colleagues (Tuttle and Basmajian 1974a, b, 1978a, b; Tuttle et al. 1972, 1983, 1992) were the first to publish EMG on nonhuman primates with their studies on forelimb muscles of gorillas, chimpanzees, and orangutans. In their first study, they used fine-wire electrodes in the forearm muscles of a habituated, captive gorilla and recorded muscle activation patterns during knuckle-walking and manual activities in an effort to understand whether hand and wrist position were stabilized through osteoligamentous means or whether muscular activity was necessary to support the postures characterizing gorilla knuckle walking (Tuttle et al. 1972; Tuttle and Basmajian 1974b). Later, they extended their sample to include chimpanzees and orangutans and found variation associated with hand position (Tuttle and Basmajian 1978a, b; Tuttle et al. 1983, 1992). Susman and Stern (1979) conducted a similar analysis on chimpanzees and concluded that the extrinsic digital flexors were not active at all during stance phase of quadrupedal stance or slow to moderate knuckle walking. They demonstrated, instead, that these muscles are highly active during suspensory locomotion and concluded that the extrinsic digital flexors, as well as their bony attachments, are adaptations for use during suspensory locomotion.

After discovering no active role in support during knuckle walking for the extrinsic flexors, Susman and Stern (1980) tested whether the intrinsic muscles of the hand served any such role during knuckle walking. They used EMG on the interossei and lumbricals of the hand, muscles that had already been demonstrated to function in manipulation in humans. In fact, the interossei and lumbricals showed very similar functions in chimpanzees and humans, and the muscles did not seem to play any necessary stabilization role in apes during knuckle walking. They later followed up these studies with an examination of the accessory interosseus muscle found in gibbons (Susman et al. 1982). This muscle was primarily used to assist with adduction during flexion because of the wide digital cleft in gibbons, but it was also used substantially during pinch grasps between the thumb and index finger.

While most of these studies have addressed topics directly related to the identification of function in the fossil record, more recent studies have used EMG to address fundamental questions regarding hand use during primate locomotion. Patel et al. (2012), following up on work described above regarding digitigrady in baboons, examined the activity of the wrist and digital flexor muscles. They found that at slow speeds, baboons can adopt a digitigrade posture and not need to recruit wrist or finger flexors to support the digits. However, at high speeds, as the hand becomes more palmigrade, active muscle recruitment is necessary. They interpreted this by suggesting that manual digitigrady in baboons and some other Old World monkeys is an energy-reducing adaptation for walking long distances on the ground. Together with palmar pressure data, they demonstrated that using more palmigrade postures at higher speeds reduces pressures under the hand when moving at higher speeds.

Studies reporting on muscle activation patterns during gripping and other manual activities are mostly limited to humans (Kaufman et al. 1999; Klein Breteler et al. 2007; Vigouroux et al. 2007; Johnston et al. 2010), and a comparative nonhuman primate dataset on muscle recruitment during manipulative behavior remains, to date, lacking. EMG has also been used on humans to address questions about early hominin behavior by looking at muscle activation patterns during tool making and use in humans (Marzke et al. 1998; Hamrick et al. 1998). Hamrick et al. (1998) sought to understand the role of the independent flexor pollicis longus muscle in humans (see Chap. 7). The human arrangement for the flexor pollicis longus, as well as the muscle markings on fossil distal phalanges, had been associated with facility for tool making in early hominins (Susman 1988, 1989, 1998; see Chaps. 11 and 18). Hamrick et al. (1998) confirmed that the flexor pollicis longus is used extensively during forceful stone toolmaking behaviors as well as during cutting/scraping using a precision grip in novice knappers, supporting suggestions that fossil hominins such as O.H. 7 were capable and did make and use tools. However, Marzke et al. (1998) found that the flexor pollicis longus is not strongly recruited during precision pinch grips in experienced stone tool knappers.

6 Medical Imaging

In addition to the experimental studies reviewed above, primate hand function has also been studied using medical imaging techniques. This can be done in a static way, using plain X-ray films of cadaveric or bone material positioned in different configurations to document internal motion of the carpal bones. Recently, Daver et al. (2012) used a static approach, comparable to the earlier approach of Jouffroy and Medina (2002), to study motion of the carpal bones in a cadaveric sample consisting of seven monkey species. They demonstrated that all studied monkeys displayed a greater midcarpal mobility than antebrachiocarpal mobility (see Chap. 3). Despite this similar general pattern, they did also observe differences in carpal kinematics between ceboids and cercopithecines that might be linked to differences in carpal morphology and locomotor hand postures.

A fully dynamic approach to study internal bone motion was developed by Jenkins and colleagues (Jenkins and Fleagle 1975; Jenkins et al. 1978; Jenkins 1981) who used 2D and 3D cineradiography to study forelimb limb joint motion during locomotion in nonhuman primates. Cineradiography proved very useful, especially for the shoulder joint and wrist joints with their complex motions in different planes. For example, they recorded spider monkeys while brachiating and, using cineradiography, they were able to describe and quantify the motion of the shoulder girdle and wrist in great detail (Jenkins et al. 1978; Jenkins 1981). Later on, 2D cineradiography or video fluoroscopy was used by Schmidt and Fischer (2000) to study forelimb kinematics of a brown lemur (*Eulemur fulvus*) during quadrupedal locomotion on a horizontal rope mill. The method also proved useful to quantify the 3D motion of the shoulder girdle. While cineradiography is a valuable technique to study motion of long or large bones, it can be more challenging for small and irregular bones with a complex 3D configuration such as the carpal bones (Jenkins and Fleagle 1975, but see Jenkins 1981).

In the last two decades, several CT-based 3D imaging techniques have been developed that are applicable to study carpal kinematics in vivo. Crisco and colleagues developed a semi-dynamic technique by calculating bone kinematics from CT scans taken at incremental steps throughout a motion cycle (Crisco et al. 1999; Wolfe et al. 2000; Moore et al. 2007; Rainbow et al. 2008, 2013). They used this technique primarily to study carpal kinematics during wrist or thumb motion in humans (Wolfe et al. 2000; Rainbow et al. 2008) and even published a full dataset of carpal kinematics during wrist motion (Moore et al. 2007). Orr et al. (2010) have extended the application of this technique to study carpal kinematics, and specifically wrist motion, in nonhuman primates (see Chap. 9). Moojen et al. (2002) also developed a semi-dynamic CT scan technique—comparable to the technique used by Crisco and colleagues—to quantify carpal bone kinematics during wrist motion in 11 human subjects. One of their findings was the broad spectrum of movement patterns of the scaphoid, with motion sometimes confined to the sagittal plane and sometimes mostly to the frontal plane.

Carelsen and colleagues developed a fully dynamic CT scanning technique to study bone motion in vivo (Carelsen et al. 2005, 2009; Foumani et al. 2009). They adapted a protocol, originally conceived for angiography (i.e., visualization of coronary arteries), to record internal bone motion of a moving wrist joint in three dimensions. More recently, Kerkhof et al. (2013) used dynamic or 4D CT to study internal motion of the basal thumb joint during thumb opposition in human cadavers. One of the big advantages of dynamic CT scanning is that you get a direct view on the moving bones and that the arthrokinematics can be studied and quantified without using markers. An additional advantage of the method developed by Kerkhof et al. (2013) is that it uses a non-gated protocol, i.e., data acquisition does not need to be synchronized with rotation of the CT scanner (contrary to the gated protocol of Carelsen and colleagues).The dynamic CT scanning technique has so far only been used in human subjects, but can also be used to study bone motion in nonhuman primates (Fig. 10.7). However, since a smooth movement at a constant speed is required, we believe this technique can only be used in cadaveric samples or anesthetized subjects while imposing a passive motion using a custom-made motion simulator.

Medical imaging techniques, mostly CT, have also been used to perform a detailed and quantitative anatomical analysis of the carpal bones in primates (Schilling et al. 2014; Tocheri et al. 2003). As these studies mainly focus on morphology and shape of the articular facets, they will not be discussed in this chapter.

7 Computer Modeling and Simulation

A further development is the use of computer modeling and simulation to get insight into the functioning of a structure, in this case, the hand. A musculoskeletal model is built using anatomical information of both osteology (e.g., shape of the bones and articular facets) and surrounding soft tissues (e.g., muscle size and path, fascicle length, tendon slack length). Next, functional data—collected using the

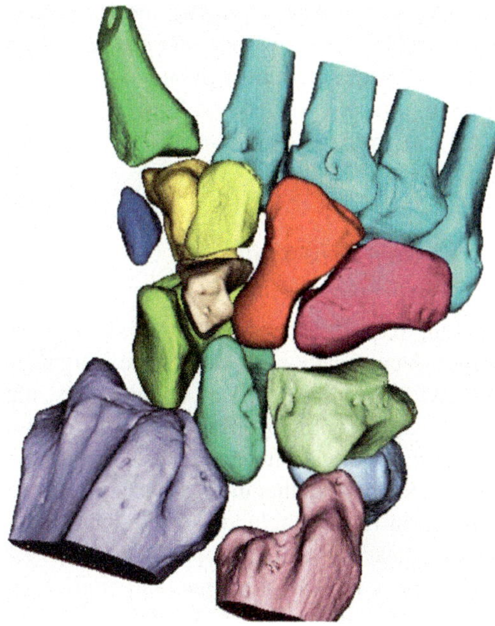

Fig. 10.7 Three-dimensional reconstruction of a baboon hand from CT images (using Mimics software, Materialise, Leuven)

methodologies discussed above—are used as input to the model. In an inverse dynamic analysis, kinematic and kinetic data are used as input, and joint torques and muscle forces are then calculated by the model. A forward dynamic analysis can also calculate joint kinematics from muscle forces. Even though validation of such models remains difficult, because actual muscle and joint forces are difficult to measure, they offer valuable insight in the functioning of specific joints or structures and allow one to investigate the impact or importance of specific morphological characteristics (e.g., muscle size or path, ligament position). In simulations, virtual modifications can be made to the anatomical configuration, offering the possibility to establish form-function relationships.

Several studies have developed a 3D model of the human hand (e.g., Buchholz and Armstrong 1992; van der Smagt and Stillfried 2008; Gustus et al. 2012; de Monsabert et al. 2014), but studies modeling the nonhuman primate hand remain scarce. Chan and Moran (2006) developed a 3D musculoskeletal model of a macaque forelimb, including the shoulder girdle, arm, forearm, hand, and 38 muscles. The model was developed using SIMM software and made freely available as supplementary data to their paper. The model can also be used in the open-source software OpenSim, making it available and accessible to the whole scientific community. Their main goal was to develop a primate forelimb model that could be used in neurophysiological studies on motor control and reaching. The model is, however, equally useful to study primate hand function. Building an accurate musculoskeletal model is certainly not straightforward as it requires an integrated dataset, including detailed anatomical information, joint kinematics, muscle activation patterns, and kinetics. The development of musculoskeletal models of the forelimb and hand of a series of primate species is thus not something that will easily be achieved. However, it would offer a powerful way to investigate primate hand function. In addition, musculoskeletal modeling also opens up possibilities for the study of fossil primates, where input from fossil remains can be complemented with functional data from extant primates.

8 Future Directions

Over the past 50 years, we have made tremendous progress in the way we study primate hand function, as well as the insights gained in hand function and adaptation. With the advent of new technologies, the possibilities for detailed study of hand function continue to increase. There are two important aspects that we want to highlight: the use of integrated data collection and the inclusion of different primate species. We believe that these aspects are key to further improve our understanding of the form-function relationships in the primate hand. With integrated data collection, we mean the simultaneous and preferably synchronized, collection of kinematics, kinetics, palmar pressures, and/or EMG in combination with an anatomical study of hand morphology. The combination of morphological and functional information is particularly important to improve our insight of the adaptive processes that shaped the primate hand and accuracy to interpret fossil remains.

Of equal importance is the inclusion of a wider range of primate species in functional studies and implementation of truly comparative studies on primate hand function. For more invasive or technically demanding setups, inclusion of different subjects or species is difficult. Probably for that reason, the scientific attention to a few easily accessible primate species (mainly macaques and chimpanzees) has been disproportional relative to the rare interest given to other species. There are, however, several studies that have included five or more primate species (e.g., O'Connor and Rarey 1979; Schmitt 1999, 2003a, b; Pouydebat et al. 2009; Orr et al. 2010; Reghem et al. 2013), and these have contributed significantly to our understanding of interspecific differences in grasping strategies, hand function, and hand morphology. The added value of including different species in one study is that data are obtained in a consistent way, using the same methodology for each species. The difficulty with single-species studies is that results from different studies, often obtained with different methodologies, must be compared to come to conclusions about interspecific differences or patterns. We believe that more effort should be made to include more species in each study or, alternatively, to standardize the methods and improve transparency about data collection and analysis. Open-access databases are certainly a highly valuable instrument in this context.

From his earliest studies, Napier pointed out the diversity in primate hand morphology. Since then, we have gained insight into how primate hands function by studying how hands grasp objects, which muscles are recruited, how they are loaded, and what pinch and grip forces they can produce. However, many questions remain, mainly pertaining to the evolution of the human hand, the functioning of hands in our early ancestors, and how we can recognize characteristics related to human-like dexterity in the fossil record. To this end, more studies focusing on hand form and function are needed, especially those that include a diversity of extant species and integrate different types of data that, together, can inform our understanding of the relationships among hand morphology, function, and ecological context.

References

Aldien Y, Welcome D, Rakheja S, Dong R, Boileau PE (2005) Contact pressure distribution at hand-handle interface: role of hand forces and handle size. Int J Ind Ergon 35:267–286

Banks JJ, Lavender SA, Buford JA, Sommerich CM (2007) Measuring pad-pad pinch strength in a non-human primate: *Macaca fascicularis*. J Electromyogr Kinesiol 17:725–730

Bishop K (2007) Aerodynamic force generation, performance and control of body orientation during gliding in sugar gliders. J Exp Biol 210:2593–2606

Buchholz B, Armstrong TJ (1992) A kinematic model of the human hand to evaluate its prehensile capabilities. J Biomech 25:149–162

Burgar C, Valero-Cuevas FJ, Hentz VR (1997) Fine-wire electromyographic recording during force generation: application to index finger kinesiologic studies. Am J Phys Med Rehabil 76:494–501

Bury SD, Plautz EJ, Liu W, Quaney BM, Luchies CW, Maletsky RA, Nudo RJ (2009) A novel device to measure power grip forces in squirrel monkeys. J Neurosci Methods 179:264–270

Carelsen B, Bakker NH, Strackee SD, Boon SN, Maas M, Sabczynski J, Grimbergen CA, Streekstra GJ (2005) 4D rotational x-ray imaging of wrist joint dynamic motion. Med Phys 32:2771–2776

Carelsen B, Jonges R, Strackee SD, Maas M, van Kemenade P, Grimbergen CA, van Herk M, Streekstra GJ (2009) Detection of in vivo dynamic 3-D motion patterns in the wrist joint. IEEE Trans Biomed Eng 56:1236–1244

Castiello U, Bennet K, Paulignan Y (1992) Does the type of prehension influence the kinematics of reaching? Behav Brain Res 50:7–15

Chan SS, Moran DW (2006) Computational model of a primate arm: from hand position to joint angles, joint torques and muscle forces. J Neural Eng 3:327–337

Chang Y, Bertram JEA, Lee DV (2000) External forces and torques generated by the brachiating white-handed gibbon (*Hylobates lar*). Am J Phys Anthropol 216:201–216

Channon AJ, Crompton RH, Günther MM, D'Août K, Vereecke EE (2010) The biomechanics of leaping in gibbons. Am J Phys Anthropol 143:403–416

Channon AJ, Usherwood JR, Crompton RH, Günther MM, Vereecke EE (2012) The extraordinary athletic performance of leaping gibbons. Biol Lett 8:46–49

Choppin S, Lane B, Wheat J (2014) The accuracy of the Microsoft Kinect in joint angle measurement. Sport Technol 7:98–105

Christel MI (1993) Grasping techniques and hand preferences in Hominoidea. In: Preuschoft H, Chivers DJ (eds) Hands of primates. Springer Verlag, Vienna, pp 91–108

Christel MI, Billard A (2002) Comparison between macaques' and humans' kinematics of prehension: the role of morphological differences and control mechanisms. Behav Brain Res 131:169–184

Christel MI, Fragaszy D (2000) Manual function in *Cebus apella*: digital mobility, preshaping, and endurance in repetitive grasping. Int J Primatol 21:697–719

Christel MI, Kitzel S, Niemitz C (1998) How precisely do bonobos (*Pan paniscus*) grasp small objects? Int J Primatol 19:165–194

Corneil BD, Goonetilleke SC, Peel TR, Green KA, Welch ID (2012) Ultrasound-guided insertion of intramuscular electrodes into suboccipital muscles in the non-human primate. J Electromyogr Kinesiol 22:553–559

Costello MB, Fragaszy DM (1988) Prehension in *Cebus* and *Saimiri*: I. Grip type and hand preference. Am J Primatol 15:235–245

Courtine G, Roy RR, Hodgson J, Mckay H, Zhong H, Yang H, Tuszynski MH, Edgerton VR, Raven J (2005) Kinematic and EMG determinants in quadrupedal locomotion of a non-human primate (Rhesus). J Neurophysiol 93:3127–3145

Crisco JJ, Mcgovern RD, Wolfe SW (1999) Noninvasive technique for measuring in vivo three-dimensional carpal bone kinematics. J Orthop Res 17:96–100

Daver G, Berillon G, Grimaud-Hervé D (2012) Carpal kinematics in quadrupedal monkeys: towards a better understanding of wrist morphology and function. J Anat 220:42–56

de Monsabert BG, Vigouroux L, Bendahan D, Berton E (2014) Quantification of finger joint loadings using musculoskeletal modelling clarifies mechanical risk factors of hand osteoarthritis. Med Eng Phys 36(2):177–184

Elftman H (1934) A cinematic study of the distribution of pressure in the human foot. Anat Rec 59:481–491

Elftman H, Manter J (1934) The axis of the human foot. Science 80:484

Elftman H, Manter J (1935) Chimpanzee and human feet in bipedal walking. Am J Phys Anthropol 20:69–79

Foumani M, Strackee SD, Jonges R, Blankevoort L, Zwinderman AH, Carelsen B, Streekstra GJ (2009) In-vivo three-dimensional carpal bone kinematics during flexion-extension and radio-ulnar deviation of the wrist: dynamic motion versus step-wise static wrist positions. J Biomech 42:2664–2671

Gustus A, Stillfried G, Visser J, Jörntell H, van der Smagt P (2012) Human hand modelling: kinematics, dynamics, applications. Biol Cybern 106:741–755

Hamrick MW, Churchill SE, Schmitt D, Hylander WL (1998) EMG of the human flexor pollicis longus muscle: implications for the evolution of hominid tool use. J Hum Evol 34:123–136

Hall C (1997) External pressure at the hand during object handling and work with tools. Int J Ind Ergon 20:191–206

Hedrick TL (2008) Software techniques for two- and three-dimensional kinematic measurements of biological and biomimetic systems. Bioinspir Biomim 3:034001

Hennig EM, Nicol K (1978) Registration methods for time-dependent pressure distribution measurements with mats working as capacitors. In: Asmussen E, Joergensen K (eds) Biomechanics VI-A. University Park Press, Baltimore, pp 361–367

Higurashi Y, Hirasaki E, Kumakura H (2010) Palmar and plantar pressure while walking on a horizontal ladder and single pole in *Macaca fuscata*. Int J Primatol 31:181–190

Inouye S (1994) Ontogeny of knuckle-walking hand postures in African apes. J Hum Evol 26:459–485

Jenkins FA Jr (1981) Wrist rotation in primates: a critical adaptation for brachiators. Symp Zool Soc Lond 48:429–451

Jenkins FA Jr, Fleagle JG (1975) Knuckle-walking and the functional anatomy of the wrists in living apes. In: Tuttle RH (ed) Primate functional morphology and evolution. Mouton, The Hague, pp 213–231

Jenkins FA Jr, Dombrowski PJ, Gordon EP (1978) Analysis of the shoulder in brachiating spider monkeys. Am J Phys Anthropol 48:65–76

Johnston JA, Bobich LR, Santello M (2010) Coordination of intrinsic and extrinsic hand muscle activity as a function of wrist joint angle during two-digit grasping. Neurosci Lett 474:104–108

Jolly CJ (1970) The seed-eaters: a new model of hominid differentiation based on a baboon analogy. Man 5:5–26

Jouffroy FK, Medina MF (2002) Radio-ulnar deviation of the primate carpus: an X-ray study. Z Morphol Anthropol 83:275–289

Jude J (1993) Manipulative behavior of hamadryas baboons. Senior thesis, Arizona State University, Tempe, AZ, USA

Jungers WL, Stern JT Jr (1981) Preliminary electromyographical analysis of brachiation in gibbon and spider monkey. Int J Primatol 2:19–33

Kaufman KR, An K-N, Litchy WJ, Cooney WP, Chao EY (1999) In-vivo function of the thumb muscles. Clin Biomech 14:141–150

Kerkhof F, Brugman E, D'Agostino P, Stockmans F, Jonkers I, Vereecke EE (2013) Determining carpal and metacarpal bone kinematics using dynamic CT: opportunities and challenges. Congress of the International Society of Biomechanics, Natal, Brazil, 4–9 August 2013 (abstract)

Kimura T, Okada M, Ishida H (1979) Kinesiological characteristics of primate walking: its significance in human walking. In: Morbeck ME, Preuschoft H, Gomberg N (eds) Environment, behavior and morphology: dynamic interactions in primates. Gustav Fischer, New York, pp 297–311

Kivell TL, Schmitt D, Wunderlich RE (2010) Hand and foot pressures in the aye-aye (*Daubentonia madagascariensis*) reveal novel biomechanical trade-offs required for walking on gracile digits. J Exp Biol 213:1549–1557

Klein Breteler MD, Simura KJ, Flanders M (2007) Timing of muscle activation in a hand movement sequence. Cereb Cortex 17:803–815

Lammers AR (2007) Locomotor kinetics on sloped arboreal and terrestrial substrates in a small quadrupedal mammal. Zoology 110:93–103

Lammers AR, Gauntner T (2008) Mechanics of torque generation during quadrupedal arboreal locomotion. J Biomech 41:2388–2395

Lemon RN (1999) Neural control of dexterity: what has been achieved? Exp Brain Res 128:6–12

Macfarlane NBW, Graziano MS (2009) Diversity of grip in *Macaca mulatta*. Exp Brain Res 197:255–268

Marzke MW (1997) Precision grips, hand morphology, and tools. Am J Phys Anthropol 102:91–110

Marzke MW, Marzke RF (1987) The third metacarpal styloid process in humans: origin and functions. Am J Phys Anthropol 73:415–431

Marzke MW, Marzke RF (2000) Evolution of the human hand: approaches to acquiring, analysing and interpreting the anatomical evidence. J Anat 197:121–140

Marzke MW, Shackley MS (1986) Hominid hand use in the Pliocene and Pleistocene: evidence from experimental archeology and comparative morphology. J Hum Evol 15:439–460

Marzke MW, Wullstein K (1996) Chimpanzee and human grips: a new classification with a focus on evolutionary morphology. Int J Primatol 17:117–139

Marzke MW, Wullstein K, Viegas S (1992) Evolution of the power (squeeze) grip and its morphological correlates in hominids. Am J Phys Anthropol 89:283–298

Marzke MW, Toth N, Schick K, Reece S, Steinberg B, Hunt K, Linscheid R, An K-N (1998) EMG study of hand muscle recruitment during hard hammer percussion manufacture of Oldowan tools. Am J Phys Anthropol 105:315–332

Matarazzo S (2013) Manual pressure distribution patterns of knuckle-walking apes. Am J Phys Anthropol 152:44–50

Metcalf CD, Robinson R, Malpass A, Bogle T, Dell T, Harris C, Demain S (2013) Markerless motion capture and measurement of hand kinematics: validation and application to home-based upper limb rehabilitation. IEEE Trans Biomed Eng 60:2184–2192

Michilsens F, D'Août K, Vereecke EE, Aerts P (2012) One step beyond: different step-to-step transitions exist during continuous contact brachiation in siamangs. Biol Open 1:411–421

Moojen T, Snel J, Ritt MJP, Kauer JM, Venema H, Bos K (2002) Three-dimensional carpal kinematics in vivo. Clin Biomech 17:506–514

Moore DC, Crisco JJ, Trafton TG, Leventhal EL (2007) A digital database of wrist bone anatomy and carpal kinematics. J Biomech 40:2537–2542

Napier JR (1955) Form and function of the carpometacarpal joint of the thumb. J Anat 89:362–369

Napier JR (1956) The prehensile movements of the human hand. J Bone Joint Surg [Br] 38B:902–913

Napier JR (1960) Studies of the hand of living primates. Proc Zool Soc Lond 134:647–657

Napier JR (1961) Prehensility and opposability in the hands of primates. Symp Zool Soc Lond 5:115–132

Napier J (1980) Hands. Panthenon Books, New York

Nicol K, Hennig EM (1976) Time dependent method for measuring force distribution using a flexible mat as a capacitor. In: Komi P (ed) Biomechanics V-B. University Park Press, Baltimore, pp 433–440

O'Connor BL, Rarey KE (1979) Normal amplitudes of radioulnar pronation and supination in several genera of anthropoid primates. Am J Phys Anthropol 51:39–43

Orr CM, Leventhal EL, Chivers SF, Marzke MW, Wolfe SW, Crisco JJ (2010) Studying primate carpal kinematics in three dimensions using a computed-tomography-based markerless registration method. Anat Rec 293:692–709

Patel BA (2009) Not so fast: speed effects on forelimb kinematics in cercopithecine monkeys and implications for digitigrade postures in primates. Am J Phys Anthropol 140:92–112

Patel BA (2010) The interplay between speed, kinetics, and hand postures during primate terrestrial locomotion. Am J Phys Anthropol 141:222–234

Patel BA, Polk J (2010) Distal forelimb kinematics in *Erythrocebus patas* and *Papio anubis* during walking and galloping. Int J Primatol 31:191–207

Patel BA, Wunderlich RE (2010) Dynamic pressure patterns in the hands of olive baboons (*Papio anubis*) during terrestrial locomotion: implications for cercopithecoid primate hand morphology. Anat Rec 293:710–718

Patel BA, Larson SG, Stern JT Jr (2012) Electromyography of wrist and finger flexor muscles in olive baboons (*Papio anubis*). J Exp Biol 215:115–123

Pizzimenti MA, Darling WG, Rotella DL, McNeal DW, Herrick JL, Ge J, Stilwell-Morecraft KS, Morecraft RJ (2007) Measurement of reaching kinematics and prehensile dexterity in nonhuman primates. J Neurophys 98:1015–1029

Pontzer H, Raichlen DA, Rodman PS (2014) Bipedal and quadrupedal locomotion in chimpanzees. J Hum Evol 66:64–82

Pouydebat E, Coppens Y, Gorce P (2006a) Évolution de la préhension chez les primates humains et non humains: la précision et l'utilisation d'outils revisitées. Anthropologie 110:687–697

Pouydebat E, Berge C, Gorce P, Coppens Y (2006b) La préhension chez les Primates : précision, outils et perspectives évolutives. C R Palevol 5:597–602

Pouydebat E, Gorce P, Coppens Y, Bels V (2009) Biomechanical study of grasping according to the volume of the object: human versus non-human primates. J Biomech 42:266–272

Rainbow MJ, Crisco JJ, Moore DC, Wolfe SW (2008) Gender differences in capitate kinematics are eliminated after accounting for variation in carpal size. J Biomech Eng 130:041003

Rainbow MJ, Kamal RN, Leventhal E, Akelman E, Moore DC, Wolfe SW, Crisco JJ (2013) In vivo kinematics of the scaphoid, lunate, capitate, and third metacarpal in extreme wrist flexion and extension. J Hand Surg [Am] 38:278–288

Reghem E, Chèze L, Coppens Y, Pouydebat E (2013) Unconstrained 3D-kinematics of prehension in five primates: lemur, capuchin, gorilla, chimpanzee, human. J Hum Evol 65:303–312

Reynolds TR (1985) Stresses on the limbs of quadrupedal primates. Am J Phys Anthropol 67:351–362

Rolian C, Lieberman DE, Zermeno JP (2011) Hand biomechanics during simulated stone tool use. J Hum Evol 61:26–41

Rose MD (1977) Positional behavior of olive baboons (*Papio anubis*) and its relationship to maintenance and social activities. Primates 18:59–116

Rudroff T (2008) Kinesiological fine wire EMG. A practical introduction to fine wire EMG applications. Version 1.0

Schick KD, Toth N, Garufi G, Savage-Rumbaugh ES, Rumbaugh D, Sevcik R (1999) Continuing investigations into the stone tool-making and tool-using capabilities of a bonobo (*Pan paniscus*). J Archaeol Sci 26:821–832

Schilling A-M, Tofanelli S, Hublin J-J, Kivell TL (2014) Trabecular bone structure in the primate wrist. J Morphol 275:572–585

Schmidt M, Fischer MS (2000) Cineradiographic study of forelimb movements during quadrupedal walking in the brown lemur (*Eulemur fulvus*, Primates: Lemuridae). Am J Phys Anthropol 262:245–262

Schmitt D (1994) Forelimb mechanics as a function of substrate type during quadrupedalism in two anthropoid primates. J Hum Evol 26:441–457

Schmitt D (1998) Forelimb mechanics during arboreal and terrestrial quadrupedalism in Old World monkeys. In: Strasser E, Fleagle J, Rosenberger AL, McHenry H (eds) Primate locomotion: recent advances. Springer, New York, pp 175–200

Schmitt D (1999) Compliant walking in primates. J Zool 248:149–160
Schmitt D (2003a) Substrate size and primate forelimb mechanics: implications for understanding the evolution of primate locomotion. Int J Primatol 24:1023–1036
Schmitt D (2003b) Mediolateral reaction forces and forelimb anatomy in quadrupedal primates: implications for interpreting locomotor behavior in fossil primates. J Hum Evol 44:47–58
Schmitt D, Lemelin P (2002) Origins of primate locomotion: gait mechanics of the woolly opossum. Am J Phys Anthropol 118:231–238
Schmitt D, Hanna JB (2004) Substrate alters forelimb to hindlimb peak force ratios in primates. J Hum Evol 46:237–252
Schwartz P, Heath A (1937) Some factors which influence the balance of the foot in walking. J Bone Joint Surg [Am] 19:431–442
Schwartz C, Lempereur M, Burdin V, Jacq J-J, Remy-Neris O (2007) Shoulder motion analysis using simultaneous skin shape registration. Conf Proc IEEE Eng Med Biol Soc. pp 533–536
Sellers WI, Hirasaki E (2014) Markerless 3D motion capture for animal locomotion studies. Biol Open 3:656–668
Solomonow M, Baratta R, Bernardi M, Zhou B, Lu Y, Zhu M, Acierno S (1994) Surface and wire EMG crosstalk in neighbouring muscles. J Electromyogr Kinesiol 4:131–142
Spinozzi G, Truppa V, Laganà T (2004) Grasping behavior in tufted capuchin monkeys (*Cebus apella*): grip types and manual laterality for picking up a small food item. Am J Phys Anthropol 125:30–41
Spinozzi G, Lagana T, Truppa V (2007) Hand use by tufted capuchins (*Cebus apella*) to extract a small food items from a tube: digit movements, hand preference, and performance. Am J Primatol 69:336–352
Stern JT Jr, Wells JP, Jungers WL, Vangor AK, Fleagle JG (1980) An electromyographic study of the pectoralis major in atelines and *Hylobates*, with special reference to the evolution of a pars clavicularis. Am J Phys Anthropol 52:13–25
Stevens NJ, Schmitt D, Cole TM, Chan L (2006) Technical note: out-of-plane angular correction based on a trigonometric function for use in two-dimensional kinematic studies. Am J Phys Anthropol 402:399–402
Susman RL (1988) Hand of *Paranthropus robustus* from Member 1, Swartkrans: fossil evidence for tool behavior. Science 240:781–784
Susman RL (1989) New hominid fossils from the Swartkrans formation (1979–1986 excavations): postcranial specimens. Am J Phys Anthropol 79:451–474
Susman RL (1998) Hand function and tool behavior in early hominids. J Hum Evol 35:23–46
Susman RL, Stern JT Jr (1979) Telemetered electromyography of flexor digitorum profundus and flexor digitorum superficialis in *Pan troglodytes* and implications for interpretation of the O.H. 7 hand. Am J Phys Anthropol 50:565–574
Susman RL, Stern JT Jr (1980) EMG of the interosseus and limbrical muscles in the chimpanzee (*Pan troglodytes*) hand during locomotion. Am J Anat 157:389–397
Susman RL, Tuttle RH (1976) Knuckling behavior in captive orangutans and a wounded baboon. Am J Phys Anthropol 45:123–124
Susman RL, Jungers WL, Stern JT Jr (1982) The functional morphology of the accessory interosseous muscle in the gibbon hand: determination of locomotor and manipulatory compromises. J Anat 134:111–120
Tocheri MW, Marzke MW, Liu D, Bae M, Jones GP, Williams RC, Razdan A (2003) Functional capabilities of modern and fossil hominid hands: three-dimensional analysis of trapezia. Am J Phys Anthropol 122:101–112
Toth N, Schick K (1993) Early stone industries and inferences regarding language and cognition. In: Gibson K, Ingold T (eds) Tools, language and cognition in human evolution. Cambridge University Press, Cambridge, pp 346–362
Toth N, Schick KD, Savage-Rumbaugh ES, Sevcik RA, Rumbaugh DM (1993) *Pan* the toolmaker: investigations into the stone tool-making and tool-using capabilities of a bonobo (*Pan paniscus*). J Archaeol Sci 20:81–91

Tuttle RH (1969) Knuckle-walking and problem of human origins. Science 166:953–961
Tuttle RH, Basmajian JV (1974a) Electromyography of brachial muscles in *Pan gorilla* and hominoid evolution. Am J Phys Anthropol 41:71–90
Tuttle RH, Basmajian JV (1974b) Electromyography of forearm musculature in *Gorilla* and problems related to knuckle-walking. In: Jenkins FA Jr (ed) Primate Locomotion. Academic Press, New York, pp 293–348
Tuttle RH, Basmajian JV (1978a) Electromyography of pongid shoulder muscles. II. Deltoid, rhomboid and "rotator cuff.". Am J Phys Anthropol 49:47–56
Tuttle RH, Basmajian JV (1978b) Electromyography of pongid shoulder muscles. III. Quadrupedal positional behavior. Am J Phys Anthropol 49:57–69
Tuttle RH, Basmajian JV, Regenos E, Shine G (1972) Electromyography of knuckle-walking: results of four experiments on the forearm of *Pan gorilla*. Am J Phys Anthropol 37:255–265
Tuttle RH, Velte MJ, Basmajian JV (1983) Electromyography of brachial muscles in *Pan troglodytes* and *Pongo pygmaeus*. Am J Phys Anthropol 61:75–83
Tuttle RH, Hollowed JR, Basmajian JV (1992) Electromyography of pronators and supinators in great apes. Am J Phys Anthropol 87:215–226
van der Smagt P, Stillfried G (2008) Using MRI data to compute a hand kinematic model. 9th Conf. Motion Vib. pp 1–10
Vereecke EE, D'Août K, De Clercq D, Van Elsacker L, Aerts P (2003) Dynamic plantar pressure distribution during terrestrial locomotion of bonobos (*Pan paniscus*). Am J Phys Anthropol 120:373–383
Vereecke EE, D'Août K, Van Elsacker L, De Clercq D, Aerts P (2005) Functional analysis of the gibbon foot during terrestrial bipedal walking: plantar pressure distributions and three-dimensional ground reaction forces. Am J Phys Anthropol 128:659–669
Vigouroux L, Quaine F, Labarre-Vila A, Amarantini D, Moutet F (2007) Using EMG data to constrain optimization procedure improves finger tendon tension estimations during static fingertip force production. J Biomech 40:2846–2856
Williams EM, Richmond BG (2012) Manual pressure distribution during stone tool use. Am J Phys Anthropol 147(Suppl 54):303 (abstract)
Williams EM, Gordon AD, Richmond BG (2010) Upper limb kinematics and the role of the wrist during stone tool production. Am J Phys Anthropol 143:134–145
Williams EM, Gordon AD, Richmond BG (2012) Hand pressure distribution during Oldowan stone tool production. J Hum Evol 62:520–532
Wolfe SW, Neu C, Crisco JJ (2000) In vivo scaphoid, lunate, and capitate kinematics in flexion and in extension. J Hand Surg [Am] 25:860–869
Wunderlich RE (1999) Pedal form and plantar pressure distribution in anthropoid primates. Ph.D. dissertation, State University of New York at Stony Brook
Wunderlich RE, Patel B (2008) Peak pressures and cheiridial postures in baboons (*Papio anubis*). Society of Integrative and Comparative Biology, e245 (abstract)
Wunderlich RE, Jungers WL (2009) Manual digital pressures during knuckle-walking in chimpanzees (*Pan troglodytes*). Am J Phys Anthropol 139:394–403
Yousef H, Boukallel M, Althoefer K (2011) Tactile sensing for dexterous in-hand manipulation in robotics: a review. Sensors Actuat A Phys 167:171–187

Chapter 11
Biomechanics of the Human Hand: From Stone Tools to Computer Keyboards

Erin Marie Williams-Hatala

1 Introduction

Evolution is a contextually driven process, meaning that the form of a feature in its present state is the synthesis of the ancestral form and function, the present function and biological role, and the ecology of the organism (Bock and von Wahlert 1965). Free of this context, features cease to convey their significance to the life of an organism, and their meaning is lost to those of us studying them. With this in mind, this chapter reviews the information we have gleaned from experimental biomechanics with respect to the evolutionary history of the modern human hand and how our relationship with technology may have impacted the form of the human hand and wrist.

Napier (1965) stressed that the significance of the modern human hand is not as much a matter of structural uniqueness, given that in many ways it remains primitive in form, but rather a matter of the novel and continually evolving functions that the hand is currently, and has been in the past, capable of performing. There is certainly some truth to this view, and Napier was not alone in regarding the human hand as primitive. Wood Jones stated that the human hand is "...the absolute bed-rock of mammalian primitiveness," (Wood Jones 1920: 17). Indeed, in many respects the human hand does retain the primitive tetrapod *Bauplan* that first appeared in the early Devonian (Napier 1993; Laurin et al. 2000; Hinchliffe 2002).

However, within the broad spectrum of overall morphological conservation in the modern human hand, there also exist numerous synapomorphic and autapomorphic features. These have evolved in response to the biological role—and the associated habitual stresses and strains—of the hand in the lives of our ancestors throughout our evolutionary history (Marzke 1997; Marzke and Marzke 2000;

E.M. Williams-Hatala (✉)
Department of Biology, Chatham University, Pittsburgh, PA 15232, USA
e-mail: ewilliamshatala@chatham.edu

T.L. Kivell et al. (eds.), *The Evolution of the Primate Hand*, Developments in Primatology: Progress and Prospects, DOI 10.1007/978-1-4939-3646-5_11

Tocheri 2007). So while it is true that the modern human hand retains an overall primitive form in many respects, it has also undergone functionally significant, though perhaps structurally small, changes that, in combination with associated neural and cognitive developments (see Chap. 6), enable us to perform a myriad of tasks that would fall out of the reach of our ancestors and closest living relatives. Consequently, the human hand could no sooner meet the locomotor demands of any of the early pentadactyl tetrapods than their forelimbs could knap a stone tool or type on a computer keyboard.

The ways in which the human hand has changed over our evolutionary history are thought to reflect our lineage's relationship with enhanced manipulative skills, especially involving Paleolithic stone tools and the biomechanics related to their manufacture and use. The rationale behind this hypothesis can be understood by considering the advantages our ancestors gained by making and using these simple, early tools. The behaviors facilitated by the use of stone tools gave our early hominin ancestors, with their relatively small, gracile bodies and reduced canine teeth, considerable advantages that would not typically be attained by a mammal with their overall anatomical form, such as increased access to, and more efficient processing of, high-quality food items (Schick and Toth 1993; Plummer 2004; Braun et al. 2010; McPherron et al. 2010). These advantages contributed to a series of adaptive changes that led to the emergence of our own species (Aiello and Wheeler 1995; Navarrete et al. 2011).

Stone tool use and manufacture required an array of manipulative motions and induced joint and muscle stresses that were different from those experienced during forms of quadrupedal and suspensory locomotion—the activities that dominated the upper limb biomechanical demands of our more distant ancestors (Marzke and Shackley 1986; Preuschoft and Chivers 1993; Hartwig and Doneski 1998; Faisal et al. 2010; Williams et al. 2010; Rolian et al. 2011). Hands adept at climbing, clinging, and/or knuckle-walking would have been ill suited for complex stone tool manufacture and use. Consequently, researchers hypothesize that our ancestors evolved adaptations in their hands and wrists that enabled them to commit to the production and manipulation of stone tools.

At its simplest, the production of early Paleolithic stone tools involves removing a series of stone flakes from a larger stone nodule (called the core), either by striking the nodule with another stone or by throwing the nodule against a hard surface. Archaeological evidence indicates that direct hard-hammer percussion was the most commonly relied upon method (Schick and Toth 1993; Toth and Schick 2009). This method involves striking a small rounded stone held in the striking (i.e., dominant) hand against the larger core in such a way that small, sharp flakes of stone are removed from the core (Fig. 11.1). These flakes, and the nodule from which they were removed, can then be used for a variety of purposes, including the quick removal of meat from animal bones (Bunn 1981). Archaeological evidence suggests that stone tools were also used to crack open nuts, to break open bones to access the marrow inside, and to help process vegetable matter for consumption (Blumenschine and Pobiner 2007; Haslam et al. 2009).

Historically, the functional significance of many of these proposed adaptations has been difficult to isolate and assess. However, advances in technology are

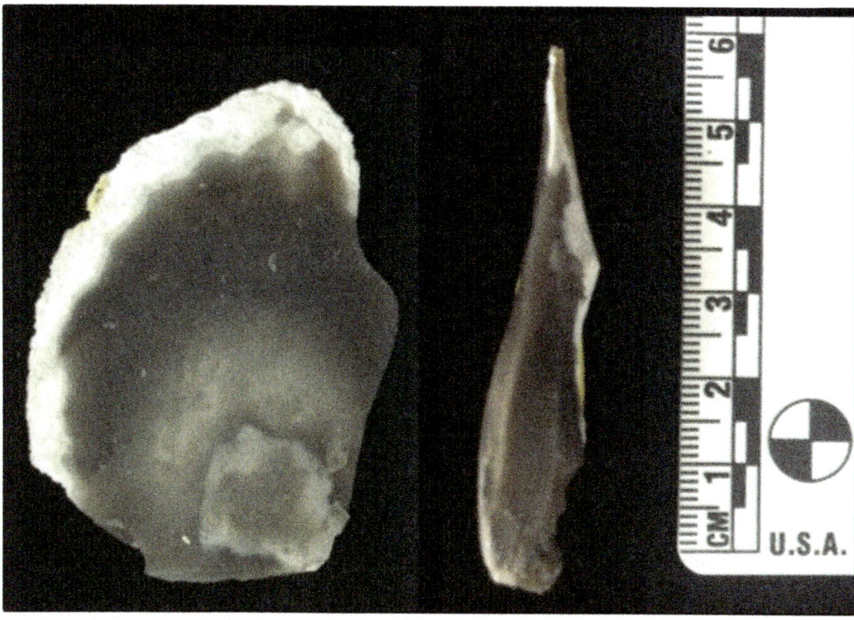

Fig. 11.1 Oldowan-style flakes produced during hard-hammer percussion

enabling researchers to quantify the biomechanics of stone tool behaviors, thereby allowing them to quantitatively test hypotheses relating modern human hand and upper limb anatomy to stone tool behaviors (i.e., use and manufacture). Systems such as high-speed motion capture devices (now even freed from the use of reflective markers and cables), in vivo computer-based imaging and modeling, and real-time pressure sensing systems generate high-resolution biomechanical and functional data that document the rapid motions associated with stone tool behaviors (Fig. 11.2). These types of data are necessary for the evaluation of hypotheses linking anatomical form to biomechanical function.

This chapter explores the relationship between those stone tool behaviors practiced by our hominin ancestors and the evolution of the modern human hand. Why do anthropologists think the two are linked? Which morphological features of the human hand may be related to stone tool behaviors? And how do we test hypotheses linking the two? To address these issues, the chapter begins by reviewing the grips performed by primates, both human and nonhuman, and discusses some of the differences in gripping behaviors among the species and the reasons behind these differences. Next, methods for evaluating whether a feature can be regarded as an adaptation are discussed, as well as the ways to determine how a feature is adapted to a specific behavior. The chapter continues with descriptions of the experiments that have attempted to evaluate the adaptive quality of some of the features of the upper limb as they pertain to stone tool behaviors. Finally, the chapter ends by looking at some of the questions that remain unanswered and how we may go about answering them.

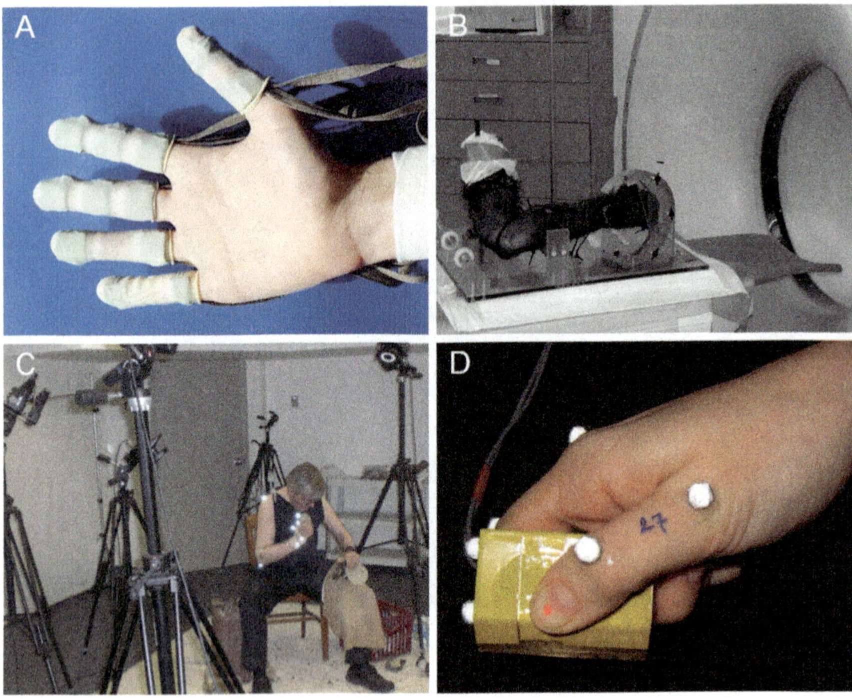

Fig. 11.2 Laboratory-based systems used to produce and capture data on the form, function, and movement patterns of upper limb anatomy. (**a**) High-speed manual pressure sensors used to capture forces and pressures acting across the hand (photo credit: E.M. Williams-Hatala); (**b**) CT scanner and polycarbonate wrist-positioning jig used to rotate and scan the region of interest (e.g., the wrist) in order to generate a visual record of the orientation of the joint through full rotation (photo credit: C. Orr); (**c**) infrared Vicon™ motion capture cameras and subject wearing reflective markers while making a stone tool, both used to record high-speed motions in X, Y, and Z coordinates (photo credit: E.M. Williams-Hatala); and (**d**) the hand fitted with reflective markers to capture motion data and holding a replica of an Oldowan flake instrumented with load cells to capture force data (photo credit: C. Rolian)

2 Grasping and Grips

Despite the complexity of the motions involved in stone tool behaviors and the technology used to document them, it appears that the vast majority of grips used during stone tool behaviors, as well as many of the grips used during contemporary manual behaviors, are variants on the two prehensile grip types originally proposed by Napier in the 1950s: the power and the precision grips (Napier 1956).

Napier observed that hands perform two types of movement—nonprehensile and prehensile. The former involves non-grasping behaviors in which objects are merely pushed or tapped. These types of movement are relatively simple and are by no means limited to humans or even to primates. Nor do they require any degree of manual dexterity; dolphins are just as capable of nonprehensile object manipulation as are humans. Prehensile movements, on the other hand, are defined as manipulative behaviors in which an object is held between the manual digits and the palm and require a degree of manual dexterity (Napier 1956, 1993; see Chap. 12).

In general, anatomical traits related to prehensile grasping behaviors may have originally evolved as an adaptation to facilitate terminal branch feeding (Rasmussen 1990; Sussman 1991; Bloch and Boyer 2002) and "hand-to-mouth" feeding techniques (Napier 1993). For example, Eocene euprimates show adaptations toward grasping behaviors in both their hands [e.g., short metacarpals relative to long proximal phalanges (Hamrick 2001) and an abducted hallux and pollex (Cartmill 1992; Dagosto 1988)] and feet [a saddle-shaped joint between the hallucal metatarsal and entocuneiform (Bloch and Boyer 2002)] (see also Chap. 14). From this beginning, extant primates have evolved extremities capable of performing a wide variety of grasping behaviors.

Humans differ from early primates and our extant relatives in that our hands do not have the dual role of acting as both a "hand and foot" (Napier 1993), which has enabled us to become uniquely adapted for one-handed forceful grasping and manipulative behaviors. Our capacity to do so hinges on our highly derived, long, and robust thumb and our ability to rotate the thumb around into a position of full opposition to the fingers. This ability has opened up the full repertoire of modern human grips, which are largely encompassed within two broad categories: power and precision grips.

2.1 Power and Precision Grips

Napier (1956) defined power grips as those in which an object is secured between the flexed fingers and the palm of the hand. The thumb acts as a buttress, further securing the object in the hand, though it is not necessarily the primary stabilizing component of the grip (Napier 1962, 1993). This is not true of precision grips, however, in which the thumb plays the main supportive role. Napier defined precision grips as those grips during which an object is pinched between the palmar aspects of the fingers and the pollical distal phalanx (Napier 1956, 1993).

Marzke (1997) expanded on Napier's precision grip definition because she did not feel that it adequately covered the range of precision behaviors used when interacting with stone tools (and by extension, during non-stone tool behaviors, as well). First, she specified the role of the palm, stating that although it may be involved in the performance of a precision grip, it is not a requirement. Second, Marzke subdivided the grips into precision finger pinch grips, precision handling movements (in which objects are maneuvered within a single hand), and precision finger/passive palm pinch grips (which involve greater contact between the object and the palmar aspects of various digits or the palm itself) [see Table 2 in Marzke (1997)]. Third, Marzke added an additional type of grip or behavior: the ability to forcefully cup the hand around objects, thereby allowing one-handed object manipulation and control (Marzke 1997, 2009).

Humans are by no means the only species capable of precision grips (in addition to power and cupping grips). A variety of Old and New World primates use such grips (and tools, too, at times) for subsistence and grooming purposes (see Chap. 12 for a comprehensive review). For example, captive and wild chimpanzees have been documented using the cup grip, as well as various power and precision grips to accomplish food-related tasks and to manipulate tools of various shapes and sizes (McGrew 1974; Jones-Engel and Bard 1996; Marzke and Wullstein 1996; Pouydebat et al. 2009). Capuchin monkeys, macaques, and orangutans also use power grips and a wide variety of precision grips during feeding and grooming behaviors (Costello and Fragaszy 1988; Westergaard and Suomi 1997; Pouydebat et al. 2009).

What makes humans unique in terms of our gripping ability is that we are the only species known to be capable of applying large forces with a single hand when using precision grips (Marzke 1997, 2009; Marzke et al. 1992). Our ability to do so hinges on our derived thumb anatomy and the motions we can perform because of that anatomy. Some of the most important features include a higher thumb-to-finger length ratio to facilitate full palmar contact between the digits, well-developed intrinsic and extrinsic pollical muscles to help increase grip strength [in particular the flexor pollicis longus muscle (Diogo et al. 2012)], and a large range of motion at the first carpometacarpal and metacarpophalangeal joints (Kuczynski 1974; Napier 1962; Marzke 1992, 1997; Marzke et al. 1998; Tocheri et al. 2008). Features outside of the thumb also facilitate the forceful application of precision grips and object control, including mobile digits, broad ungual tufts with ungual spines on the distal phalanges, a radial orientation of the third metacarpal head, and asymmetry of the second and fifth metacarpal heads (Susman 1979, 1994; Shrewsbury and Johnson 1983; Marzke 1997; Marzke et al. 1998; Mittra et al. 2007; Tocheri et al. 2008).

With a shorter or weaker thumb, we may be forced to grasp objects between flexed fingers and the palm or between the fingers (i.e., scissor grip), thereby limiting greatly our manual behavioral repertoire. The utility of our derived thumb is evident in experiments with modern experienced stone tool knappers; they naturally tend to use variants of grips that heavily utilize the thumb and that fall under the "precision" umbrella: the pad-to-side or "key" grip (thumb pad to the side of the index finger), the three-jaw chuck grip, and the cradle precision pinch grip between the pads of the thumb and all of the fingers (Marzke and Shackley 1986; Williams et al. 2012). Traditionally, the evolution of a humanlike hand has been linked to stone tool behaviors because of the loose chronological synchronicity between archaeological evidence of stone tool behaviors (Leakey 1971; Clark et al. 1994; Semaw et al. 1997; de Heinzelin et al. 1999; Roche et al. 1999) and the appearance of features such as a higher thumb-to-finger length ratio, greater thumb robusticity, and broader apical tufts in fossil hominins (Fig. 11.3) (Marzke 1983, 1997; Ricklan 1987; Susman 1988, 1994; Tocheri et al. 2003; see also Chap. 18).

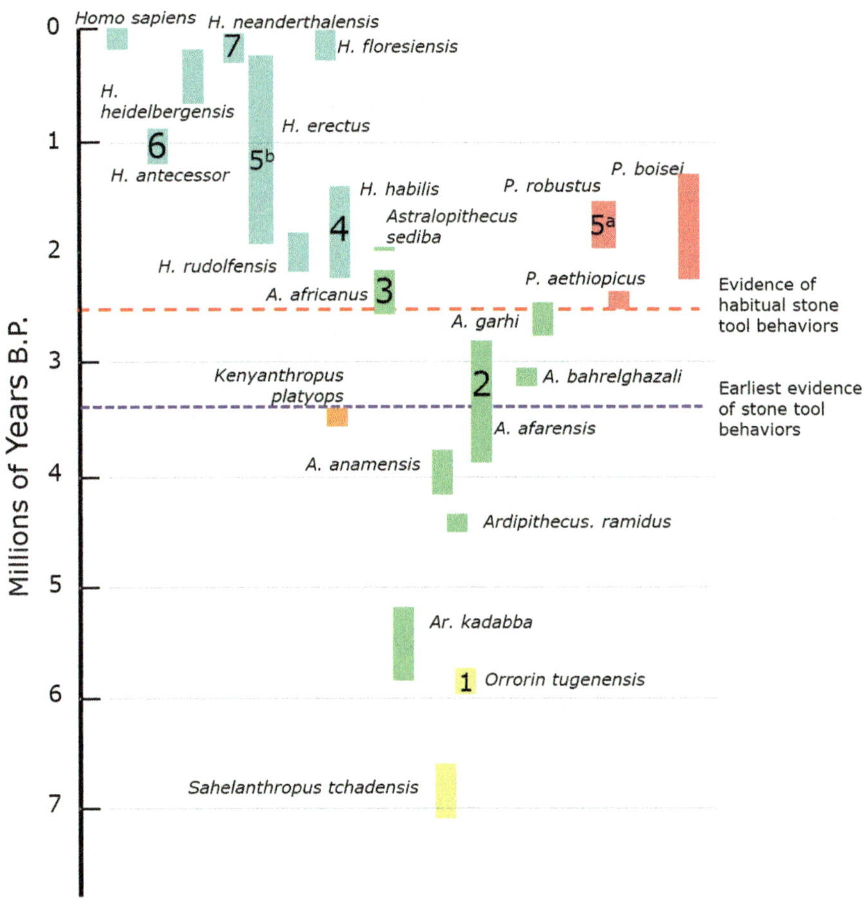

Fig. 11.3 Human fossil family tree showing the temporal relationship between major events in stone tool behaviors and the first appearance of derived features in the hand and wrist. (1) *Orrorin tugenensis*, the distal phalanx of the thumb has insertion site for the flexor pollicis longus and a broad apical tuft; (2) *Australopithecus afarensis*, possibly modern humanlike hand proportions and asymmetric metacarpal (Mc) heads on Mc2, Mc4, and Mc5 heads; (3) *Australopithecus africanus*, no radial extension on the distal radius; (4) *Homo habilis*, flat articular surface between the trapezium and Mc1; (5a) *Paranthropus robustus* (also possibly early *Homo* as specimens from Swartkrans are of uncertain species attribution), robust Mc1 relative to thumb length and relative to length of Mc2–5; (5b) *Homo erectus*, Mc3 styloid process present; (6) *Homo antecessor*, derived palmarly expanded trapezoid shape, modern humanlike capitate-Mc2 and capitate-trapezoid articulations, minimally curved proximal phalanges, and weakly developed flexor sheaths; and (7) *Homo neanderthalensis*, radial orientation of the Mc3 head. *P. robustus* and *H. erectus* are listed as 5a and 5b due to uncertainty of the taxonomic affiliation of some of the relevant fossils

3 When Is a Morphological Feature an Adaptation?

Archaeological and fossil evidence do not by any means confirm the notion that the derived traits of the human hand are adaptations for stone tool behaviors. For one thing, this evidence is incomplete, making it challenging if not impossible to draw a direct causal relationship between tools and morphology based on the apparent synchronicity of the two records. Furthermore, the functional significance of many of the derived traits of the human hand as they relate to specific stone tool behaviors has yet to be thoroughly investigated; it is possible that the hand and wrist are behaving in ways not previously considered (Williams et al. 2012; Williams and Richmond 2012) and that muscles are interacting with bones at unexpected locations or in unexpected ways (Marzke et al. 2007). Finally, hand features frequently associated with stone tool behaviors may be in fact by-products of pleiotropy or morphological integration and may result from selection acting on other parts of the body, such as the feet (Rolian 2009; Rolian et al. 2010). In such cases, derived features of the human hand may have been exapted, rather than adapted, for stone tool behaviors (Alba et al. 2003; Almécija and Alba 2014). Exaptations are "features that now enhance fitness but were not built by natural selection for their current role" (Gould and Vrba 1982: 4).

Manual digit proportions provide a good illustration of the difficulty of evaluating whether a feature may have been adapted specifically to stone tool behaviors or whether it arose for some other reason. The presence of a long thumb relative to finger length is regarded as an indicator of precision gripping abilities in fossil hominins (Napier 1960, 1962; Marzke 1997) and stone tool behaviors when multiple fossil hominins and artifacts are found at the same site (Susman 1988, 1994, 1998). However, Alba and colleagues (2003) suggest that thumb-to-finger length proportions in *Australopithecus afarensis* by 3.4 Ma may have been similar to those of modern humans [see also Almécija and Alba (2014) and Rolian and Gordon (2013, 2014) for different perspectives on this topic]. The earliest evidence of stone tool behaviors (potentially cut-marked bones from Dikika, Ethiopia) also dates to ~3.4 Ma (McPherron et al. 2010 [but see Dominguez-Rodrigo et al. 2010, 2011, 2012; McPherron et al. 2011 for different viewpoints]), and the earliest evidence for stone tool production dates to ~3.3 Ma (Harmand et al. 2015). However, the isolated nature of these tools suggests that, though hominins may have engaged in tool production, they seem to have done so sporadically. Stone tools are found a multiple sites starting after 2.6 Ma (Semaw 2000; Semaw et al. 2003), at least 0.8 Ma after the last known appearance of *Au. afarensis*. This chronological discontinuity between the first occurrence stone tool *use* and habitual stone *toolmaking* challenges the notion that modern human thumb length-to-finger length proportions were an adaptation for the latter or even stone tool use (Alba et al. 2003). Further complicating the issue, Rolian and Gordon (2013, 2014) recently reanalyzed the fossil evidence and concluded that digit proportions in *Au. afarensis* were intermediate between gorilla and modern humans, which would have diminished their ability to perform precision grips. That being said, there is a clear biomechanical benefit to having a long thumb when using a precision grip during stone tool manufacture (Rolian et al. 2011; see below), indicating that our hand proportions may have been exapted for stone tool behaviors (specifically tool manufacture and, by logical extension, stone tool use).

The presence of a true flexor pollicis longus (FPL) muscle is another important example of a feature whose designation as an adaptation toward stone tool behaviors is no longer as certain as was once thought (Susman 1988, 1994). The FPL is regarded as an important feature for stone tool behaviors because it helps to stabilize the thumb joints against high extension moments and because it helps produce high flexion forces, facilitating a strong grip on the tool in hand. In most nonhuman primates, the tendon to the pollical distal phalanx and other deep digital tendons share a common muscle belly (flexor digitorum profundus or FDP). In contrast, the pollical tendon of modern humans arises from a separate muscle belly (FPL) that is independent from the adjacent FDP (see Chap. 7).

Electromyography (EMG) data demonstrate that the FPL is heavily recruited during tool use behaviors (Hamrick et al. 1998), but its degree of recruitment during tool production appears to vary depending on skill level of the knapper (Marzke et al. 1998). Additionally, the amount of force acting on the pollical distal phalanx varies significantly across stone tool use behaviors (e.g., nutcracking versus slicing with a flake; Williams and Richmond 2012), suggesting that not all tool use behaviors would have been selectively significant. Furthermore, it is difficult to determine when a distinct FPL muscle first appears in the hominin lineage. A palmar depression on the pollical distal phalanx was once regarded as a good indication of the derived muscular condition and therefore of the potential for stone tool behaviors (e.g., Napier 1962; Ricklan 1987; Susman 1989, 2005; Marzke 1997; Moyá-Solá et al. 1999). However, comparative anatomical dissections of this morphology have shown that it is a palmar gabled ridge just distal to the depression, and not the depression itself, that is the actual attachment for the FPL (Shrewsbury et al. 2003; Marzke and Shrewsbury 2006; see Almécija et al. 2010 for a good diagram for this morphology). Furthermore, a gabled FPL attachment is also found in other primates without a separate FPL (e.g., chimpanzees, orangutans, lemurs), and the FPL often attaches in a complex manner such that the skeletal morphology does not necessarily reflect the size of the muscle (Shrewsbury et al. 2003). There is also evidence of a palmar ridge on the pollical distal phalanx of numerous hominin species that predate tool behaviors by millions of years, including *Orrorin tugenensis* (Almécija et al. 2010) and *Ardipithecus ramidus* (Almécija et al. 2010; Lovejoy et al. 2009; but see Ward et al. 2012). Thus, it is not clear whether an FPL muscle is an adaptation to stone tool production and stone tool use or whether it predates both behaviors, making it an exaptation for stone tool behaviors in general.

In light of the difficulties described above, Kay and Cartmill (1977) proposed criteria for differentiating between exaptations and adaptations that take into account the vagaries of the fossil and archaeological records:

1. Demonstrate that the feature has the same adaptive role in all living taxa that possess it.
2. Experimentally establish a functional link between the feature and behavior in question.
3. Show that the feature does not predate the behavior (e.g., by tracing character evolution on a phylogeny; see Tocheri et al. 2008).
4. Demonstrate independent evolution (in parallel or by convergence) of the feature in organisms that also practice the behavior.

Despite the intended application of these criteria to the fossil record, the nature of many of the derived features of the human hand makes it difficult to fulfill these criteria. For example, following these criteria, the development of the FPL may be an exaptation, rather than an adaptation, for stone tool manufacture specifically, given its appearance in taxa long before stone tools are recognized in the archaeological record. However, it is possible that early taxa like *Orrorin* engaged in tool use behaviors of which we have no fossil evidence, but for which the enhancement of the FPL could have evolved as an adaptation. Or, in another example, how does one demonstrate definitively that asymmetry of the joint surfaces of the second and fifth metacarpal heads was an adaptation specifically toward stone tool production? While it is true that this feature facilitates such behaviors, particularly the performance of the three-jaw chuck grip [a derived feature of humans with respect to chimpanzees (Marzke 1997) that is commonly used during stone tool production (Marzke and Shackley 1986; Marzke 1997)], is it realistic to expect scientists to inspect the metacarpals of all mammals, or even all primates, to verify that this feature is truly unique to humans? Kay and Cartmill (1977) might say that this is a reasonable and necessary expectation, while others would suggest the investigation of a more limited sample (i.e., via "phylogenetic bracketing") is sufficient (e.g., Witmer 1995). Also, how can we demonstrate that humans would be less adept at stone tool production without pronation of the index finger and supination of the little finger that this morphological configuration facilitates? It may be possible to demonstrate experimentally that the cupping behavior enabled by this metacarpal joint asymmetry is useful during stone tool behaviors, but this is not the same as demonstrating that we would be hindered without it. To be fair, no one has suggested that this feature is a specific adaptation for stone tool behaviors, but it serves to illustrate the point.

Two significant issues further complicate the application of the above criteria to humans and our fossil ancestors: the uniqueness of modern humans and the gaping holes in the archaeological and fossil records. Modern humans are unique in both our anatomy and behavior to the point that we are frequently the sole extant species in possession of the relevant trait and/or that practices the relevant behavior. Even features that may appear at first to be shared are often determined to be uniquely derived in humans upon closer inspection. For example, olive and gelada baboons also have a long thumb relative to their second digit (Etter 1973). But this is due to a decrease primarily in the length of the second digit rather than an elongation of the thumb relative to the fingers, as is the case in humans (Jolly 1970). In many instances, our anatomical and behavioral uniqueness makes it difficult, if not impossible, to fulfill the first and last criteria outlined by Kay and Cartmill (1977). Furthermore, the nature of the data is such that paleontologists will never be certain or even reasonably secure in the chronologies they construct based on fossil and cultural artifacts. How then can we reasonably evaluate whether the behavior does indeed predate the anatomy (third criterion)?

The second criterion can be addressed independent of fossil species, testing the functional link between morphology and behavior in extant species. Therefore, in some circumstances, biomechanical studies are capable of providing the robust evidence needed for evaluating the functional significance of derived features of the modern human upper limb. However, as will be seen below, biomechanical studies have their own limitations to consider.

4 Experimental Biomechanics and Stone Tool Behaviors

Experimental studies of stone tool behaviors in humans and chimpanzees concentrate on satisfying the second criterion in determining if a particular feature of the human hand is an adaptation or exaptation, establishing a functional link between the feature and behavior in question. In fulfilling this criterion, researchers have had to come up with novel methods for mimicking or modeling the proposed ancestral condition (that of the presumed last common ancestor), with physical and ethical issues to be taken into consideration. After all, it would be impossible, both physically and ethically, to surgically alter people's manual proportions or the shape of their carpals to test performance ability. Instead, scientists have developed far less invasive methods that rely on experimental biomechanics and computer modeling to investigate the functional links between the derived manual features present in modern humans and stone tool behaviors (Marzke and Shackley 1986; Hamrick et al. 1998; Marzke et al. 1998; Williams et al. 2010; Rolian et al. 2011; Orr 2012; Williams and Richmond 2012). Other researchers have used a different experimental approach by investigating the cognitive basis associated with stone tool production using brain imaging technologies. The application of imaging methods, such as magnetic resonance imaging (MRI) and positron emission tomography (PET), enables researchers to study brain activation patterns during various Paleolithic stone tool behaviors to better understand the underlying cognitive foundation of stone tool behaviors (Stout et al. 2000, 2008). The two experimental approaches—biomechanics and brain imaging—were recently combined to investigate whether increased right hemispheric activity reflects an emphasis on grasp control by the left hand, right hemisphere contributions toward motor sequences, or both (Faisal et al. 2010).

One of the earliest mentions of an experimental investigation into the biomechanics of stone tool behaviors can be found in Napier (1962: 411). Therein, he states conducting "personal experiments in the construction of 'Oldowan' pebble-tools and 'Chellean' [i.e., Acheulean] hand-axes," which convinced him that a precision grip was unnecessary for their production. Marzke and Shackley (1986) followed this up more than two decades later with a more systematic study of the grips involved in stone tool production and use. Their study was one of the first presentations of the knapping swing from an explicit biomechanics perspective and for which kinematic and kinetic aspects were directly related to upper limb anatomy. Much of what was qualitatively described in Marzke and Shackley (1986) was later quantified as new technology became available.

4.1 The Knapping Swing

Marzke and Shackley (1986) and Williams et al. (2010, 2014) described the path of the dominant arm during the knapping swing (Fig. 11.4). They described the arm as forcefully moving downward through strike until the elbow reached near-full extension poststrike. They then went on to describe the orientation of the hand around the hammerstone, and the location of peak reaction forces on the hand (see Sect. 4.3 below for a full discussion). During the downswing portion of the knapping swing, both novice and experienced knappers employed a variant of an upper limb motion strategy now commonly used in sports (e.g., when throwing a ball) known as the *proximal-to-distal (PD) joint sequence* (Williams et al. 2010, 2014), in which the limb moves like a whip, with motions starting at the shoulder, continuing down to the elbow, and finally ending at the wrist and fingers (Hirashima et al. 2002, 2007; Putnam 1991, 1993).

Strong and early muscular activity at the shoulder leads to muscular benefits further down the limb such that the need for muscular input to move the limb at the elbow and wrist joints decreases, which in turn increases muscular efficiency (Hirashima et al. 2003, 2007; Furuya and Kinoshita 2008). Instead of contributing to the overall movement of the limb, muscle activity in the forearm and hand is reserved primarily to play the less demanding role of refining movement to increase accuracy (Dounskaia et al. 1998; Dounskaia 2010). As motion proceeds from one joint to the next, each successive joint picks up speed, resulting in significantly higher velocities at the wrist than could be reached using a different motion strategy (Bunn 1955; Putnam 1991, 1993).

An increase in target accuracy is a beneficial by-product of this motion strategy due to the resulting joint separation that is inherent in such a limb joint sequence. As the limb joints become disassociated from one another, they are better able to move independently from one another. Thus, the arm replaces a rigid columnar movement pattern with a more complex joint motion pattern in which joints have the opportunity to act in concert with one another to accomplish a target-based task (Bernstein 1967; Chowdhary and Challis 1999).

Knappers have been found to use a variant of the standard PD sequence when making stone tools. Motions begin at the shoulder and continue toward the wrist, and velocity increases from shoulder to elbow to wrist. But they vary from the standard sequence in terms of the timing of peak velocities among the three joints:

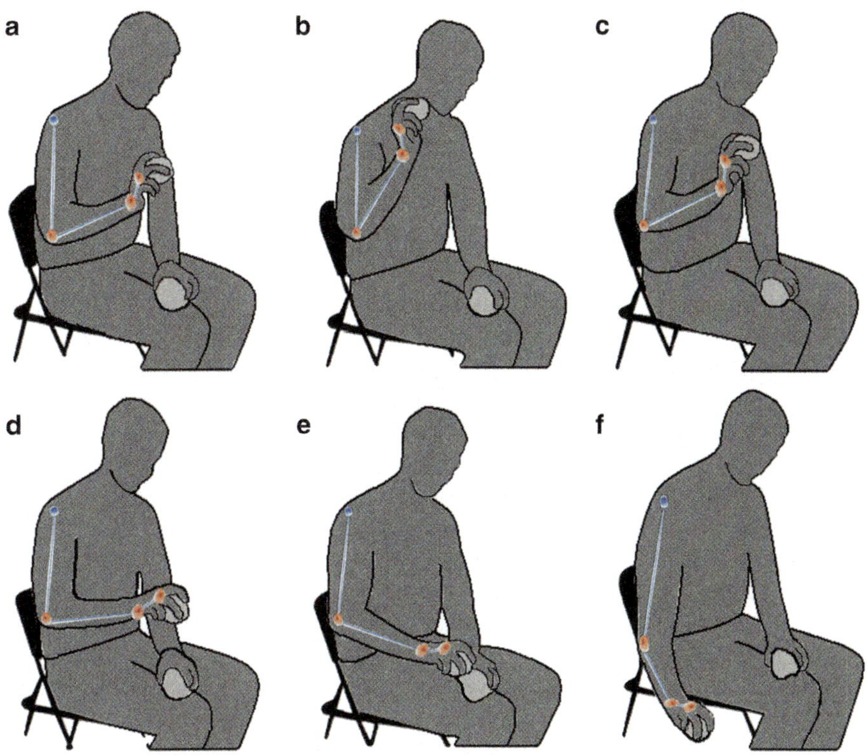

Fig. 11.4 The knapping swing: the upper limb during upswing and downswing with an overlay of the 3D model. (**a**) Mid upswing, the shoulder and elbow joints are flexing and the wrist joint is extending; (**b**) upswing apex, the shoulder begins to extend, while the elbow continues to flex and the wrist continues to extend; and (**c**) early downswing, the elbow transitions to extension and the forelimb begins to move downward. The hand also moves downward, while the wrist continues to extend. (**d**) Mid downswing, the upper limb continues to move downward and the wrist transitions into flexion; (**e**) strike, the wrist reaches peak flexion at strike and has reached it maximum velocity and acceleration; and (**f**) poststrike, the wrist is propelled out of peak flexion, back into extension, while the arm and forearm follow through to complete the swing (originally published in Williams et al. 2014)

velocity at the shoulder peaks first, followed by a peak at the wrist, with the elbow peaking last (Williams et al. 2010, 2014). Knappers that are particularly skillful are still able to achieve a velocity summation effect through the use of a distinct secondary, though smaller, peak in elbow angular velocity just prior to the peak in wrist angular velocity.

Performance differences between novice and skilled knappers resulting from control of upper limb motions, such as the ability to use the PD sequence to achieve a velocity summation effect, have also been documented in terms of their ability to control flake shape and target accuracy (Nonaka et al. 2010), as well as their ability to alter motions in response to changes in the knapping scenario (e.g., mass of the striking stone) while maintaining performance parameters [e.g., kinetic energy

(Bril et al. 2010)]. Together, these studies suggest that despite the gross similarities in the motion sequences of novice and skilled knappers (e.g., the use of the PD sequence), the motions of skilled knappers differ from novices in ways that affect efficiency and the ability to alter strategy and attain knapping goals.

4.2 The Role of the Wrist During Knapping

The modern human wrist is derived in numerous ways compared with that of extant African apes and our hominin ancestors (Tocheri et al. 2008), including its ability to attain high degrees of extension (Tuttle 1967, 1969; Jenkins and Fleagle 1975; Richmond and Strait 2000; Almquist 2001). This ability has been hypothesized to contribute to knapping efficiency and accuracy (Marzke 1971; Ambrose 2001). Williams and colleagues (2014) tested the role of wrist extension during knapping performance by having human subjects knap while wearing a wrist brace that mimicked the hypothetical ancestral condition, limiting wrist extension to approximately 35°. This is both the mean extension limit of extant chimpanzees (Tuttle 1967, 1969; Jenkins and Fleagle 1975; Richmond 2006) and the potential limit exhibited by *Australopithecus anamensis* and *Au. afarensis*, based on shape similarities of the dorsal aspect of the distal radius with extant knuckle-walkers (Richmond and Strait 2000; Richmond et al. 2001).

Extension is the dominant wrist motion used by both novice and skilled knappers, with subjects using 58–99 % of the total extension range and averaging ~47° among novices and ~59° among skilled knappers (Williams et al. 2010, 2014). None of the tested knappers, however, flex their wrists past the neutral position (i.e., when the hand is in line with the forearm). Taking advantage of the radiocarpal stability achieved by the natural coupling of extension/radial deviation and flexion/ulnar deviation, known as the “dart thrower’s arc” (Crisco et al. 2005), knappers also use high degrees of radial extension. Wrist extension peaks at the beginning of the downswing phase, cocking the wrist in preparation for rapid flexion to facilitate a forceful strike (Williams et al. 2010). This rapid flexion is also evident in the peak activity of the flexor carpi ulnaris muscle immediately prior to strike (Marzke et al. 1998). Immediately following the strike, the wrist undergoes sudden extension (~40°) by the strong reaction force created by the hammerstone and the core (Williams et al. 2014).

The ability of the modern human wrist to attain high degrees of extension contributes to knapping in three ways compared with the extension-limited hypothetical ancestral condition: (1) knappers are able to reach significantly greater angular wrist velocities due to the mechanical advantage wrist flexor muscles experience as extension increases at this joint (Pigeon et al. 1996), in combination with the whip-like motion pattern that results from the PD joint sequence; (2) target accuracy is increased when knappers are able to fully extend, thereby enabling to use the “dart thrower’s arc” of motion (Palmer et al. 1985; Wolfe et al. 2006); and (3) decreased risk of damage in the carpal region due to reaction-force-induced poststrike hyperextension (Linscheid and Dobyns 1985; Rettig 2003).

4.3 The Dominant Hand During Toolmaking and Use

Though the human hand is capable of performing numerous grips, knappers using their dominant hand (i.e., typically the right hand in most humans) to hold the hammerstone rely almost exclusively on just one—the three-jaw chuck, a type of precision grip (Marzke 1997)—during the production of early Paleolithic tools (Marzke and Shackley 1986; Williams and Richmond 2012). Marzke and Shackley (1986) noted that, while using this grip, strike occurs directly opposite the second and third metacarpal heads, implying that reaction forces would be concentrated in this region rather than at the thumb. This is contrary to the expected distribution of forces according to hypotheses linking modern human thumb robusticity to stone tool production as an adaptation to help withstand the elevated thumb joint reaction forces generated during stone tool production (Marzke 1992; Susman 1994; Marzke et al. 1998).

The location of peak force (and pressure) over the second and third metacarpal heads of the dominant hand was recently confirmed by Williams and colleagues (2012) in knapping experiments documenting manual force and pressure distributions during the production of a type of Paleolithic tool called bifacial Oldowan chopper (Fig. 11.5). Strike force and pressure, impulse, and pressure-time integrals (i.e., sum of pressure over time) were significantly higher on the second and/or third digits compared to the thumb (Williams et al. 2012). During tool manufacture the thumb acts as a buttress against the side of the hammerstone in a modified precision grip, as suggested by Marzke and colleagues (1998), rather than acting as the primary stabilizing mechanism. However, Williams et al.'s (2012) experiment was conducted using raw Texas flint, a fine-grained raw material not typical of the Oldowan industry. To test the effect of the raw material used, the same experimental procedure was replicated with novice knappers making flakes from Koobi Fora basalt and ignimbrite, two material types used by early human ancestors for stone tool manufacture (Braun et al. 2009) with different material properties than flint (e.g., toughness, grain size). Again, pressures and forces were concentrated on the second and/or third rays, despite the differences in material properties between the three materials (Fig. 11.6) (Williams-Hatala and Richmond, unpublished data). This was likely due to the maintenance of (1) the orientation of the hand around the hammerstone and (2) the location of strike on the opposite side of the hammerstone across raw material type. Specifically, knappers continued to hold the striking hammer in such a way that the second and third metacarpal heads continued to be oriented directly opposite to the location of strike, regardless of the type of material they were knapping. The pattern of pressures and forces acting across the hand was therefore maintained across raw material types because the orientation of the hand to the hammerstone was independent of the stone's material toughness.

Other experimental studies of simulated stone tool production and use found that forces were, in fact, concentrated on the thumb (Rolian et al. 2011). The seemingly contradictory results are likely due to methodological differences. Rolian and colleagues (2011) placed load cells into a triangular brass instrument,

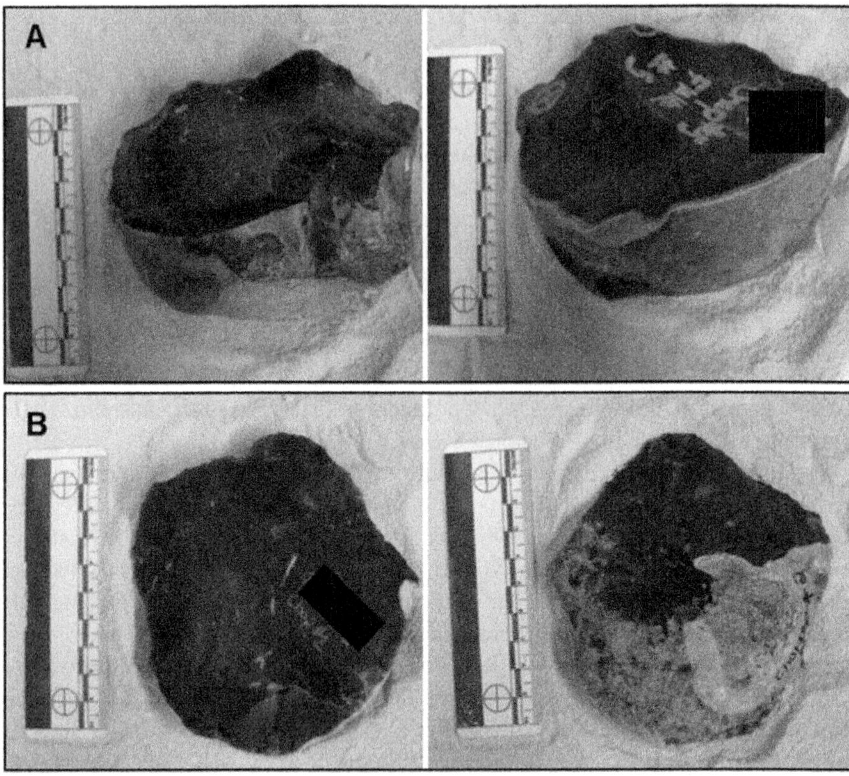

Fig. 11.5 Examples of two Oldowan-style bifacial choppers (**a**, **b**) produced by experienced knappers. Both surfaces of each chopper are shown

which required subjects to orient their hands around the devise in a predetermined manner. This orientation and the triangular shape of the "striking stone" suited the purpose of their study, which was to examine the relationship between digit length and biomechanical efficiency. However, an untended effect may have been the alteration of the subject's natural hand/striking stone orientation and consequently on pressure distribution patterns. In contrast, the work of Williams and colleagues (2012) relied on manual pressure sensing strips applied onto the palmar surface of the digits, enabling more direct quantification of the interactions between the hand and natural hammerstones.

There is a clear biomechanical benefit gained during stone tool manufacture from having relatively longer digits, including both a long thumb and relatively longer fingers (Rolian et al. 2011). However, it is not digit length per se that provides an advantage, but rather the relatively larger joint surface areas resulting from the isometric scaling relationship between digit length and articular width within modern humans (Rolian 2009; Rolian et al. 2011). Consequently, humans with longer digits have also relatively larger articular areas, which results in relatively lower

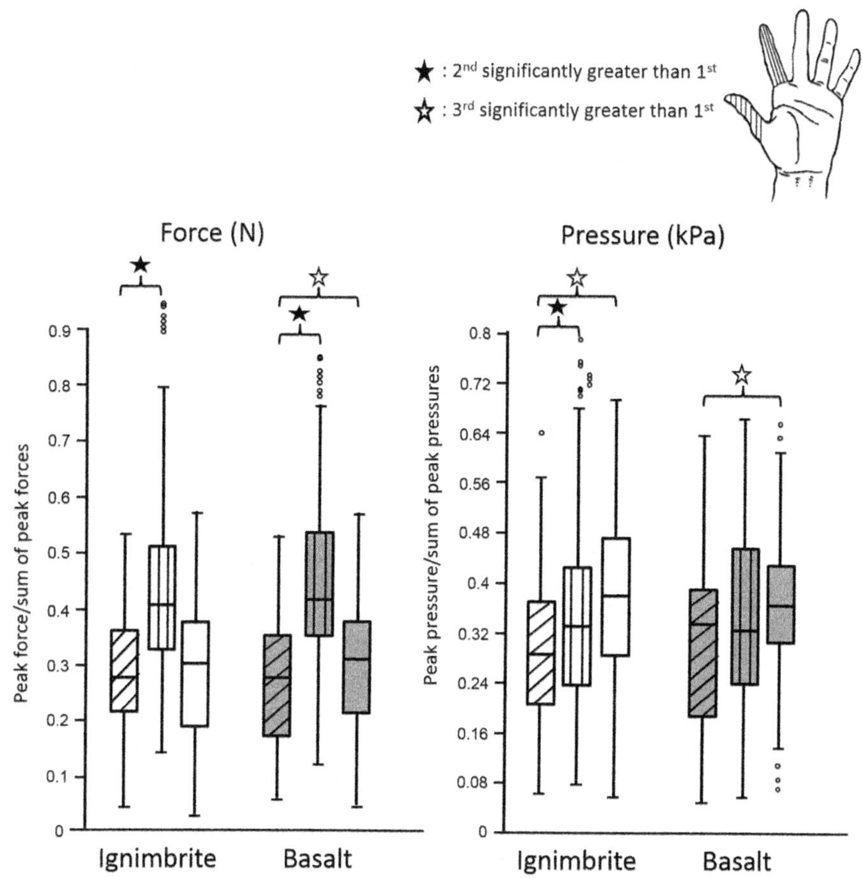

Fig. 11.6 Pressures and forces acting on the hand during flake production with Koobi Fora ignimbrite and Koobi Fora basalt

flexion forces and joint contact stresses during stone tool manufacture (Rolian et al. 2011). In chimpanzees, this relationship is negatively allometric, meaning that their long fingers are not coupled with relative increase in joint size (Rolian et al. 2011).

The pressure and force distribution pattern found by Williams and colleagues (2012) and Williams-Hatala and Richmond (unpublished data) during stone tool production seems to be reversed during tool use behaviors: pressures and forces are significantly greater on the thumb compared with the other regions of the hand during nutcracking, slicing animal tissue with a flake and a hand ax, and marrow acquisition with a hammerstone (Williams-Hatala and Richmond, unpublished data). Furthermore, the cumulative and absolute force both on the thumb specifically and across the hand as a whole is significantly greater during slicing and marrow acquisition than other Paleolithic stone tool behaviors. By extension, joint reaction forces are also higher during slicing and marrow acquisition. This evidence suggests that the evolution of robust thumbs in the hominin lineage at ~2 Ma (Trinkaus and Long 1990; Susman 1994) was the result of the intensification of tool use activities involving high pollical forces, such as slicing tissue with a flake and acquiring marrow with a hammerstone. It is worth noting that the behaviors that place the highest loads on the thumb are those using precision grips between the thumb and fingers (e.g., slicing animal tissue with a flake or a hand ax). They are not the behaviors that place the highest loads across the entire hand (e.g., accessing the marrow cavity), as may be expected.

4.4 The Nondominant Hand During Knapping

During knapping, the nondominant hand (usually the left hand in most humans) is generally tasked with holding the nodule or core steady, ready to be struck by the rock hammer held in the dominant hand. Perhaps because of this supportive role, the nondominant hand has received relatively little attention. Two studies have shown that the nondominant hand plays a significant and biomechanically complex role during the manufacture of stone tools (Marzke and Shackley 1986; Faisal et al. 2010; also see Key and Dunmore 2015[1]). Marzke and Shackley (1986) highlighted the role of the nondominant hand, with an emphasis on the fifth digit, in their description of grips used during stone tool production sequences. During the production of Oldowan and Developed Oldowan tools—the two earliest types of tools present in the archaeological record—the dominant hand relied exclusively on a three-jaw chuck, while the nondominant hand adopted many different grips as the core decreased in size, including both power and precision grips. When holding larger nodules, knappers tended to hold the nodule as one grips a book—resting one edge of the nodule against the palm with the fingers and thumb wrapped around it using a power grip. For smaller nodules, knappers used a variety of precision grip types, including grips involving the fingers and thumb without the palm and grips involving the palm and fingers without the thumb.

In a more recent study, Faisal and colleagues (2010) quantitatively described the grips used in the nondominant hand during the manufacture of Oldowan tools and of tools from the next cultural stage, the Acheulean. Despite the increased complexity of tool form seen in the Acheulean compared with the Oldowan, the results from this study showed that there were no significant differences in the types of grips used by the nondominant hand.

[1]Key and Dunmore (2015) report significantly greater forces acting on the nondominant thumb compared with the other digits tested on the nondominant hand. However, the reported forces are low relative to other tool-using and tool-making activities, including nutcracking (see Williams and Richmond 2012; Williams et al. 2012). Instead, reported forces are similar to those experienced during common, relatively low-force activities such as writing and playing the piano (Parlitz et al. 1998; Chau et al. 2006).

Given the range of motion available to the modern human thumb, variability in joint configurations was highest for this digit compared with the other digits during both Oldowan and Acheulean reduction sequences. Positional variability was also high for the fifth digit (see Fig. 2 in Faisal et al. 2010), an observation also noted by Marzke and Shackley (1986). Contrary to Marzke and Shackley (1986), Faisal and colleagues (2010) found little variation in positioning of the index finger between the two stone tool traditions and in positioning between the index and middle fingers overall (see Fig. 2 in Faisal et al. 2010).

These results demonstrated that, contrary to what may be expected, motion control and complexity in the nondominant hand do not seem to differ between Oldowan and Acheulean reduction sequences.

4.5 *Muscle Activity During Stone Tool Manufacture*

Advances in technology have also enabled researchers to gain insight into how the muscles of the upper limb are behaving during the production and the use of Paleolithic tools. Two EMG studies have investigated muscle recruitment during stone tool behaviors, one looking at 17 intrinsic (located in the palm) and extrinsic (located in the forearm) hand muscles (Marzke et al. 1998), while the other focused specifically on one extrinsic muscle, the flexor pollicis longus (FPL) (Hamrick et al. 1998).

Marzke et al. (1998) recorded EMG data from three subjects during the manufacture of stone tools, one with more than 20 years of knapping experience, one with moderate experience, and one novice knapper. The resulting data demonstrated that the FPL muscle was not consistently heavily recruited during the manufacture of stone tools relative to a baseline established prior to knapping (the baseline for each muscle was defined as the maximum isometric contraction force of a given muscle). Instead, ten muscles were found to be principally involved in hard-hammer percussion: five intrinsic muscles of the dominant hand, four intrinsic muscles of the nondominant hand (two, the small flexor of the first digit and the small flexor of the fifth digit, also peaked in the dominant hand), and three extrinsic muscles. The authors suggested that all these intrinsic muscles would be primarily responsible for securing the striking stone and core nodules against strong contact forces, which in turn cause of significant joint stress.

Of those ten muscles that were heavily recruited in both the dominant and nondominant hands, five are associated with the fifth digit. Although the little finger has received less attention than the thumb, high activity in muscles associated with the fifth digit is consistent with behavioral results reported by Faisal and colleagues (2010) and supports the hypothesis that the little finger plays an important role during the manufacture of stone tools (Marzke and Shackley 1986; Faisal et al. 2010). This hypothesis is further supported by the relative robusticity of the fifth digits of two known stone toolmakers, modern humans and Neanderthals (see Chap. 19), compared with two non-stone toolmakers: extant African apes and *Ardipithecus ramidus* (Kivell et al. 2011). However, interpretations of the role of the fifth digit are complicated by the fact that *Au. afarensis*, a hominin that likely engaged in stone tool behaviors (McPherron et al. 2010; Harmand et al. 2015), had a relatively gracile fifth digit, whereas recovered fifth digits assigned to *Australopithecus africanus* are robust, but not associated with stone tools or evidence of their use (but see Skinner et al. 2015). Clearly, further quantitative data on the use of the fifth digit during stone tool behaviors, as well as additional comparisons of fifth digit anatomy between modern humans, fossil hominins, and extant primates, are needed in order to clarify its role and adaptive significance with regard to stone tool behaviors.

Hamrick et al. (1998) recorded recruitment of the FPL muscle from nine novice knappers during tool production and tool use behaviors (Schmitt personal communication). Contrary to Marzke et al. (1998), Hamrick and colleagues reported high levels of recruitment of the FPL muscle during various types of tool use and tool production, particularly those involving a power grip and two types of precision grips. The differences in FPL recruitment between the two studies and within the subjects working with Marzke et al. (1998) may correspond to the degree of knapping experience of the subject: there appears to be a negative relationship between knapping experience and recruitment of FPL. Marzke et al. (1998) suggested that low FPL recruitment in their most experienced knapper may be a strategy to reduce muscle fatigue by avoiding firm pinch grips or to increase control of the striking stone by using the entire palmar thumb surface rather than just the surface of the distal phalanx.

The FPL muscle, coupled with a relatively robust pollex, has long been regarded an adaptation toward stone tool production to help withstand the high, repetitive forces believed to act on the pollex during hard-hammer percussion (Marzke 1992; Susman 1988, 1994; see also Sect. 4.3 above). In fact, Susman (1994) argued that *Paranthropus robustus* should be regarded as a toolmaker based solely on the robusticity of SKX 5020, a metacarpal from Swartkrans Member I that he attributed to *P. robustus* (but see Trinkaus and Long 1990). The low recruitment of the flexor pollicis longus muscle among more experienced knappers during stone tool production explains the low external pressure exerted by the distal phalanx during stone tool manufacture (Williams-Hatala and Richmond, unpublished data) and is further an evidence that stone tool use, rather than stone tool production, was more influential in selection for pollical robusticity (Hamrick et al. 1998; Rolian et al. 2011; Williams-Hatala and Richmond, unpublished data).

5 Discussion and Future Directions

What was the impetus for the transition from a grasping hand whose primary function was locomotion to the dexterous hand we posses today, capable of the fine motor control needed to perform everything from eye surgery to Joplin's Maple Leaf Rag? And why have humans in particular been endowed with such a high level of manual dexterity and such a wide diversity of grip types compared with other primates? The answers to all of these questions lie in the relaxation of one selective pressure and the intensification of another (Alba et al. 2003; Marzke 2009; see Chap. 18). As previously discussed, all primates exhibit a propensity for grasping behaviors and a relatively high degree of manual control. However, prior to the advent of bipedality, upper limb anatomy was also highly constrained by the stronger selective forces exerted by locomotor demands (Preuschoft and Chivers 1993; Richmond et al. 2001). Many of the locomotor patterns practiced by nonhuman primates call for upper limb joint stability (Hartwig and Doneski 1998; Richmond and Strait 2000), and they tend to evoke higher joint reaction forces than manipulative behaviors (Preuschoft and Chivers 1993). More arboreal behaviors select for strong digital flexor muscles and long, curved phalanges, which better distribute strain along the digits (Richmond 2007). However, many of the forceful, one-handed manipulative behaviors, particularly those associated with stone tool behaviors, call for a different suite of upper limb features. These include (1) increased thumb and wrist mobility, particularly extension in the wrist (Richmond and Strait 2000; Ambrose 2001; Williams et al. 2010), (2) a high thumb-to-finger length ratio (Marzke and Marzke 2000; Panger et al. 2002; Rolian et al. 2011), and (3) broad apical tufts on the distal phalanges (Susman and Creel 1979; Panger et al. 2002; Mittra et al. 2007). As Alba and colleagues (2003) pointed out, although nonhuman primate locomotor patterns and manual dexterity are not necessarily functionally exclusive behaviors, the selective pressures they place on hand anatomy are frequently in opposition.

Tocheri and colleagues (2008) observed [in reference to the primitive wrist anatomy of *Homo floresiensis* (Tocheri et al. 2007)] that more than one type of the hominin hand was capable of making and using stone tools; the full suite of derived traits present in the modern human hand and wrist was by no means necessary for stone tool production or use. In other words, the commitment to manual manipulation and dexterity that we display required more than the relaxation of one selective pressure and the casual application of another. It required something on the order of an adaptive shift that firmly anchored hominins to activities that induced joint and muscle stresses that were novel not only compared with those experienced during different forms of quadrupedal and/or arboreal locomotion, but novel or more intense than those experienced by early tool-associated hominins such as *Au. afarensis* and *Homo habilis*, species that are traditionally associated with unmodified or primitive tools. An adaptive shift of this magnitude is evident in the fossil and archaeological record of *Homo erectus s.l.*, making it a tempting candidate for the first hominin to display a morphological commitment to stone tool behaviors (Aiello and Wheeler 1995; Antón 2003; Tocheri et al. 2008; Williams and Richmond 2012). Unfortunately, much of the fossil data needed to evaluate manual dexterity in *H. erectus* is lacking [presently available are descriptions of two juvenile first metacarpals potentially associated with KNW WT 15000 (Walker and Leakey 1993), a partial lunate from Zhoukoudian (Weidenreich 1941), two distal phalanges from Dmanisi (Lordkipanidze et al. 2007), and a complete third metacarpal (Ward et al. 2014) (see also Chap. 18)]. Similarly, large gaps remain in the archaeological record pertaining to the timing and intensity of various stone tool behaviors. Until some of these gaps are filled, it will remain difficult to tease out the relationship between stone tool behaviors and the evolution of derived features of the human hand and wrist.

We also need further data on the biomechanics of stone tool use and production, not only from human knappers, but from chimpanzees and other tool-using primates. The burgeoning field of primate archaeology has reinforced the tool-using capacities of nonhuman primates in the wild (Mercader et al. 2007; Carvalho et al. 2009, 2012; Haslam et al. 2009), and capuchin studies have demonstrated the feasibility of capturing biomechanical tool use data in wild settings (Liu et al. 2009). Advances in motion capture technology now enable the collection of digital markerless 3D motion data, freeing researchers from highly controlled (and contrived) laboratory settings and in favor of more naturalistic conditions.

One of the most exciting new methods merges 3D imaging of movements with theoretical rigid body kinematics in order to produce high-resolution visualizations of the motion of the individual bones (e.g., carpals, metacarpals, phalanges) that comprise complex osteological units (Brainerd et al. 2010; Gatesy et al. 2010; Orr et al. 2010). These methods enable scientists to visualize and accurately quantify in 3D movements that were formerly obscured due to the complexity of both the movement itself and the anatomical makeup of the region of interest. Now scientists can view and test the functional significance of some of the subtle changes in shape and orientation that have occurred in hand and wrist bones (Tocheri et al. 2003, 2005, 2007). By registering 3D models of some of the more complete fossil individuals, such as *Australopithecus sediba* (Kivell et al. 2011), to in vivo 3D video of human movement patterns, it may become possible to create high-resolution animations of complex systems, such as the wrist, during evolutionarily significant behaviors (e.g., stone tool use and production). For example, we may soon actually be able to examine the functional contributions of asymmetrical second and fifth metacarpal heads to stone tool behaviors.

More than 2.5 Ma ago, our ancestors knocked over the proverbial technology domino that kick-started the evolution of the genus *Homo* and the tech-driven iWorld in which we now live. And as technology advances, our hands, formerly used mainly for feeding and locomotor purposes, are finding themselves subject to new technologically mediated motion patterns and joint stresses. For example, the increasing use of mobile devices (Yao et al. 2012) has been linked to thumb, hand, arm, and shoulder pain (Berolo et al. 2011). This is likely due in part to the repetition of highly stereotypical movements and in part to the novel biomechanical regimes called for by such devices: handheld mobile devices regularly elicit interphalangeal (IP) thumb flexion joint angles of approximately 90° (Yao et al. 2012). In contrast, common tasks of daily life, such as using a pen or a key, combing your hair, or brushing your teeth, elicit average IP thumb flexion joint angles of 18° and metacarpophalangeal angles of 21° (Hume et al. 1990). The joint angles assumed during stone tool behaviors are most likely similar to those reported by Hume and colleagues (1990), given the use of key pinch grips during slicing activities (Marzke 1997; Williams et al. 2012) and the knappers' tendency to slightly extend the pollical distal phalanx during tool manufacture (Marzke et al. 1998; Williams et al. 2012). It is impossible to tell whether these new biomechanical regimes will have an effect on hand anatomy; however, they are already having a measured impact on the rates of bone disease and degeneration, the rates and types of surgeries performed, and the way we use and view technology and on technology itself. Evolution is a contextually driven process; over generations, species respond to their environment and to the world in which they live. Our world is dominated by the technology born from the tools our ancestors began creating more than 2 Ma ago, and our relationship with technology has influenced our lives in ways both (literally) small (i.e., changes in hand and wrist structure) and large (i.e., the evolution of the genus *Homo* and even perhaps our own species). Technology will no doubt continue to impact the course of our lineage.

References

Aiello LC, Wheeler P (1995) The expensive tissue hypothesis. Curr Anthropol 36:199–221

Alba DM, Moyá-Solá S, Köhler M (2003) Morphological affinities of the *Australopithecus afarensis* hand on the basis of manual proportions and relative thumb length. J Hum Evol 44:225–254

Almécija S, Alba DM (2014) On manual proportions and pad-to-pad precision grasping in *Australopithecus afarensis*. J Hum Evol 73:88–92

Almécija S, Moyá-Solá S, Alba DM (2010) Early origin for human-like precision grasping: a comparative study of pollical distal phalanges in fossil hominins. PLoS One 5(7):e11727

Almquist EE (2001) Painful conditions of the forearm, wrist, and hand. In: Loeser JD, Bonica JD (eds) Bonica's management of pain, 3rd edn. Lippincott, Williams & Wilkins, Philadelphia, pp 1084–1109

Ambrose SH (2001) Paleolithic technology and human evolution. Science 291:1748–1753

Antón SC (2003) Natural history of *Homo erectus*. Yearb Phys Anthropol 46:126–170

Bernstein N (1967) The coordination and regulation of movements. Pergamon, New York

Berolo S, Wells RP, Amick BC (2011) Musculoskeletal symptoms among mobile hand-held device users and their relationship to device use: a preliminary study in a Canadian university population. Appl Ergon 42:371–378

Bloch J, Boyer D (2002) Grasping primate origins. Science 298:1606–1610

Blumenschine RJ, Pobiner BL (2007) Zooarchaeology and the ecology of Oldowan hominin carnivory. In: Ungar PS (ed) Evolution of the human diet: the known, the unknown and the unknowable. Oxford University Press, Oxford, pp 167–190

Bock WJ, von Wahlert G (1965) Adaptation and the form-function complex. Evolution 19:269–299

Brainerd EL, Baier DB, Gatesy SM, Hedrick TL, Metzger KA, Gilbert SL, Crisco JJ (2010) X-Ray reconstruction of moving morphology (XROMM): precision, accuracy and application in comparative biomechanics research. J Exp Zool 313A:262–279

Braun D, Harris J, Maina D (2009) Oldowan raw material procurement and use: evidence from the Koobi Fora Formation. Archaeometry 51:26–42

Braun DR, Harris JW, Levin NE, McCoy JT, Herries AI, Bamford MK, Bishop LC, Richmond BG, Kibunjia M (2010) Early hominin diet included diverse terrestrial and aquatic animals 1.95 Ma in East Turkana, Kenya. Proc Natl Acad Sci U S A 107:10002–10007

Bril B, Rein R, Nonaka T, Wenban-Smith F, Dietrich G (2010) The role of expertise in tool use: skill differences in functional action adaptations to task constraints. J Exp Psychol 36:825–839

Bunn JW (1955) Scientific principles of coaching. Prentice-Hall, Inc., New York

Bunn HT (1981) Archaeological evidence for meat-eating by Plio-Pleistocene hominids from Koobi Fora and Olduvai Gorge. Nature 291:574–577

Cartmill M (1992) New views on primate origins. Evol Anthropol 1:105–111

Carvalho S, Biro D, McGrew WC, Matsuzawa T (2009) Tool-composite reuse in wild chimpanzees (*Pan troglodytes*): archaeologically invisible steps in the technological evolution of early hominins? Anim Cogn 12:S103–S114

Carvalho S, Biro D, Cunha E, Hockings K, McGrew WC, Richmond BG, Matsuzawa T (2012) Chimpanzee carrying behaviour and the origins of human bipedality. Curr Biol 22:R180–R181

Chau T, Ji J, Tam C, Schwellnus H (2006) A novel instrument for quantifying grip activity during handwriting. Arch Phys Med Rehabil 87:1542–1547

Chowdhary AG, Challis JH (1999) Timing accuracy in human throwing. J Theor Biol 201:219–229

Clark JD, de Heinzelin J, Schick KD, Hart WK, White TD, WoldeGabriel G, Walter RC, Suwa G, Asfaw B, Vrba E, H-Selassie Y (1994) African *Homo erectus*: old radiometric ages and young Oldowan assemblages in the Middle Awash Valley, Ethiopia. Science 264:1907–1910

Costello MB, Fragaszy D (1988) Prehension in *Cebus* and *Saimiri*: I. Grip type and hand preference. Am J Primatol 15:235–245

Crisco JJ, Coburn JC, Moore DC, Akelman E, Arnold-Peter CW, Wolfe SW (2005) In vivo radiocarpal kinematics and the dart thrower's motion. J Bone Joint Surg Am 87:2729–2740

Dagosto M (1988) Implications of postcranial evidence for the origin of euprimates. J Hum Evol 17:35–56

de Heinzelin J, Clark JD, White T, Hart W, Renne P, WoldeGabriel G, Beyene Y, Vrba E (1999) Environment and behavior of 2.5-million-year-old Bouri hominids. Science 284:625–629

Diogo R, Richmond BG, Wood B (2012) Evolution and homologies of primate and modern human hand and forearm muscles, with notes on thumb movements and tool use. J Hum Evol 63:64–78

Dominguez-Rodrigo M, Pickering TR, Bunn HT (2010) Configurational approach to identifying the earliest hominin butchers. Proc Natl Acad Sci U S A 107:20929–20934

Dominguez-Rodrigo M, Pickering TR, Bunn HT (2011) Reply to McPherron et al.: doubting Dikika is about data, not paradigms. Proc Natl Acad Sci U S A 108:E117

Dominguez-Rodrigo M, Pickering TR, Bunn HT (2012) Experimental study of cut marks made with rocks unmodified by human flaking and its bearing on claims of ~3.4-million-year-old butchery evidence from Dikika, Ethiopia. J Archaeol Sci 39:205–214

Dounskaia NV (2010) Control of human limb movements: the leading joint hypothesis and its practical application. Ex Sports Sci Rev 28:201–208

Dounskaia NV, Swinnen SP, Walter CB, Spaepen AJ, Verschueren SMP (1998) Hierarchical control of different elbow-wrist coordination patterns. Exp Brain Res 121:239–254

Etter HF (1973) Terrestrial adaptations in the hands of cercopithecinae. Folia Primatol 20:331–350

Faisal A, Stout D, Apel J, Bradley B (2010) The manipulative complexity of Lower Paleolithic stone toolmaking. PLoS One 5:1–11

Furuya S, Kinoshita H (2008) Expertise-dependent modulation of muscular and non-muscular torques in multi-joint arm movements during piano keystroke. Neuroscience 156:390–402

Gatesy SM, Baier DB, Jenkins FA, Dial KP (2010) Scientific rotoscoping: a morphology-based method of 3-D motion analysis and visualization. J Exp Zool 313A:244–261

Gould SJ, Vrba ES (1982) Exaptation—a missing term in the science of form. Paleobiology 8:4–15

Hamrick MW (2001) Primate origins: evolutionary change in digital ray pattern and segmentation. J Hum Evol 40:339–351

Hamrick MW, Churchill SE, Schmitt D, Hylander WL (1998) EMG of the human flexor pollicis longus muscle: implications for the evolution of hominid tool use. J Hum Evol 34:123–136

Harmand S, Lewis JE, Feibel CS, Lepre CJ, Prat S, Lenoble A, Boes X, Quinn RL, Brenet M, Arroyo A, Taylor N, Clement S, Daver G, Brugal J-P, Leakey L, Mortlock RA, Wright JD, Lokorodi S, Kirwa C, Kent DV, Roche H (2015) 3.3-million-year-old stone tools from Lomekwi 3, West Turkana, Kenya. Nature 521:310–315

Hartwig WC, Doneski K (1998) Evolution of the hominid hand and tool making behavior. Am J Phys Anthropol 106:401–402

Haslam M, Hernandez-Aguilar A, Ling V, Carvalho S, de la Torre I, DeStefano A, Du A, Hardy B, Harris J, Marchant L, Matsuzawa T, McGrew W, Mercader J, Mora R, Petraglia M, Roche H, Visalberghi E, Warren R (2009) Primate archaeology. Nature 460:339–344

Hinchliffe JR (2002) Developmental basis of limb evolution. Int J Dev Biol 46:835–845

Hirashima M, Kadota H, Sakurai S, Kudo K, Ohitsuki T (2002) Sequential muscle activity and its functional role in the upper extremity and trunk during overarm throwing. J Sports Sci 20:301–310
Hirashima M, Kudo K, Ohtsuki T (2003) Utilization and compensation of interaction torques during ball-throwing movements. J Neurophysiol 89:1784–1796
Hirashima M, Kudo K, Watarai K, Ohtsuki T (2007) Control of 3D limb dynamics in unconstrained overarm throws of different speeds performed by skilled baseball players. J Neurophysiol 97:680–691
Hume MC, Gellman H, McKellop H, Brumfield RH (1990) Functional range of motion of the joint of the hand. J Hand Surg Am 15:240–243
Jenkins FA Jr, Fleagle JG (1975) Knuckle-walking and the functional anatomy of the wrists in living apes. In: Tuttle RH (ed) Primate functional morphology and evolution. Mouton, The Hague, pp 213–231
Jones-Engel LE, Bard KA (1996) Precision grip in young chimpanzees. Am J Primatol 39:1–15
Jolly CJ (1970) The seed-eaters: a new model of hominid differentiation based on a baboon analogy. Man 5:5–26
Kay RF, Cartmill M (1977) Cranial morphology and adaptations of *Palaechthon nacimienti* and other Paromomyidae (Plesiadapoidea, Primates) with a description of a new genus and species. J Hum Evol 6:19–53
Key AJM, Dunmore CJ (2015) The evolution of the hominin thumb and the influence exerted by the non-dominant hand during stone tool production. J Hum Evol 78:60–69
Kivell TL, Kibii JM, Churchill SE, Schmid P, Berger LR (2011) *Australopithecus sediba* hand demonstrates mosaic evolution of locomotor and manipulative abilities. Science 333:1411–1417
Kuczynski K (1974) Carpometacarpal joint of the human thumb. J Anat 118:119–126
Laurin M, Girondot M, Ricqlès D (2000) Early tetrapod evolution. Trends Ecol Evol 15:118–123
Leakey MD (1971) Olduvai Gorge: excavations in beds I and II, 1960-1963. Cambridge University Press, Cambridge
Linscheid RL, Dobyns JH (1985) Athletic injuries of the wrist. Curr Orthop Pract 198:141–151
Liu Q, Simpson K, Izar P, Ottoni E, Visalberghi E, Fragaszy D (2009) Kinematics and energetics of nut-cracking in wild capuchin monkeys (*Cebus libidinosus*) in Piaui. Brazil Am J Phys Anthropol 138:210–220
Lordkipanidze D, Jashashvili T, Vekua A, Ponce de Leon MS, Zollikofer CP, Rightmire GP, Pontzer H, Ferring R, Oms O, Tappen M, Bukhsianidze M, Agusti J, Kahlke R, Kiladze G, Martinez-Navarro B, Mouskhelishvili A, Nioradze M, Rook L (2007) Postcranial evidence from early *Homo* from Dmanisi, Georgia. Nature 449:305–310
Lovejoy CO, Simpson S, White T, Asfaw B, Suwa G (2009) Careful climbing in the Miocene: the forelimb of *Ardipithecus ramidus* and humans are primitive. Science 326:70e71–70e78
Marzke MW (1971) Origin of the human hand. Am J Phys Anthropol 34:61–84
Marzke MW (1983) Joint functions and grips of the *Australopithecus afarensis* hand, with special reference to the region of the capitate. J Hum Evol 12:197–211
Marzke MW (1992) Evolutionary development of the human thumb. In: Manske P (ed) Hand clinics and thumb reconstructions. W.B. Sanders, Philadelphia, pp 1–8
Marzke MW (1997) Precision grips, hand morphology, and tools. Am J Phys Anthropol 102:91–110
Marzke MW (2009) Upper-limb evolution and development. J Bone Joint Surg Am 91:26–30
Marzke MW, Marzke RF (2000) Evolution of the human hand: approaches to acquiring, analysing and interpreting the anatomical evidence. J Anat 197:121–140
Marzke MW, Shackley MS (1986) Hominid hand use in the Pliocene and Pleistocene: evidence from experimental archaeology and comparative morphology. J Hum Evol 15:439–460
Marzke MW, Shrewsbury MM (2006) The *Oreopithecus* thumb: pitfalls in reconstructing muscle and ligament attachments from fossil bones. J Hum Evol 51:213–215
Marzke MW, Wullstein KL (1996) Chimpanzee and human grips: a new classification with a focus on evolutionary morphology. Int J Primatol 17:117–139
Marzke MW, Wullstein KL, Viegas SF (1992) Evolution of the power (squeeze) grip and its morphological correlates in hominids. Am J Phys Anthropol 89:283–298

Marzke MW, Toth N, Schick K, Reece S, Steinberg B, Hunt K, Linscheid RL, An K-N (1998) EMG study of hand muscle recruitment during hard hammer percussion manufacture of Oldowan tools. Am J Phys Anthropol 105:315–332
Marzke MW, Shrewsbury MM, Horner KE (2007) Middle phalanx skeletal morphology in the hand: can it predict flexor tendon size and attachments? Am J Phys Anthropol 134:141–151
McGrew WC (1974) Tool use by wild chimpanzees in feeding upon driver ants. J Hum Evol 3:501–504
McPherron SP, Alemseged Z, Marean CW, Wynn JG, Reed D, Geraads D, Bobe R, Bearat HA (2010) Evidence for stone-tool-assisted consumption of animal tissues before 3.39 million years ago at Dikika, Ethiopia. Nature 466:857–860
McPherron SP, Alemseged Z, Marean CW, Wynn JG, Reed D, Geraads D, Bobe R, Bearat HA (2011) Tool-marked bones from before the Oldowan change the paradigm. Proc Natl Acad Sci U S A 108:E116
Mercader J, Barton H, Gillespie J, Harris J, Kuhn S, Tyler R, Boesch C (2007) 4,300-year-old chimpanzee sites and the origins of percussive stone technology. Proc Natl Acad Sci U S A 104:3043–3048
Mittra ES, Smith HF, Lemelin P, Jungers WL (2007) Comparative morphometrics of the primate apical tuft. Am J Phys Anthropol 134:449–459
Moyá-Solá S, Köhler M, Rook L (1999) Evidence of hominid-like precision grip capability in the hand of the Miocene ape *Oreopithecus*. Proc Natl Acad Sci U S A 96:313–317
Napier JR (1956) The prehensile movements of the human hand. J Bone Joint Surg Br 38:902–913
Napier JR (1960) Studies of the hands of living primates. Proc Zool Soc Lond 134:647–657
Napier J (1962) Fossil hand bones from Olduvai Gorge. Nature 196:409–411
Napier JR (1965) Evolution of the human hand. Proc R Inst G B 40:544–557
Napier J (1993) Hands. Revised edition by Russell H. Tuttle. Princeton University Press, Princeton
Navarrete A, van Schaik CP, Isler K (2011) Energetics and the evolution of human brain size. Nature 480:91–93
Nonaka T, Bril B, Rein R (2010) How do stone knappers predict and control the outcome of flaking? Implications for understanding early stone tool technology. J Hum Evol 59:155–167
Orr CM (2012) Kinematics and morphometrics of the radiocarpus in anthropoids with implications for reconstructing the evolution of hominin wrist mechanics. Am J Phys Anthropol 147(Suppl 54):229 (abstract)
Orr CM, Leventhal EL, Chivers SF, Marzke MW, Wolfe SW, Crisco JJ (2010) Studying primate carpal kinematics in three dimensions using a computed-tomography-based markerless registration method. Anat Rec 293:692–709
Palmer A, Werner F, Murphy D, Glisson R (1985) Functional wrist motion: a biomechanical study. J Hand Surg Am 10:39–46
Panger MA, Brooks AS, Richmond BG, Wood B (2002) Older than the Oldowan? Rethinking the emergence of hominin tool use. Evol Anthropol 11:235–245
Parlitz D, Peschel T, Altenmüller E (1998) Assessment of dynamic finger forces in pianists: effects of training and expertise. J Biomech 31:1063–1067
Pigeon P, Yahia L, Feldman AG (1996) Moment arms and lengths of human upper limb muscles as functions of joint angles. J Biomech 29:1365–1370
Plummer T (2004) Flaked stones and old bones: biological and cultural evolution at the dawn of technology. Yearb Phys Anthropol 47:118–164
Pouydebat E, Gorce P, Coppens Y, Bels V (2009) Biomechanical study of grasping according to the volume of the object: human versus non-human primates. J Biomech 42:266–272
Preuschoft H, Chivers DJ (1993) Introduction. In: Preuschoft H, Chivers DJ (eds) Hands of primates. Springer-Verlag, Vienna, pp 1–3
Putnam CA (1991) A segmental interaction analysis of proximal-to-distal sequential segment motion patterns. Med Sci Sports Exerc 23:130–144
Putnam CA (1993) Sequential motions of body segments in striking and throwing skills: descriptions and explanations. J Biomech 26:125–135
Rasmussen DT (1990) Primate origins: lessons from a neotropical marsupial. Am J Primatol 22:263–277

Rettig AC (2003) Athletic injuries of the wrist and hand. Part 1: traumatic injuries of the wrist. Am J Sports Med 31:1038–1048

Richmond BG (2006) Functional morphology of the midcarpal joint in knuckle-walkers and terrestrial quadrupeds. In: Ishida H, Tuttle R, Pickford M, Ogihara N (eds) Human origins and environmental backgrounds. Springer, New York, pp 105–122

Richmond BG (2007) Biomechanics of phalangeal curvature. J Hum Evol 53:678–690

Richmond BG, Strait DS (2000) Evidence that humans evolved from a knuckle-walking ancestor. Nature 404:382–385

Richmond BG, Begun DR, Strait DS (2001) Origin of human bipedalism: the knuckle-walking hypothesis revisited. Yearb Phys Anthropol 44:70–105

Ricklan DE (1987) Functional anatomy of the hand of *Australopithecus africanus*. J Hum Evol 16:643–664

Roche H, Delagnes A, Brugal JP, Feibel C, Kibunjia M, Mourre V, Texier PJ (1999) Early hominid stone tool production and technical skill 2.34 Myr ago in West Turkana, Kenya. Nature 399:57–60

Rolian C (2009) Integration and evolvability in primate hands and feet. Evol Biol 36:100–117

Rolian C, Gordon AD (2013) Reassessing manual proportions in *Australopithecus afarensis*. Am J Phys Anthropol 152:393–406

Rolian C, Gordon AD (2014) Response to Almécija and Alba (2014)—on manual proportions in *Australopithecus afarensis*. J Hum Evol 73:93–97

Rolian C, Lieberman DE, Hallgrimsson B (2010) The coevolution of human hands and feet. Evolution 64:1558–1568

Rolian C, Lieberman DE, Zermeno JP (2011) Hand biomechanics during simulated stone tool use. J Hum Evol 61:26–41

Schick KD, Toth N (1993) Making silent stones speak. Human evolution and the dawn of technology. Simon & Schuster, New York

Semaw S (2000) The world's oldest stone artefacts from Gona, Ethiopia: their implications for understanding stone technology and patterns of human evolution between 2.6-1.5 million years ago. J Archaeol Sci 27:1197–1213

Semaw S, Renne P, Harris JW, Feibel CS, Bernor RL, Fesseha N, Mowbray K (1997) 2.5-million-year-old stone tools from Gona, Ethiopia. Nature 385:333–336

Semaw S, Rogers MJ, Quade J, Renne PR, Butler RF, Dominguez-Rodrigo M, Stout D, Hart WS, Pickering T, Simpson SW (2003) 2.6 million-year-old stone tools and associated bones from OGS-6 and OGS-7, Gona, Afar, Ethiopia. J Hum Evol 45:169–177

Shrewsbury MM, Johnson RK (1983) Form, function, and evolution of the distal phalanx. J Hand Surg Am 8:475–479

Shrewsbury MM, Marzke MW, Linscheid RL (2003) Comparative morphology of the pollical distal phalanx. Am J Phys Anthropol 121:30–47

Skinner MM, Stephens NB, Tsegai ZJ, Foote AC, Nguyen NH, Gross T, Pahr DH, Hublin J-J, Kivell TL (2015) Human-like hand use in *Australopithecus africanus*. Science 347:395–399

Stout D, Toth N, Schick KD (2000) Stone tool-making and brain activation: position emission tomography (PET) studies. J Archaeol Sci 27:1215–1223

Stout D, Toth N, Schick KD, Chaminade T (2008) Neural correlates of early stone age toolmaking: technology, language and cognition in human evolution. Philos Trans R Soc B 363:1939–1949

Susman RL (1979) Comparative and functional morphology of hominoid fingers. Am J Phys Anthropol 50:215–236

Susman RL (1988) Hand of *Paranthropus robustus* from Member 1, Swartkrans: fossil evidence for tool behavior. Science 240:781–784

Susman RL (1989) New hominid fossils from the Swartkrans Formation (1979-1986 excavations): postcranial specimens. Am J Phys Anthropol 79:451–474

Susman RL (1994) Fossil evidence for early hominid tool use. Science 265:1570–1573

Susman RL (1998) Hand function and tool behavior in early hominids. J Hum Evol 35:23–46

Susman RL (2005) *Oreopithecus*: still apelike after all these years. J Hum Evol 45:405–411

Susman RL, Creel N (1979) Functional and morphological affinities of the subadult hand (O.H. 7) from Olduvai Gorge. Am J Phys Anthropol 51:311–331

Sussman RW (1991) Primate origins and the evolution of angiosperms. Am J Primatol 23:209–223

Tocheri MW (2007) Three-dimensional riddles of the radial wrist: derived carpal and carpometacarpal joint morphology in the genus *Homo* and the implications for understanding the evolution of stone tool-related behaviors in hominins. Ph.D. dissertation, Arizona State University

Tocheri MW, Marzke MW, Liu D, Bae M, Jones GP, Williams RC, Razdan A (2003) Functional capabilities of modern and fossil hominid hands: three-dimensional analysis of trapezia. Am J Phys Anthropol 122:101–112

Tocheri MW, Razdan A, Williams RC, Marzke MW (2005) A 3D quantitative comparison of trapezium and trapezoid relative articular and non-articular surface areas in modern humans and great apes. J Hum Evol 49:570–586

Tocheri MW, Orr CM, Larson SG, Sutikna T, Jatmiko, Saptomo EW, Due RA, Djubiantono T, Morwood MJ, Jungers WL (2007) The primitive wrist of *Homo floresiensis* and its implications for hominin evolution. Science 317:1743–1745

Tocheri MW, Orr CM, Jacofsky MC, Marzke MW (2008) The evolutionary history of the hominin hand since the last common ancestor of *Pan* and *Homo*. J Anat 212:544–562

Toth N, Schick K (2009) The Oldowan: the tool making of early hominins and chimpanzees compared. Annu Rev Anthropol 38:289–305

Trinkaus E, Long JC (1990) Species attribution of the Swartkrans Member 1 first metacarpals: SK 84 and SKX 5020. Am J Phys Anthropol 83:419–424

Tuttle RH (1967) Knuckle-walking and the evolution of hominoid hands. Am J Phys Anthropol 26:171–206

Tuttle RH (1969) Quantitative and functional studies on the hands of the Anthropoidea: I. The Hominoidea. J Morphol 128:309–363

Walker A, Leakey R (1993) The Nariokotome *Homo erectus* skeleton. Harvard University Press, Cambridge

Ward CV, Kimbel WH, Harmon EH, Johanson DC (2012) New postcranial fossils of *Australopithecus afarensis* from Hadar, Ethiopia (1997-2007). J Hum Evol 63:1–51

Ward CV, Tocheri MW, Plavcan JM, Brown FH, Manthi FK (2014) Early Pleistocene third metacarpal from Kenya and the evolution of modern human-like hand morphology. Proc Natl Acad Sci U S A 111:121–124

Weidenreich F (1941) The extremity bones of *Sinanthropus pekinensis*. Paleontol Sin New Ser D 5:1–150

Westergaard GC, Suomi SJ (1997) Capuchin monkey (*Cebus apella*) grips for the use of stone tools. Am J Phys Anthropol 103:131–135

Williams EMS, Richmond BG (2012) Manual pressure distribution during stone tool use. Am J Phys Anthropol 147(Suppl 54):303 (abstract)

Williams-Hatala EMS, Richmond BG (2012) The evolution of human thumb robusticity and tool use

Williams EMS, Gordon AD, Richmond BG (2010) Upper limb kinematics and the role of the wrist during stone tool production. Am J Phys Anthropol 43:134–145

Williams EMS, Gordon AD, Richmond BG (2012) Hand pressure distribution during Oldowan stone tool production. J Hum Evol 62:520–532

Williams EMS, Gordon AD, Richmond BG (2014) Biomechanical strategies for accuracy and force generation during stone tool production. J Hum Evol 72:52–63

Witmer LM (1995) The extant phylogenetic bracket and the importance of reconstructing soft tissues in fossils. In: Thomason J (ed) Functional morphology in vertebrate paleontology. Cambridge University Press, Cambridge, pp 19–33

Wolfe SW, Crisco JJ, Orr CM, Marzke MW (2006) The dart-throwing motion of the wrist: is it unique to humans? J Hand Surg Am 31:1429–1437

Wood Jones F (1920) The principles of anatomy as seen in the hand. J. & A. Churchill, London

Yao J, Park MJ, Davis D, Chang J (2012) Ideal position for thumb interphalangeal arthrodesis in the era of smartphones and text communication. Orthopedics 35:955–957

Chapter 12
Functions of the Hand in Primates

Dorothy M. Fragaszy and Jessica Crast

1 Introduction

Nearly 100 years ago, Wood Jones (1920) characterized the hand as a specialized sensory organ and the use of the hands in exploration as a defining characteristic of primates. Thus, from the beginning, researchers have investigated how primates use their hands to touch and handle objects, as well as during locomotion. Understanding how primates use their hands is fundamental to reconstructing and interpreting the evolution of the order to which humans belong to (Napier 1960, 1961, 1980). Primates were present in the Eocene, feeding and moving predominantly on thin and flexible terminal branches (Cartmill 1974; Sussman 1991; Sussman et al. 2013; see Chap. 14). They used both hands and feet to locomote in this environment and the hands to feed. Prehension of small objects (such as insects or fruits) by one hand is considered a primitive characteristic of primates (Washburn 1951; Napier 1961, 1980) and is integral to feeding in all living primates. However, this characteristic is shared with other orders as primates are not alone in using one appendage in prehension. Indeed, Iwaniuk and Whishaw (2000) reported that, of 29 tetrapod orders for which the presence or absence of "skilled forelimb movements" (which includes reaching, grasping, and manipulation with one forelimb) was examined, 13 orders were characterized by such movements. Iwaniuk and Whishaw (2000) showed that skilled movements of the forelimbs in mammals probably share a common origin in early tetrapods, appearing after the divergence of therian mammals (marsupials and placentals) from the monotremes. The prehensile actions of primates must therefore be considered against this ancient backdrop.

D.M. Fragaszy (✉)
Department of Psychology, University of Georgia, Athens, GA 30602, USA
e-mail: doree@uga.edu; dfragaszy@gmail.com

J. Crast
Yerkes National Primate Research Center, Emory University, Atlanta, GA 30322, USA
e-mail: jessicrast@emory.edu; jessicrast@gmail.com

T.L. Kivell et al. (eds.), *The Evolution of the Primate Hand*, Developments in Primatology: Progress and Prospects, DOI 10.1007/978-1-4939-3646-5_12

2 Prehension in Primates Compared to Other Therian Mammals

We start this chapter by discussing the primitive physical features of primate hands, described by Wood Jones (1920) nearly 100 years ago. All primates have pentadactyl hands with relatively long and unwebbed digits, a morphology that enables flexion, extension, and rotation of the digits relative to one another and enables the digits of one hand to close around a small object to press it against the palm. All primates have tactile pads on the palmar surface of the terminal phalanges that are richly innervated with sensory receptors, enabling the hand to serve exploratory as well as postural functions. Animals in other orders also possess unwebbed digits, long digits, or a rich supply of sensory receptors on the terminal phalanges (Lemelin and Grafton 1998), but the uniform presence across the order of all three features is characteristic only of primates.

Species lacking some of these characteristics can still use the forelimbs in surprisingly skilled ways. Rodents, for example, with nearly 2000 species inhabiting a vast range of habitats, display skilled movements with the forelimbs (Whishaw et al. 1998). Common skilled forelimb movements in rodents include grasping food using an elbow-in movement and manipulating food with the digits (Whishaw et al. 1998). Beyond these shared features, rodents present a wide variety of specialized skilled forelimb movements across species, in keeping with their ecological diversity, such as unilateral and bilateral grasping (i.e., holding two objects simultaneously). Whishaw (2005) provided an illuminating review of the range of skilled forelimb movements in the common rat (*Rattus norvegicus*). The rat's first digit can move medially toward the palm, so that it can hold an object between the terminal pad of the pollex and the terminal pads of other digits. The fifth digit moves independently, turning medially during grasping, in a thumb-like manner. The two paws can move in parallel and in complementary ways to hold or move an object. Rats make visually guided limb movements, rotate the forearm to aim, and pronate and supinate the limb during reaching. After the digits contact an object, they converge to the palmar pads to hold objects in various orientations. The rich repertoire of skilled forelimb movements in rats reminds us that prehensile skills shared by primates are also found in some members of other orders (see also Ivanco et al. 1996). However, rats, like other rodents, preferentially use olfaction to locate food and pick up food with the mouth when possible, rather than lifting it to the mouth with their paws. Primates, on the other hand, preferentially grasp food with their hands.

Research related to unraveling the evolution of morphology and prehensile skills in primates has looked to procyonid carnivores (such as raccoons and kinkajous) and didelphid marsupials as useful comparative models (Rasmussen 1990; McClearn 1992; Iwaniuk and Whishaw 1999; Lemelin 1999; Lemelin and Schmitt 2007). Kinkajous, for example, grasp objects unimanually while feeding from terminal branches (McClearn 1992). Raccoons grasp objects between the digits, between the apical and distal palmar pads, or between two paws, rather than in a grasp for which several digits in one hand converge. Raccoons manipulate objects by rolling them between the palms of both paws, with little or no digit movement. Thus, although raccoons use visually guided reaching like primates, they are more similar to other carnivores in how they use their forelimbs (bimanually and without convergent digits during grasping; Iwaniuk and Whishaw 1999).

Didelphid marsupial species exhibit a body mass range that overlaps with cheirogaleid primates (i.e., dwarf and mouse lemurs). Some are primarily terrestrial and others are primarily arboreal (reviewed in Lemelin 1999). The neotropical woolly opossum, *Caluromys*, for example, is almost exclusively arboreal (Charles-Dominique et al. 1981; Rasmussen 1990). Woolly opossums use vision more than other opossums to locate and capture mobile insect prey, collect fruits from terminal branches with their forelimbs, and frequently adopt suspensory postures using the hind limbs, as do cheirogaleid primates (Rasmussen 1990; Lemelin 1999). This is assumed to be the primitive locomotor pattern and feeding niche of early primates (Cartmill 1974; Sussman 1991; Sussman et al. 2013). We know little about skilled prehensive movements during prey capture in these animals, but on ecological grounds one could predict a primate-like pattern of actions, with both forelimbs participating in grabbing flying prey and the palm contacting the prey item first. For example, lorisoids studied by Charles-Dominique (1977) sometimes captured slow-moving invertebrate prey (caterpillars, beetles, ants, etc.) by pressing them against a branch. Ivanco et al. (1996) described prey capture movements in the gray short-tailed opossum (*Monodelphis domestica*), a more terrestrial species than *Caluromys*. The gray short-tailed opossum can capture prey on a solid substrate with one forelimb, although they exhibit less variable and simpler movements compared to rats (*Rattus norvegicus*) in the same situations (Ivanco et al. 1996). Similarly, the pygmy tree shrew (*Tupaia minor*) grasps food items in one hand and moves nimbly on small-diameter supports, whereas the more terrestrial large tree shrew (*Tupaia tana*) does not, suggesting that the former could also be a useful model of prehensive uses of the forelimb in early primates (Sargis 2001).

Additional comparative work on prehensile behavior and the neuromotor system (particularly the corticospinal tracts and their terminations) supporting these movements in rodents, tree shrews, and opossums would be useful for understanding the functional significance of variations in this system across mammals and their relation to inhabiting a small branch environment and their feeding niche. The small branch environment in which early primates lived together with feeding on mobile insect prey probably contributed in a synergistic way to the development of primate-typical use of the hands to prehend food items. Indeed, Toussaint et al. (2013) found that gray mouse lemurs (*Microcebus murinus*) used their hands (rather than the mouth) to grasp food more often while clinging to narrow substrates and used the hands to grasp moving prey regardless of the substrate to which they were clinging.

3 Manual Function in Prosimians as a Window onto Primitive Features of the Primate Hand

Some prosimians, particularly small-bodied cheirogaleids and galagids, are generally thought to be more representative of early primates than the anthropoids and, therefore, are of strong interest for the identification of primitive characters in primates (Charles-Dominique 1972; Martin 1972a). However, one must keep in mind that the extant species of prosimians (i.e., strepsirrhines and tarsiers) reflect their own long evolutionary history and the lemurs in particular have undergone remarkable radiation (Martin 1972b, 1990) and diversification of hand morphology and use (Lemelin and Jungers 2007). The case of the aye-aye (*Daubentonia madagascariensis*) comes to mind and will be discussed further in this chapter.

Bishop (1962, 1964; Jolly 1964; as a point of clarification, Alison Bishop and Alison Jolly are the same person) perceptively described manual function in several species of prosimians with an eye to understanding primitive features that presaged and supported the later elaborations of manual function in anthropoids. Bishop combined experimental studies of orientation of the hand during locomotion with observational studies of spontaneous use of the hands in daily life in unconstrained captive individuals. Thus, she was able to consider manual function in relation to exploration, play, self-care, and social behavior, as well as locomotion and feeding. Despite the variations across lemurs and lorises in hand postures during prehension of substrates during locomotion and while grasping food items, all species studied by Bishop (1964) used the hands in some way during social grooming, to explore novel objects and surfaces, in self-care, and in play. Even lemurs, which often pick up small objects from a surface using the mouth rather than the hands (Bishop 1964; Torigoe 1985), explored novel surfaces with their hands. Schöneich (1993) observed *Lemur catta* using their hands to lift, push, and slide fasteners (chains, pins, hooks, sliding panels) to open a box containing food, replicating Bishop's (1964; Jolly 1964) observations. Ring-tailed lemurs used both left and right hands singly and together and adjusted their hands quickly when the fasteners were presented in novel positions. We can surmise that using the hand for multiple purposes (in social behavior, self-care, play, and exploration, as well as in feeding and locomotion) is a primitive characteristic of primates and that the hands moved and gripped objects and touched surfaces in varied ways in these different situations. In other words, the hands enabled a primate-typical way of life.

4 Functions of the Hand

With the broad functional importance of the hands to primates in mind, we adopt a conceptual framework following Jones and Lederman (2006) that is novel in comparative treatments of manual function. Jones and Lederman (2006) cast hand function into four categories falling along a continuum from sensory functions to skilled movements that do not involve prehension. These are (a) tactile sensing, (b) active haptic sensing, (c) prehension, and (d) nonprehensile skilled movements. Definitions and examples of common human actions in each category are given in Table 12.1. **Tactile sensing** serves to effect contact between the stationary hand and a surface or an object (which may be moving). This category is likely most employed in nonhuman primates in postural and locomotor activities (e.g., maintaining a secure grip), rather than in prehension of portable objects. The ubiquitous presence of dense sensory receptors in the glabrous skin of the feet as well as the hand (Talbot et al. 1968; Hoffman et al. 2004; see Chaps. 6 and 8) supports this proposal. **Active haptic sensing** serves to effect contact between the hand and a surface as the hand moves voluntarily over a surface or object. It has an exploratory character. This is the usual and preferred activity for identifying objects and extracting information about them. **Prehension** includes reaching to grasp an object and holding it. In humans, the configuration of the hand during prehension is determined by the objective of the task and the properties of the object(s) to be held and so changes dynamically as the task progresses. Most of our treatment in this chapter concerns this category—it is by far the most studied category of manual function in primates. **Nonprehensile skilled movements**

Table 12.1 Explanation of categories of manual functions (following Jones and Lederman 2006)

Category	Function	Characteristic manual action	Common situation in humans
Tactile sensing	Effect contact between the stationary hand and a surface or an object. Affords information about certain properties (e.g., surface texture), especially if the object or surface moves across the skin	The hand is passive. Affords information about certain properties (e.g., surface texture), especially if the object or surface moves across the skin	Climbing a rocky surface using the hands, bracing a hand against a wall, touching a hand lightly on a railing while descending a staircase
Active haptic sensing	Effect contact between the hand and a surface as the hand moves voluntarily over a surface or object	The hand is active	Feeling the texture of a fabric, squeezing an object to evaluate firmness, running fingers around an outer contour to evaluate shape and size of an object
Prehension	Reaching to grasp an object and holding it	The hand is active	Picking up a cup, buttoning a shirt, using a knife and fork, washing dishes
Nonprehensile skilled movements	Pointing and aiming movements, gestures, and actions with instrumental outcomes apparent in humans	The hand is active	Gestures during speech or other symbolic activities (e.g., dancing, conducting music), pressing keys on a keyboard, using a touchscreen, flipping a light switch

include pointing and aiming movements, gestures, and actions with instrumental outcomes. This category is highly apparent in humans. Although most nonhuman primates do not gesture, they still perform a variety of nonprehensile skilled movements as we shall see.

Jones and Lederman (2006) developed this framework to organize a vast literature about normal manual function in healthy humans arising from the disciplines of anatomy, neurophysiology, cognitive science, experimental psychology, developmental psychology and gerontology, kinesiology, hand surgery and rehabilitation medicine, haptic software and robotics, and human factors. We find it useful for our comparative purpose when considering the range of manual function in the primate order, which presents diversity in this domain as in others. This framework has the advantage of treating actions without an anthropomorphic focus; it does not privilege actions particularly valued in humans. It also has the advantage of linking manual function with the sensory systems of the skin, muscles, tendons, and joints. In our treatment of the variations across primate taxa in the given categories of manual function, we consider if the evidence allows us to assign "primitive" and "derived" status to aspects of function. Our assignments in these cases are speculative; we hope that this exercise will prompt others to provide more complete analyses to flesh out these ideas.

4.1 Tactile Sensing

All primates examined to date have rich sensory mechanoreceptors (Meissner corpuscles) in the glabrous skin of the hands and feet (Hoffman et al. 2004; see Chap. 6). These receptors are particularly densely packed beneath epidermal ridges (Martin 1990) and are sensitive to friction (see Chap. 8). Thus, primates are equipped to detect the stability of their grip on the weight-supporting substrate through passive tactile means. Some marsupials have similar receptors, but most other mammal species apparently do not (Winkelmann 1964). These features of tactile sensing shared with marsupials could be linked to movement in an arboreal environment. We are not aware of derived variations among primates in this category of manual function.

4.2 Active Haptic Sensing

In contrast to tactile sensing, in which the hand is passive, active haptic sensing (using the hands to locate, identify, and explore surfaces and objects) is unlikely to be related to locomotor requirements. Exploration is a fundamental aspect of behavior in mobile vertebrates. However, most vertebrates use olfaction, vision, and audition to explore rather than contact with forelimb appendages. Exploring with the hands is a primitive behavior in primates, but derived compared to other mammals in its variety and frequency. Glickman and Sroges (1966), in a classic study comparing exploratory behavior toward novel objects in zoo animals representing several vertebrate orders, commented that carnivores and primates exhibited more frequent investigatory behavior than rodents or a group of "primitive" mammals. Primates were the only order in which grasping, visual inspection, and manipulation were common. These behaviors are associated with processing and bringing food to the mouth in primates, whereas carnivores mainly use the forelimb when subduing and eating prey (batting, steadying; see also Power 2000 for review of the scanty, more recent literature on this topic concerning animals other than primates). For example, Whishaw (2005) does not describe manual actions that fit investigatory behavior as occurring in rats, although sniffing, rearing, and locomotor exploration are typical of rats (see also Berlyne 1966). In contrast, Bishop (1964) describes lemurs as moving their fingers over the edges of a cut plastic plate, feeling this novel object in an exploratory way, and lemurs, lorises, and galagos as playing with objects. Thus, although mammals from other orders investigate objects using their forelimbs to some extent, primates do so routinely. Within primates, the consensus view is that those species with omnivorous tendencies and that obtain foods via extractive methods (e.g., tearing or pulling items out of the wood or the soil, sifting through leaf debris, and breaking open husked fruits) have the most diverse repertoires of investigatory behaviors (Glickman and Sroges 1966; Parker and Gibson 1977; Torigoe 1985; Fragaszy and Adams-Curtis 1991; Westergaard 1992).

Aye-ayes (*Daubentonia madagascariensis*) present the most specialized actions in primates used for active haptic perception. Their specializations are in accord with the notion that extractive foraging promotes active haptic sensing. Aye-ayes possess an elongated third digit, which they tap extremely rapidly against woody substrates (called tap scanning) while searching for wood-boring larvae (Erickson 1994; Fig. 12.1). Tap scanning produces both active haptic and auditory information about the density of the material, which can then be used to guide and direct ayes-ayes' gnawing and probing with the elongated third digit (Erickson 1991; Erickson et al. 1998). Younger animals tap objects they encounter apparently for general exploratory purposes (Soligo 2005).

Other movements of the aye-aye's third digit are also specialized. When probing into cavities in search of food, the third digit moves independently from the other digits (Fig. 12.1). Uniquely among primates, the third digit can be moved laterally to enter acute, obtuse, and even right-angled extensions of tunnels because of the ball-and-socket metacarpophalangeal (McP) joint of this digit (Milliken et al. 1991). The terminal phalanx can be hyperextended as much as 30° to allow the finger to follow a tunnel and to move over an encountered object, so that when flexed, the elongated claw on the terminal phalanx acts as a hook to capture what it encounters (Milliken et al. 1991). The fourth digit may also be used for probing, but not tapping, and does not have the special ball-and-socket joint at the McP joint nor the enhanced flexibility in the terminal phalanx that the third digit possesses (Milliken et al. 1991).

Other primates use a different set of movements for active haptic perception. In humans, species-typical exploratory actions with the hands have been characterized by Lederman and Klatzky (1987, 1990). Exploratory actions include, for

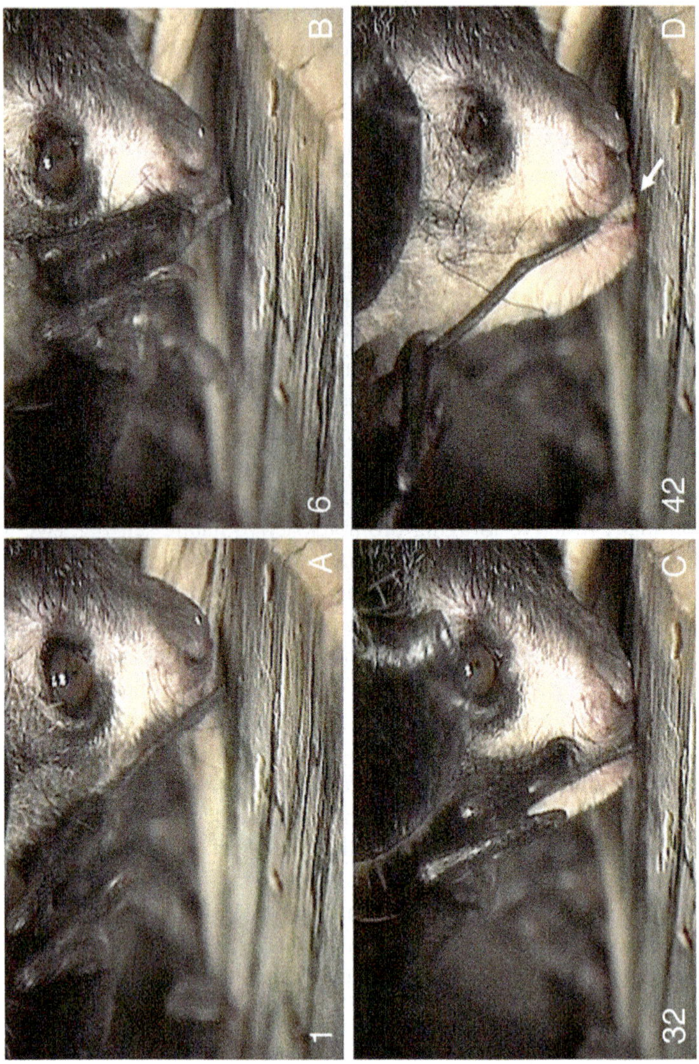

Fig. 12.1 A sequence of images of an aye-aye (*Daubentonia madagascariensis*) retrieving a grub from a hole in a tree trunk. The third digit is inserted (**a**, **b**) and then retracted (**c**), bringing the grub to the mouth (**d**). Numbers on the *lower left* refer to frame number (from 30 fps video). Images courtesy of Pierre Lemelin

example, rubbing to detect texture, probing to detect hardness, and contour following to detect shape. Actions of this kind are so commonplace in other species of primates that we are simply likely to overlook them. Researchers have looked for exploratory actions most systematically in tufted capuchin monkeys (*Sapajus apella*, formerly *Cebus apella*[1]), which have a reputation for varied and persistent manipulation of objects (reviewed in Fragaszy et al. 2004). For example, Lacreuse and Fragaszy (1997) observed humans and tufted capuchin monkeys reaching through an aperture in an opaque panel to find sunflower seeds deposited on the surface of irregularly shaped clay objects. Monkeys displayed the same classes of exploratory behavior as of humans (probe, pinch, enclosure, contour following, and lateral movement), although humans explored the objects more exhaustively, while monkeys explored them one region at a time. While foraging, capuchin monkeys probe inside holes and crevices, using their whole hands to feel for prey. They use a wide array of digital postures and movements, including movements of the index finger to scrape, pull, and tear (Figs. 12.2 and 12.3). Wild bearded capuchin monkeys (*Sapajus libidinosus*) knock nuts into pits on anvil surfaces as a means of detecting when they are in a stable position prior to cracking

Fig. 12.2 A wild bearded capuchin monkey (*Sapajus libidinosus*) illustrating several grips and independent use of the index finger in a sequence of photos taken while the monkey opened an immature cashew nut pod. (**a**) The monkey rubs the pod against the rough tree bark with its right hand. The thumb is parallel to the other digits, a common power grip in platyrrhine monkeys. (**b**) The monkey rubs the same pod in a different direction. The thumb is now at right angles to the other digits. This position of the thumb is not achieved by other platyrrhines as far as is now known. (**c**) The monkey inserts the index finger of the left hand into the breached pod of the immature nut to extract the kernel. The right hand holds the pod. The thumb on the right hand presses the nut toward the palm and the other digits. This is a second common form of a power grip in platyrrhine monkeys. Photos courtesy of Marino Junior Fonseca de Oliveira

[1]See Lynch Alfaro et al. (2012) for reclassification of the robust (tufted) species of the genus *Cebus*, including *apella* and *libidinosus*, into the genus *Sapajus*.

Fig. 12.3 A wild bearded capuchin monkey (*Sapajus libidinosus*) illustrating coordinated bimanual action and the use of the fingers to probe and scrape at a tucum nut that she has cracked open. (**a**) Scraping with an extended index finger. (**b**) The terminal phalanx of the index finger is flexed while scraping. (**c**) Pieces of the nut kernel are held beneath the index and middle fingers. Photos courtesy of Valentina Truppa

them with a stone (Fragaszy et al. 2013). They tap and lift stones in the process of determining which of the two stones is heavier (Visalberghi et al. 2009; Fragaszy et al. 2010). Wild tufted capuchins tap branches with their fingertips or fingernails while searching for invertebrate prey (Gunst et al. 2010). Captive tufted capuchins (*Sapajus* spp.) tap nuts with their fingertips in the course of choosing which ones to open, avoiding empty ones (Visalberghi and Neel 2003; Phillips et al. 2004), and use the index finger to find food in tubes (Spinozzi et al. 2007). Thus, capuchins use their hands to explore their environment in many different ways.

Torigoe's (1985) study remains the most comprehensive direct comparison of exploratory manual activity in many species of nonhuman primates. He presented a length of rope and a wood cube to members of 74 species of primates. These relatively uninteresting, inedible objects elicited exploratory behavior from all the groups he observed. In particular, guenons, mangabeys, and baboons (*Cercopithecus*, *Cercocebus*, and *Papio*, respectively), together with capuchins (*Cebus* and *Sapajus*), and lesser and greater apes (Hylobatidae and Pongidae) exhibited a wide range of exploratory actions.

Some aspects of active haptic perception have been studied in nonhuman primates using psychophysical and physiological methods (e.g., to characterize the function of the various receptors in the skin of the fingers and palm; Talbot et al. 1968). The increased density of Meissner corpuscles (mentioned earlier in relation to passive tactile sensing of friction) in the fingertips suggests that humans and other primates may have "tactile fovea" (regions of enhanced tactile sensitivity that would support active haptic sensing) at the ends of the digits (Hoffman et al. 2004). Hoffman et al. (2004) showed that of nine representative species of nonhuman primates, the more frugivorous species possessed a higher density of Meissner corpuscles in the fingertips, in accord with the hypothesis that these receptors afford perception of elastic texture of fruit (in addition to friction, as noted above), although other plausible relationships could not be disambiguated because of the small data set. However, sensitivity of the hands is largely unstudied in naturalistic tasks. It is plausible, but not confirmed, that anthropoid primates have more sensitive fingertips and engage in more active haptic perception using their fingertips than strepsirrhine primates and nonprimates. The value of active haptic sensing via the fingertips in foraging (e.g., to detect ripeness of fruit via palpation, banging, or tapping) is recognized (e.g., Dominy et al. 2004), but little studied. We lack comparative data on sensitivity to temperature, hardness, and other physical properties. Psychophysical studies of captive individuals could be extremely informative in this area.

4.3 Prehension

Prehension refers to grasping and holding an object in the hand (Napier 1956). Below we describe four types of prehension in nonhuman primates, each differentiated by hand posture and contact points when grasping an object: power grips, precision grips, in-hand movements, and compound grips. First, however, we describe variation in thumb opposition in primates, as this is the foundation of many forms of prehension.

4.3.1 Thumb Opposition

Napier recognized the importance of thumb opposition in prehension and defined "true" thumb opposition, in which the thumb rotates so that the distal pad is directly opposed to, and/or makes contact with, the distal pads of one or more of the remaining digits (Napier 1961; Napier and Napier 1967; Fig. 12.4). The ability to oppose the thumb to other digits enables a secure purchase on an object, although it is not required to hold an object, as we describe below. Primate taxa vary in the degree of thumb opposition they are able to achieve and, thus, the types of grips and movements they can execute. All Old World monkeys, apes, and humans are able to achieve true pad-to-pad thumb opposition (with the exception of the thumbless colobus monkeys) due to the saddle joint at the first carpometacarpal (CM1) joint (Napier and Napier 1967; Rose 1992). Prosimians and New World monkeys achieve what Napier (1960; Napier and Napier 1967, 1985) referred to as "pseudo-opposition," as the surface of the CM1 joint is hinge like, therefore limiting rotation of the thumb. Consequently, these primates can press the thumb to the lateral aspect of the second digit (i.e., pad to side), but cannot achieve full pad-to-pad contact. Although Napier did not consider pseudo-opposition to allow for precision handling, functionally both "true" and

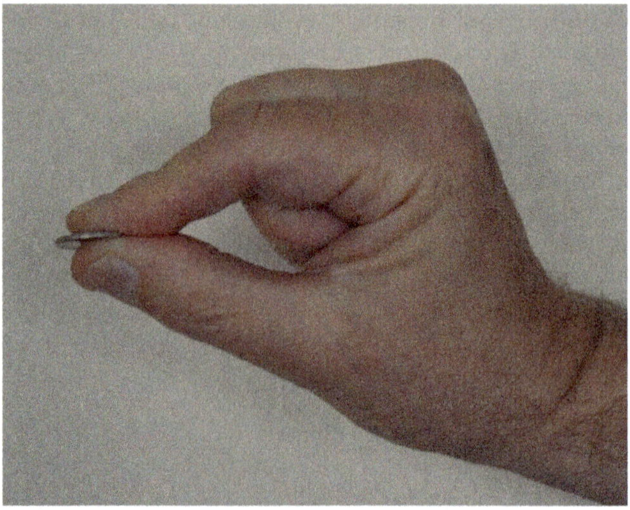

Fig. 12.4 Humans easily make full contact between pads of the thumb and index finger. This posture is what Napier (1961) identified as "true opposition." Photo by D. Fragaszy

Table 12.2 A sample of terminology used to describe power, precision, hook, and scissor grips among nonhuman primates

Category	Hominoids	Catarrhines	Platyrrhines	Prosimians
Power grip	• Power grip[a] • Squeeze grip[a] • Power grips[b]	• Hand wrap[c] • Finger-splayed wrap[c] • Hand-to-torso grip[c] • Climbing wrap[c]	• Power grip[d] • Enclosed thumb-palm[e] • Thumb/ index-palm[e] • Thumb-thenar[e] • Palm-thenar[e]	• One hand grasp[f]
Hook type	• Hook[a] • Loose grip[a] • Transverse hook grip[g] • Diagonal hook grip[g] • Extended transverse hook grip[g] • Extended diagonal hook grip[g] • Ulnar-palmar grasp[h]	• Climbing hook[c]		
Precision grip	• Pinch grip[a] • Pencil grip[a] • Tip-to-tip hold[g] • Pad-to-tip hold[g] • Pad-to-side hold[g] • Cup hold[g] • Radial-palmar grasp[h] • Imprecise grasp[h] • Pincer grip[h] • Tip-to-tip precision grip[i] • Lateral precision grip[i] • Precision grips[b] • Thumb-to-finger pad(s)[b] • Imprecise[b]	• Tip-to-tip precision grip[i] • Pad-to-pad precision grip[i] • Lateral precision grip[i] • Pad-to-side grip[c] • Pad-to-pad grip[c] • Thumb-to-second-third grip[c] • Tip-to-inside grip[c] • Three-tip grip[c] • Thumb-and-four-finger grip[c] • All-tip grip[c] • Thumb-to-outside grip[c]	• Precision grip[j] • Precision grip (types: 1-2; 1-2,3; 1-2,3,4; 1-2,3,4,5)[d] • Precision grip (types: I-II distal areas, I-II distal-to-other areas, I-II, III/distal areas, other variants)[e]	
Scissor type	• Scissor grip[a] • Index and middle finger grip[h] • Adduction grip[i] • Other grips[b]	• Between-finger grip[c]	• Precision grip (types: 2-3, 3-4, 4-5)[d]	

Note that within each category of movement defined by Napier (1956, 1980), there is a variety in which digits are used in the grasp and the contact points of the object on those digits. This table provides an overview of terms that have been used by researchers to describe various grips that fall within each broad category and highlights the need for a common lexicon that can be used across taxa

(continued)

Table 12.2 (continued)

[a]Byrne and Corp (2001); *Gorilla g. beringei*
[b]Jones-Engel and Bard (1996); *Pan troglodytes*
[c]Macfarlane and Graziano (2009); *Macaca mulatta*
[d]Costello and Fragaszy (1988); *Sapajus* spp.
[e]Spinozzi et al. (2004); *Sapajus* spp.
[f]Reghem et al. (2011); *Microcebus murinus*
[g]Marzke and Wullstein (1996); *Pan troglodytes*
[h]Tonooka and Matsuzawa (1995); *Pan troglodytes*
[i]Christel (1993); *Pan paniscus*, *Pan troglodytes*, *Gorilla g. gorilla*, *Pongo pygmaeus* and *P. abelii*, *Hylobates lar lar* and *H. l. moloch*, *Theropithecus gelada*, *Macaca silenus*, and *Cercocebus aterrimus*
[j]Christel and Fragaszy (2000); *Sapajus* spp.

"pseudo-opposition" allow for a multitude of precision grips that have been described in Old World and some New World primates (see Table 12.2). In our view, the categorical distinction between "true" and "pseudo-opposition," derived from anatomical considerations, does not contribute to our understanding of the array of functional uses of digits opposing each other and the palm that are evident in primates. For example, Pellis and Pellis (2012) described how aye-ayes use their thumb in a distinctive manner to secure food to the palm with no reference to whether they are using "true" or "pseudo-opposition."

4.3.2 Defining Power and Precision Grips

Napier (1956) coined the terms **power** and **precision** grips to describe the two most basic grasping patterns in humans. Many of the varieties of grips described for humans have also been identified in nonhuman primates (Table 12.2). With the power grip, an object is stabilized against the palm, and the digits converge around the object, as in holding the handle of a hammer or a tennis racket (Fig. 12.5a). Napier identified the **hook** grip as similar to the power grip, but without involvement of the thumb (as in holding the handle of a briefcase; Napier 1956). With the precision grip, an object is held by the digits alone (away from the palm), and in humans the thumb is abducted and rotated to face the palm, thus opposing the other digits, as in holding a tennis ball off the palm (Fig. 12.5b). Napier (1980) also defined the scissor grip, a form of precision grasping that does not involve the thumb (as in holding a pencil between the index and middle fingers). While the properties of the object to be grasped influence whether predominantly a power or precision grip is needed, often a key factor is the intended action (Napier 1956, 1980). For example, a precision grip is used to align the lid of a jar to the threads, but a power grip can be used to tighten the lid (Napier 1980). Due to varying object properties and the varying actions that the objects afford, there is a multitude of forms of power and precision grips that can be adopted (summarized below). In this chapter, we use the term "precision grip(s)"

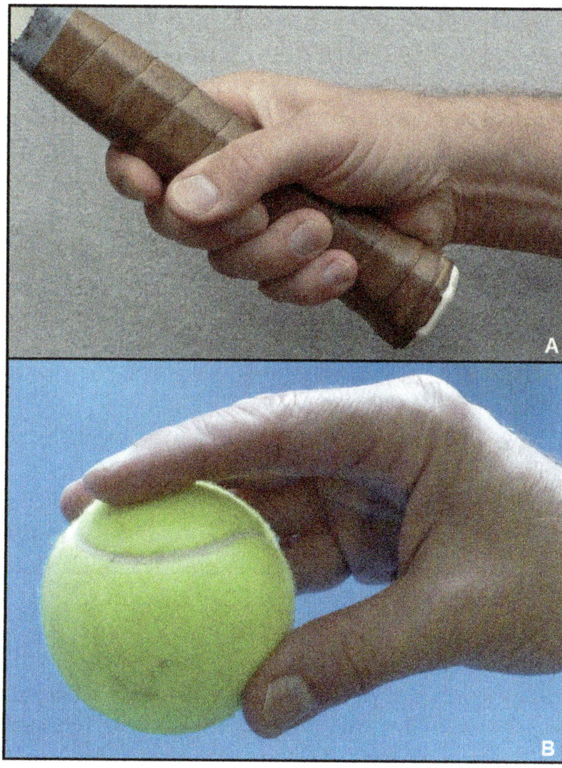

Fig. 12.5 Human power (**a**) and precision (**b**) grips, as described by Napier (1956). Photos by D. Fragaszy

in a general sense to refer to the grasping of an object with the distal aspects of the digits, acknowledging that animals can execute precision grips with different contact points on the digits, all of which can be individually recognized (e.g., Marzke et al. 2009; Pouydebat et al. 2009; Table 12.2).

4.3.3 Power Grips

All primates are able to execute a power grip quite efficiently, although there is considerable variation in the orientation of the digits during closure, depending on the degree of thumb opposition. Usually, a power grip involves contact by all five digits, each flexing and securing an object against the palm. However, a power grip can also be achieved with fewer than five digits, as long as the digits involved are gripping an object against the palm.

Lorisiform and lemuriform species are characterized by a single prehensive pattern, a power grip, with final grips determined by the shape of the object they grasp. Bishop (1964) identified general patterns of prehension in strepsirrhine taxa in terms of posture during reaching and the point of contact of the hand with an object, as well as in terms of the forms of contact of digits with the palm and with each other. Lorisoids picking up small objects from a flat surface reached with the fingers splayed, contacted the object first with the palmar pads, and then closed all the digits convergently, flexing the interphalangeal (IP) joints and pressing the object between the digital pads and the interdigital pads (Fig. 12.6a). Bishop (1964) described this pattern as precise, although without differentiated control of the digits. The lemuriform pattern differs from the lorisiform pattern in Bishop's (1964) descriptions in that the lemurs reached with roughly parallel (rather than splayed) digits and contacted objects with the digits first, rather than the palm, in common with anthropoid primates. Lemurs flex the fingers at the McP and proximal IP joints, so that an object is held between the digital pads and the proximal palmar pads (Fig. 12.6b). Lemurs also hooked small objects under the two most distal phalanges of digits 2–5 and pulled them toward the palm, with the thumb in line or at about 90° to the palm.

Aye-ayes present a unique variation on the power grip. In this species, the long third digit remains on the dorsal side of the hand when not in use for probing or tapping. During locomotion (in which a power grip is used), especially during head-first descent (which is common in aye-ayes), the third digit is often hyperextended at the McP joint, flexed at the IP joints, and held to one side (Krakauer et al. 2002; Soligo 2005; Kivell et al. 2010). In this position, it does not participate in grasping the substrate. The longer, more robust fourth digit supports a strong grip on the substrate (Soligo 2005). Although aye-ayes cannot achieve a high degree of thumb opposability, they are able to abduct the thumb enough to secure a small object against the palm without assistance from the other digits. This ability, shared only with sifakas among the lemuriforms, suggests that the relative independence of the thumb

Fig. 12.6 A slender loris (*Loris tardigradus*) and a sifaka (*Propithecus verreauxi*) illustrate the different prehensive patterns described by Bishop (1962, 1964) for lorises and lemurs. (**a**) The loris holds an insect against the palm with all interphalangeal joints flexed. (**b**) The sifaka holds plant matter with flexed metacarpophalangeal and proximal interphalangeal joints, pressing it against the proximal palmar pads with the distal pads of the digits. Images courtesy of Pierre Lemelin

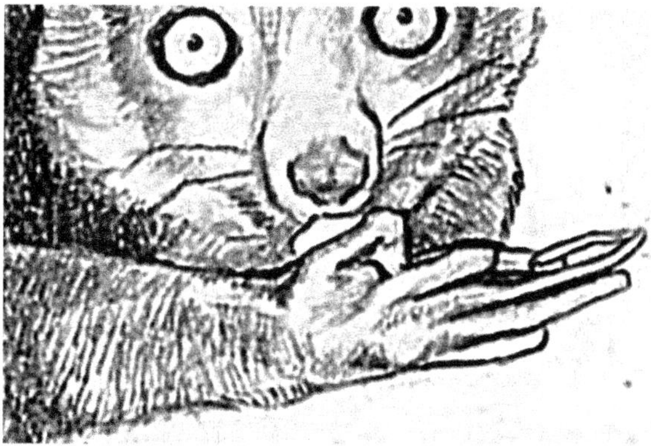

Fig. 12.7 Drawing of an aye-aye (*Daubentonia madagascariensis*) holding a food item against the palm with the thumb and with minimal support from the other digits. Reprinted from Pellis and Pellis (2012) with permission of the publisher

is a derived feature that allows the aye-aye to hold the small extracted food items away from its cumbersome, elongated third and fourth digits (Pellis and Pellis 2012; Fig. 12.7).

Tarsiers, nocturnal and small-bodied primates, also present a taxonomically unique pattern of prehension, which can be viewed as a variation on a power grip. They feed on vertebrate and invertebrate prey captured with one or both hands (Niemitz 1984). Niemitz (1984) describes the typical grip of Western tarsiers (*Tarsius* (or *Cephalopachus*) *bancanus*) in which digits 2–4 flex toward the palm, with the pollex flexing more or less parallel to digits 2–4. The fifth digit rotates at the McP joint and flexes to be perpendicular to the other digits, "much like the thumb" (Niemitz 1984, 69; Fig. 12.8). This movement is reminiscent of grasping behavior in rats (see above), which sometimes rotate the fifth digit toward the palm. This hand posture prevents captured mobile prey from escaping. No other primate rotates the fifth digit; this is apparently a derived feature present only in tarsiers. Casual observations of the Philippine tarsier (*Tarsius* (or *Carlito*) *syrichta*) reveal the typical parallel position of the thumb, but do not indicate the rotation of the fifth digit described by Niemitz (1984) (P. Lemelin, personal communication). As the genus *Tarsius* has recently been recognized to be composed of three genera (Groves and Shekelle 2010) and multiple species within each genus, these discrepancies in hand behavior parallel generic variability in tarsiers. In general, we know less about prehensive function in tarsiers than in other primate taxa. Further study of manual function and associated neuromuscular systems in this taxon will be particularly enlightening.

Platyrrhine and catarrhine species use a variety of power grips, and the terminology to describe these grips is likewise varied (Table 12.2). Preshaping of the hand to match the size of the object to be grasped or the features of the surface to be contacted has been described in macaques and tufted capuchin monkeys (Christel and Fragaszy 2000; Roy et al. 2000; Christel and Billard 2002). The precise orientation of the fingers with respect to the object held against the palm varies according to the shape and size of the object relative to the size of the hand and on the degree of radial abduction of the thumb. For example, an individual holding a long, thin

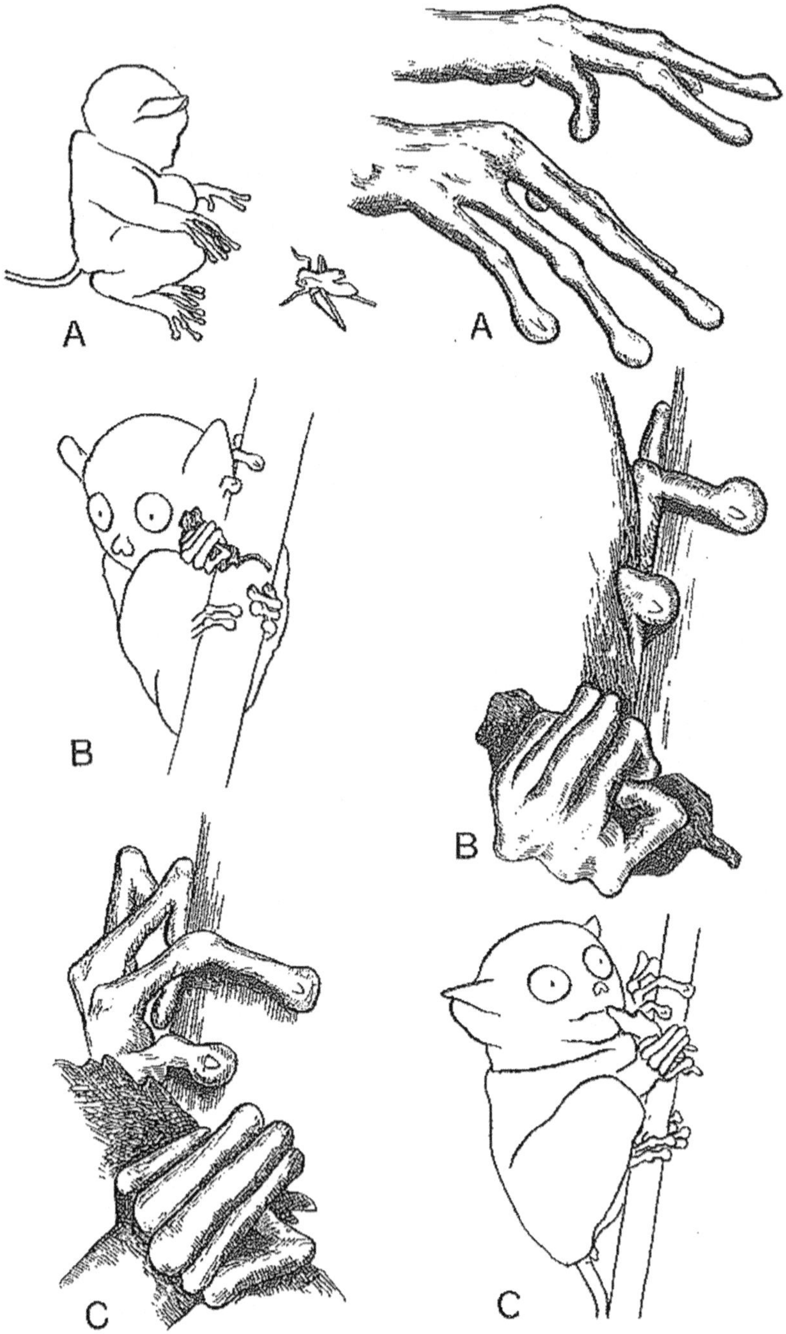

Fig. 12.8 A tarsier (*Cephalopachus bancanus*) just prior to capturing an insect, with fingers splayed (**a**), holding an insect with the fifth digit rotated to be perpendicular to the other digits (**b** and **c**), and eating it (**c**). The grip posture illustrated in **b** and **c** is observed uniquely in tarsiers. Drawn by C. Niemitz and W. von Bischoffshausen, and reproduced from Niemitz (1984) with permission of the author

object commonly abducts the thumb to wrap it over the dorsal side of the other digits. Platyrrhine monkeys can achieve this kind of power grip, although they often keep all the digits parallel during grasping due to their inability to rotate the thumb, as illustrated in the golden-handed tamarin (*Saguinus midas*) by Lemelin and Grafton (1998). Like the aye-aye and sifaka, squirrel monkeys can press the thumb laterally on the sides of an object, and the proximal pollical phalanx can flex 180° to contact the palm, providing a secure grip for small objects (Fragaszy 1983; Costello and Fragaszy 1988; Lemelin and Grafton 1998). Tufted capuchin monkeys preferentially grasp small moving targets using a power grip, but use precision grips on small stationary targets (Costello and Fragaszy 1988). Because they can abduct the thumb with respect to the palm, catarrhine primates probably achieve a greater variety of power grips compared to platyrrhine monkeys. In this regard, Macfarlane and Graziano (2009) provided an extensive description of power grips used by free-ranging rhesus monkeys (*Macaca mulatta*, Table 12.2).

4.3.4 Precision Grips

Precision handling involves the manipulation of objects with the distal surfaces of the digits used in any combination (Landsmeer 1962). The proportional length of the digits and the degree of opposability of the thumb determine the contact points between an object and the digits in a precision grip. Prosimians and many New World monkeys are limited in precision handling due to their inability to rotate the thumb. With the exception of capuchin monkeys (*Cebus* and *Sapajus*), these primates move their digits in unison and parallel to one another and are unable to achieve a precision grip between digits 1 and 2, although many New World monkeys can use a scissor grip between digits 2 and 3 or digits 3 and 4 (*Cacajao*, *Ateles*, and *Lagothrix*, Bishop 1964; *Sapajus*, Costello and Fragaszy 1988; *Saguinus*, Lemelin and Grafton 1998).

Capuchin monkeys (*Cebus* and *Sapajus*) possess some individuated control of the digits, as demonstrated in probing actions with digit 2 and in an array of precision grips executed between the thumb and other digits (Costello and Fragaszy 1988; Christel and Fragaszy 2000; Spinozzi et al. 2004, 2007; Fig. 12.2, Table 12.2). Christel and Fragaszy (2000) noted that tufted capuchin monkeys move single digits independently when the fingers rest on the surface of a board in contrast to a relatively uniform preshaping pattern when fingers are coordinated in space. It appears that contact of the hand with a substrate (a surface or tube) supports better coordination of single digits for diverse grips. Aside from capuchins, Bishop (1964) reported that uakari monkeys (genus *Cacajao*) use the sides of digit 1 against those of digit 2 when manipulating small objects, but her initial observations have not been replicated. Indeed, we are not aware of additional information about prehension in *Cacajao* other than Bishop's (1964) landmark studies; this is clearly a topic deserving further investigation.

Catarrhine primates, particularly cercopithecoids, use a variety of precision grips in foraging and grooming. For example, gelada baboons (*Theropithecus gelada*) pluck grass between the thumb and index finger while holding a bundle of grass against the palm with the other digits (Maier 1993). Japanese macaques (*Macaca fuscata*) use several variations of a delicate pad-to-pad precision grip between the thumb and index finger to remove louse eggs from the hair follicle while grooming (Tanaka 1998). All catarrhine monkeys routinely use precision grips to pick up small objects. Colobus monkeys (*Colobus*), which have a vestigial thumb, flex the index finger until it touches the pollical nub when picking up small objects (Bishop 1964). Assuming similar function across catarrhine genera, these monkeys exhibit substantial preshaping of the hand when reaching for objects of different sizes (Roy et al. 2000) and kinematic similarities to humans during reaching for prehension (Christel and Billard 2002).

In great apes, digits 2–5 are relatively long and curved, and the thumb is proportionally short compared to other primates, thus limiting pad-to-pad contact in thumb opposition (Napier and Napier 1967; Marzke and Wullstein 1996). Although all great apes are capable of pad-to-pad contact between the thumb and index finger, they have lesser contact area and pressure compared to humans and typically achieve a precision grip between the thumb pad and side of the index finger (Christel 1993; Fig. 12.9). Humans achieve the greatest area of pad-to-pad contact in a precision grip among primates due to a greater ability cup the palm and a relatively long thumb (Marzke 1983; compare Figs. 12.4 and 12.9a). Nevertheless, many different types of precision grips have been documented in apes (e.g., see Jones-Engel and Bard 1996; Marzke and Wullstein 1996; Table 12.2). Christel (1993) documented variations in precision grips in humans, chimpanzees, and bonobos (*Pan troglodytes*

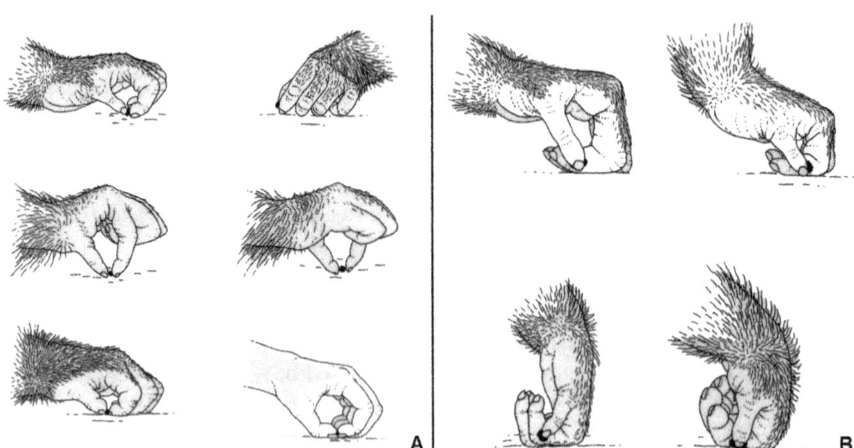

Fig. 12.9 Precision grips in hominids when picking up a small object. (**a**) Opposition of the thumb to the index finger. *Top* row, chimpanzee (*Pan troglodytes*); *middle* row, orangutan (*Pongo pygmaeus*); and *bottom* row, gorilla (*Gorilla gorilla*) (*left*) and human (*Homo sapiens*) (*right*). (**b**) Lateral opposition of the thumb to the middle phalanx of the index finger. All chimpanzees (*Pan troglodytes*). Drawings by M. Christel and H. Schulze, unpublished and reprinted from Christel (1993) with permission of the publisher and the artists

and *Pan paniscus*), Western gorillas (*Gorilla gorilla*), orangutans (*Pongo pygmaeus*), and white-handed and moloch gibbons (*Hylobates lar* and *Hylobates moloch*) picking up small pieces of food (Fig. 12.9). Pad-to-pad contact between the thumb and index finger was achieved by chimpanzees, gorillas, and orangutans (see Fig. 12.9). For this task, humans habitually preferred the pad-to-pad precision grip, whereas apes frequently used the lateral side of the index finger in precision grips.

4.3.5 In-Hand Movements

The above discussion highlights the routine use of a variety of grasping patterns in the daily lives of primates. This variety is a hallmark of the order Primates and distinguishes it from other taxonomic groups. Some primates act with objects in skillful ways beyond static grasping. They turn over, rotate, and otherwise move objects held within the hand using **in-hand movements**. An in-hand movement involves manipulation of an object within a single hand using the digits of that hand alone, such as flipping the cap off of a pen (Elliott and Connolly 1984). Human infants exhibit simple forms of in-hand movements within the first year of life, and children are able to perform all documented forms of in-hand movements by the age of 8 years, although with less proficiency than adults (Exner 1992; Manoel and Connolly 1998).

Elliott and Connolly (1984) provided a comprehensive taxonomy of the various forms of in-hand movements used by humans, which was adapted to describe in-hand movements in children (Exner 1992), in-hand movements in chimpanzees (Crast et al. 2009), and in-hand movements in wild mountain gorillas and chimpanzees (Byrne and Corp 2001; Corp and Byrne 2002) (Table 12.3). In-hand movements are categorized as simultaneous, in which an object is moved by the concurrent movement of two or more digits, or as sequential, in which the digits move sequentially to change an object's orientation in the hand (referred to as sequential movements and palmar combinations; see definitions in Table 12.3). Here we discuss which primates are known to use these forms of prehension.

In Elliott and Connolly's (1984) classification of in-hand movements, simultaneous movements are distinguished as simple synergies or reciprocal synergies. A simple synergy involves prehending an object using a static grip and then flexing and extending the digits to move the object through space, without changing the object's orientation (e.g., moving a needle through fabric using only the flexion/extension of the digits). Presumably, any primate that can use a precision grip is able to flex and extend its digits while maintaining a grip on the object (i.e., a simple synergy). This has been documented in adult chimpanzees during a movement called "turnover" (Crast et al. 2009); but even primates using whole-hand control should be able to execute simple synergies by grasping an object with the tips of all five digits and then flexing those digits.

A reciprocal synergy involves the simultaneous movement of two or more digits in opposite directions to turn an object about one of its axes, as in turning a screw. Reciprocal synergies emerge in human children around the age of 2–3 years and are

Table 12.3 Known terminology used to describe in-hand movements among nonhuman primates using Elliott and Connolly's (1984) classification system

In-hand movement	Definition	Hominoids	Catarrhines	Platyrrhines
Simple synergies	Prehending an object using a static grip and then flexing and extending the digits to move the object through space without changing the object's orientation (e.g., moving a needle through fabric using only the flexion/extension of the digits)	• Combine[a]		
Reciprocal synergies	Movement of the digits after prehension to rotate object about one axis; the thumb moves in the opposite direction of digit(s) simultaneously (e.g., turning a doorknob using only the fingers)	• Thumb abduction/adduction[b] • Roll[b]		• Roll[c]
Sequential movements	Concurrent movement of digits in opposing directions allowing for movement of an object about more than one axis (e.g., rotating a pen within the hand)	• Manipulate[a,d] • Rotation[b] • Turnover[b]		
Palmar combination	Object is stabilized in the palm, and independent digit movements are used to manipulate part of the object, as in pushing off the cap of a pen	• Thumb push[b]		
Transfer[e]	Object is grasped in a precision grip with digits 1–2, brought into the palm with a simple synergy and secured with digits 3–5 while the first two digits grasp another object	• Digit-role differentiation[a] Unimanual multitasking[d]	• Transfer[f,g]	

[a]Byrne and Corp (2001); *Gorilla g. beringei*
[b]Crast et al. (2009); *Pan troglodytes*
[c]Crast (2006); *Sapajus* spp.
[d]Corp and Byrne (2002); *Pan troglodytes*
[e]The term "transfer" was not used by Macfarlane and Graziano (2009) or Maier (1993), nor was it a category of movements described by Elliott and Connolly (1984). We include it here because it is a type of in-hand movement that is distinct from sequential movements in some ways. In this regard, Byrne and Corp (2001) considered digit-role differentiation and unimanual multitasking to be forms of sequential movements
[f]Macfarlane and Graziano (2009); *Macaca mulatta*
[g]Maier (1993); *Theropithecus gelada*

refined by the age of 7–8 years (Exner 1992; Manoel and Connolly 1998). Similarly, young chimpanzees at age 5 years are fully proficient at using reciprocal synergies to manipulate small objects, as adult chimpanzees do in a variety of ways, routinely and with ease (Crast et al. 2009; Table 12.3). Because capuchins have a relatively high level of individuated control for digits 1 and 2, we investigated whether they can use reciprocal synergies using the same experimental design that elicited these movements in chimpanzees. Two monkeys each used a rudimentary reciprocal synergy once (Crast 2006). They both placed their forelimb against the vertical panel and, while holding the object in a precision grip against the surface of the panel, flexed the index or middle finger toward the palm as the thumb simultaneously extended, thus turning the object about one of its axes. These observations, however, require replication, as the events were rare and the usual means of reorienting the object in the hand was to move it to the other hand or the mouth. This suggests that the neuromuscular anatomy required to execute more complex in-hand movements is less developed in capuchins compared to apes and humans.

Although adult chimpanzees did not execute sequential movements as finely as adult humans, they performed a variety of forms including rotations and turnovers (Crast et al. 2009). To perform a turnover, adult chimpanzees grasped the object between the index and middle fingers and then flexed those digits, bringing the object into the palm (a simple synergy); the object was then rolled over the index finger to be grasped between it and the thumb (Fig. 12.10). When rotating an object within one hand, adult chimpanzees frequently cradled the object in the palm while using the digits to readjust the object's orientation. In contrast, humans are quite capable of moving an object of comparable size without the use of the palm.

4.3.6 Compound Grips and In-Hand Movements

Macfarlane and Graziano (2009) analyzed the rich variety of spontaneous manual actions in rhesus monkeys (*Macaca mulatta*) ranging freely in a natural setting, including gripping an object with two hands concurrently and bracing an object

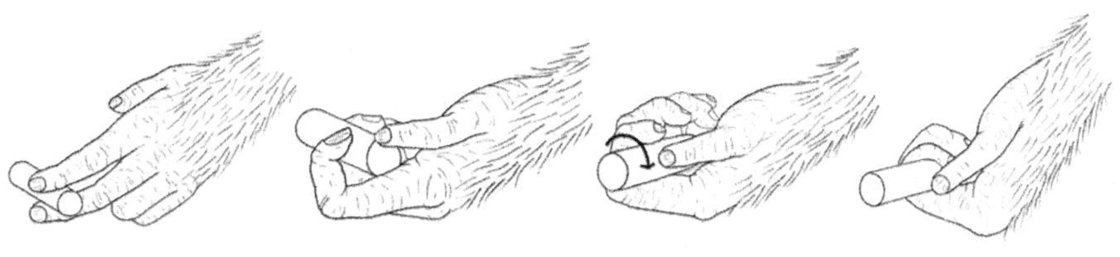

Fig. 12.10 Example of in-hand movements in the chimpanzee (*Pan troglodytes*). The drawings illustrate a "turnover" sequence used by an adult chimpanzee. This sequence was used to pick up an object from the floor and reorient the object within the hand to align it with a correspondingly shaped cutout in a transparent panel. The object is grasped between the second and third digits, which flex and bring the object toward the palm; the object is then rolled over the distal end of the index finger by reciprocal movement of the thumb and index finger; the object is finally grasped in a precision grip between the tip of the thumb and the side of the index finger (drawn by Cheryl Reese)

against the torso. In addition to the typically defined grips of one object held in one hand, these authors defined a variety of compound grips when the monkey held more than one object in one hand. For example, the monkey could hold one or more objects in a "storage grip," a power grip with the object braced against the palm using digits 4 and 5, and concurrently pick up another object with digits 1–3. Interestingly, Macfarlane and Graziano also observed the macaques shifting objects picked up with the first two digits to digits 3–5 for storage and then picking up an additional object with the first two digits (M. Graziano, personal communication). The transfer of an object within the hand using the digits of that hand alone is by definition an in-hand movement: the monkey used one grip to prehend an object (precision grip with digits 1–2), a simple synergy to bring the object into the palm, and then a second grip to hold the object (power grasp with digits 3–5).

Compound grips and transfer movements have also been identified in other primates, including wild gelada baboons (*Theropithecus gelada*) that continuously pluck grass between the thumb and index finger and transfer the grass to a bundle held against the palm with the other digits (Maier 1993). Wild mountain gorillas (*Gorilla gorilla beringei*) use the same movement with the thistle plant *Carduus nyassanus* (Byrne and Corp 2001), as do chimpanzees (*Pan troglodytes*) with the fruit *Saba florida* (Corp and Byrne 2002). Gorillas' movements were termed "digit role differentiation" and chimpanzees' movements "unimanual multitasking" (one hand carries out two actions simultaneously); both were identified as sequential in-hand movements (Byrne and Corp 2001; Corp and Byrne 2002). Transferring an object in this manner may be considered distinct from reciprocal and sequential in-hand movements because concurrent movement of digits in opposite directions is not necessarily present. "Transfer" may qualify as a new category of in-hand movement, as it requires a series of grasps and movements in order to move an object's location within the hand (Table 12.3).

Both gorillas and chimpanzees in Byrne and Corp's studies also demonstrated a more sophisticated form of in-hand movement, termed manipulate, in which an object was rearranged within one hand using the digits of that hand alone. As shown in captive chimpanzees (Crast et al. 2009), skill at manipulating and transferring items within the hand increased with age (proficiency develops by age 4–6; Byrne and Corp 2001; Corp and Byrne 2002). Notably, studies have not examined whether catarrhine monkeys use in-hand movements other than the transfer movement. Clearly, however, macaques, gorillas, and chimpanzees make use of compound grips when executing a transfer movement. Chimpanzees probably also utilize a compound grip together with a transfer movement during nutcracking, as they often store one or more nuts in the palm of the hand while preparing another nut to be hammered against an anvil (Boesch and Boesch 1993).

Compound grips and in-hand movements probably rely upon shared neuromotor abilities to activate muscles of the hand and multiple digits independently and in varied coordinative structures. These dynamic aspects of manual function, which are elaborated in humans, require further investigation by primatologists. We know far more about static grip postures than about movements of the hands while manipulating gripped objects. With respect to taxonomic variation, at present we can only say that in-hand movements and compound grips are evident in several catarrhine taxa, require validation in capuchin monkeys, and are totally uninvestigated in other groups of primates. Researchers with interests in robotics and prosthetics have developed alternative frameworks to measure movements of the hand (e.g., Bullock and Dollar 2011; Fu and Santello 2011), and we can expect that treatments of this aspect of manual function in nonhuman primates will eventually broaden to include these frameworks as well.

4.4 Nonprehensile Skilled Movements

Movements that fall within the last category of manual function in Jones and Lederman's (2006) framework, nonprehensile skilled movements, have received attention from behavioral scientists studying nonhuman primates, but not from functional anatomists or others concerned with the evolution of manual function in primates, to our knowledge. Nonprehensile skilled movements include actions such as the following: (1) urine washing (urinating into the palm of the hand and wiping the urine on the sole of the foot, a common behavior in New World monkeys and prosimians), (2) spreading the hair while grooming (most taxa), (3) sweeping loose debris off a surface (as capuchins do when cracking nuts or digging up roots; D. Fragaszy, personal observation), (4) searching in loose plant debris for animal prey (as some species of lion tamarins, *Leontopithecus*, do; Rylands 1993; Passos and Keuroghlian 1999), (5) positioning the hand to collect ants as they climb onto the hand (slender loris, *Loris*; Kumara, et al. 2005), (6) capturing small objects on the fingertip by pressing them against a hard surface (macaques and chimpanzees; Christel 1993), (7) probing into a narrow opening using a single digit (capuchins, aye-ayes, and chimpanzees), and (8) gesturing communicatively (as in pointing in chimpanzees; Leavens et al. 1996).

Humans produce a far larger variety of these movements than any other species, including manual gestures used in sign language, actions used to modify objects (e.g., molding clay), and actions used to control devices (e.g., keyboards, musical instruments). The same features of movement control that enable compound grips and in-hand movements (i.e., finely controlled and independent movements of the digits, in sequence and concurrently), together with elaborated proprioceptive and kinesthetic sensitivity, also support nonprehensile skilled movements. In this case, it is interesting to note that capuchin monkeys—which may be able to move the thumb in opposite directions from the index finger and move the index finger to the thumb in a precision grip—also move the index finger with partial independence when probing into an opening (Spinozzi et al. 2007; Figs. 12.2c and 12.3a). When they probe with the index finger, the index finger leads and the other digits are initially partially flexed. They flex further toward the palm as downward pressure is applied to the object with the tip of digit 2. That is, the full extension of digit 2 may be passive, while digits 3–5 are flexed (Fig. 12.3b). Thus, the control of digit 2 may be less independent in capuchin monkeys than in chimpanzees, which can point the extended index finger of an unsupported hand (with digits 2–5 flexed) at a distal target (Leavens et al. 1996). The use of the index finger in probing has not been described for catarrhine monkeys, to our knowledge, but they do scrape with the index finger (e.g., Tanaka 1998). Underlying the ability of capuchins and catarrhines to move the index finger independently, and to oppose the thumb and the index finger, is a relatively elaborated corticospinal tract compared to other primates (for details, see Bortoff and Strick 1993; Lemon 1993; Lemon and Griffiths 2005; see Chap. 6). No doubt this elaboration is also involved in the production of in-hand movements and compound grips.

5 Summary of Major Differences in Manual Function Among Primates

Skilled forelimb movements (aimed reaching, prehension with a single appendage using digital closure) are primitive characteristics in the order Primates that are shared with some other orders of mammals. Some manual functions are apparently derived in primates, and some are derived within certain clades of nonhuman primates. For discussion purposes only (as the data are currently inadequate for proper analysis), a tentative list of these derivations in various phylogenetic groups is given in Table 12.4.

We examined manual function using the four classes of function identified by Jones and Lederman (2006). Tactile sensing is primitive in primates and we are unaware of variations in this function across primates. Active haptic sensing is also probably primitive in primates, but we have few studies of this aspect of manual function in primates and even fewer in other orders. Several derivations have

Table 12.4 Derived characters in manual function observed to date in primates

Taxonomic group	Derived characters
Daubentonia	Specialized use of digit 3 in tapping and extraction, modified joints permitting rotation at McP[a] joint and hyperextension at distal IP[b] joint
Tarsiiformes	Rotation at McP joints of the thumb and digit 5, permitting an oppositional grip; possibly, gains in independent movements of digits (suggested by Niemitz 1984, but not confirmed)
Anthropoidea	Adjacent digits can be used in scissor grip
Cebus and *Sapajus*	(a) The thumb moves medially to provide functional opposition precision grip with digit 2 and digit 3 (b) Partial independent movement of digit 2, affording probing/extraction (c) Reciprocal movement of the thumb and digit 2 (nascent in-hand movement)
Catarrhini	(a) The thumb "fully" opposes other digits (achieves pulp-to-pulp contact) (b) Compound grips and transfer in-hand movements
Hominoidea	Full independent movement of digit 2 (extension and flexion)
Hominidae	Reciprocal and sequential in-hand movements

[a]*McP* metacarpophalangeal

[b]*IP* interphalangeal

occurred in prehensile function. Principal among these are the following: (1) the ability to press digits together to hold objects between the lateral sides of adjacent digits in a scissor grip, which apparently evolved after anthropoids diverged from prosimians; (2) the ability to oppose the thumb and other digits and individuated control of the thumb and digit 2, which evolved early in the radiation of catarrhine primates and independently and to a lesser extent in cebids; (3) the ability to move objects held within one hand and to perform concurrently more than one manipulative action with some digits and another manipulative action with the remaining digits of the same hand (in-hand movements and compound grips), which evolved early in the radiation of catarrhine primates; (4) the ability to rotate digit 5 toward the palm, reported only for tarsiers; and (5) the independent movement of digit 3 for tapping and probing and the placement to one side of this digit during some locomotor circumstances, both present only in aye-ayes. Nonprehensile skilled actions are probably derived in primates compared to other mammals, specialized across taxa, and vastly elaborated in humans. The independent use of the index finger in probing and scraping actions is the most commonly described nonprehensile behavior in nonhuman primates; this ability evolved in catarrhine primates and independently, and to a lesser extent, in cebids.

6 Future Directions

We see four topics as high priority for attention in future research. First, there is the problem of a common vocabulary to identify hand postures and movements that applies across primate taxa. Table 12.2 is intended to highlight the overlap in prehensile abilities across primates and the problem of having multiple taxonomies to describe them. We need to develop a standard terminology to describe grips involving specific contact points on the hand and sequences of movements.

Second, active haptic perception requires attention because uses of the hand in exploration are mostly unknown outside of humans. It is telling that our most comprehensive comparative studies in this area were published decades ago and were meant as initial explorations of this topic. They relied upon presentations of large and simple objects, such as wooden blocks and rope (Glickman and Sroges 1966; Torigoe 1985), and did not address the precise manner in which individuals use the body to investigate and handle objects. Bishop (1962, 1964; Jolly 1964) set the standard for more detailed studies in the early 1960s, and we have yet to approach that standard again.

Third, although much of the discussion concerning the evolution of manual function in primates has centered around grips, we suggest that in-hand movements and nonprehensile movements contribute as much or more than grips per se to the differentiation of manual function across primates. These movements underlie many species-typical skilled actions in humans, including exploratory functions of the hand. We know almost nothing about the extent to which Old World monkeys and lesser apes use these movements in their daily lives and very little about how great apes use them. The comparative approach can reveal much more about the evolution of complex and intricate manual dexterity, which is so natural and routine for humans that it goes completely unappreciated in our day-to-day activities. We see this dimension of manual function as an important target for comparative research.

Fourth, we need to assess manual actions in natural circumstances, to develop an integrated comparative understanding of manual function in daily life of various species across the order. At present we know variations exist across taxa in the details of manual function, but we do not have a systematic basis for comparing either the quality or quantity of variation (see Leca et al. 2010 for a detailed description of stone-handling patterns in Japanese macaques and Macfarlane and Graziano 2009 for an equally detailed but independent treatment of manual function in rhesus macaques). One promising direction for understanding ecological and behavioral factors associated with different types of hand movements is to explore evolutionary convergence in the forms and the kinematics of these actions (e.g., Reghem et al. 2013) in New and Old World primates. This effort will be possible once a systematic taxonomy of manual function in primates is in hand (pun intended). If we can achieve this at the centennial of the publication of Wood Jones' (1920) insightful volume, we will be able to answer more of his and John Napier's fundamental queries about this appendage that is so central to human activity and human identity.

Acknowledgments and Dedication We thank Marino Junior Oliveira and Valentina Truppa for the use of their photos, Pierre Lemelin for providing photos of lemurs and lorises, and Marianne Christel for drawings of hominid grips. Thanks to Bernard Wood for lending his copy of Wood Jones (1920) to D.F. Thanks to Myron Shekelle and Carsten Niemitz for discussing manual function in tarsiers. Thanks to Elisabetta Visalberghi, Mary Marzke, and Valentina Truppa for commenting on early drafts of the manuscript and Pierre Lemelin and two anonymous reviewers for their constructive comments on the manuscript. We dedicate this chapter to Alison Jolly, a pioneer in the behavioral study of nonhuman primates and in particular of their manual function, who passed away in early 2014. We miss her deep knowledge, good sense, and good humor.

References

Berlyne DE (1966) Curiosity and exploration. Science 153:25–33

Bishop A (1962) Control of the hand in lower primates. N Y Acad Sci 120:316–337

Bishop A (1964) Use of the hand in lower primates. In: Buettner-Janusch J (ed) Evolutionary and genetic biology of primates. Academic Press, New York, pp 133–223

Boesch C, Boesch H (1993) Different hand postures for pounding nuts with natural hammers by wild chimpanzees. In: Preuschoft H, Chivers DJ (eds) Hands of primates. Springer-Verlag, Vienna, pp 31–43

Bortoff GA, Strick PL (1993) Corticospinal terminations in two new-world primates: further evidence that corticomotoneuronal connections provide part of the neural substrate for manual dexterity. J Neurosci 13:5105–5118

Bullock IM, Dollar AM (2011) Classifying human manipulation behavior. IEEE International Conference on Rehabilitation Robotics (ICORR), Zurich, Switzerland, 29 June to 1 July 2011

Byrne RW, Corp N (2001) Manual dexterity in the gorilla: bimanual and digit role differentiation in a natural task. Anim Cogn 4:347–361

Cartmill M (1974) Rethinking primate origins. Science 184:436–443
Charles-Dominique P (1972) Écologie et vie sociale de *Galago demidovii* (Fischer 1808; Prosimii). Z Tierpsychol Suppl 9:7–41
Charles-Dominique P (1977) Ecology and behaviour of nocturnal primates. Columbia University Press, New York
Charles-Dominique P, Atramentowicz M, Charles-Dominique M, Gérard H, Hladik A, Hladik CM, Prévost MF (1981) Les mammifères frugivores arboricoles nocturnes d'une forêt guyanaise: inter-relations plantes-animaux. Rev Ecol (Terre Vie) 35:341–435
Christel M (1993) Grasping techniques and hand preferences in Hominoidea. In: Preuschoft H, Chivers DJ (eds) Hands of primates. Springer-Verlag, Vienna, pp 91–108
Christel M, Billard A (2002) Comparison between macaques' and humans' kinematics of prehension: the role of morphological differences and control mechanisms. Behav Brain Res 131:169–184
Christel M, Fragaszy DM (2000) Manual function in *Cebus apella*. Digital mobility, preshaping, and endurance in repetitive grasping. Int J Primatol 21:697–719
Corp N, Byrne RW (2002) Leaf processing of wild chimpanzees: physically defended leaves reveal complex manual skills. Ethology 108:673–696
Costello MB, Fragaszy DM (1988) Prehension in *Cebus* and *Saimiri*: 1. Grip type and hand preference. Am J Primatol 15:235–245
Crast J (2006) A comparison of dynamic hand movements in chimpanzees (*Pan troglodytes*) and capuchin monkeys (*Cebus apella*). Master's thesis, University of Georgia
Crast J, Fragaszy D, Hayashi M, Matsuzawa M (2009) Dynamic in-hand movements in adult and young juvenile chimpanzees (*Pan troglodytes*). Am J Phys Anthropol 138:274–285
Dominy NJ, Ross CF, Smith TD (2004) Evolution of special senses in primates: past, present, and future. Anat Rec A 281A:1078–1082
Elliott JM, Connolly KJ (1984) A classification of manipulative hand movements. Dev Med Child Neurol 26:283–296
Erickson CJ (1991) Percussive foraging in the aye aye, *Daubentonia madagascariensis*. Anim Behav 41:793–801
Erickson CJ (1994) Tap-scanning and extractive foraging in the aye-aye, *Daubentonia madagascariensis*. Folia Primatol 62:125–135
Erickson CJ, Nowicki S, Dollar L, Goehring N (1998) Percussive foraging: stimuli for prey location by aye-ayes (*Daubentonia madagascariensis*). Int J Primatol 19:111–122
Exner CE (1992) In-hand manipulation skills. In: Case-Smith J, Pehoski C (eds) Development of hand skills in the child. American Occupational Therapy Association, Inc., Bethesda, MD, pp 1–11
Fragaszy DM (1983) Preliminary quantitative studies of prehension in squirrel monkeys (*Saimiri sciureus*). Brain Behav Evol 23:81–92
Fragaszy DM, Adams-Curtis LE (1991) Generative aspects of manipulation in tufted capuchin monkeys (*Cebus apella*). J Comp Psychol 105:387–397
Fragaszy DM, Visalberghi E, Fedigan LM (2004) The complete capuchin. Cambridge University Press, Cambridge
Fragaszy D, Greenberg R, Visalberghi E, Ottoni EB, Izar P, Liu Q (2010) How wild bearded capuchin monkeys select stones and nuts to minimize the number of strikes per nut cracked. Anim Behav 80:205–214
Fragaszy DM, Liu Q, Wright BW, Allen A, Brown CW (2013) Wild bearded capuchin monkeys (*Sapajus libidinosus*) strategically place nuts in a stable position during nut-cracking. PLoS One 8:E56182
Fu Q, Santello M (2011) Towards a complete description of grasping kinematics: a framework for quantifying human grasping and manipulation. Proceedings, 33rd Annual International Conference of the IEEE EMBS, Boston, MA, 30 August to 3 September 2011. pp 8247–8250
Glickman SE, Sroges RW (1966) Curiosity in zoo animals. Behaviour 26:151–187
Groves C, Shekelle M (2010) The genera and species of *Tarsiidae*. Int J Primatol 31:1071–1082
Gunst N, Boinski B, Fragaszy D (2010) Development of skilled detection and extraction of embedded preys by wild brown capuchin monkeys (*Cebus apella apella*). J Comp Psychol 124:194–204

Hoffman JH, Montag AG, Dominy NJ (2004) Meissner corpuscles and somatosensory acuity: the prehensile appendages of primates and elephants. Anat Rec A 281A:1138–1147

Ivanco TL, Pellis SM, Whishaw IQ (1996) Skilled forelimb movements in prey catching and in reaching by rats (*Rattus norvegicus*) and opossums (*Monodelphis domestica*): relations to anatomical differences in motor systems. Behav Brain Res 79:163–181

Iwaniuk AN, Whishaw Q (1999) How skilled are the skilled limb movements of the raccoon (*Procyon lotor*)? Behav Brain Res 99:35–44

Iwaniuk AN, Whishaw Q (2000) On the origin of skilled forelimb movements. Trends Neurosci 23:372–376

Jolly A (1964) Prosimians' manipulation of simple object problems. Anim Behav 12:560–570

Jones LA, Lederman SJ (2006) Human hand function. Oxford University Press, New York

Jones-Engle L, Bard K (1996) Precision grips in young chimpanzees. Am J Primatol 39:1–15

Kivell TL, Schmitt D, Wunderlich RE (2010) Hand and foot pressures in the aye-aye (*Daubentonia madagascariensis*) reveal novel biomechanical trade-offs required for walking on gracile digits. J Exp Biol 213:1549–1557

Krakauer E, Lemelin P, Schmitt D (2002) Hand and body position during locomotor behavior in the aye-aye (*Daubentonia madagascariensis*). Am J Primatol 57:105–118

Kumara HN, Kumar S, Singh M (2005) A novel foraging technique observed in slender loris (*Loris lydekkerianus malabaricus*) feeding on red ants in the Western Ghats. India Folia Primatol 76:116–118

Lacreuse A, Fragaszy DM (1997) Manual exploratory procedures and asymmetries for a haptic searching task: a comparison between capuchins (*Cebus apella*) and humans. Laterality 2:247–266

Landsmeer JMF (1962) Power grip and precision handling. Ann Rheum Dis 21:164–170

Leavens D, Hopkins W, Bard K (1996) Indexical and referential pointing in chimpanzees (*Pan troglodytes*). J Comp Psychol 110:346–353

Leca J-B, Gunst N, Huffman MA (2010) Principles and levels of laterality in unimanual and bimanual stone handling patterns by Japanese macaques. J Hum Evol 58:155–165

Lederman S, Klatzky R (1987) Hand movement: a window into haptic object recognition. Cogn Psychol 22:342–368

Lederman S, Klatzky R (1990) Haptic classification of common objects: knowledge driven exploration. Cogn Psychol 22:421–459

Lemelin P (1999) Morphological correlates of substrate use in didelphid marsupials: implications for primate origins. J Zool 247:165–175

Lemelin P, Grafton BW (1998) Grasping performance in *Saguinus midas* and the evolution of hand prehensility in primates. In: Strasser E, Fleagle JG, Rosenberger AL, McHenry HM (eds) Primate locomotion: recent advances. Plenum Press, New York, pp 131–144

Lemelin P, Jungers WL (2007) Body size and scaling of the hands and feet in prosimian primates. Am J Phys Anthropol 133:828–840

Lemelin P, Schmitt D (2007) Origins of grasping and locomotor adaptations in primates: comparative and experimental approaches using an opossum model. In: Ravosa MJ, Dagosto M (eds) Primate origins, adaptations and evolution. Springer, New York, pp 329–380

Lemon RN (1993) Cortical control of the hand. Exp Physiol 78:263–301

Lemon RN, Griffiths J (2005) Comparing the function of the corticospinal system in different species: organizational differences for motor specialization? Muscle Nerve 32:261–279

Lynch Alfaro JW, Boubli JP, Olson LE, Di Fiore A, Wilson B, Gutiérrez-Espeleta GA, Chiou KL, Schulte M, Neitzel S, Ross V, Schwochow D, Nguyen MTT, Farias I, Janson CH, Alfaro ME (2012) Explosive Pleistocene range expansion leads to widespread Amazonian sympatry between robust and gracile capuchin monkeys. J Biogeogr 39:272–288

Macfarlane NBW, Graziano MSA (2009) Diversity of grip in *Macaca mulatta*. Exp Brain Res 197:255–268

Maier W (1993) Adaptations in the hands of cercopithecoids and callitrichids. In: Preuschoft H, Chivers DJ (eds) Hands of primates. Springer-Verlag, Vienna, pp 191–198

Manoel EH, Connolly KJ (1998) The development of manual dexterity in young children. In: Connolly KJ (ed) The psychobiology of the hand. Mac Keith Press, London, pp 177–198

Martin RD (1972a) A preliminary field-study of the Lesser Mouse Lemur (*Microcebus murinus* J.F. Miller, 1777). Z Tierpsychol Suppl 9:43–89
Martin RD (1972b) Adaptive radiation and behaviour of the Malagasy lemurs. Philos Trans R Soc Lond B264:295–352
Martin RD (1990) Primate origins and evolution. A phylogenetic reconstruction. Princeton University Press, Princeton
Marzke MW (1983) Joint functions and grips of the *Australopithecus afarensis* hand, with special reference to the region of the capitate. J Hum Evol 12:197–211
Marzke MW, Wullstein KL (1996) Chimpanzee and human grips: a new classification with a focus on evolutionary morphology. Int J Primatol 17:117–139
Marzke MW, Pouydebat E, Laurin M, Gorce P, Bels V (2009) A clarification of Pouydebat, et al., 2009, evolution of grasping among anthropoids. J Evol Biol 22:2554–2557
McClearn D (1992) Locomotion, posture, and feeding behavior of kinkajous, coatis, and raccoons. J Mammal 73:245–261
Milliken GW, Ward JP, Erickson CJ (1991) Independent digit control in foraging by the aye aye (*Daubentonia madagascariensis*). Folia Primatol 56:219–224
Napier JR (1956) The prehensile movements of the human hand. J Bone Joint Surg [Br] 38B:902–913
Napier JR (1960) Studies of the hands of living primates. Proc Zool Soc Lond 134:647–657
Napier JR (1961) Prehensility and opposability in the hands of primates. Symp Zool Soc Lond 5:115–132
Napier J (1980) Hands. Pantheon Books, New York
Napier JR, Napier PH (1967) A handbook of living primates. Academic Press, London
Napier JR, Napier PH (1985) The natural history of the primates. The MIT Press, Cambridge, MA
Niemitz C (1984) Synecological relationships and feeding behavior of the genus *tarsius*. In: Niemitz C (ed) Biology of tarsiers. Gustav Fischer Verlag, Stuttgart, pp 59–76
Parker ST, Gibson KR (1977) Object manipulation, tool use and sensorimotor intelligence as feeding adaptations in cebus monkeys and great apes. J Hum Evol 6:623–641
de Passos FC, Keuroghlian A (1999) Foraging behavior and microhabitats used by black lion tamarins, *Leontopithecus chrysopyqus* (Mikan) (Primates, Callitrichidae). Revta Bras Zool 16(Suppl 2):219–222
Pellis SM, Pellis VC (2012) Anatomy is important, but need not be destiny: novel uses of the thumb in aye-ayes compared to other lemurs. Behav Brain Res 231:378–385
Phillips KA, Goodchild LM, Haas ME, Ulyan MJ, Petro S (2004) Use of visual, acoustic, and olfactory information during embedded invertebrate foraging in brown capuchins (*Cebus apella*). J Comp Psychol 118:200–205
Pouydebat E, Gorce P, Coppens Y, Bels V (2009) Biomechanical study of grasping according to the volume of the object: human versus non-human primates. J Biomech 42:266–272
Power TG (2000) Play and exploration in children and animals. L. Earlbaum Associates, Hillsdale, NJ
Rasmussen DT (1990) Primate origins: lessons from a neotropical marsupial. Am J Primatol 22:263–277
Reghem E, Cheze L, Coppens Y, Pouydebat E (2013) Unconstrained 3D-kinematics of prehension in five primates: lemur, capuchin, gorilla, chimpanzee, human. J Hum Evol 65(3):303–312
Reghem E, Tia B, Bels V, Pouydebat E (2011) Food prehension and manipulation in *Microcebus murinus* (Prosimii, Cheirogaleidae). Folia Primatol 82:177–188
Rose MD (1992) Kinematics of the trapezium-1st metacarpal joint in extant anthropoids and miocene hominoids. J Hum Evol 22:255–266
Roy AC, Paulignan Y, Farne A, Jouffrais C, Boussaoud D (2000) Hand kinematics during reaching and grasping in the macaque monkey. Behav Brain Res 117:75–82
Rylands A (1993) The ecology of the lion tamarins, *Leontopithecus*: some intrageneric differences and comparisons with other callitrichids. In: Rylands A (ed) Marmosets and tamarins: systematics, behavior, and ecology. Oxford University Press, New York, pp 296–313
Sargis EJ (2001) The grasping behaviour, locomotion and substrate use of the tree shrews *Tupaia minor* and *T. tana* (Mammalia, Scandentia). J Zool 253:485–490

Schöneich S (1993) Hand usage in the ring-tailed lemur (*Lemur catta* Linnaeus 1758) when solving manipulative tasks. In: Preuschoft H, Chivers DJ (eds) Hands of primates. Springer-Verlag, Vienna, pp 7–20
Soligo C (2005) Anatomy of the hand and arm in *Daubentonia madagascariensis*: a functional and phylogenetic outlook. Folia Primatol 76:262–300
Spinozzi G, Truppa V, Laganà T (2004) Grasping behavior in tufted capuchin monkeys (*Cebus apella*): grip types and manual laterality for picking up a small food item. Am J Phys Anthropol 125:30–41
Spinozzi G, Laganà T, Truppa V (2007) Hand use by tufted capuchins (*Cebus apella*) to extract a small food item from a tube: digit movements, hand preference, and performance. Am J Primatol 69:336–352
Sussman RW (1991) Primate origins and the evolution of angiosperms. Am J Primatol 23:209–223
Sussman RW, Rasmussen DT, Raven PH (2013) Rethinking primate origins again. Am J Primatol 75:95–106
Talbot WH, Darian-Smith I, Kornhuber HH, Mountcastle VB (1968) The sense of flutter-vibration: comparison of the human capacity with response patterns of mechanoreceptor afferents from the monkey hand. J Neurophysiol 31:301–334
Tanaka I (1998) Social diffusion of modified louse egg-handling techniques during grooming in free-ranging Japanese macaques. Anim Behav 56:1229–1236
Tonooka R, Matsuzawa T (1995) Hand preferences of captive chimpanzees (*Pan troglodytes*) in simple reaching for food. Int J Primatol 16:17–35
Torigoe T (1985) Comparison of object manipulation among 74 species of non-human primates. Primates 26:182–194
Toussaint S, Reghem E, Chotard H, Herrel A, Ross CF, Pouydebat E (2013) Food acquisition on arboreal substrates by the grey mouse lemur: implication for primate grasping evolution. J Zool 291:235–242
Visalberghi E, Neel C (2003) Tufted capuchins (*Cebus apella*) use weight and sound to choose between full and empty nuts. Ecol Psychol 15:215–228
Visalberghi E, Addessi E, Truppa V, Spagnoletti N, Ottoni E, Izar P, Fragaszy D (2009) Selection of effective stone tools by wild bearded capuchin monkeys. Curr Biol 19:213–217
Washburn SL (1951) The analysis of primate evolution with particular reference to the origin of man. Cold Spring Harb Symp Quant Biol 15:67–77
Westergaard GC (1992) Object manipulation and the use of tools by infant baboons (*Papio cynocephalus anubis*). J Comp Psychol 106:398
Whishaw I (2005) Prehension. In: Whishaw I, Kolb B (eds) The behavior of the laboratory rat. Oxford University Press, Oxford, pp 162–170
Whishaw IQ, Sarna JR, Pellis SM (1998) Rodent-typical and species-specific limb use in eating: evidence for specialized paw use from a comparative analysis of ten species. Behav Brain Res 96:79–91
Winkelmann RK (1964) Nerve endings of the North American opossum (*Didelphis virginiana*): a comparison with nerve endings of primates. Am J Phys Anthropol 22:253–258
Wood Jones F (1920) The principles of anatomy as seen in the hand. J. and A. Churchill, London

Chapter 13
Patterns, Variability, and Flexibility of Hand Posture During Locomotion in Primates

Daniel Schmitt, Angel Zeininger, and Michael C. Granatosky

1 Introduction

When anthropologists stand in front of a classroom and describe to students the suite of features that distinguish primates from other mammals, they almost always mention two features related to the hand: the presence of nails instead of claws (see Chap. 8) and grasping (prehensile) hands. It is that latter feature—the ability to grasp and manipulate objects with a single hand (observed even in the claw-bearing callitrichids; see Lemelin and Grafton 1998)—that generations of students remember as a definitive characteristic of the order of mammals to which we belong.

This conception of the hand as a unique, defining characteristic of Primates was first articulated by Wood Jones (1916, 1920) and elaborated by Napier (1962, 1980). Rather than emphasizing derived anatomical specializations, Wood Jones and Napier noted that primates retained a fundamentally primitive, pentadactyl hand (see Chap. 2 for a review of the work of Wood Jones and Napier). They recognized that what makes the primate hand special is that relatively primitive anatomy is coupled with remarkable dexterity. In this context, the innovations in neuromuscular control of the hand (see Chaps. 6 and 7) can be seen as key to the invasion and exploitation of an arboreal niche by the earliest primates (see Jenkins 1974), specifically an environment with thin terminal branches [the "fine-branch environment," first coined by Charles-Dominique and Martin (1970) and later emphasized by Cartmill (1972, 1974, 1992) in his visual predation hypothesis of primate origins]. In such an environment, early primates faced a complex array of branch sizes and orientations that they needed to grip and grasp, run along, and leap from. Thus, a hand that could adjust to complex surfaces was critical to their success. This idea, articulated in detail first by Wood Jones (1916) in *Arboreal Man*, is reflected in all

D. Schmitt (✉) • A. Zeininger • M.C. Granatosky
Department of Evolutionary Anthropology, Duke University, Durham, NC 27708, USA
e-mail: daniel.schmitt@duke.edu

T.L. Kivell et al. (eds.), *The Evolution of the Primate Hand*, Developments in Primatology: Progress and Prospects, DOI 10.1007/978-1-4939-3646-5_13

later models of primate origins and evolution, regardless of the types of arboreality, feeding strategies, or body size and life history patterns that are assumed in each model (Napier and Walker 1967; Cartmill 1972, 1974, 1992; Szalay and Dagosto 1980, 1988; Sussman 1991; Schmitt and Lemelin 2002; Soligo and Martin 2006).

While all extant primates retain prehensile hands from their earliest ancestors, primates have radiated into a vast diversity of habitats (trees and ground) and are capable of a wide range of locomotor modes. This diversity in locomotor ecology is reflected in remarkable diversity in hand anatomy (see Chaps. 3, 4, 7, and 8) and hand posture during manipulation (see Chap. 12) and locomotion in living primates.

This chapter reviews hand postures used by extant primates during locomotion and explores how well those postures reflect ecology and anatomy. Interpreting the functional anatomy of the hand in light of variation in hand posture can be seen as a challenge to functional morphologists since hand anatomy does not always predict or reflect habitual ecology. Nor does hand anatomy always reflect the remarkable flexibility in hand postures primates are capable of exhibiting. However, this apparent challenge in relating form to function may reveal a more important aspect of primate adaptations that illuminate the success of the order. Variation in hand posture within individuals, species, and across Primates can be seen as a measure of mechanical flexibility, which in itself can have substantial adaptive value allowing an animal to function as a "generalist" in terms of the ability to move on a variety of substrates. This mechanical flexibility should not be confused with anatomical flexibility or range of motion. Rather, it is meant as a measure of the range of options an animal has in adopting a specific posture. This could be described as "versatility." We prefer the formal definition of mechanical flexibility proposed by Wainwright et al. (2008) that reflects the extent to which an animal can alter its behavior (in this case hand posture) in response to a stimulus.

2 Hand Posture Types

We will discuss two main hand posture orientations: vertical and horizontal manus. Each orientation includes one or more hand posture types and subtypes: (1) palmigrady (on vertical and horizontal substrates), (2) digitigrady, (3) knuckle- and fist-walking, and (4) suspensory (below-branch hook grip and inverted quadrupedal postures). The descriptions that follow are summarized in Table 13.1.

2.1 Horizontal Manus

As the most common hand posture used by primates, palmigrady (Figs. 13.1a and 13.2b) is presumably the basal hand posture of the order and is not by itself associated with any specific anatomical adaptations. There is, however, considerable variation in the details of wrist and digit posture within palmigrady that may reveal deeper ecological patterns. Palmigrade hand postures in which the fingers and palm

Table 13.1 Common hand postures used by primates during locomotion

Orientation of the manus	Posture type	Posture subtype	Substrate	Primary loading area	Primary wrist position[a]	Primary McP position[b]	Primary IP position[c]	Example groups[d]
Horizontal	Palmigrady	Non-grasping	Terrestrial or large[e] bough	Palm	Ulnar deviated or neutral[f], extended	Neutral/ulnar deviated[f]	Neutral or extended/flexed as in "bowed up"[g] can be schizodactylous[h]	Cercopithecoids, *Lemur catta*, small primate quadrupeds on relatively large branches
		Grasping	Arboreal (horizontal and vertical)	Palm	Ulnar deviated or neutral, slightly extended	Neutral	Flexed	Most arboreal quadrupeds and climbers
Vertical	Digitigrady		Terrestrial	Digits	Neutral	Hyperextended[i]	Neutral, may be extended/flexed as in "bowed up"[g]	Terrestrial cercopithecoids
	Knuckle-walking	Palm in/palm out	Terrestrial or large[e] bough	Digits (middle phalanx) and metacarpal heads	Neutral, slightly extended	Neutral or hyperextended	Flexed	*Pan* and *Gorilla*
	Fist-walking		Terrestrial	Lateral edge of hand (ray 5) or proximal phalanx	Neutral or deviated, hyper-extended	Flexed	Flexed ("curled")	*Pongo*

(continued)

Table 13.1 (continued)

Orientation of the manus	Posture type	Posture subtype	Substrate	Primary loading area	Primary wrist position[a]	Primary McP position[b]	Primary IP position[c]	Example groups[d]
	Suspensory	Grasping	Arboreal	Digits	Neutral	Neutral	Flexed	Hylobatids, atelids, and primates that use inverted quadrupedal postures (e.g., *Varecia*)

[a]Primary wrist position describes radial and ulnar deviation and then flexion or extension

[b]Primary McP position describes flexion and extension at the metacarpophalangeal (McP) joints

[c]Primary IP position describes flexion and extension at the proximal interphalangeal (IP) joints

[d]Example species can be found in Bishop (1964), Tuttle (1967), Susman (1974), Napier and Napier (1967), Whitehead (1993), Fleagle (2013), Gebo (2014), and Hunt et al. (1996)

[e]Size is expressed relative to hand length and width, a large support is one that cannot be encircled by the digits

[f]Ulnar deviated substrate-hand contact may run along an axis between digits 1 and 2 or digits 2 and 3

[g]"Bowed up" is defined as hyperextension at MP joints and flexion at proximal IP joints (see Fig. 13.3 and Nieschalk and Demes 1993)

[h]In a schizodactylous grip, substrate-hand contact may run between digits 2 and 3 or digits 3 and 4 when those digits are abducted. In this case the grip might be considered a grasping type of palmigrady

[i]Hyperextended refers to position of the McP joints beyond neutral or 180°

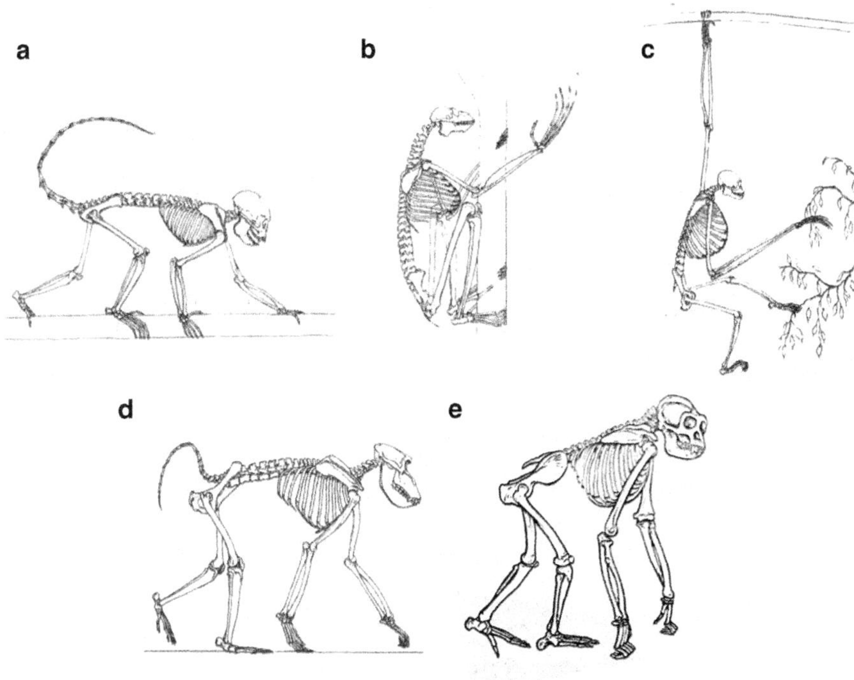

Fig. 13.1 Common hand postures among primates during positional behaviors. (**a**) Arboreal quadrupedal primate with hand in pronated, palmigrade (grasping), and ulnar-deviated position; (**b**) vertical clinging primate with partially supinated hands surrounding support and palms in contact with trunk; (**c**) suspensory primate with hand in a hook-like position; (**d**) digitigrade primate with a vertical manus and middle and distal phalanx on ground; (**e**) knuckle-walking primate with hands pronated and middle phalanx in contact with the ground. All images adapted from Fleagle (2013)

contact the substrate can be observed during movement and resting postures on vertical trunks (grasping) and horizontal branches (grasping and non-grasping) as well as on the ground (non-grasping).

The mechanics of palmigrade hand postures on vertical trunks (Figs. 13.1b and 13.2a) were first discussed in detail in a seminal paper by Cartmill (1974). More recently, Johnson et al. (2015) examined the models of hand loading during vertical clinging and grasping (VCG) by video recording hand positions and applied forces using an indwelling force plate during vertical grasping (Fig. 13.2). Primates climbing vertical tree trunks grasp the substrate with hands positioned halfway between supination and pronation (relative to the anatomical position; in a standing biped, this would be with palms facing medially) (Johnson 2012). Depending on the size of the substrate, the interphalangeal (IP) and metacarpophalangeal (McP) joints may be in neutral position (i.e., a neutral position with respect to flexion/extension or abduction/adduction would be one in which segments form a straight line with an angle at or near 180°) when the

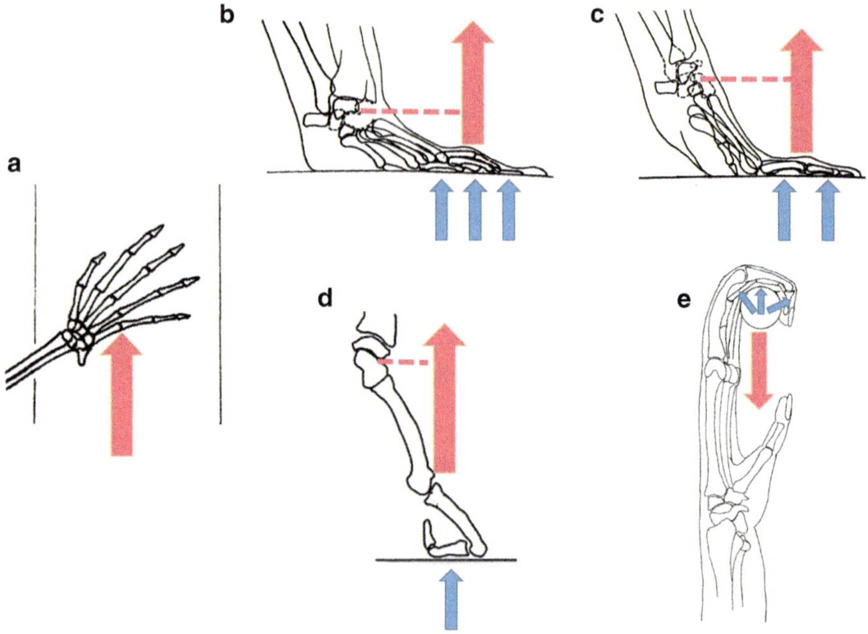

Fig. 13.2 Hand postures and schematic renditions of loading patterns. *Blue arrows* represent potential distribution and location of manual or digital pressures. *Pink arrows* represent vertical ground reaction force vectors. (**a**) Vertical grasping (modified from Fleagle 2013); (**b**) palmigrade (non-grasping) hand position on ground showing the wide contact with the palm and the relatively large moment arm of the ground reaction force (GRF) to wrist and digits (modified from Whitehead 1993); (**c**) digitigrade hand position showing pressure under proximal and middle, and distal phalanges and relatively small moment arm (modified from Whitehead 1993); (**d**) knuckle-walking digital position showing pressure under middle phalanges only and relatively small moment arm of GRF (modified from Zeininger et al. 2011); (**e**) suspensory hand posture showing loading of proximal, middle, and distal phalanges (Modified from Zeininger et al. 2011)

substrate is large and the hand is pressed against the surface or flexed so that the fingers and palm can conform around the trunk (Johnson 2012; Fig. 13.2a).

When walking on horizontal branches, palmigrade primates can either grasp around the branch with flexed McP and IP joints (most common on relatively small branches) or walk above the substrate without grasping (most common on relatively large branches) (Lemelin and Schmitt 1998; Fig. 13.2b). During walking or standing on horizontal branches and terrestrial supports, primates may use non-grasping postures in which the palmigrade hand is pronated and lies flat against the substrate with the McP and IP joints aligned in a neutral position with respect to the plane of flexion/extension (i.e., 180°).

Finally, there are palmigrade postures, reviewed in more detail later, in which the fingers and wrist may be ulnar deviated (common in grasping palmigrady, but also in non-grasping palmigrady) (Bishop 1962, 1964; Lemelin and Schmitt 1998) or in which the McP joints are extended and proximal IP joints are flexed, placing the digits in a "bowed-up" position (Bishop 1964; Day and Iliffe 1975; Nieschalk and Demes 1993; Lemelin 1996; Congdon 2014). In all forms, palmigrady provides a large contact area between the hand and substrate and distributes pressure across the volar pads more evenly, thereby reducing stress on individual hand bones (Patel 2010b; Patel and Wunderlich 2010).

2.2 *Vertical Manus*

Vertical manus postures in which the palm of the hand is elevated above the substrate (digitigrady, knuckle- and fist-walking) are restricted to catarrhines and are most often used during quadrupedal locomotion on the ground. While the palm does not contact the substrate in any vertical manus postures, the orientation of the IP and McP joints and loading of the fingers differs between digitigrady and knuckle-walking. Cercopithecoids that have relatively short fingers and spend time on the ground (e.g., baboons, patas monkeys, and grivets) walk with digitigrade hands (Weidenreich 1931; Bishop 1964; Jolly 1965; Napier and Napier 1967; Yalden 1972; Clevedon Brown and Yalden 1973; Rose 1973; Rollinson and Martin 1981; Nengo 1993; Rawlins 1993; Whitehead 1993; Hayama et al. 1994; Patel 2009, 2010a, b; Patel and Polk 2010; Patel and Wunderlich 2010). Digitigrady is similar to non-grasping palmigrady in that the hand is pronated, the IP joints may be in neutral positions, and the palmar surface of the fingers contacts the ground. However, in digitigrady, the palm is elevated such that the McP joints are hyperextended (i.e., extended beyond 180° or neutral) and the metacarpal heads and fingers are weight bearing (Fig. 13.2c). Elevating the palm may align the wrist with the ground reaction force resultant, reducing wrist moment arm lengths, wrist moments, and extensor muscle force required to support body weight (Biewener 1983, 1989; Polk 2002; Sockol et al. 2007; Pontzer et al. 2009; Foster et al. 2013; Fig. 13.2c). Compared to palmigrady, digitigrady is associated with an increased effective forelimb length (distance from the shoulder to the point of contact with the ground) in more terrestrial monkeys. This may improve locomotor cost of transport by reducing forelimb digital flexor musculature recruitment and by increasing stride length at a given speed, lowering stride frequency (Patel 2010b; Patel et al. 2012).

Knuckle-walking postures are seen exclusively in African apes (Figs. 13.1e and 13.2d). During knuckle-walking, the palm and proximal phalanges are elevated above the substrate (Tuttle 1967; Susman 1974; Inouye 1994; Figs. 13.1e and 13.2d), and the hand can be pronated or partially supinated [palm back or palm in, respectively; see Wunderlich and Jungers (2009) and Zeininger et al. (2014)]. Flexion at the IP joints positions the dorsal aspect of the middle phalanges in contact with the substrate as the main weight-bearing elements. The McP joints are held in a neutral position (i.e., 180°) or slightly hyperextended. Internal subchondral and trabecular bone mineral density (Zeininger et al. 2011) and trabecular fabric properties (Tsegai et al. 2013) of the dorsal metacarpal head suggest that this region experiences compressive McP joint reaction forces during knuckle-walking. It is thought that African apes adopt knuckle-walking postures to reduce bending loads that their fingers would experience in palmigrade or digitigrade postures (Tuttle 1967; Inouye 1994).

A less-studied variant of this functional solution to loading long digits when walking on the ground or large branches is fist-walking (Tuttle 1967; Susman 1974). There are many variations on fist-walking, observed only in orangutans, that have been described by Tuttle (1967) and Susman (1974). The hand is formed into a "fist" by flexion at the IP and McP joints. Weight is borne in most cases on the proximal phalanges rather than the middle phalanges as in knuckle-walking. The hand can be pronated or partially supinated, and the thumb can touch the ground or not. In adducted or abducted fist-walking (wrist deviated either radially or ulnarly), weight can be borne on the medial or lateral edge of the manus, on the thumb and medial edge of ray 2, or on the lateral edge of ray 5 (see description and images in Tuttle 1967).

Suspensory hand postures also involve a vertical manus. Suspensory locomotion may be defined relative to the pull of gravity. During suspensory locomotion (Stern and Oxnard 1973) and sometimes during vertical clinging (Cartmill 1974, 1985; Preuschoft 2002; Johnson et al. 2015), the gravity vector moves the body away from the support. In postures in which gravity moves the animal's body (center of mass) away from the support, the limbs are often described as "in tension" (Stern and Oxnard 1973; Swartz et al. 1989). However, it should be noted that compressive forces can and do occur along the bones and at joints because of muscular forces counteracting tensile forces of body weight leading to an overall reduction in net axial and bending stresses (Swartz et al. 1989; Richmond 2007). In primates, suspensory hand postures can be observed during both arm-swinging and inverted quadrupedal locomotion ("in tension" loading of both the forelimbs and hind limbs) (Hunt et al. 1996). Suspensory locomotion is common in primates and can be observed in strepsirrhines, New and Old world monkeys, and apes (Stern and Oxnard 1973; Hunt et al. 1996; Meldrum et al. 1997; Cant et al. 2003; Granatosky et al. 2016). During both arm-swinging and inverted quadrupedal locomotion, the hand can be described as being positioned like a hook to wrap above and around the support (Figs. 13.1c and 13.2e) (Swartz et al. 1989; Jungers et al. 1997; Richmond 2007; Fleagle 2013). The proximal phalanges are usually positioned above the support, while intermediate and distal phalanges make contact on the side and near the bottom of the support, respectively, although posture is likely to vary with support size (Richmond 2007). Commonly, the metacarpals have no contact with the support, and the pollex is thought to be of little importance during suspensory movement (Fleagle 2013). For species that utilize suspensory postures as their primary form of locomotion (e.g., *Ateles* and *Hylobates*), substantial phalangeal curvature can be observed throughout the digital portion of the ray (Susman 1979; Jungers et al. 1997; Richmond 1997, 1998, 2007; Congdon 2012, 2014; see Chap. 4). The hook style grip observed during suspensory locomotion serves as an important mechanism for mitigating bending strain while below branches. Basically, as a digit is loaded during suspensory postures, joint reaction forces load the articular ends of the phalanx in compression and dorsally, while digital flexor musculature pulls its mid-shaft palmarly (Richmond 2007). The presence of highly curved digits substantially mitigates the strains experienced during suspensory locomotion (Richmond 2007), which can be further decreased by behavioral adjustment of the hand.

Patterns of wrist morphology in suspensory species are thought to promote mobility and overall range of motion (Cartmill and Milton 1977; Mendel 1979; Lewis 1985a, b; Fleagle 2013; Kivell et al. 2013; see Chap. 3). Many species that commonly adopt suspensory postures have a reduced or absent ulnocarpal articulation and have an increased expansion of radiocarpal joint surfaces (Hamrick et al. 2000). Additionally, certain arm-swinging primates possess a more mobile midcarpal region in which the capitate and hamate rotate about more proximal carpals (Jenkins 1981; see Chap. 3). Together, these traits are thought to reflect adaptations for enhanced joint mobility that allow primates to move fluidly below branches (Cartmill and Milton 1977; Mendel 1979; Jenkins 1981).

3 Variability in Hand Posture Use

One of the most interesting aspects of hand postures in primates is not just the variety of hand postures that can be identified within the order, but also the variation within those posture categories. Most mammals place their hands in consistently pronated postures while walking and the (often relatively short) digits of nonprimate mammals are often not deviated away from the plane of locomotion. The variation in primate hand postures is wider compared to many other mammals (see Heffner and Masterton 1975 for a numerical scale devised to describe hand dexterity and the rank of various animals on that scale and Iwaniuk and Whishaw 1999 for a discussion of that analysis and exceptions to the general patterns presented there). Obviously, ungulate mammals have no manual dexterity and their limbs move primarily in a sagittal plane. Most carnivores also have limited dexterity (Heffner and Masterton 1975; Iwaniuk and Whishaw 1999), although there are some clear exceptions, such as raccoons. However, even raccoons do not show dexterity as high as that of primates (Iwaniuk and Whishaw 1999). Some rodents and marsupials also show dexterity and can handle some items with a prehensile grip, but not with the skill and flexibility of primates (Wishaw and Pellis 1990; Ivanco et al. 1996; Wishaw and Coles 1996). It is interesting to note that kinkajous (McClearn 1992), raccoons (McClearn 1992; Iwaniuk and Whishaw 1999), and woolly opossums (Lemelin 1996, 1999) have a high degree of hand mobility and dexterity and that these mammals are all active arborealists (see Chap. 12).

Just within the category of palmigrady in primates, there are many nuanced variants in hand posture. Some species use one variant with a higher frequency than another, and often individuals will use variants of palmigrady in different settings or even within a locomotor sequence. It is easy to see these variants as "noise," but we think that they represent key behavioral adjustments associated with ecological contexts and growth during ontogeny. These variants can be expressed as a measure of a primate's behavioral flexibility. Below we try to capture some of the variation observed. Discussion of all of the variation in primate hand postures during positional behavior is beyond the scope of this chapter, and thus we have simplified this by looking at variation in the context of larger categories. There are texts that provide considerably more detail. For example, we recommend that readers consult Bishop (1964) for details on variation in hand placement on dowels of different diameters to get a sense of the enormous range of options in primate hand posture.

3.1 *Variants in Horizontal Manus*

Much of the variation within hand posture types involves moving the digits out of the path of locomotion. Palmigrady represents a grip that provides maximum contact with the substrate, increases stability, and, when the substrate is relatively narrow, allows the fingers to grasp around the support. In addition to the horizontal and vertical substrate variants mentioned above, considerable variation exists within the grasping and non-grasping palmigrade postures (Bishop 1964; Lemelin and Schmitt 1998). In grasping palmigrady, the hand can be non-deviated (Fig. 13.2a) or ulnar deviated (Figs. 13.1a and 13.3a) at the wrist joint, and the fingers can be in contact with the substrate (Figs. 13.1a, 13.2a, and 13.3a–c) or bowed up (Fig. 13.3c, d) and abducted to varying degrees. When walking on large supports, primates favor non-deviated hand positions in line with the support, running along or near the axis of ray 3 (Bishop 1962, 1964; Rawlins 1993; Lemelin and Schmitt 1998; Fig. 13.2a). When walking on small supports, primates deviate their hands or align the branch between the fingers, so that bending moments on the digits are mitigated (Bishop 1962, 1964; Nieschalk and Demes 1993; Rawlins 1993; Lemelin and Schmitt 1998; Figs. 13.1a and 13.3a). There is a wide range of ulnar deviation among primates, often when walking or standing on poles (Lemelin and Schmitt 1998), that appears to facilitate a complete grasp and moderate load on the digits by shortening load arms on the rays. The extreme ulnar deviation of the spider monkey and lorises (Bishop 1962, 1964; Nieschalk and Demes 1993; Lemelin and Schmitt 1998) are clear examples of this behavior.

These deviated forms of palmigrady moderate or entirely avoid digit loading and appear especially valuable in reducing bending loads on relatively gracile digits. This is an important issue for all primates and appears to drive much of the divergence from simple palmigrady in which the entire palmar surface of the digits and metacarpals are aligned and in contact with the substrate. There are additional ways which primates can moderate high stresses on the digits. These can be anatomical (see Chap. 4) and/or postural. For example, primates such as the aye-aye (*Daubentonia*) and small-bodied galagos, lemurs, tarsiers, and New World monkeys raise the digits off the substrate either completely or in a bowed-up posture (Bishop 1962, 1964; Day and Iliffe 1975; Nieschalk and Demes 1993; Lemelin 1996; Lemelin and Schmitt 1998; Krakauer et al. 2002; Kivell et al. 2010; Fig. 13.3c, d). The aye-aye exhibits the extreme version of this non-digital palmigrady by pulling its fingers completely off the substrate (Oxnard et al. 1990; Lemelin 1996;

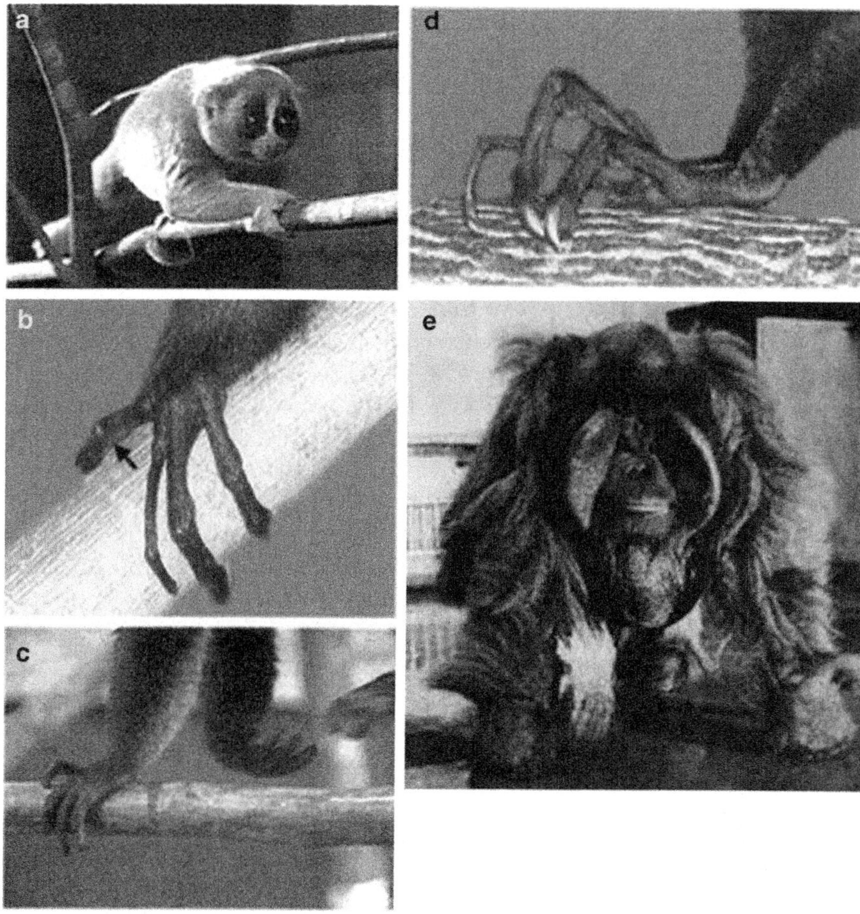

Fig. 13.3 Hand postures that pull digits out of the path locomotion and may reduce loading on the digits. (**a**) Highly deviated posture of lorisines (adapted from Lemelin and Schmitt 1998); (**b**) Schizodactylous hand in *Daubentonia* (common in many strepsirrhines and in *Alouatta*) (adapted from Krakauer et al. 2002); (**c**) Schizodactylous grip in *Eulemur* with bowed-up digit (adapted from Lemelin and Schmitt 1998); (**d**) Extremely bowed-up posture of *Daubentonia* (adapted from Krakauer et al. 2002). (**e**) Fist-walking posture in *Pongo* (Adapted from Susman 1974)

Karakauer et al. 2002; Fig. 13.3d), applying very little pressure on the tips of the digits when touching the support (Kivell et al. 2010).

Bowing up of the digits observed in primates is still poorly understood, and we do not wish to imply that a single explanation can be applied across all animals for which this hand posture is observed. It seems clear that for the aye-aye, given its specialized foraging-adapted digit morphology, reducing load on the digits is an important aspect of this hand posture (Kivell et al. 2010). This may be true for other primates as well (Nieschalk and Demes 1993), but may be less of an important issue for smaller strepsirrhines and in some small anthropoids. One could argue that locomotor stresses represent less risk of injury to very small primates. Thus, the use of bowed-up postures could relate to muscle forces generated across the joints and the need to apply a normal force, and/or proximally-directed frictional forces (Cartmill 1974; Nieschalk and Demes 1993; Lemelin 1996). Flexion of the proximal IP joint may allow the flexor digitorum superficialis to more effectively generate force along the ray.

Another option for repositioning digits from a simple palmigrade position is schizodactyly. In schizodactylous postures, a branch is grasped between the second and third digit, instead of the thumb and second digit. Thus, digits 2 and 3 (or sometimes 3 and 4) are abducted relative to each other, and the substrate (a branch in this case) runs along that space. If the branch is small enough, the digits run alongside the support, allowing the animal to grip it. This is a common grip used by howler monkeys (*Alouatta*), in which their long fingers are well-suited to this posture (Grand 1968). But, as reported in Bishop (1962, 1964) and Lemelin and Schmitt (1998), other primates, such as aye-ayes (Fig. 13.3b) and small lemurs (Fig. 13.3c), use this grip as well. However, it has never been reported in Old World monkeys or apes. Like ulnar-deviated hand postures and bowed-up finger positions, the digits are pulled away from the sagittal plane (corresponding to the line of travel of the animal) or off the substrate, presumably reducing bending loads on the digit.

3.2 *Variants of Vertical Manus Postures*

Among primates, digitigrady (Figs. 13.1c and 13.2c) is less common than palmigrady and mostly used by primates that walk frequently on the ground. However, some primates also use digitigrady while standing or walking on branches (Whitehead 1993). It is important to recognize that primates preferring digitigrady are not restricted to using this posture. Captive infant baboons (2–9 months old) use both palmigrade and digitigrade hand postures (Zeininger et al. 2007). When the hand is digitigrade at touchdown, it often lowers and becomes less digitigrade or even palmigrade from touchdown to midstance (Zeininger et al. 2007). This lowering of the hand may indicate that the mobile wrist joints of infant baboons are forced into palmigrady at midstance as the animal's body weight passes over the hand. While incidence of hand lowering and palmigrady increases with increasing speed in digitigrade adult baboons (Patel 2009), no correlations between walking speed and hand posture were observed in infant baboons (Zeininger et al. 2009). As the frequency of palmigrady decreases with age, the frequency of digitigrady increases. By about 2 years of age, juvenile baboons only use digitigrade hand postures (Zeininger et al. 2007). This flexibility in hand posture use between digitigrady and palmigrady may explain why multivariate analyses of metacarpal morphology are not able to distinguish palmigrade from digitigrade primates (Patel 2010a).

Knuckle- and fist-walking are exhibited by great apes. It is commonplace to argue that these hand postures represent a biomechanical solution to the problem of loading long digits when terrestrial. While primates with relatively short digits use digitigrady (e.g., baboons and patas monkeys), it is not an option for climbing primates with long fingers that also walk on the ground habitually (e.g., chimpanzees). Knuckle- and fist-walking are present (though fist-walking is rare and seen only on the ground) in large-bodied apes presumably because they have relatively long digits for arboreal locomotion that must be behaviorally shortened for terrestrial locomotion (as well as aligning the manus and wrist with vertical reaction forces) (Tuttle 1967, 1981; Susman 1974, 1979; Whitehead 1993; Inouye 1989, 1994; Kivell and Schmitt 2009; Wunderlich and Jungers 2009; Fig. 13.2d). This simple and intuitively sensible model is challenged by the fact that gorillas have short digits relative to their body size compared with other apes (Almécija et al. 2015). This raises two possibilities: either knuckle-walking is retained from a common ancestor of African apes that had longer fingers or an alternative explanation is required. It is likely that knuckle-walking, like digitigrady, increases effective limb length allowing greater limb excursions and postures associated with raised shoulders, a pattern reported for baboons (Larson et al. 2001; Patel et al. 2013). Kivell and Schmitt (2009) speculated that there are significant differences in the mechanics of knuckle-walking between chimpanzees and gorillas. It is also possible that different selective pressures were at work in favoring knuckle-walking over other hand postures in these two African ape lineages. But considerably more data are required to test this hypothesis.

Extensive discussions have been devoted to identifying anatomical correlates of knuckle-walking (Tuttle 1967, 1981; Susman 1979; Inouye 1989, 1992, 1994; Whitehead 1993; Richmond and Strait 2000 [and associated replies]; Richmond et al. 2001; Kivell and Schmitt 2009; Williams 2010; Begun and Kivell 2011; Zeininger et al. 2011) and attempting to match those correlates with fossil morphology (Richmond and Strait 2000; Begun and Kivell 2011). Chimpanzees and gorillas moderate bending moments on their rays by bearing weight on the middle and a vertical column made of the proximal phalanx and metacarpal (with the McP joint held more or less neutral) (Fig. 13.2c). However, based on qualitative observations from video and cineradiographic images, Kivell and Schmitt (2009) argued that gorillas fit the above description, but not chimpanzees and bonobos, which exhibit a shallower ground to proximal phalanx angle and a more hyperextended McP joint [see images in Tuttle (1967) and Whitehead (1993)]. They argued that this corresponded to anatomical differences in the phalanges, metacarpals, and carpals between gorillas and chimpanzees, both as adults and during ontogeny (Kivell and Schmitt 2009). The fact that no unequivocal features exclusive to knuckle-walking have ever been identified makes this hand posture especially intriguing (but see Richmond et al. 2001; Matarazzo 2008). This problem was discussed in detail most recently by Kivell and Schmitt (2009) [but see Williams (2010) for a recent contrary position] and focused on ontogenetic changes in African ape wrist bone morphology (see also Inouye 1992, 1994), as well as biomechanical differences in knuckle-walking postures between chimpanzees and gorillas.

Interestingly, primates that are digitigrade or knuckle-walkers as adults use palmigrade hand postures when they first begin to walk (Doran 1992, 1997; Zeininger et al. 2007). In gorillas, palmigrade crawling begins by week 9 in captivity (Groves 1970) and week 14 in the wild (Doran 1997). While independent quadrupedal walking with a palmigrade hand posture begins between 4.5 and 6 months, adult-like knuckle-walking manual postures are not used until 10–15 months of age (Doran 1997). By 21 months, gorillas are proficient knuckle-walkers, but still adopt palmigrade hand postures on uneven surfaces (Doran 1997).

Chimpanzees also transition from palmigrady to knuckle-walking as they age, but the transition occurs later in chimpanzees than in gorillas. In chimpanzees, first attempts at knuckle-walking begin between 12 and 15 months with habitual knuckle-walking not taking place until 2–3 years of age (Doran 1992). Even though chimpanzees have begun to knuckle-walk, they still use deliberate palmigrady as late as 2 years (Doran 1992). Despite sharing similar hand morphology, chimpanzees and bonobos differ significantly in their hand posture. In adult bonobos, although knuckle-walking is still the dominant quadrupedal mode of locomotion, palmigrady is used more frequently than in chimpanzees, especially when walking on arboreal substrates (Doran 1993). As such, adult quadrupedal hand posture in bonobos more closely resembles that of infant rather than adult chimpanzees (Doran 1992, 1993).

Ontogenetic studies have emphasized differences in the mechanics of knuckle-walking between gorillas and chimpanzees. From the onset of knuckle-walking, western lowland gorillas adopt a fully pronated, or palm-back, hand posture (Tuttle 1969; Inouye 1989; Zeininger et al. 2014). Chimpanzees, however, use palm-back as well as palm-in (supinated) postures (Tuttle 1967, 1969; Inouye 1994; Wunderlich and Jungers 2009; Zeininger et al. 2014). During knuckle-walking in gorillas, digits 2–5 contact the ground while only digits 2–4 contact the ground in chimpanzees (Tuttle 1967; Inouye 1994). Peak pressure is most often beneath digit 2 in gorillas, but is more variable in chimpanzees (Matarazzo 2013). Additionally, digit loading changes across age and substrate in chimpanzees. In 4–5-year-old chimpanzees, pressures beneath digits 3 and 4 are highest while walking on the ground compared to equally distributed pressures across digits 2–4 while walking on a pole (Wunderlich and Jungers 2009). Pressure distribution changes with age such that 7-year-old chimpanzees load digits 2 and 3 more while on the ground than at younger ages (Wunderlich and Jungers 2009).

Studies on the development of hand postures have shown considerable overlap in hand posture use at a given age. For example, infant baboons use palmigrade and digitigrade postures from the age of 2–9 months (Zeininger et al. 2007). Likewise, bonobos and chimpanzees still use palmigrady even after they begin knuckle-walking (Doran 1992, 1993, 1997). Moreover, ontogenetic studies have shown that primates have the dynamic flexibility of adjusting hand postures within a step (from touchdown to midstance), between steps, and throughout ontogeny (Zeininger et al. 2007). Infant primates may be able to behaviorally modify the position of the hand during walking and influence other important biomechanical variables, such as the position of the center of mass, effective forelimb length, bending moments on the digits, and wrist joint moments. This behavioral flexibility may help juvenile primates manage functional demands such as postural stability needed for a still developing nervous system in the youngest infants or increased stride length in older juveniles that have relatively short limbs compared to adults.

These ontogenetic studies reveal patterns that make mechanical sense (i.e., the use of palmigrady at an early development stage). But they also reinforce the enormous flexibility that occurs in primates throughout their lifetime and may be at its extreme when locomotor variability is at its most extreme. For example, juvenile macaques spend more time in the trees than adult macaques and their hand postures are much more variable (Rawlins 1993).

Knuckle- and fist-walking have been presented as a singular mechanical problem—avoiding loading on long digits—with two solutions: one in chimpanzees (and bonobos) and gorillas (described above) and one in orangutans. Orangutans adopt fist-walking postures (described above) that avoid loading the digits along their long axis, thus increasing bending moments (Tuttle 1967; Susman 1974; Fig. 13.3e). Orangutans have hands and feet that are very anatomically similar and often adopt foot postures that, like the hand, load the lateral border of the foot.

Kinematic and behavioral studies of hand position during suspensory locomotion have been somewhat limited (Turnquist 1975; Richmond 2007; Michilsens et al. 2010; Nyakatura et al. 2010; Fujiwara et al. 2011; Guillot 2011; Tripp et al. 2015; Granatosky 2015). The positioning of the hand that results in a functional hook during below-branch movement is ubiquitous during both arm-swinging and inverted quadrupedal locomotion. Even anatomically less specialized species (e.g., *Lemur*, *Macaca*, *Pygathrix*) can adopt suspensory positional behaviors, as long as a functional hook is achieved (Stern and Oxnard 1973). The digits wrap around the substrate with the proximal phalanx near the top of the support (or the proximal IP joint near the top), the rays perpendicular to the support, and metacarpals facing toward or away from the support (depending on the grasp; *Lagothrix*, e.g., varies this often by alternating handholds [Schmitt, unpublished data]). The position of the thumb, however, is highly variable during suspensory locomotion. In many species that commonly arm-swing (e.g., *Hylobates* and *Ateles*), the thumb shows little involvement during suspensory postures. This is especially true for *Ateles*, which lacks an external thumb. It is therefore unlikely that the thumb performs any important functional role for this behavior. In contrast, species that commonly use below-branch quadrupedal locomotion (e.g., *Varecia*, *Lemur*, *Propithecus*, and *Alouatta*) almost always wrap the pollex around the support forming an additional point of contact (McClure et al. 2012; Granatosky unpublished data) and apply some pressure unto the support to further increase grip strength (Congdon 2014).

During suspensory locomotion, the more distal portion of the hand provides a fixed point on the support, whereas its more proximal part facilitates movement of the body relative to the fixed point (Cartmill and Milton 1977; Mendel 1979; Jenkins 1981). It appears that arm-swinging primates like *Ateles* and *Hylobates* rotate under the support, spinning about the long axis their wrist joints (Jenkins 1981; see Chap. 3). Great apes can do this as well. However, great apes and less anatomically specialized arm-swingers like *Lagothrix* also progress using considerable ranges of ulnar and radial deviation of the hand. Recent studies of wrist motion during support phase of arm-swinging in three species of primates (*Hylobates*, *Ateles*, and *Pygathrix*) show intriguing patterns associated with hand positions (Tripp et al.

2015; Granatosky 2015). *Hylobates* tends to maintain slight radial deviation throughout the support phase, likely indicating a greater reliance on wrist rotary movements (although not measured in these studies) to achieve forward progression. In contrast, *Ateles* and *Pygathrix* demonstrate substantial ulnar and radial deviation throughout support phase (Fig. 13.4a). This finding indicates that primate species can adopt a range of kinematic solutions to accomplish the same intended goal (i.e., arm-swinging). Although *Ateles* has the wrist characteristics that allow for rotary wrist movements (Jenkins 1981; Rosenberger et al. 2008), it does not seem to swivel about its wrist as much as *Hylobates* during arm-swinging.

Based on patterns of wrist movements during arm-swinging in *Lagothrix* (Schmitt et al. 2009) and *Pygathrix* (Granatosky 2015), one could assume that wide

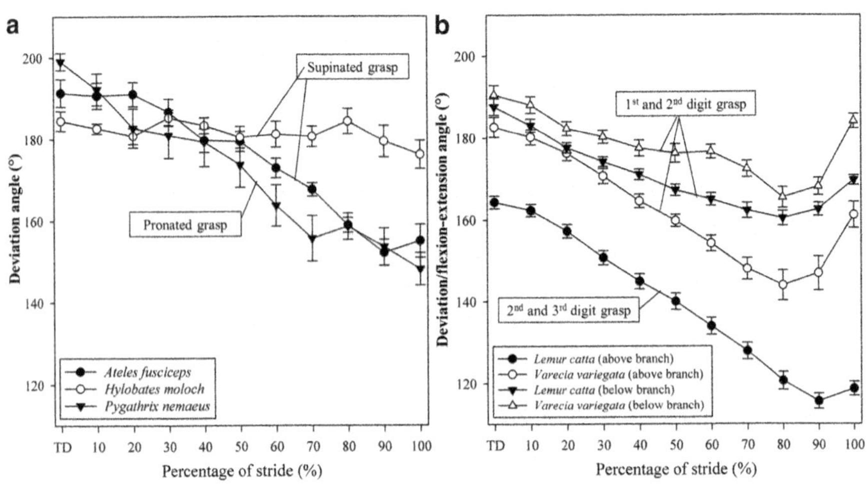

Fig. 13.4 Patterns of wrist movements (mean and standard error) from touchdown (TD) to the end of support phase during (**a**) arm-swinging in *Ateles fusciceps*, *Hylobates moloch*, and *Pygathrix nemaeus*, and (**b**) above and below-branch quadrupedal locomotion in *Lemur catta* and *Varecia variegata*. All angular measurements are reported in degrees. (**a**) During arm-swinging, animals can grasp the support with a supinated hand position (as do *A. fusciceps* and *H. moloch*) where the substrate is grasped on the same side as the arm doing the grasping and is in a "palm-in" position. But some animals (*P. nemaeus*) reach under the pole and grasp the contralateral side of the substrate relative to the arm doing the grasping and place the hand in a pronated or "palm-out" position. Note that *Hylobates* shows little wrist deviation during a stride, whereas *Ateles* and *Pygathrix* show a great deal of wrist deviation despite having different grips. In pronated (palm-out) hand positions, the wrist is first ulnarly deviated, then subsequently radially deviated throughout the remainder of the support phase. For supinated (palm-in) hand positions, the pattern is opposite. (**b**) During below-branch quadrupedal locomotion, all the animals grasped with a hook grip (grasping between digits 1 and 2 or 1–2 grasp). Above branch, *Varecia* also used a 1–2 grasp, but *Lemur* used a schizodactylous grip (2–3 grasp). For 2–3 grasping primates, wrist angle provides a measure of flexion and extension. The degrees of motion shown in the graphs should be read in this context: a 1–2 grasp results in degree changes in the radioulnar (deviation) plane, whereas a 2–3 grasp represents degree changes in the flexion-extension plane. Data reanalyzed from Tripp et al. (2015) and Granatosky (2015)

ranges of radial and ulnar deviation of the hand may be a way for anatomically generalized primates (e.g., *Lagothrix, Pygathrix,* and most below-branch quadrupeds) to achieve proficient suspensory locomotion. This does not appear to be the case. In a study by Tripp et al. (2015), wrist position was measured during below-branch quadrupedal walking in *Varecia variegata* and *Lemur catta*. During below-branch quadrupedal walking, both species grasped the support with a radially deviated wrist position and subsequently ulnarly deviated throughout most of the stance phase. Although radial deviation was greater during below-branch quadrupedal locomotion, the overall change in wrist position (excursion) was still greater during above-branch quadrupedal locomotion (Fig. 13.4). Although studies of hand and wrist positions during suspensory locomotion are currently limited (Turnquist 1975; Jenkins 1981; Granatosky 2015), it appears evident that the mechanical demands of suspensory locomotion can be solved with a broad range of behavioral solutions, even in the absence of hand anatomical patterns commonly associated with suspensory locomotion.

4 Manual Mechanical Flexibility as an Adaptive Feature of Primates

On the basis of studies reviewed above, we can conclude that many and varied behavioral solutions to the problems of loading long digits can be found within the order of Primates. These solutions include ulnar deviation, proximal IP joint flexion (bowed-up fingers), McP joint hyperextension (fingers off the substrate as in aye-ayes), schizodactyly, fist-walking, chimpanzee-style knuckle-walking, and gorilla-style knuckle-walking. One possible way to parse this further is to argue that non-hominoid solutions (hand deviation, bowing-up, schizodactyly) are solutions found only in an arboreal context, while hominoid solutions (knuckle- and fist-walking) are associated with terrestrial locomotion. This would make intuitive mechanical sense, but behavioral observations do not support that argument. Lorises and spider monkeys walking on the ground abduct their hands (Schmitt unpublished data). The aye-aye pulls its long fingers up off flat surfaces and large boughs, but wraps them when grasping smaller branches (Krakauer et al. 2002). Finally, chimpanzees exhibit knuckle-walking postures on arboreal and terrestrial supports (Doran 1992, 1993, 1997). Therefore, a common and important strategy for primates with long fingers is to moderate bending moments with behavioral mechanisms. In this way, the knuckle-walking postures of apes are not so unusual or linked to terrestrial environments, but rather one solution among many that can be used to moderate digital loads on a variety of substrates, among other advantages like increased effective limb length.

This review emphasizes flexibility in hand posture as an adaptive strategy of primates. We also emphasize the lack of straightforward relationships between anatomy and hand position during locomotion and posture, an observation made repeatedly by others (see Bishop 1962, 1964; Whitehead 1993; Lemelin and Schmitt 1998). This perspective shows that studying the functional anatomy of the hand is challenging and may be futile for reconstructing hand postures in fossil primates. This statement, that will make most readers wince, is not nearly as negative as it sounds. The hands of primates provide a perfect example of the challenges of functional anatomy and issues of mechanical design and biological role, as detailed by Bock and von Wahlert (1965). The problem of predicting hand postures from hand anatomy brings into sharp relief the important distinction between habitual postures and mechanical flexibility. Habitual postures like grasping (which is reflected in curved phalanges) or digitigrady (reflected in short phalanges) may still be inferred from anatomy. Bishop (1964) showed that while strepsirrhines and most platyrrhines are highly variable in hand posture, callitrichids and cercopithecoids are much less so. This suggests that the use of large trunks and terrestrial substrates reduces the need for and advantage of variability in hand posture.

Seen this way, anatomical features of the hand that reflect habit should not be seen as constraints. In fact, the primate hand is special not for its specialization, as is the case for many other mammal groups, but instead for its mechanical and behavioral flexibility. Flexibility in hand posture—such as the ability to grasp a substrate or place the hand on top of it, the ability to deviate or supinate the hand at the wrist, and the ability to walk on the metacarpal heads, lateral border of the hand (in a fist), or even the dorsal aspect of the middle phalanges (knuckles)—can be seen as one of the keys to the successful radiation and the diversity of locomotor and postural behavior that has evolved in our family tree.

The capacity for animals to change intrinsic aspects of locomotion (e.g., altering gaits, limb postures, body position, hand and wrist movements, etc.) fluidly and efficiently is key to locomotor success in multiple settings. This ability seems especially important in a complex arboreal milieu occupied by many mammalian species (Blanchard and Crompton 2011; Fleagle 2013). The idea that primates have a well-developed capacity for modifying aspects of their gaits to effectively adjust to particular environmental circumstances is not new (see Vilensky and Larson 1989; Schmitt 2010a, b). This mechanical flexibility (see Wainwright et al. 2008; Iriarte-Diaz et al. 2012) can be seen as reflecting underlying neuromuscular mechanisms that may have allowed for the great locomotor diversity within the primate order (Vilensky and Larson 1989; Schmitt 2010b).

For most nonprimate mammals, locomotion is thought to be controlled by central pattern generators (CPGs), which are neural networks that produce rhythmic outputs without significant cortical feedback (MacKay-Lyons 2002; Drew et al. 2004; Ijspeert 2008) and produce the basic action of stepping without higher commands from the cortex (Mori 1987; Mori et al. 1996; Golubitsky et al. 1999; MacKay-Lyons 2002; Drew et al. 2004; see also Chap. 6). Evidence for such circuits in higher order primates, including humans, is tenuous, and studies suggest that supraspinal inputs, and likely cortical inputs, have a more important role in the generation of primate locomotion (Mori 1987; Vilensky and Larson 1989; Mori et al. 1996; Schmitt 2010b). Recent neuromuscular studies suggest that humans and some nonhuman primates may have evolved a flexible coupling of thoracolumbar and cervical centers that allows humans to use the upper limbs for manipulative and skilled movements or for locomotor tasks (Dietz 2002, 2003).

We argue that palmigrady represents the primitive condition for primates and that all other hand postures likely evolved from this condition and involved moving load from the palm to the digits (Table 13.1). Throughout their evolution, even the most specialized primates have maintained a hand (and a foot for that matter) whose primitive five digits and ranges of movements compared to many other mammals has allowed for fluid movements from one substrate to another, kinematic adjustment to changes in speed and gaits, and biomechanical adjustment throughout ontogeny. This perspective reinforces the view that much of the success of the primate radiation has to do with maintaining generalized "cheiridial compromise" that allows flexibility.

5 Future Directions

This perspective on primate manual (and pedal) behavioral flexibility raises the question of where to go next? Does research on the functional anatomy of the hand simply stop at this point? The answer to that is clearly no. What is needed is dual efforts in the field and in semi-captive situations. Bishop (1964) provides a model for this. With the availability of high-resolution, light-weight digital cameras (see Schmitt 2010a; Granatosky et al. 2016) that can zoom in easily and have long enough battery life to be used in the field, researchers can observe hand positioning during locomotion and posture. Following the approach of Bishop (1964), hand axis can be quantified. In addition hand-substrate angle and joint angles can be analyzed. We need more data on what primates do every day. In her study, Bishop (1964) showed the amazing variation in strepsirrhine hand axis orientations when grasping dowels, but far less variation in callitrichids and cercopithecoids. Would that pattern bear out in the wild? We will not know without rigorous data collection in those settings.

Simple postural observations alone are not enough. Vereecke and Wunderlich (Chap. 10 in this volume) detailed the dramatic improvement in technology for studying loading and movements of the hand. Many of those techniques can be easily adopted to semi-captive situations in wildlife centers and zoos. Data collection in those conditions can allow for testing of specific ideas we presented here about palm and digital loading. Taken together, field data and data collected in semi-captive conditions will provide a more complete picture of primate hand postures and loading. These data, once collected, can also test for hypotheses about how the primate hand evolved from a fundamental organ of weight support to a flexible and versatile organ allowing for movement and foraging in a range of environments, from the most simple to the most complex, promoting the exploitation of arboreal niches in ways that other mammals have not.

Acknowledgments A review chapter like this develops only as a result of extensive conversations with colleagues, collaborators, and friends. The topic of the evolution of the hand remains a deep and profound one in our field, and we have discussed some of the ideas here with many of our colleagues in the recent and more distant past. We thank the following people (listed in a sort of temporal order) for discussions about the ideas concerning functional anatomy of the hand presented in this chapter: Susan Larson, Jack Stern, Scott McGraw, Chris Wall, Liza Shapiro, Rich Kay, Matt Cartmill, Jean Turnquist, John Cant, Jandy Hanna, David Raichlen, Chris Vinyard, Laura Johnson, Karyne Rabey, Charlotte Miller, Steve Churchill, Doug Boyer, and Anne-Claire Fabre. We owe a special debt of gratitude to Mike Rose whose many articles on this topic and wise advice and opinions helped shape much of our thinking on hand anatomy. We are especially grateful to Pierre Lemelin, Tracy Kivell, and Brian Richmond for heavy and constructive editing of this chapter. More importantly, we are grateful to them for the many and lengthy discussions about this chapter and other issues concerning the evolution of the hand that we had over the past 3 years. Much of the data described here come from studies carried out by one or more of us at SUNY Stony Brook (laboratory of Jack Stern and Susan Larson), The University of Texas at Austin (laboratory of Jody Jensen), Monkey Jungle, and the Duke Lemur Center. We thank all the faculty and staff there in general and specifically Marianne Crisci, Sian Evans, David Brewer, Bill Hess, and Erin Ehmke.

References

Almécija S, Smaers JB, Jungers WL (2015) The evolution of human and ape hand proportions. Nat Commun 6:7717

Begun DR, Kivell TL (2011) Knuckle-walking in *Sivapithecus*? The combined effects of homology and homoplasy with possible implications for pongine dispersals. J Hum Evol 60:158–170

Biewener AA (1983) Allometry of quadrupedal locomotion: the scaling of duty factor, bone curvature and limb orientation to body size. J Exp Biol 105:47–171

Biewener AA (1989) Scaling in body support in mammals: limb posture and muscle mechanics. Science 245:45–48

Bishop A (1962) Control of the hand in lower primates. N Y Acad Sci 120:316–337

Bishop A (1964) Use of the hand in lower primates. In: Buettner-Janusch J (ed) Evolutionary and genetic biology of primates. Academic Press, New York, pp 133–223

Blanchard ML, Crompton RH (2011) Field study methods for primate locomotor ecology and biomechanics. In: D'Août K, Vereecke EE (eds) Primate locomotion: linking field and laboratory research. Springer, New York, pp 183–199

Bock WJ, von Wahlert G (1965) Adaptation and the form-function complex. Evolution 19:269–299

Cant JGH, Youlatos D, Rose MD (2003) Suspensory locomotion of *Lagothrix lagothricha* and *Ateles belzebuth* in Yasuni National Park, Ecuador. J Hum Evol 44:685–699

Cartmill M (1972) Arboreal adaptations and the origin of the order Primates. In: Tuttle R (ed) The functional and evolutionary biology of primates. Aldine-Atherton, Chicago, pp 97–122

Cartmill M (1974a) Pads and claws in arboreal locomotion. In: Jenkins FA Jr (ed) Primate locomotion. Academic Press, New York, pp 45–83

Cartmill M (1985) Climbing. In: Hildebrand M, Bramble DM, Liem KF, Wake DB (eds) Functional vertebrate morphology. Harvard University Press, Cambridge, pp 73–88

Cartmill M (1992) New views on primate origins. Evol Anthropol 1:105–111

Cartmill M, Milton K (1977) The lorisiform wrist joint and the evolution of "brachiating" adaptations in the Hominoidea. Am J Phys Anthropol 47:249–272

Charles-Dominique P, Martin RD (1970) Evolution of lorises and lemurs. Nature 227:257–260

Clevedon Brown J, Yalden DW (1973) The description of mammals—2 limbs and locomotion of terrestrial mammals. Mamm Rev 3:107–134

Congdon KA (2012) Interspecific and ontogenetic variation in proximal pedal phalangeal curvature of great apes (*Gorilla gorilla*, *Pan troglodytes*, and *Pongo pygmaeus*). Int J Primatol 33:418–427
Congdon KA (2014) Grasping pressures and phalangeal curvature in primates: an experimental in vivo approach. Ph.D. dissertation, University of Missouri
Day MH, Iliffe SR (1975) The contrahens muscle layer in *Tarsius*. Folia Primatol 24:241–249
Dietz V (2002) Do human bipeds use quadrupedal coordination? Trends Neurosci 25:462–467
Dietz V (2003) Spinal cord pattern generators for locomotion. Clin Neurophysiol 114:1379–1389
Doran DM (1992) The ontogeny of chimpanzee and pygmy chimpanzee locomotor behavior: a case study of paedomorphism and its behavioral correlates. J Hum Evol 23:139–157
Doran, DM (1993) Comparative locomotor behavior of chimpanzees and bonobos: the influence of morphology on locomotion. Am J Phy Anthropol, 91(1): pp.83–98.
Doran DM (1997) Ontogeny of locomotion in mountain gorillas and chimpanzees. J Hum Evol 32:323–344
Drew T, Prentice S, Schepens B (2004) Cortical and brainstem control of locomotion. Prog Brain Res 143:251–261
Fleagle JG (2013) Primate adaptation and evolution, 3rd edn. Academic Press, San Diego
Foster AD, Raichlen DA, Pontzer H (2013) Muscle force production during bent-knee, bent-hip walking in humans. J Hum Evol 65:294–302
Fujiwara S, Endo H, Hutchinson JR (2011) Topsy-turvy locomotion: biomechanical specializations of the elbow in suspended quadrupeds reflect inverted gravitational constraints. J Anat 219:176–191
Gebo DL (2014) Primate comparative anatomy. Johns Hopkins University Press, Baltimore
Golubitsky M, Stewart I, Pietro-Luciano B, Collins JJ (1999) Symmetry in locomotor central pattern generators and animal gaits. Nature 401:693–695
Grand TI (1968) Functional anatomy of the upper limb. Bibl Primatol 7:104–125
Granatosky MC (2015) Kinematic and kinetic analysis of arm-swinging in the red-shanked douc langur (*Pygathrix nemaeus*). J Viet Primatol 2(4):33–40
Granatosky MC, Tripp CH, Schmitt D (2016) Gait kinetics of above and below branch quadrupedal locomotion in lemurid primates. J Exp Biol 219:53–63
Groves CP (1970) Gorillas. Arthur Barker, London
Guillot D (2011) Forelimb suspensory gait characteristics of wild *Lagothrix poeppigii* and *Ateles belzebuth*: developing video-based methodologies in free-ranging primates. In: D'Août K, Vereecke EE (eds) Primate locomotion: linking field and laboratory research. Springer, New York, pp 247–269
Hamrick MW, Simons EL, Jungers WL (2000) New wrist bones of the Malagasy giant subfossil lemurs. J Hum Evol 38:635–650
Hayama S, Chantani K, Nakatsukasa M (1994) The digitigrade hand and terrestrial adaptation in Japanese macaques. Anthropol Sci 102(Suppl):115–125
Heffner RS, Masterton RB (1975) Variation in form of the pyramidal tract and its relationship to digital dexterity. Brain Behav Evol 12:188–200
Hunt KD, Cant JGH, Gebo DL, Rose MD, Walker SE, Youlatos D (1996) Standardized descriptions of primate locomotor and postural modes. Primates 37:363–387
Ijspeert AJ (2008) Central pattern generators for locomotion control in animals and robots: a review. Neural Netw 21:642–653
Inouye SE (1989) Variability of hand postures in the knuckle-walking behavior of African apes. Am J Phys Anthropol 78:245 (abstract)
Inouye SE (1992) Ontogeny and allometry of African ape manual rays. J Hum Evol 23:107–138
Inouye SE (1994) Ontogeny of knuckle-walking hand postures in African apes. J Hum Evol 26:459–485
Iriarte-Diaz J, Riskin DK, Breuer KS, Swartz SM (2012) Kinematic plasticity during flight in fruit bats: individual variability in response to loading. PLoS One 7:e36665
Ivanco TL, Pellis SM, Whishaw IQ (1996) Skilled forelimb movements in prey catching and in reaching by rats (*Rattus norvegicus*) and opossums (*Monodelphis domestica*): relations to anatomical differences in motor systems. Behav Brain Res 79:163–181

Iwaniuk AN, Whishaw IQ (1999) How skilled are the skilled limb movements of the raccoon (Procyon lotor)? Behav Brain Res 99(1):35–44
Jenkins FA Jr (1974) Tree shrew locomotion and the origins of primate arborealism. In: Jenkins FA Jr (ed) Primate locomotion. Academic Press, New York, pp 85–115
Jenkins FA Jr (1981) Wrist rotation in primates: a critical adaptation for brachiators. Symp Zool Soc Lond 48:429–451
Johnson LE (2012) Biomechanics of vertical clinging and grasping in primates. Ph.D. dissertation, Duke University
Johnson LE, Hanna J, Schmitt D (2015) Single-limb force data for two lemur species while vertically clinging. Am J Phys Anthropol 158:463–474
Jolly CJ (1965) The origins and specializations of long-faced Cercopithecoidea. Ph.D. dissertation, University of London
Jungers WL, Godfrey LR, Simons EL, Chatrath PS (1997) Phalangeal curvature and positional behavior in extinct sloth lemurs (Primates, Palaeopropithecidae). Proc Natl Acad Sci U S A 94:11998–20001
Kivell TL, Schmitt D (2009) Independent evolution of knuckle-walking in African apes shows that humans did not evolve from a knuckle-walking ancestor. Proc Natl Acad Sci U S A 106:14241–14246
Kivell TL, Schmitt D, Wunderlich RE (2010) Hand and foot pressures in the aye-aye (*Daubentonia madagascariensis*) reveal novel biomechanical trade-offs required for walking on gracile digits. J Exp Biol 213:1549–1557
Kivell TL, Barros AP, Smaers JB (2013) Different evolutionary pathways underlie the morphology of wrist bones in hominoids. BMC Evol Biol 13:229
Krakauer E, Lemelin P, Schmitt D (2002) Hand and body position during locomotor behavior in the aye-aye (*Daubentonia madagascariensis*). Am J Primatol 57(3):105–118
Larson SG, Schmitt D, Lemelin P, Hamrick MW (2001) Limb excursion during quadrupedal walking: how do primates compare to other mammals? J Zool 255:353–365
Lemelin P (1996) The evolution of manual prehensility in primates. A comparative study of prosimian primates and didelphid marsupials. Ph.D. dissertation, State University of New York at Stony Brook
Lemelin P (1999) Morphological correlates of substrate use in didelphid marsupials: implications for primate origins. J Zool 247:165–175
Lemelin P, Grafton BW (1998) Grasping performance in *Saguinus midas* and the evolution of hand prehensility in primates. In: Strasser E, Fleagle J, Rosenberger A, McHenry H (eds) Primate locomotion: recent advances. Plenum Press, New York, pp 131–144
Lemelin P, Schmitt D (1998) The relation between hand morphology and quadrupedalism in primates. Am J Phys Anthropol 105:185–197
Lewis OJ (1985a) Derived morphology of the wrist articulations and theories of hominoid evolution. Part II. The lorisine joints. J Anat 140:447–460
Lewis OJ (1985b) Derived morphology of the wrist articulations and theories of hominoid evolution. Part I. The midcarpal joints of higher primates. J Anat 142:151–172
MacKay-Lyons M (2002) Central pattern generation of locomotion: a review of the evidence. Phys Ther 82:69–83
Matarazzo S (2008) Knuckle walking signal in the manual digits of *Pan* and *Gorilla*. Am J Phys Anthropol 135:27–33
Matarazzo S (2013) Manual pressure distribution patterns of knuckle-walking apes. Am J Phys Anthropol 152:44–50
McClearn D (1992) Locomotion, posture, and feeding behavior of kinkajous, coatis, and raccoons. J Mamm 73:245–261
McClure NK, Phillips AC, Vogel ER, Tocheri MW (2012) Unexpected pollex and hallux use in wild *Pongo pygmaeus wurmbii*. Am J Phys Anthropol 147(Suppl 54):203 (abstract)
Meldrum DJ, Dagosto M, White J (1997) Hindlimb suspension and hind foot reversal in *Varecia variegata* and other arboreal mammals. Am J Phys Anthropol 103:85–102

Mendel FC (1979) The wrist joint of two-toed sloths and its relevance to brachiating adaptations in the Hominoidea. J Morphol 162:413–424

Michilsens F, Vereecke EE, D'Août K, Aerts P (2010) Muscle moment arms and function of the siamang forelimb during brachiation. J Anat 217:521–535

Mori S (1987) Integration of posture and locomotion in acute decerebrate cats and in awake, freely moving cats. Prog Neurobiol 28:161–195

Mori S, Matsuyama K, Miyashita E, Nakajima K, Asanome M (1996) Basic neurophysiology of primate locomotion. Folia Primatol 66:192–203

Napier J (1962) The evolution of the hand. Sci Am 207:56–62

Napier J (1980) Hands. Pantheon Books, New York

Napier JR, Napier PH (1967) A handbook of living primates. Academic Press, London

Napier JR, Walker AC (1967) Vertical clinging and leaping—a newly recognized category of locomotor behaviour of primates. Folia Primatol 6:204–219

Nengo O (1993) Integrating mechanical and historical approaches to organismal design: the case in the hands and feet of catarrhines. Ph.D. dissertation, Harvard University

Nieschalk U, Demes B (1993) Biomechanical determinants of reduction of the second ray in Lorisinae. In: Preuschoft H, Chivers DJ (eds) Hands of primates. Springer-Verlag, Vienna, pp 225–234

Nyakatura JA, Petrovitch A, Fischer MS (2010) Limb kinematics during locomotion in the two-toed sloth (*Choloepus didactylus*, Xenarthra) and its implications for the evolution of the sloth locomotor apparatus. Zoology 113:221–234

Oxnard CE, Crompton RH, Lieberman SS (1990) Animal lifestyles and anatomies: the case of the Prosimian primates. University of Washington Press, Seattle

Patel BA (2009) Not so fast: speed effects on forelimb kinematics in cercopithecoid monkeys and implications for digitigrade postures in primates. Am J Phys Anthropol 140:92–112

Patel BA (2010a) Functional morphology of the cercopithecoid primate metacarpals. J Hum Evol 58:320–337

Patel BA (2010b) The interplay between speed, kinetics and hand postures during primate terrestrial locomotion. Am J Phys Anthropol 141:222–234

Patel BA, Polk JD (2010) Distal forelimb kinematics in *Erythrocebus patas* and *Papio anubis* during walking and galloping. Int J Primatol 31:191–207

Patel BA, Wunderlich RE (2010) Dynamic pressure patterns in the hands of olive baboons (*Papio anubis*) during terrestrial locomotion: implications for cercopithecoid primate hand morphology. Anat Rec 293:710–718

Patel BA, Larson SG, Stern JT Jr (2012) Electromyography of wrist and finger flexor muscles in olive baboons (*Papio anubis*). J Exp Biol 215:115–123

Patel BA, Horner AM, Thompson NE, Barrett L, Henzi SP (2013) Ontogenetic scaling of fore-and hind limb posture in wild chacma baboons (*Papio hamadryas ursinus*). PLoS One 8:e71020

Polk JD (2002) Adaptive and phylogenetic influences on musculoskeletal design in cercopithecine primates. J Exp Biol 205:3399–3412

Pontzer H, Raichlen DA, Sockol MD (2009) The metabolic cost of walking in humans, chimpanzees, and early hominins. J Hum Evol 56:43–54

Preuschoft H (2002) What does "arboreal locomotion" mean exactly and what are the relationships between "climbing", environment and morphology? Z Morphol Anthropol 83:171–188

Rawlins RG (1993) Locomotor and manipulative use of the hand in Cayo Santiago macaques (*Macaca mulatta*). In: Preuschoft H, Chivers DJ (eds) Hands of primates. Springer-Verlag, Vienna, pp 21–30

Richmond BG (1997) Ontogeny of phalangeal curvature and locomotor behavior in lar gibbons. Am J Phys Anthropol 104(Suppl 24):197 (abstract)

Richmond BG (1998) Ontogeny and biomechanics of phalangeal form in primates. Ph.D. dissertation, State University of New York at Stony Brook

Richmond BG (2007) Biomechanics of phalangeal curvature. J Hum Evol 53:678–690

Richmond BG, Strait DS (2000) Evidence that humans evolved from a knuckle-walking ancestor. Nature 404:382–385

Richmond BG, Begun DR, Strait DS (2001) Origin of human bipedalism: the knuckle-walking hypothesis revisited. Yearb Phys Anthropol 44:70–105

Rollinson J, Martin RD (1981) Comparative aspects of primate locomotion, with special reference to arboreal cercopithecoids. Symp Zool Soc Lond 48:377–427

Rose MD (1973) Quadrupedalism in primates. Primates 14:337–357

Rosenberger AL, Halenar L, Cooke SB, Hartwig WC (2008) Morphology and evolution of the spider monkey, genus *Ateles*. In: Campbell CJ (ed) Spider monkeys: behavior, ecology and evolution of the genus *Ateles*. Cambridge University Press, Cambridge, UK, pp 19–49

Schmitt D (2010a) Translating primate locomotor biomechanical variables from the laboratory to the field. In: D'Août K, Vereecke EE (eds) Primate locomotion: linking field and laboratory research. Springer, New York, pp 7–28

Schmitt D (2010b) Primate locomotor evolution: biomechanical studies of primate locomotion and their implications for understanding primate neuroethology. In: Platt ML, Ghazanfar AA (eds) Primate neuroethology. Oxford University Press, Oxford, pp 10–30

Schmitt D, Lemelin P (2002) Origins of primate locomotion: gait mechanics of the woolly opossum. Am J Phys Anthropol 118:231–238

Schmitt D, Turnquist JE, Rose MD (2009) The relationship between arm-swinging behavior and postcranial morphology in *Ateles* and *Lagothrix*. Am J Phys Anthropol 138(Suppl 38):230 (abstract)

Soligo C, Martin RD (2006) Adaptive origins of primates revisited. J Hum Evol 50:414–430

Sockol MD, Raichlen DA, Pontzer H (2007) Chimpanzee locomotor energetics and the origin of human bipedalism. Proc Natl Acad Sci U S A 104:12265–12269

Stern JT Jr, Oxnard CE (1973) Primate locomotion: some links with evolution and morphology. Primatologia 4:1–93

Susman RL (1974) Facultative terrestrial hand postures in an orangutan (*Pongo pygmaeus*) and pongid evolution. Am J Phys Anthropol 40:27–37

Susman RL (1979) Comparative and functional morphology of hominoid fingers. Am J Phys Anthropol 50:215–236

Sussman RW (1991) Primate origins and the evolution of angiosperms. Am J Primatol 23:209–223

Swartz SM, Bertram JE, Biewener AA (1989) Telemetered in vivo strain analysis of locomotor mechanics of brachiating gibbons. Nature 342:270–272

Szalay FS, Dagosto M (1980) Locomotor adaptations as reflected on the humerus of Paleogene primates. Folia Primatol 34:1–45

Szalay FS, Dagosto M (1988) Evolution of hallucial grasping in the primates. J Hum Evol 17:1–33

Tripp CH, Granatosky MC, Schmitt D (2015) Forelimb mechanics during below-branch quadrupedalism in primates: a comparison to above-branch quadrupedalism and bimanual locomotion. Am J Phys Anthropol 156(Suppl 60):307 (abstract)

Tsegai ZJ, Kivell TL, Gross T, Nguyen NH, Pahr DH, Smaers JB, Skinner MM (2013) Trabecular bone structure correlates with hand posture and use in hominoids. PLoS One 8:e78781

Turnquist JE (1975) The elbow and forearm of *Ateles*: an anatomical and behavioral study of locomotion, Ph.D. dissertation, University of Pennsylvania

Tuttle RH (1967) Knuckle-walking and the evolution of hominoid hands. Am J Phys Anthropol 26:171–206

Tuttle RH (1969) Knuckle-walking and the problem of human origins. Science 166:953–961

Tuttle RH (1981) Evolution of hominid bipedalism and prehensile capabilities. Phil Trans R Soc Lond B292:89–94

Vilensky JA, Larson SG (1989) Primate locomotion: utilization and control of symmetrical gaits. Annu Rev Anthropol 18:17–35

Wainwright PC, Mehta RS, Higham TE (2008) Stereotypy, flexibility and coordination: key concepts in behavioral functional morphology. J Exp Biol 211:3523–3528

Weidenreich F (1931) Der primäre Greitcharacter der menschlichen Hände und Füsse und seine Bedeutung für das Abstammungsproblem. Verbandlungen Gesellschaft Phys Anthropol 5:97–110

Whitehead PF (1993) Aspects of the anthropoid wrist and hand. In: Gebo DL (ed) Postcranial adaptation in nonhuman primates. Northern Illinois University Press, DeKalb, pp 96–102

Williams SA (2010) Morphological integration and the evolution of knuckle-walking. J Hum Evol 58:432–440

Whishaw IQ, Coles BL (1996) Varieties of paw and digit movement during spontaneous food handling in rats: postures, bimanual coordination, preferences, and the effect of forelimb cortex lesions. Behav Brain Res 77:135–148

Whishaw IQ, Pellis SM (1990) The structure of skilled forelimb reaching in the rat: a proximally driven movement with a single distal rotatory component. Behav Brain Res 41:49–59

Wood Jones F (1916) Arboreal man. Edward Arnold, London

Wood Jones F (1920) The principles of anatomy as seen in the hand. J. and A. Churchill, London

Wunderlich RE, Jungers WL (2009) Manual digital pressures during knuckle-walking in chimpanzees (*Pan troglodytes*). Am J Phys Anthropol 139:394–403

Yalden DW (1972) The form and function of the carpal bones in some arboreally adapted mammals. Acta Anat 82:383–406

Zeininger A, Shapiro LJ, Raichlen DA (2007) Ontogeny of digitigrade hand and foot postures in baboons (*Papio cynocephalus*). Am J Phys Anthropol 44:255 (abstract)

Zeininger A, Shapiro LJ, Raichlen DA (2009) The effects of digitigrade cheiridial postures on speed and gait in infant baboons. Am J Phys Anthropol 138(Suppl 48):279 (abstract)

Zeininger A, Richmond BG, Hartman G (2011) Metacarpal head biomechanics: a comparative backscattered electron image analysis of trabecular bone mineral density in *Pan troglodytes*, *Pongo pygmaeus*, and *Homo sapiens*. J Hum Evol 60:703–710

Zeininger A, Schmitt D, Wunderlich R (2014) Developmental timing of heel-strike plantigrady in chimpanzees and gorillas. Am J Phys Anthropol 153(Suppl 58):281–282 (abstract)

Part III
Paleontological Evidence

Chapter 14
Hands of Paleogene Primates

Doug M. Boyer, Gabriel S. Yapuncich, Stephen G.B. Chester, Jonathan I. Bloch, and Marc Godinot

1 Introduction

Morphological specializations of the hand that allow arboreal clinging, one-handed grasping (prehension), thumb (pollical) opposability, and object manipulation are thought to be key adaptations in primate evolution (Wood Jones 1916; Haines 1955; Napier 1960, 1961; Le Gros Clark 1971; Cartmill 1972, 1974a, b; Martin 1986; Godinot and Beard 1991, 1993; Jouffroy et al. 1991; Godinot 1992; Lemelin 1996, 1999; Hamrick 2001, 2007; Bloch and Boyer 2002; Kirk et al. 2008). However, despite considerable scholarly focus, the distribution of these specializations, their functional significance, and their novelty relative to other arboreal mammals are still poorly understood. For example, prehensility, while often cited as a primate innovation, is now known to characterize close relatives of primates and other arboreal mammals (Lemelin 1996, 1999; Kirk et al. 2008; Boyer et al. 2013a). In contrast, many argue that true pollical opposability is not a primate trait per se, but is restricted to catarrhines

D.M. Boyer (✉) • G.S. Yapuncich
Department of Evolutionary Anthropology, Duke University, Box 90383, 130 Science Drive, Durham, NC 27708, USA
e-mail: douglasmb@gmail.com

S.G.B. Chester
Department of Anthropology and Archaeology, Brooklyn College, CUNY, Brooklyn, NY, USA

Department of Anthropology, Graduate Center of the City University of New York, New York, NY 10016, USA

New York Consortium in Evolutionary Primatology, New York, NY, USA

J.I. Bloch
Florida Museum of Natural History, University of Florida, Gainesville, FL, USA

M. Godinot
UMR 7207 CR2P, École Pratique des Hautes Études, Paris, France

T.L. Kivell et al. (eds.), *The Evolution of the Primate Hand*, Developments in Primatology: Progress and Prospects, DOI 10.1007/978-1-4939-3646-5_14

(e.g., Napier 1961). Other researchers nonetheless view morpho-functional evidence for opposability to be more widespread (Jouffroy and Lessertisseur 1959; Etter 1974, 1978; Boyer et al. 2013a). More generally, some researchers view specializations of the thumb to be secondary in primates, with the ancestral form having a non-divergent, non-opposable digit, whereas others see an earlier evolution of this specialization as more likely [see discussion in Napier (1961) and Godinot and Beard (1993)]. Better documentation of the distribution of specializations (both morphological and functional) remains essential for establishing patterns of morphological change through time and for evaluating the functional/adaptive context for the evolution of primate hands. The fossil record provides the only direct evidence for testing specific hypotheses focused on the evolution of unique and functionally significant skeletal specializations that characterize the hands of the earliest primates and their descendants.

Ultimately, functional implications of hand morphology in early fossil primates can influence competing hypotheses addressing the nature of adaptive transition(s) in the evolution and radiations of stem and crown primates. Here we review the evidence from the hands of early fossil primates and their close relatives and discuss alternative ideas for some of the ecological transitions that occurred during early primate evolution. In particular, it has been suggested that primate hand morphology reflects *increasing specialization* for either: (1) locomotion and foraging in a fine-branch niche (Lemelin 1996, 1999; Hamrick 2001; 2007; Bloch and Boyer 2002; Bloch et al. 2007; Kirk et al. 2008); (2) prehensile clinging on relatively "large" and often vertical supports (e.g., Napier and Walker 1967; Godinot 1991); (3) manual predation in the fine-branch niche in certain lineages derived from a clade of terrestrial/scansorial "olfactory-guided" insectivores (e.g., Cartmill 1972, 1974a, b, 1992; Lemelin 1996); or (4) a "lunging grasp" (Godinot 2007) predation style similar to that often used by galagos and tarsiers, in which the animal "almost leaps" by rapidly extending its hind limbs while reaching out and grabbing at prey but never disengaging from the substrate.

We begin by summarizing the information currently available about the hands of early fossil primates, specifically plesiadapiforms, adapiforms, omomyiforms, and Paleogene anthropoids. We close with an attempt to synthesize these data in terms of the functional-adaptive transitions during early primate evolution. Our goal is to provide a clearer view of what is known about the hands of early fossil primates (Fig. 14.1; Tables 14.1 and 14.2), the transitions they underwent, and the functional and adaptive significance of these changes. Figure 14.2 illustrates and provides anatomical terminology used in this chapter.

Fig. 14.1 (continued) bones, replicated to give a more complete (if hypothetical) view of the hand. *Notharctus* reconstruction based on AMNH 127167 (Hamrick and Alexander 1996). Note that lengths of third and fourth intermediate phalanges and third metacarpal (Mc3) are not known. *Adapis* reconstruction primarily based on Rosières 311 (Godinot and Beard 1991). Three isolated phalanges from the Phosphorites de Quercy were scaled to represent locality mean proportions of Mc3, proximal, intermediate, and distal phalanges. All proximal phalanges illustrated are AMNH 140719, intermediate phalanges are represented by an unnumbered specimen from the Montauban Quercy collection (MaPhQ), and distal phalanges are UM ECA 1400 (Godinot 1992). Note hyperextension of the metacarpophalangeal joints on clawed *Plesiadapis*, the wide divergence between Mc1-Mc2 and Mc2-Mc3 in *Notharctus*, and that Mc2 is relatively much shorter in *Notharctus* compared to *Adapis*. Modified from Boyer et al. (2013a)

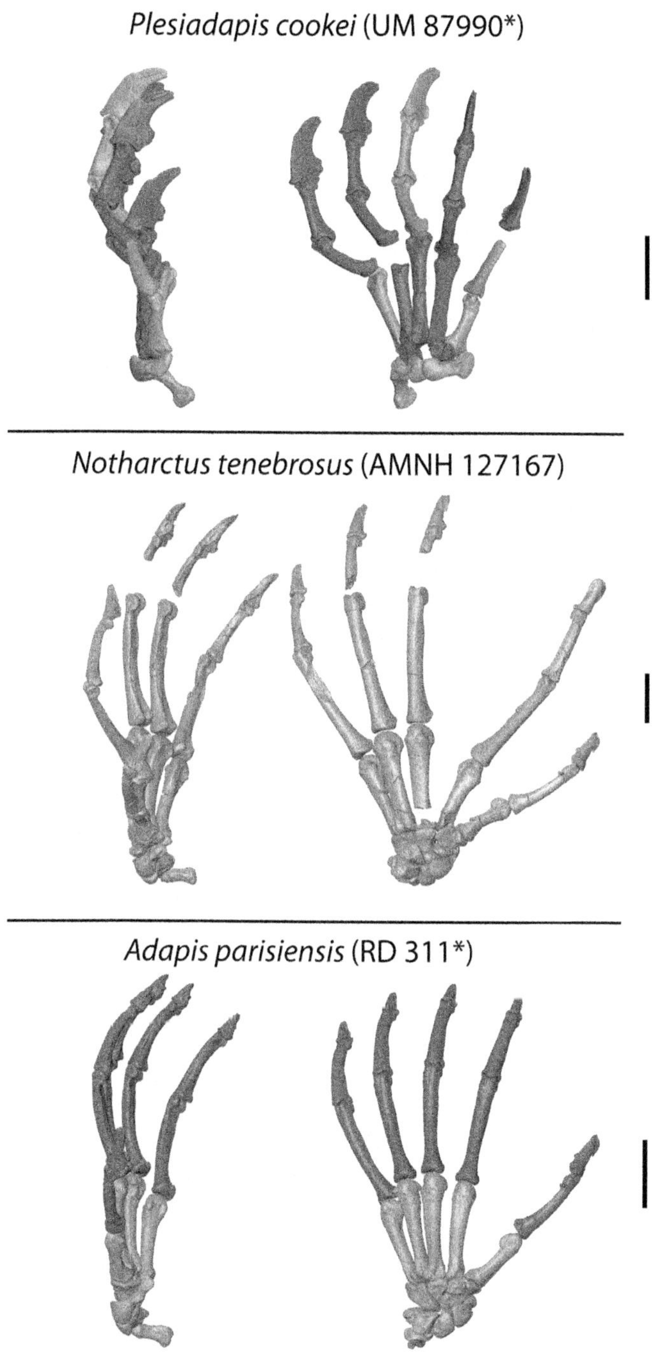

Fig. 14.1 Three fossil primate hand reconstructions (*Plesiadapis cookei*, *Notharctus tenebrosus*, *Adapis parisiensis*). Reconstructed hands depicted in radial (*left*) and dorsal (*right*) views. Both *Plesiadapis* and *Adapis* represent composites (*asterisk*). *Plesiadapis* reconstruction based on UM 87990 from Clarks Fork Basin locality SC-117 (Boyer 2009). The *dark gray* bones are those that are tentatively assigned to the specimen (see text). Phalanges on all digit rays represent the three

Table 14.1 Plesiadapiform hand fossil material

Taxon	Specimen/s	Elements represented	References
Plesiadapis cookei	UM 87990 (1 specimen)	R, U, Sc, Ln, Tr, Ps, Mc1,5, PP*s*, IP*s*, DP*s*	Hamrick (2001), Bloch and Boyer (2002, 2007), Bloch et al. (2007), Boyer and Bloch (2008), Kirk et al. (2008), Boyer (2009), Boyer et al. (2013a)
Plesiadapis tricuspidens	MNHN (~56 specimens)	4 R*s*, 8 U*s*, Hm, Tr, 4 Mc3*s*, 2 Mc4*s*, 2 Mc5*s*, 5 PP*s*, 21 IP*s*, 8 DP*s*	Godinot and Beard (1991, 1993), Beard (1989, 1993a, b), Youlatos and Godinot (2004), Boyer (2009), Boyer et al. (2013a)
Plesiadapis insignis	MNHN coll. Menat specimen	Articulated hand/forelimb	Russell (1967), Gingerich (1976), Boyer (2009), Boyer et al. (2013a)
Nannodectes intermedius	USNM 442229 (1 specimen)	R, U, Sc, Hm, Cp, Ps, Mc1,3,5, PP*s*, IP*s*, DP*s*	Beard (1989, 1990, 1993a, b), Boyer and Bloch (2008), Kirk et al. (2008), Boyer (2009), Boyer et al. (2010a, 2013a)
Nannodectes gidleyi	AMNH 17379 (1 specimen)	R, U, Mc1,3, PP*s*, IP*s*, DP*s*	Simpson (1935), Beard (1989), Boyer and Bloch (2008), Kirk et al. (2008), Boyer (2009), Boyer et al. (2013a)
Dryomomys szalayi	UM 41870 (1 specimen)	R, U, Carp, Mc*s*1-5, PP*s*, IP*s*, DP*s*	Bloch and Boyer (2007), Bloch et al. (2007), Kirk et al. (2008), Boyer and Bloch (2008), Boyer et al. (2013a)
Tinimomys graybulliensis	UM 85176 (1 specimen) USNM (2 specimens)	3 R, 3 U, Carp, Mc*s*1-5, PP*s*, IP*s*, DP*s*	Beard (1989, 1993a, b), Boyer and Bloch (2008), Boyer et al. (2010a), Boyer et al. (2013a)
Ignacius graybullianus	USNM (4 specimens)	R, PP, 2 IP*s*	Beard (1989, 1990, 1993a, b), Krause (1991), Runestad and Ruff (1995), Hamrick (1999), Boyer et al. (2013a)
Ignacius clarkforkensis	UM (3 specimens)	R, U, Sc, Mc*s*1,3-5, PP*s*, IP*s*, DP*s*	Bloch and Boyer (2007), Bloch et al. (2007), Kirk et al. (2008), Boyer and Bloch (2008)
Phenacolemur jepseni	USGS 17847 (1 specimen)	Tr	Beard (1989, 1990), Boyer et al. (2013a)
Phenacolemur praecox	UM (2 specimens)	PP, IP	Beard (1989, 1990), Boyer and Bloch (2008), Boyer et al. (2013a)
Phenacolemur simonsi	USNM (13 specimens)	3 R*s*, 2 U*s*, Mc*s*1,3, PP*s*, 4 IP*s*	Beard (1989, 1990, 1993a, b), Krause (1991), Runestad and Ruff (1995), Hamrick et al. (1999), Boyer et al. (2013a)
Carpolestes simpsoni	UM 101963 (1 specimen)	R, U, Carp, Mc*s* 1-5, PP*s*, IP*s*, DP*s*	Bloch and Boyer (2002, 2007), Bloch et al. (2007), Boyer and Bloch (2008), Kirk et al. (2008), Boyer et al. (2013a)

Abbreviations: *AMNH* American Museum of Natural History; *MNHN* Muséum National d'Histoire Naturelle; *NMB* Naturhistorisches Museum Basel; *UM* University of Michigan; *USGS* United States Geological Survey; *USNM* United States National Museum, Smithsonian Institute; *R* radius; *U* ulna; *Carp* carpals; *Cp* capitate; Ce centrale; *Sc* scaphoid; *Ln* lunate; *Tr* triquetrum; *Trp* trapezoid; *Trm* trapezium; *Hm* hamate; *Ps* pisiform; *Mc* metacarpal; *PP* proximal phalanx; *IP* intermediate phalanx; *DP* distal phalanx

Table 14.2 Early euprimate hand fossil material

Taxon	Specimen(s)	Elements represented	References
Europolemur koenigswaldi	SMF-ME 1228 (1 specimen)	R, U, Mc1-5, PP1-5, IP3,4, DP2,3	Franzen (1987, 1988, 1993), Franzen and Frey (1993), Godinot and Beard (1991, 1993), Boyer et al. (2013a)
Europolemur keileri	SMF-ME 1683 (1 specimen)	R, U, Carp, Mc1-5, PP1,2,4,5, IP2,5, DP1,2	Franzen (1988, 1993), Godinot and Beard (1991, 1993), Hamrick (1996c, 2001), Boyer et al. (2013a)
Godinotia neglecta	GMH L-2 (1 specimen)	R, U, Carp, Mc1-5, PP1-5, IP3-5, DP1,4,5	Godinot (1992), Thalmann et al. (1989), Thalmann (1994), Boyer et al. (2013a)
Darwinius masillae	PMO 214.214 (1 specimen)	R, U, Carp, Mc1,5 (full), Mc3, 4 (partial), PP1-5, IP2-5, DP1-5	Franzen et al. (2009), Boyer et al. (2013a)
Notharctus tenebrosus	AMNH (3 specimens)	3 R*s*, 3 U*s*, 2 Sc*s*, Ln, 2 Tr*s*, 2 Ps*s*, 3 Hm*s*, 2 Cp*s*, 2 Ce*s*, Trp, Trm, Px, Mc*s* 1-5, PP*s*1-5, IP*s*, DP*s*1-5	Gregory (1920), Beard and Godinot (1988), Godinot (1992), Godinot and Beard (1991, 1993), Jouffroy et al. (1991), Hamrick (1996c, 2001), Hamrick and Alexander (1996), Kirk et al. (2008), Boyer et al. (2013a)
	USNM (5 specimens)		
Notharctus robustior	CM 11910 (1 specimen)	Sc, Cp, Hm	Beard and Godinot (1988)
Smilodectes gracilis	USNM (5 specimens)	2 Sc*s*, 3 Ln*s*, 3 Hm*s*, 3 Tr*s*, 2 Ps*s*, 4 Cp*s*, Ce, Trm	Beard and Godinot (1988), Godinot and Beard (1991, 1993), Hamrick (1996c), Boyer et al. (2013a)
Adapis parisiensis	RD 311 (private coll.)	R, U, Sc, Ln, Tr, Ps, Hm, Cp, Ce, Trp, Trm, Px, Mc*s*1-5, 4 PP*s*, 3 IP*s*	Dagosto (1983), Godinot and Jouffroy (1984), Beard and Godinot (1988), Godinot (1991, 1992), Godinot and Beard (1991, 1993), Jouffroy et al. (1991), Hamrick (1996c), Boyer et al. (2013a)
	NMB (4 specimens)		
	AMNH 140719 (1 specimen)		
	MaPhQ (4 specimens)		
Leptadapis magnus	NMB (8 specimens)	Mc*s*2-5, 3 PP*s*, IP	Dagosto (1983), Boyer et al. (2013a)
Omomys carteri	UM (3 specimens)	2 Hm*s*, Ps	Hamrick (1999), Boyer et al. (2013a)
Teilhardina belgica	IRNSB (5 specimens)	U. Mc, 2 PP*s*, 5 IP*s*	Gebo et al. (2012, 2015), Boyer et al. (2013a)
	Private Coll. (4 specimens)		
Archicebus achilles	IVPP 18618 (1 specimen)	R, U, carpal impressions	Ni et al. (2013)
Aegyptopithecus zeuxis	DPC (3 specimens)	Mc3, 2 PP*s*	Hamrick et al. (1995), Rose et al. (1996), Boyer et al. (2013a)
Apidium sp.	DPC 79-513 (1 specimen)	2 PP*s*	Hamrick et al. (1995), Boyer et al. (2013a)
	YPM 25812 (1 specimen)		

Abbreviations: *CM* Carnegie Museum of Natural History; *GMH* Geiseltalmuseum Halle; *IRNSB* Institut Royal des Sciences Naturelles de Belgique; *MaPhQ* Montauban Musée d'Histoire Naturelle Victor Brun, Phosphorites de Quercy locality; *PMO* Geological Museum, Natural History Museum, University of Oslo; *RD* private collection of A. Collier; *Ros* Rosières locality, France; *SMF-ME* Messel collection, Forschungsinstitut Senckenberg; *P*x prepollex See Table 14.1 for additional abbreviations

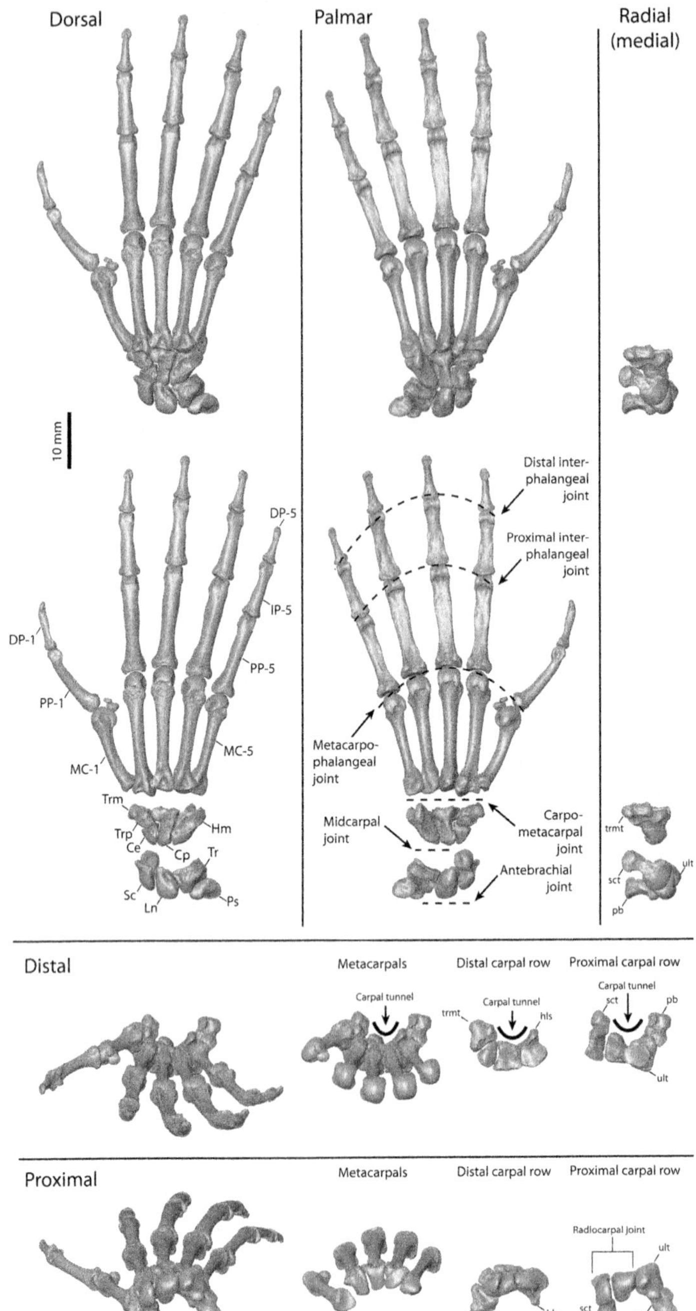

Fig. 14.2 Anatomical features and terminology for the hand. Illustrated on the hand of *Cebus apella* (EA 054, Duke Univ. Coll.). Abbreviations: *Ce* centrale, *Cp* capitate, *DP* distal phalanx, *hls* hamate hamulus, *Hm* hamate, *IP* intermediate phalanx, *Ln* lunate, *MC* metacarpal, *PP* proximal phalanx, *pb* pisiform body, *Ps* pisiform, *Sc* scaphoid, *sct* scaphoid tubercle, *Tr* triquetrum, *Trm* trapezium, *trmt* trapezium tubercle, *Trp* trapezoid, and *ult* tubercle for ulnopisiform ligament. Reproduced from Boyer et al. (2013a)

1.1 Fossil Taxa Reviewed: Systematic Considerations

To identify the key innovations in the hand skeleton that contributed to or were perpetuated by the radiation of extant primates from their common ancestor, we must examine the morphological patterns found in both stem and crown primates. **Stem primates** are taxa that are phylogenetically basal to the last common ancestor of extant primates, but closer to extant primates than to other extant taxa. **Crown primates** are taxa in the clade containing all of the descendants of the common ancestor of extant primates. Essentially, if we want to know what the common ancestor of extant primates looked like and what evolutionary steps occurred to make it that way, it is important to document as many stem primates and early members of crown primates as possible. Plesiadapiforms, a geographically widespread radiation of Paleogene [beginning at 66 million years ago (mya)] mammals (Silcox et al. 2007), are the only group of fossil mammals widely regarded as probable stem primates (Simpson 1935; Gingerich 1976; Szalay and Delson 1979; Szalay et al. 1987; Bloch and Boyer 2002; Bloch et al. 2007; Janečka et al. 2007; Chester et al. 2015). Though various researchers remain skeptical about plesiadapiforms as stem primates (see Godinot 2007) and results of some well-sampled cladistic analyses do not support this (e.g., Ni et al. 2013), plesiadapiforms *are* universally regarded as members of Euarchonta, a group whose extant members include primates, dermopterans (*Cynocephalus* and *Galeopterus*, the colugos or "flying lemurs"), and scandentians (Tupaiidae and Ptilocercidae, the treeshrews) (Szalay and Decker 1974; Szalay and Delson 1979; Szalay and Dagosto 1980; Szalay and Drawhorn 1980; Szalay et al. 1987; Cartmill 1992; Bloch and Boyer 2002; Bloch et al. 2007; Chester et al. 2015). At the very least, plesiadapiforms are the only euarchontans with a sampled postcranial fossil record just prior to the appearance of taxa usually assumed to be members of euprimates. As such, they can help evaluate hypotheses for morphological and ecological transitions involved in primate origins (Bloch and Boyer 2002). They can help provide an understanding of what aspects of euprimate hands are primitive, as well as what features are more likely to be innovations within a lineage close to the common ancestor of euprimates.

While plesiadapiforms may represent primates of the Paleocene, a new kind of primate appeared at the beginning of the Eocene (55.8 mya) (Gingerich 1986; Smith et al. 2006; Rose et al. 2011). These "new primates" are typically referable to either the Omomyiformes or Adapiformes, the two principal Eocene radiations of primates. They are often assumed to be part of the crown clade (Wible and Covert 1987): the former is assigned to haplorhines, whereas the latter is frequently aligned with extant strepsirrhines (Szalay and Delson 1979; Beard et al. 1988; Dagosto 1990; Kay et al. 1997, 2004; Godinot 1998; Seiffert et al. 2009). However, debate continues about the alternative possibility that adapiforms also have a closer affinity to haplorhines or anthropoids (Franzen et al. 2009; Seiffert et al. 2009; Boyer et al. 2010b; Gingerich et al. 2010; Williams et al. 2010; Gingerich 2012; Maiolino et al. 2012). Hoffstetter (1977) used the term **euprimates** in reference to a specific formulation of crown primates that includes omomyiforms and adapiforms as stem haplorhines and stem strepsirrhines, respectively. However, it is important to keep in mind the possibility that the Eocene radiations *also* represent stem groups (like plesiadapiforms) given that they maintain significant differences from all modern radiations and remain undersampled [see Boyer et al. (2013a) for further discussion].

1.2 Functional Anatomy

Ideally, morphological data derived from fossils can allow us to better understand functional-behavioral changes in primate hand evolution. Across extant primates, variation in digit length relative to the metacarpus has been linked to arboreality versus terrestriality (Etter 1973, 1974; Jouffroy et al. 1991; Lemelin 1996, 1999; Kirk et al. 2008; see Chap. 4). Distal phalanx morphology and shape of the metacarpophalangeal (McP) joints have been linked to the use of a clinging versus prehensile grasp (Cartmill 1974a; Boyer et al. 2013a; see Chap. 4). Proximal phalanx curvature has been linked to the use of suspensory behaviors (Jungers et al. 1997; Stern et al. 1995; Richmond 2007; Deane and Begun 2008). Finally, it has been argued that arboreal quadrupedalism, vertical clinging, and loris-like slow-climbing—all potential locomotor behaviors of early primates—each demand different and consistent functions from the wrist and hand, leading to tendencies for standard suites of morphological characteristics (Hamrick 1996a, b, c, 1997). These form-function relationships in extant primates should allow behavioral inferences in fossil taxa for which the wrist and hand bones have been recovered.

Arboreal pronograde quadrupeds tend to use extended (dorsiflexed), pronated hand postures with limited ulnar deviation and to transmit greater compressive forces through the wrist than more orthograde clinging, climbing, or suspensory primates (Hamrick 1995, 1996b, c, and references therein; see Chap. 3). Therefore, arboreal quadrupeds generally exhibit a broader carpus and flatter carpal joints (particularly a radioulnarly oriented plane of the midcarpal joint surface) compared with other functional groups, which promote pronation instead of supination and limit wrist mobility, thus requiring less muscular effort to maintain stability. Arboreal quadrupedal primates also tend to have an enlarged pisiform to enhance propulsion from a dorsiflexed posture, but only a moderately developed carpal tunnel as the flexor muscles are not as well developed as in clinging and climbing primates (Hamrick 1996c, 1997, and references therein).

During locomotion, extant vertical clinging primates habitually use a more neutral or flexed wrist with ulnar-deviated and supinated hand posture (Hamrick 1996c; Reghem et al. 2012). The wrist joints probably experience reduced compressive forces, while the metacarpals and phalanges may experience higher substrate reaction forces and intra-joint stresses during grasping than pronograde quadrupeds (Hamrick 1996c). These functional demands are reflected in the proximally facing and deeply cupped radiocarpal facets, the strongly curved (proximally convex) midcarpal joint, a proximodistally oriented triquetrohamate articulation that is also dorsally exposed and distally extended, and a small pisiform (Hamrick 1996c). The use of ulnar-deviated postures may also explain the presence of a strongly divergent thumb and an ectaxonic hand (i.e., the fourth digit is the longest) (Hamrick 1996c;

Reghem et al. 2012). Finally, long rays (if not long arms) and a large, divergent thumb (and a corresponding large trapezium) relative to body size are also common in vertical clinging primates, especially larger species (Napier and Walker 1967; Beard and Godinot 1988; Hamrick 1996c; Hamrick and Alexander 1996; Lemelin and Jungers 2007). On vertical supports, a primate must produce enough frictional force to prevent from sliding down the support (Cartmill 1974b); a larger hand span relative to the support enables the animal to wrap its fingers and thumb more completely around the support such that the digits can generate opposing forces and maintain a critical level of friction between substrate and pads to prevent slippage (as in palming a basketball).

As small branch specialists, lorises also use ulnarly deviated hand postures. They have the most prominently developed ectaxonic hands among primates, including the relatively largest and most divergent thumb and the most reduced second ray (Jouffroy et al. 1991; Lemelin 1996; Lemelin and Jungers 2007). Though they exhibit strong ectaxony like vertical clingers, their fingers differ from those of vertical clingers by being shorter relative to body size (despite having a high scaling coefficient as shown by Lemelin and Jungers 2007). Requiring more mobility at the wrist joint, they also have a greater degree of curvature at the midcarpal joint and greater reduction of the pisiform than typical vertical clingers (Hamrick 1996c).

2 The Earliest Euarchontans: Plesiadapiforms

Hand elements are known from over a dozen plesiadapiform taxa (Table 14.1). Generally, the morphologies of these specimens are strongly consistent with an arboreal habitus. Initially, most specimens including hand elements came from the family Plesiadapidae, which is now known to be one of the more derived plesiadapiform families; most plesiadapid taxa have larger body sizes (greater than 400 g and up to ~3 kg) and reduced dental formulas compared to basal members of plesiadapiforms and euprimates. *Nannodectes gidleyi* was the first plesiadapid to have manual elements attributed to it (Simpson 1935). It was shown to have radioulnarly compressed distal phalanges and a robust pollex, as would be expected for an arboreal claw-clinging mammal. Russell (1967) attributed a skeleton from Menat, France, to *Plesiadapis insignis*. This skeleton is preserved in a rock slab, and impressions of the hand skeleton show relatively long fingers (Gingerich 1976). The manual claws are obscured, but the pedal elements are preserved and claw-like. Godinot and Beard (1991) provided the first reconstructed manual ray of a plesiadapiform attributed to the plesiadapid *Plesiadapis tricuspidens* from the Cernaysian quarries near Mont de Berru, France. This was a composite ray with a large, radioulnarly compressed, hooklike claw; long, straight intermediate and proximal phalanges; and a relatively short metacarpal. The digit was reconstructed as having a habitually hyperextended McP joint and flexed interphalangeal joints. They interpreted the long fingers and claws as correlates of an arboreal habitus in which squirrel-like clinging and climbing was practiced.

Beard (1990) presented a reconstruction of a partial hand of *Nannodectes intermedius*, a smaller and more basal plesiadapid than *Plesiadapis tricuspidens*

(Gingerich 1976; Boyer et al. 2012), based on a relatively complete skeleton recovered from freshwater limestone dated to the beginning of the late Paleocene in south central Montana (Bangtail Plateau, western Crazy Mountains Basin; see Gingerich et al. 1983). The specimen includes phalanges, several carpals (scaphoid, capitate, pisiform, and hamate), and three metacarpals (Mc) assigned to rays 1, 2, and 3. Beard (1990) argued that the corresponding morphologies on the proximal articular surfaces of Mc1 and Mc2 indicated a habitual articulation between them and dictated a high degree of divergence between the digits (~75°). Beard (1989, 1990) suggested this feature could indicate functional similarities to lorises, the only extant primates exhibiting such high divergence. However, Boyer (2009) and Boyer et al. (2013a) provide evidence that what was originally interpreted as Mc2 is in fact Mc5. This is also likely to be the case for the metacarpal used in Godinot and Beard's (1991) reconstruction of *P. tricuspidens*. If the Mc2s of *N. intermedius* and *P. tricuspidens* have been incorrectly identified, there is currently no basis for estimating pollical divergence in plesiadapiforms.

Beard (1990) described the first possible (but not dentally associated) non-plesiadapid plesiadapiform phalanges: these were attributed to the Paromomyidae. Beard's analysis suggested surprisingly dermopteran-like morphology and proportions. In Beard (1993a, b) descriptions of phalanges dentally associated with yet another plesiadapiform family—the Micromomyidae—appeared to show a similar pattern, thereby supporting the earlier work. These patterns were treated as evidence of (1) dermopteran-like mitten gliding in both paromomyids and micromomyids and (2) a monophyletic relationship between extant dermopterans and all known plesiadapiforms, with the exception of the Microsyopidae. Krause (1991) and Runestad and Ruff (1995) questioned the strength of evidence for dermopteran-like proportions in paromomyids and micromomyids. However, a study by Hamrick et al. (1999) obtained results consistent with Beard's findings: their multivariate analysis showed that the phalanges described by Beard are similar to those of taxa that cling to large diameter supports (i.e., tree trunks), such as bats and dermopterans.

Starting in the early 2000s, data from new dentally associated skeletons from the Clarks Fork Basin of Wyoming began to surface. Hamrick (2001) published a ternary plot with extant euarchontan metacarpal, proximal phalanx, and intermediate phalanx proportions and included data on a previously unpublished skeleton of *Plesiadapis cookei* (UM 87990) from the Clarks Fork Basin. The plot showed that extant primates occupied a unique morphospace, whereas *P. cookei* had proportions similar to tupaiid treeshrews. From these data, he argued that there was a developmental change in digit ray patterning in the ancestral crown primate that allowed the evolution of longer fingers and more effective exploitation of the fine-branch niche.

Bloch and Boyer (2002) published a description and analysis of the first known skeleton of *Carpolestes simpsoni* (a plesiadapiform in the family Carpolestidae; Bloch and Gingerich 1998). The partial skeleton of *C. simpsoni* (UM 101963) includes a skull, jaw, vertebrae, and upper and lower limb bones including girdles. Bloch and Boyer (2002) attributed a number of bones to the hand specifically, though the specimen's hand was not preserved in articulation. Kirk et al. (2008) give a more detailed account of how digit attributions were estimated for this specimen. Bloch and Boyer (2002) refigured Hamrick's (2001) ternary diagram with data for *C. simpsoni* and other plesiadapiform skeletons (Bloch et al. 2007; Boyer and Bloch 2008) that were unpublished at the time. All included plesiadapiforms except *P. cookei* plot directly with euprimates. The digit ray of *C. simpsoni* differs from that of *P. tricuspidens* figured by Godinot and Beard (1991) in having a more curved proximal phalanx, more gracile proximal and intermediate phalanges, and distal phalanges that are proportionally shorter, less hooklike and have a distally extended volar process for an expanded apical pad (Bloch and Boyer 2002, 2007).

New data from associated and semi-articulated skeletons of the paromomyid *Ignacius clarkforkensis*, failed to confirm the presence of dermopteran-like hand proportions in that species of paromomyid (Bloch and Boyer 2007; Bloch et al. 2007; Boyer and Bloch 2008). Additionally, a new micromomyid skeleton with semi-articulated hand bones provided contextual evidence of metacarpal positions, independent of morphological assessments, as well as clear identification of phalanges as either manual or pedal (Bloch and Boyer 2007; Boyer and Bloch 2008; Boyer et al. 2013a).

Bloch et al. (2007) and Kirk et al. (2008) provided the most comprehensive assessments of the adaptive significance of euprimate hand proportions and implied grasping abilities based on comparisons with outgroups, including plesiadapiforms. They corroborated the patterns illustrated by Bloch and Boyer (2002), showing that *C. simpsoni* and *I. clarkforkensis* plot within the range of modern primates for hand proportions that reflect prehensility. Additionally, they showed that *Ptilocercus lowii*, an arboreal treeshrew not sampled in previous analyses, also overlaps extant primates with prehensile hand proportions. Kirk et al. (2008) interpreted this as evidence that no significant clade-level shifts in hand proportions occurred in the primate stem lineage, or in the common ancestor of crown primates, and suggested that improved prehensility was not a novel adaptation leading to the radiation of crown primates. A caveat to this conclusion was Kirk et al.'s (2008) finding that *Notharctus* (the only included adapiform) has proportionately very elongated digits, which would suggest a shift in hand proportions if *Notharctus* reflects the common ancestor of extant primates [as entertained by Godinot (1991)]. Kirk et al. (2008) preferred to consider the unusual proportions of *Notharctus* as derived within its own lineage. We present more data on this issue below (see Sect. 3.3).

Regarding the plesiadapiform carpus, it should be noted that when direct articulation is lacking (as it is for most preserved skeletons), questions of attribution often arise. For example, the skeleton of *Plesiadapis cookei* (UM 87990) was preserved with a skeleton of *Uintacyon*, an arboreal carnivoran of nearly identical size, and thus identification of certain carpal and metapodial elements remains tentative (Boyer et al. 2013a). Furthermore, certain carpal bones attributed to *Nannodectes intermedius* (Beard 1989) appear to have been misidentified (Boyer et al. 2013a). As indicated in Fig. 14.1, only the scaphoid, lunate, triquetrum, Mc1, and Mc5 can be identified for *Plesiadapis cookei* with relative certainty at this time (see Boyer et al. 2013a for details). Additional dental-postcranial associations or articulated specimens will be necessary to triangulate the correct attributions.

Functionally speaking, known carpal elements of *Nannodectes* and *Plesiadapis* are consistent with arboreal quadrupedal postures; the dorsally positioned radial articular surfaces of the scaphoid and lunate suggest that the wrist was habitually dorsiflexed (Boyer 2009; Boyer et al. 2010a, 2013a), and a quadrate shape to the triquetrum suggests the hand had limited ulnar deviation (Beard 1989, 1993a; Boyer 2009; Boyer et al. 2010a, 2013a). However, the unusual triangular shape of the triquetrum in *Phenacolemur jepseni* and a reduced distal ulna may suggest more ulnarly deviated postures than used by extant arboreal quadrupeds (Beard 1989, 1993a) and are consistent with vertical climbing. *Nannodectes intermedius* USNM 442229 is the only specimen complete enough to be assessed for overall hand proportions, combining data on the carpus, metacarpus and phalanges (Boyer et al. 2013a), and it plots among the extant radiation of primates, closest to a specimen of *Indri* (Fig. 14.3a). It is not clear what this means in terms of specific functional analogy since the interphalangeal proportions (not measured in Fig. 14.3a) are quite different between plesiadapiforms and modern primates due to clawed distal phalanges of plesiadapiforms.

In sum, most, if not all, plesiadapiforms have long, euprimate-like fingers indicative of an arboreal lifestyle. In all known plesiadapiforms, these long fingers are tipped with clawed (falcular) distal phalanges. The claws are radioulnarly compressed and dorsopalmarly deep, indicating arboreal rather than terrestrial activities in all taxa for which they are known (Bloch and Boyer 2007). As well, they exhibit prominent, bilateral nutrient foramina proximal to flexor tubercle and volar process. Godinot and Beard's (1991) reconstruction of joint angles for a plesiadapiform manual ray still appears to be basically accurate even considering morphology of the additional species that have been described since. Finally, the wrist provides some indication of variation in positional behavior with plesiadapids having morphology suggesting dorsiflexed hand postures indicative of pronograde locomotion on medium-to-large diameter supports, whereas paromomyids exhibit limited evidence of ulnarly deviated hand postures, possibly reflecting vertical supports (Beard 1989, 1993a; Boyer and Bloch 2008).

3 Early Euprimates: Adapiformes and Omomyiformes

While important questions remain on the broader relationships of adapiforms and omomyiforms, the intra-clade systematics of these groups can be effectively delineated. The adapiform infraorder is typically divided into three primary families: the early to middle Eocene Notharctidae, the middle Eocene Adapidae, and the Asiatic, Eocene-Miocene family Sivaladapidae (Godinot 1998, 2015; Gebo 2002). Notharctidae is conventionally split into two subfamilies: the predominantly North American Notharctinae and the predominantly European Cercamoniinae (Gebo 2002; Gunnell and Silcox 2010; Godinot 2015). Recently, a third notharctid subfamily was proposed: the Asiadapinae from the early Eocene of India (Rose et al.

Fig. 14.3 (continued) phalanx as percentages of total ray 3 length. Data for extant species from Kirk et al. (2008). Data for fossil species presented in Tables 14.3 and 14.4. Minimum convex hulls bound indicated phylogenetic groups. Modified from Boyer et al. (2013a)

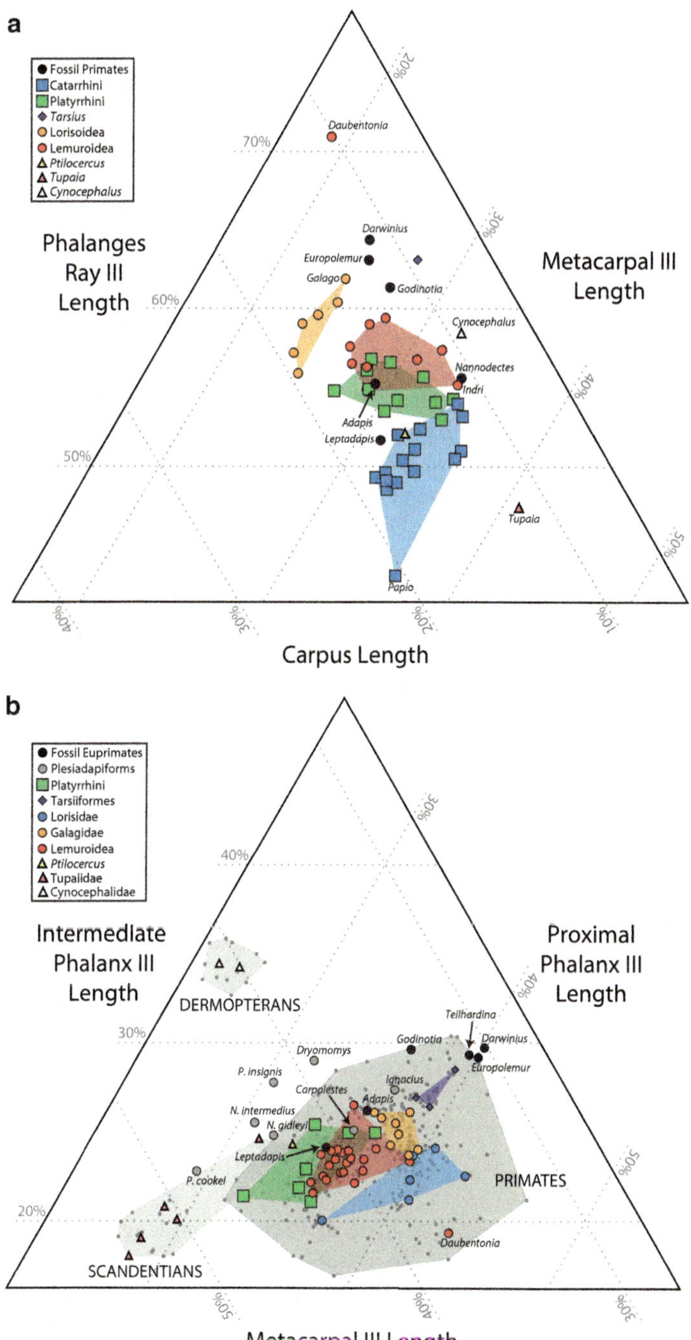

Fig. 14.3 Intrinsic hand proportions. (**a**) Lengths of the carpus, third metacarpal, and third digit ray (proximal phalanx + intermediate phalanx + distal phalanx) as percentages of total hand length. Data for extant species from Jouffroy et al. (1991). Data for fossil species are presented in Tables 14.3 and 14.4. (**b**) Length of the third ray metacarpal, proximal phalanx, and intermediate

2009; Godinot 2015). It should be noted however that the relationships among these families are not well resolved. In particular, it is quite possible that Notharctidae is paraphyletic, with cercamoniines more closely related to adapines (Seiffert et al. 2009; Patel et al. 2012; Gladman et al. 2013).

Hand fossils are preserved for seven notharctid species (Table 14.2). These include the notharctines *Notharctus tenebrosus* (Gregory 1920; Hamrick and Alexander 1996), *Notharctus robustior* (Beard and Godinot 1988), and *Smilodectes gracilis* (Covert 1985a, b, 1986; Beard and Godinot 1988; Godinot and Beard 1991; Godinot 1992; Godinot and Beard 1993; Alexander and Burger 2001) from the Bridger Formation of North America; the cercamoniines *Europolemur kelleri* (Franzen 1988, 1993, 2000; Franzen and Frey 1993), *Europolemur koenigswaldi* (Franzen 1987), and *Darwinius masillae* (Franzen et al. 2009) from the Messel oil shale; and *Godinotia neglecta* (Thalmann et al. 1989; Thalmann 1994) from the Geiseltal lignite beds (Fig. 14.4). Only the North American species are known from multiple specimens within a given species.

Hand fossils are preserved for at least two adapines: *Adapis parisiensis* and *Leptadapis magnus* (Dagosto 1983; Godinot and Jouffroy 1984; Godinot and Beard 1991, 1993; Godinot 1992). However, only *Adapis* includes articulated remains (Fig. 1: RD 311, the Rosières specimen), and these unfortunately lack associated phalanges. Nonetheless, thanks to large samples of isolated adapine phalanges from specific localities, Boyer et al. (2013a) provided tentative estimates of hand proportions, which we review here. Much less is known of omomyiform hands and a review of the systematics of the clade is therefore unnecessary. In fact, when Godinot and Beard (1991, 1993) wrote their reviews, no omomyiform hand fossils were known. The first published description of omomyid hand fossils was provided by Hamrick (1999), who described a hamate and pisiform from the Bridger Basin and attributed them to *Omomys* based on linear measures of absolute size. Since then, Gebo et al. (2012, 2015) have attributed intermediate phalanges, proximal phalanges, and a metacarpal to the "anaptomorphine" *Teilhardina belgica*. Finally, there appear to be carpal impressions for a partial skeleton of *Archicebus achilles*, a *Teilhardina*-like basal omomyiform (Ni et al. 2013).

We provide an overview of the morphological patterns observed in this early euprimate hand material and include comparisons to plesiadapiforms.

3.1 Carpus

The carpus of adapiforms ranges from 11 to 16% of hand length in taxa for which sufficient fossils are known (Fig. 14.3; Tables 14.3 and 14.4), including various notharctids and *Adapis*. No North American specimens, including the *Notharctus tenebrosus* specimen (AMNH 127167) described by Hamrick and Alexander (1996), are complete enough to reliably estimate relative carpus length (Boyer et al. 2013a). Compared to other mammals, the adapiform carpus is relatively short, which is simply a reflection of long phalanges that are typical for most primates (Tables 14.5A and 14.5B).

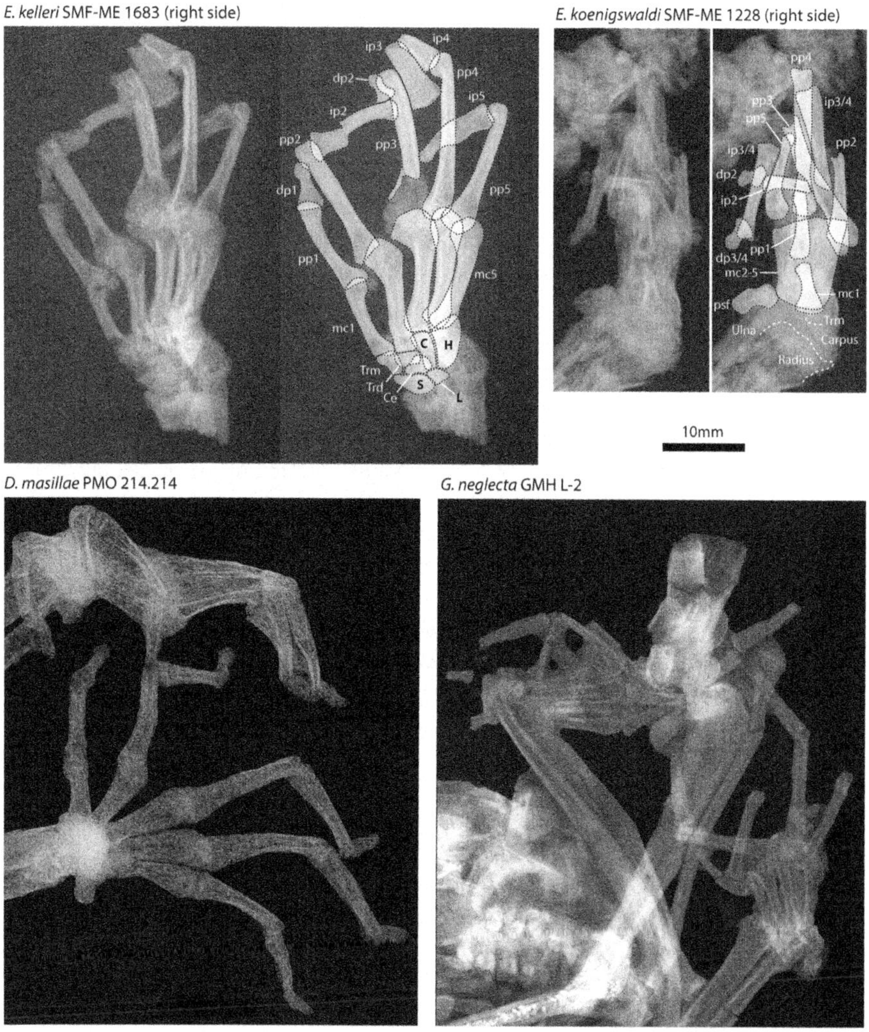

Fig. 14.4 Radiographs of adapiform hands. Two specimens of *Europolemur* (*E. kelleri*, SMF-ME 1683, *top left*; *E. koenigswaldi*, SMF-ME 1228, *top right*) were used to create a set of composite measurements for the hands of *Europolemur*. Traces and labels represent interpretations incorporated in Boyer et al. (2013a) and for this chapter. These high-resolution radiographs were provided by J. Franzen (Jörg Habersetzer made them and the copyright is with Senckenberg Forschungsinstitut Frankfurt am Main), and casts of original specimens were provided by R. F. Kay. Measurements in Boyer et al. (2013a; see their appendix Table V) are based directly off the bones as interpreted in this image. The hand of *Darwinius masillae* is a radiograph image reproduced from Franzen et al. (2009) and also made by Jörg Habersetzer. The hand of *Godinotia neglecta* is based on original radiographs used in Thalmann (1994) that were provided by U. Thalmann. Measurements of *Darwinius* and *Godinotia* were reassessed in Boyer et al. (2013a). Summary measurements and computed proportions for these specimens are provided in Tables 14.3 and 14.4. Abbreviations: *S* scaphoid, *L* lunate, *H* hamate, *C* capitate, *Ce* centrale, *Trm* trapezium, *Trd* trapezoid, *psf* pisiform, *mc* metacarpal, *pp* proximal phalanx, *ip* intermediate phalanx, *dp* distal phalanx. Modified from Boyer et al. (2013a)

The most broadly represented bones of the wrist are the hamate and pisiform (Figs. 14.5 and 14.6). The hamate is similar among all Eocene euprimates for which it is known in having a spiral facet for the triquetrum and a relatively small hamulus (Godinot and Jouffroy 1984; Godinot and Beard 1991; Hamrick 1997, 1999). The spiral facet is rare among extant non-primates, as well as plesiadapiforms, and thus appears to be a euprimate synapomorphy. The reduced hamulus suggests lesser development of the digital flexors than in extant strepsirrhines, which have relatively large hamuli (Hamrick 1997). Notharctines and *Adapis* are said to have a relatively radioulnarly oriented plane of the proximal facet similar to arboreal quadrupedal lemurs (Hamrick 1996c), whereas *Europolemur kelleri* and *Omomys* have a more proximodistally oriented, ulnarly facing facet plane also characteristic of extant vertical clingers (Hamrick 1996c, 1999). Plesiadapid plesiadapiforms have the more radioulnarly oriented condition.

Regarding the scaphoid, all known fossil euprimates exhibit a relatively large tubercle, as noted by Hamrick (1997: 114), who suggested the ancestral euprimate differed from tupaiids in having a larger scaphoid tubercle related to an expanded carpal tunnel, serving as a "windlass mechanism for the pollical branch of the *flexor digitorum profundus* tendon." This would aid the powerful pollical adduction needed for clasping onto relatively small-diameter supports. However, plesiadapiforms have a strongly developed scaphoid tubercle as well (Boyer et al. 2013a), suggesting pollical adduction was also powerful in early stem primates.

In all Eocene euprimates known, the lunate is relatively large and participated in the radial joint. Only *Notharctus*, *Smilodectes*, and *Adapis* have lunates well preserved enough to comment on more detailed aspects of morphology. In particular, *Adapis* has a very dorsally centered radial facet, whereas *Notharctus* has a more palmarly centered one [see Fig. 5 in Boyer et al. (2013a)]. This may suggest more habitually dorsiflexed hands in *Adapis*.

Relative to the other carpal bones, the pisiform is generally larger in Eocene euprimates than in extant lemuriforms, again suggesting a proclivity for arboreal quadrupedal behaviors in which the hands are dorsiflexed (Hamrick 1996c, 1999). On the other hand, the pisiform in plesiadapiforms is proportionally even bigger (Fig. 14.6), suggesting that early euprimates were less specialized than plesiadapiforms for pronograde postures that require a large and robust pisiform. It should also be noted that tarsiers also have a relatively tall pisiform (Hamrick 1997; Fig. 14.6), despite their committed use of vertical clinging and leaping, suggesting that functional implications of a tall pisiform are not always straightforward.

The relative sizes of the triquetral and ulnar facets on the pisiform vary between *Adapis* and those of other fossil taxa where the morphology can be observed (e.g., *Notharctus* and *Omomys*) (Fig. 14.6). In *Adapis*, the pisiform is strepsirrhine-like, with a large and deeply excavated ulnar facet and smaller triquetral facet. In *Notharctus* and *Omomys*, the pisiform is more haplorhine-like, with triquetral and ulnar facets of subequal size (Beard and Godinot 1988; Hamrick 1996c, 1999). Differences among taxa illustrated in Fig. 14.6 suggest that variation in pisiform morphology must be more carefully studied before systematically informative character states, and character-state combinations can be reliably identified.

Table 14.3 Measurements used in plots and body masses

Bone	*Carpolestes*	*P. insignis*	*N. intermedius*	*Teilhardina*	*E. kelleri*	*Darwinius*	*Notharctus*	*Godinotia*	*Adapis*	*Leptadapis*
Carpus	–	–	3.77	–	7.38[a]	5.49 *r*	11.58	7.39 *l*	8.6	16.24[b]
Mc1	–	–	–	–	9.96	7.30 *r*	12.48	10.43 *l*	9.53	–
Mc2	–	–	–	–	14.44	–	16.65	13.90 *l*	15.34	27.93[c]
Mc3	7.18	11.7	11.51	5.96[d]	15.33	–	23.8[e,f]	16.93 *l*	16.15	30.47[c]
Mc4	–	–	–	–	15.23	11.75 *r*	21.42	16.42 *l*	15.88	32.47[c]
Mc5	–	–	–	4.9	14.12	11.08 *r*	19.06	14.33 *l*	13.97	28.71[c]
PP1	–	–	–	–	11.96	9.07	16.39	12.31	–	–
PP2	–	–	–	–	16.56	13.35	23.98	16.41	–	–
PP3	6.61	8.6	7.76	5.6 (7.55)	19.97[g]	15.62	27.66	18.25	15.53[h]	26.33[c]
PP4	–	–	–	–	20.61	15.66	28.51	19.61	–	–
PP5	–	–	–	–	16.99	11.66	25.47	16.06	–	–
IP2	–	–	–	–	11.76	8.12	16.08	–	–	–
IP3	4.6	7.8	6.56	5.6	14.54[g]	11.56	–	14.77	11.17[h]	17.81[c]
IP4	–	–	–	–	13.78[g]	12.27	–[f]	14.87	–	–
IP5	–	–	–	–	10.61	9.44	13.94	11.47 *r*	–	–
DP1	–	–	–	–	6.12	5.59	8.98	6.49 *r*	–	–
DP2	–	–	–	–	3.82	4.13 *r*	8.14	–	–	–
DP3	2.3	–	5.22	1.99	4.30[g]	4.01	9.38	–	3.96[h]	6.5[b]
DP4	–	–	–	–	–	4.21	10.44	5.60	–	–
DP5	–	–	–	–	–	3.13 *r*	8.96	3.82 *r*	–	–
Digit 3	13.51	–	19.54	13.2 (15.1)	38.81	31.19	na	38.62	30.66	50.60
BM (g)	100	–	352	47	1485	660	2305	1325	1066	6425

Individual specimen data were modified to generate composite hands with maximum completeness. A composite for *Europolemur kelleri* was generated using proportional information from *Europolemur koenigswaldi*. A composite for *Darwinius masillae* was generated by averaging measurements from the left and

(continued)

Table 14.3 (continued)

right sides of PMO 214.214. A composite for *Godinotia neglecta* was generated by averaging measurements from left and right sides of GMH L-2 when measurements were close, but primarily the left side (which is more complete and better preserved). A composite for *Notharctus tenebrosus* was generated using AMNH 127167 and proportional information from AMNH 11478. *Adapis parisiensis* was generated using RD 311 and proportionally adjusted based on locality means, for the metacarpals and phalanges (see appendices of Boyer et al. 2013a). *Leptadapis* is represented by species means (see appendices of Boyer et al. 2013a) with some values inferred by assuming proportional equivalence with *Adapis parisiensis*. *Teilhardina* is a composite based on material presented in Gebo et al. (2012, 2015) and the distal phalanx USNM 521825 from Rose et al. (2011). We assign the metacarpals described by Gebo et al. (2015) to ray 5, based on a better morphological fit. We use two different options for proximal phalanx length in *Teilhardina* because the longest manual phalanx of Gebo et al. (2015) seems too short given the longest intermediate phalanx of Gebo et al. (2012). Values for *Carpolestes* are those reported by Boyer and Bloch (2008) for UM 101963 with the distal phalanx measurements taken on that specimen for this study. Body mass estimates were generated in the following ways: For *Carpolestes*, we used the estimate of Bloch and Gingerich (1998). For *E. kelleri*, we used the dimensions of the calcaneal cuboid facet (5.01 and 5.01 mm) preserved in HLD-ME 7430 and the equation published by Boyer et al. (2013b). For *Darwinius*, we used the mean estimate based on the postcranium reported by Franzen et al. (2009). For *Godinotia*, we used diameters of the humerus (5.7 and 6.13 mm) and femur (5.87 mm and 5.87 mm) to generate a geometric mean (5.91). This geometric mean was used in a regression for "prosimian" body mass based on the geometric mean of these four measures with data from Lemelin (1996) (as also done by Lemelin and Jungers 2007): $[\ln(BM) = 2.601 \times \ln(geomean) + 2.57]$. See figure 23 of Boyer et al. (2013a). Body masses of *Teilhardina*, *Notharctus*, *Adapis*, *Leptadapis*, and *Nannodectes* are based on calcaneal cuboid facet means from Boyer et al. (2013b). All measurements in mm

(–) No measurement available

(r) From right side

(l) From left side

[a]Based on a caliper measurement

[b]Estimated by assuming identical proportions between *Leptadapis* and *Adapis*

[c]Locality averages

[d]Estimated using strepsirrhine average proportion of Mc3/Mc5 = 1.22 (based on data of Lemelin 1996)

[e]Estimated length for AMNH 127167 assuming identical proportions to AMNH 11478

[f]Measurements given in Hamrick and Alexander (1996) are unconstrained due to breakage (see Fig. 14.1)

[g]Locality averages adjusted so that proportion to Mc3 of Rosières 2 matches proportion of raw value to locality average Mc3

[h]Estimated lengths for *E. kelleri*, using *E. koenigswaldi*, assuming identical proportions

Table 14.4 Proportions used in plots based on measures in Table 14.3

Ratio	*Carpolestes*	*P. insignis*	*N. intermedius*	*Teilhardina*	*E. kelleri*	*Darwinius*	*Notharctus*	*Godinotia*	*Adapis*	*Leptadapis*
carp/hnd	na	na	10.83	na	12.00	11.34	na	11.74	15.55	16.70
Mc3/hnd	na	na	33.06	na	24.92	24.26	na	26.90	29.12	31.30
PP3/hnd	na	na	22.29	na	32.46	32.25	na	28.74	28.04	27.06
IP3/hnd	na	na	18.84	na	23.63	23.87	na	23.47	20.16	18.30
DP3/hnd	na	na	14.99	na	6.99	8.28	na	8.90	7.15	6.64
D3/hnd	na	na	56.12	na	63.09	64.40	na	61.36	55.35	51.99
carp/Mc3	na	na	32.75	na	48.14	46.72	49–55	43.65	53.32	53.31
Mc1/Mc3	na	na	na	na	65.00	na	52–59	61.61	59.01	na
Mc2/Mc3	na	na	na	na	94.24	na	70–79	82.10	94.98	91.66
Mc4/Mc3	na	na	na	na	99.39	na	90–102	96.99	98.33	106.56
PP3/Mc3	92.06	73.50	67.42	94–127	130.27	132.94	116–129	107.80	96.30	86.41
IP3/PP3	69.60	90.70	84.54	100–74	72.81	73.96	na	90.04	71.90	67.64
DP3/PP3	34.80	na	67.27	36–26	21.53	25.67	29.43	30.68	24.55	24.54
PP3/PP4	na	na	na	na	96.89	99.74	97.02	93.06	na	na

Abbreviations: *carp* carpus, *hnd* hand, *D* digit, *Mc* metacarpal, *PP* proximal phalanx, *IP* intermediate phalanx, *DP* distal phalanx, *na* not available

Table 14.5A Hand proportions for "prosimians"

	Data from Jouffroy et al. (1991)				Data from Kirk et al. (2008)				Data computed from Lemelin (1996)		
Taxon	*n*	%Carp	%Mc3	%D3	*n*	Mc3/[a]D3	PP3/[a]D3	IP3/[a]D3	*n*	Mc1/Mc3	s.d.
Arctocebus calabarensis	14	19.1	23.7	57.2	11	43.1	36.9	20.0	11	88.47	3.76
Loris tardigradus	10	19.5	24.6	55.9	12	36.8	39.5	23.7	11	84.24	3.64
Nycticebus coucang	21	16.6	23.8	59.6	17	37.6	40.1	22.3	17	70.74	5.04
Nycticebus pygmaeus					5	35.4	40.5	24.1	5	74.83	1.98
Perodicticus potto	7	17.7	23.3	59.1	13	38.2	40.6	21.2	13	68.69	2.86
Euoticus elegantulus	20	15.1	24.5	60.4	11	36.3	39.6	24.0	11	71.74	4.23
Galago moholi	54	13.9	24.2	61.9	7	37.3	36.9	25.8	7	65.40	4.24
Galago senegalensis					14	36.7	37.8	25.5	13	69.49	3.14
Galago zanzibaricus					1	37.5	36.4	26.1			
Galagoides demidoff					11	35.7	38.2	26.1	11	76.25	4.41
Galagoides alleni					5	36.9	38.3	24.9	5	71.16	3.43
Otolemur crassicaudatus					9	37.6	38.2	24.2	9	70.91	2.87
Otolemur garnetti					7	36.9	39.3	23.7	7	73.06	2.99
Cheirogaleus major	7	16.2	27.2	56.4	8	40.9	36.4	22.7	7	64.74	3.44
Cheirogaleus medius					6	40.2	36.0	23.8	6	66.45	2.41
Microcebus murinus	16	14.0	27.0	59	17	38.3	35.2	26.5	14	59.75	2.86
Mirza coquereli					1	40.4	37.4	22.1			
Phaner sp.	2	15.8	26.6	57.6	1	39.3	37.2	23.4			
Eulemur coronatus					2	40.4	35.7	23.9			
Eulemur fulvus					13	41.2	35.7	23.1	13	56.77	1.93
Eulemur macaco					7	40.7	35.8	23.5	7	59.46	3.64
Eulemur mongoz					7	40.7	35.4	23.9	7	57.34	3.82
Eulemur rubriventer					2	40.3	36.4	23.3			
Hapalemur griseus	9	15.5	28.2	56.3	12	40.0	36.5	23.5	12	57.82	3.11
Hapalemur simus					2	39.9	36.2	23.9			
Lemur catta	39	16.1	29.0	54.9	12	41.3	35.0	23.7	12	54.55	2.45
Lepilemur leucopus	11	12.9	27.7	59.4	7	38.4	37.6	24.0	7	59.10	2.16
Lepilemur mustelinus					5	39.7	37.5	22.8	5	54.08	1.07
Lepilemur ruficaudatus					1	37.1	39.6	23.3			
Varecia variegata	9	16.2	28.9	54.9	10	40.7	36.5	22.8	10	57.53	1.40
Avahi laniger	6	10.8	31.8	57.4	10	42.6	35.3	22.1	10	43.32	2.31
Propithecus verreauxi	6	12.5	30.7	56.8	12	41.8	35.9	22.3	12	58.91	2.35
Propithecus diadema					7	41.8	35.7	22.5	7	59.59	1.74
Indri indri	7	11.1	33.7	55.2	8	42.8	35.7	21.6	8	61.11	2.24
D. madagascarensis	9	10.2	18.9	70.9	9	37.1	43.5	19.3	9	38.30	2.99
Tarsius bancanus	10	9.4	27.5	63.1	13	32.3	39.2	28.5	12	65.53	2.32
Tarsius spectrum					11	34.6	39.0	26.4	11	69.97	4.18
Tarsius syrichta					12	35.0	38.1	26.9	12	68.55	2.31

[a]Digit (D) 3 ratios do not include distal phalanx length. Abbreviations: *s.d.* standard deviation; see also Table 14.4

Mc2/Mc3	s.d.	Mc4/ Mc3	s.d.	PP3/ Mc3	s.d.	IP3/ PP3	s.d.	DP3/ PP3	s.d.	PP3/ PP4	s.d.	Digit L res.	Mc L res.
71.39	4.25	101.52	2.50	85.41	4.41	54.09	5.13	40.43	4.09	74.92	3.78	−0.249	−0.115
85.05	2.49	96.39	3.97	107.48	6.48	61.37	5.90	27.06	2.96	96.48	5.24	0.017	−0.023
82.39	3.71	97.47	4.73	106.75	5.70	55.56	4.27	26.25	2.33	91.74	2.55	−0.111	−0.177
85.23	4.23	95.66	1.42	113.95	3.77	59.63	1.69	28.16	0.95	93.56	1.41	−0.002	−0.131
71.68	3.08	98.93	1.71	106.23	4.53	52.25	14.89	31.99	3.82	86.38	3.87	−0.140	−0.244
93.53	2.52	96.50	2.49	109.16	3.78	60.36	2.41	27.52	1.98	91.05	1.19	0.146	0.048
87.83	4.37	93.03	2.02	99.03	3.26	69.79	2.07	28.53	3.60	94.02	3.45	0.002	−0.031
89.98	2.37	93.56	2.09	102.6	4.66	67.45	4.16	28.48	2.57	93.04	3.49	−0.021	−0.085
91.90	2.89	94.88	4.22	107.08	5.17	68.27	3.02	29.28	3.53	89.99	3.19	0.128	0.061
87.71	3.40	95.51	2.57	104.09	2.18	64.79	4.42	23.8	2.46	86.69	1.55	0.124	0.071
92.00	3.26	95.56	4.39	101.78	3.63	63.59	4.55	27.73	3.61	89.93	5.83	−0.072	−0.148
90.06	2.33	96.80	1.40	106.45	2.29	60.53	1.04	27.88	2.82	91.39	1.51	−0.057	−0.171
94.58	1.51	93.27	1.29	87.38	4.13	62.74	3.35	28.45	2.62	94.94	1.58	−0.074	0.033
95.68	5.08	93.60	2.31	89.50	2.28	65.98	1.56	30.21	1.94	94.22	1.60	−0.039	0.057
93.18	1.95	93.51	1.86	90.59	3.37	74.71	5.46	25.05	2.58	93.96	5.79	0.062	0.173
95.34	2.21	97.03	1.87	86.90	1.93	64.79	2.53	26.08	2.52	95.7	1.13	−0.062	−0.013
96.37	1.19	96.32	1.59	87.79	1.79	65.89	2.56	27.12	0.93	94.49	2.13	−0.082	−0.063
94.47	1.56	95.74	0.85	87.15	2.86	67.51	1.52	28.83	1.79	95.52	1.28	−0.066	−0.042
88.13	1.75	97.97	28.29	91.55	3.48	64.29	1.89	29.36	2.23	93.58	1.88	0.023	0.038
94.02	1.87	96.93	3.25	84.98	4.05	67.30	19.37	27.65	1.74	94.47	3.23	−0.104	−0.043
86.61	1.84	100.90	1.47	98.35	3.65	63.70	4.59	31.10	1.78	88.15	1.51	0.055	0.006
86.48	2.35	101.07	1.20	94.23	2.15	60.78	2.51	29.83	1.33	90.20	1.08	0.036	0.027
95.46	1.38	96.00	1.10	89.58	4.32	62.70	1.95	28.28	1.00	98.07	0.63	0.035	0.013
85.61	1.50	103.71	2.04	82.66	2.22	62.34	3.19	31.39	3.14	90.24	1.27	0.119	0.222
84.17	6.23	101.74	1.71	85.83	2.88	62.38	1.73	28.66	2.28	90.41	1.82	0.108	0.142
85.40	0.95	101.70	0.79	85.21	2.14	63.09	0.92	28.02	1.47	89.34	2.11	0.196	0.202
90.30	2.51	101.81	0.96	83.27	3.29	60.53	2.54	23.83	3.17	91.90	1.36	0.302	0.393
54.59	4.57	69.29	5.04	117.26	11.29	44.49	2.01	13.27	2.86	90.57	2.79	0.460	0.396
94.19	2.06	84.42	1.91	120.37	6.57	72.42	22.67	20.96	3.27	112.16	2.27	0.680	0533
94.28	2.88	84.90	3.82	111.85	6.96	68.10	6.59	22.29	3.00	107.99	4.15	0.657	0.471
91.99	2.55	83.25	1.23	109.30	3.13	70.41	2.40	23.76	3.00	110.92	2.09	0.578	0.488

Table 14.5B Hand proportions for anthropoids, scandentians, and dermopterans

Taxon	Data from Jouffroy et al. (1991)				Data from Kirk et al. (2008)				Data from Jouffroy et al. (1991)						
	n	%Carp	%Mc3	%D3	*n*	Mc3/[a]D3	PP3/[a]D3	IP3/[a]D3	*n*	Mc1/Mc3	s.d.	Mc2/Mc3	s.d.	Mc4/Mc3	s.d.
Alouatta sp.	8	15.0	28.2	56.8					8	62.3	2.6	88.3	3.6	96.7	7.1
Aotus sp.	8	16.2	28.9	54.9	5	39.3	35.8	24.9	8	65.5	4.1	90.2	5.8	97.2	5.5
Ateles sp.	9	13.1	33.9	53.0					9	42.3	4.5	91.1	2.9	98.0	3.5
Brachyteles sp.	1	11.8	33.9	54.3					1	34.0	--	95.2	--	100.8	--
Callicebus sp.	3	14.1	29.3	56.6	6	38.0	37.1	24.9	3	67.2	2.3	89.8	3.6	98.0	1.8
Callithrix sp.	6	12.7	31.6	55.7	8	42.5	34.6	22.9	6	66.8	1.6	92.5	6.2	96.3	5.4
Cebus sp.	11	16.0	30.5	53.5					11	73.5	5.2	100.2	10.4	97.0	3.1
Chiropotes sp.	2	18.1	27.1	54.8											
Lagothrix sp.	6	15.6	28.3	56.1					6	66.6	5.7	94.1	3.7	99.1	2.6
Leontopithecus sp.	7	12.9	33.0	54.1	7	46.3	32.3	21.4	7	67.4	9.8	86.9	4.2	95.9	3.8
Saguinus midas					8	43.4	34.8	21.8							
Saguinus oedipus					13	43.1	35.8	21.1							
Saimiri sp.	8	14.9	30.9	54.2	5	40.7	33.8	25.4	8	69.7	4.7	93.0	4.8	89.7	5.0
Cercopithecus sp.	11	18.5	32.2	49.3					11	63.6	2.5	95.1	4.0	95.7	3.1
Colobus sp.	10	16.5	33.1	50.4					10	44.9	5.0	90.6	7.2	99.0	1.9
Gorilla sp.	6	16.2	34.1	49.7					6	56.6	6.8	103.6	7.2	95.9	3
Homo sapiens	3	18.0	32.9	49.1					13	76.3	7.9	111.8	13.0	88.9	10.0
Hylobates sp.	8	11.7	34.3	54.0					8	61.1	5.0	109.1	4.4	92.0	2.1
Macaca sp.	12	18.3	33.1	48.5					12	65.7	5.7	98.0	5.4	98.3	5.6
Miopithecus sp.	7	14.6	33.0	52.4					7	61.2	6.5	92.2	5.8	93.8	3.8
Pan sp.	14	13.6	35.9	50.5					14	46.8	3.4	100.8	1.8	93.5	2.7
Papio sp.	10	20.6	36.4	43.0					10	65.5	3.5	98.7	3.8	96.8	2.4
Pongo sp.	9	13.0	36.0	51.0					9	47.4	3.6	101.5	2.5	95.7	2.7
Presbytis sp.	10	15.4	33.5	51.1					10	51.6	4.0	94.9	1.3	96.4	1.8
Rhinopithecus sp.	2	17.6	33.5	48.9					2	44.8	4.3	98.9	3.8	101.0	2.0

Semnopithecus sp.	5	17.8	32.6	49.6					5	52.0	4.1	96.6	9.1	100.9	4.0
Symphalangus sp.	3	11.9	34.9	53.2					3	58.5	2.7	106.1	0.6	92.3	2.4
Trachypithecus sp.	2	15.9	32.1	52.0											
Ptilocercus lowii	1	15.6	32.4	52.1	1	42.4	33.3	24.3							
Tupaia glis	1	11.7	40.9	47.4	10	52.6	28.4	19.0							
Tupaia longipes					2	50.5	28.7	20.8							
Tupaia minor					1	44.0	31.4	24.6							
Tupaia tana					3	50.3	29.7	20.1							
Tupaia gracilis					1	53.7	28.3	18.0							
Galeopterus variegatus					7	41.0	24.6	34.4							
Cynocephalus volans	1	9.3	32.3	58.4	5	40.1	25.7	34.2							

[a]Digit (D) 3 ratios do not include distal phalanx length. Abbreviations: *s.d.* standard deviation; see also Table 14.4

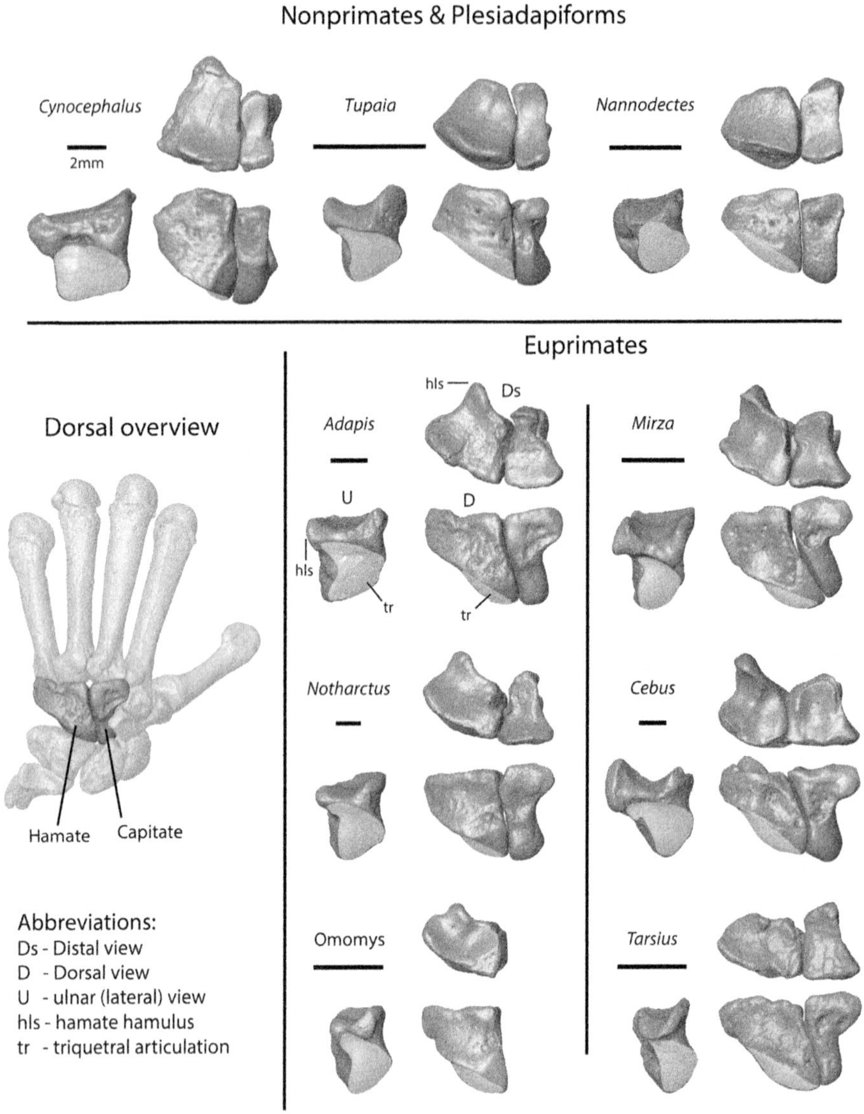

Fig. 14.5 Hamate and capitate of extant euarchontans and Paleogene primates (standardized to same proximodistal length). Views for each taxon are distal (*top*), dorsal (*bottom*), and ulnar (*left*). Scale bars equal 2 mm. Overview image is of carpus and metacarpus of *Adapis* RD 311 highlighting position of these carpals in dorsal view. Specimens depicted include *Cynocephalus volans* (UNSM 11502), *Tupaia glis* (EA 0174, Duke Univ. Coll.), *Nannodectes intermedius* (USNM 442229), *Adapis parisiensis* (RD 311), *Notharctus tenebrosus* (AMNH 127167), *Omomys carteri* (UM 32319), *Mirza coquereli* (DPC 137), *Cebus apella* (EA 054, Duke Univ. Coll.), and *Tarsius spectrum* (AMNH 109367). Reproduced from Boyer et al. (2013a)

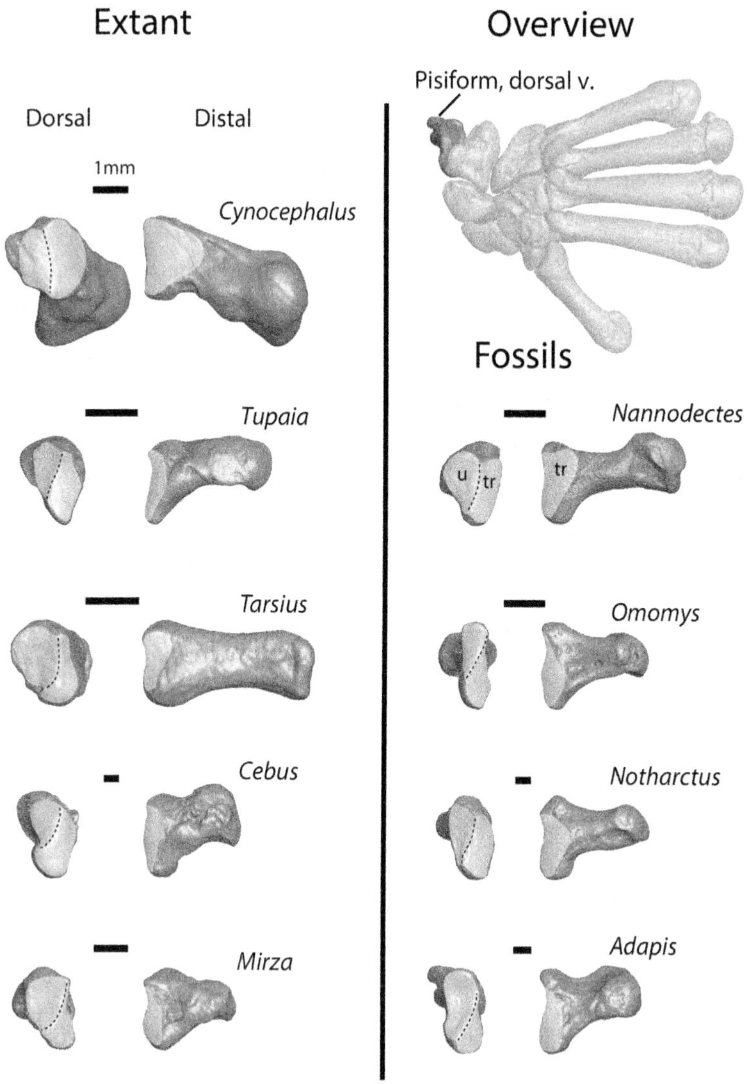

Fig. 14.6 Pisiforms of extant euarchontans and Paleogene primates (standardized to same radio-ulnar width). Ulnar and triquetral facets highlighted. Views for each taxon are distal with palmar up (*right columns*) and dorsal, with distal toward the top of the page (*left columns*), as shown in overview image of carpals and metacarpals of RD 311. A *dashed line* is used to separate ulnar and triquetral facets. Scale bars equal 1 mm. Specimens depicted include *Nannodectes intermedius* (USNM 442229), *Omomys carteri* (UM 32319), *Notharctus tenebrosus* (AMNH 127167), *Adapis parisiensis* (RD 311), *Cynocephalus volans* (UNSM 11502), *Tupaia glis* (EA 0174, Duke Univ. Coll.), *Tarsius spectrum* (AMNH 109367), *Cebus apella* (EA 054, Duke Univ. Coll.), and *Mirza coquereli* (DPC 137). Modified from Boyer et al. (2013a)

All Eocene taxa discussed here have a relatively small centrale like that of extant haplorhines and tupaiids (Beard and Godinot 1988; Godinot 1992; Hamrick and Alexander 1996). This morphology is in contrast to the larger centrale, which articulates with the hamate (and isolates the capitate from the lunate when viewed dorsally), in modern strepsirrhines, *Ptilocercus*, and *Cynocephalus* (Beard and Godinot 1988; Stafford and Thorington 1998; Sargis 2002; Boyer et al. 2013a).

The trapezium exhibits some variation among Eocene euprimates. In *Notharctus*, it is relatively large compared to that of *Adapis* (Fig. 14.1) (Hamrick 1996c). Vertical clinging taxa tend to have a larger trapezium and pollex than arboreal quadrupeds (Hamrick 1996c). The Mc1 facet of the trapezium is sellar shaped in *Notharctus* (Hamrick and Alexander 1996). It therefore joins indriids and lorisiforms (Jouffroy and Lessertisseur 1959; Etter 1974, 1978) and possibly *Mirza* and *Tarsius* (Boyer et al. 2013a) on the list of non-catarrhines that exhibit Napier's (1961) morphological correlates of "true opposability" of the thumb (Fig. 14.7). In contrast to *Notharctus*, *Adapis* appears to have a flat Mc1 articular surface (Godinot and Beard 1993) (Fig. 14.7), as argued to be typical of extant strepsirrhines and platyrrhines (Napier 1961).

The orientation of the Mc1 facet on the trapezium can contribute to pollical divergence, which in turn affects the prehensile capability of the hand (Napier 1961; Boyer et al. 2013a). *Notharctus*, *Adapis*, and *Darwinius masillae* have around 30–40° of divergence between first and second digits (though in *D. masillae*, this interpretation is based on in situ postmortem digit postures) (Fig. 14.8a; Table 14.6). Modern strepsirrhines and hominoids have a similar degree of divergence as these Eocene forms, whereas tarsiers, platyrrhines, and non-primate euarchontans have less pollical divergence (Fig. 14.8b; Table 14.6).

3.2 Metacarpus

Detailed morphology, proportions, and articular configurations of the metacarpus of Eocene euprimates can be best assessed from fossils of *Notharctus* and *Adapis* (Fig. 14.1). Metacarpals reported for *Smilodectes* (Beard and Godinot 1988; Godinot 1991, 1992; Godinot and Beard 1991) apparently lack associated phalanges. Messel specimens of cercamoniines usually preserve several metacarpals in articulation with digit and carpal elements (Fig. 14.4). In some cases joint angles between bones of Messel specimens may even represent somewhat natural postures (Franzen et al. 2009; Fig. 14.4). Unfortunately, the proximal ends and epiphyses of the metacarpals tend to be obliterated, making assessment of detailed morphology and proportions difficult (Fig. 14.4).

Well-preserved metacarpals known for Eocene adapiforms and omomyiforms look quite similar to those of modern strepsirrhines and platyrrhines (Fig. 14.9). Generally speaking, primate metacarpals are distinctive among mammals, most notably in the globular morphology of their distal articular surfaces. Compared to known plesiadapiforms, the metacarpals of adapiforms and omomyiforms are dissimilar in several respects. First, adapiform metacarpals are relatively shorter and

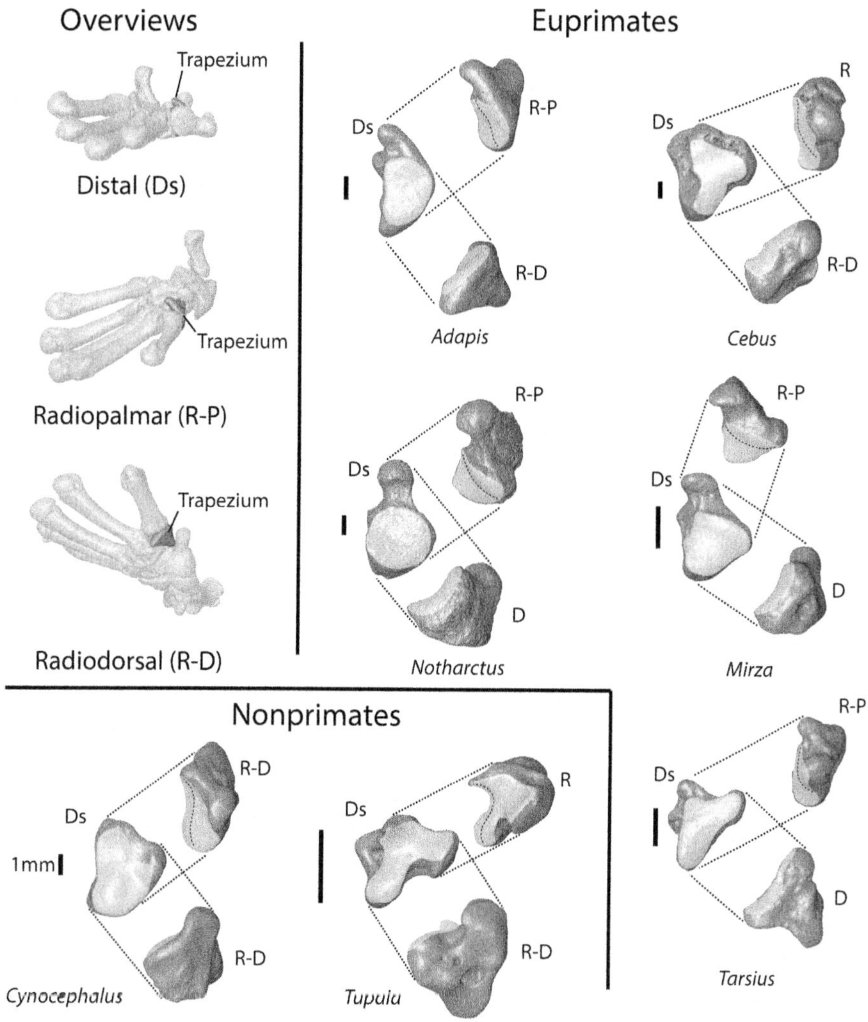

Fig. 14.7 Trapezia of extant euarchontans and Paleogene primates (standardized to same articular surface size). Note that saddle-shaped metacarpal facets are quite common among euarchontans (*contra* Napier 1961). The facet for the pollical metacarpal is highlighted. The main view is distal with palmar toward the top of the page, so that the metacarpals would be pointing out of the page and the pisiform would be pointing toward the top in an articulated specimen, as shown in the overviews using the carpals and metacarpals of *Adapis* RD 311. Other views illustrate principal curvatures of the Mc1 facet. The dashed black line on the surface in the radial and radiopalmar views shows the arc of the facet surface. No such line is necessary for the dorsal views as the arc is visible in profile if present. Note that *Notharctus*, *Tarsius*, and *Mirza* have two pronounced curvatures that constitute a "saddle shape" (which are very similar to catarrhines) and should permit opposition movements at the carpometacarpal joint. Non-primate euarchontans have a dorsal "ball-like" convexity for unrestricted mobility in rotation, abduction, and adduction during dorsiflexed postures, but recurvatures palmarly and radially restricting any mobility in palmar-flexed postures. Abbreviations: *D* dorsal, *Ds* distal, *R-D* radiodorsal, *R-P* radiopalmar, *R* radial. Scale bars equal 1 mm. Specimens depicted include *Adapis parisiensis* (RD 311), *Cebus apella* (EA 54, Duke Univ. Coll.), *Notharctus tenebrosus* (AMNH 127167), *Mirza coquereli* (DPC 137), *Tarsius spectrum* (AMNH 109367), *Tupaia glis* (EA 0174, Duke Univ. Coll.), and *Cynocephalus volans* (UNSM 15502). Reproduced from Boyer et al. (2013a)

more robust than those of plesiadapiforms. Additionally, the distal ends are radioulnarly narrower relative to their dorsopalmar depth. Finally, the distal articular surfaces for the proximal phalanges face distally and have a relatively large radius of curvature in adapiforms, whereas in known plesiadapiforms, this articular surface is palmarly restricted and therefore leads to a dorsal-facing articulation (Figs. 14.1 and 14.9). The differences in orientation and shape of the phalangeal facets suggest a limited capacity for stable (and probably less frequent) hyperextension of the McP joint in adapiforms relative to plesiadapiforms. Instead, palmar flexion was probably more effective in adapiforms. Furthermore, the larger radius of curvature and greater proportional depth of this surface equate to greater articular surface area relative to the overall size of the metacarpal, suggesting a greater capacity for transmitting force while maintaining both low joint stress and greater mobility (Hamrick 1996a). Despite these differences, both plesiadapiforms and Eocene euprimates share globular metacarpal heads, suggesting shared capacities for mobility in abduction, adduction, and axial rotation; all of which are expected to be beneficial for arboreality (Figs. 14.1 and 14.9).

Metacarpal proportions for available adapiforms show Mc1 to be the shortest, Mc3 to be the longest, and Mc4 to be the second longest [AMNH 127167 has a broken Mc3, so the length reported by Hamrick and Alexander (1996) is an estimate], with the exception of *Leptadapis*, in which Mc4 may have been the longest (Boyer et al. 2013a). Furthermore, notharctids appear to have a fairly short Mc2 relative to Mc3, making them similar to lorisids, *Daubentonia*, and *Tupaia* in relative metacarpal proportions (Tables 14.3, 14.4, and 14.6). *Adapis* has a longer Mc2, which is more typical among euprimates. In all Eocene euprimates with sufficient preservation, there is a high degree of pollical divergence similar to strepsirrhines and hominoids (Fig. 14.8b). However, *Notharctus* and *Darwinius* appear to have pronounced divergence between Mc2 and Mc3 as well (Figs. 14.1 and 14.4), possibly indicating a habitual schizodactylous grasp (in which digits 1 and 2 oppose digits 3–5) in these taxa.

3.3 Phalanges

Though there is important functional information in the detailed morphology of the phalanges, the relative proportions (Godinot and Beard 1991; Jouffroy et al. 1991; Lemelin 1996, 1999; Kirk et al. 2008) and degree of curvature (Jungers et al. 1997; Richmond 2007) are most frequently discussed. Phalanges are known for most adapiform taxa represented by other hand elements, but no associated accumulations of metacarpals and phalanges have been recovered for adapines. Prehensility has thus been difficult to estimate for these taxa (Godinot and Beard 1991, 1993; Godinot 1992; Boyer et al. 2013a). In this section, we begin by discussing morphological details of the phalanges of Eocene euprimates and then consider possible intrinsic hand proportions.

The proximal phalanges of Eocene adapiforms (Fig. 14.10) are superficially somewhat similar to those of some large-bodied plesiadapiforms, particularly

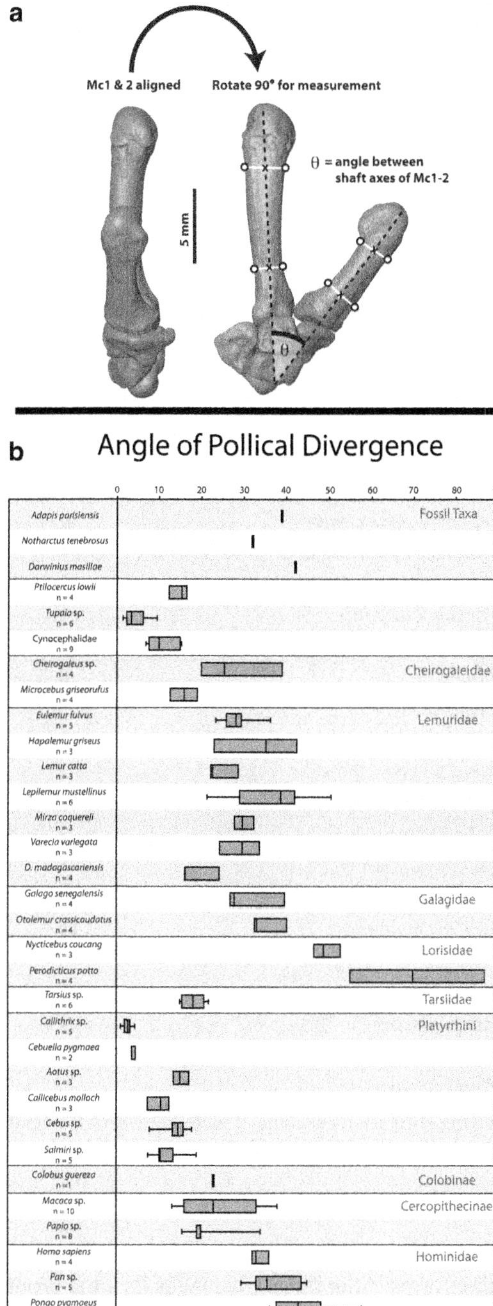

Fig. 14.8 Pollical divergence in extant euarchontans and Paleogene primates. (**a**) Method for calculating pollical divergence angle. First, it must be verified that corresponding facets on Mc1, trapezium, trapezoid, and Mc2 (and preferably the capitate and centrale) are in closest packed positions. Next, the bones are viewed and photographed (or a screenshot is obtained) with the plane of the Mc1-2 perpendicular to the viewing plane to ensure that the maximum angle is recorded. Then, the shaft axes are approximated by taking the midpoint of the shaft at two points along its length ("x's") and connecting a line through these points. The pollical divergence angle is the angle between the approximate axes of Mc1-2. (**b**) Box-and-whisker plot for pollical divergence (see Table 14.7 for sample statistics). Boxes encompass 50 % of data, and whiskers 75–100 %. *Vertical lines* represent the median. Note that the pattern of variation in divergence makes it difficult to reconstruct the degree of divergence in the euprimate ancestor. C. Orr provided access to catarrhine scans for these measurements. Modified from Boyer et al. (2013a)

Table 14.6 Pollical divergence

Taxon	*n*	Mean	s.d.
Adapis parisiensis	1	36.18	–
Notharctus tenebrosus	2	34.74	–
Darwinius masillae	1	43.00	–
Tupaia sp.	6	4.36	3.00
Ptilocercus lowii	4	15.07	1.91
Cynocephalus sp.	9	10.66	3.09
Daubentonia madagascariensis	4	19.06	3.85
Microcebus griseorufus	4	16.52	2.97
Cheirogaleus sp.	4	27.58	8.07
Mirza coquereli	3	30.03	2.27
Varecia variegata	3	29.32	4.66
Eulemur fulvus	5	28.65	4.82
Hapalemur griseus	3	33.61	9.82
Lemur catta	3	24.60	3.57
Lepilemur mustelinus	6	36.58	10.18
Otolemur crassicaudatus	4	35.31	3.48
Galago senegalensis	4	32.37	6.08
Perodicticus potto	4	71.38	13.11
Nycticebus coucang	3	49.47	3.17
Tarsius sp.	6	18.21	2.75
Cebus sp.	5	13.85	3.76
Saimiri sp.	5	12.07	4.37
Aotus sp.	3	15.19	1.83
Callicebus moloch	3	10.10	2.54
Cebuella pygmaea	2	4.11	–
Callithrix jacchus	5	2.67	1.31
Colobus guereza	1	23.00	–
Macaca mulatta	10	23.23	9.26
Papio sp.	5	21.69	7.12
Homo sapiens	5	33.78	1.67
Pan paniscus	1	33.00	–
Pan troglodytes	8	37.81	5.57
Pongo pygmaeus	5	44.62	8.76

See Fig. 14.8 for measurement method. C. Orr provided access to catarrhine scans for these measurements. Abbreviation: *s.d.* standard deviation

P. cookei (UM 87990; Boyer 2009), though clearly they are more elongated in adapiforms. Adapiform phalanges lack the pronounced flexor sheath ridges of paromomyids, micromomyids, and *C. simpsoni* (Bloch and Boyer 2002; Boyer and Bloch 2008). *C. simpsoni* also differs from notharctines in the greater curvature of its proximal phalanges (Bloch and Boyer 2002). The most salient differences in the proximal phalanges of known plesiadapiforms compared with Eocene euprimates appear in the morphology of the proximal end and its metacarpal articular surface.

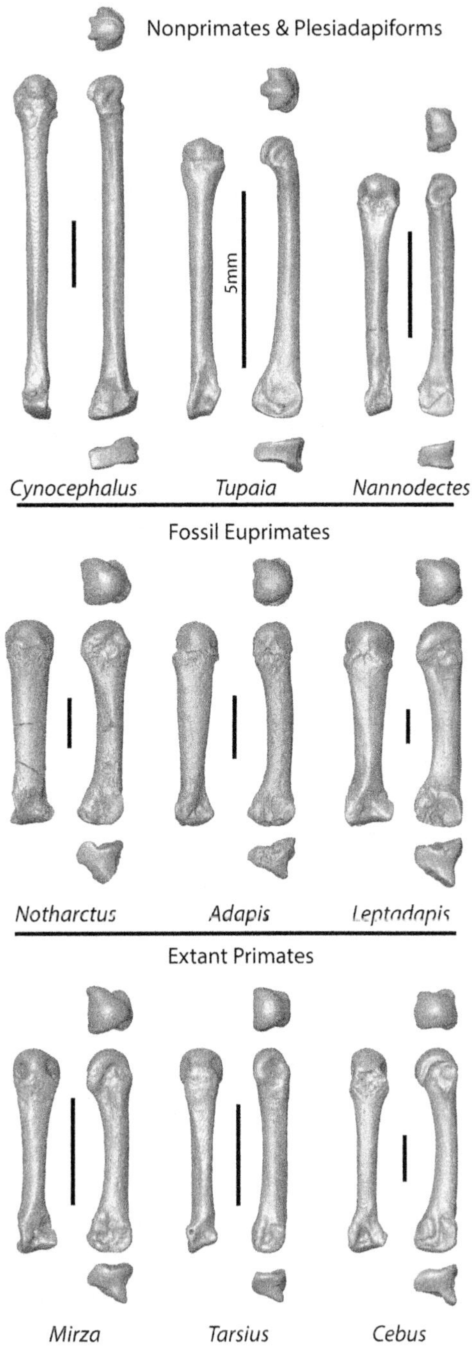

Fig. 14.9 Third metacarpal of extant euarchontans and Paleogene primates (*top row* is standardized to same radioulnar width of distal end; *bottom rows* are depicted with a similar distal end width, but standardized to same proximodistal length). Note plesiadapiforms and non-primate euarchontans have distal articular surfaces that face primarily dorsad, are dorsopalmarly shallower, and have more pronounced palmar keels. Scale bars equal 5 mm. Specimens depicted include *Cynocephalus volans* (UNSM 15502; Mc4 depicted due to image availability), *Tupaia glis* (EA 0174, Duke Univ. Coll.), *Nannodectes intermedius* (USNM 442229), *Notharctus tenebrosus* (AMNH 131950), *Adapis parisiensis* (RD 311), *Leptadapis magnus* (MaPhQ no#), *Mirza coquereli* (DPC 137), *Tarsius spectrum* (AMNH 109367), and *Cebus apella* (EA 54, Duke Univ. Coll.). Reproduced from Boyer et al. (2013a)

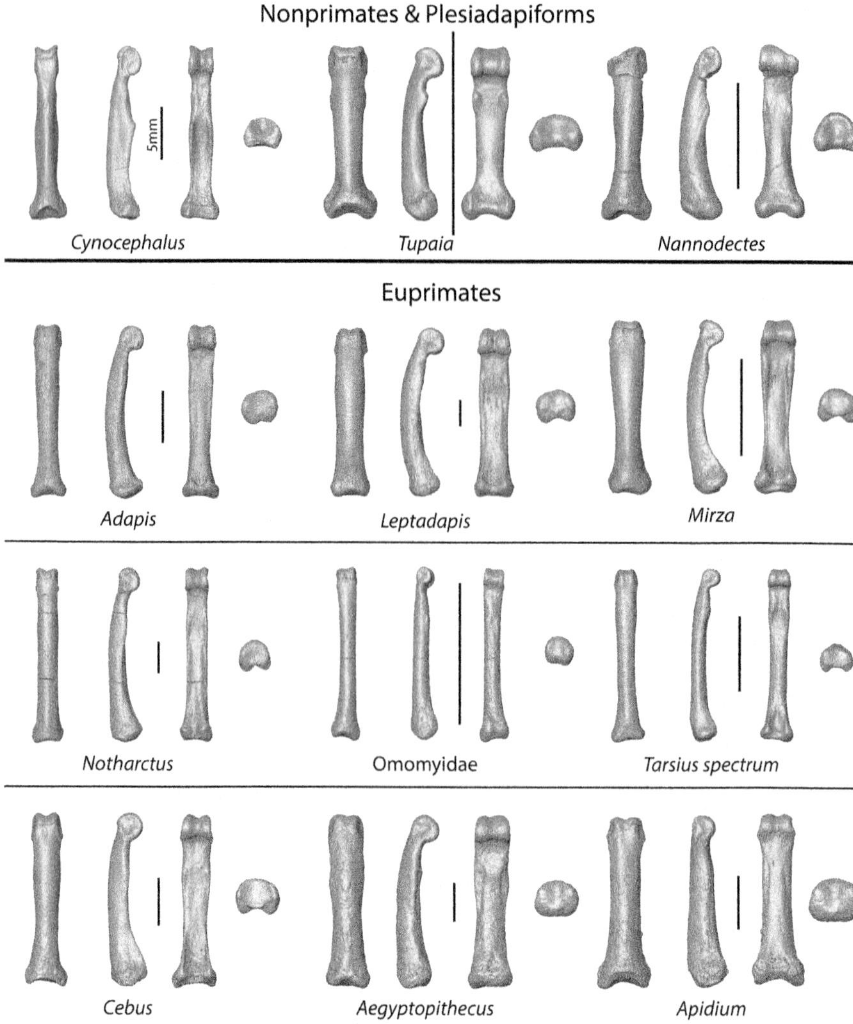

Fig. 14.10 Proximal phalanx of the third ray of extant euarchontans and Paleogene primates (standardized to same proximodistal length). Note plesiadapiforms and non-primate euarchontans have palmar tubercles on the proximal articular surface that extend proximally and give the proximal articular surface a dorsal cant. Scale bars equal 5 mm. Specimen information: *Cynocephalus volans* (UNSM 15502), *Tupaia glis* (EA 0174, Duke Univ. Coll.), *Nannodectes intermedius* (USNM 442229), *Adapis parisiensis* (AMNH 140719), *Leptadapis magnus* (NMB Q.L. 255: this is probably a pedal element, but we did not have access to scan imagery of manual proximal phalanges for *Leptadapis*), *Mirza coquerli* (DPC 137), *Northarctus tenebrosus* (AMNH 127167), Omomyidae (UCMP 218417), *Tarsius spectrum* (AMNH 109367), *Cebus sp.* (EA 54, Duke Univ. Coll.), *Aegyptopithecus zeuxis* (DPC 1005), *Apidium phiomense* [DPC 1294 (80-272)]. Reproduced from Boyer et al. (2013a)

Specifically, in plesiadapiforms, the metacarpal articular surfaces appear to face more dorsally due to proximally projecting palmar tubercles, whereas in adapiforms and omomyiforms, the facets are more proximally facing (Fig. 14.10). The morphology of plesiadapiform proximal phalanges enhances the pattern dictated by the distal articular surface of the metacarpals and promotes hyperextension of the McP joint (Figs. 14.1 and 14.9). These patterns hold for other Eocene euprimates examined here, except among adapids, which have phalangeal curvature more comparable to *C. simpsoni*, and among omomyiforms, which have reduced curvature similar to *Tarsius*, parapithecids, and terrestrial cercopithecoids (Table 14.7).

Whereas proximal phalanges of plesiadapiforms and euprimates are superficially similar, there is no mistaking the intermediate phalanges of Eocene euprimates for those of plesiadapiforms (Fig. 14.11). While plesiadapiform intermediate phalanges are characterized by radioulnarly narrow and dorsopalmarly deep articular ends and shafts (Boyer and Bloch 2008; Boyer et al. 2013a), the intermediate phalanges of Eocene euprimates are radioulnarly broad and dorsopalmarly flattened. Additionally, plesiadapiforms exhibit relatively straight intermediate phalanges, whereas Eocene euprimates exhibit more dorsal convexity in their intermediate phalanges (Boyer and Bloch 2008). The degree of curvature is difficult to evaluate in cercamoniines due to distortion of the Messel specimens, but these contrasts with plesiadapiforms otherwise hold for all known early euprimates.

When evaluating overall hand proportions, it is clear that notharctids and at least *Teilhardina* among omomyiforms have exceptionally long digits, with the third digit phalanges of *Darwinius* and *Europolemur* making up 63–64 % of the length of the entire hand (calculated as the sum length of carpus, metacarpus, and phalanges; Fig. 14.3). *Teilhardina* lacks preservation of carpal elements, but its proportions would likely have been similar. Among extant primates, only *Tarsius* and *Daubentonia* match these proportions (Tables 14.3, 14.4, and 14.6), though *Galago* and *Euoticus* [species means of 61.2 % and 60.4 %, respectively, as reported by Jouffroy et al. (1991)] come close (Fig. 14.3). We use the term "hyper-prehensility" for taxa with hands that are similar to *Daubentonia* and *Tarsius* in the ratio of digit length to hand length. Though prehensility is fundamentally a behavioral capacity for grasping an object of a given size securely in one hand (Napier 1960, 1961), Kirk et al. (2008) identified elongated phalanges relative to metacarpus as likely correlating with this capacity. The term hyper-prehensility is not a reference to the behavioral capacity but to the skeletal proportions. As mentioned above and discussed in detail in Boyer et al. (2013a), the hand of *Notharctus* (AMNH 127167) is still too poorly known to report such proportions. Nevertheless, the preserved elements demonstrate distinctive similarities between *Notharctus* and the Messel adapiforms. In particular, the estimated ratio of the third proximal phalanx to the third metacarpal (a "proximal prehensility" index; Tables 14.5A and 14.5B) in *Notharctus* (~1.16–1.30) is similar to that of the Messel species (1.31). Though values greater than 1.16 are seen in some individuals of *Loris tardigradus* and *Nycticebus pygmaeus*, the only extant primates in which individuals often have a third digit proximal prehensility index of 1.30 or more are *Daubentonia madagascariensis* (species average = 1.18) and *Tarsius bancanus* (species average = 1.21) (Tables 14.5A and 14.5B).

Table 14.7 Phalangeal curvature (included angle method following Jungers et al. 1997)

Taxon	H/F	*n*	Mean	s.d.	Source
Fossil					
Notharctus tenebrosus	H	9	34.0	9.0	[a]
Notharctus tenebrosus	F	10	25.8	6.2	[a]
cf. *Adapis* sp.	H	19	48.3	7.2	[a]
cf. *Adapis* sp.	F	13	48.1	6.7	[a]
cf. *Leptadapis magnus*	H	15	57.9	6.5	[a]
cf. *Leptadapis magnus*	F	20	53.4	9.0	[a]
Omomyid	?	2	25.9	5.8	[a]
Aegyptopithecus	H	2	55.5	6.4	[b]
Aegyptopithecus	F	1	48.7	–	[b]
Apidium	H	2	27.6	3.5	[b]
Apidium	F	1	23.5	–	[b]
Babakotia radofilai	H	27	58.3	6.3	[c]
Babakotia radofilai	F	11	60.9	6.4	[c]
Archaeolemur edwardsi	H-F	28	27.9	6.3	[c]
Palaeopropithecus kelyus	H-F	11	73.3	7.3	[c]
Palaeopropithecus ingens	H-F	45	60.3	9.9	[c]
Palaeopropithecus maximus	H-F	13	57.2	9.8	[c]
Mesopropithecus dolichobrachion	H-F	5	65.8	6.7	[c]
Megaladapis edwardsi	H-F	18	49.2	7.8	[c]
Megaladapis madagascariensis	H-F	23	46.1	8.1	[c]
Extant					
Tarsius spp.	H	16	28.3	5.2	[a]
Tarsius spp.	F	16	27.3	5.6	[a]
Varecia variegata	H	12	51.6	5.4	[c]
Varecia variegata	F	12	51.2	5.8	[c]
Propithecus diadema	H	20	35.8	4.4	[c]
Propithecus diadema	F	28	27.7	6.0	[c]
Indri indri	H	66	35.0	6.0	[c]
Indri indri	F	64	31.2	6.5	[c]
Lagothrix spp.	H	12	62.4	6.0	[c]
Lagothrix spp.	F	12	53.1	3.2	[c]
Ateles spp.	H	27	55.2	6.5	[c]
Ateles spp.	F	28	53.2	5.0	[c]
Nasalis larvatus	H	12	38.0	5.6	[c]
Nasalis larvatus	F	14	35.3	3.8	[c]
Papio spp.	F	19	11.1	6.7	[c]
Hylobates syndactylus	H	20	53.1	5.6	[c]
Hylobates lar	H-F	68	47.8	5.4	[c]
Pongo pygmaeus (both subspecies)	H	88	64.9	6.6	[c]
Pongo pygmaeus (both subspecies)	F	24	85.1	7.8	[c]

(continued)

Table 14.7 (continued)

Taxon	H/F	*n*	Mean	s.d.	Source
Gorilla gorilla	H	88	37.2	4.2	c
Gorilla gorilla	F	31	33.0	4.3	c
Pan paniscus	H	38	44.8	4.2	c
Pan paniscus	F	54	39.2	6.8	c
Pan troglodytes	H	63	42.4	4.8	c
Pan troglodytes	F	37	41.2	6.9	c

Data from Hamrick et al. (1995)[b], Jungers et al. (1997)[c], and Boyer et al. (2013a)[a]. Abbreviations: *H* hand, *F* foot, *s.d.* standard deviation. T. Clarke generated data in Boyer et al. (2013a) for an independent study project at Brooklyn College

Although the Messel adapiforms, *Daubentonia* and *Tarsius*, all have third digits that are unusually long relative to the metacarpus and carpus, these two extant taxa differ from the fossils in having digits that are also unusually long for their respective body masses (Fig. 14.12). Notharctid adapiforms examined here appear to have only slightly longer than expected digits relative to their estimated body masses [see Table 14.3 legend Boyer et al. (2013a) for a description of methods for computing residuals of digit and metacarpus length relative to body mass]. The unusual, hyper-prehensile intrinsic proportions of the Messel adapiforms appear to stem from a combination of slightly elongated digits and slightly shorter-than-expected metacarpals relative to their body mass (Fig. 14.12). Therefore, similar intrinsic ray proportions (Fig. 14.3) of the Messel adapiforms (and probably *Notharctus*), *Tarsius bancanus*, and *Daubentonia* may have different functional/adaptive purposes. Though the adapid *Godinotia* shares the pattern of slight digit elongation combined with metacarpal reduction, it is slightly less extreme in many of its intrinsic proportions than those of the Messel adapiforms (Fig. 14.3; Tables 14.3 and 14.4). Its digits make up 61.5 % of its hand length, and the third digit proximal prehensility index is 1.07. Nonetheless, *Godinotia* is fairly unusual in at least one respect: the intermediate phalanges are closer in length to the proximal phalanges (~90 %) than any other extant primate or any other adapiform (Tables 14.3, 14.4, 14.5A, 14.5B, and 14.6). Interestingly, many plesiadapiforms approach *Godinotia* in this respect (Tables 14.3, 14.4, 14.5A, 14.5B, and 14.6). *Adapis* and *Leptadapis* probably had shorter digits relative to their palm lengths (and body mass), resulting in intrinsic hand proportions most similar to *Perodicticus* and various platyrrhines (Figs. 14.3 and 14.12; Tables 14.3, 14.4, 14.5A, 14.5B, and 14.6). The plesiadapiforms *Carpolestes* and *Nannodectes* have digit and metacarpal lengths that are almost perfectly predicted by the regression line based on extant primates using body mass (Fig. 14.12; Tables 14.3, 14.4, 14.5A, 14.5B, and 14.6). Interestingly, the composite *Teilhardina* digit ray presented here [based on data in Gebo et al. (2012, 2015) and Rose et al. (2011)] suggests it matches tarsiers in both intrinsic (Fig. 14.3b) and extrinsic (Fig. 14.12) proportions. That is, it exhibits hyper-prehensility as well as extremely long digits and metacarpals relative to body mass (we consider the possible adaptive implications for this pattern below).

Relative digit lengths can be reconstructed for just two adapiforms among known early Eocene forms: *Europolemur* and *Godinotia* (Fig. 14.13). *Europolemur* has a

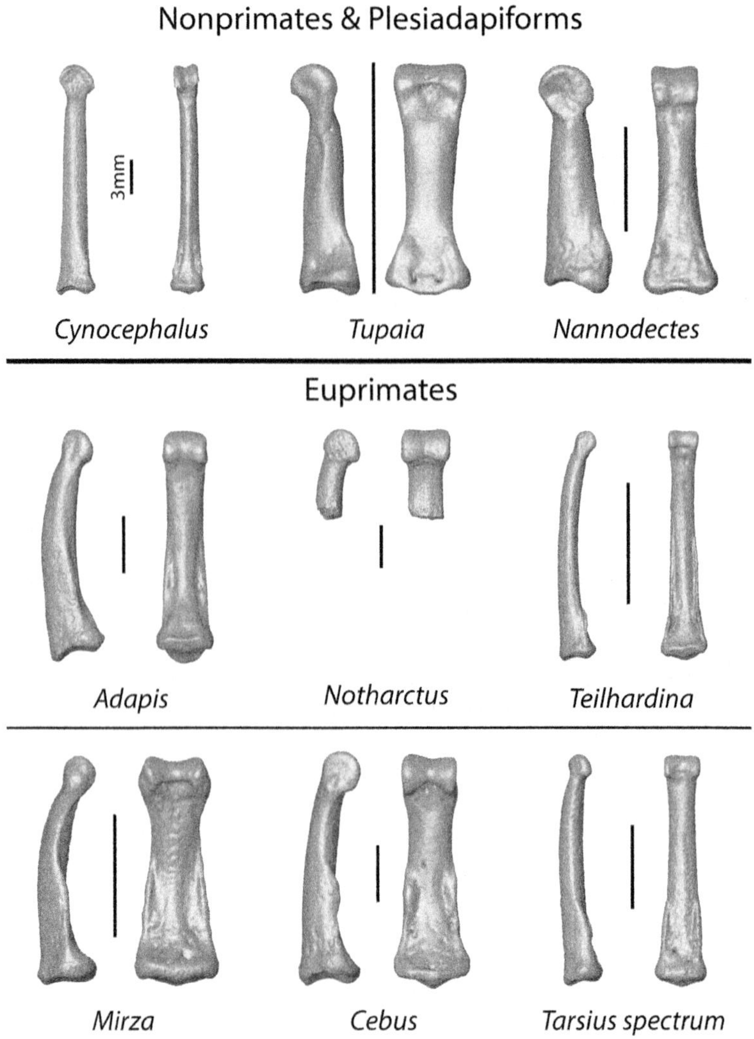

Fig. 14.11 Intermediate phalanges of extant euarchontans and Paleogene primates (standardized to same proximodistal length). Scale bars equal 3 mm. Note plesiadapiforms have radioulnarly narrow, dorsopalmarly deep shafts, whereas euprimates exhibit the opposite dimensions. Specimen information: *Cynocephalus volans* (UNSM 15502), *Tupaia glis* (EA 0174, Duke Univ. Coll.), *Nannodectes intermedius* (USNM 442229), *Adapis parisiensis* (MaPhQ no#), *Notharctus tenebrosus* (AMNH 127167), *Teilhardina belgica* (IRSNB M 1266), *Mirza coquerli* (DPC 137), *Cebus sp.* (EA 54, Duke Univ. Coll.), *Tarsius spectrum* (AMNH 109367). Reproduced from Boyer et al. (2013a)

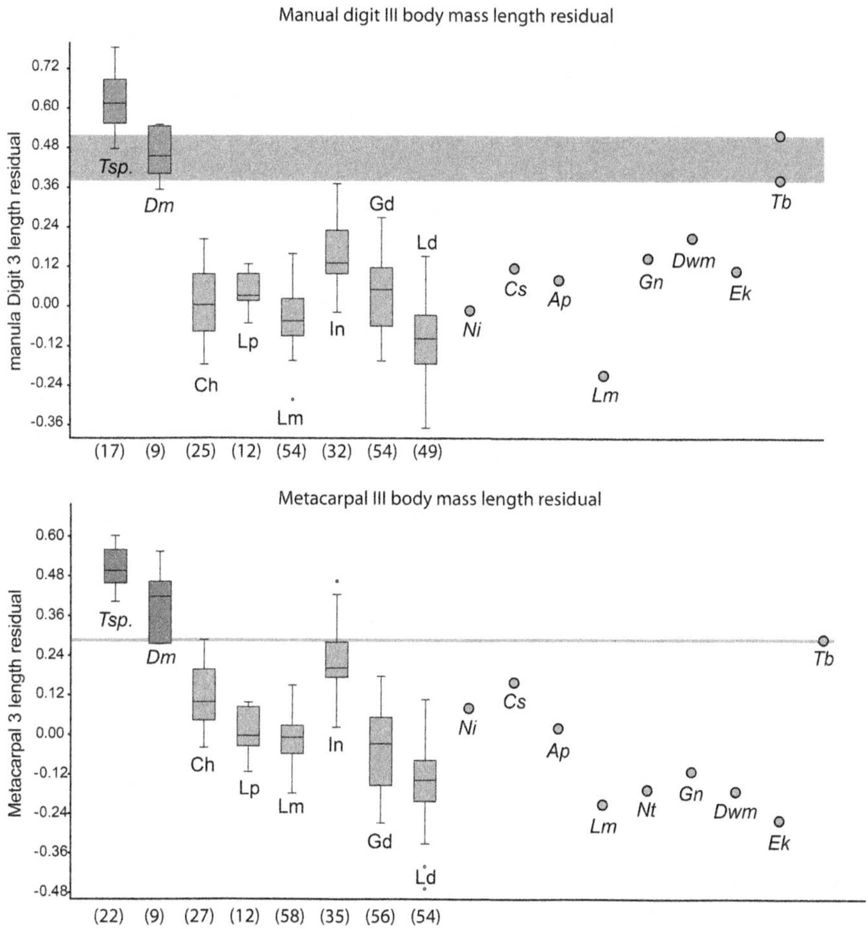

Fig. 14.12 Hand proportions relative to body mass. Box plot of residuals from a regression of digit length versus estimated body mass (**a**) and residuals from a regression of third metacarpal length versus estimated body mass (**b**) for extant strepsirrhines, tarsiers, and Paleogene primates. Note that extant data were analyzed at the level of individual specimens rather than species means to improve comparability of results to fossil taxa that are only available as single individuals or composites. The *number in parentheses* under each box plot is the number of individuals included in each group, which may include a single species or several species depending on the group. Extant species mean residual values are presented in Table 14.5A. Ordinary least squares (OLS) regression analysis was run on log-transformed digit and metacarpal lengths separately against log-transformed body mass estimates generated from data of Lemelin (1996) (which is also analyzed by Lemelin and Jungers 2007). Mass estimates for fossils were generated in various ways as described in Table 14.3. See Boyer et al. (2013a) for more details on regression method. Note that notharctids have slightly shorter than expected metacarpals, but longer than expected digits, yielding "tarsier-like" manual *intrinsic* proportions without "tarsier-like" *extrinsic* proportions. On the other hand, *Teilhardina* is clearly in the range of tarsiers and *Daubentonia*. *Teilhardina* is represented by two values in part **a** because there is uncertainty about the most appropriate proximal phalanx length to use as discussed in the Table 14.3 legend. Abbreviations for extant taxa: *Tsp. Tarsius* sp., *Dm Daubentonia madagascariensis*, *Ch* Cheirogaleidae, *Lp* Lepilemuridae, *Lm* Lemuridae, *In* Indriidae, *Gd* Galagidae, *Ld* Lorisidae. Abbreviations for fossil taxa: *Ni Nannodectes intermedius*, *Cs Carpolestes simpsoni* (Plesiadapiformes), *Ap Adapis parisiensis*, *Lm Leptadapis magnus*, *Gn Godinotia neglecta*, *Dwm Darwinius masillae*, *Ek Europolemur kelleri* (Adapiformes), *Nt Notharctus tenebrosus*, *Tb Teilhardina belgica* (Omomyiformes)

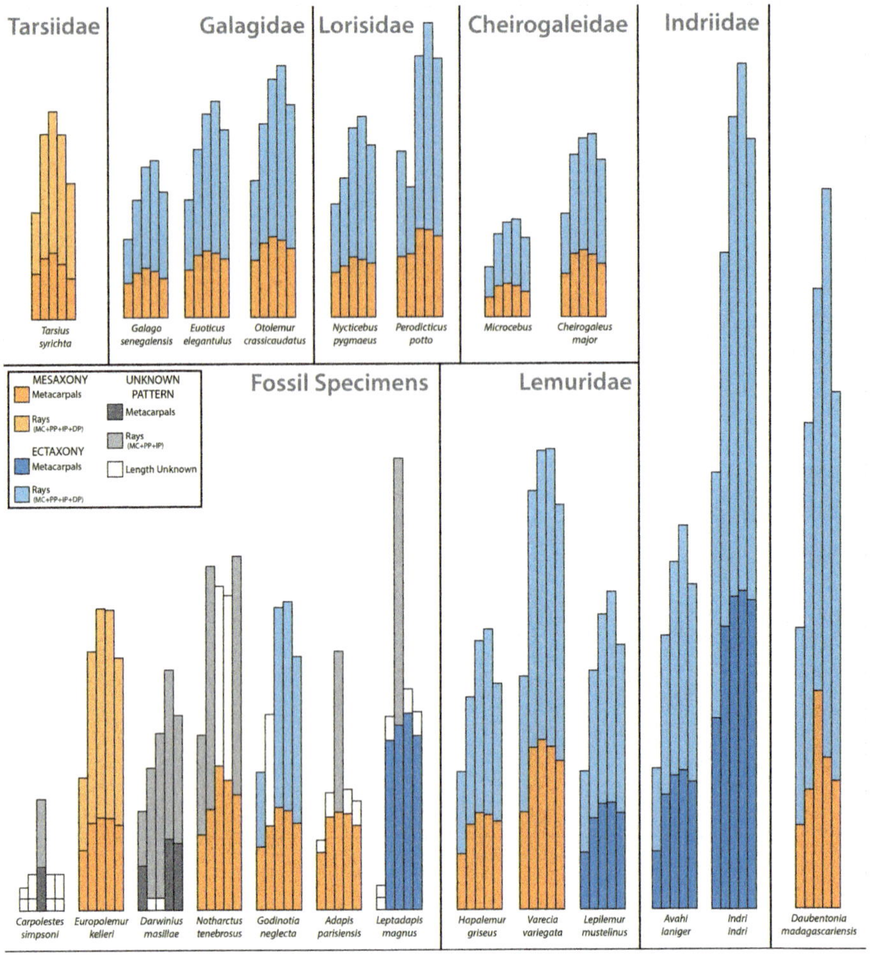

Fig. 14.13 Axonic patterning of extant strepsirrhines, tarsiers, and Paleogene primates. Relative lengths of all five metacarpals and digit rays (from *left* to *right*). Mesaxonic (third ray longest and forms the axis) and ectaxonic (fourth ray longest and forms the axis) patterning are indicated for both metacarpals and digits. For clarity, paraxonic (third and fourth rays subequal) patterning is not considered. All lengths are proportionally displayed and generated from species means. Note that most strepsirrhines have mesaxonic metacarpals within ectaxonic hands (with the exceptions of indriids, *Lepilemur*, and *Arctocebus*). Fossil taxa, with the exception of Leptadapis, have mesaxonic metacarpals. Godinotia follows the strepsirrhine pattern of ectaxonic digits, whereas *Europolemur* displays slight mesaxony. Modified from Boyer et al. (2013a)

haplorhine pattern of overall mesaxony (third digit is the longest), and *Godinotia* is like most non-indriid strepsirrhines with overall ectaxony, but metacarpal mesaxony. It seems likely that *Leptadapis* would have exhibited indriid-like overall ectaxonic proportions. No plesiadapiform hands are sufficiently preserved to estimate overall axony, though micromomyids exhibit metacarpal mesaxony (Boyer et al. 2013a).

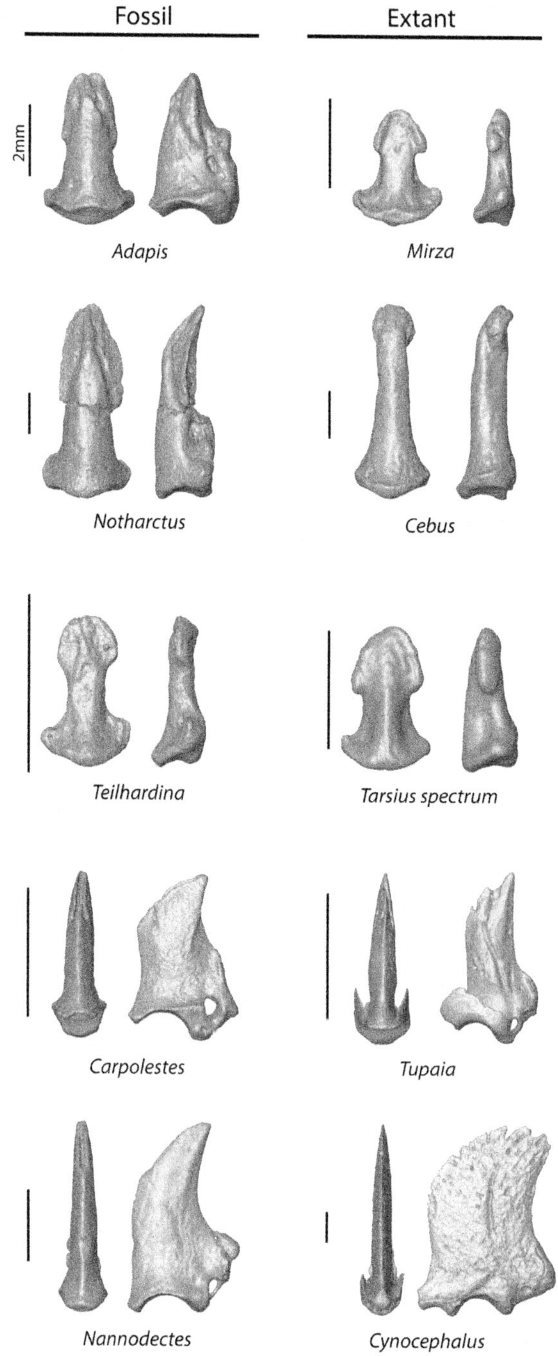

Fig. 14.14 Distal phalanges of extant euarchontans and Paleogene primates (standardized to same radioulnar width of the proximal end). Note plesiadapiforms have radioulnarly narrow, dorsopalmarly deep shafts, whereas euprimates exhibit the opposite dimensions. Scale bars equal 2 mm. Specimen information: *Adapis parisiensis* (UM ECA 1400), *Notharctus tenebrosus* (AMNH 127167), *Teilhardina brandti* (USNM 540587), *Carpolestes simpsoni* (UM 101963), *Nannodectes intermedius* (USNM 442229), *Mirza coquerli* (DPC 137), *Cebus apella* (EA 054, Duke Univ. Coll.), *Tarsius spectrum* (AMNH 109367), *Tupaia glis* (EA 0174, Duke Univ. Coll.), *Cynocephalus volans* (UNSM 15502). Modified from Boyer et al. (2013a)

Distal phalanges are known for several Eocene euprimates, but can only be confidently assigned to the hand in *Notharctus*, the Messel adapiforms (Franzen 1993; Hamrick and Alexander 1996; Franzen et al. 2009), and *Godinotia* (Thalmann et al. 1989; Thalmann 1994). Though all manual distal phalanges identified for Eocene euprimates are unguliform, possessing a flattened apical tuft (which indicates the presence of a flattened nail rather than a claw; see Chap. 8), there appears to be substantial variation (Fig. 14.14) (Gregory 1920; Dagosto 1988; Gebo et al. 1991; Godinot 1991; 1992; Franzen 1993; Hamrick and Alexander 1996; Rose et al. 2011; Ni et al. 2013). *Notharctus* (AMNH 127167) has long, narrow, curved manual distal phalanges, possibly suggesting a more claw-like nail or functional tegulae (Godinot 1991). In contrast, those of *Cantius*, *Smilodectes* (Gebo et al. 1991; Godinot 1992; Bloch et al. 2010), Messel adapiforms (Franzen 1993; Thalmann 1994; Franzen et al. 2009), and *Godinotia* (Thalmann et al. 1989; Thalmann 1994) appear proportionally shorter, broader, and flatter. Most adapiform distal phalanges retain a feature more typically characteristic of falcular (claw-bearing) phalanges: bilateral, well-developed nutrient foramina, as also seen in plesiadapiform distal phalanges, though these foramina face laterally in euprimates rather than palmarly as in plesiadapiforms (Maiolino et al. 2012). Known omomyiform distal phalanges are very similar to those of tarsiers and tend to lack well-developed nutrient foramina (Dagosto 1988; Rose et al. 2011). The adaptive significance of nails instead of claws is long debated (e.g., Cartmill 1974b, 1985; Soligo and Müller 1999; Soligo and Martin 2006), and much promising work has been the focus of recent dissertation by Maiolino (2015), which incorporates into its assessment the patterns observed in omomyiforms and adapiforms.

4 Conclusions and Future Directions

The available fossil evidence on proportions and morphology suggests dramatic transitions in functional capability and adaptive utility of the hands of early euprimates. One of the most impressive innovations is hyper-prehensility of the hands. This is somewhat surprising because it is now clear from comparative studies [e.g., Kirk et al. (2008)] that—except in *Daubentonia*, *Tarsius*, and rarely certain lorisiforms—extant primates do not have unusual levels of prehensility compared to certain plesi-adapiforms and many arboreal non-primates (such as the treeshrew *Ptilocercus* and many arboreal marsupials, carnivorans, and rodents). Therefore, the most distinctive proportional features of early euprimates (adapiforms and omomyiforms) appear to have been lost by most extant primate lineages. The near ubiquity of hyper-prehensility in early euprimates suggests this morphological innovation occurred near the euprimate node. As such, Hamrick's (2001, 2007) hypotheses that a change in digit ray development allowed natural selection to produce a dramatic change in digit proportions may well be correct, and it is possible these prehensile features provided access to niche space not available to other arborealists, contributing to the initial success of euprimates.

Although extant primates do not differ from plesiadapiform stem primates in prehensility, they differ from plesiadapiforms and resemble early euprimates in features reflecting grasping mechanics. For example, based on the form of the McP joints and distal phalanges, known plesiadapiforms appear specialized for claw clinging that entailed hyperextended McP joints and palmar-flexed proximal phalangeal joints (convergently similar to callitrichine primates). The earliest known euprimates are already similar to extant primates in ways suggesting they must have more frequently used grips in which the McP joints were more palmarly flexed and wrapped around the substrate. This hand posture maximizes contact surface area and frictional resistance to torque generated by body mass.

It is difficult to identify the biological role that hyper-prehensile intrinsic proportions may have served in the earliest euprimates. A more secure grasp seems the most likely explanation, but this explanation does not suggest novel positional behaviors. Extrinsic proportions appear to explain more about behavior. Among extant primates, those with the longest fingers relative to body mass [and taking allometry into account; Lemelin and Jungers (2007)] tend to be more active vertical clingers and leapers (Napier and Walker 1967; Boyer et al. 2013a). Thus, with moderately long digits relative to body mass, notharctids were likely more committed to vertical postures and acrobatic behaviors than adapines. However, the relative digit lengths of notharctids are not particularly long compared to "prosimian" euprimates generally and would not be considered specialized leapers from that perspective. This interpretation of notharctid positional behavior is therefore compatible with Hamrick's observations (1996b, c, 1997) that notharctid carpal structure suggests arboreal quadrupedalism.

Extrinsic proportions of *Teilhardina* reveal quite a different picture: unusually large hands that are uniquely similar to those of tarsiers and *Daubentonia*. The large hands of tarsiers have been proposed as adaptations for insect apprehension (Niemitz 1984; Godinot 1991; Lemelin 1996; Lemelin and Jungers 2007), so the proportional similarities of tarsiers and *Teilhardina* make a compelling case for predatory utility in *Teilhardina* as well. Alternatively, the finger proportions of *Teilhardina* may be functionally related to vertical clinging and leaping as recent postcranial evidence of this taxon (Rose et al. 2011; Gebo et al. 2012, 2015) and its close relative *Archicebus* (Ni et al. 2013) reveal clear leaping specializations in other parts of the skeleton. Furthermore, Boyer et al. (2013b) found evidence for consistent and pervasive trends toward increased leaping specialization in the stem lineage, leading to the ancestral euprimate and in basal haplorhine and strepsirrhine lineages. So finger elongation could be part of this trend toward greater leaping specialization.

In sum, it is possible that *Teilhardina*-like extrinsic proportions and hyper-prehensility developed primarily for a selective advantage in acrobatic leaping that frequently served a critical role in prey capture by the ancestral euprimates. Hyper-prehensility would enhance the "netting" function of the hands by allowing an absolutely greater spread of the fingers for a given hand length and angle of abduction at the McP joint. Hyper-prehensility in adapiforms would then be interpreted as a retention of the ancestral condition, while reduced extrinsic proportions developed as euprimate diets diversified and became less insectivorous overall. Another possibility is that demands from leaping and prey capture were not actually involved in this process; if clawless prehensile hands are a synapomorphy of living primates, this trait may have critically limited substrate availability by hand length. If true, there would have been a selective advantage toward proportionally larger hands in absolutely smaller primates. This would explain why *Teilhardina*—as a small bodied taxon—has high extrinsic proportions, but known adapiforms (that are all larger in body size) do not. It could be argued that this trend persists among extant primates as overall negative allometry in ray lengths (Lemelin and Jungers 2007). However, under this scenario, the need for hyper-prehensility is not explained except as a by-product. More fossils and future comparative data analyses will further test these hypotheses. Of particular relevance to the scenarios above, greater sampling of early fossil primates could help reveal whether tarsier-like proportions are truly primitive for the clade. Likewise, a better sampling and morphometric characterization of early primate distal phalanges will help establish the morphological and functional spectrum they encompass; a better description of the evolutionary transformations in distal phalanx shape in early primates will help us better understand whether claw clinging or prehensile grasping is primitive for euprimates.

Acknowledgments We wish to acknowledge P. Gingerich, G. Gunnell, J. Franzen, U. Thalmann, L. Costeur, C. Argot, P. Tassy, W. Jungers, A. Bergeret, N. Simmons, E. Westwig, D. Lunde, L. Gordon, and P. Holroyd and their respective institutions for access to specimens, measurements, and casts that were critical for this publication. J. Thostenson, M. Hill, I. Wallace, J. Lovoi, J. Butler, A. Garberg, A. Freeman, G. Almor, and S-H Kim provided help acquiring and processing scan data. T. Clarke generated the measurements for phalangeal curvature. C. Orr provided data scans for divergence measurements in catarrhine primates. We thank D. Schmitt, P. Lemelin, T. Kivell, B. Richmond, and three anonymous reviewers for useful feedback on earlier versions of this paper. Support from NSF BCS 1317525, 1440742, and 1440558 and Duke University helped DMB's work on this project. The Graduate School of Duke University supported GSY's work on this project. NSF SBE-1028505 (to EJ Sargis and SGBC) and a Leakey Foundation Research Grant supported SGBC's work on this project. Support from NSF BCS 1440558 helped JIB's involvement. Data provided by P. Lemelin was essential for the perspectives presented in this work, and we are greatly indebted to him for this resource.

References

Alexander JP, Burger BJ (2001) Stratigraphy and taphonomy of Grizzly Buttes, Bridger Formation, and the Middle Eocene of Wyoming. In: Gunnell GF (ed) Eocene biodiversity: unusual occurrences and rarely sampled habitats. Kluwer Academic/Plenum Publishers, New York, pp 165–196

Beard KC (1989) Postcranial anatomy, locomotor adaptations, and palaeoecology of early Cenozoic Plesiadapidae, Paromomyidae, and Micromomyidae (Eutheria, Dermoptera). Ph.D. dissertation, The Johns Hopkins University

Beard KC (1990) Gliding behavior and palaeoecology of the alleged primate family Paromomyidae (Mammalia, Dermoptera). Nature 345:340–341

Beard KC (1993a) Origin and evolution of gliding in Early Cenozoic Dermoptera (Mammalia, Primatopmorpha). In: MacPhee RDE (ed) Primates and their relatives in phylogenetic perspective. Plenum Press, New York, pp 63–90

Beard KC (1993b) Phylogenetic systematics of the Primatomorpha, with special reference to Dermoptera. In: Szalay FS, Novacek MJ, McKenna MC (eds) Mammal phylogeny: placentals. Springer-Verlag, New York, pp 129–150

Beard KC, Godinot M (1988) Carpal anatomy of *Smilodectes gracilis* (Adapiformes, Notharctinae) and its significance for lemuriform phylogeny. J Hum Evol 17:71–92

Beard KC, Dagosto M, Gebo DL, Godinot M (1988) Interrelationships among primate higher taxa. Nature 331:712–714

Bloch JI, Boyer DM (2002) Grasping primate origins. Science 298:1606–1610

Bloch JI, Boyer DM (2007) New skeletons of Paleocene-Eocene Plesiadapiformes: a diversity of arboreal positional behaviors in early primates. In: Dagosto M, Ravosa MJ (eds) Primate origins: adaptations and evolution. Springer, New York, pp 535–582

Bloch JI, Gingerich PD (1998) *Carpolestes simpsoni*, new species (Mammalia, Proprimates) from the late Paleocene of the Clarks Fork Basin, Wyoming. Contributions from the Museum of Paleontology University of Michigan 30:131–162

Bloch JI, Silcox MT, Boyer DM, Sargis EJ (2007) New Paleocene skeletons and the relationship of plesiadapiforms to crown-clade primates. Proc Natl Acad Sci U S A 104:1159–1164

Bloch JI, Randall Z, Silcox MT, Chester S, Boyer DM (2010) Distal phalanges of Eocene North American notharctines (Mammalia, Primates): implications from primate and anthropoid origins. J Vert Paleont 30(Suppl 2):60A (abstract)

Boyer DM (2009) New cranial and postcranial remains of late Paleocene Plesiadapidae ("Plesiadapiformes," Mammalia) from North America and Europe: description and evolutionary implications. Ph.D. dissertation, Stony Brook University

Boyer DM, Bloch JI (2008) Evaluating the mitten-gliding hypothesis for Paromomyidae and Micromomyidae (Mammalia, "Plesiadapiformes") using comparative functional morphology of new Paleogene skeletons. In: Sargis EJ, Dagosto M (eds) Mammalian evolutionary morphology. A tribute to Frederick S. Szalay. Springer, New York, pp 233–284

Boyer DM, Chester SGB, Bloch JI (2010a) New fossil evidence on the structure of the plesiadapiform hand skeleton: implications for the evolution of grasping in primates. Am J Phys Anthropol 141(Suppl 50):72 (abstract)

Boyer DM, Seiffert ER, Simons EL (2010b) Astragalar morphology of *Afradapis* a large adapiform primate from the earliest late Eocene of Egypt. Am J Phys Anthropol 143:383–402

Boyer DM, Scott CS, Fox RC (2012) New craniodental material of *Pronothodectes gaoi* Fox (Mammalia, "Plesiadapiformes") and relationships among members of Plesiadapidae. Am J Phys Anthropol 147:511–550

Boyer DM, Yapuncich GS, Chester SGB, Bloch JI, Godinot M (2013a) Hands of early primates. Yearb Phys Anthropol 57:33–78

Boyer DM, Seiffert ER, Gladman JT, Bloch JI (2013b) Evolution and allometry of calcaneal elongation in living and extinct primates. PLoS One 8:e67792

Cartmill M (1972) Arboreal adaptations and the origin of the order Primates. In: Tuttle R (ed) The functional and evolutionary biology of primates. Aldine, Chicago, pp 97–122

Cartmill M (1974a) Pads and claws in arboreal locomotion. In: Jenkins FA Jr (ed) Primate locomotion. Academic Press, New York, pp 45–83

Cartmill M (1974b) Rethinking primate origins. Science 184:436–443

Cartmill M (1985) Climbing. In: Hildebrand M, Bramble D, Liem K, Wake D (eds) Functional vertebrate morphology. Harvard University Press, Cambridge, pp 73–88

Cartmill M (1992) Primate origins, human origins, and the end of higher taxa. Evol Anthropol 21:208–220

Chester SGB, Bloch JI, Boyer DM, Clemens WA (2015) Oldest known euarchontan postcrania and affinities of Paleocene *Purgatorius* to Primates. Proc Natl Acad Sci U S A 112:1487–1492

Covert HH (1985a) Adaptations and evolutionary relationships of the Eocene primate family Notharctidae. Ph.D. dissertation, Duke University

Covert HH (1985a) The skeleton of *Smilodectes gracilis*. Am J Phys Anthropol 66:159 (abstract)

Covert HH (1986) Biology of early Cenozoic primates. In: Swindler DR, Erwin J (eds) Comparative primate biology. Alan R. Liss, New York, pp 335–359

Dagosto M (1983) Postcranium of *Adapis parisiensis* and *Leptadapis magnus* (Adapiformes, Primates). Folia Primatol 41:49–101

Dagosto M (1988) Implications of postcranial evidence for the origin of euprimates. J Hum Evol 17:35–56

Dagosto M (1990) Models for the origin of the anthropoid postcranium. J Hum Evol 19:121–139

Deane AS, Begun DR (2008) Broken fingers: retesting locomotor hypotheses for fossil hominoids using fragmentary proximal phalanges and high-resolution polynomial curve fitting (HR-PCF). J Hum Evol 55:691–701

Etter HF (1973) Terrestrial adaptations in the hands of cercopithecinae. Folia Primatol 20:331–350

Etter HF (1974) Morphologisch- und metrisch-vergleichende untersuchung am handskelet rezenter Primaten. Gegenbaurs Morphol Jahrb 120:1–484

Etter HF (1978) Lorisiform hands and their phylogenetic implications: a preliminary report. In: Chivers DH, Joysey KA (eds) Recent advances in primatology. Academic Press, London, pp 161–170

Franzen JL (1987) Ein neuer Primate aus dem Mitteleozän der Grube Messel (Deutschland, S-Hessen). Cour Forsch-Inst Senckenberg 91:151–187

Franzen JL (1988) Ein weiterer Primatenfund aus der Grube Messel bei Darmstadt. Cour Forsch-Inst Senckenberg 107:275–289

Franzen JL (1993) The oldest primate hands: additional remarks and observations. In: Preuschoft H, Chivers DJ (eds) Hands of primates. Springer-Verlag, Vienna, pp 379–386

Franzen JL (2000) *Europolemur kelleri* n. sp. von Messel und ein Nachtrag zu *Europolemur koenigswaldi* (Mammalia, Primates, Notharctidae, Cercamoniinae). Senckenbergiana lethaea 80:275–287

Franzen JL, Frey CJ (1993) *Europolemur* completed, Kaupia. Darmstädter Beiträge Naturgeschichte 3:113–130

Franzen JL, Gingerich PD, Habersetzer J, Hurum JH, von Koenigswald W, Smith BH (2009) Complete primate skeleton from the middle Eocene of Messel in Germany: morphology and paleobiology. PLoS One 4:e5723

Gebo DL (2002) Adapiformes: phylogeny and adaptation. In: Hartwig WC (ed) The primate fossil record. Cambridge University Press, Cambridge, pp 21–82

Gebo DL, Dagosto M, Rose KD (1991) Foot morphology and evolution in early Eocene *Cantius*. Am J Phys Anthropol 86:51–73

Gebo DL, Smith T, Dagosto M (2012) New postcranial elements for the earliest Eocene fossil primate *Teilhardina belgica*. J Hum Evol 65:305–218

Gebo DL, Smith R, Dagosto M, Smith T (2015) Additional postcranial elements of *Teilhardina belgica*, the oldest European primate. Am J Phys Anthropol 156:388–406

Gingerich PD (1976) Cranial anatomy and evolution of Early Tertiary Plesiadapidae (Mammalia, Primates). University of Michigan Papers on Paleontology 15:1–141

Gingerich PD (1986) Early Eocene *Cantius torresi*—oldest primate of modern aspect from North America. Nature 319:319–321

Gingerich PD (2012) Primates in the Eocene. Palaeobiodiv Palaeoenviron 92:649–663

Gingerich PD, Houde P, Krause DW (1983) A new Earliest Tiffanian (Late Paleocene) mammalian fauna from Bangtail Plateau, western Crazy Mountain Basin, Montana. J Paleontol 57:957–970

Gingerich PD, Franzen JL, Habersetzer J, Hurum JH, Smith BH (2010) *Darwinius masillae* is a haplorhine—reply to Williams et al. (2010). J Hum Evol 59:574–579

Gladman JT, Boyer DM, Seiffert ER, Simons EL (2013) Morphology and phenetic affinities of a calcaneus attributable to the primitive late Eocene anthropoid *Proteopithecus sylviae*. Am J Phys Anthropol 151:372–397

Godinot M (1991) Approches fonctionnelles des mains de primates paléogenes. Geobios 13:161–173

Godinot M (1992) Early euprimate hands in evolutionary perspective. J Hum Evol 22:267–283

Godinot M (1998) A summary of adapiform systematics and phylogeny. Folia Primatol 69:218–249

Godinot M (2007) Primate origins: a reappraisal of historical data favoring tupaiid affinities. In: Dagosto M, Ravosa MJ (eds) Primate origins: adaptations and evolution. Springer, New York, pp 83–142

Godinot M (2015) Fossil record of primates from the Paleocene to the Oligocene. In: Henke W, Tattersall I (eds) Handbook of paleoanthropology. Springer, Berlin, pp 1137–1259

Godinot M, Beard KC (1991) Fossil primate hands: a review and an evolutionary inquiry emphasizing early forms. Hum Evol 6:307–354

Godinot M, Beard KC (1993) A survey of fossil primate hands. In: Preuschoft H, Chivers DJ (eds) Hands of primates. Springer-Verlag, Vienna, pp 335–377

Godinot M, Jouffroy FK (1984) La main d'*Adapis* (Primate, Adapidé). In: Buffetaut E, Mazin JM, Salmion E (eds) Actes du symposium paléontologique Georges Cuvier. Montbéliard, Paris, pp 221–242

Gregory WK (1920) On the structure and relations of *Notharctus*, an American Eocene primate. Mem Am Mus Nat Hist 3:45–243

Gunnell GF, Silcox MT (2010) Primate origins: the early Cenozoic fossil record. In: Larsen CS (ed) A companion to biological anthropology. Wiley-Blackwell, Chichester, pp 275–289

Haines RW (1955) The anatomy of the hand of certain insectivores. Proc Zool Soc Lond 125:761–776

Hamrick MW (1995) Carpal joint morphology and function in the strepsirhine primates. Ph.D. dissertation, Northwestern University

Hamrick MW (1996a) Articular size and curvature as determinants of carpal joint mobility and stability in strepsirhine primates. J Morphol 230:113–127

Hamrick MW (1996b) Functional morphology of the lemuriform wrist joints and the relationship between wrist morphology and positional behavior in arboreal primates. Am J Phys Anthropol 99:319–344

Hamrick MW (1996c) Locomotor adaptations reflected in the wrist joints of early Tertiary primates (Adapiformes). Am J Phys Anthropol 100:585–604

Hamrick MW (1997) Functional osteology of the primate carpus with special reference to Strepsirhini. Am J Phys Anthropol 104:105–116

Hamrick MW (1999) First carpals of the Eocene primates family Omomyidae. Contributions from the Museum of Paleontology, University of Michigan 30:191–198

Hamrick MW (2001) Primate origins: evolutionary change in digit ray patterning and segmentation. J Hum Evol 40:339–351

Hamrick MW (2007) Evolvability, limb morphology, and primate origins. In: Dagosto M, Ravosa MJ (eds) Primate origins: adaptations and evolution. Springer, New York, pp 381–401

Hamrick MW, Alexander JP (1996) The hand skeleton of *Notharctus tenebrosus* (Primates, Notharctidae) and its significance for the origin of the primate hand. Am Mus Novit 3182:1–20

Hamrick MW, Meldrum DJ, and Simons EL (1995) Anthropoid phalanges from the Oligocene of Egypt. J Hum Evol 28:121–145.

Hamrick MW, Rosenman BA, Brush JA (1999) Phalangeal morphology of the Paromomyidae (?Primates, Plesiadapiformes): the evidence for gliding behavior reconsidered. Am J Phys Anthropol 109:397–413

Hoffstetter R (1977) Phylogénie des primates. Confrontation des résultats obtenus par les diverses voies d'approche du problème. Bulletins et mémoires de la Société d'Anthropologie de Paris t. 4, Série XIII, 327–346

Janečka JE, Miller W, Pringle TH, Wiens F, Zitzmann A, Helgen KM, Springer MS, Murphy WJ (2007) Molecular and genomic data identify the closest living relative of primates. Science 318:792–794

Jouffroy FK, Lessertisseur J (1959) La main des lémuriens malgaches comparée à celle des autres primates. Mémoires de l'Institut Scientifique de Madagascar Série A, Tome XIII, 195–219

Jouffroy FK, Godinot M, Nakano Y (1991) Biometrical characteristics of primate hands. Hum Evol 6:269–306

Jungers WL, Godfrey LR, Simons EL, Chatrath PS (1997) Phalangeal curvature and positional behavior in extinct sloth lemurs (Primates, Palaeopropithecidae). Proc Natl Acad Sci U S A 94:11998–12001

Kay RF, Ross C, Williams BA (1997) Anthropoid origins. Science 275:797–804

Kay RF, Williams BA, Ross CF, Takai M, Shigehara N (2004) Anthropoid origins—a phylogenetic analysis. In: Ross CF, Kay RF (eds) Anthropoid origins: new visions. Kluwer Academic/ Plenum Publishers, New York, pp 91–135

Kirk EC, Lemelin P, Hamrick MW, Boyer DM, Bloch JI (2008) Intrinsic hand proportions of euarchontans and other mammals: implications for the locomotor behavior of plesiadapiforms. J Hum Evol 55:278–299

Krause DW (1991) Were paromomyids gliders? Maybe, maybe not. J Hum Evol 21:177–188

Le Gros Clark WE (1971) The antecedents of man. Quadrangle Books, New York

Lemelin P (1996) The evolution of manual prehensility in primates: a comparative study of prosimians and didelphid marsupials. Ph.D. dissertation, State University of New York at Stony Brook

Lemelin P (1999) Morphological correlates of substrate use in didelphid marsupials: implications for primate origins. J Zool 247:165–175

Lemelin P, Jungers WL (2007) Body size and scaling of the hands and feet of prosimian primates. Am J Phys Anthropol 133:828–840

Maiolino S, Boyer DM, Bloch JI, Gilbert CS, Groenke J (2012) Evidence for a grooming claw in a North American adapiform primate: implications for anthropoid origins. PLoS One 7:e29136

Maiolino SA (2015) Comparative morphology of primate distal phalanges: implications for early primate evolution and the origins of the primate nail. Ph.D. dissertation, Stony Brook University

Martin RD (1986) Primates: a definition. In: Wood B, Martin L, Andrews P (eds) Major topics in primate and human evolution. Cambridge University Press, Cambridge, pp 1–31

Napier JR (1960) Studies of the hands of living primates. Proc Zool Soc Lond 134:647–656

Napier JR (1961) Prehensility and opposability in the hands of primates. Symp Zool Soc Lond 5:115–132

Napier JR, Walker AC (1967) Vertical clinging and leaping—a newly recognized category of locomotor behaviour of primates. Folia Primatol 6:204–219

Ni X, Gebo DL, Dagosto M, Meng J, Tafforeau P, Flynn JJ, Beard KC (2013) The oldest known primate skeleton and early haplorhine evolution. Nature 498:60–64

Niemitz C (1984) Synecological relationships and feeding behavior of the genus *tarsius*. In: Niemitz C (ed) Biology of tarsiers. Gustav Fischer Verlag, Stuttgart, pp 59–76

Patel BA, Seiffert ER, Boyer DM, Jacobs RL, St Clair EM, Simons EL (2012) New primate first metatarsals from the Paleogene of Egypt and the origin of the anthropoid big toe. J Hum Evol 63:99–120

Reghem E, Byron C, Bels V, Pouydebat E (2012) Hand posture in the grey mouse lemur during arboreal locomotion on narrow branches. J Zool 288:76–81

Richmond BG (2007) Biomechanics of phalangeal curvature. J Hum Evol 53:678–690

Rose KD, Rana RS, Sahni A, Kumar K, Missiaen P, Singh L, Smith T (2009) Early Eocene primates from Gujarat, India. J Hum Evol 56:366–404

Rose KD, Chester SGB, Dunn RH, Boyer DM, Chew AE, Bloch JI (2011) New fossils of the oldest North American euprimate *Teilhardina brandti* (Omomyidae) from the Paleocene-Eocene Thermal Maximum. Am J Phys Anthropol 146:281–305

Runestad JA, Ruff CB (1995) Structure adaptations for gliding in mammals with implications for locomotor behavior in paromomyids. Am J Phys Anthropol 98:101–119

Russell DE (1967) Sur *Menatotherium* et l'âge Paléocène du gisement de Menat (Puy-de-Dôme). Problèmes actuels de paléontologie. Évolution des vertébrés. Colloques internationaux du C.N.R.S. 163, 483–490

Sargis EJ (2002) Functional morphology of the forelimb of tupaiids (Mammalia, Scandentia) and its phylogenetic implications. J Morphol 253:10–42

Seiffert ER, Perry JMG, Simons EL, Boyer DM (2009) Convergent evolution of anthropoid-like adaptations in Eocene adapiform primates. Nature 461:1118–1121

Silcox MT, Sargis EJ, Bloch JI, Boyer DM (2007) Primate origins and supraordinal relationships: morphological evidence. In: Henke W, Tattersall I (eds) Handbook of paleoanthropology, vol 2, Primate evolution and human origins. Springer-Verlag, Berlin, pp 831–859

Simpson GG (1935) The Tiffany fauna, Upper Paleocene II. Structure and relationships of *Plesiadapis*. Am Mus Novit 816:1–30

Smith T, Rose KD, Gingerich PD (2006) Rapid Asia–Europe–North America geographic dispersal of earliest Eocene primate *Teilhardina* during the Paleocene-Eocene Thermal Maximum. Proc Natl Acad Sci U S A 103:11223–11227

Soligo C, Martin RD (2006) Adaptive origins of primates revisited. J Hum Evol 50:414–430

Soligo C, Müller AE (1999) Nails and claws in primate evolution. J Hum Evol 36:97–114

Stafford BJ, Thorington RW Jr (1998) Carpal development and morphology in archontan mammals. J Morphol 235:135–155

Stern JT Jr, Jungers WL, Susman RL (1995) Quantifying phalangeal curvature: an empirical comparison of alternative methods. Am J Phys Anthropol 97:1–10

Szalay FS, Dagosto M (1980) Locomotor adaptations as reflected on the humerus of Paleogene primates. Folia Primatol 34:1–45

Szalay FS, Decker RL (1974) Origins, evolution, and function of the tarsus in late Cretaceous Eutheria and Paleocene primates. In: Jenkins FA Jr (ed) Primate locomotion. Academic Press, New York, pp 239–259

Szalay FS, Delson E (1979) Evolutionary history of the primates. Academic Press, New York

Szalay FS, Drawhorn G (1980) Evolution and diversification of the Archonta in an arboreal milieu. In: Luckett WP (ed) Comparative biology and evolutionary relationships of tree shrews. Plenum Press, New York, pp 133–169

Szalay FS, Rosenberger AL, Dagosto M (1987) Diagnosis and differentiation of the order Primates. Yearb Phys Anthropol 30:75–105

Thalmann U (1994) Die Primaten aus dem eozänen Geiseltal bei Halle/Saale (Deutschland). Cour Forsch-Inst Senckenberg 175:121–123

Thalmann U, Haubold H, Martin RD (1989) *Pronycticebus neglectus*—an almost complete adapid primate specimen from the Geiseltal (GDR). Palaeovertebrata 19:115–130

Wible JR, Covert HH (1987) Primates: cladistic diagnosis and relationships. J Hum Evol 16:1–22

Williams BA, Kay RF, Kirk EC, Ross CF (2010) *Darwinius masillae* is a strepsirrhine—a reply to Franzen et al. (2009). J Hum Evol 59:567–573

Wood Jones F (1916) Arboreal man. Edward Arnold, London

Youlatos D, Godinot M (2004) Locomotor adaptations of *Plesiadapis tricuspidens* and *Plesiadapis n.* sp. (Mammalia, Plesiadapiformes) as reflected in selected parts of the postcranium. J Anthropol Sci 82:103–118

Chapter 15
The Hands of Subfossil Lemurs

Laurie R. Godfrey, Michael C. Granatosky, and William L. Jungers

1 Introduction

The "giant" extinct (subfossil) lemurs were endemic to Madagascar and persisted late into the Holocene (Burney and Ramilisonina 1999; Godfrey and Jungers 2003; Burney et al. 2004; Godfrey et al. 2008; Crowley 2010; Goodman and Jungers 2014). They belonged to three extinct families: the Archaeolemuridae (or "monkey lemurs"), the Palaeopropithecidae (or "sloth lemurs"), and the Megaladapidae (or "koala lemurs"). In addition, two still-extant families of lemurs have recently extinct, large-bodied representatives, the Daubentoniidae (aye-ayes) and Lemuridae. There was a large aye-aye (*Daubentonia robusta*) that was not very different from the living *Daubentonia madagascariensis* but at least three times the body size (Lamberton 1934; Simons 1994; Jungers et al. 2008). The large-bodied extinct lemurid (*Pachylemur*) was a close relative of the ruffed lemur, *Varecia*, but again around three times its body size and more robust (Jungers et al. 2008).

In this chapter we focus on the dominant modes of locomotion of the extinct lemurs and how they relate to variation in the morphology of their hands. We review each of the three families of extinct lemurs (Archaeolemuridae, Palaeopropithecidae, and Megaladapidae). We stress that, while variation in hand morphology clearly

L.R. Godfrey (✉)
Department of Anthropology, University of Massachusetts,
240 Hicks Way, Amherst 01003, MA, USA
e-mail: lgodfrey@anthro.umass.edu

M.C. Granatosky
Department of Evolutionary Anthropology, Duke University, Durham, NC, USA

W.L. Jungers
Department of Anatomical Sciences, Association Vahatra, Antananarivo, Madagascar,
Stony Brook University, Stony Brook, NY, USA

Association Vahatra, BP 3972, Antananarivo 101, Madagascar,
Stony Brook University, Stony Brook, NY, USA

T.L. Kivell et al. (eds.), *The Evolution of the Primate Hand*, Developments
in Primatology: Progress and Prospects, DOI 10.1007/978-1-4939-3646-5_15

relates to important ways hands are used in locomotion, it is also influenced by non-locomotor functions such as procuring food, grooming, or securing objects (including other individuals, in mating or fighting). Our main focus here is locomotion. First, however, we provide a general overview of their body size, life history, and phylogeny, focusing on shared characteristics as well as morphological heterogeneity and the general manner in which the families distinguish themselves.

2 Locomotion, Body Size, and Life History

None of the giant lemurs show adaptations for leaping. This can be ascertained from their hind-limb and vertebral anatomy. Some were highly suspensory, others were slow climbers, yet others were arboreal quadrupeds, and still others were highly terrestrial. All can be characterized as having had relatively short limbs for their body mass (Jungers et al. 2002; Godfrey et al. 2006a). Animals with relatively short forelimbs and hind limbs are biomechanically poorly adapted for speedy locomotion. Thus, on the basis of their basic postcranial morphology alone, we can infer that these animals were neither cursorial on the ground nor especially agile in the trees. In a word, they were all likely slow, some more so than others.

Several additional lines of evidence support the inference that the giant lemurs were slow. Walker et al. (2008) showed that subfossil lemurs have semicircular canals with small radii of curvature, as is typical of animals with poor locomotor agility (Spoor et al. 2007). Such a relationship derives from the role that the semicircular canals play in maintaining locomotor balance; the size of the radius of the semicircular canal is correlated with locomotor agility. Semicircular canal anatomy also indicates a range of variation among subfossil lemurs, with *Palaeopropithecus* and *Megaladapis* the least agile and *Archaeolemur* the most agile (but still not cursorial) (Walker et al. 2008).

Histological sections of the teeth of subfossil lemurs may also provide information regarding activity levels of extinct lemurs; among primates in general, the number of cross striations (Retzius periodicity) between "long period" lines (or striae of Retzius) in the enamel is correlated with body size, brain size, the radius of the semicircular canals, and other variables related to energy consumption and life history (Bromage et al. 2012; Hogg et al. 2015). Giant lemurs are peculiar in that they have low values for Retzius periodicities. With the exception of the archaeolemurids (*Archaeolemur* and *Hadropithecus*), which have Retzius periodicities of four, subfossil lemurs have Retzius periodicities of two or three. These values, even those for the archaeolemurids, are considerably lower than those of anthropoids of comparable body size. For example, the chimpanzee-sized *Palaeopropithecus* has a Retzius periodicity of two, while the chimpanzee has a periodicity of six; female gorilla-sized *Megaladapis* has a Retzius periodicity of three, while orangutans and gorillas have periodicities of nine to ten (Hogg et al. 2015). Low Retzius periodicity is correlated with small brain size and may also reflect hypometabolism in the giant lemurs. This, and the low values for the radius of curvature of the semicircular canals, may relate to the energy constraints experienced by the giant lemurs in their unpredictable habitats.

Finally, of course, the "giant" extinct lemurs of Madagascar were all large in body size; the smallest was larger than the largest of the still-extant lemurs. The largest of the subfossil lemurs was *Archaeoindris*, a palaeopropithecid about which we know little, rivaled adult male gorillas in body mass (~160 kg, Jungers et al. 2008). It has some peculiar features of its long bones that suggest that it climbed trees, but, because of its enormous size and the fairly open habitat in which it lived, we can surmise that this lemur spent most of its time on the ground and may have been the most terrestrial of the subfossil lemurs. Other giant lemurs likely descended regularly to the ground. Indeed, all of the extinct lemurs, no matter how specialized for arboreality, likely spent some time on the ground, if only to get from tree to tree or to a pool of drinking water. Direct evidence for at least some ground locomotion for giant lemurs exists, in that, in virtually all regions of Madagascar for which we have subfossil sites, giant lemurs fell victim to crocodiles (Meador and Godfrey 2014). This included highly specialized suspensory species such as *Palaeopropithecus*, which would have been especially vulnerable when moving terrestrially. Score marks on the bones of the prey provide evidence of the crocodile's "death roll" and prove that the lemurs were not merely scavenged on the ground after death.

3 Phylogeny and the Heterogeneity of Hand Morphology in Subfossil Lemurs

3.1 Phylogenetic Relationships

Relationships of extinct lemur families (Megaladapidae, Archaeolemuridae, and Palaeopropithecidae) to extant lemur families (Cheirogaleidae, Lepilemuridae, Indriidae, Lemuridae, and Daubentoniidae) have been reconstructed on the basis of both morphological and molecular data. To a large extent, morphological and genetic data support similar phylogenetic topologies. Thus, for example, there is no question that palaeopropithecids and indriids are sister taxa, despite large differences in body size and their hind-limb morphology. These taxa share a host of craniodental and developmental traits (Tattersall and Schwartz 1974; Schwartz and Tattersall 1985; Schwartz et al. 2002; Godfrey and Jungers 2003), but differ dramatically in postcranial characteristics. Indriids (including *Indri*, *Propithecus*, and *Avahi*) have specialized leaping anatomy, while *Mesopropithecus*, *Babakotia*, and *Palaeopropithecus* all have, to varying degrees, specialized suspensory adaptations (Godfrey et al. 1995, 2008; Jungers et al. 1997; Hamrick et al. 2000). Monophyly for the palaeopropithecids and indriids is supported by molecular data (Yoder et al. 1999; Karanth et al. 2005; Orlando et al. 2008; Kistler et al. 2015). It is noteworthy that the larger-bodied indriids, despite being skilled leapers, also regularly adopt suspensory postures.

Molecular support for monophyly exists for the Archaeolemuridae (*Archaeolemur* and *Hadropithecus*) (Orlando et al. 2008), which share numerous postcranial and craniodental traits (Tattersall 1973; Godfrey 1988). Certain dental traits link the Archaeolemuridae to the palaeopropithecid/indriid clade (e.g., the loss of the lower canines, aspects of molar shape), but these animals differ tremendously in their postcranial anatomy. Orlando et al. (2008) demonstrated molecular support for the traditional position that the archaeolemurids are sister taxa to the palaeopropithecids and indriids, and this has been affirmed by Kistler et al. (2015).

The relationship of the daubentoniids to other lemurs has provoked vigorous debate, as has the relationship of the Megaladapidae to other lemurs. Molecular data support a distant relationship of aye-ayes to other lemurs; *Daubentonia* is phylogenetically basal to all other Malagasy lemurs, extinct and extant (Kistler et al. 2015). With regard to morphology, *Daubentonia* is also the most autapomorphic of all lemurs (Soligo 2005). This greatly complicates any attempt to use them as "primitive" morphotypes despite their early branching. The "giant" aye-aye (*D. robusta*) was much larger and more robust than its extant congener (*D. madagascariensis*); differences in limb proportions also distinguish the two (Jungers 1980). In other ways, including specializations of the hand, the two were quite alike (Lamberton 1934), confirming similar (and very peculiar) feeding adaptations. As best as we can tell, *D. robusta* competitively excluded *D. madagascariensis* from the southwest sector of the island, and there is no evidence that the latter has invaded this region since the demise of the former.

Megaladapis had been reconstructed as the sister taxon to the sportive lemurs, or Lepilemuridae, on the basis of certain craniodental characteristics (including the molar morphology, the loss of the upper permanent incisors, and the development of an expanded facet on the neck of the mandibular condyle that articulates with the postglenoid process) (e.g., Schwartz and Tattersall 1985; Wall 1997). This sister taxon relationship was initially supported by molecular data (Montagnon et al. 2001), but this result turned out to be a likely product of contamination. New molecular data (Karanth et al. 2005; Orlando et al. 2008; Kistler et al. 2015) have consistently supported a sister taxon relationship for the Megaladapidae and the Lemuridae, which is supported by a limited number of morphological traits (including some features of the carpus, as described below, and dental eruption sequences; see Schwartz et al. 2005). Kistler et al. (2015) were able to sequence the complete mitochondrial genome for *Megaladapis edwardsi*, which affirmed its sister taxon relationship to the Lemuridae.

Kistler et al. (2015) also provided the first direct evidence of the time since divergence of all three extinct lemur families from their respective sister taxa. According to this study, the Megaladapidae, Lemuridae, Archaeolemuridae, Palaeopropithecidae, Indriidae, Cheirogaleidae, and Lepilemuridae belong to a monophyletic clade that began diversifying around 31 million years ago (Ma), a good 20 Ma after the initial split of the lemurs into two lineages: an aye-aye lineage and all the rest. If accurately timed, the diversification at ~31 Ma would have followed a precipitous drop in temperature at the Eocene/Oligocene boundary (the "Grande Coupure"). First, according to Kistler et al. (2015), a branch ancestral to a megaladapid-lemurid clade diverged from a branch ancestral to the remaining lemur families. The Megaladapidae and Lemuridae themselves split at around 27 Ma, while, almost simultaneously, a branch ancestral to the extant cheirogaleids and lepilemurids separated from an archaeolemurid/palaeopropithecid/indriid clade. At around 24 Ma, the Archaeolemuridae diverged from the palaeopropithecid/indriid clade. Finally, at around 21 Ma, the palaeopropithecids and indriids diverged (Kistler et al. 2015).

We can now ask: does phylogeny help us to understand variation in the hand morphology of subfossil lemurs? The answer is yes. There are some traits that link closely related lemur families, but these are relatively few, and there is remarkable heterogeneity across closely related families and even within families. We will return to this question after reviewing variation in basic hand proportions and simple traits and then summarizing key characteristics of the hands of archaeolemurids, palaeopropithecids, and megaladapids.

3.2 *Relative Hand Size*

Relative hand length (i.e., hand length ÷ [humerus + radius + hand length]) and manual digit proportions are extremely variable among lemurs; indeed, when the subfossils and extant lemurs are included, the range for relative hand length index values surpasses that of most other primates (Fig. 15.1, Table 15.1). Most living lemurs

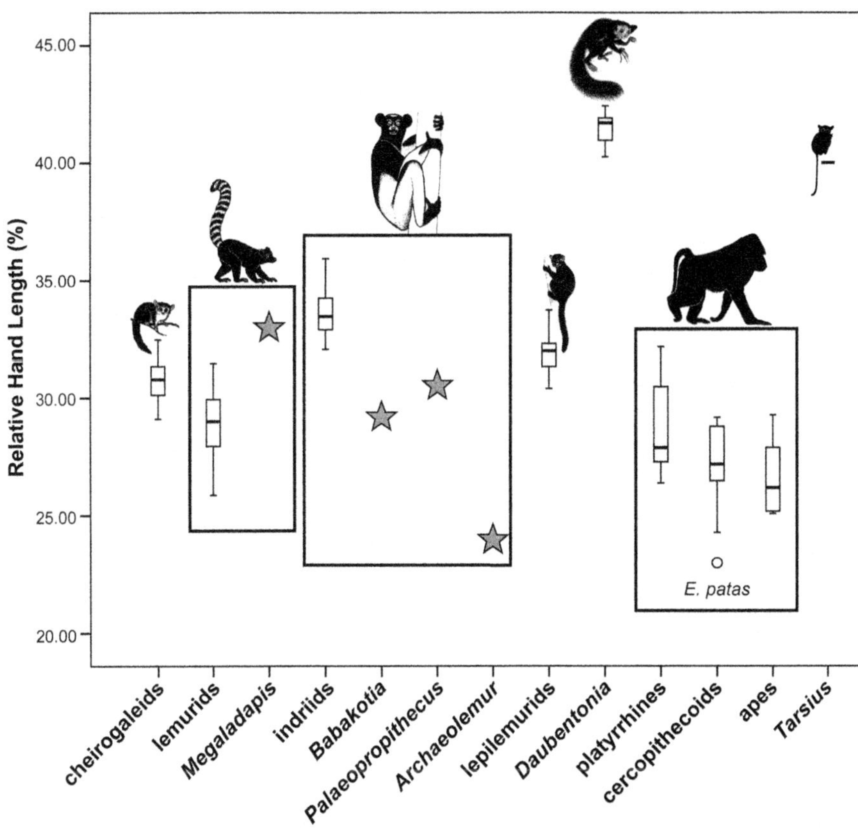

Fig. 15.1 Variation in relative hand length (100 × hand length ÷ [humerus + radius + hand length]) in subfossil lemurs and other primates. Drawings by Luci Betti-Nash

Table 15.1 Relative hand length indices (100 × hand length ÷ [hand + humerus + radius length]) in subfossil lemurs and other primates[a]

Species or species group	*N*	Mean ± std. deviation
Archaeolemur sp. cf. *edwardsi*	1	24.0
Lesser and great apes	5	26.7 ± 1.8
Cercopithecoidea	14	27.1 ± 1.8
Platyrrhines	14	28.9 ± 2.0
Lemuridae	55	28.9 ± 1.4
Babakotia radofilai	1	29.2
Palaeopropithecus kelyus	1	30.5
Cheirogaleidae	23	30.8 ± 0.8
Lepilemuridae	22	32.0 ± 0.9
Megaladapis edwardsi	1	33.0
Indriidae	30	33.6 ± 1.0
Tarsius	1	40.0
Daubentonia madagascariensis	9	41.5 ± 0.7
Total	177	30.8 ± 3.6

[a]For all strepsirrhines, *N* is the number of individuals in each group (data from Jungers 1976; Jungers et al. 2005; Jungers unpublished). For all haplorhines, *N* is the number of genera in each group (data from Jouffroy et al. 1991; Napier and Napier 1967)

have relatively long hands (see also Chap. 4). Aye-ayes have extremely long hands; this was likely the case for the giant extinct aye-aye (whose known hand bones are quite similar to those of living aye-ayes; Lamberton 1934). The exceptionally long digits of *Daubentonia* relate to the special manner in which they use their digits in extractive foraging. Only tarsiers, among haplorhines, have hands approaching the relative length of those of aye-ayes (Jouffroy and Lessertisseur 1979; Jouffroy et al. 1991; Lemelin and Jungers 2007), and they too use their hands in a specialized fashion in procuring food. Thus, the remarkable morphological heterogeneity of hand morphology in lemurs reflects their foraging and locomotor diversity.

Suspensory species among the extinct lemurs do not have the relatively longest hands (when calculated as a percentage of total forelimb length). This is because, as might be expected, they also have long humeri and radii. The same is true of suspensory haplorhines, including spider monkeys, gibbons, siamangs, and chimpanzees. Among extant lemurs, indriids and lepilemurids (both vertical clingers and leapers, or VCLs) have hands that are long relative to their forelimbs, perhaps for a variety of reasons (Napier and Walker 1967). Small-bodied VCLs such as the lepilemurids secure their grips on relatively large trunks. The larger-bodied indriids (*Indri* and *Propithecus*) engage in suspensory postures, sometimes arm swinging for short distances (e.g., Roberts and Davidson 1975; LRG and WLJ pers. obs.) (Fig. 15.2). Indriids and lepilemurids use hind limbs for propulsion and landing, and have relatively short forelimbs, so their elongated fingers are relatively longer than those of the more fully suspensory palaeopropithecids (including the most specialized *Palaeopropithecus*). Palaeopropithecids have moderate relative hand length

Fig. 15.2 Indriid *Propithecus diadema* in suspensory position at Andasibe, Madagascar (a). Arm swinging in *Propithecus coquereli*, drawn from video footage taken at Ankarafantsika in northwestern Madagascar. The whole sequence involves seven handholds and was taken at 18 frames/s. Frames shown are separated by about 1.4 s intervals. Reprinted from Godfrey (1977, p. 87). (**b**). Photo credit for (**a**): Travis Steffens. Video credit for (**b**): Paul Godfrey. Drawing by Laurie Godfrey

indices by virtue of the fact that their humeri and radii are elongated. The megaladapids, as slow climbers assuming vertical support postures, are like indriids in having relatively long hands (and feet; Wunderlich et al. 1996). In this manner they differ from their closest relatives, the lemurids.

At the other extreme, with extremely short hands relative to forelimb length, are the Archaeolemuridae. *Archaeolemur* had the relatively shortest hands of all strepsirrhines, rivaling highly terrestrial patas monkeys (Jungers et al. 2005). Shortened digits occur in highly terrestrial animals, not merely in primates but many orders of mammals. As is discussed below, the archaeolemurids are believed to have spent considerable time on the ground.

3.3 Distal Phalangeal Morphology

Variation in structures such as the distal volar pads and the associated distal phalanges may also relate to locomotion. Expanded apical pads might increase the stability of nails and thus aid in increasing contact area with the substrate, such as in climbing with the hands grasping supports. But variation here, as with relative hand length, is confounded by other factors.

Two indices capture variation in distal phalangeal shape: the distal phalanx expansion index (100 × apical tuft breadth ÷ distal phalangeal base breadth) and the distal phalanx robusticity index (100 × apical tuft breadth ÷ phalangeal length). In general, apical tufts are more expanded in strepsirrhines than in haplorhines, but there is considerable overlap in index values (Mittra et al. 2007) (Fig. 15.3, Tables 15.2 and 15.3). There are some patterns, however. The two indices are positively correlated, and the mean values for strepsirrhines and haplorhines differ significantly in both. Both also significantly correlate (but negatively) across primates with the mass-adjusted humeral length index (humeral length ÷ the cube root of body mass). In other words, species whose humeri are short relative to body mass tend to have wide apical tufts and robust distal phalanges. In general, strepsirrhines have shorter, more robust humeri than haplorhines; they also tend to have more robust distal phalanges and expanded apical tufts (Fig. 15.3).

The extinct lemurs fit the pattern displayed by the smaller-bodied extant strepsirrhines in this respect. Even highly suspensory species (such as *Palaeopropithecus*) have humeri that are relatively short when adjusted for estimated body mass (Jungers et al. 2002; Godfrey et al. 2006a). The same is true of the femora. Basically, the limbs of all giant lemurs are relatively short and robust and their trunks are relatively large. Among the giant lemurs, the species with the relatively longest and most gracile humeri (*Palaeopropithecus* spp.) have the least robust distal phalanges and the least expanded apical tufts (converging in shape on *Pongo*), and the stocky and more terrestrial *Archaeolemur* has very robust distal phalanges with expanded apical tufts. It is tempting to interpret manual phalangeal robusticity as an indicator of the magnitude of biomechanical loads operating on the phalanges and apical tuft expansion as related to this. However, some extinct lemurs deviate from the expected relationships. The sloth lemur, *Babakotia*, has robust distal phalanges with expanded apical tufts, but moderately gracile humeri. *Megaladapis* has robust humeri, but its distal phalanges are similar to *Palaeopropithecus* in robusticity, and its apical tufts are only moderately expanded (Figs. 15.4, 15.5, and 15.6).

Another factor that might influence apical tuft function and dimensions is manual touch sensitivity (mechanoreception) in the fingertips. The primary motor cortex of the brain has been shown to be critical in sensorimotor integration and control (Sessle et al. 2005; see also Chap. 6). It plays a major role in the planning and execution of limb movements, but a relatively minor role in semiautomatic movements involved in locomotion (but see Vilensky and Larson 1989). It also plays a role in orofacial sensorimotor function, which has been shown to relate to diet (Muchlinski 2010; Muchlinski et al. 2011). It is possible that the use of the hands in tactile exploration of foods correlates with the use of the oral region for the same purpose and that both

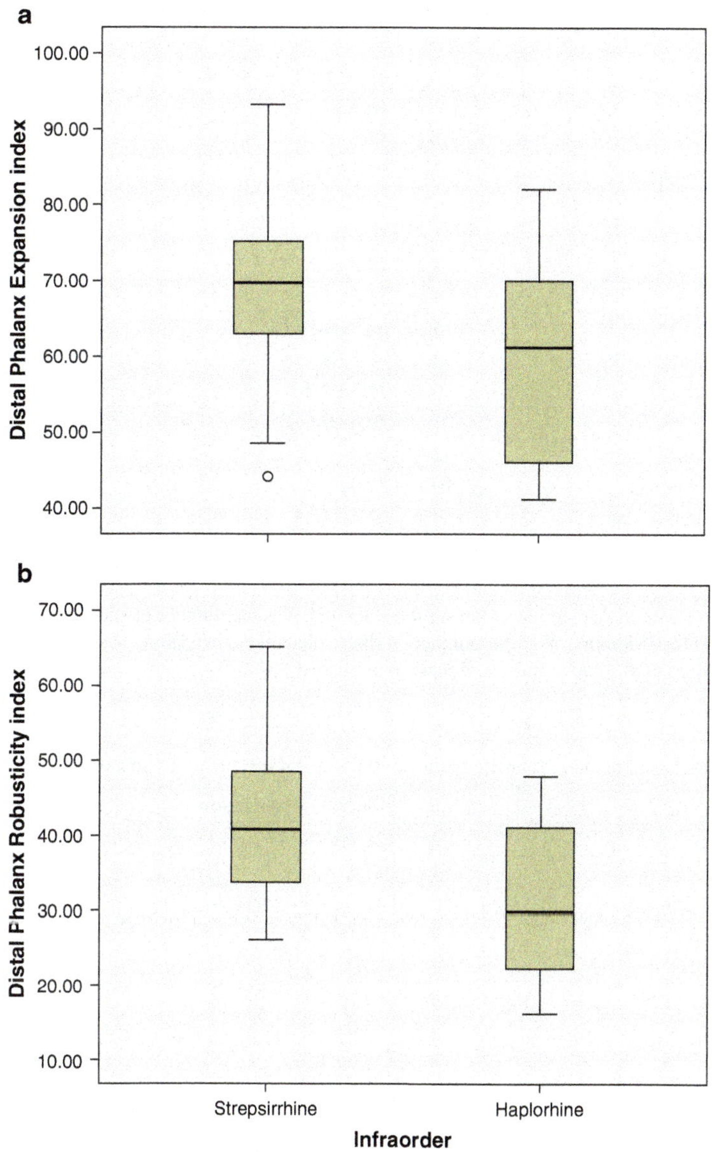

Fig. 15.3 Box and whisker plots showing differences between strepsirrhines and haplorhines in the distal phalanx expansion index (**a**) 100 × apical tuft breadth ÷ distal phalangeal base breadth) and distal phalanx robusticity index (**b**) 100 × apical tuft breadth ÷ phalangeal length). Strepsirrhines and haplorhines differ significantly in both. For the distal phalanx expansion index, the strepsirrhine mean is 68.8 ± 10.4, and the haplorhine mean is 59.5 ± 13.8; $t = 2.8$, df = 112, $p < 0.01$. For the distal phalanx robusticity index, the strepsirrhine mean is 41.7 ± 9.7, and the haplorhine mean is 31.3 ± 10.9; $t = 3.5$, df = 112, $p = 0.001$. Both indices are significantly negatively correlated with mass-adjusted humeral length. For the distal phalanx expansion index, $r = -0.40$, $N = 35$, $p < 0.05$. For the distal phalanx robusticity index, $r = -0.64$, $N = 35$, $p < 0.001$

Table 15.2 Distal phalanx expansion index (100 × apical tuft breadth ÷ distal phalangeal base breadth)[a]

Taxon	*N*	Mean ± std. deviation	Range
Atelidae	2	46.7 ± 7.9	41.2–52.3
Cebidae	1	42.5	42.5
Cercopithecidae	2	69.3 ± 7.9	63.8–74.9
Hominidae	4	60.0 ± 9.9	46.5–67.6
Hylobatidae	1	45.5	45.5
Tarsiidae	2	77.3 ± 6.8	72.5–82.1
Lorisidae	4	51.3 ± 5.1	44.2–55.7
Indriidae	4	66.1 ± 1.9	64.3–68.5
Cheirogaleidae	3	63.5 ± 8.1	56.6–72.5
Lemuridae	6	71.3 ± 2.3	68.2–74.3
Galagidae	7	59.0 ± 9.6	49.6–76.3
Lepilemuridae	1	66.7	66.7
Archaeolemuridae	10	81.0 ± 6.0	73.7–89.3
Palaeopropithecidae	37	67.7 ± 12.4	48.6–93.3
Megaladapidae	30	71.2 ± 4.3	63.9–81.5
Total	114	67.8 ± 11.1	41.2–93.3

[a]For all taxa with the exception of the subfossil lemurs, *N* is the number of species in each family. For the subfossil lemurs, *N* is the number of distal phalanges measured. *Sources*: Jungers et al. (2005), Mittra et al. (2007), and previously unpublished

Table 15.3 Distal phalanx robusticity index (100 × apical tuft breadth ÷ phalangeal length)[a]

Taxon	*N*	Mean ± std. deviation	Range
Atelidae	2	21.4 ± 7.4	16.1–26.6
Cebidae	1	17.4	17.4
Cercopithecidae	2	41.1 ± 1.4	40.2–42.1
Hominidae	4	30.1 ± 5.3	23.9–36.8
Hylobatidae	1	20.4	20.4
Tarsiidae	2	46.1 ± 2.7	44.2–48.0
Lorisidae	4	35.0 ± 7.3	26.2–42.7
Indriidae	4	38.6 ± 1.3	37.3–40.1
Cheirogaleidae	3	53.2 ± 5.6	49.4–59.6
Lemuridae	6	55.6 ± 4.3	48.4–59.6
Galagidae	7	41.3 ± 7.5	32.7–55.7
Lepilemuridae	1	45.1	45.1
Archaeolemuridae	10	47.1 ± 4.4	42.1–55.4
Palaeopropithecidae	37	42.5 ± 11.2	26.0–65.2
Megaladapidae	30	36.4 ± 6.3	27.3–47.4
Total	114	40.6 ± 10.3	16.1–65.2

[a]For all taxa with the exception of the subfossil lemurs, *N* is the number of species in each family. For the subfossil lemurs, *N* is the number of distal phalanges measured. *Sources*: Jungers et al. (2005), Mittra et al. (2007), and previously unpublished

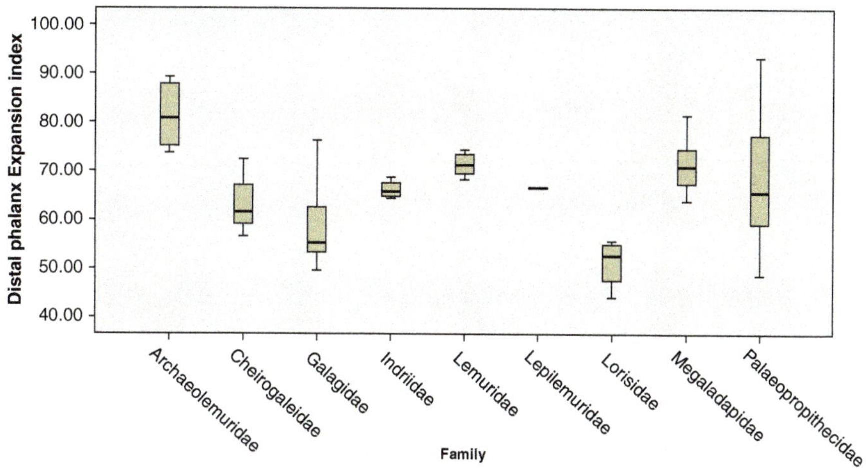

Fig. 15.4 Variation in the distal phalanx expansion index (100 × apical tuft breadth ÷ distal phalangeal base breadth) in strepsirrhines. Note the extreme variance in the Palaeopropithecidae resulting from *Babakotia* falling near the top of the range of variation in strepsirrhines and *Palaeopropithecus* falling near the bottom

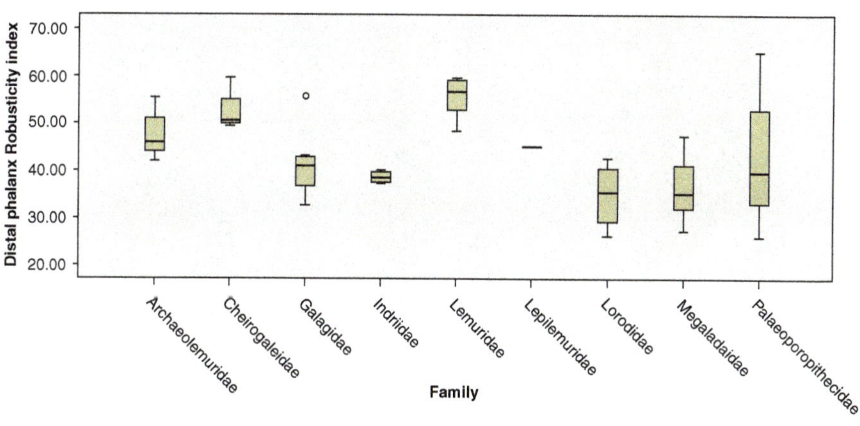

Fig. 15.5 Variation in the distal phalanx robusticity index in strepsirrhines. Note the extreme variance in the Palaeopropithecidae resulting from *Babakotia* falling at the top of the range of variation in strepsirrhines and *Palaeopropithecus* falling at the bottom

are tied to diet. There is some evidence that fingertip anatomy varies with diet in anthropoids; Hoffmann et al. (2004) demonstrate that Meissner corpuscle density in the fingertips of anthropoids increases with frugivory. To date, however, there has been no comprehensive study of how apical pad dimensions relate to mechanoreception in the fingertips of primates and to diet, if at all. Many complicating factors exist; species differ in the relative importance of tactile cues (either manual or facial) as opposed to visual, chemical, gustatory, acoustic, or cognitive assessment of potential food items (e.g., Dominy et al. 2004; Laska et al. 2007).

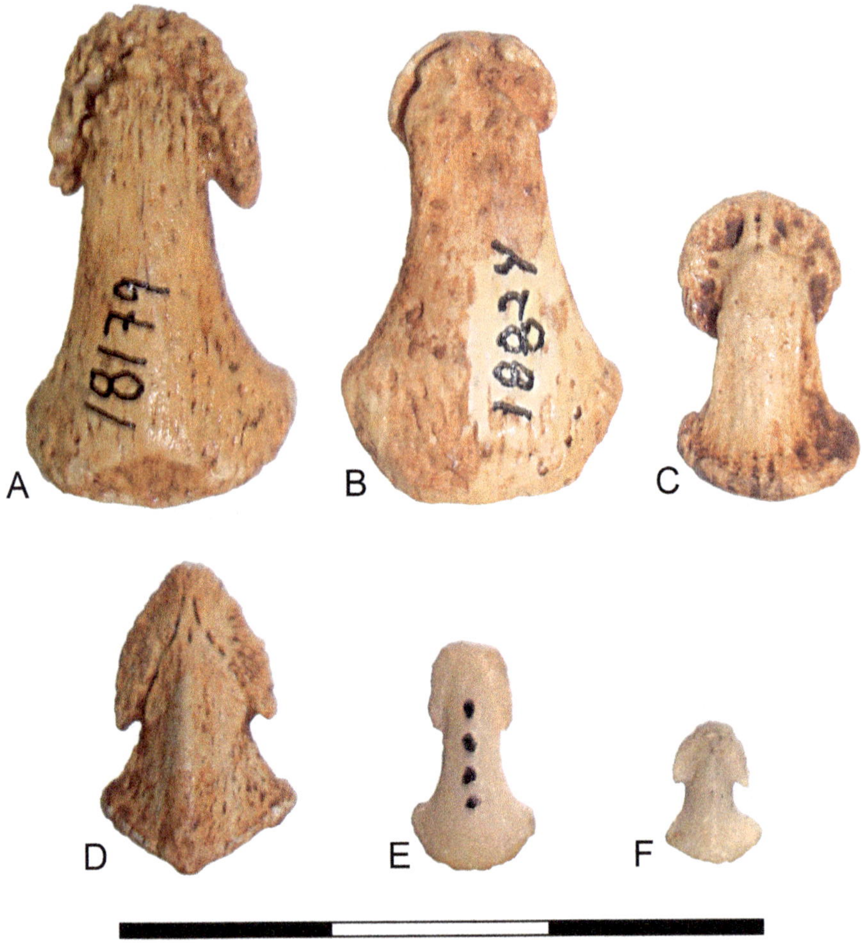

Fig. 15.6 Distal phalanges (dorsal view) of *Megaladapis madagascariensis* (**a**), *Palaeopropithecus ingens* (**b**), *Archaeolemur edwardsi* (**c**), *Babakotia radofilai* (**d**), *Indri indri* (**e**), and *Varecia variegata* (**f**). Photo credit: William Jungers

3.4 *Phalangeal Curvature*

Proximal phalangeal curvature (included angle; see Stern et al. 1995) does carry a locomotor signal. The degree of phalangeal curvature within the proximal phalanx is thought to be an accurate predictor of the importance and proportion of arboreality and suspensory positional behaviors (Stern and Susman 1983; Stern et al. 1995; Rein 2011; Congdon 2012). It has been proposed that within primates, greater phalangeal curvature of the proximal phalanx is a mechanism to mitigate high bending stresses involved in highly flexed joints used during below-branch locomotion (Richmond 2007). Among living primates, phalangeal curvature appears to be greatest in highly suspensory species, such as gibbons, orangutans, and spider monkeys compared with low curvature in pronograde quadrupedal species, with the most terrestrial species tending to show the least curvature (Jungers et al. 1997, 2005; Rein 2011; Rein et al. 2011; Congdon 2012). There is also a signal in the intermediate phalanges (and in the relationship between proximal and intermediate phalangeal curvature; Matarazzo 2008; see also Chap. 4), but this has not been studied in strepsirrhines.

Giant lemurs fall at both ends of spectrum for included angle of the proximal phalanges (Table 15.4). Sloth lemurs have high values, regardless of body size. *Archaeolemur* has straight proximal phalanges, but not as flat as those of baboons. The proximal phalanges of *Megaladapis* exhibit moderate curvature. In this character, phylogenetic signals, particularly for the archaeolemurid/palaeopropithecid/indriid clade, are weak. The palaeopropithecids differ markedly from the indriids and especially from the archaeolemurids. One might note that the megaladapids do not differ markedly from the single lemurid, *Varecia*, in our sample, but *Varecia* is one of the most suspensory of lemurids. The palaeopropithecids resemble suspensory atelids, orangutans, and siamangs in their included angles. *Archaeolemur* resembles most closely cercopithecoids such as mangabeys, vervets, and some colobines, but its proximal phalanges are not as straight as those of baboons.

3.5 Fusion of the Scaphoid and Os Centrale

Variation in the fusion of two carpal bones, the scaphoid and the os centrale, is of interest because it is extremely rare in primates, and because its functional significance remains unclear. Among primates, only hominoids and lemurs (including the giant extinct lemurs) show significant variation in this trait. The lemurs display a striking phylogenetic signal (Table 15.5), as fusion regularly occurs in only three of the eight families, including the extinct palaeopropithecids and their extant sister taxon, the indriids. This signal was not initially recognized (e.g., Hamrick et al. 2000) because some palaeopropithecids and indriids show mixed fusion, and because one of the extant lemurs (i.e., *Hapalemur*, a lemurid) that appears now to regularly display the unfused condition was initially thought to display scaphoid-centrale fusion (Kivell and Begun 2007). If indeed these two carpal bones are regularly unfused in *Hapalemur*, then scaphoid-centrale fusion is manifested among lemurs only in members of the palaeopropithecid/indriid clade, as stated above, and in the Lepilemuridae. *Lepilemur* is, like indriids, a vertical clinger and leaper; unlike indriids and palaeopropithecids, it rarely engages in suspensory postures or locomotion. In hominoids, scaphoid-centrale fusion is common in African apes (knuckle walkers) and humans, but not the more suspensory orangutans (Sarmiento 1988; Lewis 1989; Whitehead 1993; Kivell and Begun 2007). Sarmiento (1994) thought fusion might relate to increased loading of the radial side of the hand in vertical climbing and

Table 15.4 Proximal phalangeal curvature (included angle in degrees)[a]

Taxon	Mean ± std. deviation	*N*	Source
Papio anubis	14.7 ± 5.0	19	Rein (2011)
Chlorocebus aethiops	24.2 ± 7.7	20	Rein (2011)
Presbytis melalophos	25.4 ± 4.8	11	Rein (2011)
Archaeolemur edwardsi	27.9 ± 6.3	28	Jungers et al. (1997)
Lophocebus albigena	29.5 ± 10.1	6	Rein (2011)
Colobus guereza	30.1 ± 5.7	20	Rein (2011)
Cercopithecus mitis	31 ± 7.6	21	Rein (2011)
Trachypithecus obscurus	33.3 ± 7.3	20	Rein (2011)
Indri indri	35 ± 6	66	Jungers et al. (1997)
Propithecus diadema	35.8 ± 4.4	20	Jungers et al. (1997)
Sapajus apella	36 ± 3.9	15	Matarazzo (2008)
Alouatta seniculus	36.9 ± 6.3	15	Rein (2011)
Gorilla gorilla	37.2 ± 4.2	88	Jungers et al. (1997)
Nasalis larvatus	38 ± 5.6	12	Jungers et al. (1997)
Macaca nemestrina	39 ± 4.9	16	Matarazzo (2008)
Pan troglodytes	42.4 ± 4.8	63	Jungers et al. (1997)
Pan paniscus	44.8 ± 4.2	38	Jungers et al. (1997)
Macaca fascicularis	45 ± 6.6	20	Matarazzo (2008)
Hylobates lar	45.5 ± 4.8	20	Rein (2011)
Megaladapis madagascariensis	46.1 ± 8.1	23	Jungers et al. (1997)
Procolobus badius	46.1 ± 9.1	14	Rein (2011)
Hylobates spp.	47.8 ± 5.4	68	Jungers et al. (1997)
Megaladapis edwardsi	49.2 ± 7.8	18	Jungers et al. (1997)
Ateles geoffroyi	50.8 ± 8.3	7	Rein (2011)
Varecia variegata	51.6 ± 5.4	12	Jungers et al. (1997)
Symphalangus syndactylus	53.1 ± 5.6	20	Jungers et al. (1997)
Ateles sp.	55.2 ± 6.5	27	Jungers et al. (1997)
Palaeopropithecus maximus	57.2 ± 9.8	13	Jungers et al. (1997)
Babakotia radofilai	58.3 ± 6.3	27	Jungers et al. (1997)
Palaeopropithecus ingens	60.3 ± 9.9	45	Jungers et al. (1997)
Lagothrix sp.	62.4 ± 6	12	Jungers et al. (1997)
Pongo spp.	64.9 ± 6.6	88	Jungers et al. (1997)
Mesopropithecus dolichobrachion	65.8 ± 6.7	5	Jungers et al. (1997)
Palaeopropithecus kelyus	73.3 ± 7.3	11	Jungers et al. (1997)

[a]Extinct lemurs are bolded; for extant primates, sample is entirely manual; for extinct lemurs (with the exception of *Babakotia* for which samples are only manual), phalanges from the hand and foot are pooled.

quadrupedalism (i.e., knuckle-walking in hominoids), but quadrupedalism of any sort is rare or nonexistent in the lemurs showing scaphoid-centrale fusion. Indeed, a lack of scaphoid-centrale fusion is the norm for the quadrupedal lemurs, including species belonging to the Lemuridae, Megaladapidae, Cheirogaleidae, and Daubentoniidae, as well as the sister taxon to the palaeopropithecid/indriid

Table 15.5 Fusion of the scaphoid and os centrale in giant lemurs and their relatives

Genus	Family	Fusion of scaphoid and os centrale	Sources
Propithecus[a]	Indriidae	Mixed	Jouffroy (1975), Sarmiento (1985), Kivell and Begun (2007)
Indri	Indriidae	Fused	Kivell and Begun (2007)
Avahi	Indriidae	Fused	Kivell and Begun (2007)
Palaeopropithecus	Palaeopropithecidae	Mixed	Hamrick et al. (2000), Kivell and Begun (2007)
Babakotia	Palaeopropithecidae	Fused	Hamrick et al. (2000)
Archaeolemur	Archaeolemuridae	Unfused	Jungers et al. (2005)
Hadropithecus	Archaeolemuridae	Unfused	Lemelin et al. (2008)
Hapalemur[b]	Lemuridae	Unfused	Kivell and Begun (2007)
Varecia	Lemuridae	Unfused	Kivell and Begun (2007)
Lemur	Lemuridae	Unfused	Kivell and Begun (2007)
Megaladapis	Megaladapidae	Unfused	Hamrick et al. (2000)
Lepilemur	Lepilemuridae	Mixed	Sarmiento (1985)
Cheirogaleus	Cheirogaleidae	Unfused	Jouffroy (1975)
Microcebus	Cheirogaleidae	Unfused	Jouffroy (1975)
Daubentonia	Daubentoniidae	Unfused	Jouffroy (1975)
Galago[c]	Galagidae	Unfused	Kivell and Begun (2007), Yalden (1972)

[a]Sarmiento (1985) found fusion of the scaphoid and os centrale in some *Propithecus*

[b]Kivell and Begun (2007) found that, contrary to previous reports, the scaphoid and os centrale are "firmly bound," but not fused, in *Hapalemur*. If prior reports of fusion are accurate for some individuals, then this taxon would exhibit a mixed signal

[c]On the basis of a sample of 51 individuals, Kivell and Begun (2007) concluded that the scaphoid and os centrale are not fused in *Galago*, although rare fusion was reported by Yalden (1972)

clade, the Archaeolemuridae. Begun (2004) thought that scaphoid-centrale fusion might correlate with having a large and divergent thumb, which again is contradicted by the lemur evidence. There is no difference in mean relative thumb length ($100 \times Mc1 \div Mc4$) between extant lemurs with and without scaphoid-centrale fusion, and subfossil lemurs with at least some fusion (e.g., *Palaeopropithecus*), as well as others with no fusion (e.g., *Archaeolemur*, *Hadropithecus*), had very short or possibly vestigial thumbs (see below). Thus, scaphoid-centrale fusion does not predict exceptional thumb use in lemurs.

4 The Archaeolemuridae

The archaeolemurids, commonly referred to as "monkey lemurs," are represented by two genera: *Archaeolemur* (two species) and *Hadropithecus* (one species). Body mass estimates for monkey lemurs range from ~18 to ~35 kg (Jungers et al. 2008). First described shortly after discovery in the late 1890s as monkeys, the monkey lemurs were quickly recognized as lemurs with anthropoid convergences (see review by Godfrey and Jungers 2002). A papionin model was embraced first by Lamberton (1938) and later by Jouffroy (1963). Walker (1974) likened *Archaeolemur* to *Papio* and *Hadropithecus* to patas monkeys or gelada baboons; the latter analogue was preferred by Jolly (1970), mainly on the basis of its cranial anatomy. This view was endorsed by Tattersall (1973), but Godfrey (1988), pointing out ways in which the archaeolemurids differ from the more cursorial and terrestrial cercopithecoids, defended a mixed arboreal and terrestrial quadrupedal model for both *Archaeolemur* and *Hadropithecus*. She reconstructed these animals as short-limbed, stocky, deliberate, and non-cursorial. She also cautioned that the then-current hindlimb attributions for *Hadropithecus* may be incorrect. This turned out to be the case, as was proven when the first associated hind- and forelimb bones of *Hadropithecus* were discovered at Andrahomana Cave in 2003 (see Godfrey et al. 1997, 2006b).

The postcrania of *Archaeolemur* are very well known, including the hand (Jungers et al. 2005). Hand bones are known from a nearly complete specimen from Anjohikely Cave in northwestern Madagascar (Fig. 15.7) and from multiple individuals from the Ankarana Massif in the extreme north. The only known hand bones of *Hadropithecus* are the scaphoid, hamate, first metacarpal (Mc1), and two Mc5s, all from a single individual found at Andrahomana Cave in southeastern Madagascar (Godfrey et al. 2006b; Fig. 15.8).

It is perhaps in features of the carpus that *Archaeolemur* and *Hadropithecus* converge most strongly with baboons and diverge most strongly from other lemurs, and indeed most platyrrhines, other cercopithecoids, and hominoids (Jungers et al. 2005; Lemelin et al. 2008). As in baboons, the digits of *Archaeolemur* are very short relative to the length of the carpus (Jungers et al. 2005). However, in *Archaeolemur*, the limbs are also short relative to body mass (Jungers et al. 2002; Godfrey et al. 2006a), which makes its hands very short relative to body mass, like the paws of many non-primate mammals. The carpus is long only in comparison to the more distal parts of the hand; in actuality, the carpus is short, but the distal elements are even shorter. In the ratios of the lengths of the proximal and intermediate phalanges to the corresponding metacarpal for digits 2–5, *Archaeolemur* falls between the very terrestrial baboons (*Theropithecus* and *Papio*) and the forest baboons (*Mandrillus*) (Jungers et al. 2005). In other words, these relationships (called phalangeal indices and used to measure ability to grip small substrates or superstrates or to grip objects in single hand) are low for all digits in *Archaeolemur*, but not nearly as extreme as in *Theropithecus* or *Papio*. The manual digital proportions are nevertheless much more like those of semiterrestrial cercopithecoids than like those of other lemurs, which have much longer phalanges on all rays, particularly the fourth. The weak disparity among phalangeal indices across the manus of

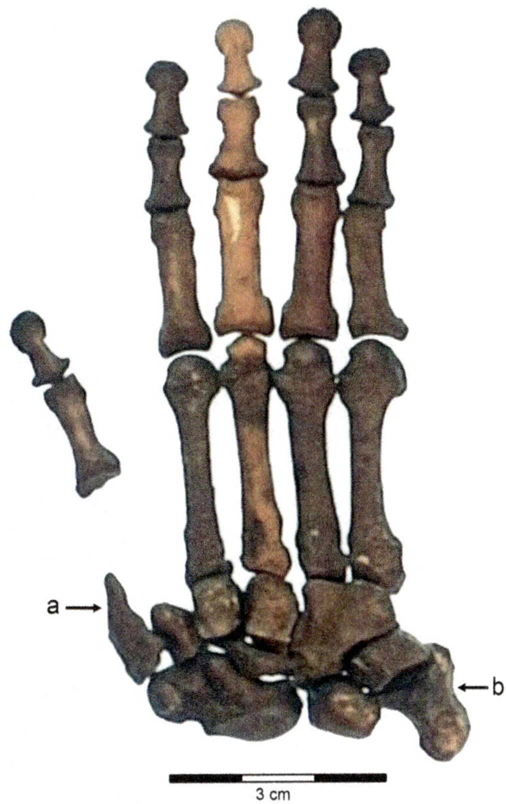

Fig. 15.7 Hand of *Archaeolemur edwardsi* from Anjohikely, northwestern Madagascar, with labeled prepollex (**a**) and pisiform (**b**)

Archaeolemur, combined with the universal reduced length of the individual rays, distinguishes *Archaeolemur* from other strepsirrhines and aligns this animal more with semiterrestrial quadrupeds, such as *Mandrillus*.

Other cercopithecoid likenesses relate to the relative lengths of the rays. As in cercopithecoids, the third and fourth digits are quite similar in total length. Conspicuously absent is the pattern found in strepsirrhine slow climbers and VCLs, wherein the second digit is reduced in length and the fourth elongated so that it far exceeds the length of any other digit. The exaggerated ectaxonic digit pattern found in strepsirrhine slow climbers and VCLs is associated with strong ulnar deviation. The digit pattern of *Archaeolemur* converges toward the anthropoid condition of mesaxony (Jungers et al. 2005).

The pollex is exceptionally short in *Archaeolemur*, suggesting limited pollical grasping capabilities (Jungers et al. 2005). The pollical metacarpal of *Archaeolemur* is less than 50 % as long as the Mc4. The trapezium is also small; its position within the carpus implies that the thumb was less abducted than is typical for lemurs. While we cannot construct the same comparisons for *Hadropithecus* (as we lack its Mc4 and trapezium), we do know that the pollical metacarpal of *Hadropithecus* was short in comparison to the length of the Mc5, which in turn should have been shorter than the fourth (Lemelin et al. 2008).

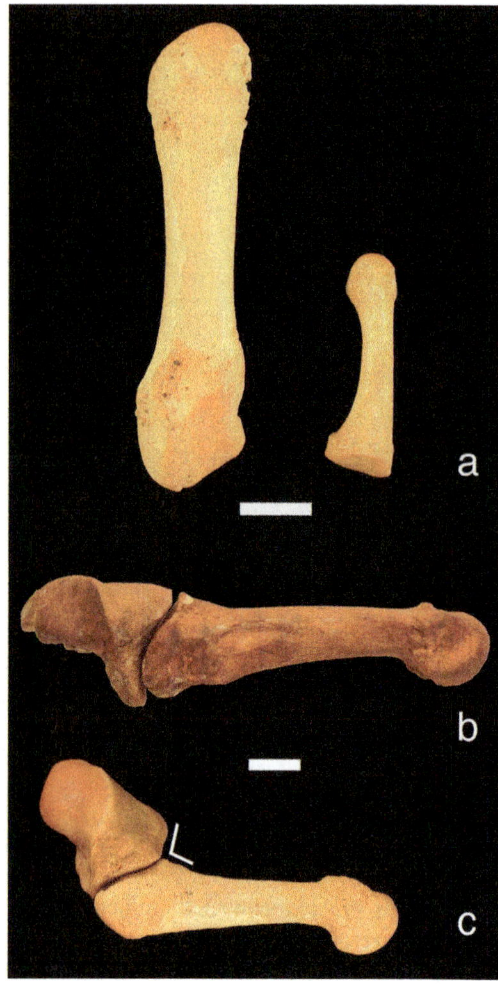

Fig. 15.8 Dorsal view of left Mc5 (*top left*) and Mc1 (*top right*) of *Hadropithecus stenognathus* from Andrahomana (AHA-I) (**a**). Ulnar view of right hamate and right Mc5 of *Archaeolemur edwardsi* (USNM 447012) from Anjohikely (**b**) and the same two bones of *Hadropithecus stenognathus* (AHA-I) from Andrahomana (**c**). Scale bars are 5 mm. Photo credits: Pierre Lemelin and Brian Richmond

The pisiform bone, the heel of the hand to which the tendon of the flexor carpi ulnaris (FCU) muscle attaches, is long and robust in *Archaeolemur* (Fig. 15.7). Hamrick and Alexander (1999) showed that the length of the pisiform distinguishes pronograde quadrupeds from slow climbers, which have reduced pisiforms. This feature underscores the quadrupedal signal in the rest of the archaeolemurid post-cranium and is perhaps amplified by an increase in terrestriality. Electromyography of FCU in baboons reveals that this muscle is recruited at low levels during standing and increases in amplitude during walking on the ground and on arboreal supports to prevent collapse of the wrist into dorsiflexion due to the substrate reaction force (Patel et al. 2012). As speed increases during running, so does the level of muscle activity in this "antigravity" wrist flexor muscle. A long pisiform provides mechanical advantage to FCU in a dedicated quadruped like *Archaeolemur*.

Two features allow one to reconstruct the depth of the carpal tunnel in extinct animals. They are the lengths of the hamulus (or hook of the hamate), which forms part of the ulnar border of the carpal tunnel, and the size of the scaphoid tubercle, which forms part of its radial border (see also Chap. 3). In suspensory taxa and vertical climbers, the carpal tunnel is deep, the hamulus well developed, and the scaphoid tubercle projects palmarly. In sharp contrast, in both *Archaeolemur* and, to a greater extent, *Hadropithecus*, the hamulus is small, and the scaphoid tubercle reduced, indicating a shallow carpal tunnel for the digital flexor muscles. Indeed, in *Hadropithecus*, the hamulus is a diminutive protuberance and the scaphoid tubercle is very small (Lemelin et al. 2008).

The orientation, in *Archaeolemur*, of the hamate's proximal ("spiral") facet for articulation with the triquetrum resembles more those of pronograde arboreal quadrupeds than those of VCLs or slow climbers (Hamrick et al. 2000; Jungers et al. 2005). It is structured to limit mobility and increase the stability of the midcarpal joint. This is the case, to a greater degree, for the proximal spiral facet of the hamate in *Hadropithecus*, which most closely resembles those of baboons or gorillas (Lemelin et al. 2008). In this manner, also, the archaeolemurids converge most closely, among lemurs, to the most terrestrial lemurid, *Lemur catta* (Lemelin et al. 2008). They would have used more extended, pronated, and neutral hand postures, unlike VCLs or slow climbers, which regularly employed greater ulnar deviation.

The tubercle on the base of the Mc5 (for attachment of the extensor carpi ulnaris muscle) is robust in *Archaeolemur*, but somewhat smaller in *Hadropithecus*. The metacarpal heads of *Archaeolemur* have expanded dorsal articular surfaces with some ridging (indicative of an extended set to the digits) as in some African apes and cercopithecoids (Walker 1974; Whitehead 1993). An expanded dorsal articular surface on the head of the Mc5 is lacking in *Hadropithecus*.

As discussed earlier, the proximal phalanges of *Archaeolemur* are straighter in comparison to most lemurs, including quadrupedal ones (but more curved in comparison to those of baboons) (Table 15.4). The included angle of the proximal phalanges of *Archaeolemur* averages 27.9°, while those of the manual phalanges of baboons average 14.7°. Other primates with similar degrees of proximal manual phalangeal curvature are colobines (such as *Presbytis melalophos*, *Colobus guereza*) and cercopithecines (*Chlorocebus aethiops*, *Lophocebus albigena*, and *Cercopithecus mitis*). These are largely arboreal quadrupeds but some are quite comfortable in disturbed habitats (such as plantations) and in savanna habitats (Fleagle 2013). Of the extant lemur species, the indriids have the closest values; the proximal phalanges of lemurids are more curved.

A feature of the hand of *Archaeolemur* that has defied explanation is its enormous prepollex, an enlarged sesamoid bone that articulated with the palmar surface of the scaphoid bone (Jungers et al. 2005) (Fig. 15.7). Judging from the presence of a large, ovoid articular facet on the palmar surface of the scaphoid tubercle in *Hadropithecus*, this animal likely also had a large prepollex (Lemelin et al. 2008). The prepollex may have been covered by a fatty pad, which would have served to broaden the wrist and distribute the load in a plantigrade posture. A fleshy covering of the enlarged manual "heel bone" (pisiform) might have also increased the overall size of the palmar pad.

The function of apical tufts in subfossil lemurs also remains uncertain, but it is clear that the apical tufts of *Archaeolemur* were very broad [and bear a striking resemblance to the fingertips of Neanderthals (Mittra et al. 2007)]. It is likely that the pulp of the palmar digital pad was also broad and that *Archaeolemur* had voluminous apical pads. They may have functioned like the prepollex and pisiform in distributing the load during terrestrial locomotion. There is also the possibility that these pads functioned in some unique and currently unknown processing of tactile information (see above). Alternatively, they might simply be a secondary, correlated effect of overall forelimb and manual robusticity.

Another trait which defies explanation, this time relating to *Hadropithecus*, is the peculiar hyperextended set of the carpometacarpal joint for the fifth digit (Lemelin et al. 2008) (Fig. 15.8). This morphology may compensate for a lack of extended metacarpophalangeal joint postures. However, a hyperextended set to the carpometacarpal joints has no analogue in primates, and it is not yet clear whether it is related to its locomotor repertoire or to other aspects of hand use, such as an unusual feeding behavior.

5 The Palaeopropithecidae

The palaeopropithecids, the "sloth lemurs," are represented by four genera: *Mesopropithecus* (three species), *Babakotia* (one species), *Palaeopropithecus* (three species), and *Archaeoindris* (one species) (Godfrey et al. 1990; Simons et al. 1992, 1995; Jungers et al. 1997; Godfrey and Jungers 2003; Gommery et al. 2009). Body mass estimates for sloth lemurs range from ~11 to ~160 kg (Jungers et al. 2008), and when *Archaeoindris* is excluded, the largest of the sloth lemurs was around 45 kg (Palaeopropithecus). Along with an extensive collection of skulls and dental remains, there have been a large number of postcranial elements collected and attributed to three of the four genera of sloth lemurs (i.e., *Mesopropithecus*, *Babakotia*, and *Palaeopropithecus*), which has allowed for thorough and reliable reconstructions of positional behavior (Godfrey et al. 1995; Jungers et al. 1997; Hamrick et al. 2000). *Mesopropithecus*, *Babakotia*, and *Palaeopropithecus* have all been interpreted as relatively large-bodied antipronograde primates with varying degrees of suspensory locomotion. This interpretation has led many to believe that these genera filled a similar ecological role to living sloths (Carleton 1936; Godfrey et al. 1995, 2008; Jungers 1980; Jungers et al. 1997; Hamrick et al. 2000; Godfrey and Jungers 2003; Shapiro et al. 2005; Granatosky et al. 2014). However, in contrast to extant sloths, sloth lemurs lack claws and have prehensile hands more typical to those of other primates, a pattern recently quantified (Granatosky et al. 2012).

Indicators of suspensory behavior are present throughout the hand (Napier 1960, 1961; Jouffroy and Lessertisseur 1979; Cartmill and Milton 1977; Sarmiento 1988; Lewis 1989; Jouffroy et al. 1991; Whitehead 1993; Richmond 2007) (Fig. 15.9). Phalangeal curvature of *Mesopropithecus*, *Babakotia*, and *Palaeopropithecus* converges with that of orangutans, atelids, and other highly

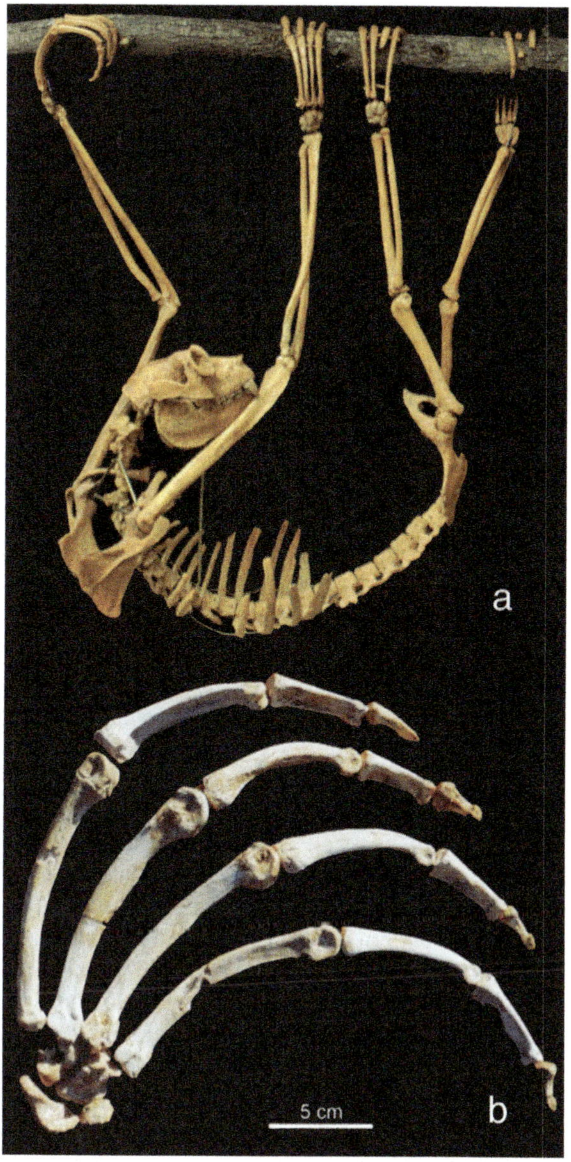

Fig. 15.9 *Palaeopropithecus kelyus* reconstruction, skeleton from Anjohibe in northwestern Madagascar (**a**). Field reconstruction of the hand of *Palaeopropithecus ingens* from Ankilitelo Cave in SW Madagascar (**b**). Photo credit: Don DeBlieux

suspensory species (Jungers et al. 1997) (Table 15.4). Wrist morphology of the smaller sloth lemur genera is similar in many respects to their extant relatives, the indriids, but certain details support other evidence that the sloth lemurs were highly suspensory quadrupedal mammals. The most suspensory was *Palaeopropithecus*, which is derived among strepsirrhines, and all other primates, in having a mortar-and-pestle type of articulation between the ulnar styloid process and the triquetrum (Hamrick et al. 2000). This pattern would have allowed an ulnar-deviated wrist to rotate around the ulna during pronation and supination, which in turn would have promoted a wider range of motion at the radioulnar joint; this is a unique and alternative solution to retreat of the ulnar styloid and development of a fibrocartilaginous triangular disk in hominoid wrists (Cartmill and Milton 1977; Mendel 1979; O'Connor and Rarey 1979; Hamrick et al. 2000).

A distal articulation between the pisiform and triquetrum is also evident in *Palaeopropithecus*, and this may be indicative of very limited contact between the pisiform and the ulna (Cartmill and Milton 1977; Mendel 1979; Sarmiento 1988; Lewis 1985, 1989; Hamrick et al. 2000). These traits together are characteristic of animals associated with antipronograde positional behaviors, such as deliberate climbing and suspension (Sarmiento 1988; Hamrick 1996a).

A deep carpal tunnel has been observed in VCL and suspensory mammals (Hamrick 1996a). This allows for the passage of large, well-developed digital flexor tendons (Napier 1961; Mendel 1979; Hamrick 1996a). In many species, ulnar deepening of the carpal tunnel is accomplished by a palmarly elongated and more curved hamate hamulus, which serves as an attachment site of the flexor retinaculum (transverse carpal ligament). *Babakotia* and *Palaeopropithecus* both possess an elongated hamate hamulus, relatively similar in length to that of the vertical clinger *Propithecus* (Hamrick et al. 2000). This finding suggests that suspensory and VCL positional behaviors were common in *Babakotia* and *Palaeopropithecus*. In contrast, *Mesopropithecus* does not have an elongated hamulus, but instead has a hamulus with proportions more similar to pronograde lemurs (Hamrick et al. 2000). This supports other studies (Godfrey 1988; Jungers et al. 1991; Shapiro et al. 2005; Granatosky et al. 2014) that suggest *Mesopropithecus* practiced above-branch locomotion more often than *Babakotia* or *Palaeopropithecus*.

Various lines of evidence suggest that *Palaeopropithecus* moved in a manner similar to living sloths (Lamberton 1947; Godfrey et al. 1995; Shapiro et al. 2005; Granatosky et al. 2014). However, the hands of *Palaeopropithecus* and living sloths are by no means perfect analogues (Table 15.6). The sloth wrist is the more extreme of the two, being considerably modified from the primitive mammalian pattern in order to house large and complex flexor musculature and allow a great deal of wrist flexion (100°), extension (81°), and ulnar deviation (90–110°) (Mendel 1979). Two-toed sloths are incapable of moving on large-diameter (i.e., >91 mm) branches (Mendel 1981a). This is a likely consequence of their specialization for small-diameter (i.e., 13–31 mm) superstrates. Sloth lemur hands were more dexterous and superstrate use was not nearly as constrained by branch size. The metacarpal and phalangeal morphology of sloths is characterized by anatomical specializations that reduce muscular effort during suspensory locomotion, but at the cost of manual dexterity and versatility (Mendel 1981b). The metacarpals of sloths are generally straight and show minimal curvature compared to suspensory primates (but their claws are long and very curved). The two-toed sloth has only two functional digital rays and the Mc1 and Mc4 are nothing more than immobile splints tightly bound from their distal ends to the heads of the Mc2 and Mc3, respectively (Mendel 1979). This arrangement, with the combination of tightly bound ligaments between adjacent metacarpals, allows for little independent movement of the individual metacarpal

Table 15.6 Function and structure in the hands of *Palaeopropithecus* and two-toed sloths (*Choloepus*)

Functional similarity	*Palaeopropithecus*	*Choloepus*
Hook-like hand	The elongated and strongly curved metacarpals, proximal phalanges, and intermediate phalanges of digits 2–5 form a hook	Long and highly curved claws of the distal phalanges of digits 2 and 3 form a hook. The proximal and intermediate phalanges are short and straight (Beard 1990; Hamrick et al. 1999; Boyer and Bloch 2008)
Digital reduction	Digit 1 is likely reduced or absent. Digits 2–5 are functional	Only digits 2 and 3 are functional
Ulnar deviation at the wrist and associated suspensory adaptations	A mortar-and-pestle type of articulation between the ulnar styloid process and the triquetrum allows the wrist to rotate around the ulna under strong ulnar deviation (Hamrick et al. 2000)	In the neutral position, no carpal bone articulates with the ulna. During extreme ulnar deviation, both the scaphoid and lunate are pushed radially, and the triquetrum, which is held in place by loose ligamentous connections, abuts against the ulna, for which it has a discrete facet. Even greater ulnar deviation is achieved through the hamate moving proximally and sliding along the now-stationary triquetrum (Mendel 1979)
Deep carpal tunnel	An elongated, palmarly directed scaphoid tubercle forms a deep lateral border of the carpal tunnel (Hamrick et al. 2000)	Scaphoid tubercle is long and palmarly directed
Phalanges have deep, well-protected channels for digital flexor muscles	The palmar surfaces of the phalanges are deeply excavated and the ridges for the flexor sheaths are well developed	The proximopalmar surface of the proximal phalanx is the site of fusion for two spur-like sesamoid bones that provide a deep and well-defined channel through which the digital flexors run (Mendel 1981b)
Metacarpophalangeal joints sacrifice mobility for stability in flexion and extension	Metacarpal heads are grooved; the reciprocal surfaces of the proximal phalanges are keeled (Lamberton 1947; Walker 1974). This "tongue and groove" morphology limits adduction, abduction, and torsion of the digits, sacrificing these for increased stability in flexion and extension	Large vertical splines extend dorsally and palmarly from the Mc2 and Mc3 and on the corresponding intermediate phalanges, and deeply grooved articular surfaces of the proximal and intermediate phalanges fit neatly into the pockets created by them. These splines act as bony stops that provide passive resistance to adduction, abduction, and torsion at the metacarpal and interphalangeal joints, allowing only flexion and extension (Mendel 1979, 1981b)

bones. Each metacarpal is capable of a few degrees of flexion or extension at best, but in most instances, this portion of the hand moves as an individual unit (Mendel 1981b). Additionally, in sloths the proximal phalanx is greatly reduced in length compared to both the intermediate and distal phalanges (Mendel 1981b; Beard 1990; Boyer and Bloch 2008). This contrasts strongly with the configuration in sloth lemurs (and in primates generally) in which the proximal phalanx is the longest, the intermediate phalanx is shorter, and the distal phalanx is the shortest (see also Chap. 4). *Palaeopropithecus* is unique among primates in possessing a "tongue and groove" metacarpophalangeal joint in digits 2–5 (Lamberton 1947; Walker 1974). The metacarpal head is grooved dorsopalmarly, and the articulating proximal phalanx has a midline keel; this configuration presumably limited abduction-adduction of these digits and limited the joints of a hooklike hand to flexion-extension.

Among highly suspensory living primates, the orangutan converges on *Palaeopropithecus* in various other ways (Lamberton 1947; Walker 1974). They share not only strongly curved proximal (and intermediate) phalanges, but their distal phalanges have small, knobby apical tufts. They also share very high intermembral indices (Jungers et al. 2002), high femoral neck-shaft angles, and globular femoral heads lacking a fovea capitis (implying convergent loss of the ligamentum teres). The olecranon process of *Palaeopropithecus* is also very reduced as in *Pongo*, and the humeral heads in both project above the greater and lesser tuberosities. Although no first metacarpals have yet been recovered for *Palaeopropithecus*, its first metatarsal is greatly reduced and almost vestigial (even more than in orangutans; Patel et al. 2013); in view of the genetic link between reduced first rays in the hand and foot of *Pongo* (Tuttle and Rogers 1966), we predict that the first metacarpal of this sloth lemur was also greatly reduced.

6 The Megaladapidae

The megaladapids, or "koala lemurs," are represented by one genus (*Megaladapis*) and two subgenera (*Megaladapis* and *Peloriadapis*) (Vuillaume-Randriamanantena et al. 1992). A taxonomic reassessment of the northern and northwestern members of this genus is overdue, and for now we refer these specimens to the intentionally ambiguous taxon *M. grandidieri/madagascariensis* (Goodman and Jungers 2014). Body mass estimates for species of *Megaladapis* range from 45 kg (*M. madagascariensis*) to 85 kg (*M. edwardsi*), but there is a good deal of variation in body size within each species (Jungers et al. 2008). All species of megaladapids were much larger than the living koala; the marsupial analogy is most often invoked to describe the unusual cranial anatomy of the genus (Tattersall 1972). All postcranial evidence points to arboreality and climbing adaptations in the genus (Walker 1974; Wunderlich et al. 1996; Jungers et al. 2002), but large body size probably offered some protection from predators when on the ground. Leaping was not part of the locomotor repertoire.

The wrist of *M. edwardsi* and *M. madagascariensis* resembles those of extant pronograde arboreal primates, displaying a dorsopalmarly expanded pisiform and a well-developed "spiral" facet on the hamate (Lorenz von Liburnau 1905; Walker 1967); "conjunct ulnar deviation and pronation were frequent movements at the triquetrohamate articulation" (Hamrick et al. 2000: 643). The hamulus of the hamate is short, thick, and blunt; it projects slightly palmarward, distally, and ulnarly; this condition is more similar to *Varecia* than to any indriid. The distal articular surfaces for Mc4 and Mc5 are gently undulating surfaces without distinct facets. The scaphoid bears a facet for a free os centrale (Walker 1967; Hamrick et al. 2000). As is typical of most lemurs, the scaphoid dominates the radiocarpal joint in *Megaladapis*. It has a modestly developed tubercle, but one that is not as elongated as that of sloth lemurs. Rodlike pisiforms are known for the smallest megaladapids, and they also differ from palaeopropithecids in that the body is very tall dorsopalmarly, and the base bears a distinct facet for the ulnar styloid process (Hamrick et al. 2000). The carpal tunnel of *Megaladapis* is shallow in comparison to sloth lemurs and suggests less reliance on extrinsic digital flexors for antipronograde postures and locomotion.

The non-pollical metacarpals of *Megaladapis* are slightly curved in the dorsopalmar plane. Metapodials attributed to *Megaladapis edwardsi* as Mc1s (Wunderlich et al. 1996) are probably better attributed to *M. madagascariensis* as very small first metatarsals (Walker 1967). Accordingly, no Mc1s of the genus are described in the literature. The Mc4 of *M. edwardsi* is the longest, a finding consistent with a typically lemuroid ectaxony. Proximal phalanges are moderately curved, intermediate in degree between chimpanzees and *Varecia*. This is consistent with frequent use of the hand in climbing and other antipronograde behaviors (but not to the extreme seen in some of the sloth lemurs). The hand is long relative to the rest of the upper extremity (Fig. 15.1, Table 15.1) and reflects its important role in arboreal grasping. The apical tufts of megaladapids are broad relative to their base as in living lemurids, but are less "robust" due to the relatively greater length of the distal phalanges. Lacking information on the thumb, it is difficult to access the manipulative capabilities of the hand of the megaladapids. The hands of koalas are very different in most respects from *Megaladapis*: the digits are all clawed except for the thumb, the radial two digits are divergent and opposable to the other three digits (schizodactyly), and the scaphoid and lunate are fused (Young 1880; Yalden 1972; Jungers 1976).

7 Discussion and Conclusions

Subfossil lemurs exhibit a huge amount of manual morphological heterogeneity, both within and across families. Some have broad, expanded apical tufts; some have narrow apical tufts. Some have long hands, others short hands. Some have highly curved phalanges, others have straight phalanges. Some displayed highly unusual adaptations—morphologies literally unknown elsewhere and certainly not in order Primates. We refer here to the adaptations for carpometacarpal hyperextension at Mc5 in *Hadropithecus* and to the modifications of the heads of the metacarpals—i.e., their tongue and groove morphology—in *Palaeopropithecus*. Giant lemurs are also remarkable for their convergences on other primates (e.g., palaeopropithecids on orangutans and atelids, archaeolemurids on cercopithecoids) and on animals outside of the primate order (e.g., palaeopropithecids on sloths, megaladapids on koalas, *Daubentonia* on marsupials such as *Dactylopsila* with similar feeding adaptations). While we have highlighted some of the ways in which the hands of sloth lemurs differ from those of sloths, it is nevertheless clear that the hands of both have traits that serve to maximize the effective range of ulnar deviation, flexion, and extension under suspension. The structural modifications of the hands of sloths may make them energetically more efficient than sloth lemurs for suspensory locomotion, but sloths achieve this by sacrificing manual dexterity and versatility, both of which are preserved in sloth lemurs.

None of the subfossil lemurs can be described as exclusively arboreal or exclusively terrestrial. The archaeolemurids were likely highly terrestrial, but had retained adaptations for climbing. The palaeopropithecids were strongly arboreal, but were able to navigate terrestrial terrains, especially the gorilla-sized *Archaeoindris*. Because of their large body size and inability to leap, we can surmise that giant lemurs would have had to have come down to the ground, if only to get from patch to patch of trees. One can only imagine how awkward an animal such as *Palaeopropithecus* may have been on the ground; nevertheless, we know that it did descend to the ground. Perhaps it crawled and dragged its trunk on the ground like living sloths (*Choloepus* and *Bradypus*) do when terrestrial (Mendel 1981a, 1985).

In many ways, the Megaladapidae show the fewest surprises: we see here adaptations for slow, vertical climbing. Oddly, *Megaladapis* is the giant lemur with the least primate-like skull and the largest foot, but apart from its relative length, it had a fairly standard, lemur-like hand (e.g., dominance of the ulnar side of the carpus and digits and no peculiar autapomorphies)—at least as far as we can surmise from the relative paucity of hand bones attributed to this genus.

At the beginning of this chapter, we asked whether phylogeny helps us to understand morphological variation in the hands of subfossil lemurs. We are now in a position to address this question in some detail.

It is difficult to identify any synapomorphy of the hands of members of the archaeolemurid/palaeopropithecid/indriid clade, as archaeolemurids diverge so dramatically in hand form and function from palaeopropithecids and indriids. The archaeolemurid/palaeopropithecid/indriid clade includes the lemurs with the greatest and the least curvature of the proximal phalanges, the greatest and least expansion of the distal phalangeal apical tufts, and the greatest and least robusticity of the distal phalanges. Members of this clade also fall near the high and at low ends of the lemur spectrum of variation in relative hand length. Indriids have relatively long hands, palaeopropithecids fall in the middle of the spectrum but closer to indriids, and *Archaeolemur* has the relatively shortest hands of any strepsirrhine, approaching patas monkeys in relative hand length. *Archaeolemur* and *Babakotia* have expanded apical tufts (but which differ in overall shape, with *Babakotia* more similar to living indriids), and *Palaeopropithecus* has relatively narrow tufts.

In fact, the hand of *Archaeolemur* is more like those of lemurids than indriids or palaeopropithecids (Hamrick 1996a, b), which prompted King et al. (2001) to question their apparently closer relationship to indriids and palaeopropithecids than to lemurids. However, molecular data strongly support an archaeolemurid/palaeopropithecid/indriid clade. The lemurid-likenesses of the hand of *Archaeolemur* probably reflect the convergent use of quadrupedalism by archaeolemurids and lemurids (Hamrick et al. 2000; Jungers et al. 2005). Alternatively, in their most basic form, they may be primitive retentions, as some form of arboreal quadrupedalism may have characterized the common ancestor of the seven families (including archaeolemurids and lemurids) that began diversifying ~31 Ma.

As documented above, *Archaeolemur* and *Hadropithecus* share features of the hand that cannot have been primitive for lemurs. These include an extreme shortening of the metacarpals (including Mc1), reduction of the hamulus of the hamate, a change in the orientation of the hamate's proximal (or "spiral") facet for articulation with the triquetrum, and enlargement of the prepollex (inferred for *Hadropithecus* from the facet on the scaphoid). Adaptations for extended, pronated, and neutral hand postures distinguish the archaeolemurids from most other lemurs, which regularly employ ulnar deviation in climbing and suspension.

The palaeopropithecids and indriids share some adaptations of the hand, in this case features of the wrist that serve to increase the effective range of flexion, extension, and ulnar deviation. However, there is considerable variation in hand morphology in this clade, particularly in the curvature of the proximal phalanges; the hand of *Palaeopropithecus* is loaded with unique specializations for suspension. One apparent palaeopropithecid/indriid synapomorphy is scaphoid-centrale fusion, which occurs in high frequency in this clade. On the basis of our current understanding of the distribution of this trait in lemurs, one might suggest a correlation with vertical climbing and suspension, but until we gain a better understanding of the biomechanical consequences of increasing radial-side midcarpal stability, we cannot make a definitive statement regarding its functional significance. The Lepilemuridae, which (like the indriids and palaeopropithecids) display scaphoid-centrale fusion, engage in considerable antipronograde behaviors. However, unlike the indriids and palaeopropithecids, they are not suspensory.

Members of the Megaladapidae and Lemuridae share certain features of the carpus associated with arboreal quadrupedal locomotion. The hand of *Megaladapis* resembles that of *Varecia* in the morphology and orientation of the hamulus of the hamate, the form of the "spiral" facet on the hamate, the form and orientation of the pisiform, and several metric traits, such as the curvature of the proximal phalanges and the degree of expansion of the apical tufts of the distal phalanges. The carpal tunnel was not as deep as in specialized VCLs and more suspensory taxa. The os centrale and scaphoid were not fused. The hand of *Megaladapis* was relatively longer than those of lemurids, which is related to forelimb dominance (e.g., higher intermembral indices) and suggests greater use of vertical climbing and quadrumanous clambering.

In general, much of what we "know" about the hands of lemurs (e.g., Jouffroy 1975) is violated by the subfossil lemurs. For example, we "know" that the morphological axis of the hands of lemurs passes through the fourth digit (this is classic ectaxony); the hands are elongated, with high phalangeal indices (i.e., the summed lengths of the proximal and middle phalanges are considerably greater than those of the corresponding metacarpals); the Mc1 is robust and divergent; and the wrist is well adapted for ulnar deviation. These generalities do not apply across the board once giant lemurs are included in the mix. As detailed above, we see instead remarkable heterogeneity in hand form and function.

New molecular evidence indicates that virtually all lemurs (including the Archaeolemuridae, Palaeopropithecidae, and Megaladapidae and excluding only living and extinct *Daubentonia*) share a common ancestor ~31 Ma. The descendant clade arguably shows greater morphological diversity in hand structure than the crown catarrhines (hominoids and cercopithecoids), whose common ancestor dates roughly to the same period of time. In each case, we know that tremendous heterogeneity in hand form and function emerged in only ~30 million years, which speaks volumes to the evolvability of the primate hand and is consistent with lemur evolution as a classic adaptive radiation.

8 Future Directions

Knowledge of the hands of extinct primates is of course always limited by available specimens. At present, due to limitations in our recovery of hand elements, we know little about the hands of *Daubentonia robusta*, *Pachylemur* spp., *Megaladapis* spp., *Hadropithecus stenognathus*, *Mesopropithecus* spp., and of course *Archaeoindris fontoynontii*. From the few elements of the hand of *Daubentonia robusta* that are known, we can be confident that it shared with its extant congener remarkable autapomorphies, but we know too little to document subtle differences in hand form or function within this genus. We still also have much to learn about extant strepsirrhine hand function, including experimental information on electromyography and kinetics that is currently lacking (but see Kivell et al. 2010), which might further illuminate the functional anatomy of subfossil hands.

We close with the familiar paleontological mantra: "we need more fossils." Fortunately, new fossil sites haves been discovered in Madagascar (e.g., Rosenberger et al. 2015), and they show tremendous promise of yielding new hand bones. As we fill in missing parts of the puzzle and increase the resolution of our phylogenetic reconstruction, we should be able to say much more about the evolution of the hands of lemurs, and the extinct lemurs should play an increasingly important role in revealing that story.

References

Beard KC (1990) Gliding behaviour and palaeoecology of the alleged primate family Paromomyidae (Mammalia, Dermoptera). Nature 345:340–341

Begun DR (2004) Knuckle-walking and the origin of human bipedalism. In: Meldrum DJ, Hilton CE (eds) From biped to strider: the emergence of modern human walking, running and resource transport. Kluwer Academic/Plenum Publishers, New York, pp 9–33

Boyer DM, Bloch JI (2008) Evaluating the mitten-gliding hypothesis for Paromomyidae and Micromomyidae (Mammalia, "Plesiadapiformes") using comparative functional morphology of new Paleogene skeletons. In: Sargis EJ, Dagosto M (eds) Mammalian evolutionary morphology: a tribute to Frederick S. Szalay. Springer, New York, pp 233–284

Bromage TG, Hogg RT, Lacruz RS, Hou C (2012) Primate enamel evinces long period biological timing and regulation of life history. J Theor Biol 305:131–144

Burney DA, Ramilisonina (1999) The kilopilopitsofy, kidoky, and bokyboky: accounts of strange animals from Belo-sur-mer, Madagascar, and the megafaunal "extinction window". Am Anthropol 100:957–966

Burney DA, Burney LP, Godfrey LR, Jungers WL, Goodman SM, Wright HT, Jull AJT (2004) A chronology for late prehistoric Madagascar. J Hum Evol 47:25–63

Carleton A (1936) The limb-bones and vertebrae of the extinct lemurs of Madagascar. Proc Zool Soc Lond 106:281–307

Cartmill M, Milton K (1977) The lorisiform wrist joint and the evolution of "brachiating" adaptations in the Hominoidea. Am J Phys Anthropol 47:249–272

Congdon KA (2012) Interspecific and ontogenetic variation in proximal pedal phalangeal curvature of great apes (*Gorilla gorilla*, *Pan troglodytes*, and *Pongo pygmaeus*). Int J Primatol 33:1–10

Crowley BE (2010) A refined chronology of prehistoric Madagascar and the demise of the megafauna. Q Sci Rev 29:2591–2603

Dominy NJ, Ross CF, Smith TD (2004) Evolution of the special senses in primates: past, present, and future. Anat Rec A 281A:1078–1082

Fleagle JG (2013) Primate adaptation and evolution, 3rd edn. Elsevier, New York

Godfrey LR (1977) Structure and function in *Archaeolemur* and *Hadropithecus* (subfossil Malagasy lemurs): the postcranial evidence. Ph.D. dissertation, Harvard University

Godfrey LR (1988) Adaptive diversification of Malagasy strepsirrhines. J Hum Evol 17:93–134

Godfrey LR, Jungers WL (2002) Quaternary fossil lemurs. In: Hartwig WC (ed) The primate fossil record. Cambridge University Press, Cambridge, pp 97–121

Godfrey LR, Jungers WL (2003) The extinct sloth lemurs of Madagascar. Evol Anthropol 12:252–263

Godfrey LR, Simons EL, Chatrath PS, Rakotosamimana B (1990) A new fossil lemur (*Babakotia*, Primates) from northern Madagascar. C R Acad Sci II 310:81–87

Godfrey LR, Sutherland MR, Paine RR, Williams FL, Boy DS, Vuillaume-Randriamanantena M (1995) Limb joint surface areas and their ratios in Malagasy lemurs and other mammals. Am J Phys Anthropol 97:11–36

Godfrey LR, Jungers WL, Wunderlich RE, Richmond BG (1997) Reappraisal of the postcranium of *Hadropithecus* (Primates, Indroidea). Am J Phys Anthropol 103:529–556

Godfrey LR, Jungers WL, Schwartz GT (2006a) Ecology and extinction of Madagascar's subfossil lemurs. In: Gould L, Sauther ML (eds) Lemurs: ecology and adaptation. Springer, New York, pp 41–64

Godfrey LR, Jungers WL, Burney DA, Vasey N, Ramilisonina, Wheeler W, Lemelin P, Shapiro LJ, Schwartz GT, King SJ, Ramarolahy MF, Raharivony LL, Randria GFN (2006b) New discoveries of skeletal elements of *Hadropithecus stenognathus* from Andrahomana Cave, southeastern Madagascar. J Hum Evol 51:395–410

Godfrey LR, Jungers WL, Schwartz GT, Irwin MT (2008) Ghosts and orphans: Madagascar's vanishing ecosystems. In: Fleagle JG, Gilbert CC (eds) Elwyn Simons: a search for origins. Springer, New York, pp 361–395

Goodman SM, Jungers WL (2014) Extinct Madagascar: picturing the island's past. University of Chicago Press, Chicago

Gommery D, Ramanivosoa B, Tombomiadana-Raveloson S, Randrianantenaina H, Kerloc'h P (2009) A new species of giant subfossil lemur from the North-West of Madagascar (*Palaeopropithecus kelyus*, Primates). C R Palevol 8:471–480

Granatosky MC, Miller CE, Lemelin P, Schmitt D (2012) Phalangeal morphology of suspensory mammals: implications for the locomotion of Malagasy subfossil sloth lemurs (Primates: Palaeopropithecidae). J Vert Paleontol S32:103 (abstract)

Granatosky MC, Miller CE, Boyer DM, Schmitt D (2014) Lumbar vertebral morphology of flying, gliding, and suspensory mammals: implications for the locomotor behavior of the subfossil lemurs *Palaeopropithecus* and *Babakotia*. J Hum Evol 75:40–52

Hamrick MW (1996a) Functional morphology of the lemuriform wrist joints and the relationship between wrist morphology and positional behavior in arboreal primates. Am J Phys Anthropol 99:319–344

Hamrick MW (1996b) Locomotor adaptations reflected in the wrist joints of early Tertiary primates (Adapiformes). Am J Phys Anthropol 100:585–604

Hamrick MW, Alexander JP (1999) The hand skeleton of *Notharctus tenebrosus* (Primates, Notharctidae) and its significance for the origin of the primate hand. Am Mus Novit 3182:1–20

Hamrick MW, Rosenman BA, Brush JA (1999) Phalangeal morphology of the Paromomyidae (? Primates, Plesiadapiformes): the evidence for gliding behavior reconsidered. Am J Phys Anthropol 109:397–413

Hamrick MW, Simons EL, Jungers WL (2000) New wrist bones of the Malagasy giant subfossil lemurs. J Hum Evol 38:635–650

Hoffmann JN, Montag AG, Dominy NJ (2004) Meissner corpuscles and somatosensory acuity: the prehensile appendages of primates and elephants. Anat Rec A 281A:1138–1147

Hogg RT, Godfrey LR, Schwartz GT, Dirks W, Bromage TG (2015) Lemur biorhythms and life history evolution. PLoS One 10(8):e0134210

Jolly CJ (1970) *Hadropithecus*: a lemuroid small-object feeder. Man 5:619–626

Jouffroy FK (1963) Contribution à la connaissance du genre *Archaeolemur* Filhol, 1895. Ann Paléontol 49:129–155

Jouffroy FK (1975) Osteology and myology of the lemuriform postcranial skeleton. In: Tattersall I, Sussman RW (eds) Lemur biology. Plenum Press, New York, pp 149–192

Jouffroy FK, Lessertisseur J (1979) Relationships between limb morphology and locomotor adaptations among prosimians: an osteometric study. In: Morbeck ME, Preuschoft H, Gomberg N (eds) Environment, behavior, and morphology: dynamic interactions in primates. Gustav Fischer, New York, pp 143–181

Jouffroy FK, Godinot M, Nakano Y (1991) Biometrical characteristics of primate hands. Hum Evol 6:269–306

Jungers WL (1976) Osteological form and function: the appendicular skeleton of *Megaladapis*, a subfossil prosimian from Madagascar (Primates, Lemuroidea). Ph.D. dissertation, University of Michigan

Jungers WL (1980) Adaptive diversity in subfossil Malagasy prosimians. Z Morphol Anthropol 71:177–186

Jungers WL, Godfrey LR, Simons EL, Chatrath PS, Rakotosamimanana B (1991) Phylogenetic and functional affinities of *Babakotia* (Primates), a fossil lemur from northern Madagascar. Proc Natl Acad Sci U S A 88:9082–9086

Jungers WL, Godfrey LR, Simons EL, Chatrath PS (1997) Phalangeal curvature and positional behavior in extinct sloth lemurs (Primates, Palaeopropithecidae). Proc Natl Acad Sci U S A 94:11998–12001

Jungers WL, Godfrey LR, Simons EL, Wunderlich RE, Richmond BG, Chatrath PS (2002) Ecomorphology and behavior of giant extinct lemurs from Madagascar. In: Plavcan JM, Kay RF, Jungers WL, van Schaik CP (eds) Reconstructing behavior in the primate fossil record. Kluwer Academic/Plenum Publishers, New York, pp 371–411

Jungers WL, Lemelin P, Godfrey LR, Wunderlich RE, Burney DA, Simons EL, Chatrath PS, James HF, Randria GFN (2005) The hands and feet of *Archaeolemur*: metrical affinities and their functional significance. J Hum Evol 49:36–55

Jungers WL, Demes B, Godfrey LR (2008) How big were the "giant" extinct lemurs of Madagascar? In: Fleagle JG, Gilbert CC (eds) Elwyn Simons: a search for origins. Springer, New York, pp 343–360

Karanth KP, Delefosse T, Rakotosamimanana B, Parsons TJ, Yoder AD (2005) Ancient DNA from giant extinct lemurs confirms single origin of Malagasy primates. Proc Natl Acad Sci U S A 102:5090–5095

King SJ, Godfrey LR, Simons EL (2001) Adaptive and phylogenetic significance of ontogenetic sequences in *Archaeolemur*, subfossil lemur from Madagascar. J Hum Evol 41:545–576

Kistler L, Ratan A, Godfrey LR, Crowley BE, Hughes CE, Lei RH, Cui YQ, Wood ML, Muldoon KM, Andriamialison H, McGraw JJ, Tomsho LP, Schuster SC, Miller W, Louis EE, Yoder AD, Malhi RS, Perry GH (2015) Comparative and population mitogenomic analyses of Madagascar's extinct, giant 'subfossil' lemurs. J Hum Evol 79:45–54

Kivell TL, Begun DR (2007) Frequency and timing of scaphoid-centrale fusion in hominoids. J Hum Evol 52:321–340

Kivell TL, Schmitt D, Wunderlich RE (2010) Hand and foot pressures in the aye-aye (*Daubentonia madagascariensis*) reveal novel biomechanical trade-offs required for walking on gracile digits. J Exp Biol 213:1549–1557

Lamberton C (1934) Contribution à la connaissance de la faune subfossile de Madagascar. Lémuriens et ratites Chiromys robustus sp nov Lamb Mém. Acad Malgache 17:40–46

Lamberton C (1938) Contribution à la connaissance de la faune subfossile de Madagascar. Note III. Les Hadropithèques. Bull Acad Malgache (ns) 20:127–170, plus plates

Lamberton C (1947) Contribution à la connaissance de la faune subfossile de Madagascar. Note XVI. *Bradytherium* ou Palaeopropithèque? Bull Acad Malgache (ns) 26:89–140

Laska M, Freist P, Krause S (2007) Which senses play a role in nonhuman primate food selection? A comparison between squirrel monkeys and spider monkeys. Am J Primatol 69:282–294

Lemelin P, Jungers WL (2007) Body size and scaling of the hands and feet of prosimian primates. Am J Phys Anthropol 133:828–840

Lemelin P, Hamrick MW, Richmond BG, Godfrey LR, Jungers WL, Burney DA (2008) New hand bones of *Hadropithecus stenognathus*: implications for the paleobiology of the Archaeolemuridae. J Hum Evol 54:405–413

Lewis OJ (1985) Derived morphology of the wrist articulations and theories of hominoid evolution. Part I.The lorisine joints. J Anat 140:447–460

Lewis OJ (1989) Functional morphology of the evolving hand and foot. Clarendon Press, Oxford

Lorenz von Liburnau L (1905) *Megaladapis edwardsi* G. Grandidier Denkschriften der Kaiserlichen Akademie der Wissenschaften. Math Nat Classe 77:451–490

Matarazzo S (2008) Knuckle walking signal in the manual digits of *Pan* and *Gorilla*. Am J Phys Anthropol 135:27–33

Meador LR, Godfrey LR (2014) Evidence of crocodile predation on the giant lemurs of Madagascar. Am J Phys Anthropol 153(Suppl 58):182–183 (abstract)

Mendel FC (1979) The wrist joint of two-toed sloths and its relevance to brachiating adaptations in the Hominoidea. J Morphol 162:413–424

Mendel FC (1981a) Use of hands and feet of two-toed sloths (*Choloepus hoffmanni*) during climbing and terrestrial locomotion. J Mammal 62:413–421

Mendel FC (1981b) The hand of two-toed sloths (*Choloepus*): its anatomy and potential uses relative to size of support. J Morphol 169:1–19

Mendel FC (1985) Use of hands and feet of three-toed sloth (*Bradypus variegatus*) during climbing and terrestrial locomotion. J Mammal 66:359–366

Mittra ES, Smith HF, Lemelin P, Jungers WL (2007) Comparative morphometrics of the primate apical tuft. Am J Phys Anthropol 134:449–459

Montagnon D, Ravaoarimanana B, Rakotosamimanana B, Rumpler Y (2001) Ancient DNA from *Megaladapis edwardsi* (Malagasy subfossil): preliminary results using partial cytochrome b sequence. Folia Primatol 72:30–32

Muchlinski MN (2010) Ecological correlates of infraorbital foramen area in primates. Am J Phys Anthropol 141:131–141

Muchlinski MN, Godfrey LR, Muldoon KM, Tongasoa L (2011) Evidence for dietary niche separation based on infraorbital foramen size variation among subfossil lemurs. Folia Primatol 81:330–345

Napier JR (1960) Studies of the hands of living primates. Proc Zool Soc Lond 134:647–657

Napier JR (1961) Prehensility and opposability in the hands of primates. Symp Zool Soc Lond 5:115–132

Napier JR, Napier PH (1967) A handbook of living primates: morphology, ecology and behaviour of nonhuman primates. Academic Press, New York

Napier JR, Walker AC (1967) Vertical clinging and leaping—a newly recognized category of locomotor behaviour of primates. Folia Primatol 6:204–219

O'Connor BL, Rarey KE (2005) Normal amplitudes of radioulnar pronation and supination in several genera of anthropoid primates. Am J Phys Anthropol 51:39–43

Orlando L, Calvignac S, Schnebelen C, Douady CJ, Godfrey LR, Hänni C (2008) DNA from extinct giant lemurs links archaeolemurids to extant indriids. BMC Evol Biol 8:121

Patel BA, Larson SG, Stern JT Jr (2012) Electromyography of wrist and finger flexor muscles in olive baboons (*Papio anubis*). J Exp Biol 215:115–123

Patel BA, Goodenberger KE, Boyer DM, Jungers WL (2013) Hallucal reduction in sloth lemurs and morphological convergence on orang-utans by *Palaeopropithecus*. Am J Phys Anthropol 150(Suppl 56):217 (abstract)

Rein TR (2011) The correspondence between proximal phalanx morphology and locomotion: implications for inferring the locomotor behavior of fossil catarrhines. Am J Phys Anthropol 146:435–445

Rein TR, Harrison T, Zollikofer CPE (2011) Skeletal correlates of quadrupedalism and climbing in the anthropoid forelimb: implications for inferring locomotion in Miocene catarrhines. J Hum Evol 61:564–574

Richmond BG (2007) Biomechanics of phalangeal curvature. J Hum Evol 53:678–690

Roberts D, Davidson I (1975) The lemur scapula. In: Tattersall I, Sussman RW (eds) Lemur biology. Plenum Press, New York, pp 125–147

Rosenberger AL, Godfrey LR, Muldoon KM, Gunnell GF, Andriamialison H, Ranivoharimanana L, Ranaivoarisoa JF, Rasoamiaramanana AH, Randrianasy J, Amador FE (2015) Giant subfossil lemur graveyard discovered, submerged, in Madagascar. J Hum Evol 81:83–87

Sarmiento EE (1985) Functional differences in the skeleton of wild and captive orang-utans and their adaptive significance. Ph.D. dissertation, New York University

Sarmiento EE (1988) Anatomy of the hominoid wrist joint: its evolutionary and functional implications. Int J Primatol 9:281–345

Sarmiento EE (1994) Terrestrial traits in the hands and feet of gorillas. Am Mus Novit 3091:1–56

Schwartz GT, Samonds KE, Godfrey LR, Jungers WL, Simons EL (2002) Dental microstructure and life history in subfossil Malagasy lemurs. Proc Natl Acad Sci U S A 99:6124–6129

Schwartz GT, Mahoney P, Godfrey LR, Cuozzo FP, Jungers WL, Randria GFN (2005) Dental development in *Megaladapis edwardsi* (Primates, Lemuriformes): implications for understanding life history variation in subfossil lemurs. J Hum Evol 49:702–721

Schwartz JH, Tattersall I (1985) Evolutionary relationships of living lemurs and lorises (Mammalia, Primates) and their potential affinities with European Eocene Adapidae. Anthropol Pap Am Mus Nat Hist 60:1–100

Sessle BJ, Yao D, Nishiura H, Yoshino K, Lee J-C, Martin RE, Murray GM (2005) Properties and plasticity of the primate somatosensory and motor cortex related to orofacial sensorimotor function. Clin Exp Pharmacol Physiol 32:109–114

Shapiro LJ, Seiffert CV, Godfrey LR, Jungers WL, Simons EL, Randria GFN (2005) Morphometric analysis of lumbar vertebrae in extinct Malagasy strepsirrhines. Am J Phys Anthropol 128:823–839

Simons EL (1994) The giant aye-aye *Daubentonia robusta*. Folia Primatol 62:14–21

Simons EL, Godfrey LR, Jungers WL, Chatrath PS, Rakotosamimanana B (1992) A new giant subfossil lemur, *Babakotia*, and the evolution of the sloth lemurs. Folia Primatol 58:197–203

Simons EL, Godfrey LR, Jungers WL, Chatrath PS, Ravaoarisoa J (1995) A new species of *Mesopropithecus* (Primates, Palaeopropithecidae) from Northern Madagascar. Int J Primatol 15:653–682

Soligo C (2005) Anatomy of the hand and arm in *Daubentonia madagascariensis*: a functional and phylogenetic outlook. Folia Primatol 76:262–300

Spoor F, Garland T, Krovitz G, Ryan TM, Silcox MT, Walker A (2007) The primate semicircular canal system and locomotion. Proc Natl Acad Sci U S A 104:10808–10812

Stern JT Jr, Susman RL (1983) The locomotor anatomy of *Australopithecus afarensis*. Am J Phys Anthropol 60:279–317

Stern JT Jr, Jungers WL, Susman RL (1995) Quantifying phalangeal curvature: an empirical comparison of alternative methods. Am J Phys Anthropol 97:1–10

Tattersall I (1972) The functional significance of airorhynchy in *Megaladapis*. Folia Primatol 18:20–26

Tattersall I (1973) Cranial anatomy of the Archaeolemurinae (Lemuroidea, Primates). Anthropol Pap Am Mus Nat Hist 52:1–110

Tattersall I, Schwartz JH (1974) Craniodental morphology and the systematics of the Malagasy lemurs (Primates, Prosimii). Anthropol Pap Am Mus Nat Hist 52:139–192

Tuttle RH, Rogers CM (1966) Genetic and selective factors in reduction of the hallux in *Pongo pygmaeus*. Am J Phys Anthropol 24:191–198

Vilensky JA, Larson SG (1989) Primate locomotion: utilization and control of symmetrical gaits. Annu Rev Anthropol 18:17–35

Vuillaume-Randriamanantena M, Godfrey LR, Jungers WL, Simons EL (1992) Morphology, taxonomy, and distribution of *Megaladapis*—giant subfossil lemurs from Madagascar. C R Acad Sci II 315:1835–1842

Walker A (1967) Locomotor adaptations in living and extinct Madagascan lemurs. Ph.D. dissertation, University of London

Walker A (1974) Locomotor adaptation in past and present prosimian primates. In: Jenkins FA Jr (ed) Primate locomotion. Academic Press, New York, pp 543–565

Walker A, Ryan TM, Silcox MT, Simons EL, Spoor F (2008) The semicircular canal system and locomotion: the case of extinct lemuroids and lorisoids. Evol Anthropol 17:135–145

Wall CE (1997) The expanded mandibular condyle of the Megaladapidae. Am J Phys Anthropol 103:263–276

Whitehead PF (1993) Aspects of the anthropoid wrist and hand. In: Gebo DL (ed) Postcranial adaptation in nonhuman primates. Northern Illinois University Press, DeKalb, pp 96–120

Wunderlich RE, Simons EL, Jungers WL (1996) New pedal remains of *Megaladapis* and their functional significance. Am J Phys Anthropol 100:115–138

Yalden DW (1972) The form and function of the carpal bones in some arboreally adapted mammals. Acta Anat 82:383–406

Yoder AD, Rakotosamimanana B, Parsons TJ (1999) Ancient DNA in subfossil lemurs: methodological challenges and their solutions. In: Rakotosamimanana B, Rasamimanana H, Ganzhorn JU, Goodman SM (eds) New directions in lemur studies. Kluwer Academic/Plenum Publishers, New York, pp 1–17

Young AH (1880) Intrinsic muscles of the marsupial hand. J Anat Physiol 14:149–165

Chapter 16
The Hands of Fossil Non-hominoid Anthropoids

Terry Harrison and Thomas R. Rein

1 Introduction

Much has been written about the evolution of the hand in humans and great apes, from both the perspectives of comparative morphology and the fossil record, but little attention has been given to the hands of fossil non-hominoid anthropoids. This is partly because the hands of New World- and Old World monkeys have less direct bearing on our understanding of the evolutionary changes that have taken place in the human hand and partly because the pertinent fossil evidence is more scanty or not studied in sufficient detail. However, comparative anatomical studies of the hand in non-hominoid anthropoids provide a broader phylogenetic perspective for understanding the evolution of hominoids and allow a more informed basis for inferring the features of the hand that characterize the ancestral catarrhine morphotypes.

The present contribution offers an overview of the morphology of the hand of fossil non-hominoid anthropoids (Fig. 16.1). The authors have carried out firsthand studies of some of this material (Harrison 1982, 1989, 2011; Harrison and Gu 1999; Rein 2010, 2011; Rein et al. 2011), but the account presented here relies heavily on the literature. It is not intended to be a comprehensive survey of the fossil record, but is instead meant to briefly describe the most complete or most informative specimens and to discuss their implications for understanding the evolution of hand form and function and positional behavior in anthropoid primates.

T. Harrison (✉)
Center for the Study of Human Origins, Department of Anthropology, New York University, 25 Waverly Place, New York, NY 10003, USA
e-mail: terry.harrison@nyu.edu

T.R. Rein
Department of Anthropology, Central Connecticut State University, New Britain, CT, USA

T.L. Kivell et al. (eds.), *The Evolution of the Primate Hand*, Developments in Primatology: Progress and Prospects, DOI 10.1007/978-1-4939-3646-5_16

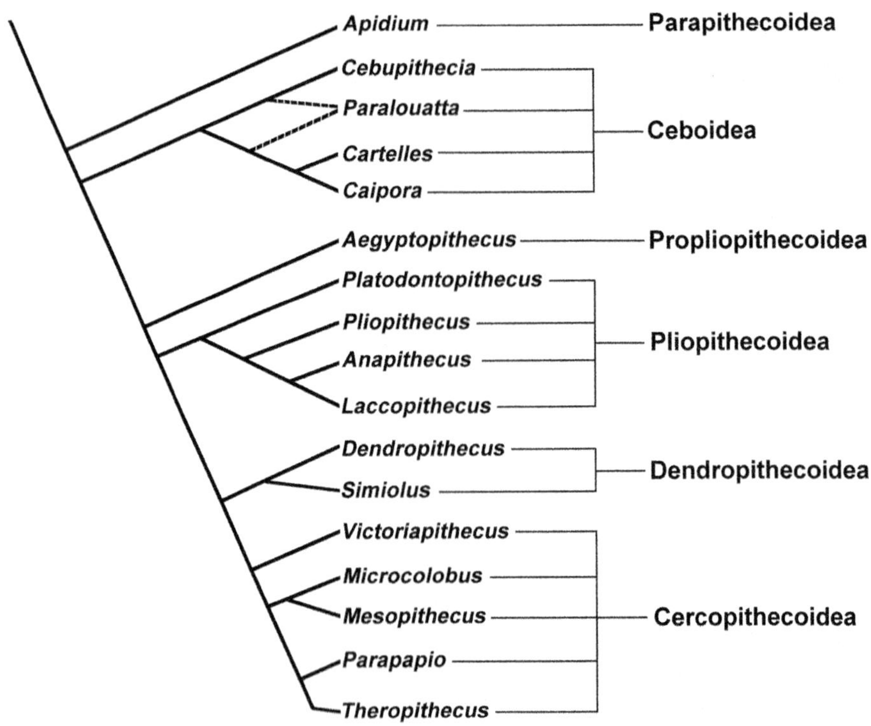

Fig. 16.1 Cladogram showing the relationships and taxonomic placement of the fossil anthropoids discussed in this chapter

2 The Fossil Evidence

2.1 Ceboidea

The oldest postcranial specimens of platyrrhines are from the early Miocene of Argentina (~20–16.5 Ma), associated with *Dolichocebus*, *Soriacebus*, and *Carlocebus*, but no elements of the hand are known (Meldrum 1993; Youlatos and Meldrum 2011). *Cebupithecia* from the middle Miocene of Colombia is the oldest platyrrhine preserving elements of the hand (Stirton and Savage 1950). The late Pleistocene platyrrhines from Brazil, *Cartelles* and *Caipora*, are known from relatively complete skeletons preserving portions of the manus (Cartelle and Hartwig 1996; Hartwig and Cartelle 1996), and a number of isolated hand elements have been referred to the contemporary *Paralouatta* from Cuba (MacPhee and Meldrum 2006).

2.1.1 *Cebupithecia sarmientoi*

The holotype of *Cebupithecia sarmientoi*, from the middle Miocene of La Venta in Colombia, consists of a partial skeleton (Stirton and Savage 1950; Stirton 1951) dating to ~13.5–11.8 Ma (Youlatos and Meldrum 2011). *Cebupithecia* is a medium-sized platyrrhine, with an estimated body mass of 1.8 kg (Rosenberger et al. 2009; Youlatos and Meldrum 2011). The craniodental anatomy provides evidence of a close relationship with extant Pitheciinae (Kay 1990; Horovitz 1999; Hartwig and Meldrum 2002; Rosenberger 2011), and the postcranial skeleton appears to approximate the ancestral pitheciine morphotype (Jones 2008; Youlatos and Meldrum 2011). The published description of the partial skeleton includes only a brief account of the hand bones (Stirton and Savage 1950). The skeleton preserves the right second metacarpal (Mc2), the right and left Mc3, and the distal ends of two other metapodials (probably Mc4 and Mc5). There are also two partial phalanges—a distal end of a proximal phalanx (PP), probably of the fourth ray, and a proximal end of a middle phalanx (MP), probably from the third ray—but these are fragmentary and uninformative (Fleagle and Meldrum 1988; Hartwig and Meldrum 2002).

The metacarpals are shorter and more robust than in *Pithecia*, which has relatively more slender metacarpals and longer digits compared with other pitheciines (Fleagle and Meldrum 1988). The metacarpals of *Cebupithecia* also differ from *Pithecia* in having slightly more expanded and bulbous distal articular surfaces and more pronounced palmar keels (Stirton and Savage 1950). Comparative studies of the postcranial skeletal, including the hand, indicate that *Cebupithecia* was a quadrupedal arboreal primate with a moderate degree of specialization for leaping and vertical clinging (Fleagle and Meldrum 1988; Ford 1990; Meldrum et al. 1990; Meldrum and Lemelin 1991; Youlatos and Meldrum 2011).

2.1.2 *Cartelles coimbrafilhoi*

Cartelles, from the late Pleistocene of Brazil, is known from a nearly complete skeleton from Toca da Boa Vista (Hartwig and Cartelle 1996). With an estimated body mass of about 25–28 kg (Halenar 2011a, 2012), it was about twice the size of the largest extant platyrrhine. *Cartelles* is evidently an atelid platyrrhine, but its relationships within this clade have proved difficult to resolve (Guedes and Salles 2005; Rosenberger et al. 2009; Rosenberger 2011; Halenar 2011b, 2012; Halenar and Rosenberger 2013). Although it is similar to *Alouatta* in cranial anatomy, the postcranial skeletal is most similar to *Ateles* and *Brachyteles*, but more robust, with forelimb specializations for suspensory postures and brachiation (Hartwig and Cartelle 1996; Jones 2008; Halenar 2011b). Carpals, metacarpals, and phalanges of the hand are preserved (Hartwig and Cartelle 1996), but these have not been described. The phalanges are long and curved, with prominent, distally-positioned flexor sheath ridges, indicating that the hand was specialized for strong grasping and suspensory postures (Halenar 2011b; Halenar and Rosenberger 2013). The first metacarpal of *Cartelles* is not preserved (Halenar 2011b), so it remains unknown if the pollical ray was reduced as in extant atelins.

2.1.3 *Caipora bambuiorum*

A nearly complete skeleton of a subadult individual of *Caipora bambuiorum* is known from the late Pleistocene of Toca da Boa Vista, Brazil (Cartelle and Hartwig 1996). *Caipora* was a large atelin platyrrhine, about 20 kg in body mass. Only a preliminary description of the postcranial morphology has been published, suggesting that it was a specialized suspensory climber and clamberer, similar to extant *Ateles* and *Brachyteles*, but much larger with relatively more robust limb bones (Cartelle and Hartwig 1996). The hand skeleton preserves carpals, metacarpals, and phalanges, but these have not been described in detail. The metacarpals are relatively long, subequal in length to the metatarsals, and indicate a hand that was specialized for suspensory postures and brachiation, as in extant *Ateles* and *Brachyteles* (Cartelle and Hartwig 1996).

2.1.4 *Paralouatta varonai*

Postcranial specimens of *Paralouatta varonai* have been recovered from two cave sites (Cueva del Mono Fósil and Cueva Alta) in western Cuba (MacPhee and Meldrum 2006). The age of the material is uncertain, but it is likely to be late Pleistocene or Holocene. *Paralouatta* is a large platyrrhine monkey, with an estimated body mass of 9–10 kg, similar in size to *Brachyteles arachnoides*. The phylogenetic relationships of *Paralouatta* remain contentious. Some authors regard it as a specialized alouattin (Rivero and Arredondo 1991; Rosenberger 2002, 2011), while others (Horovitz and MacPhee 1999; MacPhee and Horovitz 2002) contend that it is part of an endemic clade of Greater Antillean stem pitheciines. Comparisons of the postcranium indicate that *Paralouatta* was, in most respects, a generalized quadrupedal primate that lacked particular specializations for suspensory, leaping, or slow climbing behaviors (MacPhee and Meldrum 2006). Based on the morphology of the elbow and the relative length of the phalanges, MacPhee and Meldrum (2006) have tentatively suggested that *Paralouatta* might have been a semiterrestrial primate, but its overall postcranial morphology points to it being a more fully committed arboreal quadruped.

Several hand bones have been identified, including the Mc1 and Mc2 (MacPhee and Meldrum 2006). Compared to extant large-bodied New World monkeys, the Mc1 has a straighter shaft and less bulbous distal articular surface. The distal end of the shaft bears well-developed crests on the ulnar and radial sides that are absent in extant platyrrhines (MacPhee and Meldrum 2006). The proximal articular surface for the trapezium is a uniaxial and semicylindrical hinge joint, comparable to that in most extant New World monkeys. Proximally, there are small ulnar and radial facets for paired sesamoids. The Mc2 has a radioulnarly narrow distal articular surface. The shaft is short in relation to its radioulnar diameter. The proximal articular surface for the trapezoid is concave, with a well-developed radial crest that separates it from the facet for the Mc3. There is a small triangular facet for articulation with the trapezium. The Mc1 is 77.0 % the length of the Mc2, which implies that the pollical ray was well developed, unlike in extant short-thumbed atelins (*Ateles*, mean=45.1 %; *Brachyteles*, mean=34.0 %). The relative length of the Mc1 in *Paralouatta* falls at the upper end of the range for extant anthropoids (Jouffroy et al. 1993; Sarlo 1996; Tague 1997; Harrison unpublished data), being most comparable to *Cebus* (mean=73.6 %) and *Theropithecus* (mean=77.7 %), which have well-developed thumbs associated with enhanced pollical opposability and manual dexterity (Jolly 1972; Torigoe 1985).

A small collection of isolated phalanges is known for *Paralouatta*, but it is not possible to assign them to a manual or pedal digit or to a particular ray, and in some cases they are not definitively identifiable as primate. The phalanges are stout and short relative to the length of the metacarpals (MacPhee and Meldrum 2006). The proximal end of the PP has an expanded base and a concave facet for the metapodial head. The palmar aspect of the shaft of the MPs and PPs bear well-developed flexor sheath ridges. The DPs are relatively short with expanded apical tufts (unlike extant platyrrhines, which have narrow apical tufts; Mittra et al. 2007). If the phalanges are correctly assigned to this species, the average length of the digits relative to the length of the Mc2 is shorter than in all platyrrhines (but most closely approaches *Chiropotes*) and arboreal cercopithecids and resembles the short digits of terrestrial and semiterrestrial Old World monkeys. Given the short digits and the relatively long first metacarpal, the hand of *Paralouatta* resembled that of *Theropithecus*, which is specialized for terrestrial locomotion and a high degree of pollical opposability.

2.2 *Parapithecoidea*

The parapithecoids represent a diverse group of small anthropoids from the late Eocene and early Oligocene of Egypt, Algeria, and Kenya (Seiffert et al. 2010; Ducrocq et al. 2011). They are best represented in the collections from the Fayum in Egypt, dating from ~29 to 37 Ma. Parapithecoids lack several key synapomorphies of extant anthropoids that indicate that they are stem anthropoids and the sister group to platyrrhines + catarrhines (Harrison 1987; Fleagle and Kay 1987; Kay et al. 1997; Ross et al. 1998; Simons 2001; Simons et al. 2001; Seiffert et al. 2010) (Fig. 16.1). The best-known species is *Apidium phiomense*, which is represented by a large sample of craniodental and postcranial specimens (Fleagle and Simons 1983, 1995; Fleagle and Kay 1987).

2.2.1 *Apidium phiomense*

The anatomy and functional morphology of the postcranium of *Apidium phiomense* from the Oligocene of the Fayum in Egypt (~30 Ma) have been well documented (Fleagle and Simons 1995). *Apidium phiomense* is a relatively small anthropoid with an estimated body mass of 1.6 kg. The morphology of the limb bones indicate that it was a fast-moving and agile arboreal quadruped specialized for leaping and above-branch running and walking, most similar to the smaller non-callitrichid platyrrhines, such as *Saimiri* and *Callicebus* (Fleagle and Simons 1995).

The published hand material is limited to two PPs from Fayum Quarry I (Hamrick et al. 1995). The combination of a dorsopalmarly compressed distal articular surface, straight and relatively stout shaft, and weakly developed flexor sheath ridges indicates that the phalanges of *Apidium* were frequently extended, most likely during quadrupedal locomotion on horizontal or large-diameter substrates. Overall, the manual phalanges most resemble those of extant arboreal small-bodied platyrrhines and cercopithecids, such as *Saimiri*, *Cebus*, and *Cercopithecus* (Hamrick et al. 1995).

2.3 Propliopithecoidea

The propliopithecoids are primarily known from the early Oligocene (~29–32 Ma) of Fayum, Egypt (Seiffert et al. 2010), but more fragmentary material has been recovered from contemporary localities in Oman and Angola. The best-known species is *Aegyptopithecus zeuxis* from the early Oligocene of Egypt (Simons et al. 2007; Seiffert et al. 2010). Craniodental features indicate that propliopithecoids are stem catarrhines that represent the sister taxon of all other fossil and extant catarrhines (Fleagle and Kay 1987; Harrison 1987; Seiffert et al. 2010).

2.3.1 *Aegyptopithecus zeuxis*

Aegyptopithecus is the largest of the fossil catarrhines from the Fayum, with an estimated body mass of 6.7 kg (Fleagle 2013). It was a heavily built arboreal quadruped, specialized for cautious climbing and clambering, with highly prehensile cheiridia. It appears to have been most similar in its locomotor behavior to *Alouatta* and lemurids (Fleagle et al. 1975; Conroy 1976; Fleagle and Simons 1982; Fleagle 1983).

Two PPs from the third and second manual digits are attributed to *Aegyptopithecus* (Preuschoft 1974; Hamrick et al. 1995). The palmar surface has well-developed ridges for the attachment of the flexor sheath. Strong digital flexors are associated with forelimb suspension and bridging, as well as powerful manual grasping during cautious above-branch locomotion on relatively small diameter supports (Preuschoft 1974; Hamrick et al. 1995; Richmond 2007). The distal articular surface narrows dorsally, with a sloping palmar surface and a deep trochlear groove. These features imply greater stability of the interphalangeal joint during digital flexion. The phalanges are moderately long and dorsopalmarly curved as in arboreal anthropoids. The phalangeal robusticity index (midshaft breadth × 100/length) of the phalanges is 17.8 and 19.6, which falls within the upper end of the range of platyrrhines and arboreal cercopithecids (Harrison 1989; Hamrick et al. 1995). The included angle of curvature for the two phalanges is 51° and 60°, which corresponds well with the range of curvature in large platyrrhine monkeys, such as *Lagothrix*, and is greater than in arboreal cercopithecids

(Susman et al. 1984; Rein 2011; Rein et al. 2011). The proximal end has prominent palmar tubercles, and the articular surface extends onto the dorsal surface of the shaft and is slightly proximodistally canted. This is found in cercopithecids and some platyrrhines in which there is hyperextension of the metacarpophalangeal joint in palmigrade postures (Hamrick et al. 1995; Rose et al. 1996; Rein and McCarty 2012).

2.4 *Pliopithecoidea*

The pliopithecoids represent a taxonomically diverse clade of stem catarrhines (Harrison 2013). They are more derived than the propliopithecoids but more primitive than the dendropithecoids and crown catarrhines (Harrison 1987, 2013; Andrews et al. 1996; Begun 2002). They had a wide geographical distribution throughout much of Eurasia during the Miocene (~18–7 Ma), extending from northern Spain to eastern China (Andrews et al. 1996; Harrison and Gu 1999; Begun 2002; Alba et al. 2010; Harrison 2013). Pliopithecoids are small- to medium-size catarrhines with estimated average body weights ranging from ~5 to ~15 kg (Fleagle 2013). The Pliopithecoidea are divided into two families, the Dionysopithecidae and Pliopithecidae; the latter is divided into two subfamilies, the Pliopithecinae and Crouzeliinae (Harrison and Gu 1999; Begun 2002; Harrison 2013). The earliest known and most primitive pliopithecoids are the dionysopithecids from Sihong, China (~17–18 Ma) (Harrison and Gu 1999). Several partial skeletons of *Pliopithecus vindobonensis* from Slovakia have been described (Zapfe 1960), but the postcranial skeleton of other pliopithecoids are represented by a few isolated elements only (Harrison and Gu 1999; Begun 2002; Senut 2012).

2.4.1 *Platodontopithecus jianghuaiensis*

Platodontopithecus is known from the early Miocene (~18–17 Ma) of Sihong in China (Harrison and Gu 1999), and, along with *Dionysopithecus*, represents the oldest and most primitive pliopithecoid. It is a large species with an estimated average body mass of ~15 kg (Harrison and Gu 1999). The species is mostly known from isolated teeth, but several postcranial specimens have been attributed to the taxon based on their size. These include a calcaneus, a pedal phalanx, and a pollical PP (Harrison and Gu 1999). The few postcranial remains indicate that *Platodontopithecus* was a generalized quadrupedal primate that favored above-branch walking and running on relatively large-diameter supports.

The pollical PP is complete. The length of the phalanx in relation to the estimated body weight demonstrates that the fossil phalanx is relatively longer than in all extant nonhuman catarrhines, with the exception of hylobatids, and most closely resembles those of platyrrhines and lemurids. In terms of its degree of robusticity, the midshaft diameter of the pollical phalanx is 21 % of the total length of the bone, falling within the lower end of the range for extant catarrhines. The shaft exhibits dorsopalmar curvature comparable to that found in extant African apes and large platyrrhines. The shaft is radioulnarly expanded midway along its length to accommodate well-developed flexor sheath ridges. The distal end of the phalanx is broad, with widely diverging condyles and a deep trochlear groove. The proximal end of the phalanx has a saddle-shaped depression for articulation with the head of the Mc1.

2.4.2 *Pliopithecus vindobonensis*

Pliopithecus vindobonensis, a pliopithecine from Devínská Nová Ves, Slovakia (~15.0–15.5 Ma), is known from partial skeletons of at least three individuals, as well as some isolated bones of additional individuals (Zapfe 1958, 1960). It was a medium-sized catarrhine with an estimated average body mass of ~7 kg (Fleagle 2013). Overall, the morphology of the skeleton is most similar to that of extant colobines, atelids, and lemurids and indicates that *Pliopithecus* was an agile above-branch quadruped, with a positional repertoire that included climbing, leaping, bridging, and suspensory postures (Zapfe 1960; Fleagle 1983; Rose 1993; Rein et al. 2011). The Mc3 is known from Individual I; a scaphoid, lunate, trapezium, and Mc1 are known from Individual II; and a scaphoid, lunate, hamate, and the Mc2-5 (proximal part of the Mc5 only) are known from Individual III (Fig. 16.2). Several phalanges of all three individuals are also preserved.

The *Pliopithecus* scaphoid is generally similar in morphology to those of extant platyrrhines and hylobatids. The os centrale is unfused. The lunate is proportionally similar to cercopithecids among extant anthropoids, being subequal in proximodistal length and dorsopalmar height, and relatively narrow radioulnarly. It lacks the expanded radial articular surface seen in extant hominoids. The hamate is narrow proximally, with a moderately developed hamulus and a proximodistally-oriented spiral facet (Godinot and Beard 1991). The trapezium has a relatively shallow saddle-shaped articular surface for the Mc1. The shape of the facet would have allowed extensive flexion-extension and abduction-adduction of the thumb, but a restricted degree of axial rotation. This

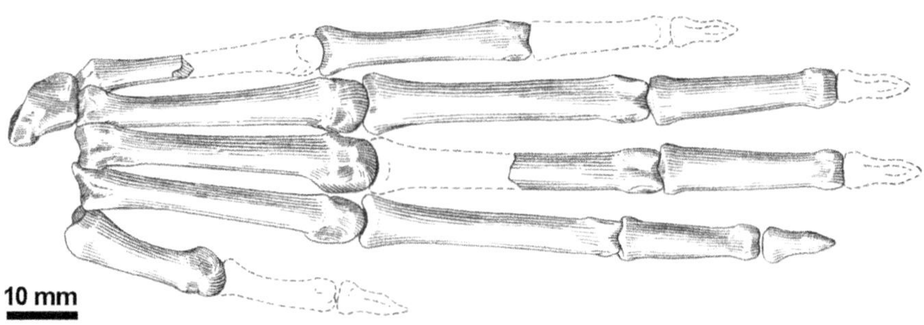

Fig. 16.2 Dorsal view of the partial left hand of *Pliopithecus vindobonensis* from Devínská Nová Ves, Slovakia. The first metacarpal is from Individual II (in mirror image) and the remaining bones are from Individual III (holotype). Adapted from Zapfe (1960)

is comparable to the pattern seen in strepsirrhines and many platyrrhines (Rose 1992) and contrasts with the more derived sellar joint typical of extant catarrhines, which permits axial rotation at the trapezium-metacarpal joint and true opposition (Napier 1961).

The Mc1 is long and stout, indicating a well-developed thumb. The length of the Mc1 (Individual II) is 56.2 % of the length of the Mc3 from Individual III and 52.2 % of the length of the Mc3 from Individual I (Zapfe 1960). Thus, the Mc1 of *Pliopithecus* is relatively shorter than that of most strepsirrhines and platyrrhines, but falls within the upper range for Asian colobines and in the lower range for cercopithecines (Zapfe 1960; Tague 1997). The proximal articular surface is saddle-shaped, with a strong dorsopalmar concavity and a shallow radioulnar convexity. This confirms that the range of flexion and extension was relatively great, while the ranges of abduction-adduction and axial rotation were more limited. The Mc2-Mc4 are moderately long, with a slight dorsopalmar curvature.

The PPs of the non-pollical digits are long and slender, with a strong degree of curvature. The shaft bears moderately well-developed flexor sheath ridges. The distal articular surface is relatively narrow and tapers dorsally. There are deep pits on the radial and ulnar sides of the head for attachment of strong collateral ligaments. The MPs are long and moderately curved, with well-developed flexor sheath ridges. The proximal articular surface has a distinct median ridge that ends palmarly and dorsally at small protuberances. The distal articular surface is broad and low, with a shallow trochlear groove and slight depressions for attachment of the collateral ligaments. A single DP is known from the second digit of the left hand of Individual III. It is relatively long and slender, with a narrow apical tuft (Zapfe 1960).

2.4.3 *Anapithecus hernyaki*

The best-known crouzeliine in Europe is *Anapithecus hernyaki* from Rudabánya in Hungary and other localities in central Europe (~10–11 Ma). It is a large pliopithecoid with an estimated body mass of ~15 kg (Begun 2002). Although considerably larger than *Pliopithecus vindobonensis*, comparisons of the postcranial material indicate that *Anapithecus* is generally similar in morphology, and like *Pliopithecus* can be inferred to have been an agile arboreal above-branch quadruped. However, *Anapithecus* differs in a number of features that suggest it was more specialized for forelimb and hind limb suspension (Kordos 2000; Begun 2002).

Several fragmentary phalanges from Rudabánya have been attributed to *Anapithecus*, some of which can be reasonably assigned to the hand (Begun 1988, 1993). These include two PPs (RUD 31 and 34) and two MPs (RUD 75 and 115). Several DPs (RUD 43, 59, and 60) are also attributed to *Anapithecus*, but it is not possible to determine if these are from the hand or foot. The PPs are long and slender. The distal end has a deep trochlear groove and prominent pits for strong collateral ligaments. The shaft exhibits a moderate degree of curvature and prominent flexor sheath ridges. All of these features resemble those of extant non-hominoid arboreal quadrupedal primates, especially colobines and large platyrrhines (Begun 1988). The

MPs have a broad proximal end, and the proximal facet has a well-developed median ridge that ends dorsally and palmarly in well-developed protuberances. Prominent tubercles on the palmar aspect of the proximal end are the sites of attachment for the collateral ligaments. The shaft is slightly curved and bears prominent flexor sheath ridges. The MPs are most similar to those of extant colobines and atelins (Begun 1988). The DPs preserve the proximal ends only. The base is broad, with a pair of concave facets separated by a weakly developed articular keel. The palmar surface of the shaft has an ill-defined tubercle for insertion of the flexor tendon.

The morphology of the phalanges indicates that *Anapithecus* had moderately long and curved manual digits, capable of powerful grasping on arboreal supports of variable diameter. It is likely that *Anapithecus* was primarily an above-branch quadrupedal climber and clamberer, but was also highly capable of adopting suspensory postures.

2.4.4 *Laccopithecus robustus*

Laccopithecus robustus, a late-surviving crouzeliine from the late Miocene (~7 Ma) of Shihuiba, Yunnan, China, is represented by a partial cranium, numerous jaw fragments, isolated teeth, and a single manual phalanx (Wu and Pan 1985; Meldrum and Pan 1988).

The incomplete PP, which preserves the distal end and much of the shaft, has been identified as coming from the fifth manual digit (Meldrum and Pan 1988). The shaft is strongly curved in the dorsopalmar plane, with an estimated included angle of curvature of 57°, which is comparable to that in extant hylobatids and atelins (Meldrum and Pan 1988), as well as in *Pliopithecus vindobonensis*. The flexor sheath ridges are well-developed. The distal articular condyles are separated by a moderately deep trochlear groove. The lateral condyle is elliptical and bears a roughened pit radially for the attachment of the collateral ligament. Both features are associated with increasing the stability of the interphalangeal joint in flexed positions. The overall morphology and the curvature of the shaft indicate that powerful manual grasping and forelimb suspension were important components of the positional repertoire.

2.4.5 Other Pliopithecoids

Isolated postcranial elements of other pliopithecids are known from Sansan in France (~14 Ma), attributable either to *Pliopithecus antiquus* or the slightly smaller *Plesiopliopithecus auscitanensis* (Senut 2012). These indicate that the Sansan pliopithecids were arboreal quadrupeds that were specialized for suspensory behaviors (Senut 2012). The only hand element is a complete and well-preserved Mc2. The triangular proximal articular surface for the trapezoid is radioulnarly compressed and bordered on the palmar margin by a well-developed tuberosity for a strong trapezium-metacarpal ligament and insertion of the flexor carpi radialis tendon. Two facets are present ulnarly for articulation with the Mc3. The shaft is slender and dorsopalmarly curved. The distal articular surface narrows dorsally and bears a pair of well-developed tubercles on the palmar surface. Overall, the Mc2 is similar to the corresponding bone in *Pliopithecus vindobonensis*, but is slightly smaller, more slender, and more strongly curved. In addition, a pliopithecid Mc1 is known from La Grive-Saint-Alban in France (~11–13 Ma) (Depéret 1887; Zapfe and Hürzeler 1957; Senut 2012), which could be referable to either *Pliopithecus antiquus* or *Plesiopliopithecus rhodanica*.

2.5 *Dendropithecoidea*

The dendropithecoids are stem catarrhines of modern aspect from the early to late Miocene of East Africa (Harrison 2002, 2010, 2013). They are more derived than the propliopithecids and pliopithecoids and represent the sister taxon of crown catarrhines (i.e., hominoids + cercopithecoids). The superfamily currently includes *Dendropithecus*, *Micropithecus*, and *Simiolus* (Harrison 2010, 2013). All are relatively small catarrhines, with an estimated average body mass ranging from 4 to 8 kg. Postcranial elements of *Dendropithecus macinnesi* and *Simiolus enjiessi* have been described (Le Gros Clark and Thomas 1951; Harrison 1982; Rose et al. 1992), but hand elements are known only for *Simiolus*. As noted by Rose et al. (1992), dendropithecoids do not closely resemble any extant anthropoids in their postcranial morphology or inferred positional behavior, but the best analogs are found among the platyrrhines. Dendropithecoids were predominantly arboreal above-branch quadrupeds, capable of scrambling, climbing, bridging, and suspensory behaviors in small branch settings (Harrison 1982; Rose et al. 1992).

2.5.1 *Simiolus enjiessi*

Postcranial remains of *Simiolus enjiessi* have been recovered from the early Miocene locality of Kalodirr in northern Kenya, dating to 16.7–17.7 Ma (Rose et al. 1992). These include a Mc3 and Mc2, as well as a PP and MP, although the phalanges cannot be determined as belonging to a hand or foot or to a particular digit.

The Mc2 consists of the proximal half only. The base and shaft are radioulnarly narrow. The trapezoid facet is concave with a pronounced ulnar lip. The radial side of the base has a small elliptical facet for the trapezium, which is similar in its shape and orientation to that of cercopithecids (Rose et al. 1992). Only the proximal end of the Mc3 is preserved. The base is radioulnarly narrow and the radial side bears a small facet dorsally for the Mc2. Just distal to this facet is a crescent-shaped tubercle for attachment of the intermetacarpal ligament. The two isolated phalanges preserve the proximal ends only. The base of the PP bears moderately well-developed palmar tubercles. The proximal articular surface of the MP is dorsopalmarly compressed, with a slight median ridge and a well-developed dorsal lip. The shaft has moderately well-developed flexor sheath ridges.

Overall, the metacarpals and phalanges indicate that the hand of *Simiolus* was narrow, with digits capable of powerful grasping of variable diameter supports, and most similar to those of extant arboreal above-branch quadrupedal cercopithecids and platyrrhines (Rose et al. 1992).

2.6 Other Small Catarrhines from the Miocene of East Africa

Several isolated hand elements from early Miocene localities in western Kenya (~17–20 Ma) have been provisionally referred to *Limnopithecus* and *Kalepithecus* (Harrison 1982). These are both small stem catarrhines of uncertain taxonomic status (Harrison 2010, 2013). The carpals include a lunate and several capitates (Harrison 1982). These are smaller than those of *Proconsul heseloni* (Napier and Davis 1959; Beard et al. 1986), but are otherwise similar in their general morphology and proportions (Harrison 1982; Daver 2009). A few metacarpals preserving only the distal end are also provisionally attributed to *Kalepithecus*, but they are too fragmentary to assign to particular rays (Harrison 1982). They are generally similar in morphology to those of *Simiolus* and *Proconsul*. The distal articular surfaces are heart-shaped in distal view, being broadest palmarly and tapering dorsally, suggesting that the metacarpophalangeal joints were most stable in semi-flexed positions. The palmar aspect of the metacarpal head has faint grooves separated by a low median keel for paired sesamoids. A similar pattern is seen in arboreal quadrupedal primates (Harrison 1982), whereas extant great apes and humans are derived in having lost the sesamoids on Mc2-5 (Le Minor 1988).

A complete Mc1 of an indeterminate small catarrhine is known from the early Miocene of Moruorot in northern Kenya (16.8–17.5 Ma) (Rose et al. 1992). It is proximodistally long relative to the diameter of the shaft, indicating that the thumb was well-developed (Rose et al. 1992). The shaft is dorsopalmarly compressed and only slightly curved. The trapezium facet is saddle-shaped, with a moderate radio-ulnar convexity and a relatively weak dorsopalmar concavity. This implies a wide range of abduction-adduction of the thumb, but that flexion-extension and axial rotation were somewhat limited. The head is radioulnarly narrow and tapers dorsally, with a slight degree of torsion relative to the base. There is also a large area of contact on the palmar surface of the distal articulation for the sesamoids of the metacarpophalangeal joint, associated with powerful grasping.

A number of isolated phalanges of indeterminate small catarrhine primates are known from the early Miocene of East Africa (Harrison 1982). Since the basic morphology is relatively uniform it is reasonable to infer that the following anatomical features characterized the digits of early Miocene small catarrhines in general. The PPs and MPs are long and slender with a slight to moderate degree of curvature, as in extant arboreal anthropoids. The distal articulation is narrow dorsally and flares palmarly, most similar to the morphology in extant arboreal quadrupedal primates. The DPs are long and slender with an apical tuft that is dorsopalmarly compressed and terminates in a relatively sharp point, as in extant hylobatids, cercopithecids, and some platyrrhines. The tips of the non-pollical digits would have borne a narrow and longitudinally curved nail or tegula. Overall, the phalangeal morphology indicates that the East African small catarrhines were arboreal quadrupeds, with relatively long, slender, and prehensile hands, that were capable of locomotion along supports of variable diameter (Harrison 1982).

2.7 *Cercopithecoidea*

Hand fossils have been described for a number of fossil cercopithecoids. These include *Victoriapithecus* and *Microcolobus* from the middle and late Miocene of East Africa, *Mesopithecus* from the late Miocene of Europe, and *Parapapio* and *Theropithecus* from the Plio-Pleistocene of Africa. Hand bones are also associated with an undescribed partial skeleton of the large terrestrial colobine, *Paracolobus mutiwa*, from the Lomekwi Member, West Turkana, Kenya (Harris et al. 1988; Jablonski and Frost 2010). Isolated hand bones of other Plio-Pleistocene cercopithecids are known, but these await formal description (Jolly 1972; Delson 1973; Frost and Delson 2002; Harrison 2011).

2.7.1 *Victoriapithecus macinnesi*

Victoriapithecus macinnesi, from the middle Miocene locality of Maboko Island in Kenya (~15–16 Ma; Harrison 1989), is the best-known representative of the stem cercopithecoid family, the Victoriapithecidae. It was relatively small, about the size of the extant vervet monkey, with an average body mass of 3.0–4.5 kg (Harrison 1989; Blue et al. 2006). Comparisons of the appendicular skeleton indicate that *Victoriapithecus* was well adapted for locomotion in both arboreal and terrestrial settings, but was predominantly a semiterrestrial quadruped (Harrison 1989; Blue et al. 2006). The hand material includes a complete right Mc1, a poorly preserved proximal Mc3, an almost complete shaft of a possible Mc3, two distal metacarpal fragments, and several phalanges from digits 2 to 5 not assigned to the hand or foot (Harrison 1989).

The Mc1 is long compared with the non-pollical metacarpals and quite stout, as in terrestrial and semiterrestrial Old World monkeys. The shaft exhibits moderate dorsopalmar curvature. The trapezium facet is elliptical and sellar in shape with a small dorsal lip and palmar projection. The radial aspect of the trapezium facet has an angular buttress along its margin and is less extensive, with a reduced degree of curvature relative to the medial side. This morphology implies a large range of abduction-adduction, flexion-extension, and a limited degree of axial rotation at the trapezium-Mc1 joint, with enhanced stability when the thumb is fully abducted. Overall, the morphology of the pollical carpometacarpal joint is most similar to that of arboreal cercopithecids.

The Mc3 is long and slender, as in extant arboreal and semiterrestrial cercopithecids. Proximally, the facet for the capitate narrows palmarly, as in extant non-hominoid anthropoids. The isolated distal metacarpals have articular heads that narrow slightly dorsally and have distinct gutters on the palmar aspect to support paired sesamoids, as in extant cercopithecids.

The PPs are short and stout. The robusticity index of the PPs is most similar to that seen in extant terrestrial cercopithecids (Harrison 1989). The end is radioulnarly broad with a dorsally canted proximal facet as in extant quadrupedal non-hominoid anthropoids. The shaft of the PPs is dorsopalmarly compressed, with moderately well-developed flexor sheath ridges and slight dorsal curvature. The MPs are short and stout, as in terrestrial and semiterrestrial cercopithecids. The distal articular surface is low and broad with a shallow trochlear groove. Rounded ridges for attachment of the flexor sheaths border the palmar surface of the shaft. The DPs are moderately long and slender, as in semiterrestrial Old World monkeys, being shorter and wider than in arboreal cercopithecids, and more slender than in highly terrestrial monkeys. The narrow apical tuft would have supported a slender nail or tegula.

The hand morphology, especially that of the phalanges, corroborates the inference drawn from the other postcranial elements that *Victoriapithecus* was a semiterrestrial cercopithecid, similar to *Chlorocebus aethiops*, that spent time moving and foraging on the ground, as well as in the trees.

2.7.2 *Microcolobus* sp.

Nakatsukasa et al. (2010) recently described two partial skeletons provisionally assigned to *Microcolobus* from the late Miocene locality of Nakali, Kenya (9.8–9.9 Ma). These skeletons represent the oldest known postcranial material of a colobine monkey. Body mass is estimated to be 4–6 kg (Nakatsukasa et al. 2010) and comparisons of the postcranial morphology with extant cercopithecids indicate that *Microcolobus* was an agile arboreal monkey.

The following hand elements are represented in the KNM-NA 47916 partial skeleton: Mc1, Mc2, and Mc4, three PPs, and two MPs (Figs. 16.3 and 16.4). The Mc1 lacks the head and part of the neck, so its total length cannot be measured or estimated. The dimensions of the proximal epiphysis and the shaft are similar to those of extant cercopithecines of similar body size, with no evidence of the pollical reduction seen in extant colobines. The proximal articular surface is saddle-shaped and relatively large compared with the small and relatively flat articular surface seen in extant African colobines. The Mc2 also lacks the distal end, but its total length can be estimated, and the Mc4 is complete. Both metacarpals are comparable in length to extant cercopithecines of similar body size (Nakatsukasa et al. 2010).

The PPs are moderately dorsopalmarly curved, with strong flexor sheath ridges, a concave palmar surface of the shaft distally, deep pits for the interphalangeal collateral ligaments, and a tall distal articular surface and deep trochlear groove, as in arboreal cercopithecids. The robusticity of the PPs

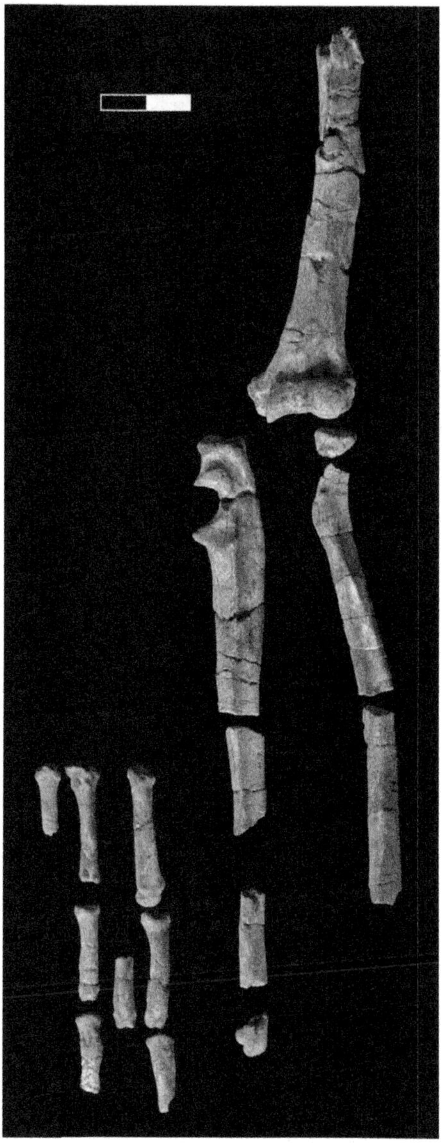

Fig. 16.3 Left forelimb of *Microcolobus* sp. (KNM-NA 47916), from Nakali, Kenya. Humerus in anterior view (*top*), radius in anterior view (*bottom, right*), ulna in radial view (*bottom, middle*), and partial manus in dorsal view (*bottom, left*). Scale bar = 2 cm. See Fig. 16.4 for detail of hand bones. Courtesy of Masato Nakatsukasa

(midshaft breadth × 100/maximum length) falls in the range of *Procolobus badius* (Harrison 1989). The two larger phalanges have been assigned to the third and fourth digits. If this assignment is correct, then the proximal phalanx/metacarpal index for the third ray indicates that *Microcolobus* had relatively long fingers, like that of arboreal cercopithecids. The two MPs are long, with a robusticity index that falls in the upper end of the range for arboreal and semi-terrestrial cercopithecids. The shafts are strongly curved dorsopalmarly as in arboreal Old World monkeys with strong flexor sheath ridges. The degree of

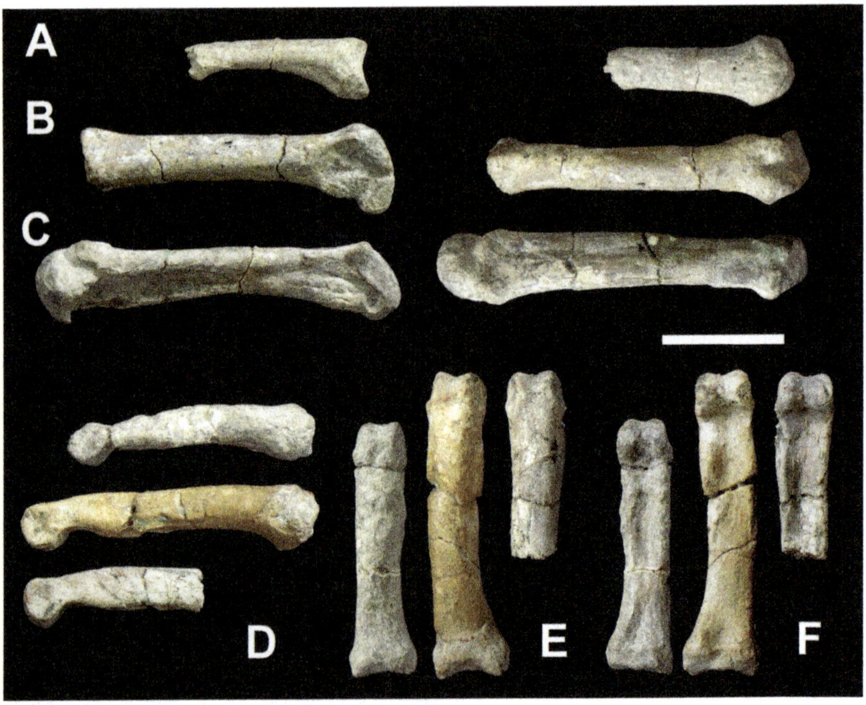

Fig. 16.4 Partial left manus of *Microcolobus* sp. (KNM-NA 47916) from Nakali, Kenya. First (**a**), second (**b**), and fourth (**c**) metacarpals, each in radial (*left*) and dorsal (*right*) views; manual proximal phalanges in radial (**d**), dorsal (**e**), and palmar (**f**) views. Scale bar = 10 mm. Courtesy of Masato Nakatsukasa

phalangeal curvature and the phalanx-metacarpal index are higher than those in semiterrestrial cercopithecids, indicating that *Microcolobus* was most likely a highly arboreal quadruped (Nakatsukasa et al. 2010).

The skeletal remains of *Microcolobus*, especially those of the hand, are important for several reasons that pertain to the evolutionary history of colobines (Nakatsukasa et al. 2010). First, if it is assumed, as seems most parsimonious, that cercopithecoids were primitively semiterrestrial, like the stem cercopithecoid *Victoriapithecus* (Andrews and Aiello 1984; Benefit 1987; Strasser and Delson 1987; Strasser 1988; Harrison 1989), then the evidence provided by *Microcolobus* shows that at least one lineage of small-bodied colobines secondarily became fully adapted to an arboreal lifestyle by 10 million years ago. Second, the retention of a well-developed and fully functional Mc1 (also seen in *Mesopithecus* from the late Miocene of Europe—see below) suggests that *Microcolobus* is more primitive than extant Asian and African colobines (with highly reduced thumbs), and that it probably represents the sister taxon of crown colobines (unless one posits the less parsimonious evolutionary scenario that extant Asian and African colobines reduced the size of their thumbs independently).

2.7.3 *Mesopithecus pentelicus*

Mesopithecus pentelicus is a fossil colobine, known mainly from the late Miocene of southeast Europe (but with a distribution that extends across Eurasia from England to Afghanistan) (Szalay and Delson 1979; Andrews et al. 1996). It has been suggested that *Mesopithecus* may be a member of the Asian colobine clade (subtribe Presbytina) (Jablonski 1998), but the consensus view is that the phylogenetic relationships are not adequately resolved, and in several respects it appears to be more primitive than all crown colobines. *Mesopithecus* is a medium-sized cercopithecid, with an estimated average body mass of 9 kg and 13 kg for females and males, respectively (Delson et al. 2000). Comparative studies of the postcranium indicate that *Mesopithecus* was a semiterrestrial monkey, similar to *Semnopithecus entellus*. The best material is from the type locality of Pikermi in Greece (~8.5 Ma), including several isolated hand elements, as well as three partial hand skeletons (Gaudry 1862; Zapfe 1991) (Fig. 16.5).

The carpals of *Mesopithecus* exhibit the characteristic morphology of extant cercopithecids (Zapfe 1991). The os centrale is unfused (i.e., is a separate carpal). As in all extant cercopithecids, the scaphoid has a well-developed, distally directed beak-like tubercle, which encompasses the proximal aspect of the os centrale and extends toward the angle between the head of the capitate and the trapezoid. The tubercle morphology allows the scaphoid to maintain greater contact with the os centrale during flexion and extension of the wrist, especially during dorsiflexion, and thus increases the stability of the midcarpal region (Napier and Davis 1959; Harrison 1982). The palmar aspect of the scaphoid bears a relatively weak tubercle compared

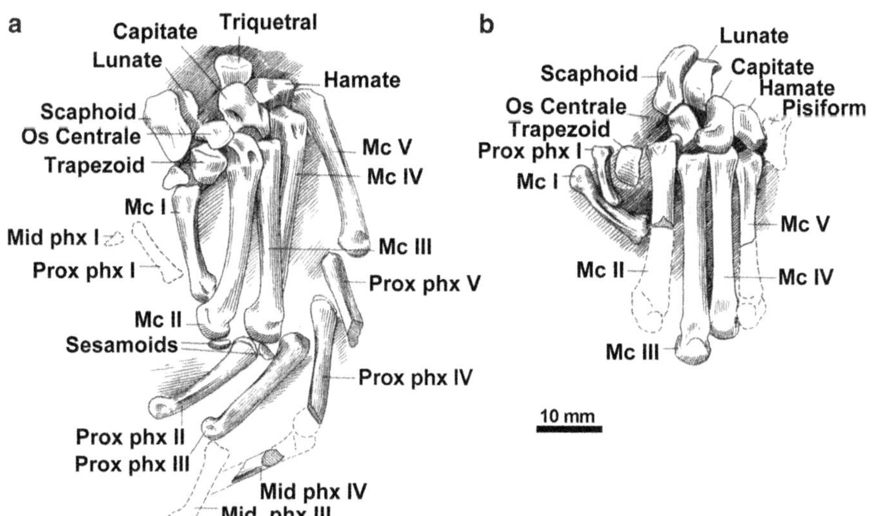

Fig. 16.5 Partially articulated hands of *Mesopithecus pentelicus* from Pikermi, Greece. (**a**) Left manus of a female individual; (**b**) left manus of a possible male individual. Abbreviations: *Mc* metacarpal, *Mid* middle, *phx* phalanx, *Prox* proximal. Adapted from Gaudry (1862)

to many extant anthropoids, but it is generally comparable to other cercopithecids. The lunate is morphologically and proportionally similar to those of extant colobines, differing from cercopithecines in having a radioulnarly narrower radial facet. The capitate has a radioulnarly wide body compared to the size of the head (Godinot and Beard 1991). The pisiform is relatively stout as in extant colobines (Zapfe 1991).

The Mc2-5 are short and robust, most similar to those of extant cercopithecines (Zapfe 1991). The length of the Mc1 is 57.2 % of the length of the Mc3, which is higher than the mean values for extant Asian (44.8–55.2 %) and African colobines (44.9–47.1 %), and falls in the lower end of the range for cercopithecines (55.9–77.7 %). This indicates that *Mesopithecus* (like *Microcolobus*) had not undergone the same degree of pollical reduction typical of extant colobines. The PPs and MPs of *Mesopithecus* are moderately short and only slightly curved, most comparable to those of semiterrestrial cercopithecids.

The well-developed thumb of *Mesopithecus* implies either that it is the primitive sister taxon of crown colobines (the inference preferred here) or that reduction of the thumb occurred independently in Asian and African colobine clades (Nakatsukasa et al. 2010). The overall hand morphology of *Mesopithecus*, which is most similar to extant terrestrial and semiterrestrial cercopithecoids, is consistent with the hypothesis that extant colobines were primitively semiterrestrial and that they secondarily became adapted to life in the trees.

2.7.4 *Parapapio ado*

Parapapio ado is known primarily from the mid-Pliocene Upper Laetolil Beds (3.65–3.8 Ma) in Tanzania. It is a medium-sized papionin with an estimated average body mass of 12 kg for females and 21 kg for males (Delson et al. 2000). Several postcranial elements have been provisionally attributed to this species (Leakey and Delson 1987; Harrison 2011), and they indicate that *Pp. ado* was a relatively slender and agile semiterrestrial monkey, generally similar in its positional behavior to extant *Cercocebus* and *Macaca*. It was likely adept in the trees but probably spent a good deal of time on the ground foraging and traveling between woodland and forest patches (Harrison 2011). A number of trails of cercopithecid footprints are known from Laetoli (Leakey 1987), and, given the presence of a sizeable thumb impression and overall similarity to those of extant baboons, these prints were most likely made by *Pp. ado*. If so, they confirm that *Pp. ado* spent time moving quadrupedally on the ground (Harrison 2011).

Harrison (2011) described several hand elements of *Pp. ado* (Fig. 16.6). The capitate is generally similar in morphology to that of *Lophocebus*. It differs primarily in being relatively proximodistally longer and having a less globular scaphoid facet and less concave Mc3 facet. The palmar tubercle is weakly developed and the hamate facet is only slightly convex. The Mc1 is long and stout, indicating that the pollex was well developed. The proximal articular surface is broad with a strong radioulnar convexity, most similar to that of *Lophocebus*, suggesting a wide range of abduction-adduction of the thumb. A fragment of the proximal end of the Mc5 is

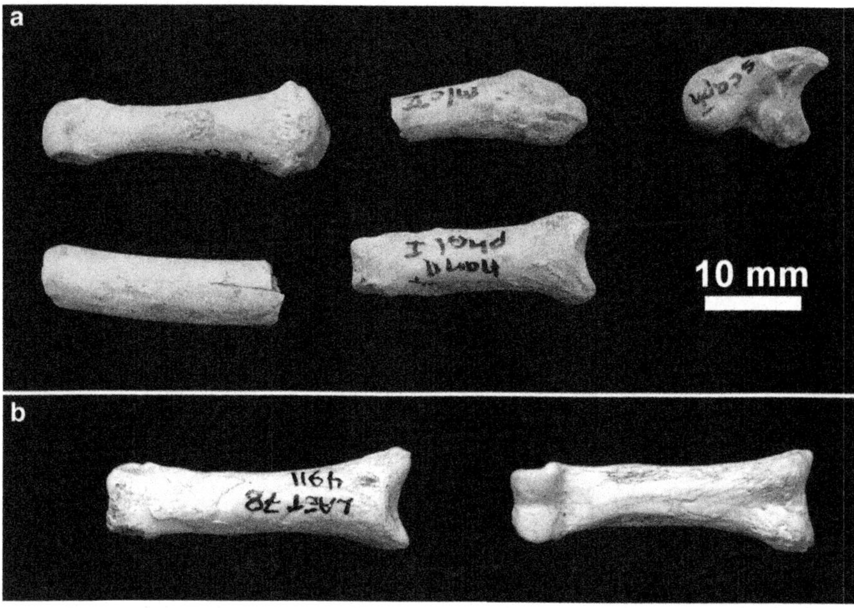

Fig. 16.6 Hand bones of *Parapapio ado* from Upper Laetolil Beds, Laetoli, Tanzania. (**a**) Associated bones of left manus (LAET 76-3904). *Top row*, from *left* to *right*, first metacarpal, fifth metacarpal, and capitate. *Bottom row*, from *left* to *right*, a metacarpal shaft fragment and proximal phalanx (probably ray 4). (**b**) Proximal manual phalanx (probably left ray 4) in dorsal (*left*) and palmar (*right*) views. After Harrison (2011)

preserved. It differs from extant African papionins in having a relatively broad proximal articular surface, a greater radioulnar convexity of the hamate facet, and a relatively longer facet for articulation with the Mc4. The manual PPs are short and robust, suggesting a high degree of terrestriality (Jolly 1967, 1972; Harrison 1989). The phalangeal robusticity index of 19.7–25.1 corresponds most closely to the range seen in terrestrial cercopithecids. The trochlea does not narrow dorsally, being most similar to *Papio* in this respect and implying joint stability in both flexion and extension. The broad and dorsopalmarly flattened shaft, the slight degree of curvature, and strongly developed and distally-placed flexor ridges are all similar to those of *Papio*.

The morphology of the hand fits well with the general interpretation of the postcranium that *Pp. ado* was a semiterrestrial papionin, similar in its locomotor behavior to extant *Cercocebus* or *Macaca*.

2.7.5 *Theropithecus brumpti*

Almost complete left and right hands of *Theropithecus brumpti* are preserved in association with a partial, adult male skeleton (KNM-WT 39368) from the Nachukui Formation of West Turkana (~3.3 Ma; Jablonski et al. 2002) (Fig. 16.7). The left

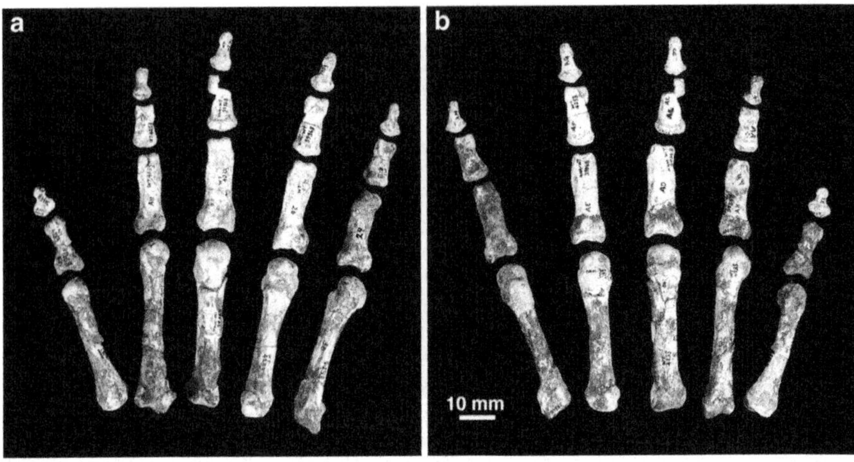

Fig. 16.7 Metacarpals and phalanges of left manus of *Theropithecus brumpti* associated with a partial skeleton (KNM-WT 39368) from Lomekwi, West Turkana, Kenya, in (**a**) palmar and (**b**) dorsal views. Courtesy of Nina Jablonski

hand preserves the phalanges, metacarpals, and most of the carpal bones. The right hand preserves, in a partially articulated state, complete second to fourth rays and partial first and fifth rays. The right lunate and scaphoid are visible as part of the articulated carpus. In addition, a partial right hand of *T. brumpti* is known from the Shungura Formation, Member E, Omo, Ethiopia (2.3–2.4 Ma; Jablonski 1986), in association with a partial skeleton of an aged male individual (L865-2) (Fig. 16.8). The hand comprises the metacarpals and PPs of all five rays, the MPs of digits 2–4, and a number of sesamoid bones.

The postcranial material indicates that *T. brumpti* was a large terrestrial papionin (males are estimated to have exceeded 40 kg), with several anatomical features that point to it being somewhat more adept at arboreal climbing than the extant *T. gelada* (Jablonski et al. 2002). The hand proportions are similar to those of extant *T. gelada*, with relatively short non-pollical digits and a long thumb. The opposability index (thumb length × 100/ray 2 length) is 69.2 in KNM-WT 39368 and estimated to be 70.6 in L865-2. These values are similar to the average opposability index of *T. gelada* (69.7) and much greater than those found in all other extant cercopithecoids (Napier and Napier 1967; Etter 1973; Jablonski 1986; Jouffroy et al. 1993). The elongated thumb and relatively short second digit in extant geladas is a specialization of the hand associated with a foraging strategy in which the thumb and index finger are used in a pincer-like opposition to harvest grass stems and blades (Napier and Napier 1967; Etter 1973). It is evident from the hand of *T. brumpti* that this specialized morphology had already been fully developed by at least the mid-Pliocene.

In addition to being short, the non-pollical metacarpals are robust and straight. They are also similar in length to one another; the length of the shortest metacarpal (Mc5) is 87–96 % the length of the longest metacarpal (Mc3). The non-pollical PPs have a robust shaft, a flattened palmar surface, a radioulnarly wide base,

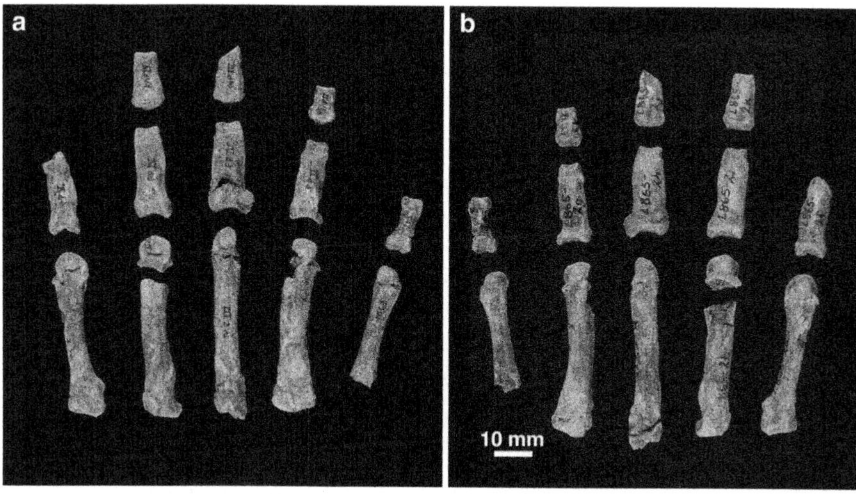

Fig. 16.8 Metacarpals and phalanges of right manus of *Theropithecus brumpti* associated with a partial skeleton (L865-2) from Shungura Formation, Omo, Ethiopia, in (**a**) palmar and (**b**) dorsal views. Courtesy of Nina Jablonski

and prominent flexor sheath ridges. They are slightly more curved than those of extant geladas, suggesting a greater facility for arboreal climbing (Jablonski 1986; Godinot and Beard 1991). However, the PPs and MPs are relatively short in relation to their midshaft breadth, with a robusticity index of 34.1 and 45.7, respectively. This exceeds the ranges of all extant cercopithecoids, except *T. gelada* (Harrison 1989), suggesting that the short, stout phalanges of *T. brumpti* would have been suitably adapted for terrestrial digitigrady. Since the phalanges are shorter and more robust than in *Papio* and *Mandrillus*, *T. brumpti* may have been more committed to terrestrial locomotion.

Overall, the evidence demonstrates that *T. brumpti* had a hand like that of modern geladas, specialized for terrestrial locomotion and for manual foraging that required precise opposition of the thumb and index finger.

3 Summary and Conclusions

Although our appreciation of the hand morphology of extinct anthropoids is limited by the paucity of the material, it is possible to make a number of observations about the evolutionary history of the hand in anthropoid primates. The only hand elements of a fossil platyrrhine older than the Pleistocene is associated with the partial skeleton of *Cebupithecia* from the late Miocene of Colombia. Hand elements are also associated with the skeletons of *Cartelles* and *Caipora* from the late Pleistocene of Brazil, and isolated hand bones are provisionally attributed to the contemporary *Paralouatta* from Cuba. These demonstrate that the diverse hand morphologies exemplified by modern-day platyrrhines were already a hallmark of platyrrhine communities in the later Tertiary of South America and the Caribbean islands. The hand of *Cebupithecia* generally resembles those of extant small platyrrhines, such as *Aotus* and *Callicebus*. This is consistent with the inference that *Cebupithecia* is a primitive pitheciin and an arboreal quadruped with specializations for leaping and vertical clinging. The hands of *Protopithecus* and *Caipora* were adapted for suspensory postures and locomotion as in extant atelins. If the hand material attributed to *Paralouatta* is correctly assigned, then it indicates that at least one lineage of platyrrhines may have successfully developed specializations for terrestrial locomotion, a locomotor behavior not seen in any extant New World monkeys. In addition, the hand of *Paralouatta* may have had a relatively long thumb and short non-pollical digits, a highly derived condition similar to that seen in the extant gelada.

In the early Oligocene of Egypt, the hands of *Apidium* and *Aegyptopithecus* are most similar to those of extant arboreal platyrrhines and cercopithecids. The stem anthropoid *Apidium* can be inferred to be an arboreal above-branch quadruped that was specialized for running along and leaping between large-diameter supports, whereas the early stem catarrhine *Aegyptopithecus* was more specialized for cautious climbing and clambering among supports of variable diameter using its powerful grasping hands.

Hand elements are known for several pliopithecoids, including the dionysopithecid *Platodontopithecus* from the early Miocene of China, *Pliopithecus* from the middle Miocene of Europe, and the crouzeliines *Anapithecus* and *Laccopithecus* from the late Miocene of Eurasia. They had long and narrow hands with well-developed thumbs and enhanced grasping capabilities. Overall, the postcranial morphology indicates that pliopithecoids were predominantly above-branch quadrupeds, specialized for climbing, clambering, and leaping, but exhibiting varying degrees of capability for forelimb (and hind limb) suspensory postures.

The hand morphology of dendropithecoids and other small stem catarrhines from the Miocene of East Africa is quite similar to that of pliopithecoids. The closest modern analogs are the hands of extant platyrrhines, rather than the more derived hand morphologies of extant cercopithecoids and hominoids. The thumb and non-pollical digits were relatively long with well-developed capabilities for powerful grasping. The evidence from the hand fits well with interpretations based on the rest of the postcranial skeleton that dendropithecoids were predominantly arboreal above-branch quadrupeds, adept at climbing, clambering, and bridging in small branch settings, but also capable of suspensory behaviors. The hand morphology in pliopithecoids and dendropithecoids demonstrates that Miocene stem catarrhines primitively retained a well-developed thumb, unlike the relatively abbreviated thumbs seen in extant colobines and most hominoids (Straus 1942).

The fossil record documenting the hand morphology of cercopithecoids is much better known than for other anthropoids. Hand elements of stem cercopithecoids (i.e., *Victoriapithecus*), early colobines (i.e., *Microcolobus* and

Mesopithecus), and Plio-Pleistocene papionin cercopithecines (i.e., *Parapapio* and *Theropithecus*) have been described previously in some detail. The short and robust phalanges of *Victoriapithecus* provide support for the conclusion, based on other elements of the postcranium, that the earliest known Old World monkeys were specialized for semiterrestrial locomotion. It has been inferred that this represents the ancestral locomotor behavior of cercopithecoids, and that crown Old World monkeys have subsequently reinvaded arboreal niches (Andrews and Aiello 1984; Benefit 1987; Strasser 1988; Harrison 1989; Blue et al. 2006). This view is supported by the inferred locomotor behavior of *Mesopithecus*, which indicates that the earliest Eurasian colobine was a semiterrestrial quadruped. However, the interpretation of *Microcolobus* from late Miocene East Africa as fully arboreal may support the alternative possibility that stem cercopithecoids were primitively arboreal (Napier 1967, 1970; Delson and Andrews 1975; Rollinson and Martin 1981). In this latter scenario, different lineages of cercopithecoids took advantage of exploiting the open woodlands and grasslands that expanded across Africa and Eurasia from the middle Miocene onward by independently developing specializations for semiterrestrial and terrestrial locomotion. However, comparative evidence and parsimony imply that this hypothesis is less plausible than the alternative scenario in which the ancestral locomotor behavior is inferred to be semiterrestrial. All fossil and extant cercopithecoids share a suite of derived features of the postcranial skeleton, in comparison to hominoids and stem catarrhines that are functionally associated with increasing the stability of the limb joints for movements in the parasagittal plane (Harrison 1989; Richmond et al. 1998). These are best interpreted as adaptations for more effective travel in terrestrial settings, implying that ancestral cercopithecoids were semiterrestrial.

The late Miocene colobines *Mesopithecus* and *Microcolobus* appear to have retained a well-developed thumb, similar to the primitive catarrhine condition, and had not undergone the degree of pollical abbreviation typical of extant Asian and African colobine monkeys. The most parsimonious assumption to draw from this evidence is that *Mesopithecus* and *Microcolobus* both represent stem colobines that diverged prior to the most recent common ancestor of crown colobines.

Lastly, the morphology of the hand of *T. brumpti*, from the mid-Pliocene of East Africa, clearly shows that by ~3.3 Ma fossil geladas had already developed the unique dietary and foraging specializations of modern *T. gelada* (as also evident from dentognathic adaptations and stable isotope data), associated with harvesting grass stems and blades (Jablonski and Frost 2010; Cerling et al. 2014).

4 Future Directions

Understanding of the anatomy, functional morphology, and evolution of the hand in non-hominoid anthropoid primates is constrained by the lack of adequate fossil material. Few taxa have a good representation of the hand, and in the majority of species, hand bones are unknown or are represented only by a few isolated elements. Clearly, additional fossil material is needed. However, many museum collections do include postcranial remains that are awaiting detailed analysis and description, and much could be learned about the anatomy and evolution of the hand from a more comprehensive study of material that is already available.

Beyond the limitations of the material, there are several key problems that relate to the evolution of the hand in anthropoids that still need to be resolved. These have been touched upon in the preceding descriptive account and discussion, but can be summarized here: (1) the hands of extant New World monkeys exhibit a remarkable diversity of form and function, so it would be important to elucidate whether platyrrhine communities in the past exhibited similar (or greater) levels of adaptive diversity and whether the morphology of the hand can contribute to current debates surrounding the phylogenetic relationships of fossil platyrrhines; (2) a better understanding of the hand morphology in fossil cercopithecoids could potentially help resolve the enduring question of whether ancestral Old World monkeys were arboreal or semiterrestrial; and (3) a shortened thumb is a characteristic derived feature of extant colobines, but it is uncertain whether the relatively well-developed thumbs in later Miocene colobines represent the retention of a more primitive condition or whether pollical reduction has occurred independently in different lineages of crown colobines.

Acknowledgments We would like to thank the editors for inviting us to prepare a contribution for this volume and for their many helpful comments on an earlier draft manuscript. We thank the following institutions and their staff for allowing access to the fossil specimens and comparative material in their care: National Museums of Kenya; National Museum of Tanzania; The Natural History Museum, London; Naturhistorisches Museum, Basel; American Museum of Natural History; and Institute of Vertebrate Paleontology and Paleoanthropology, Beijing. Masato Kunimatsu, Nina Jablonski, and Tess Wilson kindly provided photographs. Numerous colleagues have contributed to the research and ideas presented here, but the following deserve special mention: P. Andrews, D. Begun, E. Delson, J. Fleagle, Y. Gu, L. Halenar, N. Jablonski, C. Jolly, M.G. Leakey, M. Nakatsukasa, M. Rose, L. Sarlo, B. Senut, A. Walker, and H. Zapfe. Research was supported by grants from the National Geographic Society, the Leakey Foundation, and the National Science Foundation (Grants BCS-9903434 and BCS-0309513).

References

Alba DM, Moyà-Solà S, Malgosa A, Casanovas-Vilar I, Robles JM, Almécija S, Galindo J, Rotgers C, Mengual JVB (2010) A new species of *Pliopithecus* Gervais, 1849 (Primates: Pliopithecidae) from the Middle Miocene (MN8) of Abocador de Can Mata (els Hostalets de Pierola, Catalonia, Spain). Am J Phys Anthropol 141:52–75

Andrews P, Aiello L (1984) An evolutionary model for feeding and positional behaviour. In: Chivers DJ, Wood BA, Bilsborough A (eds) Food acquisition and processing in primates. Plenum Press, New York, pp 429–466

Andrews P, Harrison T, Delson E, Bernor RL, Martin L (1996) Distribution and biochronology of European and Southwest Asian Miocene catarrhines. In: Bernor RL, Fahlbusch V, Mittmann H-W (eds) The evolution of Western Eurasian Neogene mammal faunas. Columbia University Press, New York, pp 168–207

Beard K, Teaford MF, Walker A (1986) New wrist bones of *Proconsul africanus* and *P. nyanzae* from Rusinga Island, Kenya. Folia Primatol 47:97–118

Begun DR (1988) Catarrhine phalanges from the Late Miocene (Vallesian) of Rudabánya, Hungary. J Hum Evol 17:413–438

Begun DR (1993) New catarrhine phalanges from Rudabánya (Northeastern Hungary) and the problem of parallelism and convergence in hominoid postcranial morphology. J Hum Evol 24:373–402

Begun DR (2002) The Pliopithecoidea. In: Hartwig WC (ed) The primate fossil record. Cambridge University Press, Cambridge, pp 221–240

Benefit BR (1987) The molar morphology, natural history, and phylogenetic position of the Middle Miocene monkey, *Victoriapithecus*, and their implications for understanding the evolution of the Old World monkeys. Ph.D. dissertation, New York University

Blue KT, McCrossin ML, Benefit BR (2006) Terrestriality in a Middle Miocene context: *Victoriapithecus* from Maboko, Kenya. In: Ishida H, Tuttle R, Pickford M, Ogihara N, Nakatsukasa M (eds) Human origins and environmental backgrounds. Springer, Dordrecht, pp 45–58

Cartelle C, Hartwig WC (1996) A new extinct primate among the Pleistocene megafauna of Bahia, Brazil. Proc Natl Acad Sci U S A 93:6405–6409

Cerling TE, Chritz KL, Jablonski NG, Leakey MG, Manthi FK (2014) Diet of *Theropithecus* from 4 to 1 Ma in Kenya. Proc Natl Acad Sci U S A 110:10507–10512

Conroy GC (1976) Primate postcranial remains from the Oligocene of Egypt. Contrib Primatol 8:1–134

Daver G (2009) The articular wrist complex in Miocene and Plio-Pleistocene African hominoids: anatomofunctional and morphometric comparative approach. Bull Mém Soc Anthropol Paris 21:233–239

Delson E (1973) Fossil colobine monkeys of the circum-Mediterranean region and the evolutionary history of the Cercopithecidae (Primates, Mammalia). Ph.D. dissertation, Columbia University

Delson E, Andrews P (1975) Evolution and interrelationships of the catarrhine primates. In: Luckett WP, Szalay FS (eds) Phylogeny of the primates: a multidisciplinary approach. Plenum Press, New York, pp 405–446

Delson E, Terranova CJ, Jungers WL, Sargis EJ, Jablonski NG, Dechow PC (2000) Body mass in Cercopithecidae (Primates, Mammalia): estimation and scaling in extinct and extant taxa. Anthropol Pap Am Mus Nat Hist 83:1–159

Depéret C (1887) Recherches sur la succession des faunes de vertébrés miocènes de la vallée du Rhône. Arch Mus Hist Nat Lyon 4:1–269

Ducrocq S, Manthi FK, Lihoreau F (2011) First record of a parapithecid primate from the Oligocene of Kenya. J Hum Evol 61:327–331

Etter HF (1973) Terrestrial adaptations in the hands of Cercopithecinae. Folia Primatol 20:331–350

Fleagle JG (1983) Locomotor adaptations of Oligocene and Miocene hominoids and their phyletic implications. In: Ciochon RL, Corruccini RS (eds) New interpretations of ape and human ancestry. Plenum Press, New York, pp 301–324

Fleagle JG (2013) Primate adaptation and evolution, 3rd edn. Academic Press, San Diego

Fleagle JG, Kay RF (1987) The phyletic position of the Parapithecidae. J Hum Evol 16:483–532

Fleagle JG, Meldrum DJ (1988) Locomotor behavior and skeletal morphology of two sympatric pitheciine monkeys, *Pithecia pithecia* and *Chiropotes satanus*. Am J Primatol 16:227–249

Fleagle JG, Simons EL (1982) The humerus of *Aegyptopithecus zeuxis*: a primitive anthropoid. Am J Phys Anthropol 59:175–193

Fleagle JG, Simons EL (1983) The tibio-fibular articulation in *Apidium phiomense*, an Oligocene anthropoid. Nature 301:238–239

Fleagle JG, Simons EL (1995) Limb skeleton and locomotor adaptations of *Apidium phiomense*, an Oligocene anthropoid from Egypt. Am J Phys Anthropol 97:235–289

Fleagle JG, Simons EL, Conroy GC (1975) Ape limb bone from Oligocene of Egypt. Science 189:135–137

Ford S (1990) Locomotor adaptations of fossil platyrrhines. J Hum Evol 19:141–174

Frost SR, Delson E (2002) Fossil Cercopithecidae from the Hadar Formation and surrounding areas of the Afar Depression, Ethiopia. J Hum Evol 43:687–748
Gaudry A (1862) Animaux Fossiles et Géologie de l'Attique. F. Savy, Paris
Godinot M, Beard KC (1991) Fossil primate hands: a review and an evolutionary inquiry emphasizing early forms. Hum Evol 6:307–354
Guedes PG, de Salles LO (2005) New insights on the phylogenetic relationships of the two giant extinct New World monkeys (Primates, Platyrrhini). Arq Mus Nac Rio de Janeiro 63:147–159
Halenar LB (2011a) Reconstructing the locomotor repertoire of *Protopithecus brasiliensis*. I Body size. Anat Rec 294:2024–2047
Halenar LB (2011b) Reconstructing the locomotor repertoire of *Protopithecus brasiliensis*. II Forelimb morphology. Anat Rec 294:2048–2063
Halenar LB (2012) Paleobiology of *Protopithecus brasiliensis*, a plus-size Pleistocene platyrrhine from Brazil. Ph.D. dissertation, City University of New York
Halenar LB, Rosenberger AL (2013) A closer look at the "Protopithecus" fossil assemblages: new genus and species from Bahia, Brazil. J Hum Evol 65:374–390
Hamrick MW, Meldrum DJ, Simons EL (1995) Anthropoid phalanges from the Oligocene of Egypt. J Hum Evol 28:121–145
Harris JM, Brown FH, Leakey MG (1988) Stratigraphy and paleontology of Pliocene and Pleistocene localities west of Lake Turkana, Kenya. Contr Sci Mus Nat His Los Angeles 399:1–128
Harrison T (1982) Small-bodied apes from the Miocene of East Africa. Ph.D. dissertation, University of London
Harrison T (1987) The phylogenetic relationships of the early catarrhine primates: a review of the current evidence. J Hum Evol 16:41–80
Harrison T (1989) New postcranial remains of *Victoriapithecus* from the middle Miocene of Kenya. J Hum Evol 18:3–54
Harrison T (2002) Late Oligocene to middle Miocene catarrhines from Afro-Arabia. In: Hartwig WC (ed) The primate fossil record. Cambridge University Press, Cambridge, pp 311–338
Harrison T (2010) Dendropithecoidea, Proconsuloidea and Hominoidea. In: Werdelin L, Sanders WJ (eds) Cenozoic mammals of Africa. University of California Press, Berkeley, pp 429–469
Harrison T (2011) Cercopithecids (Cercopithecidae, Primates). In: Harrison T (ed) Paleontology and geology of Laetoli: human evolution in context, vol 2, Fossil Hominins and the associated fauna. Springer, Dordrecht, pp 83–139
Harrison T (2013) Catarrhine origins. In: Begun DR (ed) A companion to paleoanthropology. Blackwell, Hoboken, pp 376–396
Harrison T, Gu Y (1999) Taxonomy and phylogenetic relationships of early Miocene catarrhines from Sihong, China. J Hum Evol 37:225–277
Hartwig WC, Cartelle C (1996) A complete skeleton of the giant South American primate *Protopithecus*. Nature 381:307–311
Hartwig WC, Meldrum J (2002) Miocene platyrrhines of the northern Neotropics. In: Hartwig WC (ed) The primate fossil record. Cambridge University Press, Cambridge, pp 175–188
Horovitz I (1999) A phylogenetic study of living and fossil platyrrhines. Am Mus Novit 3269:1–40
Horovitz I, MacPhee RDE (1999) The Quaternary Cuban platyrrhine *Paralouatta varonai* and the origin of the Antillean monkeys. J Hum Evol 36:33–68
Jablonski NG (1986) The hand of *Theropithecus brumpti*. In: Else JG, Lee PC (eds) Primate evolution. Cambridge University Press, Cambridge, pp 173–182
Jablonski NG (1998) The evolution of the doucs and snub-nosed monkeys and the question of the phyletic unity of the odd-nosed colobines. In: Jablonski NG (ed) The natural history of the Douc and Snub-nosed monkeys. World Scientific Press, Singapore, pp 13–52
Jablonski NG, Frost S (2010) Cercopithecoidea. In: Werdelin L, Sanders WJ (eds) Cenozoic mammals of Africa. University of California Press, Berkeley, pp 393–428
Jablonski NG, Leakey MG, Kiarie C, Antón M (2002) A new skeleton of *Theropithecus brumpti* (Primates, Cercopithecidae) from Lomekwi, West Turkana, Kenya. J Hum Evol 43:887–923
Jolly CJ (1967) The evolution of the baboons. In: Vagtborg H (ed) The baboon in medical research, vol 2. University of Texas Press, Austin, pp 427–457

Jolly CJ (1972) The classification and natural history of *Theropithecus* (*Simopithecus*) (Andrews, 1916), baboons of the African Plio-Pleistocene. Bull Br Mus Nat Hist Geol 22:1–122

Jones AL (2008) The evolution of brachiation in ateline primates, ancestral character states and history. Am J Phys Anthropol 137:123–144

Jouffroy FK, Godinot M, Nakano Y (1993) Biometrical characteristics of primate hands. In: Preuschoft H, Chivers DJ (eds) Hands of primates. Springer, Vienna, pp 133–171

Kay RF (1990) The phyletic relationships of extant and fossil Pitheciinae (Platyrrhini, Anthropoidea). J Hum Evol 19:175–208

Kay RF, Ross C, Williams BA (1997) Anthropoid origins. Science 275:797–803

Kordos L (2000) New results of hominoid research in the Carpathian Basin. Acta Biol Szegediensis 44:71–74

Leakey MD (1987) Animal prints and trails. In: Leakey MD, Harris JM (eds) Laetoli: a Pliocene Site in Northern Tanzania. Clarendon, Oxford, pp 451–489

Leakey MG, Delson E (1987) Fossil Cercopithecidae from the Laetolil Beds. In: Leakey MD, Harris JM (eds) Laetoli: a Pliocene Site in Northern Tanzania. Clarendon, Oxford, pp 91–107

Le Gros Clark WE, Thomas DP (1951) Associated jaws and limb bones of *Limnopithecus macinnesi*. Fossil Mammals of Africa No. 3. British Museum (Natural History), London

Le Minor J-M (1988) The ventral metacarpo- and metatarso-phalangeal sesamoid bones: comparative anatomy and evolutionary aspects. Gegenbaurs Morphol Jahrb 134:693–731

MacPhee RDE, Horovitz I (2002) Extinct Quaternary platyrrhines of the Greater Antilles and Brazil. In: Hartwig WC (ed) The primate fossil record. Cambridge University Press, Cambridge, pp 189–200

MacPhee RDE, Meldrum J (2006) Postcranial remains of the extinct monkeys of the Greater Antilles, with evidence for semiterrestriality in *Paralouatta*. Am Mus Novit 3516:1–65

Meldrum J (1993) Postcranial adaptations and positional behavior in fossil platyrrhines. In: Gebo DL (ed) Postcranial adaptation in nonhuman primates. Northern Illinois University Press, DeKalb, pp 235–251

Meldrum DJ, Pan Y (1988) Manual proximal phalanx of *Laccopithecus robustus* from the latest Miocene site of Lufeng. J Hum Evol 17:719–731

Meldrum DJ, Lemelin P (1991) Axial skeleton of *Cebupithecia sarmientoi* (Pitheciinae, Platyrrhini) from the Middle Miocene of La Venta, Columbia. Am J Primatol 25:69–90

Meldrum DJ, Fleagle JG, Kay RF (1990) Partial humeri of two Miocene Colombian primates. Am J Phys Anthropol 81:413–422

Mittra ES, Smith HF, Lemelin P, Jungers WL (2007) Comparative morphometrics of the primate apical tuft. Am J Phys Anthropol 134:449–459

Nakatsukasa M, Mbua E, Sawada Y, Sakai T, Nakaya H, Yano W, Kunimatsu Y (2010) Earliest colobine skeletons from Nakali, Kenya. Am J Phys Anthropol 143:365–382

Napier JR (1961) Prehensility and opposability in the hands of primates. Symp Zool Soc Lond 5:115–132

Napier JR (1967) Evolutionary aspects of primate locomotion. Am J Phys Anthropol 27:333–341

Napier JR (1970) Paleoecology and catarrhine evolution. In: Napier JR, Napier PH (eds) Old world monkeys: evolution, systematics, and behavior. Academic Press, New York, pp 53–95

Napier JR, Davis PR (1959) The fore-limb skeleton and associated remains of *Proconsul africanus*. Fossil Mammals of Africa, No. 16. British Museum (Natural History), London

Napier JR, Napier PH (1967) A handbook of living primates. Academic, London

Preuschoft H (1974) Body posture and mode of locomotion in fossil primates—method and example: *Aegyptopithecus zeuxis*. In: Kondo S, Kawai M, Ehara A, Kawamura S (eds) Proceedings from the symposia of the fifth congress of the International Primatological Society. Japan Science Press, Tokyo, pp 345–359

Rein TR (2010) Locomotor function and phylogeny: implications for interpreting the hominoid fossil record. Ph.D. dissertation, New York University

Rein TR (2011) The correspondence between proximal phalanx morphology and locomotion: implications for inferring the locomotor behavior of fossil catarrhines. Am J Phys Anthropol 146:435–445

Rein TR, McCarty LA (2012) Metacarpophalangeal joint orientation in anthropoid manual phalanges. Anat Rec 295:2057–2068

Rein TR, Harrison T, Zollikofer CPE (2011) Skeletal correlates of quadrupedalism and climbing in the anthropoid forelimb: implications for inferring locomotion in Miocene catarrhines. J Hum Evol 61:564–574

Richmond BG (2007) Biomechanics of phalangeal curvature. J Hum Evol 53:678–690

Richmond BG, Fleagle JG, Kappelman J, Swisher CC (1998) First hominoid from the Miocene of Ethiopia and the evolution of the catarrhine elbow. Am J Phys Anthropol 105:257–277

Rivero M, Arredondo O (1991) *Paralouatta varonai*, a new Quaternary platyrrhine from Cuba. J Hum Evol 21:1–11

Rollinson J, Martin RD (1981) Comparative aspects of primate locomotion with special reference to arboreal cercopithecines. Symp Zool Soc Lond 48:377–427

Rose MD (1992) Kinematics of the trapezium-1st metacarpal joint in extant anthropoids and Miocene hominoids. J Hum Evol 22:255–266

Rose MD (1993) Locomotor anatomy of Miocene hominoids. In: Gebo DL (ed) Postcranial adaptation in nonhuman primates. Northern Illinois University Press, DeKalb, pp 252–272

Rose MD, Leakey MG, Leakey REF, Walker AC (1992) Postcranial specimens of *Simiolus enjiessi* and other primitive catarrhines from the early Miocene of Lake Turkana, Kenya. J Hum Evol 22:171–237

Rose MD, Nakano Y, Ishida H (1996) *Kenyapithecus* postcranial specimens from Nachola, Kenya. Afr Stud Monogr 24:3–56

Rosenberger AL (2002) Platyrrhine paleontology and systematics: the paradigm shifts. In: Hartwig WC (ed) The primate fossil record. Cambridge University Press, Cambridge, pp 151–159

Rosenberger AL (2011) Evolutionary morphology, platyrrhine evolution, and systematics. Anat Rec 294:1955–1974

Rosenberger AL, Tejedor MF, Cooke SB, Pekar S (2009) Platyrrhine ecophylogenetics in space and time. In: Garber PA, Estrada A, Bicca-Marques JC, Heymann EW, Strier K (eds) South American primates: comparative perspectives in the study of behavior, ecology, and conservation. Springer, New York, pp 69–113

Ross CF, Williams B, Kay RF (1998) Phylogenetic analysis of anthropoid relationships. J Hum Evol 35:221–306

Sarlo LM (1996) Hand dimensions, body size, and substrate utilization among living primates. M.A. thesis, New York University

Seiffert ER, Simons EL, Fleagle JG, Godinot M (2010) Paleogene anthropoids. In: Werdelin L, Sanders WJ (eds) Cenozoic mammals of Africa. University of California Press, Berkeley, pp 369–391

Senut B (2012) Les restes post-crâniens des Pliopithecidae (Primates) de Sansan. In: Peigné S, Sen S (eds) Mammifères de Sansan. Mémoires du Muséum national d'Histoire naturelle 203. Muséum national d'Histoire naturelle, Paris, pp 535–558

Simons EL (2001) The cranium of *Parapithecus grangeri*, and Egyptian Oligocene anthropoidean primate. Proc Natl Acad Sci U S A 98:7892–7897

Simons EL, Seiffert ER, Chatrath PS, Attia Y (2001) Earliest record of a parapithecid anthropoid from the Jebel Qatrani Formation, northern Egypt. Folia Primatol 72:316–331

Simons EL, Seiffert ER, Ryan TM, Attia Y (2007) A remarkable female cranium of the early Oligocene anthropoid *Aegyptopithecus zeuxis* (Catarrhini, Propliopithecidae). Proc Natl Acad Sci U S A 104:8731–8736

Stirton RA (1951) Ceboid monkeys from the Miocene of Colombia. Univ Calif Publ Geol Sci 28:315–355

Stirton RA, Savage DE (1950) A new monkey from the La Venta Miocene of Colombia. Compilacion de los Estudios Geologicos Oficiales en Colombia 8:347–356

Strasser E (1988) Pedal evidence for the origin and diversification of cercopithecoid clades. J Hum Evol 17:225–245

Strasser E, Delson E (1987) Cladistic analysis of cercopithecid relationships. J Hum Evol 16:81–99

Straus WL (1942) Rudimentary digits in primates. Q Rev Biol 17:228–243

Susman RL, Stern JT, Jungers WL (1984) Arboreality and bipedality in the Hadar hominids. Folia Primatol 43:113–156

Szalay FS, Delson E (1979) Evolutionary history of the primates. Academic Press, New York

Tague RG (1997) Variability of a vestigial structure: first metacarpal in *Colobus guereza* and *Ateles geoffroyi*. Evolution 51:595–605

Torigoe T (1985) Comparison of object manipulation among 74 species of non-human primates. Primates 26:182–194

Wu R, Pan Y (1985) Preliminary observation on the cranium of *Laccopithecus robustus* from Lufeng, Yunnan with reference to its phylogenetic relationship. Acta Anthropol Sinica 4:7–12

Youlatos D, Meldrum J (2011) Locomotor diversification in New World monkeys: running, climbing, or clawing along evolutionary branches. Anat Rec 294:1991–2012

Zapfe H (1958) The skeleton of *Pliopithecus* (*Epipliopithecus*) *vindobonensis* Zapfe and Hürzeler. Am J Phys Anthropol 16:441–457

Zapfe H (1960) Die Primatenfunde aus der miozänen Spaltenfüllung von Neudorf an der March (Devínská Nová Ves), Tschechoslowakei. Schweizerische paläontologische Abhandlungen 78:1–293

Zapfe H (1991) *Mesopithecus pentelicus* Wagner aus dem Turolien von Pikermi bei Athen, Odontologie und Osteologie (Eine Dokumentation). Neue Denkschriften des Naturhistorischen Museums Wien 5:1–203

Zapfe H, Hürzeler J (1957) Die Fauna der miozänen Spaltenfüllung von Neudorf an der March (ČSR), Primates. Sitzungsberichte der Österreichischen Akademie der Wissenschaften, Wien, Mathematisch-Naturwissenschaftliche Klasse 166. pp 114–123

Chapter 17
The Hands of Miocene Hominoids

Masato Nakatsukasa, Sergio Almécija, and David R. Begun

1 Introduction

Hands of extant hominoids are highly derived compared with those of non-hominoid catarrhines. Tracing the evolutionary history of the hand in Miocene apes depends on understanding hominoid (sensu lato) positional behavioral evolution and clarifying the ancestral hand morphology from which human manual dexterity evolved. Unfortunately, the hominoid fossil record is spotty and biased in time and space. Preservation issues sometimes cause differences in functional interpretations and even in the anatomical and taxonomic identification of specimens. Recurrent postcranial homoplasy for suspensory adaptation (e.g., Tuttle 1975; Begun 1993; Ward 2007) spurs controversies over phylogenetic interpretation. This chapter reviews the status of our understanding (and debate) of Miocene hominoid hand anatomy and evolution.

The ape vs. monkey (i.e., cercopithecid) dichotomy is clear in extant catarrhine postcranial anatomy. Apes are specialized for forelimb-dominated orthograde behaviors such as suspension and vertical climbing (e.g., Ward 2007). In relation to this, their hands have greater capabilities for ulnar deviation, midcarpal supination, and stable hooklike grips (Napier 1960; Tuttle 1967; Lewis 1969,

M. Nakatsukasa (✉)
Laboratory of Physical Anthropology, Kyoto University, Kyoto 6068502, Japan
e-mail: nakatsuk@anthro.zool.kyoto-u.ac.jp

S. Almécija
Center for the Advanced Study of Human Paleobiology, Department of Anthropology, The George Washington University, Washington, DC 20052, USA

Department of Anatomical Sciences, Stony Brook University, Stony Brook, NY, USA

Institut Català de Paleontologia Miquel Crusafont, Universitat Autònoma de Barcelona, Cerdanyola del Vallès, Barcelona, Spain

D.R. Begun
Department of Anthropology, University of Toronto, Toronto, ON, Canada, M5S 2S2

T.L. Kivell et al. (eds.), *The Evolution of the Primate Hand*, Developments in Primatology: Progress and Prospects, DOI 10.1007/978-1-4939-3646-5_17

1972; Corruccini et al. 1975; Morbeck 1975; O'Connor 1975; Cartmill and Milton 1977; Susman 1979; Rose 1988; Sarmiento 1988). Hooklike grips are facilitated in the most suspensory hominoids (i.e., hylobatids, *Pongo*, and *Pan*) by the possession of long non-pollical rays and curved fingers (e.g., Susman 1979) (Fig. 17.1). The hands of the knuckle-walking (KW) apes are further specialized to limit dorsiflexion and radial deviation of the wrist and to enable effective load transmission through the median (III/IV) rays (in *Pan*) or all non-pollical rays (in *Gorilla*) (Tuttle 1967; Inouye 1994; Richmond et al. 2001; Begun 2004). Therefore, the midcarpal joint is significantly broader to resist proximodistal

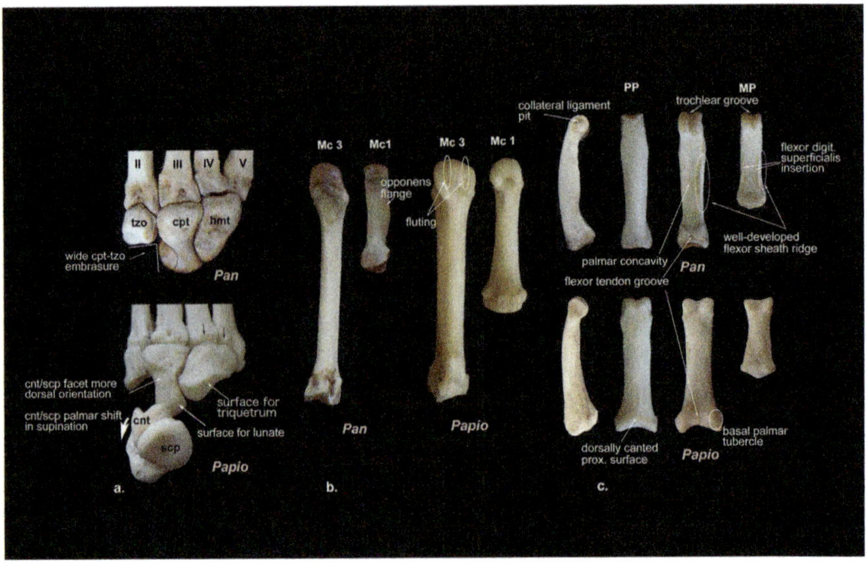

Fig. 17.1 Anatomical keys of hand bones in the suspensory/vertical climbing *Pan* and the terrestrial quadrupedal *Papio*. (**a**) Articulated right trapezoid (tzo), capitate (cpt), hamate (hmt), and II–V metacarpals (Mcs). Scaled to an approximately same width. For *Papio*, the scaphoid/centrale (scp/cnt) is shown in a proximally disarticulated position to explain joint movement. In midcarpal supination, the scp/cnt palmarly shifts around the cpt head and the triquetrum (not shown) does dorsally on the hmt. In *Pan*, the cpt-tzo embrasure is wide to allow a greater degree of scp palmar shift; cpt head is wide and the scp surface largely faces radially. In *Papio*, the scp/cnt surface is more dorsally oriented. Note dorsally extended Mc4/5 proximal surfaces in *Papio* (*short arrows*). (**b**) Right Mc3 and Mc1 in palmar view. Note the slenderness of Mc3 and elongation relative to Mc1 in *Pan*. The Mc3 head of *Papio* shows well-developed gutters in between the central articular and sesamoid areas (fluting) to form the glenoid plate/sesamoid complex, which restricts the metacarpophalangeal joint movements essentially to flexion-extension. In *Pan*, corresponding proximal articular extension is narrow and is not fluted. Also note the dorsopalmarly guttered proximal surface of Mc3 in relation to the formation of a high ridge on the cpt distal surface. (**c**) Proximal phalanx (PP) and middle phalanx (MP) of a median digit. Note well-developed secondary shaft features (i.e., fibrous flexor sheath ridges, flexor digitorum superficialis insertion (only for MP), longitudinal shaft curvature, median palmar groove or concavity), tall trochlea with a palmar bent, parallel-sided trochlear surface, and deep collateral ligament pit as well as elongation in the *Pan* phalanges. In *Papio*, the proximal surface of the PP is dorsally extended (dorsal canting). Well-developed basal palmar tubercles are related to palmigrade hand use

load transmission (convergently with terrestrial cercopithecids; Jenkins and Fleagle 1975; Richmond 2006).

In pronograde cercopithecids, the hand is used in a relatively stereotyped manner. It is typically only slightly ulnarly deviated, either in palmigrade or digitigrade postures, with the forearm pronated and forelimb excursion largely restricted to the parasagittal plane (Schmitt 1994). The wrist joint is primarily hinge-like, with the ulnocarpal articulation restricting ulnar deviation while enhancing stability (Morbeck 1975; O'Connor 1975). While midcarpal pronation is relatively free, enhancement of midcarpal supination is absent (Jenkins 1981). Proximal and middle phalanges (PP and MP, respectively) of the non-pollical rays are relatively short and lack well-developed secondary shaft features (SSFs), such as flexor sheath ridges (see Fig. 17.1), which are most pronounced in suspensory primates (Fig. 17.1). In comparison to modern apes, the pollex of pronograde cercopithecids is robust and long relative to the other digits in cercopithecines. This ape-monkey dichotomy, however, may be a relatively late event in the evolution of the apes, given the evidence of quite well-developed thumbs in fossil apes (e.g., Almécija et al. 2012, 2014, and references therein). It is unclear in the fossil record when and where extant ape-like hands evolved, although unambiguous below-branch suspensory and orthogrady adaptations are not recorded until the late Miocene (Begun 1988, 1993; Moyà-Solà and Köhler 1996; Almécija et al. 2007). Furthermore, recent evolutionary modeling suggests that *Pan*, *Pongo* and hylobatids (in this increasing order) elongated their fingers independently (Almécija et al. 2015).

2 Early Miocene

The early Miocene (23–16 Ma) is the earliest epoch from which abundant remains of fossil hominoids have been recovered. The fossils are mostly from East Africa. The hand anatomy of three hominoids has been studied so far: *Proconsul heseloni*, *Proconsul nyanzae*, and *Afropithecus turkanensis*. The hominoid status of *Proconsul* is not universally accepted (e.g., Harrison 2010). However, *Proconsul*'s position near the base of the hominoid clade makes it critical to our understanding of the hand anatomy of the stem group from which later indisputable hominoids have evolved. Hands of these early Miocene hominoids are comparatively primitive, lacking suspensory or orthograde specialization but exhibiting a suite of features that are related to powerful pollical-assisted grasping. Most likely, these apes were above-branch palmigrade quadrupeds without a tail as a balancing organ (Kelley 1997).

2.1 Proconsul

Proconsul (or *Ekembo*: see McNulty et al. (2015)) is postcranially the best-known fossil ape (Walker 1997; Ward 1998). Although the genus *Proconsul* includes several species (Begun 2007; Harrison 2010), the majority of the hand specimens are known from the 18.5 to 17 Ma old *P. heseloni*. The hypodigm of *P. heseloni* includes several partial skeletons such as KNM-RU 2036 and Kaswanga Primate Site (KPS) individuals (Napier and Davis 1959; Walker and Pickford 1983; Beard et al. 1986; Begun et al. 1994). The often-used *P. heseloni* body mass (BM) of 10 kg is based primarily on a subadult female (KNM-RU 2036) (Ruff et al. 1989; Rafferty et al. 1995; Ruff 2003). Rafferty et al. (1995) estimated BM of a subadult male KPS 1 as 13.6 kg. Considering its young growth stage (Begun et al. 1994), we estimate that this individual would likely have reached about 20 kg after maturation. Fortunately, the hand morphology of the contemporaneous and larger *P. nyanzae* is known from a male hand skeleton KNM-RU 15100. The BM of male *P. nyanzae* is estimated to be 35–40 kg (Ruff et al. 1989; Rafferty et al. 1995; Ruff 2003). The following description of the *Proconsul* hand is based on the better-represented *P. heseloni* unless otherwise indicated.

The hand of *Proconsul* was originally described as representing a mosaic of "monkey-like" and "ape-like" traits (Napier and Davis 1959), emphasizing the complexity and mosaic nature of the hand of *Proconsul*. Over the years its hand shape affinities have shifted from the opposite views of a suspensory-like ape (e.g., Lewis 1972) to those of a palmigrade primate (e.g., Morbeck 1975). Currently, it is a general agreement that the hand is overall more "cercopithecid-like" than "ape-like" (Corruccini et al. 1975; Morbeck 1975; Beard et al. 1986; Begun et al. 1994). This is due to the comparatively minimal divergence of the hand in some cercopithecid taxa from the primitive catarrhine condition. As an example, Morbeck (1975) observed that the distal surface of the radius is rather primitive, the ulnar styloid process is prominent (and thus weight bearing), and the ulnar head is narrow (thus probably limiting the range of pronation-supination). However, *Proconsul* does have some features of the hand skeleton that are reminiscent of extant apes, which may represent hominoid shared derived characters.

Typical features of the *Proconsul* hand include the articular surface for the lunate on the radius that is mildly concave and faces distally. In this regard, the radius of *Proconsul* contrasts with that of the derived KW apes that exhibits a wider lunate-scaphoid surface angle and prominent dorsal ridge (Tuttle 1967; Richmond and Strait 2000; Begun 2004).

The ulnar styloid is long and articulates with the triquetrum and pisiform (Corruccini et al. 1975; Morbeck 1975; O'Connor 1975; Beard et al. 1986). The ulnar styloid probably checked ulnar deviation at the antebrachiocarpal joint as in cercopithecids (O'Connor 1975). However, the articular morphology on the *Proconsul* triquetrum differs from that of cercopithecids. The articular surface for the styloid is essentially flat in *Proconsul* (Fig. 17.2) lacking a pronounced beak on

Fig. 17.2 (Continued) extant great apes (and *Pierolapithecus*). The distopalmar part of this specimen is broken. The trq facet on the hmt shows *Proconsul*-like orientation but is uniquely short (*16*). *Remarks for* Rudapithecus. The scp is elongated (*17*) compared with *Proconsul*. The cpt is quite narrow (*18*: head is partially broken) but has a distally extensive cnt facet (*19*: on the palmar side also). The Mc3 surface is highly keeled (*20*). The lnt dominates the radiocarpal joint (*21*). The carpal tunnel (*22*) is deep with the palmarly projecting scp tubercle and pis. *Remarks for* Oreopithecus. The proximal articular surface of the lnt is narrow radioulnarly (*23*), suggesting predominantly radial-side force transmission of the wrist like extant great apes. The cpt head is uniquely small (*24*), while the proximal hmt is very well developed (*25*)

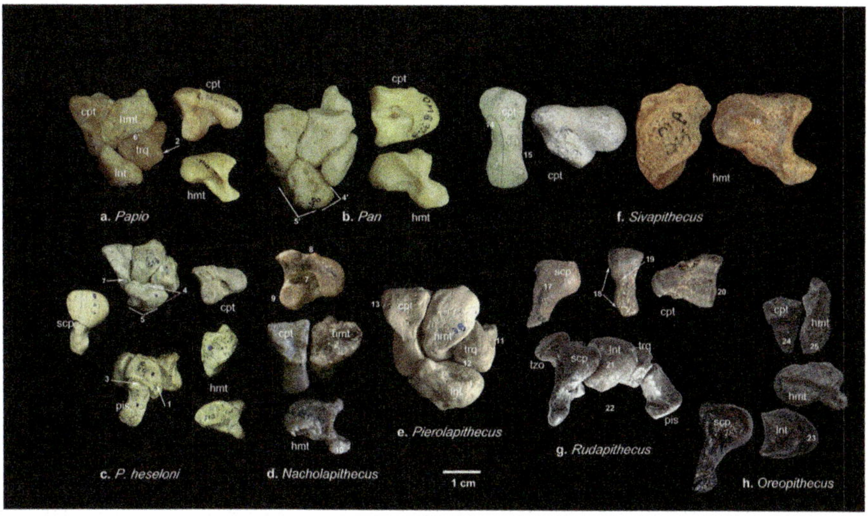

Fig. 17.2 Right carpus of extant catarrhines and fossil apes. Left-side bones are reversed for comparative purpose. Where side view is shown, it is radial for the capitate (cpt) and ulnar for the hamate (hmt). (**a**) *Papio* articulated cpt, hmt, lunate (lnt), and triquetrum (trq) in dorsal view, cpt, and hmt. (**b**) Carpus of *Pan* displayed in the same arrangement as (**a**). (**c**) *Proconsul heseloni* (KNM-KPS 3). *Clockwise* from the *top left*: articulated ulnar carpus in dorsal view; cpt and hmt (these hmt images are reversed ones of the left counterpart since the hamulus is broken in the right bone); articulated ulnar carpals in ulnar view [pisiform (pis) is also shown]; scaphoid (scp) in proximoradial view. (**d**) *Nacholapithecus* cpt (KNM-BG 40799B, reversed) and hmt (KNM-BG 42759B). These are from different individuals. (**e**) *Pierolapithecus* (IPS 21350) articulated ulnar carpals (reversed image). (**f**) *Sivapithecus* cpt (GSP 17119) and hmt (NG 940; reversed). (**g**) *Rudapithecus* scp (RUD 202), cpt (RUD 167), and associated carpus (scp, centrale, lunate, pisiform, trapezoid) in articulation (unpublished specimens). (**h**) *Oreopithecus* IGF11778 wrist bones (all reversed images). Courtesy of S. Moyà-Solà (*Pierolapithecus*), T. Kivell (*Sivapithecus*, RUD 167, 202), and A. S. Hammond and L. Rook (*Oreopithecus*). *Remarks for* Proconsul. The styloid articular surface on the trq is essentially flat and distally extensive (*1*) unlike cercopithecids in which the dorsal border of this surface forms a pronounced beak (*2*). In the pis, the styloid facet (*3*), like the trq facet, faces proximally unlike a more ulnarly faced facet in cercopithecids. The lnt is compressed proximodistally, having a narrow angle between the radial surface and the midcarpal joint surface (*4*) and a wide radial facet-cnt/scp facet angle (*5*). Compare these angles with those of *Pan* (*4′* and *5′*). The trq facet on the hmt is more proximodistally oriented (*6*) like *Pan*. *Remarks for* Nacholapithecus. In the cpt, the attachment area of the capitate-trapezoid interosseous ligament is remarkably wide and deep (*7*), the head has a dorsal stop ridge (*8*), and the palmar process of the Mc surface is longer than that of *Proconsul*, which probably provides greater stability with the cpt-Mc3 joint in dorsiflexion (*9*). The hamulus of the hmt is robust and long (*10*) though not as thick and prominent as that of extant hominoids (see *Pan*). *Remarks for* Pierolapithecus. The trq lacks the articular surface for the ulnar styloid but has a depressed porous area for insertion of the meniscus (*11*) like *Pan*. However, the trq is radioulnarly thick and retains a wide lnt contact like *P. heseloni* (*12*). The lnt is less flattened compared with *P. heseloni*. Note intermediate angles between the radial surface and the midcarpal joint surface and between the radial surface and the cnt/scp surface. The distal body of the cpt is expanded radially (*13*) and probably provided a wide embrasure for the cnt/scp. *Remarks for* Sivapithecus. The cpt is narrow. The joint surface for the cnt/scp quite distally encroaches the dorsal surface of the body (*14*; the boundary of the scp/cnt and lnt surfaces is drawn). This expansion locks the cnt when the midcarpal joint assumes maximum pronation and dorsiflexion. The hmt is tall and wide. The border of the hmt articular surface draws a single modest curvature in dorsal view (15), unlike a sigmoid one in

the dorsal border (Fig. 17.2). Therefore, the flat articular surface probably permitted a greater degree of ulnar deviation in *Proconsul* than in cercopithecid monkeys (Beard et al. 1986; Godinot and Beard 1991).

The lunate is uniquely compressed in the direction of the articular surfaces for the radius and the midcarpal joint (Ward 1998; Fig. 17.2). This condition is more similar to that in cercopithecids than to that in extant hominoids (Beard et al. 1986). The flat lunate of *Proconsul* is probably suitable for effective proximodistal load transmission when the wrist is neutral with respect to radioulnar deviation (Ward 1998).

In *Proconsul*, the long axis of the triquetral facet on the hamate is more proximodistally aligned compared with cercopithecids, where it faces more proximally (Beard et al. 1986; Fig. 17.2). This condition is reminiscent of extant apes (except *Gorilla*) and, together with the moderately long articular surface (McHenry and Corruccini 1983), probably enabled a greater magnitude of ulnar deviation and midcarpal rotation. In this position, the joint is less well suited for force transmission through the ulnocarpal articulation than in monkeys (Jenkins and Fleagle 1975; Cartmill and Milton 1977; Sarmiento 1988).

Suspensory primates are capable of wide ranges of midcarpal supination (Jenkins 1981). Midcarpal supination is initiated and maintained by a palmar shift of the centrale/scaphoid around the capitate head and a dorsal shift of the triquetrum on the hamate (Fig. 17.1). Therefore, in extant suspensory primates (including *Ateles*), the centrale surface on the capitate head faces more radially (rather than dorsally), and the embrasure between the capitate and opposing trapezoid is wider palmarly, to enable a greater degree of palmar shift of the centrale/scaphoid (Jenkins and Fleagle 1975; Jenkins 1981; Fig. 17.1). *Proconsul* lacks this suite of characters (Corruccini et al. 1975; Jenkins and Fleagle 1975; McHenry and Corruccini 1983; Beard et al. 1986). The centrale/scaphoid facet on the capitate head faces dorsoradially like cercopithecids (Fig. 17.2). In *Proconsul*, this embrasure is wide dorsally, and thus midcarpal pronation is relatively free like *Macaca* (Jenkins 1981).

The centrale is not fused to the scaphoid (this is true of all fossil apes whose scaphoid is known). The scaphoid tubercle projects radially without a palmar deflection (Beard et al. 1986), suggesting that the depth of the carpal tunnel was not particularly great, unlike the extant African apes (Napier and Davis 1959).

Proconsul had a relatively mobile trapezium-first metacarpal (CMc1) joint like extant apes (Preuschoft 1973; Rose 1992; Ward 1998). The Mc1 surface on the trapezium has a relatively distal orientation like extant apes and ceboids rather than a palmar orientation like cercopithecids (Beard et al. 1986). The saddle-shaped distal articular surface has a strong dorsopalmar curvature and is extensive palmarly and dorsally, more so than in cercopithecids. This configuration, together with a greater degree of incongruence with the CMc1 articular surface, enables wide ranges of movements, particularly for abduction-adduction of the thumb (Rafferty 1990; Rose 1992). The opponens flange is unusually well developed for a primate of this size (Walker and Pickford 1983), which suggests quite powerful pollical opposition.

At the carpometacarpal (CMc) joint, the hamulus of the *Proconsul* hamate is relatively small, resembling that of cercopithecids (Napier and Davis 1959; Corruccini et al. 1975; O'Connor 1975; Beard et al. 1986; Fig. 17.2). In extant apes, it is long and massive (Schön and Ziemer 1973; O'Connor 1975; Lewis 1989; Fig. 17.2) to restrict the hamate-Mc5 joint in hyperdorsiflexion (Tuttle 1967; Corruccini et al. 1975). The proximal articular surface of the Mc4/5 extends onto the dorsum, suggesting frequent hyperdorsiflexion at the CMc joints (O'Connor 1975; Ward 1998). In this regard, *Proconsul* resembles cercopithecids and differs from the extant apes, in which these surfaces face directly proximally.

The length formula of the Mc2-4 is ($4 \geq 3 > 2$) like many colobines, differing from ($2 \geq 3 > 4$) in extant apes (Jouffroy et al. 1991). The Mc4 is moderately short relative to BM, like cercopithecines, and much shorter compared to the extant apes (Lovejoy et al. 2009a). On the palmar aspect of the Mc head, the sesamoid articular surfaces extend more proximally along the radial and ulnar borders than in the central part (Napier and Davis 1959; Rose et al. 1996). In cercopithecids (particularly terrestrial taxa), non-pollical Mc heads are palmarly fluted (Fig. 17.1b) due to the presence of large sesamoid bones. This morphology is related to restricted movements (essentially to flexion-extension) at the metacarpophalangeal (McP) joint (Lewis 1989). In extant great apes, the sesamoids are absent and the palmar surface is flat (Le Minor 1988). In *Proconsul*, the fluting appears weak, and sesamoids were probably reduced (Napier and Davis 1959), suggesting a greater range of movement beyond simple hinge-like movements as in cercopithecids. Pits for the McP joint collateral ligaments encroach the dorsal surface of the Mc heads, approaching the midline. This eccentric attachment permits more abduction-adduction and axial rotation when the joint is in extension, with increased stability during flexed postures (Rose et al. 1996).

The non-pollical proximal phalanges (PPs) have well-developed palmar tubercles on either side of a groove for the flexor tendons (Fig. 17.3). The proximal surface is dorsally canted and mediolaterally concave, suggesting hyperdorsiflexion at the McP joints during push-off (Begun et al. 1994). Secondary shaft features (SSFs) are less developed than even some arboreal cercopithecids (Begun et al. 1994; Fig. 17.3). The shaft curvature is comparable to that of *Nasalis* (Fig. 17.4). However, median values for *Gorilla*, *Alouatta*, *Trachypithecus*, and a number of fossil taxa all fall within the 50 % interquartile range of *Proconsul*, indicating a relatively generalized degree of curvature (Deane and Begun 2008). The trochlea is moderately canted palmarly and is moderately tall with a ventral flare and a shallow trochlear groove. The palmarly wider (and proximally wider on the palmar side) trochlear articular surface, unlike that of extant apes (Fig. 17.1), suggests greater stability when the joint is flexed. The trochlea of *Proconsul* non-pollical PPs is moderately asymmetric with the radial condyle projecting more palmarly, which suggests the presence of great shearing force. In the middle phalanges (MPs), the proximal articular end bears deep concavities with a pronounced median ridge in between, which may be an adaptation to allow rotation of the MP on the PP to maintain maximum contact during torque moments. The shaft bears a well-developed insertion for the digital flexor, while the flexor sheath ridges are weakly developed (Fig. 17.5). The non-pollical PPs

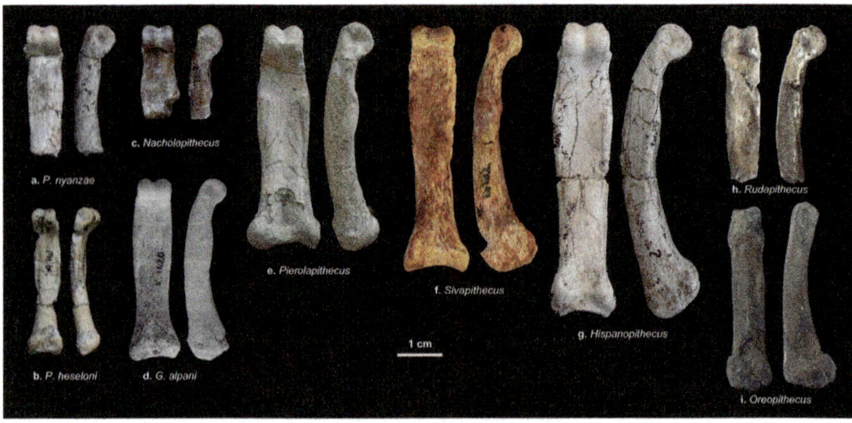

Fig. 17.3 Manual proximal phalanges of fossil apes in palmar and side views. (**a**) *Proconsul nyanzae* KNM-RU 15100H. (**b**) *Proconsul heseloni* KNM-KPS8 ph69/131. (**c**) *Nacholapithecus* KNM-BG 35250BB. (**d**) *Griphopithecus alpani* K1420. (**e**) *Pierolapithecus* IPS21350. (**f**) *Sivapithecus* GSP 19700. (**g**) *Hispanopithecus* IPS18800. (**h**) *Rudapithecus* RUD 31 (side view is reversed). (**i**) *Oreopithecus* Bac 148. Median (III/IV) ray except KNM-RU 15100H and potentially Bac 148. Courtesy of J. Kelley (*G. alpani*) and R. Susman (*Sivapithecus*)

and MPs are only modestly long (relative to Mc length and to BM) and similar to those in non-papionin cercopithecids and *Gorilla* (Lovejoy et al. 2009a), being as long as or only modestly longer than the pedal counterparts (Begun et al. 1994; Lovejoy et al. 2009a).

Proconsul has a well-developed pollex, unlike the extant great apes (Begun et al. 1994; Almécija et al. 2012; Fig. 17.6). The PP1 is about half as long as the PP3 (Napier and Davis 1959). The trochlear articular surface is proximally extensive, and the distal end exhibits a marked asymmetrical condition, suggesting the presence of strong pollical opposition. The pollical distal phalanx (DP1) has intermediate proportions between the wide and flat phalanx of humans and the "conical" phalanx of suspensory great apes and exhibits an attachment for a long flexor tendon on its palmar surface (Begun et al. 1994; Almécija et al. 2014).

There are some morphological differences between the hand morphology of the smaller *P. heseloni* and the larger *P. nyanzae*. In *P. nyanzae*, the non-pollical PPs and MPs show a stronger manifestation of SSFs (Begun et al. 1994; Nakatsukasa et al. 2003; Figs. 17.3 and 17.5). The MPs are wide, but they are absolutely long in relation to the body size. Manual and pedal phalanges are more easily distinguishable than in *P. heseloni*. Nonetheless, these specimens share basic functional features regarding articular shape. Both species probably utilized their hands in a similar manner despite body size differences.

In summary, the hand anatomy of *Proconsul* suggests arboreal palmigrade quadrupedalism with hyperdorsiflexion at the wrist and McP joints as well as powerful pollical-assisted grasping. The hand is adapted for enhanced ulnar deviation compared with cercopithecids, which are most likely related to climbing capabilities (Kelley 1997; Walker 1997; Ward 1998). The wide ulnocarpal articulation in *Proconsul* may have been present to ensure the stability of the

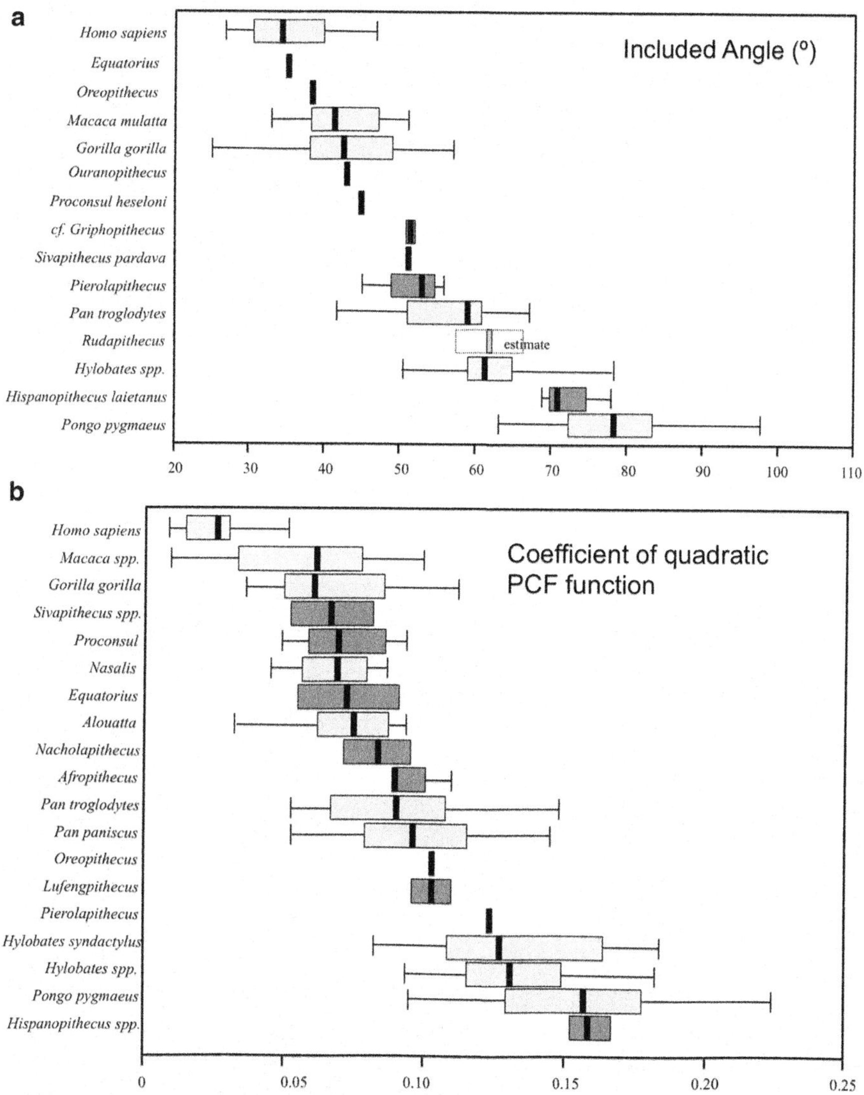

Fig. 17.4 Box and whisker plots of (**a**) included angle (IA) and (**b**) first coefficient of the (quadratic) polynomial curve fitting (PCF) function in non-pollical proximal phalanges. While the IA is calculated only from an intact bone, the PCF function can be calculated from fragmentary bones. Data source of IA: for extant taxa, Richmond and Jungers (2008); for fossil taxa, Richmond and Whalen (2001), Ersoy et al. (2008), Almécija et al. (2007, 2009), Bonis and Koufos (2014). Data of *Equatorius* (KNM-TH 28860U) and *Oreopithecus* (IGF 11778 PP5) is unpublished data of S.A. Data of PCF are from Deane and Begun (2008)

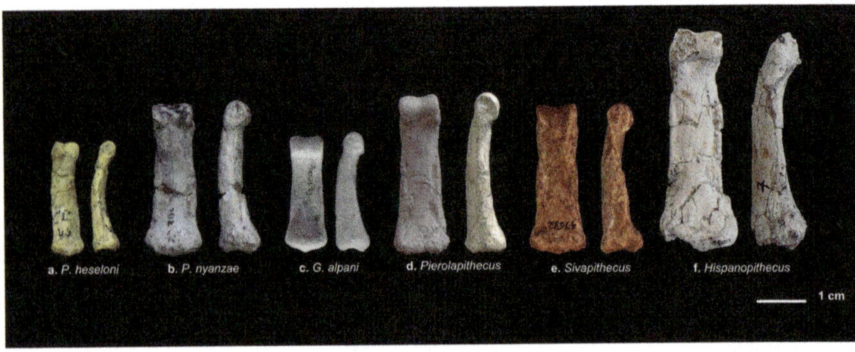

Fig. 17.5 Manual middle phalanges of fossil apes in palmar and side views. (**a**) *Proconsul heseloni* PH 222 (KPS 3). (**b**) *Proconsul nyanzae* KNM-RU 15100I. (**c**) *Griphopithecus alpani* G1004. (**d**) *Pierolapithecus* IPS21350. (**e**) *Sivapithecus* GSP 47582. (**f**) *Hispanopithecus* IPS18800. All from a median ray. The curvature of PH 222 is exaggerated by deformation. Note that MP of *P. nyanzae* is much more robust than that of *P. heseloni*. Courtesy of J. Kelley (*G. alpani*), S. Moyà-Solà (*Pierolapithecus*), and R. Susman (*Sivapithecus*)

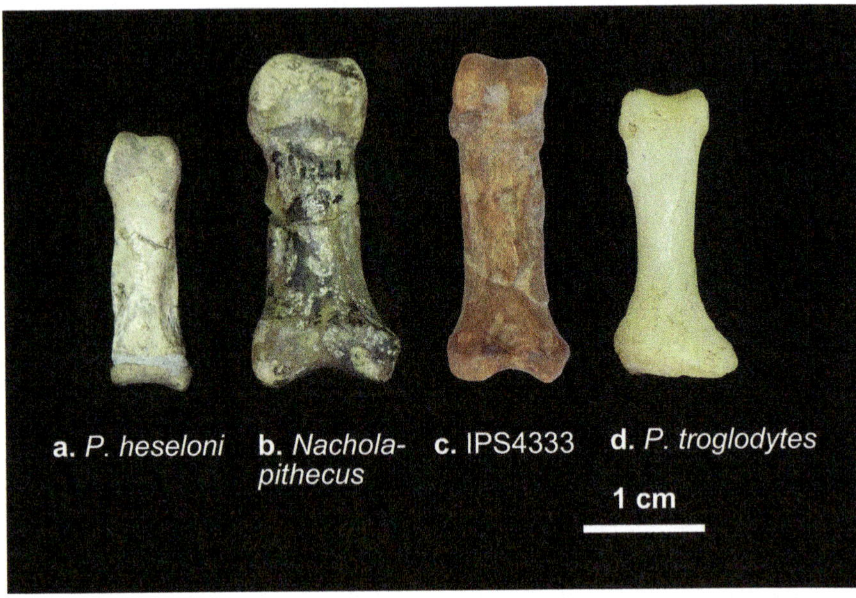

Fig. 17.6 Right pollical proximal phalanges of fossil apes and *Pan* in palmar view. (**a**) *Proconsul heseloni* KPS 1 (reversed). (**b**) *Nacholapithecus* KNM-BG 17813. (**c**) Unassigned specimen from Castell de Barberà (11.2–10.5 Ma) IPS 4333. (**d**) Male *Pan troglodytes*. The fossil phalanges resemble each other in shaft robusticity, proximally extensive palmar articular surface of the trochlea, and marked asymmetrical condition of the distal end (e.g., radially tilted head, wider ulnar portion of the trochlear palmar surface). This asymmetry suggests the presence of strong pollical opposition. In *P. heseloni* and *Nacholapithecus*, the trochlear articular surface is palmarly flared and becomes wider toward its proximal limit on the palmar side

wrist joint in exaggerated ulnar deviation, rather than to ensure a greater capacity for proximodistal load transfer during quadrupedalism. The wide range of movements at the trapezium-Mc1 joint is reminiscent of extant great apes (Rose 1992). It is probably related to pollical grasp for positional behavior. Compared with ancestral catarrhines (Hamrick et al. 1995), PPs are elongated, suggesting an enhancement for secure grasping, which would have been crucial for a larger-sized tailless primate engaged in above-branch activities (Kelley 1997).

2.2 Afropithecus turkanensis

Afropithecus is known from several 17.5–17 Ma localities in northern Kenya (Leakey et al. 1988; Leakey and Walker 1997). A talus (KNM-WK 18120) gives an estimated BM of about 35 kg (Rafferty et al. 1995). Most of the hand bones described here are close in size to those of KNM-RU 15100, which is attributed to *P. nyanzae*. The known hand elements include a lunate; a fragmentary scaphoid; a trapezium; proximal fragments of the Mc1, Mc2, and Mc4; a DP1; and several non-pollical phalanges (Leakey et al. 1988). General postcranial resemblance between *Afropithecus* and *Proconsul* has been noted (Leakey et al. 1988; Rose 1992, 1993; Leakey and Walker 1997; Ward 1998, 2007). Several hand characters in *Proconsul* (e.g., an unfused os centrale, a palmarly narrow capitate-trapezoid embrasure, flat form of the lunate, extensive Mc1 surface of the trapezium, dorsally canted Mc4 proximal surface) are also present in *Afropithecus* (Rose 1992; Ward 1998). The phalanges have not been studied in detail. The longitudinal curvature of manual PP is slightly stronger than in *Proconsul* (Fig. 17.4). However, the curvature is still within the range of variation of pronograde quadrupeds, and this is most likely the positional behavior of *Afropithecus*.

3 Early Middle Miocene

As early as 17 Ma, the geographic range of apes expanded into Eurasia (Heizmann and Begun 2001; Böhme et al. 2011.) The postcranial remains of several fossil apes have been documented from 16 to 15 Ma localities in East Africa and Europe. Aspects of hand morphology are known in three species: *Equatorius africanus*, *Griphopithecus alpani*, and *Nacholapithecus kerioi*. Compared to early Miocene apes, middle Miocene apes exhibit a greater diversity of positional behavior (Ward et al. 1999; Ishida et al. 2004), which might be related to environmental changes and their expanded geographic distribution. *Equatorius* and *Griphopithecus* might have included palmigrade terrestrial quadrupedalism in their locomotor repertoire. However, their hands are less specialized than those of terrestrial cercopithecids. *Nacholapithecus*, in contrast, with increased forelimb size, was probably specialized for more frequent orthograde behaviors. However, the hands of these middle Miocene apes are primitive and more closely resemble the hand of *Proconsul* than those of any extant catarrhine. Terrestrial signals have been reported more strongly in regions of the limbs other than the hand. This may imply that early Miocene hominoid hands were originally versatile and allowed novel terrestrial locomotion with relatively minor modifications.

3.1 Equatorius africanus

Equatorius is known from two 15 Ma Kenyan sites: the Tugen Hills (Ward et al. 1999; Sherwood et al. 2002) and Maboko Island (McCrossin and Benefit 1994; McCrossin et al. 1998). An adult male skeleton, KNM-TH 28860, from the Tugen Hills, preserves four carpals, four non-pollical Mcs, and five manual phalanges. Isolated Mcs and phalanges have also been recovered from Maboko. Body mass has been estimated as 28 kg based on various dental and postcranial elements from Maboko (McCrossin 1994). A similar value was obtained for KNM-TH 28860 (Ward et al. 1999).

A distal radius from Maboko is reported to show several features that are similar to those of extant African apes (McCrossin et al. 1998). The scaphoid/lunate surface exhibits a distal projection of the dorsal ridge and is dorsopalmarly relatively concave. Although such morphology has been interpreted to limit dorsiflexion of the wrist in the extant African apes (Tuttle 1967), these character states have not been established quantitatively in *Equatorius*, and their behavioral implications are unclear.

As reported in descriptions of the fossil material, carpal and Mc morphology of *Equatorius* is essentially similar to that of *Proconsul* (Ward et al. 1999; Sherwood et al. 2002). The scaphoid tubercle is more robust than in *P. heseloni*, but this is probably a size effect. The hamate has a relatively proximodistally aligned triquetral surface. The Mcs are slender. The Mc heads are not fluted, and pits for the McP joint collateral ligaments encroach the dorsal surface. A notable difference is a well-developed dorsal transverse ridge (DTR) of the Mc3 head reported on a Maboko specimen (McCrossin et al. 1998), which is absent in *Proconsul* (and three Mcs from the Tugen Hills; Patel et al. 2009). In the extant African apes, a similar dorsal ridge stabilizes the McP joints in hyperextension during KW (Tuttle 1967). The DTR in *Equatorius* has been reported to be comparable to that in extant African apes in its expression (Allen and McCrossin 2007), but DTR is also observed in pronograde terrestrial quadrupeds such as large-bodied *Papio* (Richmond et al. 2001). This putative DTR recalls the morphology of the Mc of *Hispanopithecus* and *Rudapithecus*, where the pits for the collateral ligaments almost meet in the midline, creating a DTR-like continuous beveling of the proximal portion of the pits (Almécija et al. 2007; DRB personal observation). The proximal joint surface of the *Equatorius* PP is dorsally canted and not suitable for maintaining the McP joint fully extended while the Mc and PP are in line (Patel et al. 2009). Taking all these attributes into account, this ridge is unlikely to be related to KW but consistent with terrestrial palmigrade hand postures.

Manual phalanges of *Equatorius* are slender with weak expression of SSFs like those of *P. heseloni* (Sherwood et al. 2002; Fig. 17.4). Phalangeal curvature is essentially the same as *Proconsul* (Deane and Begun 2008.) However, the median PP is short relative to the median Mc (80 %; see Patel et al. 2009). In this measure, *Equatorius* falls within the range of semiterrestrial cercopithecids, but near the lower end of the range of arboreal cercopithecids and below *P. heseloni* (93 %). This may imply a certain reliance on terrestriality with palmigrade quadrupedalism, though not habitual terrestriality with digitigrady (Patel et al. 2009). Terrestriality in *Equatorius* has been proposed from various postcranial signals other than the hand (McCrossin and Benefit, 1997; McCrossin et al. 1998; Ward et al. 1999).

3.2 Griphopithecus alpani

Twenty-one ape phalanges have been recovered from a middle Miocene locality of Paşalar, Turkey (Ersoy et al. 2008). Although two apes, *Griphopithecus alpani* and *Kenyapithecus kizili*, are known from Paşalar (Kelley et al. 2008), these phalanges were tentatively assigned to *G. alpani*, whose presence is much more common in the fossil assemblage (Ersoy et al. 2008). No pollical bones are known, and a reliable BM estimate is unavailable. The phalanges are slightly larger than those of male *Equatorius* but smaller than those of male *P. nyanzae*.

Morphologically, PPs of *G. alpani* are intermediate in overall morphology between habitually terrestrial taxa like *Papio* and predominantly suspensory taxa (Ersoy et al. 2008; Fig. 17.3). The base is relatively wide and palmar tubercles are well developed. The proximal articular surface is canted dorsally. The longitudinal curvature is stronger than in *P. heseloni* (Fig. 17.4). The midshaft is wider than in *P. heseloni*, but narrower than *Papio* or *Gorilla*. Proportions of the MP are similar to those in *P. nyanzae* and *Sivapithecus* (Fig. 17.5). The relative length of PP and MP is similar to *P. heseloni* or to *Sivapithecus* (for the MP) (Almécija et al. 2009). *G. alpani* probably retained a generally primitive hand morphology like *Proconsul*. Ersoy et al. (2008) concluded that *G. alpani* employed primarily pronograde quadrupedal behaviors, mostly in arboreal settings, and inferred that terrestrial activity might have been common in moving between trees, based in part on the reconstruction of the Paşalar paleoenvironment.

3.3 Nacholapithecus kerioi

Nacholapithecus is known from the 15 Ma locality of Nachola, northern Kenya (Ishida et al. 1999). The hypodigm includes the adult male skeleton KNM-BG 35250. Adult male BM is estimated at 20–23 kg (Ishida et al. 2004). *Nacholapithecus* has unique body proportions, wherein the forelimb is extremely large (but not elongated as in extant apes) compared with other body parts, suggesting a shift to more forelimb-dominated positional behavior. Although published hand elements were scant (two carpal fragments, two Mc heads, and several phalanges), several additional carpals have been recovered and published recently (Ogihara et al. 2016).

The radius has a pronounced styloid process, which is more prominent than in *P. heseloni*. The scaphoid is closely similar to that of *Proconsul* (the tubercle morphology is unknown), but its size exceeds that of male *P. nyanzae* (Ishida et al. 2004). The recently identified pisiform of KNM-BG 35250 exhibits the articular morphology similar to that of *Proconsul*. The head morphology of the non-pollical Mcs is also *Proconsul*-like (Rose et al. 1996).

The recently published carpal specimens are briefly discussed here. These carpals are straightforwardly "hyper-*Proconsul*-like," where the basic joint design is similar to that of *Proconsul*, but structures related to joint rigidity are far more heightened, reflecting the enhanced forelimb dominance. They include, for example, wide and deep attachment areas of the interosseous carpal ligaments, a long palmar process of the capitate, and a robust hamulus (Ogihara et al. 2016; Fig. 17.2).

The *Nacholapithecus* PPs lack specialization for suspension (Figs. 17.3 and 17.4). However, there are differences from *Proconsul* that suggest more pronounced general grasping ability (Nakatsukasa et al. 2003). The distal shaft is relatively wider due to an expansion at the flexor sheath ridges (Fig. 17.3). In the PP5, the asymmetry of the bone is heightened compared with *Proconsul*, suggesting a greater degree of shearing force. The pollex is very well developed. Male PP1 is (absolutely) longer and more robust than that of *Pan* (Almécija et al. 2012; Fig. 17.6).

In summary, the hand of *Nacholapithecus* retains the same basic design as that of *Proconsul*. However, it is specialized for more powerful general grasping, probably in relation to increased forelimb dominance and more frequent orthograde behaviors compared with *Proconsul* (Nakatsukasa and Kunimatsu 2009).

4 Late Middle and Late Miocene

During the period from 12 to 9 Ma, hominoids flourished in Eurasia. While some of these apes were probably still primarily above-branch quadrupeds, others exhibit different degrees (or modes) of suspensory and orthograde adaptations. In contrast, the fossil record is sparse in Africa for this time period. Although three large apes have been documented from ca. 10–8 Ma in East Africa, their postcranial adaptation is unknown, leaving the hand anatomy and positional behavior of the last common ancestor of the extant African ape-human (AAH) clade debatable.

4.1 Pierolapithecus catalaunicus

Pierolapithecus is a 12 Ma ape from Els Hostalets de Pierola, Catalonia (Moyà-Solà et al. 2004, 2009). BM of the holotype, IPS-21350, an adult male, was estimated at 30–35 kg based on various dimensions of a lumbar vertebra and dentition (Moyà-Solà et al. 2004). Begun (2009) argued that *Pierolapithecus* is a junior synonym of *Dryopithecus*. However, the taxonomic argument is not crucial here since no *Dryopithecus* fossils are treated in this chapter.

The hand of IPS-21350 includes all the carpal bones with the exception of the pisiform. *Pierolapithecus* is the earliest ape in which the retreat of the ulna is confirmed (Moyà-Solà et al. 2004; Fig. 17.2). Still, the relative triquetrum size is larger than in extant great apes, and the lunate facet is large and flat and faces proximally like *Proconsul* (Fig. 17.2). For the other carpal elements, in-depth studies await. However, brief comments may be useful for some clarification. The lunate is not strongly flattened, being intermediate between *Proconsul* and *Pan*. The body of the capitate is expanded radially and probably provided a wide embrasure for the centrale/scaphoid (Fig. 17.2). The *Pierolapithecus* wrist certainly allowed greater degrees of ulnar deviation at the antebrachiocarpal joint and midcarpal supination. However, the ulnar portion of the carpus probably played a greater role for weight transfer than in extant great apes.

The median PPs of *Pierolapithecus* resemble those of *Proconsul* and *Sivapithecus*, with strongly developed features related to palmigrady such as large basal palmar tubercles and a wide and dorsally canted proximal surface (Almécija et al. 2009; Fig. 17.3). SSFs are better developed than in *Proconsul*. The longitudinal curvature is greater than in *Proconsul* and marginally overlaps with the range of *Pan* (Fig. 17.4), although Deane and Begun (2008), using polynomial curve fitting (PCF), report a curvature close to the median for hylobatids. The overall shape of MP is similar to that in *P. nyanzae* and *G. alpani* (Fig. 17.5). The pollical DP is large and wide at the base in relation to the non-pollical DPs, suggesting that the pollex was not reduced (Almécija et al. 2012).

A long hand is correlated with arboreal locomotion, particularly suspension (Straus 1940; Erikson 1963). However, the hand of *Pierolapithecus* is significantly shorter than that of the similarly sized *Hispanopithecus* (see below), suggesting a lack of specialization for a hooklike grip (Moyà-Solà et al. 2004, 2005a; Almécija et al. 2009 also noted a relatively long pollex). The conclusion that *Pierolapithecus* was less specialized for suspension or orthogrady than extant great apes and late Miocene apes such as *Hispanopithecus* and *Rudapithecus* is also supported by a comparatively weak expression of SSFs in the PPs (Fig. 17.3).

4.2 Sivapithecus

Sivapithecus is known from 12.5 to 8.5 Ma localities in the Siwaliks and three species are recognized (Kelley 2005; Begun 2007). One capitate from the Chinji Formation is attributed to *Sivapithecus indicus* while the other bones from the later Nagri Formation are to *Sivapithecus parvada* (Fig. 17.2). BM of male *S. indicus* is slightly less than that of male *Pan* (~50 kg), and that of females is about half that of males (DeSilva et al. 2010). The largest postcranial specimen from the Chinji Formation surpasses that of female *Gorilla* (~95 kg) in size (Pilbeam et al. 1980).

Kelley (1988) estimated BM of *S. parvada* at ~69 kg based on molar size. Specimens of *S. parvada* discussed here are in the size range between female *Pan* and male *Pongo* (32–80 kg).

The capitate features are described here following Rose (1984, 1993). The capitate is radioulnarly narrow. A large part of the proximal joint surface is for the articulation with the lunate (Fig. 17.2). The proximal surface is tightly and evenly curved dorsopalmarly and less tightly but evenly curved radioulnarly. This head morphology is generally similar to that in extant African apes (Begun and Kivell 2011). The lunatocapitate joint probably accommodated more mobility for abduction-adduction than for flexion-extension. The joint surface for the centrale/scaphoid is unusually wide and encroaches the dorsal surface of the body. This joint morphology may be related to dorsiflexion stability in relatively large hominoids relying on palmigrady, but this is also one of the reasons Begun and Kivell (2011) suggest that *Sivapithecus* may have adopted KW postures (see below). The orientation of the radial surface of the head suggests relatively free supination. The articular border for the hamate is evenly and only modestly curved in dorsal view, like *Proconsul* and unlike extant great apes, *Pierolapithecus* and *Hispanopithecus*, in which this curve is sigmoid (plausible adaptation to resist a greater degree of proximodistal shearing force in relation to knuckle- or fist-walking). The distal articular surface is deeply concave dorsopalmarly. On the Mc3 surface, the dorsopalmar ridge is well developed. Rose (1984, 1993, 1994) interpreted the morphology of the capitate as reflecting generalized capabilities for palmigrade quadrupedalism, climbing, and some suspension.

The *S. parvada* hamate is dorsopalmarly tall and radioulnarly wide (Spoor et al. 1991). The triquetral facet is uniquely short with a truncation of the distal portion. The Mc articular surface is rectangular, as in extant great apes. However, the hamulus is not enlarged. The hamate of *S. parvada* is adapted for effective weight transmission through the ulnar side of the wrist, limiting ulnar deviation and extension at the triquetrohamate joint. *S. parvada* might have employed predominantly quadrupedalism rather than climbing or suspension (Spoor et al. 1991).

The proximal end of the non-pollical PP is dorsally canted, wide, and equipped with large basal palmar tubercles (Rose 1986; Fig. 17.3). The dorsopalmarly thick and radioulnarly narrow proximal shaft is reminiscent of some suspensory primates (Rose 1986). General proportion suggests moderate elongation (Almécija et al. 2009). SSFs are modestly developed. The shaft curvature (Fig. 17.4) has been interpreted as similar to *Proconsul* and other anthropoids with generalized phalanges based on PCF, or moderate, like *Pierolapithecus* and *Pan*, based on included angle (Richmond and Whalen 2001; Almécija et al. 2007; Deane and Begun 2008). The distal shaft is moderately wide, like *Pan*. The MP is radioulnarly wide like *P. nyanzae* or *Pierolapithecus* (Fig. 17.5). The magnitude of the development of SSFs resembles *Hispanopithecus*/*Rudapithecus* (Madar et al. 2002). The distal end is tall, and the articular surface is extensive proximally both dorsally and palmarly, suggesting a wide range of movement at the distal interphalangeal (IP) joint (Madar et al. 2002). The pollical PP is absolutely and relatively large in relation to other PPs, indicating that the thumb was long and capable of opposing the non-pollical digits (Madar et al. 2002). This is concordant with the morphology of the Mc1 (Spoor et al. 1991). The proximal surface of the pollical PP is weakly oval, resembling the extant African apes, but the radioulnar concavity is cercopithecine-like (Madar et al. 2002). The basal palmar tubercles (especially ulnar ones) are better developed than in extant great apes, suggesting the presence of two sesamoid bones associated with the pollical McP joint. In extant great apes, these sesamoids tend to be lost or reduced (Le Minor 1988). The flexor sheath ridges are extremely well developed. Phalangeal features suggest palmigrade quadrupedalism was an important activity in *S. parvada*. The large and powerful pollex suggests the importance of pollical-assisted grasp in arboreal activities (Rose 1986, 1994; Madar et al. 2002). The dorsopalmarly thick proximal shaft of the PP suggests the adoption of suspensory activities to a certain extent (Rose 1986; 1994).

Recently, Begun and Kivell (2011) proposed that the morphology and its functional inference of the *Sivapithecus* capitate and hamate are compatible with KW. In testing this hypothesis, Begun and Kivell (2011) argue that other aspects of the postcranial morphology of *Sivapithecus*, such as the knee and elbow joints as well as the shoulder region, are consistent, or at least not contradictory, with the hypothesis of some form of KW. However, given the variability in the expression of features biomechanically thought to be related to KW in extant knuckle-walkers, as well as their presence in non-knuckle-walkers (e.g., DTRs on the Mc heads), this hypothesis remains controversial.

4.3 Hispanopithecus laietanus

Hispanopithecus laietanus, formerly *Dryopithecus laietanus*, is known from Can Llobateres (9.65 Ma) in southern Spain (Moyà-Solà and Köhler 1996; Moyà-Solà et al. 2009; Casanovas-Vilar et al. 2011). The hypodigm of *H. laietanus* includes a partial adult male skeleton (IPS18800), which preserves most of the manual II–V ray elements (Almécija et al. 2007). BM of IPS18800 is estimated as 34 kg (Moyà-Solà and Köhler 1996). Additionally, a lunate and hamate have been partially described (Begun 1994).

The *Hispanopithecus* hand fossils exhibit a mixture of *Proconsul*-like primitive and derived features. The wide angle between the radial facet and the centrale/scaphoid facet of the lunate is similar to that of *Proconsul*. The proximal articular surface of the Mc5 is extended dorsally. However, the palmar projection of the hamate's hamulus is more *Pongo*-like and unlike the more distal orientation found in African apes. The triquetral surface on the hamate is very long proximodistally and somewhat enlarged and globular proximally. These features could be interpreted as those of a generalized quadruped with enhanced mobility on the ulnar side for orthogrady and suspension (Begun 1994; Almécija personal observation).

Hispanopithecus is unique in that the non-pollical Mcs are short and stout, intermediate between extant great apes and *Homo*, while the PPs and MPs are elongated like those of *Pongo* (Almécija et al. 2007, 2009; Lovejoy et al. 2009a).

The Mc2 head is slightly tilted radially so that the palmar side of the head faces radiopalmarly. The McP joint collateral ligament pits are very well developed, so they almost meet dorsally. The palmar surface is not fluted and proximally extensive (Moyà-Solà and Köhler 1996; Almécija et al. 2007). Besides the elongation, the PPs are dorsopalmarly strongly curved (Fig. 17.4) and show very well-developed SSFs (Fig. 17.3). The trochlea is dorsopalmarly tall and palmarly bent and bears a deep, narrow groove. The proximal articular surface of the non-pollical PP is almost circular as in *Pongo*. However, it is dorsally canted and large basal tubercles exist palmarly. The MP is wide with well-developed SSFs, and the trochlea is markedly bent palmarly (Fig. 17.5). Whereas the intermediate degree of Mc elongation and the orientation of the PPs' proximal surface are suggestive of palmigrady, many other features strongly suggest that *Hispanopithecus* was committed to suspension and orthogrady (Almécija et al. 2007), as it is also indicated by its interlimb proportions and other postcranial elements (Moyà-Solà and Köhler 1996; Alba et al. 2012; Pina et al. 2012; Tallman et al. 2013).

Hispanopithecus probably employed a combination of suspension, climbing, and above-branch palmigrade quadrupedalism with powerful grasping. The degree to which *Hispanopithecus* was adapted to palmigrady is debated (Almécija et al. 2007, 2009; Deane and Begun 2008; Alba et al. 2010). Among the authors of this chapter, DRB considers that the potential palmigrade characters of the hand (carpal morphology, short Mcs, palmar tubercles, and dorsally canted proximal articular surface on ulnar PPs) represent primitive retentions, while the features suggestive of suspension and orthogrady (e.g., remarkable elongation and pronounced curvature/axial torsion of the PPs, which is arguably inconsistent with palmigrady) are the result of dynamic biomechanical forces strongly indicative of habitual postures. Alternatively, the other authors (MN, SA) view the palmigrade characters as legitimate indicators of habitual positional behavior. In the end, *Hispanopithecus* is not like any living taxon, and it is not surprising that it is difficult using modern analogues to definitively characterize its positional behavior. Interestingly, the recent reconstruction of the IPS-18800 hand reveals that the ray IV was longer than the III, as in most male *Pongo*, indicating an ulnar shift of the main axis of the hand, which enables a better grasping of slender vertical supports (Susman 1979; Napier 1993; Almécija et al. 2007). All in all, the evidence from *Hispanopithecus* is consistent with the long-held interpretation that the Miocene apes of Europe provide the earliest unambiguous evidence of modern ape anatomy (Begun 1988, 1992, 1993, 1994), possibly coupled with behaviors involving above-branch palmigrady (see above), a combination that if real is without any extant analogue.

4.4 Rudapithecus hungaricus

Rudapithecus hungaricus (formerly called *Dryopithecus brancoi*) is a 10 Ma ape from Rudabánya, Hungary (Begun 2002, 2009). Estimates of BM range from 20 to 40 kg, suggesting a large degree of sexual dimorphism (Kivell and Begun 2009).

Within the carpus, the morphology of the scaphoid and capitate is known (Kivell and Begun 2009; Fig. 17.2). The scaphoid is elongated like that of *Pongo*. The facet for the lunate is uniquely broad, expanding to the base of the tubercle. This *Pongo*-like condition suggests a large range of extension-flexion at the wrist. Although the capitate is radioulnarly narrow and lacks broadening of the head and the distal body (Kivell and Begun 2009), the articular surface for the centrale is remarkably expansive distally. This feature suggests a greater capacity of radial deviation at the midcarpal joint. Although *Rudapithecus* did not have an enhanced potential for midcarpal rotation as evidenced from the form of the head and body, it had a relatively mobile scaphoid and a heightened capacity for radial deviation at the midcarpal joint.

A recently discovered set of associated carpals is currently under analysis, but some general preliminary impressions have been published (Begun and Kordos 2011). The articulated configuration (Fig. 17.2) is broadly most like that of *Pongo*. The carpal tunnel is extremely deep. The lunate dominates the radiocarpal joint, as in *Pongo* and unlike the more even distribution of radial joint surfaces between the lunate and scaphoid as in extant African apes.

Non-pollical phalanges exhibit a suite of features related to suspensory and other orthograde arboreal activities (Begun 1988, 1993; Fig. 17.3). The PPs have relatively large and rounded condyles, dorsally and palmarly extensive distal articular surfaces, well-developed pits for the IP joint collateral ligaments, and relatively strong SSFs (Fig. 17.3h). They are among the most strongly curved of any ape, with a median value very close to that of *Pongo* (Deane and Begun 2008). The MPs have well-developed attachment areas for the IP joint collateral ligaments, strong SSFs, pronounced dorsopalmar thickness of the proximal shaft, and a tall trochlea. Trochlear asymmetry suggests some rotation of the DP when grasping. The pollical PP has a relatively tall distal joint like *Proconsul*. The pollex must have been more developed than in extant great apes (Begun 1993; Begun and Kordos 2011; Almécija et al. 2012).

4.5 Ouranopithecus macedoniensis

Ouranopithecus macedoniensis is known from 9.3 to 9.2 Ma localities in northern Greece (Bonis and Koufos 1994). The size of large (=male) jaw/dental remains is comparable to female gorillas. However, BM estimate of ca. 50 kg, based on a male cranium, is reliable given its megadontic nature (Kelley 2001: Kappelman et al. 2003). The postcranium is represented only by two isolated phalanges from Ravin de la Pluie [RPI 86 (PP) and RPI 87 (MP)], which are considered by Bonis and Koufos (2014) to most likely to be manual.

RPI 86 is close to human manual PP3 and a chimpanzee pedal PP3 in length. It is robust (radioulnarly wide relative to the proximodistal length). SSFs are intermediate between humans (weak) and chimpanzees (stronger). The included angle (43°) is low, as is the trochlea, and best matches with PPs of terrestrial cercopithecines, according to Bonis and Koufos (2014). The proximal joint surface is small, mildly concave, mostly limited to the dorsal half of the proximal end, and oriented proximo-dorsally. The basal tubercles are very large, more closely reminiscent of pedal than manual PPs. The shaft virtually lacks a radioulnar flare distally, unlike manual PPs in other fossil and living apes. Radioulnar asymmetry of the joints and axial torsion suggest RPI 86 is either PP2 or 4. If this phalanx is manual, then there is very little indication of any arboreal behavior. Even if this specimen comes from a foot, the behavioral signal remains a mainly terrestrial one, though the length and curvature of the specimen suggests a degree of arboreality perhaps most similar to gorillas.

RPI 87 is also robust, being similar to MPs of *Homo* in relative radioulnar dimensions (Bonis and Koufos 2014). It is approximately the length of a chimpanzee manual MP5, but more robust transversely. The foveae of the proximal surface (for the condyles of the PP) are deep and separated from each other by a prominent central ridge, as is typical of manual phalanges compared with pedal ones. Radioulnar asymmetry of the joints is also strong, suggesting a fifth ray manual MP.

Together, these phalanges contain the weakest arboreal signals of all fossil ape phalanges discussed in this chapter. *O. macedoniensis* must have spent a significant amount of time on the ground, being more adapted to terrestriality than any other Miocene ape. The paleoenvironment of RPl is depicted as a savanna-like landscape with some woodland elements, yet dominated by C3 plants (Bonis and Koufos 2014). This is consistent with terrestriality and/or a gorilla-like combination of terrestrial and arboreal positional behavior.

4.6 Oreopithecus bambolii

Oreopithecus is an 8–7 Ma ape known from Tuscany and Sardinia, Italy (Begun 2002). A number of postcranial characters support the conclusion that *Oreopithecus* was highly specialized for orthograde behavior (Straus 1963; Harrison 1986, 1991; Sarmiento 1987; Rose 1993; Begun 2002). Adult male BM was estimated at 32 kg based on the partial skeleton IGF 11778 (Jungers 1987). In addition to this skeleton, there are several hand skeletons as well as a number of isolated phalanges (de Terra 1956; Moyà-Solà et al. 1999; Almécija et al. 2014). Several carpal bones (scaphoid, lunate, capitate, hamate) have been recovered (Harrison 1986; Fig 17.2). The scaphoid is elongated and the facet for the lunate is tall as in *Pongo* (Kivell and Begun 2009). The proximal articular surface of the lunate is relatively narrow radioulnarly, suggesting predominantly radial-side force transmission of the wrist like extant great apes (Rose 1993; contra Harrison 1986). The head of the capitate is relatively small, whereas the proximal portion of the hamate is very well developed. The non-pollical Mcs are slender and exhibit a slight degree of dorsopalmar curvature. Their heads appear to be relatively tall and narrow, as in *Proconsul* and unlike extant apes (Harrison 1986). The non-pollical PPs (and MPs) are gracile and the proximal articular surface of the PP faces proximally (Moyà-Solà et al. 1999). They display quite weak curvature, with an included angle close to the average of *Macaca* (Fig. 17.4a). However, based on PCF the PPs group with *Pan paniscus* and *Lufengpithecus* (Deane and Begun 2008.)

There are opposing views concerning the capability of precision grips in *Oreopithecus*. Moyà-Solà et al. (1999, 2005b) proposed that *Oreopithecus* had hominin-like precision grip capabilities based on the relative lengths of the hand and pollex, along with muscular insertion and articular features. This view was challenged by Susman (2004, 2005), who questioned the assignment of the pollical elements and functional interpretations of the morphology shared with *Oreopithecus* and hominins exclusive of extant apes. One of the most important points is whether the eroded and deformed pollical PP in IGF 11778 as assigned by Moyà-Solà et al. (1999, 2005b) is in fact a MP (Susman 2004, 2005; DRB personal observations). Moyà-Solà et al. (1999, 2005b) also compared the lengths of this PP1 against PP2 of another individual (BA140), finding that the PP1 was relatively long, as in *Papio* or *Homo*. However, since *P. heseloni* (KPS 3) is similar to *Oreopithecus* in this regard [measurements from Begun et al. (1994)], and since most Miocene apes seems to have had longer thumbs than extant apes (Almécija et al. 2012), this debate may be equivocal. Be that as it may, the pollical DPs of *Oreopithecus* seem to display better-developed muscular insertions than any other Miocene or extant ape (Moyà-Solà et al. 1999, 2005b; Almécija et al. 2014).

The hand skeleton of *Oreopithecus* appears generally slender (Szalay and Delson 1979). However, the hand is not remarkably elongated as in *Pan*, *Pongo*, or even *Hispanopithecus* when scaled to BM (Moyà-Solà et al. 2005b). *Oreopithecus* falls between these apes and the shorter-handed group (*Gorilla*, *Papio*, and *Proconsul*). Although it is almost identical to *Pierolapithecus* in this regard (Moyà-Solà et al. 2005a), the hand of *Oreopithecus* is much more gracile. While a long list of postcranial characters supports significant forelimb-dominated orthograde in *Oreopithecus*, its hand differs greatly from those of extant great apes, *Rudapithecus* or *Hispanopithecus*. Positional behavior of *Oreopithecus* might have been rather different from that of these apes. Alternatively, its postcranium might exhibit a stage in transition or a different solution for similar positional behavior.

4.7 Lufengpithecus lufengensis

Lufengpithecus lufengensis is a 7–6 Ma ape known from Shihuiba, Yunnan Province, China (Kelley 2002; Begun 2007). Although three species are recognized for *Lufengpithecus*, postcranial materials are known only for *L. lufengensis*. Hand elements are limited to two PPs (Wu et al. 1986) that are figured in Meldrum and Pan (1988). DRB has examined the original specimens. If these PPs belong to the same individual, PA 1057 would belong to a paramedian digit and PA 1056 a median digit, judging from the overall size, robusticity, and asymmetry. Although the proximal articular surface is radioulnarly wide and concave and the basal palmar tubercles are well developed, the articular surface is proximally oriented. The shaft is broader radioulnarly than dorsopalmarly and exhibits strong SSFs (also see Fig. 17.4b), suggesting that *L. lufengensis* was a specialized orthograde and probably suspensory great ape (DRB, personal observations). Deane and Begun (2008) found a degree of proximal phalangeal curvature most similar to *Pan*, *Oreopithecus*, and *Pierolapithecus* (again, based on PCF).

5 Conclusions

After this brief review of the fossil record, we cannot help but recognize that although isolated hand bones are available for several Miocene hominoids, only a few taxa preserve relatively complete hands, making it difficult to trace the evolutionary history of the ape hand (Fig. 17.7). The only certainty is that the evolution of the ape hand started from an appendage very well suited for powerful pollical-assisted grasping that supplied a balancing function in response to the loss of tail (Kelley 1997). This hand was most probably used during above-branch grasping quadrupedalism and other general arboreal activities such as palmigrady and climbing. These adaptations lasted until early middle Miocene.

Though less outstanding, a potentially functionally important sign may be hinted at in the 15 Ma *Nacholapithecus*, whose hand was adapted for a primitive pollical-assisted grasping function, but absolutely large and more powerful. This may be an initial change toward forelimb-dominated positional behaviors while retaining the generally plesiomorphic postcrania. Alternatively, it might have been an aberrant offshoot without phylogenetic significance for the adaptations seen in extant great apes. Although this question is unanswerable with currently available fossil data, it is a hypothesis worth exploring.

Pierolapithecus stands at a very important position because it is clearly derived and may represent an early form evolving toward the suspension/orthogrady seen in extant great apes, though in a mosaic way: enhanced ulnar deviation and midcarpal supination, yet lacking complete radial-side loading of the wrist and specializations for a hooklike grip. Although its phylogenetic position is yet to be resolved (Moyà-Solà et al. 2004, 2005b; Begun and Ward 2005; Alba 2012), the "*Pierolapithecus* stage" seems to fit an evolutionary link between the primitive *Proconsul*-like ape and later "modern" Eurasian great apes.

Late Miocene Eurasian great apes discussed in this chapter exhibit varying degrees (or modes) of suspension and orthograde adaptations. However, the hand anatomy that invokes such behaviors is not always morphologically identical across these apes, suggesting independent specialization for those behaviors. Nor is any late Miocene great ape hand fully "modern." It is notable that a reduction of the pollex, where fossil elements are available, is not observed in any of these species. This may suggest that above-branch

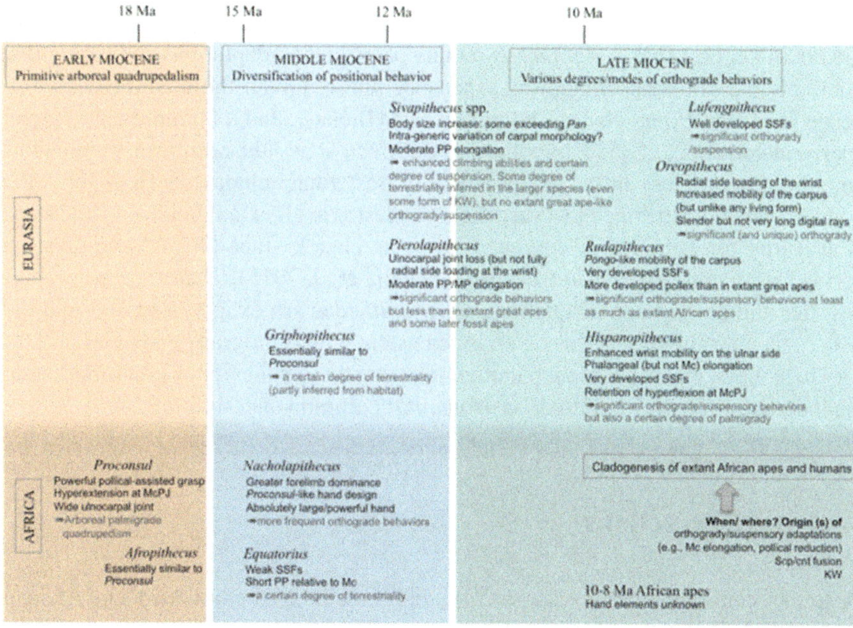

Fig. 17.7 Summary figure of the evolution of the hand of Miocene apes. See text for abbreviations

quadrupedalism/climbing with power grasping was an indispensable locomotor behavior, even in these apes (Almécija et al. 2012).

Whereas this functional/behavioral inference of the late Miocene apes is, at least broadly, a consensus, opinions are divided as to how to relate it to the evolution of AAH (see Cote 2004; Kunimatsu et al. 2007; Begun 2009, 2010; Begun et al. 2012). If the root of the AAH hand was represented by any of these apes or a close relative, it would support the interpretation of a Eurasian origin of the AAH clade (Begun 2010; Begun et al. 2012). If the Eurasian origin is refuted for any reason, then we face a complete lack of Miocene fossil clues of the origin of manual modernity after 15 Ma in Africa.

Furthermore, there are competing hypotheses regarding the original adaptation of the hand in the AAH clade (see Chap. 18). If the KW hypothesis (Begun 1992, 2004; Richmond and Strait 2000; Richmond et al. 2001) is supported, the hand of the ancestral lineage of this clade must have passed through a specialization for suspension (and/or orthogrady) as in *Hispanopithecus* or *Rudapithecus*, prior to a terrestrial adaptation. Alternatively, the arboreal palmigrady hypothesis (Lovejoy et al. 2009a,b; also see Schmitt 2003) posits that *Proconsul*-like careful above-branch quadrupedalism and clambering persisted until the last common ancestor of the *Pan*-human clade. In this view, the hand of this hypothetical ancestor had a capability for an extreme midcarpal dorsiflexion and was adapted for general grasping with elongated phalanges but relatively short non-pollical Mcs. Lovejoy et al. (2009a,b) suggest that *Ardipithecus* retains palmigrade adaptations indicating that hominins did not pass through a suspensory phase. Given the unambiguous evidence for a *Pan-Homo* clade, a hominine (AAH) clade, and a hominid clade (great apes and humans), if their hypothesis is correct, it would require a tremendous amount of homoplasy in the development of postcranial characters throughout the body universally attributed to a suspensory ancestry in all extant hominids (the same is also true for *Pongo* if *Sivapithecus* is in its clade), since all of these taxa are derived relative to *Proconsul* (see also Almécija et al. 2015). Therefore, some suggest that the palmigrade adaptations of *Ardipithecus* are exaggerated and that the bulk of the morphology of this taxon is consistent with orthogrady and even suspensory behaviors, making it more parsimonious that these behaviors were important in the locomotor repertoire of the *Pan-Homo* last common ancestor.

6 Future Directions

Despite recent discoveries of dental/gnathic remains of the late Miocene African purported hominines *Chororapithecus* and *Nakalipithecus* (Kunimatsu et al. 2007; Suwa et al. 2007), postcrania of these taxa have not been found. There are differing interpretations of the evolutionary relations of these taxa (Kunimatsu et al. 2007; Suwa et al. 2007; Begun 2009, 2010; Begun et al. 2012), and the establishment of a robust phylogenetic hypothesis of these apes is essential. Without a doubt, the recovery of more postcrania from these and other middle and late Miocene taxa will serve to test current hypotheses of the ancestral morphotype of the AAH clade.

It is also desirable to understand evolutionary context(s) of suspensory/orthograde adaptations in Eurasian Miocene apes. What factors drove several ape lineages toward those behaviors presumably independently after 10 Ma? Why do none of those fossil apes have a thumb short relative to the fingers, as in all extant great apes? The answers to these questions and many others are indispensable to refine the scenario of the evolutionary history of the AAH clade.

Acknowledgments We thank the editors for inviting us to contribute to this volume and providing us with thoughtful comments on an early version of the manuscript. We are grateful to the National Museums of Kenya for permission to study original specimens under their care. We thank Ashley Hammond, Jay Kelley, Tracy Kivell, Salvador Moyà-Solà, Lorenzo Rook, and Randy Susman for providing us original fossil photographs. This work is supported by grants from the JSPS Grant-in-Aid (#22255006, 25257408) to M.N., from the Spanish Ministerio de Economía y Competitividad (CGL2014-54373-P) and National Science Foundation (NSF-BCS 1316947) to S.A., from the American Association of Physical Anthropologists (Professional Development Program) to S.A., and from the Natural Sciences and Engineering Research Council of Canada, the National Geographic Society, and the University of Toronto to D.R.B.

References

Alba DM (2012) Fossil apes from the Vallès-Penedès Basin. Evol Anthropol 21:254–269

Alba DM, Almécija S, Moyà-Solà S (2010) Locomotor inferences in *Pierolapithecus* and *Hispanopithecus*: reply to Deane and Begun (2008). J Hum Evol 59:143–149

Alba DM, Almécija S, Casanovas-Vilar I, Méndez JM, Moyà-Solà S (2012) A partial skeleton of the fossil great ape *Hispanopithecus laietanus* from Can Feu and the mosaic evolution of crown-hominoid positional behaviors. PLoS One 7:e39617

Allen KL, McCrossin ML (2007) Functional morphology of the *Kenyapithecus* hand from Maboko Island (Kenya). Am J Phys Anthropol 132(Suppl 44):62

Almécija S, Alba DM, Moyà-Solà S, Köhler M (2007) Orang-like manual adaptations in the fossil hominoid *Hispanopithecus laietanus*: first steps towards great ape suspensory behaviors. Proc R Soc B 274:2375–2384

Almécija S, Alba DM, Moyà-Solà S (2009) *Pierolapithecus* and the functional morphology of Miocene ape hand phalanges: Paleobiological and evolutionary implications. J Hum Evol 57:284–297

Almécija S, Alba DM, Moyà-Solà S (2012) The thumb of Miocene apes: new insights from Castell de Barberà (Catalonia, Spain). Am J Phys Anthropol 148:436–450

Almécija S, Shrewsbury M, Rook L, Moyà-Solà S (2014) The morphology of *Oreopithecus bambolii* pollical distal phalanx. Am J Phys Anthropol 153:582–597

Almécija S, Smaers JB, Jungers WL (2015) The evolution of human and ape hand proportions. Nat Commun 6:7717

Beard KC, Teaford MF, Walker A (1986) New wrist bones of *Proconsul africanus* and *P. nyanzae* from Rusinga Island, Kenya. Folia Primatol 47:97–118

Begun DR (1988) Catarrhine phalanges from the late Miocene (Vallesian) of Rudabánya, Hungary. J Hum Evol 17:413–438

Begun DR (1992) Miocene fossil hominids and the chimp-human clade. Science 257:1929–1933

Begun DR (1993) New catarrhine phalanges from Rudabánya (Northeastern Hungary) and the problem of parallelism and convergence in hominoid postcranial morphology. J Hum Evol 24:373–402

Begun DR (1994) Relations among the great apes and humans: new interpretations based on the fossil great ape *Dryopithecus*. Yearb Phys Anthropol 37:11–63

Begun DR (2002) European hominoids. In: Hartwig W (ed) The primate fossil record. Cambridge University Press, Cambridge, pp 231–253

Begun DR (2004) Knuckle-walking and the origin of human bipedalism. In: Meldrum DJ, Hilton CE (eds) From biped to strider: the emergence of modern human walking, running, and resource transport. Kluwer Academic/Plenum Publishers, New York, pp 9–33

Begun DR (2007) Fossil record of Miocene hominoids. In: Henke W, Tattersall I (eds) Handbook of paleoanthropology. Springer, Berlin, pp 921–977

Begun DR (2009) Dryopithecins, Darwin, de Bonis, and the European origin of the African apes and human clade. Geodiversitas 31:789–816

Begun DR (2010) Miocene hominids and the origins of the African apes and humans. Annu Rev Anthropol 39:67–84

Begun DR, Kivell TL (2011) Knuckle-walking in *Sivapithecus*? The combined effects of homology and homoplasy with possible implications for pongine dispersals. J Hum Evol 60:158–170

Begun DR, Nargolwalla MC, Kordos L (2012) European Miocene hominids and the origin of the African ape and human clade. Evol Anthropol 21:10–23

Begun DR, Kordos L (2011) New postcrania of *Rudapithecus hungaricus* from Rudabánya (Hungary). Am J Phys Anthropol 144(Suppl 52):86 (abstract)

Begun DR, Teaford MF, Walker A (1994) Comparative and functional anatomy of *Proconsul* phalanges from the Kaswanga Primate Site, Rusinga Island, Kenya. J Hum Evol 26:89–165

Begun DR, Ward CV (2005) Comment on "*Pierolapithecus catalaunicus*, a new Middle Miocene great ape from Spain". Science 308:203c

Böhme M, Aziz HA, Prieto J, Bachtadse V, Schweigert G (2011) Bio-magnetostratigraphy and environment of the oldest Eurasian hominoid from the Early Miocene of Engelswies (Germany). J Hum Evol 61:332–339

De Bonis L, Koufos GD (1994) Our ancestors' ancestor: *Ouranopithecus* is a Greek link in human ancestory. Evol Anthropol 3:75–83

De Bonis L, Koufos GD (2014) First discovery of postcranial bones of *Ouranopithecus macedoniensis* (Primates, Hominoidea) from the late Miocene of Macedonia (Greece). J Hum Evol 74:21–36

Cartmill M, Milton K (1977) The lorisiform wrist joint and the evolution of "brachiating" adaptations in the Hominoidea. Am J Phys Anthropol 47:249–272

Casanovas-Vilar I, Alba DM, Garcés M, Robles JM, Moyà-Solà S (2011) Updated chronology for the Miocene hominoid radiation in Western Eurasia. Proc Natl Acad Sci U S A 108:5554–5559

Corruccini RS, Ciochon RL, McHenry HM (1975) Osteometric shape relationships in the wrist joint of some anthropoids. Folia Primatol 24:250–274

Cote SM (2004) Origins of the African hominoids: an assessment of the palaeobiogeographical evidence. C R Palevol 3:323–340

Deane AS, Begun DR (2008) Broken fingers: retesting locomotor hypotheses for fossil hominoids using fragmentary proximal phalanges and high-resolution polynomial curve fitting (HR-PCF). J Hum Evol 55:691–701

DeSilva JM, Morgan ME, Barry JC, Pilbeam D (2010) A hominoid distal tibia from the Miocene of Pakistan. J Hum Evol 58:147–154

de Terra H (1956) New approach to the problem of man's origin. Science 124:1282–1285

Erikson GE (1963) Brachiation in New World monkeys and in anthropoid apes. Symp Zool Soc Lond 10:135–164

Ersoy A, Kelley J, Andrews P, Alpagut B (2008) Hominoid phalanges from the middle Miocene site of Paşalar, Turkey. J Hum Evol 54:518–529

Godinot M, Beard KC (1991) Fossil primate hands: a review and an evolutionary inquiry emphasizing early forms. Hum Evol 6:307–354

Hamrick MW, Meldrum DJ, Simons EL (1995) Anthropoid phalanges from the Oligocene of Egypt. J Hum Evol 28:121–145

Harrison T (1986) A reassessment of the phylogenetic relationships of *Oreopithecus bambolii* Gervais. J Hum Evol 15:541–583

Harrison T (1991) The implications of *Oreopithecus bambolii* for the origins of bipedalism. In: Coppens Y, Senut B (eds) Origine(s) de la bipédie chez les hominidés. CNRS, Paris, pp 235–244

Harrison T (2010) Dendropithecoidea, Proconsuloidea and Hominoidea (Catarrhini, Primates). In: Werdelin L, Sanders WJ (eds) Cenozoic mammals of Africa. University of California Press, Berkeley, pp 429–469

Heizmann EPJ, Begun DR (2001) The oldest European hominoid. J Hum Evol 41:463–481

Inouye SE (1994) Ontogeny of knuckle-walking hand postures in African apes. J Hum Evol 26:459–485

Ishida H, Kunimatsu T, Nakatsukasa M, Nakano Y (1999) New hominoid genus from the middle Miocene of Nachola, Kenya. Anthropol Sci 107:189–191

Ishida H, Kunimatsu Y, Takano T, Nakano Y, Nakatsukasa M (2004) *Nacholapithecus* skeleton from the Middle Miocene of Kenya. J Hum Evol 46:67–101

Jenkins FA Jr (1981) Wrist rotation in primates: a critical adaptation for brachiators. Symp Zool Soc Lond 48:429–451

Jenkins FA Jr, Fleagle JG (1975) Knuckle-walking and the functional anatomy of the wrists in living apes. In: Tuttle RH (ed) Primate functional morphology and evolution. Mouton, The Hague, pp 213–231

Jouffroy FK, Godinot M, Nakano Y (1991) Biometrical characteristics of primate hands. Hum Evol 6:269–306

Jungers WL (1987) Body size and morphometric affinities of the appendicular skeleton in *Oreopithecus bambolii* (IGF11778). J Hum Evol 16:445–456

Kappelman J, Richmond BG, Seiffert ER, Maga AM, Ryan TM (2003) Hominoidea (Primates). In: Fortelius M, Kappelman J, Sen S, Bernor RL (eds) Geology and paleontology of the Miocene Sinap Formation, Turkey. Columbia University Press, New York, pp 90–124

Kelley J (1997) Paleobiological and phylogenetic significance of life history in Miocene hominoids. In: Begun DR, Ward CV, Rose MD (eds) Function, phylogeny, and fossils: Miocene hominoid evolution and adaptations. Plenum Press, New York, pp 173–208

Kelley J (2001) Phylogeny and sexually dimorphic characters: canine reduction in *Ouranopithecus*. In: Koufos GD, Andrews P, de Bonis L (eds) Phylogeny of the Neogene hominoid primates of Eurasia. Cambridge University Press, Cambridge, pp 269–283

Kelley J (1988) A new large species of *Sivapithecus* from the Siwaliks of Pakistan. J Hum Evol 17:305–324

Kelley J (2002) The hominoid radiation in Asia. In: Hartwig WC (ed) The primate fossil record. Cambridge University Press, Cambridge, pp 369–384

Kelley J (2005) Twenty-five years contemplating *Sivapithecus* taxonomy. In: Lieberman DE, Smith RH, Kelley L (eds) Interpreting the past: essays on human, primate, and mammal evolution. Brill, Boston, pp 123–143

Kelley J, Andrew P, Alpagut B (2008) A new hominoid species from the middle Miocene site of Paşalar, Turkey. J Hum Evol 54:455–479

Kivell TL, Begun DR (2009) New primate carpal bones from Rudabánya (late Miocene, Hungary): taxonomic and functional implications. J Hum Evol 57:697–709

Kunimatsu Y, Nakatsukasa M, Sawada Y, Sakai T, Hyodo M, Hyodo H, Itaya T, Nakaya H, Saegusa H, Mazurier A, Saneyoshi M, Tsujikawa H, Yamamoto A, Mbua E (2007) A new Late Miocene great ape from Kenya and its implications for the origins of African great apes and humans. Proc Natl Acad Sci U S A 104:19220–19225

Leakey M, Walker A (1997) *Afropithecus*: function and phylogeny. In: Begun DR, Ward CV, Rose MD (eds) Function, phylogeny, and fossils: Miocene hominoid evolution and adaptation. Plenum Press, New York, pp 225–239

Leakey RE, Leakey MG, Walker AC (1988) Morphology of *Afropithecus turkanensis* from Kenya. Am J Phys Anthropol 76:289–307

Le Minor JM (1988) The ventral metacarpo- and metatarsophalangeal sesamoid bones: comparative and evolutionary aspects. Gebenbaurs Morphol Jahrb 134:693–731

Lewis OJ (1969) The hominoid wrist joint. Am J Phys Anthropol 30:251–268

Lewis OJ (1972) Osteological features characterizing the wrists of monkeys and apes, with a reconsideration of this region in *Dryopithecus* (*Proconsul*) *africanus*. Am J Phys Anthropol 36:45–58

Lewis OJ (1989) Functional morphology of the evolving hand and foot. Clarendon Press, Oxford

Lovejoy CO, Simpson SW, White TD, Asfaw B, Suwa G (2009a) Careful climbing in the Miocene: the forelimbs of *Ardipithecus ramidus* and humans are primitive. Science 326:70e71–70e78

Lovejoy CO, Suwa G, Simpson SW, Matternes JH, White TD (2009b) The great divides: *Ardipithecus ramidus* reveals the postcrania of our last common ancestors with African apes. Science 326:100–106

Madar SI, Rose MD, Kelley J, MacLatchy L, Pilbeam D (2002) New *Sivapithecus* postcranial specimens from the Siwaliks of Pakistan. J Hum Evol 42:705–752

McCrossin ML (1994) The phylogenetic relationships, adaptations, and ecology of *Kenyapithecus*. Ph.D. dissertation, University of California, Berkeley

McCrossin ML, Benefit BR (1994) Maboko Island and the evolutionary history of Old World monkeys and apes. In: Corruccini RS, Ciochon RL (eds) Integrative paths to the past: paleoanthropological advances in honor of F. Clark Howell. Prentice Hall, Englewood Cliffs, pp 95–122

McCrossin ML, Benefit BR, Gitau SN, Palmer AK, Blue KT (1998) Fossil evidence for the origins of terrestriality among Old World higher primates. In: Strasser E, Fleagle J, Rosenberger A, McHenry H (eds) Primate locomotion. Plenum Press, New York, pp 353–396

McHenry HM, Corruccini RS (1983) The wrist of *Proconsul africanus* and the origin of hominoid postcranial adaptations. In: Ciochon RL, Corruccini RS (eds) New interpretations of ape and human ancestry. Plenum Press, New York, pp 353–366

McNulty KP, Begun DR, Kelley J, Manthi FK, Mbua EN (2015) A systematic revision of *Proconsul* with the description of a new genus of early Miocene hominoid. J Hum Evol 84:42–61

Meldrum DJ, Pan Y (1988) Manual proximal phalanx of *Laccopithecus robustus* from the Latest Miocene site of Lufeng. J Hum Evol 17:719–731

Morbeck ME (1975) *Dryopithecus africanus* forelimb. J Hum Evol 4:39–46
Moyà-Solà S, Köhler M (1996) A *Dryopithecus* skeleton and the origins of great-ape locomotion. Nature 379:156–159
Moyà-Solà S, Köhler M, Rook L (1999) Evidence of hominid-like precision grip capability in the hand of the Miocene ape *Oreopithecus*. Proc Natl Acad Sci U S A 96:313–317
Moyà-Solà S, Köhler M, Alba DM, Casanovas-Vilar I, Galindo J (2004) *Pierolapithecus catalaunicus*, a new Middle Miocene great ape from Spain. Science 306:1339–1344
Moyà-Solà S, Köhler M, Alba DM (2005a) Response to comment on "*Pierolapithecus catalaunicus*, a new Middle Miocene great ape from Spain". Science 308:203d
Moyà-Solà S, Köhler M, Rook L (2005b) The *Oreopithecus* thumb: a strange case in hominoid evolution. J Hum Evol 49:395–404
Moyà-Solà S, Köhler M, Alba DM, Casanovas-Vilar I, Galindo J, Robles JM, Cabrera L, Garcés M, Almécija S, Beamud E (2009) First partial face and upper dentition of the Middle Miocene hominoid *Dryopithecus fontani* from Abocador de Can Mata (Vallès-Penedès Basin, Catalonia, NE Spain): taxonomic and phylogenetic implications. Am J Phys Anthropol 139:126–145
Nakatsukasa M, Kunimatsu Y (2009) *Nacholapithecus* and its importance for understanding hominoid evolution. Evol Anthropol 18:103–119
Nakatsukasa M, Kunimatsu Y, Nakano Y, Takano T, Ishida H (2003) Comparative and functional anatomy of phalanges in *Nacholapithecus kerioi*, a Middle Miocene hominoid from northern Kenya. Primates 160:469–482
Napier JR (1960) Studies of the hands of living primates. Proc Zool Soc Lond 134:647–657
Napier J (1993) Hands [revised by Russell H. Tuttle]. Princeton University Press, Princeton
Napier JR, Davis PR (1959) The fore-limb skeleton and associated remains of *Proconsul africanus*. Fossil mammals of Africa, No. 16. British Museum (Natural History), London
O'Connor BL (1975) The functional morphology of the cercopithecoid wrist and inferior radioulnar joints, and their bearing on some problems in the evolution of the Hominoidea. Am J Phys Anthropol 43:113–122
Ogihara N, Almécija S, Nakatsukasa M, Nakano Y, Kikuchi Y, Kunimatsu Y, Makishima H, Shimizu D, Takano T, Tsujikawa H, Kagaya M, Ishida H, (2016) Carpal bones of *Nacholapithecus kerioi*, a middle Miocene hominoid from Northern Kenya. Am J Phys Anthropol 160:469–482
Patel BA, Susman RL, Rossie JB, Hill A (2009) Terrestrial adaptations in the hands of *Equatorius africanus* revisited. J Hum Evol 57:763–772
Pilbeam DR, Rose MD, Badgley C, Lipschutz B (1980) Miocene hominoids from Pakistan. Postilla 181:1–94
Pina M, Alba DM, Almécija S, Fortuny J, Moyà-Solà S (2012) Brief communication: Paleobiological inferences on the locomotor repertoire of extinct hominoids based on femoral neck cortical thickness: the fossil great ape *Hispanopithecus laietanus* as a test-case study. Am J Phys Anthropol 149:142–148
Preuschoft H (1973) Body posture and locomotion in some East African Miocene Dryopithecinae. In: Day M (ed) Human evolution. Taylor and Francis, London, pp 13–46
Rafferty KL (1990) The functional and phylogenetic significance of the carpometacarpal joint of the thumb in anthropoid primates. M.A. thesis, New York University
Rafferty KL, Walker A, Ruff CB, Rose MD, Andrews PJ (1995) Postcranial estimates of body weight in *Proconsul*, with a note on a distal tibia of *P. major* from Napak, Uganda. Am J Phys Anthropol 97:391–402
Richmond BG (2006) Functional morphology of the midcarpal joint in knuckle-walkers and terrestrial quadrupeds. In: Ishida H, Tuttle R, Pickford M, Ogihara N, Nakatsukasa M (eds) Human origins and environmental backgrounds. Springer, New York, pp 105–122
Richmond BG, Jungers WL (2008) *Orrorin tugenensis* femoral morphology and the evolution of hominin bipedalism. Science 319:1662–1665
Richmond BG, Strait DR (2000) Evidence that humans evolved from a knuckle-walking ancestor. Nature 404:382–385
Richmond BG, Whalen M (2001) Forelimb function, bone curvature and phylogeny of *Sivapithecus*. In: de Bonis L, Koufos GD, Andrews P (eds) African and Eurasian Miocene Hominoids and the origins of the Hominidae. Cambridge University Press, Cambridge, pp 326–348

Richmond BG, Begun DR, Strait DS (2001) Origin of human bipedalism: the knuckle-walking hypothesis revisited. Yearb Phys Anthropol 44:70–105
Rose MD (1984) Hominoid postcranial specimens from the Middle Miocene Chinji Formation, Pakistan. J Hum Evol 13:503–516
Rose MD (1986) Further hominoid postcranial specimens from the Late Miocene Nagri Formation of Pakistan. J Hum Evol 15:333–367
Rose MD (1988) Functional anatomy of the cheiridia. In: Schwartz JH (ed) Orangutan biology. Oxford University Press, New York, pp 299–310
Rose MD (1992) Kinematics of the trapezium-1st metacarpal joint in extant anthropoids and Miocene hominoids. J Hum Evol 22:255–266
Rose MD (1993) Locomotor anatomy of Miocene hominoids. In: Gebo DL (ed) Postcranial adaptation in nonhuman primates. Northern Illinois University Press, DeKalb, pp 252–272
Rose MD (1994) Quadrupedalism in some Miocene catarrhines. J Hum Evol 26:387–411
Rose MD, Nakano Y, Ishida H (1996) *Kenyapithecus* postcranial specimens from Nachola, Kenya. Afr Stud Monogr Suppl 24:3–56
Ruff CB (2003) Long bone articular and diaphyseal structure in Old World monkeys and apes. II: Estimation of body mass. Am J Phys Anthropol 120:16–37
Ruff CB, Walker A, Teaford MF (1989) Body mass, sexual dimorphism and femoral proportions of *Proconsul* from Rusinga and Mfangano Islands, Kenya. J Hum Evol 18:515–536
Sarmiento EE (1987) The phylogenetic position of *Oreopithecus* and its significance in the origin of the Hominidae. Am Mus Novit 2881:1–44
Sarmiento EE (1988) Anatomy of the hominoid wrist joint: its evolutionary and functional implications. Int J Primatol 9:281–345
Schmitt D (1994) Forelimb mechanics as a function of substrate type during quadrupedalism in two anthropoid primates. J Hum Evol 26:441–457
Schmitt D (2003) Insights into the evolution of human bipedalism from experimental studies of humans and other primates. J Exp Biol 206:1437–1448
Schön MA, Ziemer LK (1973) Wrist mechanism and locomotor behavior of *Dryopithecus (Proconsul) africanus*. Folia Primatol 20:1–11
Sherwood RJ, Ward S, Hill A, Duren DL, Brown B, Downs W (2002) Preliminary description of the *Equatorius africanus* partial skeleton (KNM-TH 28860) from Kipsaramon, Tugen Hills, Baringo District, Kenya. J Hum Evol 42:63–73
Spoor CF, Sondaar PY, Hussain ST (1991) A hominoid hamate and first metacarpal from the Late Miocene Nagri Formation of Pakistan. J Hum Evol 21:413–424
Straus WL (1963) The classification of *Oreopithecus*. In: Washburn SL (ed) Classification and human evolution. Aldine Publishing Co., Chicago, pp 146–177
Straus WL Jr (1940) The posture of the great ape hand in locomotion and its phylogenetic implications. Am J Phys Anthropol 27:199–207
Susman RL (1979) Comparative and functional morphology of hominoid fingers. Am J Phys Anthropol 50:215–236
Susman RL (2004) *Oreopithecus bambolii*: an unlikely case of hominidlike grip capability in a Miocene ape. J Hum Evol 46:105–117
Susman RL (2005) *Oreopithecus*: still apelike after all these years. J Hum Evol 49:405–411
Suwa G, Kono RT, Katoh S, Asfaw B, Beyene Y (2007) A new species of great ape from the late Miocene epoch in Ethiopia. Nature 448:921–924
Szalay FS, Delson E (1979) Evolutionary history of the primates. Academic Press, New York
Tallman M, Almécija S, Reber SL, Alba DM, Moyà-Solà S (2013) The distal tibia of *Hispanopithecus laietanus*: more evidence for mosaic evolution in Miocene apes. J Hum Evol 64:319–327
Tuttle RH (1967) Knuckle-walking and the evolution of hominoid hands. Am J Phys Anthropol 26:171–206
Tuttle RH (1975) Parallelism, brachiation, and hominoid phylogeny. In: Luckett WP, Szalay FS (eds) Phylogeny of the primates. Plenum Press, New York, pp 447–480

Walker A (1997) *Proconsul*: function and phylogeny. In: Begun DR, Ward CV, Rose MD (eds) Function, phylogeny, and fossils: Miocene hominoid evolution and adaptations. Plenum Press, New York, pp 209–224

Walker AC, Pickford M (1983) New postcranial fossils of *Proconsul africanus* and *Proconsul nyanzae*. In: Ciochon RL, Corruccini RS (eds) New interpretations of ape and human ancestry. Plenum Press, New York, pp 325–351

Ward CV (1998) *Afropithecus*, *Proconsul*, and the primitive hominoid skeleton. In: Strasser E, Fleagle J, Rosenberger A, McHenry H (eds) Primate locomotion: recent advances. Plenum Press, New York, pp 337–352

Ward CV (2007) Postcranial and locomotor adaptations of hominoids. In: Henke W, Tattersall I (eds) Handbook of paleoanthropology. Springer Verlag, Berlin, pp 1011–1030

Ward S, Brown B, Hill A, Kelley J, Downs W (1999) *Equatorius*: a new hominoid genus from the middle Miocene of Kenya. Science 285:1382–1386

Wu R, Xu Q, Lu Q (1986) Relationship between Lufeng *Sivapithecus* and *Ramapithecus* and their phylogenetic position. Acta Anthropol Sinica 5:1–30

Chapter 18
Evolution of the Early Hominin Hand

Brian G. Richmond, Neil T. Roach, and Kelly R. Ostrofsky

1 Introduction

The hand is among the most primitive anatomical regions in the human body. The basic five-digit pattern of our manus, including the number of bones in each ray and the retention of many extrinsic and intrinsic muscles (see Chap. 7), is likely retained from the ancestral condition for all mammals. While primate hands are generally characterized by two major novelties, namely, nails instead of claws and enhanced thumb divergence and opposability during grasping, human hands are not particularly derived compared with our primate relatives. The fact that our hands are so little modified from those of early mammals is a testament to the astonishing versatility of the hand as a functional organ (e.g., Wood Jones 1916, 1920; Napier 1993).

What makes the primitive gestalt of the human hand surprising is that it has been the object of major changes in function. Since the Devonian, when our ancient sarcopterygian ancestors made the transition from swimming to weight support on land some 390 million years (Ma) or so ago (Niedźwiedzki et al. 2010), hands have functioned as organs of weight support. Since then, our ancestors went through a variety of locomotor transitions involving weight support on land and in trees, including

B.G. Richmond (✉)
Division of Anthropology, American Museum of Natural History, New York, NY 10024, USA

Department of Human Evolution, Max Planck Institute for Evolutionary Anthropology, Deutscher Platz 6, Leipzig 04103, Germany
e-mail: brichmond@amnh.org

N.T. Roach
Division of Anthropology, American Museum of Natural History, New York, NY, USA

Department of Human Evolutionary Biology, Harvard University, Cambridge, MA, USA

K.R. Ostrofsky
Center for the Advanced Study of Human Paleobiology, Department of Anthropology, The George Washington University, Washington, DC, USA

T.L. Kivell et al. (eds.), *The Evolution of the Primate Hand*, Developments in Primatology: Progress and Prospects, DOI 10.1007/978-1-4939-3646-5_18

using hands in arboreal settings in a variety of ways (Chaps. 14, 16, and 17). However, after the earliest hominins split from our last common ancestor with bonobos and chimpanzees, two fundamental changes in function occurred. First, bipedality and reduced arboreality removed the need for weight support through the forelimb during locomotion. Second, intensification of hand function in manipulation, tool use, and tool making in hominins required more precise and powerful fine motor actions. These functional changes did not necessarily take place at the same time or in the same context. Indeed, the timing and context of these two major changes in function are the subject of active debate and fertile ground for new research.

This chapter explores the evolution of early hominin hand anatomy, from the *Pan-Homo* last common ancestor (LCA) through *Homo erectus*, in the context of these two major changes in function. For clarity, we treat early hominins here as three discrete groups: basal hominins, australopiths, and early *Homo*. The term "basal hominins" is used to describe *Sahelanthropus*, *Orrorin*, and *Ardipithecus* (Fig. 18.1), with the caveat that the systematics of some of these taxa are uncertain due to the limited anatomical evidence. "Australopith" is used to describe species in *Australopithecus* and *Paranthropus*, and "early *Homo*" refers to members of the genus *Homo* that are more primitive than, and sister taxa to, *H. erectus* (Fig. 18.1).

Here we explore the functional significance and evolutionary history of seven anatomical features in the early hominin hand. These are thumb/finger proportions, thumb robusticity, thumb musculature, distal tuberosities (apical tufts), carpal architecture, wrist mobility, and finger curvature. Three of these are features of the thumb that appear at different times and contexts, highlighting the perceived importance of thumb function in human evolution (Napier 1956, 1962; Marzke 1997; Susman 1998). We attempt to bring together this evidence to summarize the current picture of how hand function and anatomy changed over the course of early human evolution, concluding with a discussion of fruitful areas of future research.

2 Anatomy: Function and Evidence from the Fossil Record

2.1 *Thumb/Finger Proportions*

Hand proportions arguably influence manipulative skills more than any other anatomical characteristic. In particular, a relatively long thumb that more closely approximates finger length allows the thumb to more effectively oppose the fingers. Too much difference between thumb and finger lengths limit pad-to-pad contact in

Fig. 18.1 (continued) share numerous derived characteristics related to bipedalism, as well as hand morphology that includes short fingers relative to some basal hominins. (3) *Homo* is a clade and grade of hominins that share derived human characteristics such as enlarged brain size and aspects of mandible shape. As a clade, *Homo* shows no evidence to date of sharing derived hand anatomy relative to earlier hominins because the hands of earliest *Homo* resemble those of australopiths. Derived hand anatomy only appears later in the *Homo* clade

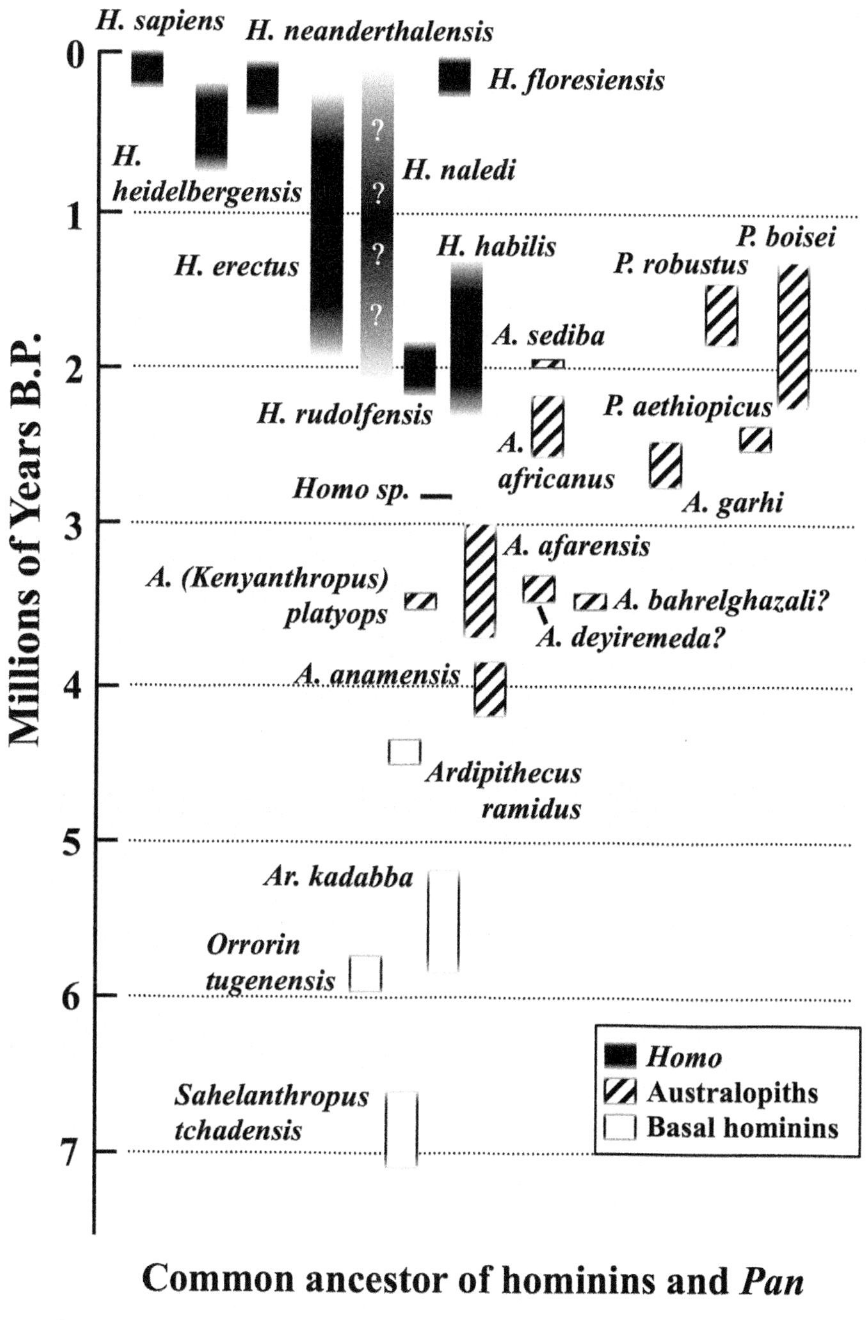

Fig. 18.1 Hominin phylogram. Hominins are organized into three groups or "grades." (1) "Basal hominins" refer the to most primitive known hominins such as *Ardipithecus ramidus* and probable hominins that are poorly known to date; basal hominins share anatomy related to orthograde posture and derived dental morphology, but it is unclear what aspects of hand anatomy are derived. (2) "Australopiths" refer to gracile (*Australopithecus*) and robust (*Paranthropus*) australopiths that

thumb opposition (Marzke and Wullstein 1996; Chap. 12). These proportions are so critical that Napier and Napier (1967) define an "opposability index" as the ratio between thumb and index finger lengths. Definitions of opposability vary widely, but typically refer to Napier's (1956) description of the ability to rotate the thumb into pad-to-pad contact with other digits or Napier's (1961) description of the position of the thumb via movements of abduction, flexion, and medial rotation of the first metacarpal (Mc1) at the first carpometacarpal (CM1) joint (see Chap. 2).

Gelada baboons have notable manual dexterity including the ability to use pad-to-pad precision grips during foraging in the wild (Jolly 1970; Rose 1977). It is thought that their high thumb/finger length proportions, while so useful for precision grips, are largely made possible by the evolution of short fingers for terrestrial locomotion. However, some anatomical features, such as short index fingers and hyper-extendable distal interphalangeal joints that facilitate pad-to-pad grip (Etter 1973; Marzke 1997), may enhance their precision grip capabilities.

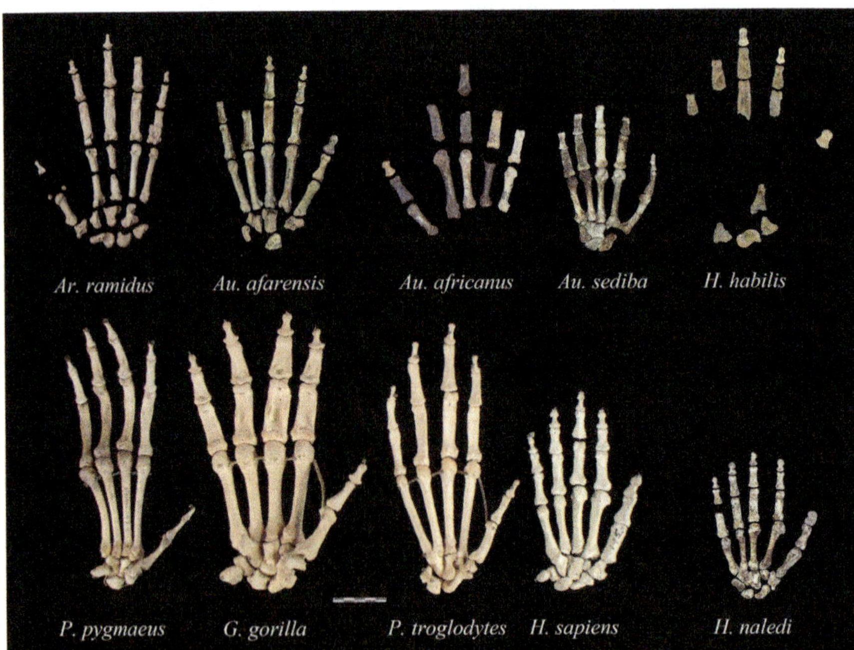

Fig. 18.2 Hand skeletons of extant hominoids and early hominins, illustrating variation in hand proportions. Extant hominoid hands (*bottom left*), all adult, include female *Pongo pygmaeus*, male *Gorilla gorilla gorilla*, female *Pan troglodytes*, and female *Homo sapiens*. Fossil hominin hands, clockwise from *top left*, attributed to *Ardipithecus ramidus* [ARA-VP-6/500 composite of right and left elements, unreversed; photo © and courtesy of T. White (2009)], *Australopithecus afarensis* (A.L. 333 site composite of left and right elements, unreversed; adapted from photos courtesy of T. Kivell), *Australopithecus africanus* (Sterkfontein composite of left and right elements, unreversed), *Australopithecus sediba* (photo © and courtesy of T. Kivell), *Homo habilis* (OH 7, adapted from photos courtesy of T. Kivell), and *Homo naledi* (photo © and courtesy of P. Schmid). All images approximately to scale

The fingers of some apes, such as chimpanzees and orangutans (Fig. 18.2), are so long that when flexed they tend to contact the tip of the thumb rather than its pad. When handling small objects, chimpanzees tend to hold objects in the manner humans typically grip a key, namely, a "two-jaw pad-to-side" grip with the thumb flexed against the side of the index finger (Christel 1993; Marzke and Wullstein 1996; Marzke 1997; see Chap. 12).

In contrast to apes, humans excel in their ability to use pad-to-pad precision grips (e.g., Napier 1956). Marzke (1997) argues that what makes human dexterity unique is the ability to firmly pinch objects, and precisely handle objects, with one hand. Humans have long thumbs relative to their fingers, proportions that are critical to these capabilities (Napier and Napier 1967; see Figs. 18.2 and 18.3).

Evidence of the evolution of hominin thumb/finger proportions is incomplete because associated hand skeletons are rare in the fossil record. *Ardipithecus ramidus* is the earliest hominin taxon with good evidence of hand proportions and the sole data point for such proportions in the basal hominins (Fig. 18.1). The partial

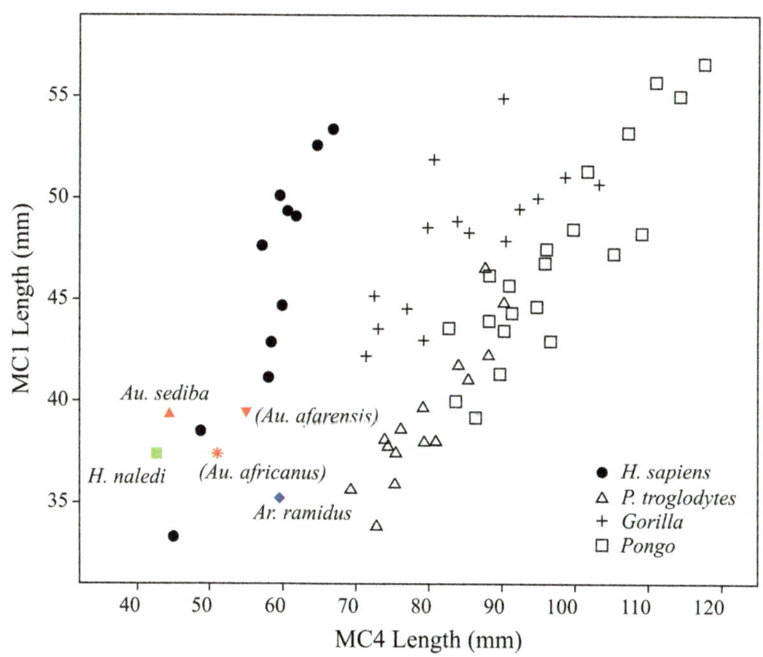

Fig. 18.3 Thumb/finger length proportions in hominoids and fossil hominins. Relative to the fourth metacarpal (Mc4) length, the first metacarpal (Mc1) length is intermediate in gorillas and much longer in modern humans. The australopiths approximate modern humans in relative Mc1 length, whereas *Ardipithecus ramidus* has thumb proportions more similar to gorillas. The long ulnar metacarpals in *Pan* and *Pongo* contribute to their very low relative Mc1 lengths. ARA-VP-6/500 represents *Ar. ramidus* (Lovejoy et al. 2009a), MH2 represents *Australopithecus sediba* (Kivell et al. 2011), and *Australopithecus afarensis* and *Au. africanus* are represented by composites (indicated by *parentheses*) of isolated hand elements from A.L. 333 and Sterkfontein, respectively

skeleton ARA-VP-6/500 has a short thumb. Relative to fifth metacarpal (Mc5) length, the Mc1 length in ARA-VP-6/500 is much shorter than in humans and appears to be slightly longer than those of great apes (Lovejoy et al. 2009a). Similarly, relative to its fourth metacarpal (Mc4) length, ARA-VP-6/500 has a short Mc1 compared with those of modern humans and several hominin species (Almécija et al. 2015; Fig. 18.3). The intermediate position of ARA-VP-6/500 between apes and later hominins appears to reflect both a short Mc1 and the lack of elongated ulnar (2–5) metacarpals seen in chimpanzees and orangutans (Drapeau et al. 2005; Almécija et al. 2015). When compared with a composite measure of "hand size," Mc1 length in *Ar. ramidus* is apelike and shorter than those of australopiths and modern humans (Kivell et al. 2011).

The non-pollical phalanges of ARA-VP-6/500 are longer than those of *Gorilla* and shorter than those of *Pan* relative to estimated body size (Lovejoy et al. 2009a). However, when taking into account confidence intervals on estimated body mass (Lovejoy et al. 2009b; Grabowski et al. 2015), relative finger length in *Ar. ramidus* could be equivalent to *Pan* or *Gorilla*. Thus, *Ar. ramidus* appears to have a thumb that is great ape-like in length and, relative to modern human hands, long non-pollical phalanges. While some of this anatomy has been interpreted as revealing the condition of the *Pan-Homo* LCA (Lovejoy et al. 2009b), *Ar. ramidus* occurs millions of years after the likely LCA divergence. Considering evolutionary changes that occur in other hominins over similar spans of time, such as those between *Australopithecus* and *H. erectus* or *H. erectus* and modern humans, it is premature to interpret *Ar. ramidus* as representing the anatomy of the LCA. It is impossible to know the primitive condition without evidence of the earliest basal hominins and the LCA. The metacarpal proportions and somewhat long phalanges suggest that precision grip capabilities in *Ar. ramidus* would more closely resemble those of modern African apes rather than modern humans.

Relative thumb lengths more similar to those of modern humans are first seen in australopiths (Figs. 18.2 and 18.3), consistent with hypotheses that members of this genus were adapted for greater manual dexterity (e.g., Marzke 1997; Panger et al. 2002; Alba et al. 2003). Although no associated hand skeletons of *Australopithecus afarensis* or *Australopithecus africanus* have yet been reported, the large fossil assemblages from Hadar, Ethiopia, and Sterkfontein cave, South Africa, are informative. Using resampling procedures, it is possible to examine the distribution of hand proportions that could produce the fossil assemblage. The results of these analyses show that *Au. afarensis* had a relatively long thumb compared with *Pan* (Alba et al. 2003; Rolian and Gordon 2013). Thumb length relative to other digits appears to be intermediate between modern humans and gorillas (Rolian and Gordon 2014) or essentially the same as in modern humans (Almécija and Alba 2014), depending on whether many of the A.L. 333 hand bones represent one individual and the ray identity of phalanges can be reliably determined based on morphology. This evidence suggests that *Au. afarensis* would have been capable of pad-to-pad precision grips, although likely with less proficiency and force than those of modern humans due to relatively limited thumb mobility and robusticity (Almécija and Alba 2014; Rolian and Gordon 2014; see below).

Later *Australopithecus* species show evidence of humanlike thumb length proportions. No associated hand is yet known for *Au. africanus*, unless StW 573 ("Little Foot") represents this taxon (Clarke 1999). However, resampling analyses of isolated bones suggest that the length of the Mc1 relative to that of the third metacarpal (Mc3) in *Au. africanus* was similar to that of modern humans, albeit with a more gracile thumb (Green and Gordon 2008). The unusually well-preserved, associated hand skeleton (MH2) of *Australopithecus sediba* shows that the hand proportions were autapomorphic, with longer thumbs than seen in any other higher primate (Jouffroy et al. 1993; Kivell et al. 2011). The great thumb length is due primarily to an elongated Mc1 in a hand that otherwise had digit length proportions similar to those of modern humans (Kivell et al. 2011).

Currently, insufficient evidence is available to know the thumb/finger length proportions in *Paranthropus*.

Relative thumb length in early *Homo* is poorly understood owing to a sparse fossil record. For example, OH 7 does not preserve enough elements to estimate hand proportions for *Homo habilis* (Napier 1962). The only primitive member of the genus *Homo* that preserves sufficient evidence is *Homo naledi*, recently discovered in South Africa (Kivell et al. 2015). It has a long thumb, at the upper end of the modern human range and longer (relative to third ray length) than any other known extinct hominin except for MH2 (*Au. sediba*).

Taken together, the evidence to date shows that, relative to *Ar. ramidus*, *Australopithecus* is characterized by relatively shorter fingers and longer thumbs. This suggests that as *Australopithecus* shifted to more open habitats and derived bipedal locomotion, the importance of arboreal locomotion decreased and that of manual dexterity increased. Furthermore, variation in relative thumb length among *Australopithecus* species tentatively suggests that there may be differences among them in the biomechanics of manipulation.

2.2 Thumb Robusticity

Greater robusticity is functionally related to bone strength and the capacity to withstand external loads (e.g., Ruff et al. 2006). Modern humans have expanded thumb musculature (see below and Chap. 7) that generates strong applied loading forces. Moreover, muscle recruitment creates joint reaction forces that are several times higher than the external applied force. For example, a pinch grip generates forces at the metacarpophalangeal joint that are about five to six times higher than the external force at the thumb tip, and the transarticular forces on the saddle joint (CM1) are about 12 times higher (Cooney and Chao 1977). Stress at these thumb joints can be reduced by lowering muscle force or increasing joint surface area (as stress = force/area). In modern humans, the joints of the thumb are significantly larger than those of apes (Fig. 18.4), particularly when compared with the joint sizes of non-pollical digits. Thus, modern humans have robust thumbs thought to be an adaptation to generating and resisting high forces relative to forces experienced by other regions of the hand (Susman 1994; Williams et al. 2012). Alternatively, it is possible that greater pollical robusticity may also reflect integration with a hallux that is evolving to withstand greater forces from bipedalism (Rolian et al. 2010); in this scenario, the increased robusticity of the thumb has nonetheless the functional consequence of providing enlarged attachment areas for musculature and enhanced resistance to joint stresses.

Twenty years ago, the picture of how thumb robusticity evolved appeared to be relatively simple: australopiths (represented only by A.L. 333w-39, attributed to *Au. afarensis*) resembled apes in having a small Mc1 head breadth relative to length, whereas fossils attributed to *Paranthropus* and *Homo* (at Swartkrans) appeared more humanlike (Susman 1988a, 1994).

Since then, discoveries of numerous isolated and associated hand fossils now show that the evolution of thumb robusticity and morphology might be more complex. Although Mc1s are gracile in all *Australopithecus* taxa (*Au. afarensis*, *Au. africanus*, and *Au. sediba*) for which this bone is known (Bush et al. 1982; Green

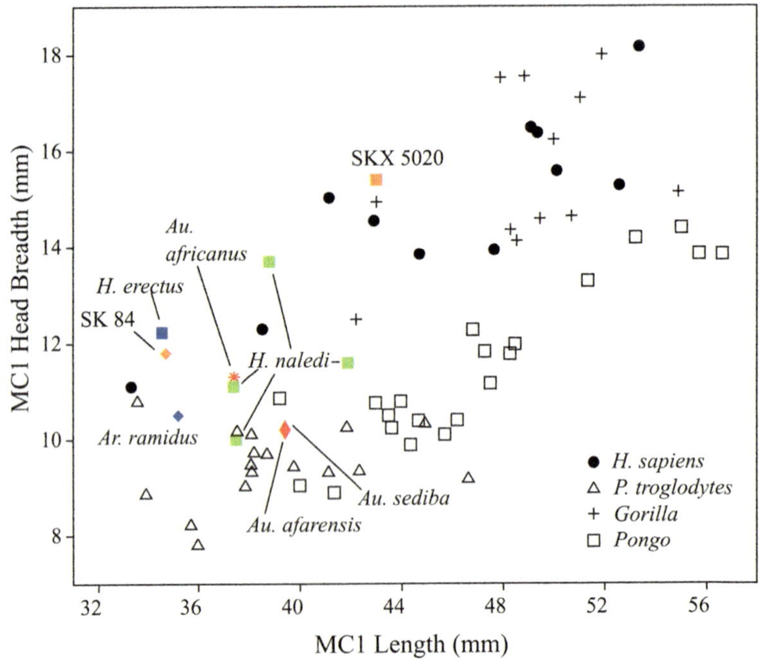

Fig. 18.4 Thumb robusticity in hominoids and fossil hominins. Relative to first metacarpal (Mc1) length, Mc1 head breadth is wider in modern humans and gorillas, although all metacarpals are more robust in the latter. *Australopithecus afarensis* (A.L. 333-39; *red inverted triangle*) and *Australopithecus sediba* (UW 88-119; *red triangle*) have gracile Mc1 heads; *Ardipithecus ramidus* (ARA-VP-6/500-015; *blue diamond*) and *Australopithecus africanus* (StW 418; *red asterisk*) are somewhat intermediate; and *Homo erectus* (KNM-WT 15000BU; *blue square*), SKX 5020 (head breadth estimated due to preservation; *orange square*) and SK 84 (both attributed to either *Paranthropus robustus* or early *Homo*; *orange diamond*) have wide Mc1 heads similar to modern humans. *H. naledi* (UW 101-007, UW 101-270, UW 101-1282, UW 101-1321) shows variability, overlapping with the distributions of modern humans and chimpanzees

and Gordon 2008; Kivell et al. 2011), *Ar. ramidus* has Mc1 head breadth/length proportions that appear to be slightly greater than in some australopiths (Fig. 18.4). However, *Ar. ramidus* also has a relatively short Mc1, so it would be informative to compare Mc1 head breadth against other variables to assess its relative size. If humanlike Mc1 joints represent the primitive condition for *Australopithecus* (Lovejoy et al. 2009a), then thumb robusticity decreased during the evolution of australopiths, such as *Au. afarensis*. More research is needed to understand this transition and its implications for early hominin hand evolution.

All known fossils potentially attributable to *Homo* have robust thumbs, with relatively broad Mc1 heads and broad shafts, typically with well-developed opponens pollicis insertions (note that no Mc1 of *H. habilis* is known). The oldest of these are two Swartkrans fossils, SK 84 and SKX 5020 (Fig. 18.4), at roughly 2 Ma. There is some debate about the taxonomic attribution of these fossils (Susman 1988a; Trinkaus and Long 1990). Comparative evidence (e.g., from Mc1s of the early *H. erectus* KNM-WT15000 skeleton and MH2; Richmond et al. 2009; Kivell et al. 2011) now suggests that SKX 5020 likely represents *Homo* and SK 84 might represent *Paranthropus robustus*, rather than the reverse (Susman 1988a, 1994). However, this does not diminish Susman's (1994: 1570) conclusion that *P. robustus* had hand anatomy capable of "advanced precision grasping" for making and using tools. It is notable that humanlike thumb robusticity evolved well after the first appearance of making and using stone tools (McPherron et al. 2010; Harmand et al. 2015).

The well-preserved *H. naledi* fossils have unique Mc1 morphology, represented by six individuals, with unusually small proximal bases and saddle joints for the trapezium, combined with humanlike broad Mc1 heads and shafts with pronounced attachments for the opponens pollicis and dorsal interossei (Kivell et al. 2015). Thus, *H. naledi* appears to have most of the derived anatomy for forceful precision pinch grips, but would not have been capable of withstanding as much force compared with the thumbs of later *Homo*.

2.3 Thumb Musculature

Modern humans have elaborate thumb musculature. Humans do not have unique muscles, as every major muscle attached to the human thumb can be found in other primate taxa. However, more muscles attach to the thumb in humans than in almost all other primates (Diogo et al. 2012; see also Chap. 7), and thumb muscles constitute a much larger proportion of total hand muscle mass in humans compared with other hominoids (Tuttle 1969).

Humans have two major extrinsic thumb muscles that are normally absent in the great apes: the flexor pollicis longus and extensor pollicis brevis (e.g., Diogo et al. 2012 and references therein). These two muscles are present in hylobatids, raising questions about whcther these muscles were independently lost in three great ape lineages, were lost at the Hominidae node (great ape and human clade) and re-evolved in hominins, or were independently gained in hylobatids and hominins, perhaps associated with their relatively long thumbs (Chap. 7). Humans and African apes also share a third, intrinsic thumb muscle almost never observed in other primates, a structure sometimes called the first volar interosseous muscle of Henle (Susman 1994; Susman et al. 1999), more accurately named the accessory adductor pollicis muscle (Diogo et al. 2012).

The most important muscle for force production during thumb flexion is the flexor pollicis longus (FPL). In most nonhuman primates, this muscle (when present) is not fully separated from the flexor digitorum profundus (FDP) muscles attaching to the other digits, meaning that the thumb would flex during forceful flexion of the fingers. In humans [and hylobatids (Diogo et al. 2012)], it is more fully separated from other muscles, so that the thumb can be forcefully flexed independently from the other digits and vice-versa. This capability confers obvious advantages during manipulation. However, to date, there is no way to detect the relative independence of the FPL from skeletal evidence alone. Because of this, a visible insertion site on the thumb's distal phalanx merely provides evidence of the attachment of the long tendon of an extrinsic flexor muscle, but cannot distinguish whether it is independent (FPL) or not (i.e., belly connected to the rest of FDP). It is interesting to note that even in humans, about 31 % of individuals lack a completely independent FPL (Lindburg and Comstock 1979).

Because great apes generally lack an extrinsic thumb flexor muscle, parsimony suggests that the LCA lacked it as well (Tocheri et al. 2008). However, the insertion site is present in most, if not all, fossil hominins to date (Fig. 18.5), including the distal phalanges (DP1s) attributed to *Orrorin* (Gommery and Senut 2006; Almécija et al. 2010), *Ar. ramidus* (Lovejoy et al. 2009a), *Au. afarensis* (Ward et al. 2012), *Au. africanus* (Ricklan 1987), *Au. sediba* (Kivell et al. 2011), *P. robustus* or early *Homo* at Swartkrans (Susman 1988b), *H. naledi* (Kivell et al. 2015), and *H. habilis* (Napier 1962). Thus, an extrinsic thumb flexor muscle is present very early in the hominin lineage, if not in the LCA, but there is no evidence regarding when the FPL became independent from the FDP muscle to the other digits.

The morphology of the FPL attachment in humans (Fig. 18.5) is distinctive in ways that may influence its function. For example, the FPL tendon inserts in a gable-shaped pattern with a consistently longer radial side. Shrewsbury et al. (2003) argue that this contributes to opposition against the fingers, together with an asymmetric interphalangeal joint that slightly pronates the distal phalanx during flexion. The distal phalanx BAR 1901'01 displays these characteristics and appears remarkably humanlike (Almécija et al. 2010). If it is correctly attributed to *Orrorin*, this suggests that a humanlike FPL attachment was present at or near the base of the hominin lineage. Australopiths (post *Au. afarensis*) and early *Homo* DP1s tend to have much larger palmar fossae and more distal FPL attachments (e.g., Kivell et al. 2011), although the specific morphology of the FPL attachment and overall morphology of the DP1 (see below) are quite variable across australopiths and early *Homo*.

Fossil Mc1s attributed to *Homo* often bear a prominent opponens pollicis muscle attachment site, developing into an opponens "flange" in later archaic *Homo* (Chap. 19). While it may seem logical to interpret the prominent attachment as indicating a large and/or intensively used opponens pollicis muscle, efforts to examine the relationships between muscle size, activity, and insertion sites have found no association (Zumwalt 2006; Rabey et al. 2015), including those specifically examining hand muscle size and

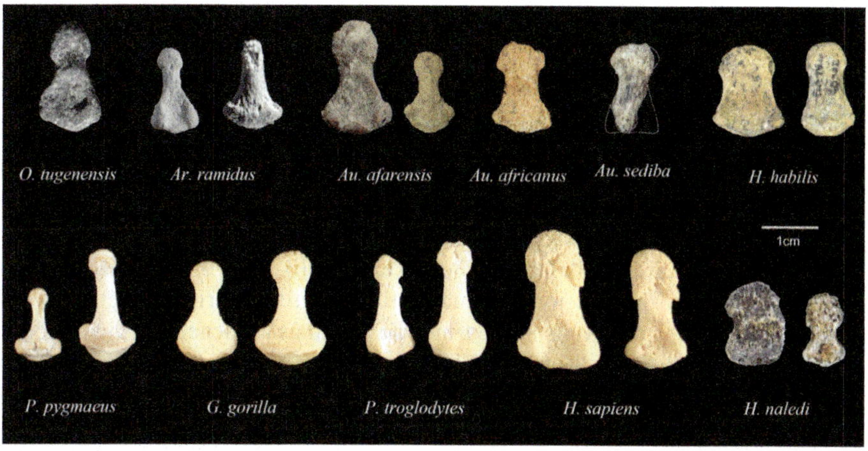

Fig. 18.5 Images of pollical (DP1) (*left*) and non-pollical (*right*) distal phalanges of humans, apes, and fossil hominins. Fossil phalanges attributed to *Orrorin tugenensis* (DP1 only, BAR 1901'01; adapted from Gommery and Senut 2006), *Ardipithecus ramidus* (ARA-VP-6/500-049, ARA-VP-6/500-050; adapted from Lovejoy et al. 2009a; photo © and courtesy of T. White [2009]), *Australopithecus afarensis* (A.L. 333-159, A.L. 333w-50), *Australopithecus africanus* (DP1 only, StW 294), *Australopithecus sediba* (DP1 only, UW 88-124, adapted from Kivell et al. 2011), *Homo habilis* (OH 7: FLK NN-A, FLK NN-B), and *Homo naledi* (DH1: UW 101-1351, UW 101-1329)

insertion morphology (Marzke et al. 2007; Williams-Hatala et al. 2016). A radial extension of the opponens crest does, however, have the biomechanical consequence of increasing the rotational moment arm of the muscle (Maki and Trinkaus 2011), and the more distally extended dorsal interosseous attachment in humans provides a longer moment arm for adduction (Tocheri et al. 2008).

2.4 Broad Distal Tuberosities

Fingertips are responsible for precision grips and handling, so it is no surprise that humans have derived distal phalanges. Humans have broad fingertips, nails, and underlying distal tuberosities (also called "apical tufts") compared with other hominoids (Mittra et al. 2007; Chaps. 4 and 8 and references therein). Shrewsbury and colleagues (Shrewsbury and Johnson 1983; Shrewsbury et al. 2003) describe a "functional compartmentalization" of the thumb tip, with a larger, thick, mobile proximal portion and a smaller, more stable distal portion. The border between these "compartments" lies at the proximal end of the tuberosity. The distal portion has more fibers and less fat content and is anchored to the tuberosity and supported by the nail, whereas the proximal portion has greater fat content and is supported by the lateral intraosseous ligaments that attach to the ungual spines on either side of the tuberosity. The proximal portion also contains neurovascular bundles and corpuscles that provide sensory information using during manipulation (Shrewsbury and Johnson 1983; see Chaps. 6 and 8).

Broader fingertips increase the surface area for precision grips between the thumb and other digits. The breadth of the soft tissue of the fingertip is correlated with the breadth of the underlying DP distal tuberosity (Mittra et al. 2007), reflecting the functional role played by the distal tuberosities in supporting the stability of the nail dorsally and fleshy fingertip palmarly.

In his analysis of the OH 7 hand, part of the *H. habilis* holotype, Napier (1962) highlighted the broad terminal phalanges as a key requirement for precision grip capabilities, but noted that the "set" of the trapezium left open doubt about whether thumb opposition was exactly like that of modern humans. Finding a hand with functional signatures of precision grips, together with some of the oldest stone tools at the time, was cited as evidence for the hypothesis that OH 7 belonged in the genus *Homo* (Leakey et al. 1964). Since then, discoveries have shown that stone tools occur substantially earlier, at 3.3–3.4 Ma (McPherron et al. 2010; Harmand et al. 2015), and broad distal tuberosities may occur even earlier.

The c. 5.8 Ma DP1 attributed to *Orrorin tugenensis*, BAR 1901'01, has remarkably humanlike anatomy, much more so than the morphology of fossils securely attributed to *Ardipithecus* and some *Australopithecus* taxa. Like human DP1s, BAR 1901'01 has a broad, distinct tuberosity with ungual spines and well-developed basal tubercles that form the attachments of the lateral intraosseous ligaments (Almécija et al. 2010) and are functionally related to the mobile proximal pulp compartment (Shrewsbury et al. 2003). The DP1 of *Ar. ramidus* (ARA-VP-6/500-049), on the other hand, appears much more like those of great apes, with a narrow distal tuberosity without ungual spines (Lovejoy et al. 2009a). Given the modern humanlike morphology of BAR 1901'01, it is surprising that the morphology in *Ar. ramidus* appears so much more primitive and potentially raises concern over the reliability of the attribution of BAR 1901'01 to *Orrorin*. More fossil evidence is needed to clarify the evolution of this anatomy near the base of the hominin clade.

Aside from the DP of *Au. afarensis*, the DPs of all gracile and robust australopiths and most members of the genus *Homo* have substantially broader distal tuberosities than the DPs of modern humans (Fig. 18.5). These taxa include *Au. africanus*, *Au. sediba*, *H. habilis*, and *H. naledi*, as well as *Homo heidelbergensis* and *Homo neanderthalensis* (Fig. 18.5; Chap. 19). This raises questions about the functional significance of the broader apical tufts in earlier hominins and why modern humans have evolved narrower apical tufts. Are modern humans less dextrous in this respect?

2.5 Carpal Architecture

One of the few osteological synapomorphies of humans and African apes occurs in the wrist, namely, the fusion of the os centrale to the scaphoid (Weinert 1932; Corruccini 1978; Richmond et al. 2001; Kivell and Begun 2007). Among anthropoids, this feature is occasionally observed in *Pongo*, but only in African apes and humans does it occur in virtually all individuals and develop very early in ontogeny (Kivell and Begun 2007). Wood Jones (1916) proposed that os centrale fusion was functionally related to stability at the base of the index finger in African apes and in humans, two groups considered to be distantly related at the time. Lovejoy et al. (2009a) proposed a nonfunctional hypothesis, positing that fusion of the os centrale may have been a pleiotropic effect of an elongation of the scaphoid tubercle and deepened carpal tunnel.

Currently, the predominant functional hypothesis for its evolution posits that the fused os centrale/scaphoid decreases mobility and improves the ability to transmit forces between the manual rays and radius during knuckle-walking (Marzke 1971; Tuttle 1975; Sarmiento 1988; Gebo 1996; Richmond et al. 2001). This is particularly important in African apes, which lack a weight-bearing articulation between the carpals and ulna and use a hand posture in which ground reaction forces travel along a "column" of metacarpals, carpals, and the radius. Some support for this hypothesis can be found in the convergent evolution of carpal fusion in knuckle-walking giant anteaters (Orr 2005), as well as chalicotheres (Richmond et al. 2001; Begun 2004). Of course, it is well known that some lemurs, including suspensory subfossil lemurs such as *Palaeopropithecus*, also show convergence in os centrale fusion (Jouffroy 1975; Hamrick et al. 2000; Chap. 15). This observation, along with other factors, leads some to question a knuckle-walking hypothesis for os centrale fusion in the African ape-human clade (e.g., Kivell and Schmitt 2009), while others argue that the wrist structure and function in lemurs is distinct enough to have evolved for different reasons (e.g., Begun 2004). This area deserves more research.

The presence of a fused os centrale/scaphoid in all known hominin fossils to date, including the *Ar. ramidus* partial skeleton ARA-VP-6/500 (Lovejoy et al. 2009a), suggests that it did not evolve independently in hominins and African apes. In contrast, Miocene apes such as *Rudapithecus* (Kivell and Begun 2009) and *Pierolapithecus* (Moyà-Solà et al. 2004) have an independent os centrale. The growing fossil record suggests that os centrale fusion most likely evolved once at, or near, the base of the African ape and human clade, begging an alternative explanation if the knuckle-walking hypothesis is incorrect.

In addition to os centrale fusion, modern humans have a carpal architecture that is quite distinct from other primates. In most anthropoids, the trapezoid is wedge shaped, with the narrow end of the wedge projecting palmarly. In humans, the palmar portion of the trapezoid is substantially expanded (Tocheri et al. 2007). This expansion effectively supinates the trapezium and brings the distal carpal row into greater radioulnar alignment (Tocheri 2007). This realignment influenced the morphology of many of the intercarpal and carpometacarpal joints on the radial side and has been hypothesized to improve the biomechanics of resisting radioulnarly oriented forces acting across the radial side of the wrist during forceful contraction of the enlarged thenar musculature (Tocheri 2007; Tocheri et al. 2008).

All non-*Homo* hominin taxa and the most primitive members of the genus *Homo*, including *H. habilis* and *Homo floresiensis*, show evidence of the primitive anthropoid wrist configuration (Tocheri et al. 2007). In contrast, later members of the genus *Homo*, including *H. neanderthalensis* and early modern humans, show evidence of a shared, derived wrist configuration (Tocheri et al. 2008). Interestingly, the wrist anatomy of

H. naledi shares the same derived anatomy with Neanderthals and modern humans (Kivell et al. 2015). Unfortunately, wrist anatomy is poorly preserved for *H. erectus*, so it remains unclear exactly when and in what context this anatomy evolved.

Humans also have a prominent styloid process on the base of the Mc3, a feature that might be linked with this radial carpal reconfiguration and was present at least 1.4 Ma in KNM-WT 51260 from West Turkana, Kenya (Ward et al. 2014). Marzke and Marzke (1987) proposed that the styloid process evolved to stabilize the carpometacarpal joint to prevent subluxation, especially due to palmarly oriented external forces experienced during stone tool use. The functional significance of this feature deserves more attention.

2.6 Wrist Mobility

Humans have more mobile wrists in some directions than do African apes and in this regard more closely resemble the Asian apes (Bradley and Sunderland 1953; Tuttle 1969; Richmond 2006). Schreiber (1936) was the first to report limited wrist mobility in chimpanzees, and Tuttle (1969) demonstrated this based on the passive mobility of upper limb joints in a large sample of extant apes. He noted that African apes, particularly *Pan troglodytes*, have much lower ranges of mobility in wrist extension and proposed that this limited range of motion represented a functional adaptation to knuckle-walking in order to provide a stable, weight-bearing column between the hand and arm.

Range of motion is difficult to predict from skeletal anatomy alone because of the critical roles played by ligaments and other soft tissues. However, the anatomy of joints also plays a crucial role in joint mobility and stability (Hamrick 1996). Systematic assessment of the movements of individual carpals when cadaveric wrists are flexed and extended shows that the morphology of the distal radius and midcarpal joint (e.g., lunate-capitate arc curvature) correlates significantly with range of motion in wrist extension (Orr et al. 2010; see Chap. 9).

Morphology of the radiocarpal and midcarpal joints suggests that early hominins had less wrist mobility in extension compared with modern humans. *Ar. ramidus*, *Australopithecus anamensis*, and *Au. afarensis* have distal radii with palmar inclinations formed in part by distally projecting dorsal ridges, or margins, similar to the morphology in modern African apes (Richmond and Strait 2000; Richmond et al. 2001; Orr 2013). The midcarpal joint morphology, such as the curvature of the lunate-capitate joint, shows that *Ar. ramidus* had restricted wrist extension (Orr 2013; *contra* Lovejoy et al. 2009a). This restricted wrist extension supports the hypothesis that the LCA practiced knuckle-walking as well as climbing (Washburn 1967; Richmond et al. 2001; Williams 2010). Nevertheless, the locomotor behavior of the LCA is still a matter of debate (e.g., Schmitt 2003; Begun et al. 2007; Thorpe et al. 2007; Kivell and Schmitt 2009; Lovejoy et al. 2009b; Williams, 2010) and will remain so until forelimb fossils of the *Pan-Homo* LCA are found.

By the late Pliocene, hominin taxa such as *Au. africanus* and *P. robustus* had radii with distally oriented ends without projecting dorsal margins (Richmond et al. 2001). Further analyses of early hominin wrist joint anatomy will help to clarify the timing and pattern of changes in wrist joint mobility during human evolution (Chap. 9). It is not clear what led to the changes in wrist mobility, but two hypotheses propose that the changes are influenced by relaxed selection for the use of the upper limb in locomotion and/or selection for manual manipulation or throwing (Marzke 1971; Ambrose 2001; Richmond et al. 2001; Roach et al. 2013). In humans today, many manipulative activities involve a high degree of wrist mobility. Mobility in wrist extension has been shown to play an important role in stone toolmaking (Williams et al. 2010, 2014; see Chap. 11) and throwing (Wolfe et al. 2006; Roach and Lieberman, 2014).

2.7 Phalangeal Curvature

There is a well-documented association between longitudinal curvature of proximal and middle phalanges and arboreal and, especially, suspensory behavior among primate taxa (Stern and Susman 1983; Jungers et al. 1997; Matarazzo 2008; Rein 2011). Biomechanical modeling based on in vivo data show that phalangeal curvature substantially reduces bending stresses during highly flexed finger postures, such as those involved in grasping branches during suspensory locomotion (Preuschoft 1970; Richmond 2007; Nguyen et al. 2014). This seems to be attributable to the counteracting influences of the palmarly oriented force of the extrinsic flexor tendons that tend to bend "open" (in the direction of straightening) the proximal phalanx and the proximo-distal component of the joint reaction forces that tend to bend the phalanx "closed" (in the direction of increasing the curvature). In contrast, straighter phalanges reduce bending when fingers use more extended postures, such as those used by primates when traveling on the ground (Richmond 1998). Longitudinal curvature is sensitive to changes in behavior during growth and, therefore, serves as a good indicator for degree of arboreal behavior during primate development (Richmond 1998; Jungers et al. 2001; Congdon 2012).

The functional morphology of phalangeal curvature is still not completely understood. Some taxa (e.g., gibbons) that are highly suspensory have lower levels of curvature than others (e.g., orangutans; Fig. 18.6). Size and scaling may play a role, especially regarding how hand size and arboreal support size influence finger postures during locomotion; for example, small primates may not use very flexed finger postures during arboreal locomotion because the supports are large relative to their hands. Very little is currently known about hand and finger postures during locomotion and during ontogeny (but see Sarringhaus 2013). At present, evidence shows that curved phalanges occur in primate taxa that climb and suspend from arboreal supports and that curvature is somewhat sensitive to changes in locomotor behavior during growth.

Among modern hominoids, curvature in proximal phalanges varies from extremely curved in orangutans to lower levels, in decreasing order, in siamangs, gibbons, bonobos, chimpanzees, and gorillas, with the straightest phalanges in modern humans (Fig. 18.6). Finger curvature is moderately high, more or less like that of chimpanzees, in basal hominins such as *Orrorin* (Richmond and Jungers 2008) and early australopiths such as *Au. afarensis* (Stern and Susman 1983). Enough fossils are now known to see that phalangeal curvature varied among early hominin taxa (Fig. 18.6). For example, it appears that phalangeal curvature is greater in *Au. afarensis* than in *Au. africanus* and *Au. sediba* (Kivell et al. 2015). The latter taxa have moderate phalangeal curvature, and more fossils will be needed to assess how some of these taxa differed from one another and how the degree of phalangeal curvature fits with the remaining morphology of the hand.

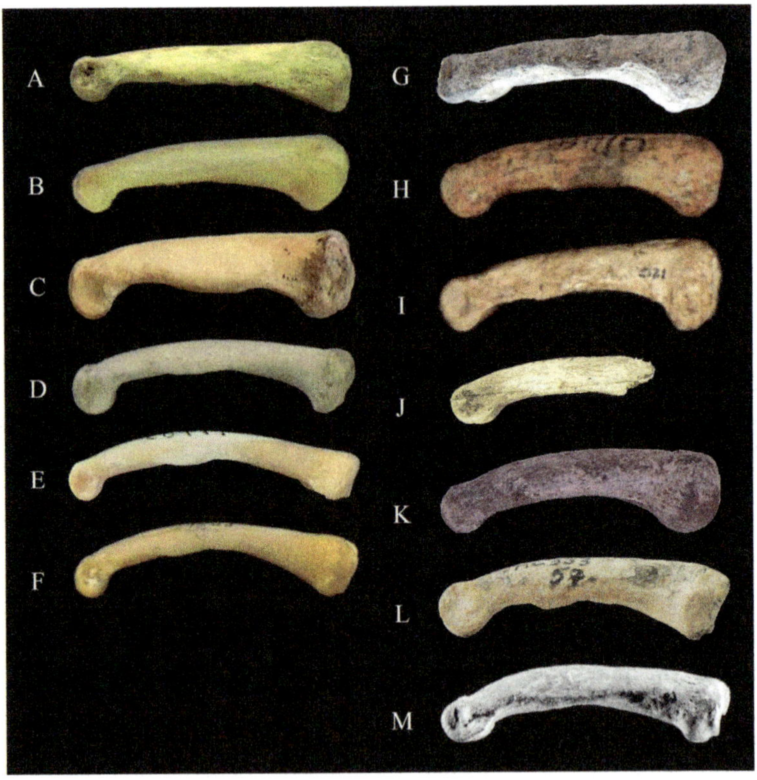

Fig. 18.6 Extant anthropoid (*left*) and early fossil hominin (*right*) manual proximal phalanges (PPs), illustrating variation in phalangeal curvature. PPs are arranged roughly from straighter at the top to those with greater curvature toward the bottom, scaled to equivalent lengths. PPs of extant taxa include (**a**) modern human, (**b**) baboon, (**c**) gorilla, (**d**) chimpanzee, (**e**) gibbon, and (**f**) orangutan. Fossil PPs include (**g**) Hominini gen. et sp. indet. (OH 86; adapted from Domínguez-Rodrigo et al. 2015), (**h**) *Australopithecus africanus* (StW 293; adapted from Kivell et al. 2011), (**i**) *Australopithecus sediba* (UW 88-120 of MH2; adapted from Kivell et al. 2011), (**j**) *Homo habilis* (OH 7, cast), (**k**) *Homo naledi* (UW 101-1327; adapted from Kivell et al. 2015), (**l**) *Australopithecus afarensis* (A.L. 333-57), and (**m**) *Ardipithecus ramidus* (ARA-VP-6/500-022; adapted from Lovejoy et al. 2009a; photo © and courtesy of T. White [2009])

Differences between species in phalangeal curvature and other characteristics may reflect slight but significant differences in positional behavior (Green et al. 2007), not unlike the differences in anatomy and behavior observed between species of *Pan* (Doran 1993).

The oldest evidence of humanlike low phalangeal curvature is seen in a manual proximal phalanx (OH 86) from Olduvai found in deposits over 1.84 Ma (Domínguez-Rodrigo et al. 2015). This phalanx differs substantially from the OH 7 phalanges, attributed to *H. habilis*, pointing to diversity in hand function at that time. Low phalangeal curvature is seen in later *Homo* fossils, such as ATE9-2 from c. 1.2 to 1.3 Ma deposits at Sima del Elefante (Lorenzo et al. 2015). Very few hand bones are securely attributed to *H. erectus*. However, based on the loss of other apelike, "arboreal" traits and the presence of a humanlike Mc3 styloid process (Ward et al. 2014), it would not be surprising if future discoveries show that *H. erectus* had straight proximal and middle phalanges, but this anatomy will remain unknown until fossil evidence is recovered.

How to interpret the behavioral and adaptive implications of phalangeal curvature has been a matter of considerable debate, despite having a better understanding of its functional significance than that of many anatomical characteristics. Primitive anatomy can be retained without the continuation of primitive behaviors (e.g., a highly mobile, apelike shoulder in modern humans), complicating the interpretation of primitive anatomy (Latimer, 1990; Stern 2000; Ward 2002). However, comparative and developmental evidence suggest that the *Pan*-like levels of manual (and pedal) phalangeal curvature in *Orrorin*, *Ardipithecus*, *Australopithecus*, and early *Homo* are evidence that climbing trees, probably to obtain food and avoid predation, was a significant component of the behavioral repertoire of these taxa (e.g., Stern and Susman 1983; Richmond 1998; Kivell et al. 2015).

3 Changes in Hominin Hand Functions: A Process, Not an Event

This review of changes in hominin hand anatomy drives home the point that the two major functional changes—the loss of hand use for locomotion and the intensification of hand use in manipulation—are better characterized as "processes" than "events." What do we mean by "process"? We use the term process in the same sense that Mike Rose (1991) used it to argue that the origin of bipedalism was a process rather than an event. In that example, it is extremely unlikely that the earliest hominins switched, across a single species transition, from a locomotor repertoire involving a small proportion of bipedalism (comparable to that seen in modern great apes) to one that was exclusively bipedal like that seen in modern humans. Instead, it is far more likely that bipedalism became a more important component in the locomotor repertoires of some hominin descendants of the *Pan*-*Homo* LCA. However, the increase in the importance of bipedalism was not necessarily linear, nor was the decrease in arboreality and increase in manual manipulation capabilities. Different species, even different populations, may have varied in the importance of bipedalism and arboreality as components of their locomotor repertoires, just as African ape species, subspecies, and populations differ in the relative proportions of suspension, climbing, knuckle-walking, and bipedalism in their repertoires (e.g., Doran 1993, 1996) and in related functional anatomy. An example can be seen in the variation in medial cuneiform morphology and locomotor behavior among gorilla species and subspecies (Tocheri et al. 2011). The many factors that influence modern primate locomotion would also have influenced the repertoires of the earliest hominin taxa. Therefore, we argue that the "null hypothesis" for the pattern of early hominin locomotor evolution should be one in which different species (or smaller divisions if we could accurately reconstruct them) differ in important, and perhaps subtle, ways from one another depending on their ecological settings, diets, body sizes, and other factors.

The mosaic pattern observed in the changes in functional complexes of the hand, including the seven reviewed above, supports the hypothesis that the evolution of locomotor and manipulative hand function was complex.

3.1 Changes in Locomotion

As locomotor mode changed over the course of human evolution, so too must have the selective forces acting on the hand. The locomotor behavior of the LCA is a matter of considerable debate (e.g., Richmond and Strait 2000; Begun et al. 2007; Thorpe et al. 2007; Kivell and Schmitt 2009; Lovejoy et al. 2009b; Williams 2010). However, there is broad consensus that arboreal climbing and suspension were very important components of the LCA's behavior and that bipedalism was likely no more important than that seen in the repertoires of modern great apes. There is some evidence that bipedality was a relatively greater component of the repertoires of the earliest basal hominins, including *Sahelanthropus*, *Orrorin*, and *Ardipithecus kadabba*, but the evidence is frustratingly fragmentary (reviewed in Richmond and Hatala 2013). By 4.4 Ma, there is much more abundant evidence of the anatomy of *Ar. ramidus* (White et al. 2009). Arboreal locomotion was likely a very important part of the repertoire in *Ar. ramidus*, which had curved manual and pedal phalanges and a divergent, grasping hallux (Lovejoy et al. 2009a, c). However, evidence of a stiff foot and short pelvis suggests that bipedalism was also an important component of the locomotor adaptation of *Ar. ramidus*, likely more so than in any extant great ape (Lovejoy et al. 2009c, d).

Derived anatomy in *Australopithecus* shows that bipedalism played a greater role in the repertoires of australopiths than those of basal hominins. However, the role of arboreality continues to be a matter of enduring debate based mainly on differences in interpretation of the functional significance of primitive traits (e.g., Ward 2002). Regardless, some species of *Australopithecus* differ in these apelike traits, showing that they underwent evolutionary changes. Although one could invoke genetic drift to explain interspecific differences, some characteristics (phalangeal curvature, limb joint size proportions) differ among closely related anthropoid taxa in ways that reflect significant differences in locomotor repertoires (Fleagle 1977; Doran 1993). Therefore, some of the differences between *Australopithecus* species may reflect differences in the relative degrees of arboreal behavior in the repertoire (Green et al. 2007). Similarly, there is compelling evidence of different locomotor repertoires between neighboring hominins living at c. 3.5 Ma, with greater emphasis on bipedalism in *Au. afarensis* at Hadar and greater emphasis on arboreal locomotion in another hominin taxon (*Ardipithecus* sp. or *Australopithecus deyiremeda*?) at nearby Burtele (Haile-Selassie et al. 2012, 2015).

The oldest evidence from hand fossils that hominins abandoned arboreal behavior comes from the over 1.84 Ma straight proximal phalanx at Olduvai (Domínguez-Rodrigo et al. 2015). However, based on the loss of morphology and cross-sectional geometry related to arboreality from other anatomical regions, there is broad consensus that by the early Pleistocene the locomotor repertoire of *H. erectus* comprised only bipedalism with no significant arboreal component (e.g., Jungers 1988; Ruff and Walker 1993; Bramble and Lieberman 2004; Ruff 2009). Current evidence suggests that the role of arboreality decreased from *Ar. ramidus* to *Australopithecus* to *H. erectus*. However, the pattern is certainly more complex, including broadly contemporaneous lineages that differ in their repertoires (e.g., hominins at Hadar and Burtele) and likely diversity among australopiths and early *Homo* in the degree of arboreality.

While the details of the evolutionary history must await more fossil discoveries, it is clear that hominins essentially stopped using their hands for locomotion "for good" by the time of *H. erectus* or possibly earlier.[1] This represents one of two fundamental changes in the functional and biological roles (sensu Bock and von Wahlert 1965) of the hominin hand.

3.2 Changes in Hand Manipulation

The review of changes in hominin hand anatomy also highlights the mosaic pattern in which the human hand has been shaped by evolution for greater manual dexterity, particularly in the anatomy and function of the thumb. However, investigating *how* the hand might be adapted for dexterity is challenging. In addition to a sparse fossil record, the uniqueness of human hand function limits the ability to use the comparative approach, arguably the most potent tool for investigating adaptation in extinct species. For example, long, curved fingers are good examples of adaptations to arboreal locomotion because this morphology evolved independently in South American atelines and Asian apes (Erickson 1963; Jungers et al. 1997), taxa that share the functional similarity of engaging in relatively high frequencies of suspensory behavior in their locomotor repertoires. In contrast, since humans are the only

[1] We note, of course, that hands are used for crawling in humans during a brief period of development, and some people occasionally use their hands to climb trees in natural settings (Kraft et al. 2014). However, these instances of hand-based locomotion are rare.

natural stone toolmakers, no other taxa are available to examine structure-function relationships specific to this behavior (Susman 1998). Furthermore, the unique morphology in the hands of early hominins complicates attempts to infer their hand function based on experiments replicating primitive tools, because the human subjects in these experiments obviously have modern, not early hominin, hand morphology and cognitive abilities.

Despite these challenges, there are several lines of evidence regarding the evolution of manual dexterity and tool use. First, the fossil record demonstrates where and when hand morphologies existed, including characteristics that have functional consequences. Second, in some cases the fossil record preserves evidence of hand function, such as the pattern of trabecular bone density in metacarpal heads from Sterkfontein attributed to *Au. africanus* (Skinner et al. 2015). Third, the archeological record of tool use, such as cut marks and percussion marks on bones, and tools themselves demonstrates the products of hominin hand function. Fourth, experimental evidence of the manipulative skills needed to make and use tools, and the mechanical demands involved, provide an important context for understanding the functional morphology of making and using tools (see Chap. 11). Finally, studies of living human and nonhuman primate tool use, toolmaking, and carrying in natural settings provide important perspectives on the behavioral and ecological contexts of hand manipulation (e.g., Carvalho et al. 2013; Koops et al. 2014; Chap. 12).

These various lines of evidence suggest the following hypothesis for the evolution of hominin hand manipulation and tool use. Basal hominins likely used and made tools or at least had the capacity to do so in the right ecological settings; modern great apes in the wild make and use tools, and parsimony suggests that this capability was present in the *Pan-Homo* LCA as well (Panger et al. 2002). Basal hominins in the late Miocene may have had slightly better manipulative abilities than modern great apes, as suggested by distal pollical anatomy (Almécija et al. 2010) and potentially by hand proportions less derived than in great apes (Almécija et al. 2015), but this hypothesis remains very tentative until more fossil evidence is recovered. The oldest direct evidence of manipulation and tool use comes from the archeological record, including cut marks and percussion marks on large mammal bones from 3.4 Ma sediments at Dikika, Ethiopia (McPherron et al. 2010), and c. 3.3 Ma stone tools from Lomekwi, Kenya (Harmand et al. 2015), predating the earliest evidence of the genus *Homo* (Villmoare et al. 2015). This Pliocene archeological record is consistent with evidence that australopiths had greater manipulative capabilities than modern apes, mainly because australopiths had fairly humanlike thumb/finger proportions, broad distal tuberosities on fingertips, and an extrinsic flexor muscle to the thumb. Australopiths are derived relative to *Ar. ramidus* in having more humanlike thumb/finger proportions.

In the early Pleistocene, the hand of *H. habilis* is similar to that of australopiths, despite the name "handy man." *H. habilis* is found in association with primitive Oldowan stone tools, but so is *Paranthropus*, leaving open the possibility that both engaged in this behavior. Wrist mobility appears to have increased by the early Pleistocene in *Au. africanus* and *P. robustus* (Richmond and Strait 2000; Richmond et al. 2001). Wrist mobility improves the accuracy of throwing and accuracy and force production during stone toolmaking (Williams et al. 2010, 2014; Roach et al. 2013; Roach and Lieberman 2014), suggesting that these australopiths had greater manual capabilities than their predecessors. Rearrangement of carpal architecture is seen in *H. naledi* but not in *H. habilis* or *H. floresiensis*, pointing to major changes in force transmission patterns through the hand after the origin of *Homo*.

More sophisticated Acheulean stone tools date to c. 1.7 Ma (Lepre et al. 2011) and are first found in association with *H. erectus* (and *Paranthropus*). These tools require greater manipulative and cognitive skills than earlier Oldowan tools, and a hand anatomy suited to make and use them. Hand fossil elements that can be reliably attributed to *H. erectus* are scarce, but Mc1s associated with the Nariokotome skeleton indicate a robust thumb capable of generating higher forces and resisting greater stresses associated with strong thumb flexion (Fig. 18.4). The oldest evidence of a hand with essentially modern human anatomy comes after a substantial chronological gap, later in the early Pleistocene at 0.8 Ma (see Chap. 19). Much would be learned about the timing, context, and potential selective pressures responsible for more humanlike morphology from the recovery of hand fossils securely attributed to *H. erectus* and *Paranthropus*.

It is clear that morphology related to forceful precision grips evolved in a mosaic fashion, with derived features appearing in multiple species of *Australopithecus*, possibly *Paranthropus*, and multiple species of *Homo*. It is notable that several modern human features, such as a robust thumb, greater wrist mobility, reorganized carpal architecture, and straight phalanges, appear well after stone tools first appear. This supports the hypothesis that *intensification* of making and using tools, rather than the origin of these behaviors, drove much of the evolution of derived human hand anatomy.

4 Conclusions and Future Directions

In conclusion, the fossil record preserves evidence of hominin hand anatomy spanning about 6 million years. When examined in a comparative and functional context, it is clear that the anatomy of the hominin hand evolved in a complex, mosaic fashion.

This review highlights several major conclusions:

1. Two major functional changes occurred during human evolution: the loss of a locomotor role and the intensification of manipulation.
2. These two changes in function were not abrupt. The abandonment of arboreal locomotion and rise of manipulative capabilities evolved over considerable spans of time, in varied contexts, and in a nonlinear manner.
3. Differences in hand morphology among early hominin species suggest differences in their locomotor repertoires and manipulative abilities, not unlike differences in behavior seen among closely related species today (e.g., bonobos and chimpanzees).

4. Evidence supports the hypothesis that the *intensification* of manipulation, rather than the origin of stone tool making, drove the evolution of a number of derived aspects of human hand anatomy.

4.1 Future Directions I: Is the Human Thumb "Hyper-opposable"?

In the process of reviewing what we currently know about the early evolution of the human hand, it became clear that there is considerable variation in what is meant by an "opposable" thumb, and this topic deserves further attention. There is broad consensus that a thumb capable of opposing the other digits is a key characteristic of primates (e.g., see Chap. 14) and that humans are derived in the ability to oppose the thumb for precision gripping and handling. Definitions of opposability vary widely, but typically defer to Napier's traditional definitions describing the ability to rotate the thumb into pad-to-pad contact with other digits (Napier 1956) or position the thumb via movements at the CM1 joint (Napier 1961). It would be useful to have a separate term that describes the derived ability in humans to produce an extensive area of contact between the thumb and other digits and forcefully secure and precisely handle objects between the thumb and other digits through pad-to-pad contact (Marzke 1997). We propose the term *hyper-opposable* to describe this derived human condition.

4.2 Future Directions II: Other Areas of Research

It goes without saying that more fossils are needed to document the evolution of human hand anatomy, although some fossils would be more valuable than others. Efforts to reconstruct the *Pan-Homo* LCA have preoccupied scholars for well over a century, so fossil evidence of the LCA and its close relatives (including fossil *Pan* and *Gorilla*) would be particularly enlightening, despite the challenges in reaching consensus among scientists regarding the systematics and phylogenetic position of fossils near the LCA once they are found. Other notable gaps in the fossil record include the lack of hand fossils, especially associated elements, that are securely attributed to *Paranthropus*, *H. erectus*, and other early *Homo* species. In addition to finding fossils of taxa for which we currently know virtually nothing about their hand morphology, a more complete fossil and archeological record will improve our knowledge about the timing and context of the changes in hominin hand function.

A second area of research that deserves more attention is primate functional morphology. While primate hand morphology is relatively well characterized, our understanding of hand function, especially in natural settings, is still in its infancy.

Very little is known about how primates use their hands during locomotion and how hand postures and forces vary with support size and orientation, types of locomotion, speed, body size, and other factors (see Chaps. 10 and 13). Categories used to describe locomotion in the wild, such as "arboreal" and "terrestrial," are limited in their utility for understanding hand biomechanics. For example, for a small primate, the biomechanics of the hand can differ more between locomotion on large versus small arboreal supports than between a large arboreal support and the ground. What we need is a "hands-eye" view of a primate's world to understand the functional demands that hands face.

Similarly, too little is known about the biomechanics of the human hand when engaged in functions that were likely to be important during human evolution. A number of studies have examined aspects of biomechanics involved in making primitive stone tools, and research in this area is growing in sophistication (see Chap. 11). More research is needed to examine the biomechanics of other activities, such as the use of spears, digging sticks, food processing, and throwing, as well as the use of various forms of stone tools and flakes in different settings. In a similar vein, more research is needed on the ecological and behavioral context of making and using tools, and carrying, in humans as well as nonhuman primates in natural settings.

Finally, a better understanding of the factors that influence skeletal morphology during growth and in adulthood will improve our ability to make robust inferences about behavior from external morphology, cortical geometry, and trabecular structure.

Acknowledgments We thank Sergio Almécija, Erin Marie Williams-Hatala, Marisa Macias, Erik Trinkaus, Tracy Kivell, Pierre Lemelin, and Dan Schmitt for their helpful discussions and suggestions that improved this manuscript. We thank Tracy Kivell for *H. naledi* hand data and images of a number of early hominin fossils and Tim White for images of the hand of *Ardipithecus ramidus* (ARA-VP-6/500). We are grateful to the following collection managers and curators for access to skeletal remains and fossils in their care: Bernhard Zipfel and Lee Berger, University of the Witswatersrand; Kyalo Manthi and Emma Mbua, National Museums of Kenya; Kristofer Helgen, Richard Thorington, and Darrin Lunde, National Museum of Natural History, Smithsonian Institution; and Eileen Westwig, American Museum of Natural History. This work was supported by the Wenner-Gren Foundation for Anthropological Research, the George Washington University, and NSF DGE-0801634, BCS-0924476, BCS-1128170, and BCS-1515054.

References

Alba DM, Moyà-Solà S, Köhler M (2003) Morphological affinities of the *Australopithecus afarensis* hand on the basis of manual proportions and relative thumb length. J Hum Evol 44:225–254

Almécija S, Alba DM (2014) On manual proportions and pad-to-pad precision grasping in *Australopithecus afarensis*. J Hum Evol 73:88–92

Almécija S, Moyà-Solà S, Alba DM (2010) Early origin for human-like precision grasping: a comparative study of pollical distal phalanges in fossil hominins. PLoS One 5:e11727

Almécija S, Smaers JB, Jungers WL (2015) The evolution of human and ape hand proportions. Nat Commun 6:7717
Ambrose SH (2001) Paleolithic technology and human evolution. Science 291:1748–1753
Begun DR (2004) Knuckle-walking and the origin of human bipedalism. In: Meldrum DJ, Hilton CE (eds) From biped to strider: the emergence of modern human walking, running, and resource transport. Springer, New York, pp 9–33
Begun DR, Richmond BG, Strait DS (2007) Comment on "Origin of human bipedalism as an adaptation for locomotion on flexible branches." Science 318:1066, author reply 1066
Bock WJ, von Wahlert G (1965) Adaptation and the form-function complex. Evolution 19:269–299
Bradley KC, Sunderland S (1953) The range of movement at the wrist joint. Anat Rec 116:139–145
Bramble DM, Lieberman DE (2004) Endurance running and the evolution of *Homo*. Nature 432:345–352
Bush ME, Lovejoy CO, Johanson DC, Coppens Y (1982) Hominid carpal, metacarpal, and phalangeal bones recovered from the Hadar Formation: 1974-1977 collections. Am J Phys Anthropol 57:651–677
Carvalho S, Biro D, Cunha E, Hockings K, McGrew WC, Richmond BG, Matsuzawa T (2013) Chimpanzee carrying behaviour and the origins of human bipedality. Curr Biol 22:180–181
Christel M (1993) Grasping techniques and hand preferences in Hominoidea. In: Preuschoft H, Chivers DJ (eds) Hands of primates. Springer-Verlag, Vienna, pp 91–108
Clarke RJ (1999) Discovery of complete arm and hand of the 3.3 million-year-old *Australopithecus* skeleton from Sterkfontein. S Afr J Sci 95:477–480
Cooney WP, Chao EY (1977) Biomechanical analysis of static forces in the thumb during hand function. J Bone Joint Surg Am 59:27–36
Congdon KA (2012) Interspecific and ontogenetic variation in proximal pedal phalangeal curvature of great apes (Gorilla gorilla, Pan troglodytes, and Pongo pygmaeus). Int J Primatol 33:418–427
Corruccini RS (1978) Comparative osteometrics of the hominoid wrist joint, with special reference to knuckle-walking. J Hum Evol 7:307–321
Diogo R, Richmond BG, Wood B (2012) Evolution and homologies of primate and modern human hand and forearm muscles, with notes on thumb movements and tool use. J Hum Evol 63:64–78
Domínguez-Rodrigo M, Pickering TR, Almécija S, Heaton J, Baquedano E, Mabulla A, Uribelarrea D (2015) Earliest modern human-like hand bone from a new >1.84-million-year-old site at Olduvai in Tanzania. Nat Commun 6:7987
Doran DM (1993) Comparative locomotor behavior of chimpanzees and bonobos: the influence of morphology and locomotion. Am J Phys Anthropol 91:83–98
Doran DM (1996) Comparative positional behavior of the African apes. In: McGrew W, Marchant L, Nishida T (eds) Great ape societies. Cambridge University Press, Cambridge, pp 213–224
Drapeau MS, Ward CV, Kimbel WH, Johanson DC, Rak Y (2005) Associated cranial and forelimb remains attributed to *Australopithecus afarensis* from Hadar, Ethiopia. J Hum Evol 48:593–642
Erickson GE (1963) Brachiation in the New World monkeys and in anthropoid apes. Symp Zool Soc Lond 10:135–164
Etter HF (1973) Terrestrial adaptations in the hands of Cercopithecinae. Folia Primatol 20:331–350
Fleagle JG (1977) Locomotor behavior and skeletal anatomy of sympatric Malaysian leaf-monkeys. Yearb Phys Anthropol 20:440–453
Gebo DL (1996) Climbing, brachiation, and terrestrial quadrupedalism: historical precursors of hominid bipedalism. Am J Phys Anthropol 101:55–92
Gommery D, Senut B (2006) The terminal thumb phalanx of *Orrorin tugenensis* (Upper Miocene of Kenya). Geobios 39:372–384
Grabowski MW, Hatala KG, Jungers WL, Richmond BG (2015) Body size estimates of the earliest possible hominins. J Hum Evol 85:75–93

Green DJ, Gordon AD (2008) Metacarpal proportions in *Australopithecus africanus*. J Hum Evol 54:705–719

Green DJ, Gordon AD, Richmond BG (2007) Limb-size proportions in *Australopithecus afarensis* and *Australopithecus africanus*. J Hum Evol 52:187–200

Haile-Selassie Y, Saylor BZ, Deino A, Levin NE, Alene M, Latimer BM (2012) A new hominin foot from Ethiopia shows multiple Pliocene bipedal adaptations. Nature 483:565–569

Haile-Selassie Y, Gibert L, Melillo SM, Ryan TM, Alene M, Deino A, Levin NE, Scott G, Saylor BZ (2015) New species from Ethiopia further expands Middle Pliocene hominin diversity. Nature 521:483–488

Hamrick MW (1996) Articular size and curvature as determinants of carpal joint mobility and stability in strepsirhine primates. J Morphol 230:113–127

Hamrick MW, Simons EL, Jungers WL (2000) New wrist bones of the Malagasy giant subfossil lemurs. J Hum Evol 38:635–650

Harmand S, Lewis JE, Feibel CS, Lepre CJ, Prat S, Lenoble A, Boës X, Quinn RL, Brenet M, Arroyo A, Taylor N, Clément S, Daver G, Brugal J-P, Leakey L, Mortlock RA, Wright JD, Lokorodi S, Kirwa C, Kent DV, Roche H (2015) 3.3-million-year-old stone tools from Lomekwi 3, West Turkana, Kenya. Nature 521:310–315

Jolly CJ (1970) The large African monkeys as an adaptive array. In: Napier JR, Napier PH (eds) Old world monkeys. Academic Press, New York, pp 139–174

Jouffroy FK (1975) Osteology and myology of the lemuriform postcranial skeleton. In: Tattersall I, Sussman RW (eds) Lemur biology. Plenum Press, New York, pp 149–192

Jouffroy FK, Godinot M, Nakano Y (1993) Biometrical characteristics of primate hands. In: Preuschoft H, Chivers DJ (eds) Hands of primates. Springer-Verlag, Vienna, pp 133–173

Jungers WL (1988) Relative joint size and hominoid locomotor adaptations with implications for the evolution of hominid bipedalism. J Hum Evol 17:247–265

Jungers WL, Godfrey LR, Simons EL, Chatrath PS (1997) Phalangeal curvature and positional behavior in extinct sloth lemurs (Primates, Paleopropithecidae). Proc Natl Acad Sci U S A 94:11998–12001

Jungers WL, Godfrey LR, Simons EL, Wunderlich RE, Richmond BG, Chatrath PS, Rakotosamimanana B (2001) Ecomorphology and behavior of giant extinct lemurs from Madagascar. In: Plavcan JM, Kay RF, Van Schaik CP, Jungers WL (eds) Reconstructing behavior in the primate fossil record. Kluwer Academic/Plenum Publishers, New York, pp 371–411

Kivell TL, Begun DR (2007) Frequency and timing of scaphoid-centrale fusion in hominoids. J Hum Evol 52:321–340

Kivell TL, Begun DR (2009) New primate carpal bones from Rudabánya (late Miocene, Hungary): taxonomic and functional implications. J Hum Evol 57:697–709

Kivell TL, Schmitt D (2009) Independent evolution of knuckle-walking in African apes shows that humans did not evolve from a knuckle-walking ancestor. Proc Natl Acad Sci U S A 106:14241–14246

Kivell TL, Kibii JM, Churchill SE, Schmid P, Berger LR (2011) *Australopithecus sediba* hand demonstrates mosaic evolution of locomotor and manipulative abilities. Science 333:1411–1417

Kivell TL, Deane AS, Tocheri MW, Orr CM, Schmid P, Hawks J, Berger LR, Churchill SE (2015) The hand of *Homo naledi*. Nat Commun 6:8431

Koops K, Visalberghi E, van Schaik CP (2014) The ecology of primate material culture. Biol Lett 10:20140508

Kraft TS, Venkataraman VV, Dominy NJ (2014) A natural history of human tree climbing. J Hum Evol 71:105–118

Latimer BM (1990) Locomotor adaptations in *Australopithecus afarensis*: the issue of arboreality. In: Coppens Y, Senut B (eds) Origine(s) de la Bipédie chez les Hominidés. Centre National de la Recherche Scientifique, Paris, pp 169–176

Leakey LSB, Tobias PV, Napier JR (1964) A new species of the genus *Homo* from Olduvai Gorge. Nature 202:7–9

Lepre CJ, Roche H, Kent DV, Harmand S, Quinn RL, Brugal J-P, Texier P-J, Lenoble A, Feibel CS (2011) An earlier origin for the Acheulian. Nature 477:82–85

Lindburg RM, Comstock BE (1979) Anomalous tendon slips from the flexor pollicis longus to the flexor digitorum profundus. J Hand Surg Am 4:79–83

Lorenzo C, Pablos A, Carretero JM, Huguet R, Valverdú J, Martinón-Torres M, Arsuaga JL, Carbonell E, Bermúdez de Castro JM (2015) Early Pleistocene human hand phalanx from the Sima del Elefante (TE) cave site in Sierra de Atapuerca (Spain). J Hum Evol 78:114–121

Lovejoy CO, Simpson SW, White TD, Asfaw B, Suwa G (2009a) Careful climbing in the Miocene: the forelimbs of *Ardipithecus ramidus* and humans are primitive. Science 326:70e1–70e8

Lovejoy CO, Suwa G, Simpson SW, Matternes JH, White TD (2009b) The great divides: *Ardipithecus ramidus* reveals the postcrania of our last common ancestors with African apes. Science 326:100–106

Lovejoy CO, Latimer B, Suwa G, Asfaw B, White TD (2009c) Combining prehension and propulsion: the foot of *Ardipithecus ramidus*. Science 326:72e1–72e8

Lovejoy CO, Suwa G, Spurlock L, Asfaw B, White TD (2009d) The pelvis and femur of *Ardipithecus ramidus*: the emergence of upright walking. Science 326:71e1–71e6

Maki J, Trinkaus E (2011) Opponens pollicis mechanical effectiveness in Neandertals and early modern humans. PaleoAnthropology 2011:62–71

Marzke MW (1971) Origin of the human hand. Am J Phys Anthropol 34:61–84

Marzke MW (1997) Precision grips, hand morphology, and tools. Am J Phys Anthropol 102:91–110

Marzke MW, Marzke RF (1987) The third metacarpal styloid process in humans: origin and functions. Am J Phys Anthropol 73:415–431

Marzke MW, Wullstein KL (1996) Chimpanzee and human grips: a new classification with a focus on evolutionary morphology. Int J Primatol 17:117–139

Marzke MW, Shrewsbury MM, Horner KE (2007) Middle phalanx skeletal morphology in the hand: can it predict flexor tendon size and attachments? Am J Phys Anthropol 134:141–151

Matarazzo S (2008) Knuckle walking signal in the manual digits of *Pan* and *Gorilla*. Am J Phys Anthropol 135:27–33

McPherron SP, Alemseged Z, Marean CW, Wynn JG, Reed D, Geraads D, Bobe R, Bearat HA (2010) Evidence for stone-tool-assisted consumption of animal tissues before 3.39 million years ago at Dikika, Ethiopia. Nature 466:857–860

Mittra ES, Smith HF, Lemelin P, Jungers WL (2007) Comparative morphometrics of the primate apical tuft. Am J Phys Anthropol 134:449–459

Moyà-Solà S, Köhler M, Alba DM, Casanovas-Vilar I, Galindo J (2004) *Pierolapithecus catalaunicus*, a new Middle Miocene great ape from Spain. Science 306:1339–1344

Napier JR (1956) The prehensile movements of the human hand. J Bone Joint Surg Br 38B:902–913

Napier JR (1961) Prehensility and opposability in the hands of primates. Symp Zool Soc Lond 5:115–132

Napier JR (1962) Fossil hand bones from Olduvai Gorge. Nature 196:409–411

Napier JR (1993) Hands. Revised edition by Russell H. Tuttle. Princeton University Press, Princeton

Napier JR, Napier PH (1967) A handbook of living primates. Academic Press, London

Nguyen NH, Pahr DH, Gross T, Skinner MM, Kivell TL (2014) Micro-finite element (μFE) modeling of the siamang (*Symphalangus syndactylus*) third proximal phalanx: the functional role of curvature and the flexor sheath ridge. J Hum Evol 67:60–75

Niedźwiedzki G, Szrek P, Narkiewicz K, Narkiewicz M, Ahlberg PE (2010) Tetrapod trackways from the early Middle Devonian period of Poland. Nature 463:43–48

Orr CM (2005) Knuckle-walking anteater: a convergence test of adaptation for purported knuckle-walking features of African Hominidae. Am J Phys Anthropol 128:639–658

Orr CM (2013) The earliest hominins were rigid wristed. Am J Phys Anthropol 150(Suppl 56):211 (abstract)

Orr CM, Leventhal EL, Chivers SF, Marzke MW, Wolfe SW, Crisco JJ (2010) Studying primate carpal kinematics in three dimensions using a computed-tomography-based markerless registration method. Anat Rec 293:692–709

Panger MA, Brooks AS, Richmond BG, Wood BA (2002) Older than the Oldowan: rethinking the emergence of hominin tool use. Evol Anthropol 11:234–245

Preuschoft H (1970) Functional anatomy of the lower extremity. In: Bourne GH (ed) The chimpanzee, vol 3. Karger, Basel, pp 221–294

Rabey KN, Green DJ, Taylor AB, Begun DR, Richmond BG, McFarlin SC (2015) Locomotor activity influences muscle architecture and bone growth but not muscle attachment site morphology. J Hum Evol 78:91–102

Rein TR (2011) The correspondence between proximal phalanx morphology and locomotion: implications for inferring the locomotor behavior of fossil catarrhines. Am J Phys Anthropol 146:435–445

Richmond BG (1998) Ontogeny and biomechanics of phalangeal form in primates. Ph.D. dissertation, State University of New York at Stony Brook

Richmond BG (2006) Functional morphology of the midcarpal joint in knuckle-walkers and terrestrial quadrupeds. In: Ishida H, Tuttle R, Pickford M, Ogihara N, Nakatsukasa M (eds) Human origins and environmental backgrounds. Springer, New York, pp 105–122

Richmond BG (2007) Biomechanics of phalangeal curvature. J Hum Evol 53:678–690

Richmond BG, Harris JWK, Mbua E, Braun DR, Bamford MK, Bobe R, Green DJ, Griffin NL, McCoy JT, Merritt S, Pante M, Pobiner BL, Carter-Menn H, Chirchir H, Kiura P, Kibunjia M (2009) Divergence in hominin upper limb anatomy in the early Pleistocene. Am J Phys Anthropol 136(Suppl 48):341 (abstract)

Richmond BG, Hatala KG (2013) Origin and evolution of human postcranial anatomy. In: Begun DR (ed) A companion to paleoanthropology. Blackwell Publishing Ltd., Oxford, pp 183–202

Richmond BG, Jungers WL (2008) *Orrorin tugenensis* femoral morphology and the evolution of hominin bipedalism. Science 319:1662–1665

Richmond BG, Strait DS (2000) Evidence that humans evolved from a knuckle-walking ancestor. Nature 404:382–385

Richmond BG, Begun DR, Strait DS (2001) Origin of human bipedalism: the knuckle-walking hypothesis revisited. Yearb Phys Anthropol 116:70–105

Ricklan DE (1987) Functional anatomy of the hand of *Australopithecus africanus*. J Hum Evol 16:643–664

Roach NT, Lieberman DE (2014) Upper body contributions to power generation during rapid, overhand throwing in humans. J Exp Biol 217:2139–2149

Roach NT, Venkadesan M, Rainbow MJ, Lieberman DE (2013) Elastic energy storage in the shoulder and the evolution of high-speed throwing in *Homo*. Nature 498:483–486

Rolian C, Gordon AD (2013) Reassessing manual proportions in *Australopithecus afarensis*. Am J Phys Anthropol 152:393–406

Rolian C, Gordon AD (2014) Response to Almécija and Alba (2014)—on manual proportions in *Australopithecus afarensis*. J Hum Evol 73:93–97

Rolian C, Lieberman DE, Hallgrimsson B (2010) The coevolution of human hands and feet. Evolution 64:1558–1568

Rose MD (1977) Positional behavior of olive baboons (*Papio anubis*) and its relationship to maintenance and social activities. Primates 18:59–116

Rose MD (1991) The process of bipedalization in hominids. In: Coppens Y, Senut B (eds) Origine(s) de la Bipédie chez les Hominidés. Centre National de la Recherche Scientifique, Paris, pp 37–48

Ruff CB (2009) Relative limb strength and locomotion in *Homo habilis*. Am J Phys Anthropol 138:90–100

Ruff CB, Walker AC (1993) Body size and body shape. In: Walker AC, Leakey RE (eds) The Nariokotome Homo Erectus Skeleton. Harvard University Press, Cambridge, pp 234–265

Ruff CB, Holt B, Trinkaus E (2006) Who's afraid of the big bad Wolff?: "Wolff's law" and bone functional adaptation. Am J Phys Anthropol 129:484–498

Sarmiento EE (1988) Anatomy of the hominoid wrist joint: its evolutionary and functional implications. Int J Primatol 9:281–345

Sarringhaus LA (2013) Positional and morphological development of wild chimpanzees, *Pan troglodytes*. Ph.D. dissertation, University of Michigan

Schmitt D (2003) Insights into the evolution of human bipedalism from experimental studies of humans and other primates. J Exp Biol 206:1437–1448

Schreiber H (1936) Die Extrembewegungen der Schimpansenhand. 2. Mitteilung zu: Zur Morphologie der Primatenhand. Gegenbaurs Morphol Jahrb 77:22–60

Shrewsbury MM, Johnson RK (1983) Form, function, and evolution of the distal phalanx. J Hand Surg Am 8:475–479

Shrewsbury MM, Marzke MW, Linscheid RL, Reece SP (2003) Comparative morphology of the pollical distal phalanx. Am J Phys Anthropol 121:30–47

Skinner MM, Stephens NB, Tsegai ZJ, Foote AC, Nguyen NH, Gross T, Pahr DH, Hublin J-J, Kivell TL (2015) Human-like hand use in *Australopithecus africanus*. Science 347:395–399

Stern JT Jr (2000) Climbing to the top: a personal memoir of *Australopithecus afarensis*. Evol Anthropol 9:113–133

Stern JT Jr, Susman RL (1983) The locomotor anatomy of *Australopithecus afarensis*. Am J Phys Anthropol 60:279–317

Susman RL (1988a) Hand of *Paranthropus robustus* from Member I, Swartkrans: fossil evidence for tool behavior. Science 240:781–784

Susman RL (1988b) New postcranial remains from Swartkrans and their bearing on the functional morphology and behavior of *Paranthropus robustus*. In: Grine FE (ed) Evolutionary history of the "robust" australopithecines. Aldine de Gruyter, New York, pp 149–172

Susman RL (1994) Fossil evidence for early hominid tool use. Science 265:1570–1573

Susman RL (1998) Hand function and tool behavior in early hominids. J Hum Evol 35:23–46

Susman RL, Nyati L, Jassal MS (1999) Observations on the pollical palmar interosseous muscle (of Henle). Anat Rec 254:159–165

Thorpe SK, Holder RL, Crompton RH (2007) Origin of human bipedalism as an adaptation for locomotion on flexible branches. Science 316:1328–1331

Tocheri MW (2007) Three-dimensional riddles of the radial wrist: derived carpal and carpometacarpal joint morphology in the genus *Homo* and the implications for understanding the evolution of stone tool-related behaviors in hominins. Ph.D. dissertation, Arizona State University

Tocheri MW, Orr CM, Larson SG, Sutikna T, Jatmiko, Saptomo EW, Due RA, Djubiantono T, Morwood MJ, Jungers WL (2007) The primitive wrist of *Homo floresiensis* and its implications for hominin evolution. Science 317:1743–1745

Tocheri MW, Orr CM, Jacofsky MC, Marzke MW (2008) The evolutionary history of the hominin hand since the last common ancestor of *Pan* and *Homo*. J Anat 212:544–562

Tocheri MW, Solhan CR, Orr CM, Femiani J, Frohlich B, Groves CP, Harcourt-Smith WE, Richmond BG, Shoelson B, Jungers WL (2011) Ecological divergence and medial cuneiform morphology in gorillas. J Hum Evol 60:171–184

Trinkaus E, Long JC (1990) Species attribution of the Swartkrans Member 1 first metacarpals: SK 84 and SKX 5020. Am J Phys Anthropol 83:419–424

Tuttle RH (1969) Quantitative and functional studies on the hands of the Anthropoidea: I. The Hominoidea. J Morphol 128:309–363

Tuttle RH (1975) Parallelism, brachiation, and hominoid phylogeny. In: Luckett WP, Szalay FS (eds) Phylogeny of the primates: a multidisciplinary approach. Plenum Press, New York, pp 447–480

Villmoare B, Kimbel WH, Seyoum C, Campisano CJ, DiMaggio E, Rowan J, Braun DR, Arrowsmith JR, Reed KE (2015) Early *Homo* at 2.8 Ma from Ledi-Geraru, Afar, Ethiopia. Science 347:1352–1355

Ward CV (2002) Interpreting the posture and locomotion of *Australopithecus afarensis*: where do we stand? Yearb Phys Anthropol 35:185–215

Ward CV, Kimbel WH, Harmon EH, Johanson DC (2012) New postcranial fossils of *Australopithecus afarensis* from Hadar, Ethiopia (1990–2007). J Hum Evol 63:1–51

Ward CV, Tocheri MW, Plavcan JM, Brown FH, Manthi FK (2014) Early Pleistocene third metacarpal from Kenya and the evolution of modern human-like hand morphology. Proc Natl Acad Sci U S A 111:121–124

Washburn SL (1967) Behavior and the origin of man. Proc R Anthropol Inst Great Britain Ireland 3:21–27

Weinert H (1932) Ursprung der Menschheit. Üeber den engeren Anschluss des Menschengeschlechts an die Menschenaffen. Ferdinand Enke Verlag, Stuttgart

White TD, Asfaw B, Beyene Y, Haile-Selassie Y, Lovejoy CO, Suwa G, WoldeGabriel G (2009) *Ardipithecus ramidus* and the paleobiology of early hominids. Science 326:75–86

Williams SA (2010) Morphological integration and the evolution of knuckle-walking. J Hum Evol 58:432–440

Williams EM, Gordon AD, Richmond BG (2010) Upper limb kinematics and the role of the wrist during stone tool production. Am J Phys Anthropol 143:134–145

Williams EM, Gordon AD, Richmond BG (2012) Hand pressure distribution during Oldowan stone tool production. J Hum Evol 62:520–532

Williams EM, Gordon AD, Richmond BG (2014) Biomechanical strategies for accuracy and force generation during stone tool production. J Hum Evol 72:52–63

Williams-Hatala EM, Hatala KG, Hiles S, Rabey KN (2016) Morphology of muscle attachment sites in the modern human hand does not reflect muscle architecture. Scientific Reports 6:28353

Wolfe SW, Crisco JJ, Orr CM, Marzke MW (2006) The dart-throwing motion of the wrist: is it unique to humans? J Hand Surg Am 31:1429–1437

Wood Jones F (1916) Arboreal man. Edward Arnold, London

Wood Jones F (1920) The principles of anatomy as seen in the hand. J. & A. Churchill, London

Zumwalt A (2006) The effect of endurance exercise on the morphology of muscle attachment sites. J Exp Biol 209:444–454

Chapter 19
The Evolution of the Hand in Pleistocene *Homo*

Erik Trinkaus

1 Introduction

Fine manipulative abilities and a dependence on tool use are hallmarks of modern humans, and these aspects emerged primarily throughout the evolution of the genus *Homo*. They appear among the earliest members of the genus and their probable association with the Oldowan technocomplex, and they continue through the Pleistocene to the manipulative patterns evident in the ethnohistorical present. While aspects of this behavioral and adaptive evolutionary process are well documented in the Paleolithic archeological record, the anatomical bases for it are less well known. The fossil record of the Late Pleistocene, especially of the last ~100 Ka, preserves sufficient numbers of associated hand skeletons to document the basic anatomical patterns and provide some functional inferences for both late archaic humans (especially those of western Eurasia—the Neandertals) and early modern humans. For most of the Early and Middle Pleistocene, hand skeletal remains are rare, localized in time and space, and primarily consist of isolated bones rather than associated hand skeletons (see Chap. 18). Yet, it is becoming possible through the available remains to provide an outline of the morphological and, by inference, functional implications of the manual elements representing the evolution of the hand within the genus *Homo*. This contribution is a summary of such evidence.

E. Trinkaus (✉)
Department of Anthropology, Washington University, Saint Louis, MO 63130, USA
e-mail: trinkaus@wustl.edu

T.L. Kivell et al. (eds.), *The Evolution of the Primate Hand*, Developments in Primatology: Progress and Prospects, DOI 10.1007/978-1-4939-3646-5_19

2 General Considerations

2.1 Manipulative Inferences

The inference of human (i.e., members of the genus *Homo*) manipulative evolution from the fossil record (leaving aside the Paleolithic archeological record) involves several interweaving aspects. A first step is the documentation of morphological patterns and assessments of the ancestral versus derived aspects of *Homo* hand skeletal anatomy. In addition to any taxonomic or phylogenetic implications that they might have, which are not under consideration here, the polarities of hand features provide a framework for determining the directionality of any inferred functional changes through the Pleistocene.

The second step is the functional assessment of relevant aspects of the hand skeleton. They are here divided into those concerned with (a) overall hand proportions; (b) carpometacarpal articular morphology; (c) proportional aspects of the thumb; and (d) reflections of carpal, metacarpal, and phalangeal hypertrophy. Each of these aspects has implications for habitual manipulative loading patterns and levels.

The third step, and the most difficult one, is the putting together of these particular elements into a coherent scenario for the evolution of the functional morphology of the *Homo* hand. This step is limited given that most of these Pleistocene *Homo* hand elements are basically "modern human" in contrast with those of living nonhuman primates and *Australopithecus* (Chap. 18). These assessments therefore tend to be focused on analogies with recent human hand skeletons, although some employ basic biomechanical principles of joints and muscles (see Chaps. 9–11).

All of these assessments, however, have to bear in mind that (1) a variety of morphological configurations can provide the same manipulative capabilities, and (2) human manual dexterity is based not only on musculoskeletal capabilities but also on neurological architecture (see Chap. 6). The human hand, and especially digits 1–3, has expanded neurological control with expansion of the lower cervical spinal cord for efferent and afferent synapses for nerve fibers to the brachial plexus and expansions of the motor and sensory portions of the cerebral cortex for the hand (MacLarnon 1993; Nakamura et al. 1998). However, these neurological aspects are largely invisible paleontologically. Enlargement of the cervical vertebral canal for the expanded brachial plexus synapses is found in early *Homo* (Weaver et al. 2001; Meyer 2005), implying modern humanlike hand innervation from the beginning of the genus. Despite some encephalization of early *Homo* relative to earlier hominins (Begun and Walker 1993), a marked increase in encephalization occurred throughout the evolution of the genus *Homo*, especially through the Middle Pleistocene (Ruff et al. 1997). One can only wonder how much of this cerebral increase was due to expanding manual innervation.

2.2 A Comment on Hand Functions

A substantial proportion of the literature on the evolution of the human hand is concerned with identifying anatomical features associated with manipulation (i.e., tool use) versus locomotion (e.g., Marzke and Shackley 1986; Susman 1998; Kivell et al. 2011, and references therein). This dichotomy is not a substantial concern here. Although members of the genus *Homo* likely foraged in trees and would have used trees or other protected areas for sleeping until the full domestication of fire by the earlier Middle Pleistocene (cf. Roebroeks and Villa 2011), all of these humans were fully dedicated terrestrial bipeds. Their upper limbs would have been used predominantly for manipulation and burden carrying, with only intermittent support of body weight [but see Dinaledi remains below and Chap. 18]. The morphological variation observed should therefore be seen in the context of changing technology and manipulative postures and of changing load levels and patterns associated with that varying behavior.

3 Samples and Comparisons

3.1 The Fossil Record

The paleontological samples of concern here, working backwards in time, include remains of (1) early modern humans, (2) Neandertals, (3) Middle Pleistocene humans (often included within *Homo heidelbergensis*), and (4) Early Pleistocene humans (*Homo erectus sensu lato*). The complete hand (missing only the pisiform) plus isolated bones from Dinaledi (*Homo naledi*) is also discussed here, as it is considered to represent an early member of the genus *Homo*, although its geological age remains unknown (Berger et al. 2015). The mix of australopith-like and *Homo*-like morphological features provides important evidence concerning the ancestral condition of the hand in the genus *Homo* (Kivell et al. 2015). It also provides a functional mosaic relative to our perspectives from the hands of Middle Pleistocene to recent *Homo* hands versus australopith ones. The early modern human sample is separated into those associated with Middle Paleolithic versus Upper Paleolithic archeological assemblages, given the contrasts in technology between these two groups. There is insufficient agreement on species designations within Pleistocene *Homo* to use such taxonomic terms for these fossil samples, so more general designations of Early Pleistocene, Middle Pleistocene, Neandertals, Middle Paleolithic modern humans (MPMH), and Upper Paleolithic modern humans (UPMH) are employed here.

The **Early Pleistocene** southern African remains from Swartkrans (specimen numbers with SK, SKW, and SKX) have been variously attributed to *Australopithecus*/*Paranthropus robustus* versus *Homo* (Susman 1989; Trinkaus and Long 1990;

Richmond et al. 2011; see below and Chap. 18). In addition, the Olduvai Hominid (OH) 7 juvenile hand remains have been variously attributed to *Homo* or *Australopithecus/Paranthropus boisei* (Napier 1962; Susman 1998; Moyà-Solà et al. 2008). Both sets of remains, as well as the Dinaledi hand bones, exhibit derived *Homo* features and help to establish the ancestral *Homo* pattern.

Following on the Swartkrans and Olduvai hand bones, the later Early, Middle, and Late Pleistocene samples are securely attributed to the genus *Homo*, given their associations with diagnostic craniofacial remains. The remaining Early Pleistocene sample includes a proximal phalanx from the fifth ray (OH 86) (Domínguez-Rodrigo et al. 2015), a metacarpal from Kaitio (KNM-KP 51260) (Ward et al. 2014), a proximal phalanx from Sima del Elefante (ATE9-2) (Lorenzo et al. 2015), and a small sample from the late Early Pleistocene Atapuerca-TD6 (ATD6) (Lorenzo et al. 1999). The **Middle Pleistocene** sample is composed of the very abundant (>500) sample of hand bones from Sima de los Huesos (>~430 Ka) sample at Atapuerca (Atapuerca-SH) (Lorenzo 2007; Arsuaga et al. 2015). Both of the Atapuerca samples have mixed hand bones that cannot be associated by individual.

The **Neandertal** sample derives principally from the earlier Late Pleistocene (~130 to ~50 Ka) western Eurasian sites of Amud, La Chapelle-aux-Saints, Feldhofer, La Ferrassie, Kebara, Kiik-Koba, Krapina, Palomas, Regourdou, Saint-Césaire, Shanidar, Spy, and Tabun. Most of these remains derive from associated skeletons; the exceptions are the Krapina, Spy, and some Palomas hand bones. Only two late archaic human distal phalanges (from Denisova and Sea Harvest) are known from outside of the Neandertal range.

The **early modern humans** are divided into the Marine Isotope Stage 5c (~100 Ka) southwest Asian MPMH sample from Qafzeh and Skhul, and an earlier UPMH sample (~40–20 Ka), from the sites of Arene Candide, Barma Grande, Caviglione, Dolní Věstonice, Grotte-des-Enfants (Fanciulli), Minatogawa, Mladeč, Nazlet Khater, Obłazowa, Ohalo, Paglicci, Pataud, Pavlov, Předmostí, Sunghir, and Tianyuan, from north Africa and across Eurasia. The Qafzeh, Skhul, and most of the UPMH hand bones derive from associated skeletons; the exceptions are the Mladeč and Obłazowa phalanges.

Because most of these hand remains derive from western Eurasia, it is necessary to assume that they are generally representative of archaic and early modern humans through these time periods. There is regional differentiation in human craniofacial morphology during this time period (Cartmill and Smith 2009), but the growing Pleistocene *Homo* postcranial fossil record indicates few regional contrasts within each of the samples defined here, other than in overall body size and proportions (cf. Trinkaus 2009; Shang and Trinkaus 2010; Trinkaus and Ruff 2012; Bermúdez de Castro et al. 2012). Consequently, these hand remains are taken to be representative of these Pleistocene human groups generally.

3.2 *Quantitative Comparisons*

Assessing the anatomical changes in the hand during the evolution of the genus *Homo* involves morphometric as well as qualitative observations. A set of these has been included here to illustrate some aspects of *Homo* hand evolution, chosen in part to maximize comparative sample sizes. They are compared graphically and with indices against the morphology of recent humans.

For comparisons of individual bones, isolated specimens are included. However, many aspects of hand morphology require size scaling across different hand elements; for the Late Pleistocene samples (Neandertals, MPMH, and UPMH), this can be done for a number of associated hand skeletons, but it is only possible to do so for the Dinaledi ones among the earlier remains. Consequently, for the Middle Pleistocene sample in particular, mean values for individual measurements across the samples are compared. It is also not possible to assign distal phalanges of rays 2–4 reliably to digit or side, so the available measurements have been averaged within Late Pleistocene individuals to avoid sample inflation. The isolated Atapuerca-SH and Krapina elements have been included separately, although they may duplicate some individuals.

The data are principally from personal observation, supplemented by personal communications from B. Maureille, I. Crevecoeur, J.H. Musgrave, W.H. Niewoehner, C.V. Ward, and especially C. Lorenzo, plus published data from Verneau (1906), Matiegka (1938), Susman (1989), Lorenzo et al. (1999), Smith et al. (2006), and Crevecoeur (2008). To provide a general recent human comparative framework, pooled data are included for samples of Native Americans, Europeans, Euro-Americans, and Afro-Americans (referred to as "recent" humans throughout this chapter).

4 Overall Hand Proportions

Modern human hands differ markedly from those of extant and fossil Miocene apes in the relative proportions of the fingers, especially in having a long thumb relative to the fingers (Alba et al. 2003). The hand proportions of *Australopithecus*, or at least for *A. afarensis*, *A. africanus*, and *A. sediba* (Alba et al. 2003; Green and Gordon 2008; Kivell et al. 2011; Rolian and Gordon 2013), are (or appear to be) generally more similar to those of recent humans than to great apes (Chap. 18). The relative length of the Dinaledi thumb is also similar to the proportions in modern humans (Kivell et al. 2015). The associated Late Pleistocene hand metacarpophalangeal skeletons (Trinkaus et al. 2014b), plus a composite for the Atapuerca-SH hand remains (Fig. 19.1), indicate proportions like those of recent humans.

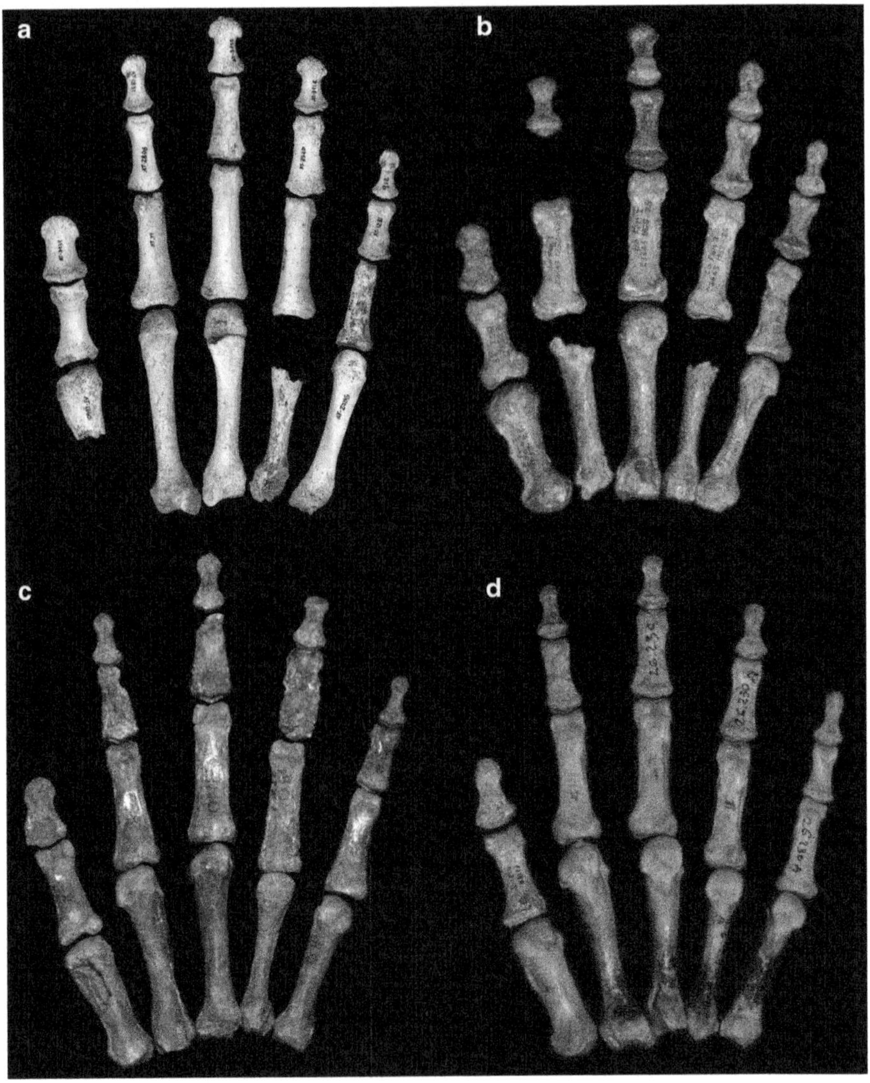

Fig. 19.1 Articulated metacarpal and phalangeal hand skeletons of (**a**) Middle Pleistocene archaic humans (Atapuerca-SH composite), (**b**) a Late Pleistocene Neandertal (La Ferrassie 1), (**c**) a Middle Paleolithic modern human (Qafzeh 9), and (**d**) an Upper Paleolithic modern human (Pataud 3). The Atapuerca "hand" in dorsal view is a composite of bones of similarly sized individuals from the mixed assemblage at the site. The other three hands are palmar views of hand remains from single individuals, photographically reversed for the Qafzeh and Pataud hands. All are approximately to the same scale

The more detailed proportions, and especially those related to relative pollical length, can be assessed directly for Middle and Late Pleistocene humans using indices of their first metacarpal (Mc1) and third metacarpal (Mc3) lengths, which are

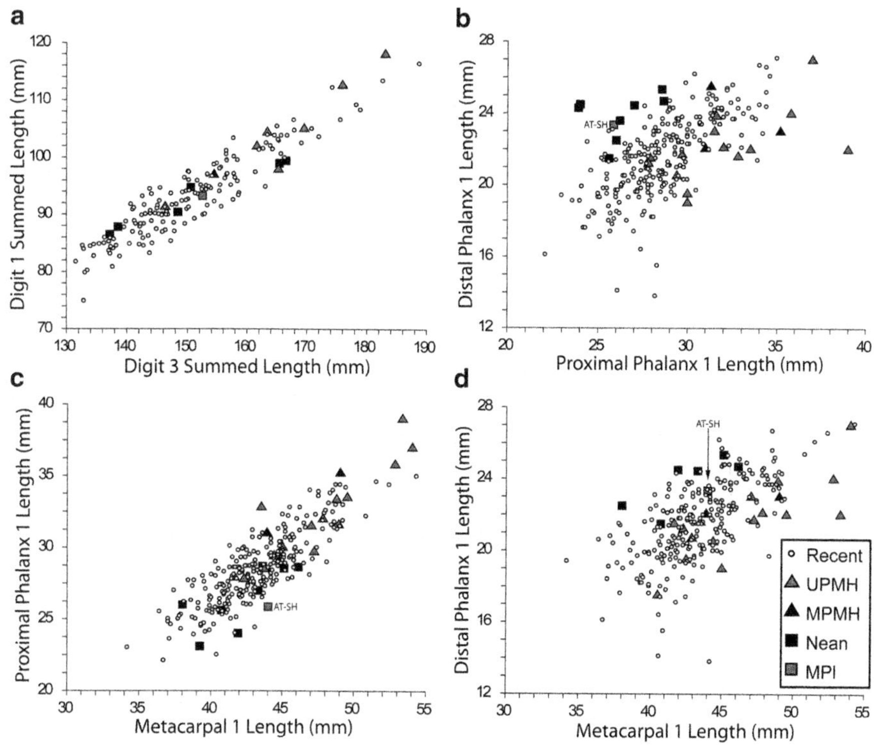

Fig. 19.2 Relative lengths of the pollex and middle finger and of pollical segments in Pleistocene and recent humans: (**a**) summed pollical articular lengths (Mc1 + PP1 + DP1) versus summed digit 3 articular lengths (Mc3 + PP3 + MP3 + DP3), (**b**) DP1 versus PP1 lengths, (**c**) PP1 versus Mc1 lengths, (**d**) DP1 versus Mc1 lengths. Recent: modern human sample comprised of Native American, Afro-American, and Euro-American modern humans; UPMH, Upper Paleolithic modern humans; MPMH, Middle Paleolithic modern humans; Nean, Late Pleistocene Neandertals; MPl, Middle Pleistocene depicted as the sum of the averages for each of the Atapuerca-SH metacarpals and phalanges, since the Atapuerca-SH remains are unassociated. For the Atapuerca-SH pollical points, the average lengths of the proximal and distal phalanges (25.9 ± 0.5 mm, N = 8; 23.4 ± 1.0 mm, N = 7), plus the length of the AT-5565 Mc1 (44.0 mm), are employed. There are no associated Early Pleistocene hands for these comparisons. See text for specimens included in the samples. Given its small size, the Dinaledi hand is not included

indistinguishable from those of recent humans [Neandertals, 68.9 ± 3.6 (n = 8); early modern humans, 69.3 ± 3.2 (n = 21); recent humans, 69.1 ± 2.8 (n = 187); Atapuerca-SH composite, 68.4; p = 0.983]. To provide a more reliable estimate of overall digit proportions, it is possible to assess summed metacarpal and phalangeal lengths for digits 1 and 3 for recent humans and a more limited number of Late Pleistocene humans (Fig. 19.2a). The single associated Dinaledi hand, although much smaller, also falls within the recent human distribution; the six Neandertal and nine early modern human hand skeletons are similar to those of recent humans. As with the metacarpal comparison, the Atapuerca-SH means place them among the more recent humans. These results should not be surprising, given the generally similar hand proportions seen in several species of *Australopithecus* (see Chap. 18).

5 Carpometacarpal Articulations

There is little of note with respect to variation in radiocarpal and intercarpal articulations in the available sample of fossils attributed to *Homo*, at least after *H. habilis* (Chap. 18). At least through the Middle and Late Pleistocene, the anatomy appears to fall well within the ranges of variation of recent humans, forming a compact carpal mass. The noted differences during the evolution of *Homo* have principally concerned the carpometacarpal (CMc) articulations, especially the CMc1 (trapezium and Mc1) and those between the capitate and the Mc2 and Mc3 (Musgrave 1971; Trinkaus 1983, 1989; Niewoehner et al. 1997; Lorenzo et al. 1999, 2012; Niewoehner 2000, 2001; Marzke et al. 2010). In addition, minor differences (condyloid versus sellar carpal facets) have been noted for the proximal Mc5 (Trinkaus 2006a), but the functional significance of this variation is unclear [see also Day and Scheuer (1973) regarding SKW 14147].

5.1 Carpometacarpal 1 Articulation

Given the prominence of the human thumb in manipulation, there has been continued focus on the shape of the pollical CMc articulation. With respect to *Homo* hands, this has focused principally on variation in the degree of dorsopalmar convexity of the trapezial Mc1 facet and of dorsopalmar concavity of the complementary saddle joint on the Mc1 for the trapezium (Musgrave 1971; Trinkaus 1983, 1989; Niewoehner 2000, 2001; Marzke et al. 2010; Walker et al. 2011; Lorenzo et al. 2012).

Most of the Neandertal trapezia and that of OH 7 have dorsopalmarly flatter saddle joints compared with those of early modern humans. One Neandertal trapezium—Palomas 92—has a markedly convex facet, and the Atapuerca-SH specimens fall within the recent human distribution. In the proximal Mc1, the Early Pleistocene Mc1 SKX 5020 has proximal articular dorsopalmar concavity similar to many recent humans (Susman 1989), but proximal Mc1s are unknown in the Middle Pleistocene. Among Neandertals there is considerable variation, with some Neandertals (e.g., La Ferrassie 1) being indistinguishable from recent humans and others (e.g., Amud 1, Regourdou 1, Shanidar 4) lacking a dorsopalmar concavity (Trinkaus 1983, 1989). The early modern human proximal Mc1 variation falls between those of the Neandertals and recent humans (Niewoehner 2001).

The Dinaledi Mc1s have relatively flat CMc1 joints, like those of the Neandertals and recent humans (and unlike *Pan*) but differ from these taxa in having an unusually small (both dorsopalmarly and radioulnarly) CMc1 articulations relative to the lengths of the Mc1s (Kivell et al. 2015). This small proximal articulation is accentuated by the radioulnarly broad distal Mc1 epiphysis and well-developed radial and ulnar flanges for muscle attachments that suggest powerful grasping by the thumb. Further research is needed to understand why it demonstrates this unusual thumb morphology and how such a small CMc1 joint could functionally cope with the large loads implied by its robust thumb.

The significance of these CMc1 variations remains unclear, since all of the paired articulations would permit three degrees of movement and any restrictions in CMc1 excursion could be compensated with intercarpal movement (Stark et al. 1977). It may well be that the relative flatness of the OH 7 and some Neandertal trapezia is largely produced by a palmar extension of the articular facet (Niewoehner 2000), which would increase the surface area perpendicular to the predominantly axial joint reaction forces across the articulation [Cooney and Chao (1977); see also Trinkaus (1989) and Richmond et al. (2011)]. Given the relative flatness of the saddle joints in OH 7, the Dinaledi CMc1s, and the Neandertals, it is nonetheless interesting that the Middle Pleistocene Atapuerca-SH sample does not share this pattern, making assessments of its polarity difficult.

5.2 Capitate: Metacarpal Articulations

The genus *Homo* has a derived joint complex between the capitate and the proximal Mc2 and Mc3, one that has been inferred (Lewis 1973; Marzke 1983; Marzke and Marzke 1987; cf. Ward et al. 2014) to improve resistance to oblique joint reaction forces through the second and third digits, resist dorsiflexion of the CMc3 articulation, and permit some rotation of the Mc2 on the capitate. These anatomical features include oblique and proximoulnarly convex facets between the proximoulnar Mc2 and distoradial capitate and a styloid process on the Mc3 that overlaps the dorsodistoradial capitate (Fig. 19.3). *Australopithecus* and the Dinaledi Mc3s lack styloid processes, and the Mc2 facets on their capitate bones tend to be parasagittally oriented (Bush et al. 1982; McHenry 1983; Ward et al. 2001). However, both *Australopithecus* and *Homo* capitates appear to have Mc2 facets that permit some Mc2 pronation (Niewoehner et al. 1997; Lorenzo et al. 1999; Ward et al. 2001).

The orientation of the Mc2 facet on the capitate shows variation in recent humans, but the angle between the Mc2 and Mc3 facets, in the coronal plane and across the dorsopalmar midline of the facet, averages around 40–45° (41.3° ± 8.1°, N=49). The MPMH and UPMH capitates have variable angles that are generally higher than those of recent human (42° and 50.3° ± 13.9°, N=7, respectively), indicating a more parasagittal Mc2 facet. The Neandertals and the Middle Pleistocene sample have even higher angles on average (53.7° ± 11.8°, N=7 and 59.7° ± 6.7°, N=6). The one Early Pleistocene ATD6 capitate has a less parasagittal Mc2 facet (48°), but nonethe-

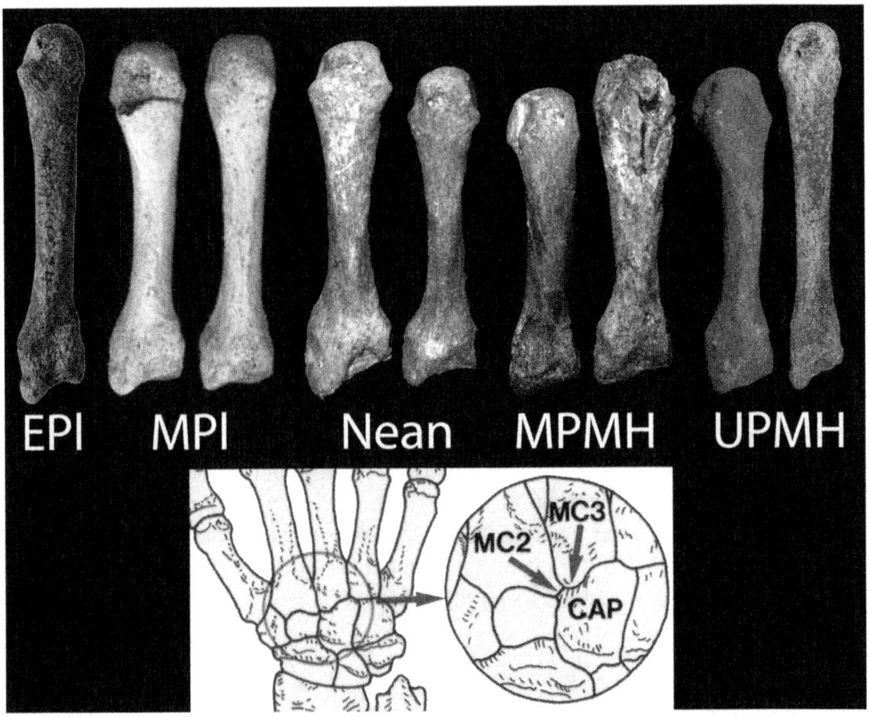

Fig. 19.3 The capitate-metacarpal articulations of Pleistocene *Homo*. *Above*: dorsal views of fossil *Homo* third metacarpals (Mc3s), all right or photographically reversed to appear as right. *Below*: dorsal view drawing of a recent human carpometacarpal skeleton, showing the capitate facet on the Mc2, the styloid process on the Mc3, and the capitate bone (CAP). For the third metacarpals: EPl, Early Pleistocene (KNM-KP 51260); MPl, Middle Pleistocene (Atapuerca-SH 2492 and 1271); Nean, Neandertals (Shanidar 4 and 6); MPMH, Middle Paleolithic modern humans (Qafzeh 8 and 9); UPMH, Upper Paleolithic modern humans (Sunghir 1, Dolní Věstonice 58). Note the variation in the projection of the styloid process [on the radial (*left*) Mc3 base], both within samples and across the fossil bones. All Pleistocene styloid processes are shorter than the majority of recent human ones

less one whose orientation overlaps those of the other samples. The Neandertal capitate Mc2 facets are joined by more parasagittal capitate facets on their proximal Mc2s (Neandertals, 26.3°±15.4°, N=12; UPMH, 47.3°±10.5°; *N*=9; recent humans, 53.0°±9.5°, *N*=48) (cf. Niewoehner et al. 1997; Trinkaus et al. 2010). There are currently no sufficiently intact Early and Middle Pleistocene *Homo* Mc2s.

At the same time, it has been noted (Niewoehner et al. 1997) that Neandertals on average have relatively shorter Mc3 styloid processes than those of recent humans. Further assessment with the addition of other Pleistocene humans (Fig. 19.3), however, shows that this pattern of variably short styloid processes is characteristic of all Pleistocene humans, from the Early Pleistocene through Late Pleistocene early modern humans (Trinkaus et al. 2010; Ward et al. 2014). It is principally recent humans who differ from earlier *Homo* in having a relatively long styloid process.

As originally noted (Niewoehner et al. 1997), the more parasagittal Neandertal capitate-Mc2 facets and smaller Mc3 styloid processes, now variably joined by other Pleistocene *Homo* remains, suggest less oblique loading of the mid-CMc region. Yet, since these articulations experience little movement, and all of them except the Dinaledi ones possess these derived *Homo* facets, any functional implications would only be in terms of the levels and patterns of manipulative loading.

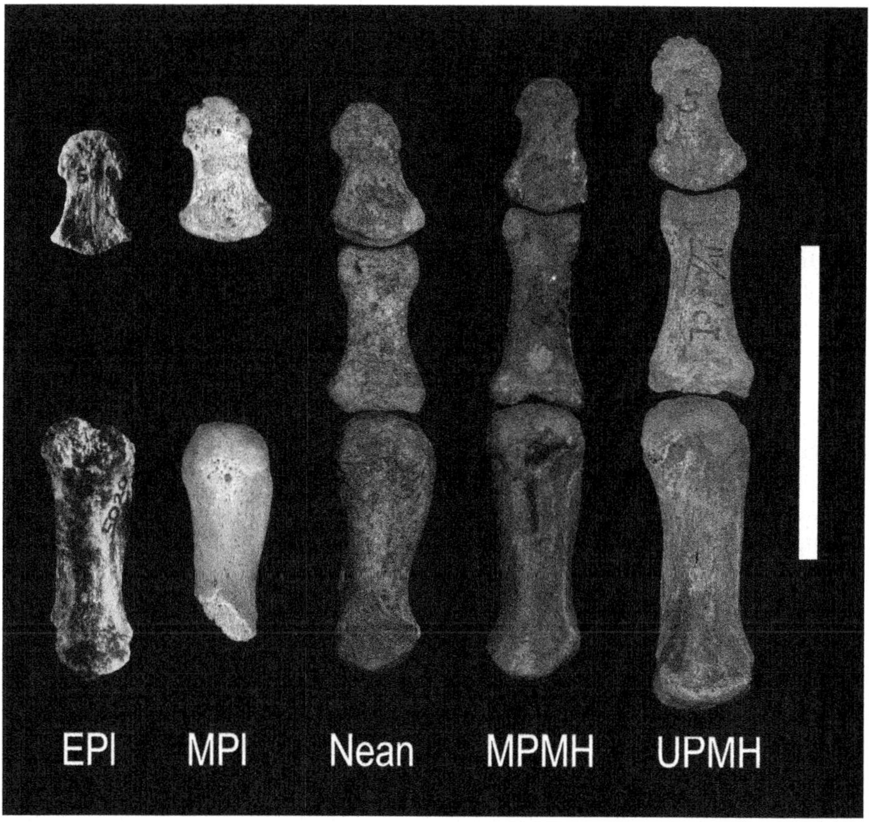

Fig. 19.4 Palmar views of pollical metacarpals and phalanges. EPl, Swartkrans SKX 5020 and SKX 5016 (attributed to either *Au. robustus* or early *Homo*); MPl, AT-1363 (reversed) and AT-1316; Nean, Regourdou 1 (reversed); MPMH, Qafzeh 9; UPMH, Sunghir 1; Scale, 5 cm. Note that the Swartkrans and Atapuerca distal phalanges are not associated with the first metacarpals (Mc1s); the three Late Pleistocene thumbs derive from associated skeletons. Note the relatively shorter distal phalanx (DP1) of the Neandertal, the large opponens pollicis flanges on the Middle Pleistocene and Neandertal Mc1s, the more pronounced ulnar deviation of Middle Pleistocene and Neandertal distal phalanges, and the absence of ulnar deviation in the Early Pleistocene DP1. Abbreviations the same as in Fig. 19.3

6 Pollical Patterns

Changes in thumb anatomy also occur across Pleistocene *Homo*, including differences in the degree of ulnar deviation in distal phalanges (DP), as well as pollical phalangeal length proportions (Fig. 19.4). The fossil record preserves sufficient evidence to assess both of these anatomical characteristics across most of the samples.

Recent humans are characterized by an ulnar deviation of the DP1 (relative to the proximal facet), which brings the distal pulp more directly toward the ulnar digits in prehension (Shrewsbury et al. 2003; Trinkaus et al. 2014a). Early modern humans tend to have relatively modest angles that cluster on the low end of the range seen in recent humans, whereas the Neandertal DP1 angles are generally much higher (recent humans, 4.4° ± 1.6°, $N = 69$; UPMH, 3.3° ± 1.2°, N = 6; MPMH: 1°, 2°; Neandertals, 7.3° ± 3.0°, $N = 14$) (Fig. 19.4; see also Fig. 19.7). The Middle Pleistocene specimens have angles (5.4° ± 1.7°, $N = 7$) that are between those of the recent human and Neandertal samples. However, the Early Pleistocene SKX 5016 has a proximal facet, which is essentially perpendicular to its long axis (angle of 0°), placing it 2.75 standard deviations from the recent human mean and further from the Middle and Late Pleistocene archaic human samples. If SKX 5016 is representative of Early Pleistocene *Homo* DP1s, it would suggest that ulnar deviation evolved later in the Pleistocene. Yet, the Dinaledi DP1 exhibits modest ulnar deviation (Kivell et al. 2015), similar to later *Homo*.

It has also been noted (Trinkaus 1983; Villemeur 1994) that Neandertal thumbs have a long DP1 relative to proximal phalanx 1 (PP1) length (Fig. 19.4). In combination with their relatively greater dorsopalmar dimension of the proximal facet, this has been inferred to indicate greater effective mechanical advantages (EMAs) in power grips across the interphalangeal (IP) articulation, but a reduction in EMA for distal pollical prehension (Trinkaus and Villemeur 1991). A reassessment of interphalangeal pollical length proportions, including early modern humans and mean values for the Middle Pleistocene Atapuerca-SH sample (Fig. 19.2b), supports this pattern. The Neandertal pollical phalanges, plus the Atapuerca-SH mean values, are at or beyond the upper limits of the recent humans, and they are even further from the early modern human thumb proportions.

This comparison, however, does not indicate whether the differences are a result of the lengths of the proximal or the distal phalanges or both. Comparing each phalangeal length to Mc1 length (Fig. 19.2c, d) provides little separation in relative PP1 lengths, yet more of a separation of the archaic versus early modern Pleistocene specimens in relative DP1 length. The Neandertal PP1s cluster among the recent humans with the relatively shorter PP1s, and the Atapuerca-SH mean value is among the lowest. However, it is important to note that only one Atapuerca-SH Mc1 provides a length, although there are eight sufficiently intact PP1s and seven DP1s from the site. The proportional difference between the phalanges is therefore driven mostly by the longer DP1s of the archaic *Homo* specimens. Interestingly, a comparison of the SKX 5016 DP1 length to that of the SKX 5020 Mc1, both from

Member 1 of Swartkrans, places it among the early modern humans, but this comparison should be interpreted with caution as the specimens are unassociated.

7 Carpal, Metacarpal, and Phalangeal Hypertrophy

There are a variety of reflections of relative hypertrophy (or robustness) in Pleistocene *Homo* hand anatomy, involving particularly epiphyseal dimensions, muscle attachments, muscle moment arms, and distal phalangeal enlargement. The discussion here of this hypertrophy is limited by the differential preservation of elements, particularly for the Early and Middle Pleistocene samples, and it focuses on carpal palmar tubercles, pollical epiphyseal breadth, opponens muscle attachments, and distal phalangeal tuberosities.

The principal issue in assessing the robustness (or gracility) of these remains is appropriate scaling to body size or the actual "robusticity" (cf. Ruff et al. 1993) of the elements. When scaled to overall arm dimensions such as humerus length [body mass would be preferable (Ruff 2000), but it is not possible to estimate it for many hand bones], some of the apparent differences between specimens become reduced (e.g., Shang and Trinkaus 2010; Walker et al. 2011). These factors are further complicated by different length proportions within digits (Villemeur 1994; Trinkaus et al. 2014a; see above), making scaling by bone length alone sometimes misleading. The measures of manual robustness considered here are nonetheless scaled mostly within bones, given the incomplete preservation of the *Homo* hand bone fossil record.

7.1 Carpal Tuberosities

The Neandertals have been noted for the large dimensions of their palmar carpal tuberosities (Trinkaus 1983), especially those of the trapezium and the hamate (Fig. 19.5), but the dimensions of the pisiform bone and the scaphoid tubercle are also included. In this hypertrophy, they contrast with the majority of recent humans and early modern humans, especially in the palmar projections of the tuberosities (Shang and Trinkaus 2010; Trinkaus et al. 2014b). Among Early and Middle Pleistocene *Homo*, there is one hamate from the late Early Pleistocene and several of each carpal bone from the Middle Pleistocene, all from Atapuerca (Arsuaga et al. 2015), plus three hamates from Dinaledi. Considerations of carpal tuberosities concern both their overall dimensions and their palmar projection. The former reflects general hypertrophy of the extrinsic and intrinsic muscles that attach on them and/or on the flexor retinaculum between them. The palmar projection forms one component of the EMA of the wrist flexors/stabilizers (especially flexor carpi ulnaris) and of the intrinsic thenar and hypothenar muscles. Tubercle projection would also increase the size of the carpal tunnel, to accommodate hypertrophied tendons for the extrinsic flexor muscles to the palm and the digits and result in greater EMA for wrist flexion during contraction of those muscles.

Comparisons of the general sizes of the trapezial tubercle and the hamulus to carpal body or articular size (Lorenzo et al. 1999, 2012; Lorenzo 2007; Trinkaus et al. 2014b) place the Middle Pleistocene and Neandertal samples at the edge of or beyond the distributions of the recent humans. In contrast, the early modern human specimens all fall within the range of recent human values, as do the Dinaledi (Kivell et al. 2015) and late Early Pleistocene ATD6-23 hamate bones. A substantial portion of the Middle and Late Pleistocene archaic *Homo* tuberosity size is a result of palmar projection. Indeed, comparisons of projection alone (Fig. 19.5) place these Middle and Late Pleistocene archaic *Homo* further from the recent and early modern humans, but less distinct from the earlier ATD6 hamate bone.

There is also a tendency for the earlier Pleistocene humans to have a distal devia-

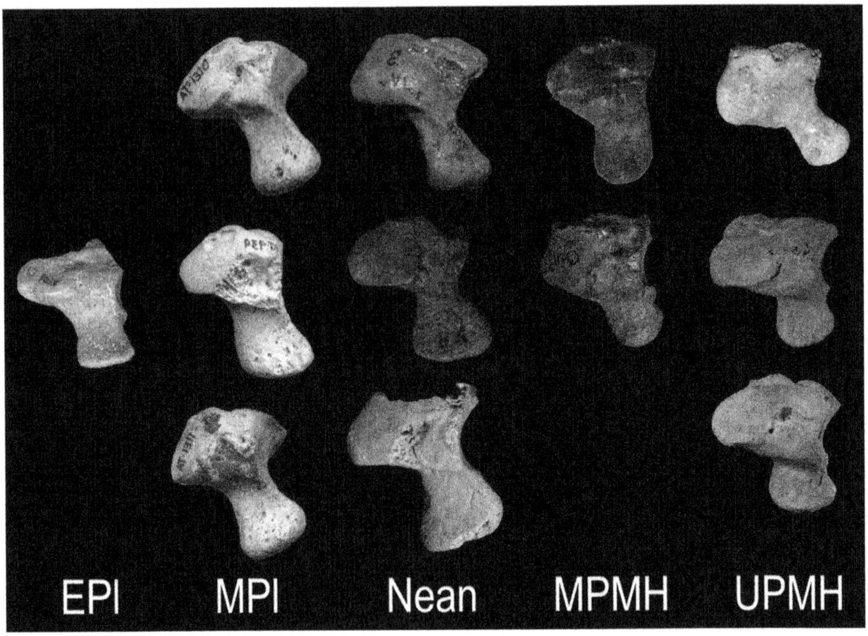

Fig. 19.5 Radial views of *Homo* hamate bones, not to scale. Each bone is oriented relative to the palmar surface of the hamate body. For each sample, from *top* to *bottom*: EPl (Atapuerca-TD6), ATD6-23; MPl (Atapuerca-SH), AT-SH: AT-1310, AT-939 (reversed), and AT-1311; Nean, Tabun 1 ("Tabun 3"), Regourdou 1 (reversed) and Amud 1; MPMH, Qafzeh 9 and Qafzeh 3 (both reversed); UPMH, Dolní Věstonice 3, Sunghir 1, and Tianyuan 1. Note variation in the degree of hamulus palmar projection, proximodistal length, and distal deviation. There is little difference across the samples in the last feature, but the Middle Pleistocene and Neandertal specimens have the largest hamuli. Abbreviations the same as in Fig. 19.3

tion of the hamulus [an apparently ancestral hominin pattern (Kivell et al. 2011)], and it is especially apparent in the Middle and Late Pleistocene archaic *Homo* specimens (Fig. 19.5). However, the early modern human samples show considerable variation in this aspect, with some specimens (Qafzeh 9 and Tianyuan 1) having palmarly oriented hamuli but others (Qafzeh 3 and Dolní Věstonice 3) evincing the more "archaic" distal deviation. It does not appear that variation in this feature would alter the effectiveness of either the extrinsic or intrinsic muscles inserting into the hamulus, but a more distal deviation might better resist bending moments from flexor carpi ulnaris.

7.2 Opponens Muscle Attachments

The Neandertals have been noted for the large flanges on their Mc1s for the insertion of the opponens pollicis muscle, which produce a waisted appearance to the bones in dorsal or palmar view and extend the distal epiphysis radially (Sarasin 1932; Musgrave 1971; Vlček 1975; Kimura 1976; Trinkaus 1983; Lorenzo et al. 2012). This pattern is evident on all but one (Tabun 1) of the Neandertal Mc1s and is ubiquitous in the Middle Pleistocene sample (Fig. 19.4). It also appears to varying degrees among early and recent modern humans but seldom to the extent that is observed among most Neandertals (Trinkaus et al. 2014b). It is also evident on the Mc1s of immature Neandertals, appearing in infancy and becoming prominent by adolescence (Vlček 1975; Heim 1982). The seven Dinaledi Mc1s (from five individuals) all demonstrate an extreme version of this morphology; the opponens pollicis (and first dorsal interosseous) muscle attachments are on well-developed flanges, which flare to create a "pinched" appearance of the palmar surface of the shaft (Kivell et al. 2015). The Swartkrans SKX 5020 Mc1 has extensive rugosity for the opponens pollicis insertion, but the area does not develop into the flange seen in later archaic *Homo* (Fig. 19.4).

It is difficult to assess to what extent the opponens pollicis flanges reflect muscle hypertrophy since the actual insertion area is principally along the radiopalmar margin of the region rather than across the palmar flange. The radial extension, however, does have the consequence of increasing the muscle's rotational moment arm during opposition, but the contrast in these terms is principally between Middle Paleolithic Mc1s (Neandertal and MPMH) and those of the UPMH and recent humans (Maki and Trinkaus 2011).

Interestingly, distinct crests for the insertion of the opponens digiti minimi muscle are present on three-quarters of the Neandertal Mc5s (Trinkaus 1983), and similar crests are present on half of the early modern human Mc5s; they are also known among recent humans (Trinkaus 2006a; Trinkaus et al. 2014b). Opponens digiti minimi crests appear to be absent from the Atapuerca-SH Middle Pleistocene sample, but a distinct one is present on the Early Pleistocene SKW 14147 Mc5 (Day and Scheuer 1973). Since opponens digiti minimi serves to rotate the fifth digit radially, a common role in prehension (Napier 1956; Marzke and Shackley 1986), any muscle hypertrophy indicated by these crests should indicate stronger or more frequent use of those hand postures.

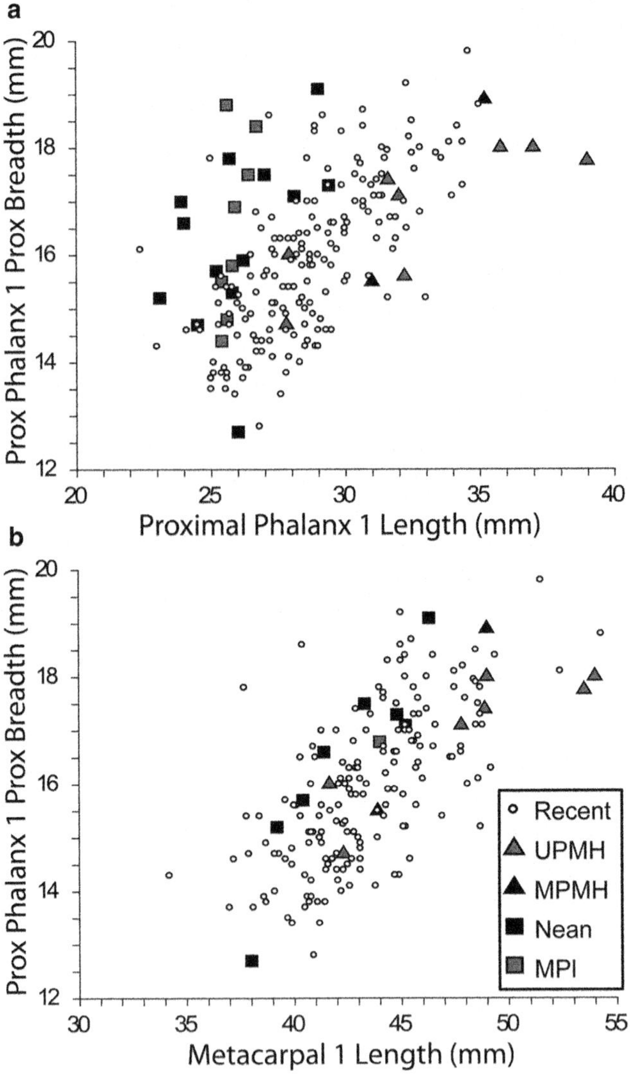

Fig. 19.6 Bivariate plots of pollical proximal phalanx (PP1) proximal epiphyseal breadth versus phalangeal length (**a**) and versus Mc1 length (**b**). Recent sample comprised of Native American, Afro-American, and Euro-American modern humans. In (**b**), the Atapuerca-SH data point is the average of the PP1 proximal breadth (16.8 ± 1.8, *N* = 10) and the AT-5565 Mc1 length (44.0 mm). Abbreviations the same as in Fig. 19.3

7.3 Phalangeal Epiphyseal Dimensions

Many researchers have noted that Neandertal phalanges have relatively broad epiphyses, involving both the distal trochlea on the proximal and middle phalanges and the articular facets and lateral tubercles on the bases of all phalanges (e.g., Musgrave 1973; Trinkaus 1983; Trinkaus et al. 2007; Semal et al. 2009; Lorenzo et al. 2012). This pattern is reflected particularly in the proximal breadths of the PP1s (Fig. 19.4), which include both the articular dimensions and the hypertrophy of the tubercles for the attachment of most of the intrinsic pollical muscles. When proximal maximum breadth is compared to phalangeal length (Fig. 19.6a), all except one (Palomas 96) of the Middle and Late Pleistocene archaic humans stand out as robust relative to the recent humans, whereas the MPMH and UPMH specimens appear gracile. Yet, there is a tendency for the Middle and Late Pleistocene archaic humans to have modestly shorter PP1s relative to their Mc1 lengths (Fig. 19.2); an alternative scaling of PP1 proximal breadth to Mc1 length (Fig. 19.6b) maintains the separation of all but one of the Neandertals and the Middle Pleistocene (Atapuerca-SH) mean value from the early modern human ones. Yet, all of these PP1s (except a couple of gracile UPMH specimens) fall within the recent human scatter. These comparisons therefore suggest that there was an average reduction in the epiphyseal (articular and musculoligamentous) dimensions from archaic to modern humans, but that the apparent metric differences are driven as much by length proportions as by phalangeal epiphyseal hypertrophy.

7.4 Distal Tuberosities (Apical Tufts)

Primates are distinctive in having flat terminal phalanges with apical tufts (or distal tuberosities) that support the nail bed and the palmar pad (Chap. 4). Moreover, there is a correlation between the radioulnar size of the skeletal tuft and the soft tissue adherent to it (Mittra et al. 2007). Among primates, humans are notable for their relatively large apical tufts, usually with distinct, proximally projecting, radial and ulnar ungual spines among recent humans (Susman 1998; Shrewsbury et al. 2003). Apical tufts that are expanded relative to those of nonhuman primates appear in *Australopithecus* (Bush et al. 1982; Kivell et al. 2011; Ward et al. 2012; see Chap. 18), but broader ones appear at Olduvai Bed I (OH 7) and at Swartkrans (SKX 5016) in the initial Early Pleistocene. Among Middle and Late Pleistocene archaic humans, the apical tufts tend to be rounded, extend radially and ulnarly well beyond the minimum diaphyseal breadth, and have small but variably prominent ungual spines (Fig. 19.7). The ungual spines are variably present or distinct in all of the samples and are not unique to human apical tufts (Susman 1998).

Metrical assessments of apical tuft breadths are hampered by the differential length proportions of the phalanges (see Sect. 6 above; also Kivell et al. 2011;

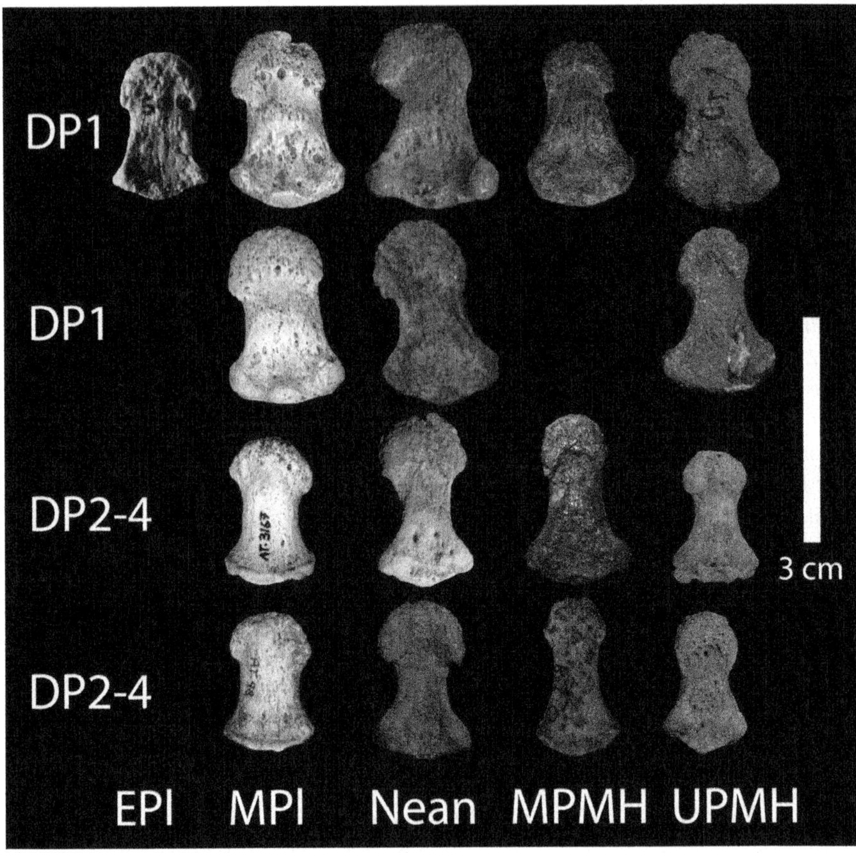

Fig. 19.7 Manual distal phalanges. From *top* to *bottom* for each group: EPl, SKX 5016 (attributed to either *Au. robustus* or early *Homo*); MPl (Atapuerca-SH), AT-SH, AT-8, AT-3129, AT-3167, and AT-98; Nean, Krapina 203.1, Shanidar 6, Amud 1, Shanidar 4; MPMH, Qafzeh 9, 7, and 8; UPMH, Sunghir 1, Caviglione 1, Tianyuan 1, and Sunghir 1. Abbreviations the same as in Fig. 19.3

Trinkaus et al. 2014a). In the thumb, a comparison of distal (tuft) breadth to length of the DP1 provides little proportional separation of the Pleistocene and recent human samples (Fig. 19.8a). However, the Neandertal and Middle Pleistocene specimens have absolutely broader tufts than most recent humans, whereas the early modern humans have narrower tufts. SKX 5016 stands out as having among the broadest apical tufts relative to DP1 length. However, the Dinaledi DP1s ($n=2$ individuals) are even broader (relative to DP1 length) than any other known fossil hominin (Kivell et al. 2015), which may be partly due to their relatively short DP1s. To correct for the relatively long DP1s of the Middle and Late Pleistocene archaic human specimens, distal breadth is also compared to summed PP1 and DP1 length (Fig. 19.8b); this analysis highlights the broad apical tufts seen in the archaic human thumb distal phalanges. It is not possible to include the OH 7 DP1 in the metric

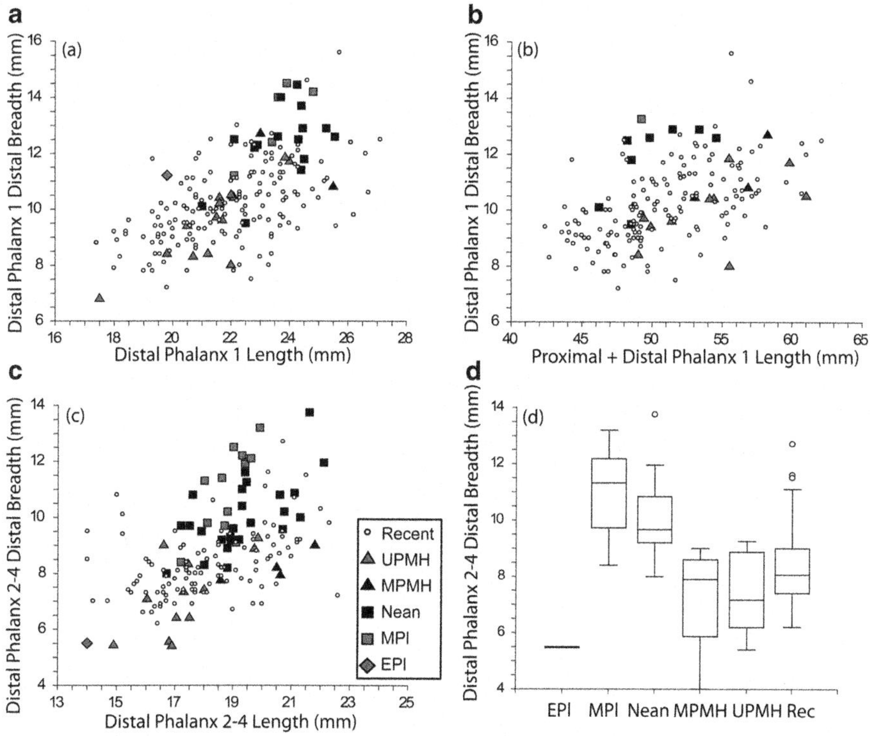

Fig. 19.8 Dimensions of the distal phalangeal apical tufts (tuberosities). (**a**) DP1 tuft breadth versus DP1 length; (**b**) DP1 tuft breadth versus the sum of the PP1 and DP1 lengths; (**c**) apical tuft breadth versus phalanx length for the middle three digits (2–4); (**d**) box plot of apical tuft breadth for distal phalanges 2–4. In (**b**) the MPl (Atapuerca-SH) data point is average value of the DP1 distal breadth of 13.3±1.3 (N=6), DP1 length is 23.4±1.0 (N=7), and PP1 length is 25.9±0.4 (N=8). Recent/Rec sample comprised of Native American, Afro-American, and Euro-American modern humans. All other abbreviations are the same as in Fig. 19.3. For the Late Pleistocene specimens with multiple distal phalanges preserved, the average of each distal breadth and each length by individual was employed in (**c**) and (**d**) to minimize duplication of individuals. Some duplication may exist in the Krapina and Atapuerca-SH samples. The small EPl specimen in (**d**) is SKX 27504, whose attribution to the genus *Homo* is uncertain

comparisons, given its immature status, but it clearly has the broad apical tuft evident in SKX 5016 and the later archaic human DP1s (Susman 1998).

The same difficulty arises is assessing apical tuft dimensions of the ulnar digits [all except the DP5, which exhibits narrow tufts in all of the Pleistocene samples, probably due to the lesser degree of involvement of the distal fifth digit in many human hand postures (Marzke and Shackley 1986)]. Although variation in ulnar phalangeal length proportions is not as pronounced as in the pollex, Neandertals tend to have relatively longer ulnar distal phalanges (Villemeur 1994). As a result, the Middle Pleistocene (Atapuerca-SH) and Neandertal DP2-4s appear (Fig. 19.8c) to follow the recent human pattern of tuft breadth to length proportion. Yet, their distal breadths are absolutely larger than those of most modern humans (Fig. 19.8d), which occurs in the context of similar overall hand size (see Mc lengths in Fig. 19.2). The MPMH and UPMH distal breadths resemble those of recent humans, in both relative and absolute tuft breadth.

There are two other Late Pleistocene archaic human ulnar distal phalanges: one from Sea Harvest, which has a small tuft and is probably from the fifth digit (Grine and Klein 1993), and one from Denisova, which is from one of the middle digits and shows the same apical tuft expansion as the Neandertal distal phalanges (Mednikova 2013). The unusually narrow Early Pleistocene specimen in Fig. 19.8d is the SKX 27504 distal phalanx, from Member 3 of Swartkrans (Susman 1989), whose narrow distal breadth does not follow the pattern evident in the SKX 5016 DP1; it is possible that these specimens represent different taxa. The apical tufts of the Dinaledi non-pollical distal phalanges are similar in breadth (relative to DP length) to those of the Neandertals (Kivell et al. 2015).

Therefore, a pattern is emerging in which archaic *Homo* distal phalanges, from the Early Pleistocene to the Late Pleistocene, had expanded apical tufts and by inference nails, nail beds, and palmar pulps. It is only with modern humans (early and recent) that there is a secondary reduction, to the smaller dimensions evident in *Australopithecus*.

8 Putting the Pieces Together

8.1 Morphological Polarities

The general pattern that emerges from Pleistocene *Homo* hand remains is that there is a general "archaic *Homo*" pattern, evident in the Middle Pleistocene (Atapuerca-SH) and Neandertal samples, that contrasts in a number of aspects with that of "modern" humans, including most of those of Middle and Upper Paleolithic early modern humans and of recent humans. Many of these aspects are related to the hypertrophy of the hand remains, and they include carpal palmar tuberosity size and projection, muscular insertion development (e.g., the opponens muscles), phalangeal epiphyseal dimensions, and broad apical tufts. Depending on how they are scaled to reflections of body size, all of these features reduce to variable degrees among early and recent humans.

As noted with respect to the fossil samples, it remains an open question as to what extent the Middle Pleistocene and Neandertal hand remains represent "archaic *Homo*" generally or might be derived in some aspects, as with other features of the Neandertal lineage (cf. Trinkaus 2006b). The late Early Pleistocene ATD6 remains, in two available comparisons (capitate-Mc2 articulation and hamulus size), are less distinct from modern humans than the Middle Pleistocene (Atapuerca-SH) and Neandertal samples. Yet, the earlier Early Pleistocene KNM-KP 51260 Mc3 has the distinctively human, though relatively short, styloid process found in other later

Pleistocene *Homo*. The SKX 5020 Mc1 lacks the opponens pollicis flange, despite marked rugosity for the muscle insertion on the radial side of the diaphysis, yet large ones are present on the Dinaledi Mc1s. All of the earlier (or apparently earlier) *Homo* DP1s (SKX 5016, OH 7 and Dinaledi DP1s) have the marked apical tuft expansion of later Eurasian archaic humans. And the OH 7 and Dinaledi trapezia exhibit the Mc1 articular "flattening" of most Neandertals, but the Middle Pleistocene Atapuerca-SH sample does not exhibit it.

Without a geological age, it is difficult to assess where Dinaledi remains may fit within the evolutionary history of *Homo*, and when and why particular aspects of its hand morphology may have evolved. The presence of some australopith-like features in the hand (e.g., curved phalanges, absence of a Mc3 styloid process) and the remainder of the postcranial skeleton suggests that the sample may stem from the base of *Homo* lineage (Berger et al. 2015; Kivell et al. 2015). However, the presence of derived Neandertal and recent humanlike changes to radial carpals, including a palmarly expanded, "boot-shaped" trapezoid that reorients the carpal articulations to cope with large loads from the thumb during tool-related behaviors, suggests several potential scenarios: (1) the derived later *Homo*-like features of the radial carpals evolved much earlier than previously thought, (2) the Dinaledi sample may be much later than its australopith-like features suggest, and/or (3) some of the ancestral or derived features evolved convergently in response to similar functional demands (e.g., curved phalanges in response to climbing or a palmarly expanded trapezoid in response to powerful loading of the thumb). At the very least, the sample demonstrates a combination of morphological features that had not previously been observed within one hand.

Therefore, the suite of manual skeletal changes between the later archaic *Homo* remains and early modern humans are relatively straightforward and are largely associated with greater robustness in Middle and Late Pleistocene archaic humans. In contrast, the scattered remains of Early Pleistocene hand bones (some of which may not belong in the genus *Homo*) provide a complex mosaic of polarity assessments. It should be apparent that the resolution of these polarity issues, therefore, will require both more complete Early Pleistocene hand remains and ones that are securely assigned to a genus. It also means that assessments of the polarities of still earlier hominins cannot merely be compared to the configurations of extant humans and nonhuman primates.

8.2 Behavioral Implications

Ultimately, the concern with the evolution of the *Homo* hand is the relationship between the hand anatomy and human manipulative abilities and behaviors. Given the extensive changes in human technology through the Paleolithic, in terms of both the manufacture and use of implements, and the hand as the interface between the body and that technology, one would expect some degree of correlation between the evolution of human hand functional anatomy and the mechanical properties of that technology. Yet, none of the aspects highlighted here would indicate any differences in basic manipulative capabilities or manual dexterity, and none of the recent assessments [ignoring earlier twentieth-century interpretations (e.g., Boule 1911-13; Bonch-Osmolovskij 1941)] have provided any evidence that the hand anatomy of these humans would have prevented them from producing the technologies with which they are associated. The only reflections of a behavioral difference may concern the CMc1 and the CMc2-3 articular facet shapes and orientations (which are nonetheless quite variable within and between the samples) and the multitude of ways that the hands of non-modern *Homo* were more robust.

These considerations therefore beg the question of the functional correlations between hand anatomy, and especially hand hypertrophy, and the archeological complexes with which they are associated. Given the general nature of the Oldowan and Acheulian (*sensu lato*) technocomplexes, with their emphasis on expedient flakes and core tools, upper limb hypertrophy is to be expected. However, the Middle Paleolithic (*sensu lato*) provides abundant evidence for the hafting of lithic cutting edges, which would have increased the mechanical effectiveness of implements and thereby reduced the upper limb strength needed to accomplish habitual tasks. Yet, associated with the Middle Paleolithic, there are both robust Neandertal hand remains and the generally more gracile ones of the Middle Paleolithic modern humans. In the past, it is the relative gracility of the latter that has generated comment (e.g., Trinkaus 1992). Yet, it may be the absence of decreases in hand robustness among the Neandertals relative to Middle Pleistocene humans that is functionally anomalous, given the multiple changes in Middle Paleolithic technology that should have reduced mechanical loads on the upper limb. Given the increasing documentation that distinctive features of Neandertal hand remains, vis-à-vis modern humans, were both present in earlier (Early and especially Middle Pleistocene) *Homo* and emerged early in Neandertal development, it is likely that many of them had become phylogenetic baggage of little direct behavioral valence by the time of the Neandertals. Indeed, for features that are clearly plastic (e.g., the rugosities for the opponens muscles), the variations through the Pleistocene extensively overlap.

The further technological elaborations of the Upper Paleolithic, especially with the development of a diversity of composite tools using both lithic and organic working edges, appear to correlate with the general further reduction in their manual hypertrophy. The greater variation in robustness in the recent human sample, and the apparently more gracile nature of some early modern human phalanges, may relate to the more linear dimensions of those early modern humans (Trinkaus 1981; Holliday 1997, 2000), if their phalangeal proportions parallel those of their long bones and need to be scaled against body mass as well as bone length (cf. Ruff 2000; Trinkaus and Ruff 2012).

In addition, the discovery of a semi-articulated, complete hand from Dinaledi demonstrates that derived Neandertal and recent humanlike features of the hand, traditionally considered adaptations for tool use and toolmaking, can occur in combination with features that clearly indicate use of the hand for locomotion. Thus the Dinaledi sample demonstrates that at least one member of the *Homo* lineage was able to retain ancestral patterns of hand (and upper limb) use for climbing while evolving morphology to facilitate forceful precision grip and tool use. Future discoveries of more complete hand skeletons that can be attributed to the earliest members of the genus *Homo* (i.e., *H. habilis*) may demonstrate alternative morphological solutions for the functional demands of both locomotion and manipulation within the hand.

9 Conclusions

A growing paleontological knowledge of hand anatomy for the Pleistocene genus *Homo*, from both new discoveries of Early and Middle Pleistocene humans and reassessments of the Late Pleistocene fossil record, has highlighted a series of changes in the details of human hand morphology through the Pleistocene (and the Paleolithic). Although there is little reason to argue for differences in manual dexterity, other than those that might be due to expanding neurological control of the upper limb, there is a series of changes in phalangeal proportions, tuberosity development, muscular attachments, and distal dimensions that argue for little long-term stasis in the details of the human hand. The anatomy of the hand continued to evolve throughout the genus *Homo*. Moreover, these changes point out that assessments of both preceding (*Australopithecus sensu lato*) and succeeding (extant human) hand morphology needs to be placed in the context of these changes through the Pleistocene.

10 Future Directions

These comments and comparisons have highlighted as many uncertainties in our knowledge of Pleistocene *Homo* hand anatomy as they have permitted the identification of key manual patterns through the genus. It goes without saying that many of the issues would be resolved with more fossil remains, in particular for the Early Pleistocene but especially of largely complete, associated hand skeletons prior to the Late Pleistocene; among hominins, there are only two prior to ~100 Ka BP, the MH2 *Au. sediba* hand (Kivell et al. 2011) and the Dinaledi DH1 hand (Kivell et al. 2015). It is only with such remains that complex issues of proportions can be resolved. It is also likely that further insight into the complex functional interplay of especially carpometacarpal articulations will come with three-dimensional modeling of those joint surfaces (Chap. 9), but in each case both (or all) sides of the joint need to be considered. Finally, it is essential that all phases of *Homo* hand evolution be considered in these evaluations, and not merely recent humans, since they [as in other features (cf. Trinkaus 2006b)] are more derived than Middle and Late Pleistocene archaic humans relative to early members of the genus *Homo*.

Acknowledgments This summary has only been possible through the willingness of C. Lorenzo to share unpublished data and images on the Atapuerca human hand remains. I. Crevecoeur, B. Maureille, J.H. Musgrave, W.A. Niewoehner, and C.V. Ward provided comparative data; F.K. Manthi and C.V. Ward furnished the KNM-KP 51260 image; R.L. Susman provided the Swartkrans images; T.L. Kivell helped with the integration of the Dinaledi hand remains into the discussion; and curators and colleagues too numerous to mention individually have permitted access to the original human remains that make up the core of the comparisons here. Portions of this work have been funded by the National Science, Wenner-Gren and Leakey Foundations. To all I am grateful.

References

Alba DM, Moyà-Solà S, Köhler M (2003) Morphological affinities of the *Australopithecus afarensis* hand on the basis of manual proportions and relative thumb length. J Hum Evol 44:225–254

Arsuaga JL, Carretero J-M, Lorenzo C, Gómez-Olivencia A, Pablos A, Rodríguez L, García-González R, Bonmatí A, Quam RM, Pantoja-Pérez A, Martínez I, Aranburo A, Gracia-Téllez A, Poza-Rey E, Sala N, Garcia N, Alcázar de Velasco A, Cuenca-Bescós G, Bermúdez de Castro JM, Carbonell E (2015) Postcranial morphology of the Middle Pleistocene humans from Sima de los Huesos, Spain. Proc Natl Acad Sci U S A 112:11524–11529

Begun D, Walker A (1993) The endocast. In: Walker A, Leakey R (eds) The Nariokotome *Homo erectus* skeleton. Harvard University Press, Cambridge, MA, pp 326–358

Berger LR, Hawks J, de Ruiter DJ, Churchill SE, Schmid P, Delezene LK, Kivell TL, Garvin HM, Williams SA, DeSilva JM, Skinner MM, Musiba CM, Cameron N, Holliday TW, Harcourt-Smith W, Ackerman RR, Bastir M, Bogin B, Bolter D, Brophy J, Cofran ZD, Congdon KA, Deane AS, Dembo M, Drapeau M, Elliot MC, Feuerriegel EM, Garcia-Martinez D, Green DJ, Gurtov A, Irish JD, Kruger A, Laird MF, Marchi D, Meyer MR, Nalla S, Negash EW, Orr CM, Radovčić D, Schroeder L, Scott JE, Throckmorton Z, Tocheri MW, VanSickle C, Walker CS, Wei P, Zipfel B (2015) *Homo naledi*, a new species of the genus *Homo* from the Dinaledi Chamber, South Africa. eLife 4:e09560

Bermúdez de Castro JM, Carretero JM, García-Gonzáles R, Rodríguez-García L, Martinón-Torres M, Rosell J, Blasco R, Martín-Francés L, Modesto M, Carbonell E (2012) Early Pleistocene human humeri from the Gran-Dolina-TD6 site (Sierra de Atapuerca, Spain). Am J Phys Anthropol 147:604–617

Bonch-Osmolovskij GA (1941) The hand of the fossil man from Kiik-Koba (in Russian with French summary). Paleolit Kryma 2:1–172

Boule M (1911–13) L'homme fossile de La Chapelle-aux-Saints. Ann Paléontol 6:111–172; 7, 21–56, 85–192; 8, 1–70.

Bush ME, Lovejoy CO, Johanson DC, Coppens Y (1982) Hominid carpal, metacarpal, and phalangeal bones recovered from the Hadar formation: 1974–1977 collections. Am J Phys Anthropol 57:651–677

Cartmill M, Smith FH (2009) The human lineage. Wiley-Blackwell, New York

Cooney WP III, Chao EYS (1977) Biomechanical analysis of static forces in the thumb during hand function. J Bone Joint Surg Am 59:27–36

Crevecoeur I (2008) Étude anthropologique du squelette du Paléolithique Supérieur de Nazlet Khater 2 (Égypte). Leuven University Press, Leuven

Day MH, Scheuer JL (1973) SKW 14147: a new hominid metacarpal from Swartkrans. J Hum Evol 2:429–438

Domínguez-Rodrigo M, Pickering TR, Almécija S, Heaton JL, Baquedano E, Mabulla A, Uribelarrea D (2015) Earliest modern human-like hand bone from a new >1.84-million-year-old site at Olduvai in Tanzania. Nat Commun 6:7987

Green DJ, Gordon AD (2008) Metacarpal proportions in *Australopithecus africanus*. J Hum Evol 54:705–719

Grine FE, Klein RG (1993) Late Pleistocene human remains from the Sea Harvest site, Saldanha Bay, South Africa. S Afr J Sci 89:145–152

Heim JL (1982) Les Enfants Néandertaliens de La Ferrassie. Masson, Paris

Holliday TW (1997) Body proportions in Late Pleistocene Europe and modern human origins. J Hum Evol 32:423–447

Holliday TW (2000) Evolution at the crossroads: modern human emergence in Western Asia. Am Anthropol 102:54–68

Kimura T (1976) Correction to the *Metacarpale I* of the Amud man. A new description especially on the insertion area of the *M. opponens pollicis*. J Anthropol Soc Nippon 84:48–54

Kivell TL, Kibii JM, Churchill SE, Schmid P, Berger LR (2011) *Australopithecus sediba* hand demonstrates mosaic evolution of locomotor and manipulative abilities. Science 333:1411–1417

Kivell TL, Deane AS, Tocheri MW, Orr CM, Schmid P, Hawks J, Berger LR, Churchill SE (2015) The hand of *Homo naledi*. Nat Commun 6:8431

Lewis OJ (1973) The hominoid os capitatum, with special reference to the fossil bones from Sterkfontein and Olduvai Gorge. J Hum Evol 2:1–11

Lorenzo C (2007) Evolución de la mano en los homínidos. Análisis morfológico de los fósiles de la Sierra de Atapuerca. Tesis doctoral, Universidad Complutense de Madrid

Lorenzo C, Arsuaga JL, Carretero JM (1999) Hand and foot remains from the Gran Dolina Early Pleistocene site (Sierra de Atapuerca, Spain). J Hum Evol 37:501–522

Lorenzo C, Carretero JM, Arsuaga JL, Martínez I, Gracia A, Quam R (2012) Hands, laterality and language: hand morphology in the Sima de los Huesos site (Sierra de Atapuerca, Spain). Am J Phys Anthropol 147(Suppl 54):195–196 (abstract)

Lorenzo C, Pablos A, Carretero JM, Huguet R, Valverdú J, Martinón-Torres M, Arsuaga JL, Carbonell E, Bermúdez de Castro JM (2015) Early Pleistocene human hand phalanx from the Sima del Elefante (TE) cave site in Sierra de Atapuerca (Spain). J Hum Evol 78:114–121

MacLarnon A (1993) The vertebral canal. In: Walker A, Leakey R (eds) The Nariokotome *Homo erectus* skeleton. Harvard University Press, Cambridge, MA, pp 359–390

Maki J, Trinkaus E (2011) Opponens pollicis mechanical effectiveness in Neandertals and early modern humans. PaleoAnthropology 2011:62–71

Marzke MW (1983) Joint functions and grips of the *Australopithecus afarensis* hand, with special reference to the region of the capitate. J Hum Evol 12:197–211

Marzke MW, Marzke RF (1987) The third metacarpal styloid process in humans: origins and functions. Am J Phys Anthropol 73:415–431

Marzke MW, Shackley MS (1986) Hominid hand use in the Pliocene and Pleistocene: evidence from experimental archaeology and comparative morphology. J Hum Evol 15:439–460

Marzke MW, Tocheri MW, Steinberg B, Femiani JD, Reece SP, Linscheid RL, Orr CM, Marzke RF (2010) Comparative 3D quantitative analyses of trapeziometacarpal joint surface curvatures among living catarrhines and fossil hominins. Am J Phys Anthropol 141:38–51

Matiegka J (1938) Homo předmostensis. Fosilní člověk z Předmostí na Moravě II. Ostatní Části Kostrové. Česká Akademie Věd a Umění, Prague

McHenry HM (1983) The capitate of *Australopithecus afarensis* and *A. africanus*. Am J Phys Anthropol 62:187–198

Mednikova MB (2013) Distal manual phalanx of *Homo* from layer 12 of Denisova Cave: an attempt of identification (in Russian). Archaeol Ethnol Anthropol Eurasia 54:146–155

Meyer MR (2005) Functional biology of the *Homo erectus* axial skeleton from Dmanisi, Georgia. Ph.D. thesis, University of Pennsylvania

Mittra ES, Smith MF, Lemelin P, Jungers WL (2007) Comparative morphometrics of the primate apical tuft. Am J Phys Anthropol 134:449–459

Moyà-Solà S, Köhler M, Alba DM, Almécija S (2008) Taxonomic attribution of the Olduvai Hominid 7 manual remains and the functional interpretation of hand morphology in robust australopithecines. Folia Primatol 79:215–250

Musgrave JH (1971) How dextrous was Neanderthal man? Nature 233:538–541

Musgrave JH (1973) The phalanges of Neanderthal and Upper Palaeolithic hands. In: Day MH (ed) Human evolution. Taylor & Francis, London, pp 59–85

Nakamura A, Yamada T, Goto A, Kato T, Ito K, Abe Y, Kachi T, Kakigi R (1998) Somatosensory homunculus as drawn by MEG. Neuroimage 7:377–386

Napier JR (1956) The prehensile movements of the human hand. J Bone Joint Surg Br 38:902–913

Napier J (1962) Fossil hand bones from Olduvai Gorge. Nature 196:409–411
Niewoehner WA (2000) The functional anatomy of Late Pleistocene and recent human carpometacarpal and metacarpophalangeal articulations. Ph.D. thesis, University of New Mexico
Niewoehner WA (2001) Behavioral inferences from Skhul/Qafzeh early modern human hand remains. Proc Natl Acad Sci U S A 98:2979–2984
Niewoehner WA, Weaver AH, Trinkaus E (1997) Neandertal capitate-metacarpal articular morphology. Am J Phys Anthropol 103:219–233
Richmond BG, Green DJ, Braun DR, Mbua E, Griffin NL, Chirchir H, Harris JWK (2011) New fossils from Ileret, Kenya, and the evolution of hominin hand function. Am J Phys Anthropol 144(Suppl 52):253 (abstract)
Roebroeks W, Villa P (2011) On the earliest evidence for habitual use of fire in Europe. Proc Natl Acad Sci U S A 108:5209–5214
Rolian C, Gordon AD (2013) Reassessing manual proportions in *Australopithecus afarensis*. Am J Phys Anthropol 152:393–406
Ruff CB (2000) Body size, body shape, and long bone strength in modern humans. J Hum Evol 38:269–290
Ruff CB, Trinkaus E, Walker A, Larsen CS (1993) Postcranial robusticity in *Homo*, I: temporal trends and mechanical interpretations. Am J Phys Anthropol 91:21–53
Ruff CB, Trinkaus E, Holliday TW (1997) Body mass and encephalization in Pleistocene *Homo*. Nature 387:173–176
Sarasin F (1932) Die Variationen im Bau des Handskeletts verschiedener Menschenformen. Z Morphol Anthropol 30:252–316
Semal P, Rougier H, Crevecoeur I, Jungels C, Flas D, Hauzeur A, Maureille B, Germonpré M, Bocherens H, Pirson S, Cammaert L, De Clerck N, Hambucken A, Higham T, Toussaint M, van der Plicht J (2009) New data on the late Neandertals: direct dating of the Belgian Spy fossils. Am J Phys Anthropol 138:421–428
Shang H, Trinkaus E (2010) The early modern human from Tianyuan Cave, China. Texas A&M University Press, College Station
Shrewsbury MM, Marzke MW, Linscheid RL, Reece SP (2003) Comparative morphology of the pollical distal phalanx. Am J Phys Anthropol 121:30–47
Smith FH, Smith MO, Schmitz RW (2006) Human skeletal remains from the 1997 and 2000 excavations of cave deposits derived from Kleine Feldhofer Grotte in the Neander Valley, Germany. In: Schmitz RW (ed) Neanderthal 1856–2006. Philipp von Zabern, Mainz, pp 187–246
Stark HH, Moore JF, Ashworth CR, Boyes JH (1977) Fusion of the first metacarpotrapezial joint for degenerative arthritis. J Bone Joint Surg Am 59:22–26
Susman RL (1989) New hominid fossils from the Swartkrans formation (1979–1986 excavations): postcranial specimens. Am J Phys Anthropol 79:451–474
Susman RL (1998) Hand function and tool behavior in early hominids. J Hum Evol 35:23–46
Trinkaus E (1981) Neanderthal limb proportions and cold adaptation. In: Stringer CB (ed) Aspects of human evolution. Taylor & Francis, London, pp 187–224
Trinkaus E (1983) The Shanidar Neandertals. Academic Press, New York
Trinkaus E (1989) Olduvai Hominid 7 trapezial metacarpal 1 articular morphology: contrasts with recent humans. Am J Phys Anthropol 80:411–416
Trinkaus E (1992) Morphological contrasts between the Near Eastern Qafzeh-Skhul and late archaic human samples: grounds for a behavioral difference? In: Akazawa T, Aoki K, Kimura T (eds) The evolution and dispersal of modern humans in Asia. Hokusen-Sha, Tokyo, pp 277–294
Trinkaus E (2006a) The upper limb remains. In: Trinkaus E, Svoboda JA (eds) Early modern human evolution in Central Europe: the people of Dolní Věstonice and Pavlov. Oxford University Press, New York, pp 327–372
Trinkaus E (2006b) Modern human versus Neandertal evolutionary distinctiveness. Curr Anthropol 47:597–620
Trinkaus E (2009) The human tibia from Broken Hill, Kabwe, Zambia. PaleoAnthropology 2009:145–165

Trinkaus E, Long JC (1990) Species attribution of the Swartkrans Member 1 first metacarpals: SK 84 and SKX 5020. Am J Phys Anthropol 83:419–424

Trinkaus E, Ruff CB (2012) Femoral and tibial diaphyseal cross-sectional geometry in Pleistocene *Homo*. PaleoAnthropology 2012:13–62

Trinkaus E, Villemeur I (1991) Mechanical advantages of the Neandertal thumb in flexion: a test of an hypothesis. Am J Phys Anthropol 84:249–260

Trinkaus E, Maki J, Zilhão J (2007) Middle Paleolithic human remains from the Gruta da Oliveira (Torres Novas), Portugal. Am J Phys Anthropol 134:263–273

Trinkaus E, Svoboda J, Wojtal P, Nývltová Fišákova M, Wilczyński J (2010) Human remains from the Moravian Gravettian: morphology and taphonomy of additional elements from Dolní Věstonice II and Pavlov I. Int J Osteoarchaeol 20:645–669

Trinkaus E, Haduch E, Valde-Nowak PW, Wojtal P (2014a) The Obłazowa 1 early modern human pollical phalanx and Late Pleistocene distal thumb proportions. Homo 65:1–12

Trinkaus E, Buzhilova AP, Mednikova MB, Dobrovolskaya MV (2014b) The people of Sunghir: burials, bodies and behavior in the earlier Upper Paleolithic. Oxford University Press, New York

Verneau R (1906) Les Grottes de Grimaldi (Baoussé-Roussé), Tome II—Fascicule I. Anthropologie. Imprimerie de Monaco, Monaco

Villemeur I (1994) La main des Néandertaliens: comparaison avec la main des hommes de type moderne. Morphologie et mécanique. Cahiers de paléoanthropologie, C.N.R.S. Éditions, Paris

Vlček E (1975) Morphology of the first metacarpal of Neanderthal individuals from the Crimea. Bull Mém Soc Anthropol Paris Sér 13(2):257–276

Walker MJ, Ortega J, López MV, Parmová K, Trinkaus E (2011) Neandertal postcranial remains from the Sima de las Palomas del Cabezo Gordo, Murcia, southeastern Spain. Am J Phys Anthropol 144:505–515

Ward CV, Leakey MG, Walker A (2001) Morphology of *Australopithecus anamensis* from Kanapoi and Allia Bay, Kenya. J Hum Evol 41:255–368

Ward CV, Kimbel WH, Harmon EH, Johanson DC (2012) New postcranial fossils of *Australopithecus afarensis* from Hadar, Ethiopia (1990–2007). J Hum Evol 63:1–51

Ward CV, Tocheri MW, Plavcan JM, Brown FH, Manthi FK (2014) An Early Pleistocene third metacarpal from Kenya and the evolution of modern human-like hand morphology. Proc Natl Acad Sci U S A 111:121–124

Weaver AH, Holliday TW, Ruff CB, Trinkaus E (2001) The fossil evidence for the evolution of human intelligences in Pleistocene *Homo*. In: Nowell A (ed) In the mind's eye, vol 13, Multidisciplinary perspectives on the evolution of human cognition. International Monographs in Prehistory Archaeological Series, Ann Arbor, pp 154–172

Index

T.L. Kivell et al. (eds.), *The Evolution of the Primate Hand*, Developments in Primatology: Progress and Prospects, DOI 10.1007/978-1-4939-3646-5

B

G

N

O

P

CPSIA information can be obtained
at www.ICGtesting.com
Printed in the USA
LVOW02*1616210816
501269LV00001B/17/P

9 781493 936441